T0363899

GERD

Reflux to Esophageal Adenocarcinoma

GERD

Reflux to Esophageal Adenocarcinoma

PARAKRAMA T. CHANDRASOMA
KECK School of Medicine, University of Southern California, Los Angeles, California, USA

TOM R. DeMEESTER
KECK School of Medicine, University of Southern California, Los Angeles, California, USA

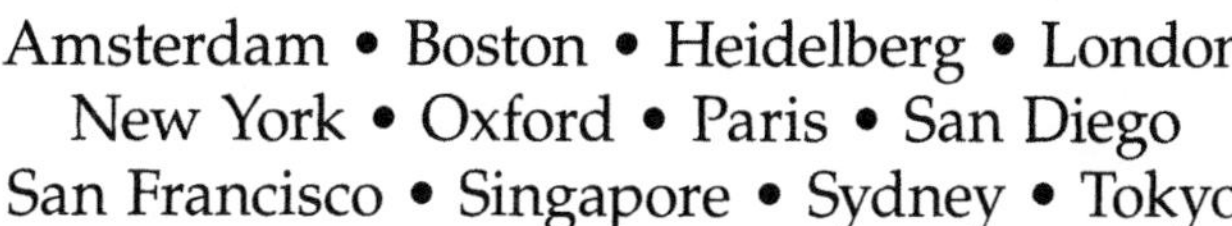
Amsterdam • Boston • Heidelberg • London
New York • Oxford • Paris • San Diego
San Francisco • Singapore • Sydney • Tokyo

Academic Press is an imprint of Elsevier
30 Corporate Drive, Suite 400, Burlington, MA 01803, USA
525 B Street, Suite 1900, San Diego, California 92101-4495, USA
84 Theobald's Road, London WC1X 8RR, UK

This book is printed on acid-free paper. 

Library of Congress Cataloging-in-Publication Data
Application Submitted

British Library Cataloguing-in-Publication Data
A catalogue record for this book is available from the British Library.

ISBN 13: 978-0-12-369416-4
ISBN 10: 0-12-369416-7

For information on all Academic Press publications visit our Web site at www.books.elsevier.com

Printed and bound by CPI Group (UK) Ltd, Croydon, CR0 4YY
Transferred to Digital Print 2011

DEDICATION

To all those patients with gastroesophageal reflux disease who paid the ultimate price for our lack of understanding.

DEDICATION

Contents

CHAPTER

3

Fetal Development of the Esophagus and Stomach

CHAPTER

4

Normal Anatomy; Present Definition of the Gastroesophageal Junction

CHAPTER

5

Histologic Definitions and Diagnosis of Epithelial Types

CHAPTER

6

Cardiac Mucosa

CHAPTER

7

New Histologic Definitions of Esophagus, Stomach, and Gastroesophageal Junction

CHAPTER

8

Pathology of Reflux Disease at a Cellular Level: Part 1—Damage to Squamous Epithelium and Transformation into Cardiac Mucosa

CHAPTER

9

The Pathology of Reflux Disease at a Cellular Level: Part 2—Evolution of Cardiac Mucosa to Oxyntocardiac Mucosa and Intestinal Metaplasia

CHAPTER

10

Pathology of Reflux Disease at a Cellular Level: Part 3—Intestinal (Barrett) Metaplasia to Carcinoma

CHAPTER

11

Pathology of Reflux Disease at an Anatomic Level

CHAPTER

12

Reflux Disease Limited to the Dilated End-Stage Esophagus: The Pathologic Basis of NERD

CHAPTER

13

Definition of Gastroesophageal Reflux Disease and Barrett Esophagus

CHAPTER

14

Diagnosis of Gastroesophageal Reflux Disease, Barrett Esophagus, and Dysplasia

CHAPTER

15

Research Strategies for Preventing Reflux-Induced Adenocarcinoma

CHAPTER

16

Rationale for Treatment of Reflux Disease and Barrett Esophagus

CHAPTER

17

Treatment Strategies for Preventing Reflux-Induced Adenocarcinoma

Preface

Failure to understand a disease has consequences to the patient. In the sixteenth century, belief that ill-humors in the body were responsible for disease led to many patients being purged and bled as part of their treatment while their diseases progressed to their natural outcome. The purging and bleeding often caused harm rather than good. For patients with reflux disease, the naïve and false belief that all facets of their disease are caused by acid has resulted in every person with reflux disease being prescribed acid suppressive drugs. Physicians declare this treatment to be a success of modern medicine, yet the mortality of patients with reflux disease has increased exponentially. This is the result of reflux-induced adenocarcinoma. While we know that esophageal adenocarcinoma is caused by reflux, we do not understand this relationship.

To be correctly understood, a disease must be characterized by its cellular and ultimately its molecular and biochemical events. In gastroesophageal reflux disease, we are far from this ideal. The present management of patients with reflux is confusing and full of contradictions. We define a patient as having reflux when they have symptoms, but recognize that many asymptomatic patients develop reflux induced complications such as Barrett's esophagus and adenocarcinoma. We define reflux by the presence of erosions but recognize the entity of "non-erosive reflux disease". We define reflux by the amount of reflux or acid exposure by impedance and 24-hour pH tests but recognize that no disease should be defined by its etiology rather than its cellular change. There is no histologic definition of reflux. In our ignorance, all we do is feed the patient with reflux disease acid suppressive drugs and hope in vain for a cure. Are acid suppressive drugs the equivalent of purging and bleeding of the sixteenth century? This book provides a theoretical basis of how acid suppressive drugs promote reflux-induced adenocarcinoma. We have unwittingly been guilty of violating the primary edict in medicine of: "Physician, do no harm."

Our work is based on the careful clinical study of over 10,000 patients over the past 15 years at the Keck School of Medicine at the University of Southern California. Patients have been examined clinically and by a range of tests, always including biopsy by a standard protocol of every patient who undergoes endoscopy. This has permitted us to describe the changes of reflux disease at a histologic level from cell to cell and millimeter to millimeter. Our translational molecular research on these patients has given us the ability to characterize the observed cellular events in their most fundamental molecular and genetic terms.

It is hoped that the increased understanding of reflux disease generated by this book will result in improved patient care, measured by a reversal of the incidence of reflux-induced cancer. Reflux induced adenocarcinoma is preceded by histologically recognizable precursor lesions for many decades. We presently ignore these and simply wait for cancer to occur, often too late to prevent death. Recognizing these early precursor changes, and aggressively detecting and treating them *can prevent cancer.* We have provided practical methods of achieving this. Our hope is that the information contained within this book will prevent misery and death caused by reflux induced esophageal adenocarcinoma, at least in the patients whose treatment will be changed by people who read it.

CHAPTER

1

Overview of Gastroesophageal Reflux Disease

Gastroesophageal reflux disease is the classical good news–bad news story. The availability of amazingly effective acid suppressive drugs has improved the quality of life for patients by controlling pain, curing erosive esophagitis, and dramatically reducing the incidence of serious ulcers and strictures. The bad news, however, is that the incidence of adenocarcinoma induced by reflux is rapidly increasing in the United States, Western Europe, Australia, and New Zealand (1). Since the mid-1970s, the incidence of esophageal adenocarcinoma has shown the greatest increase when compared with all other cancer types (Fig. 1.1).

The detection, diagnosis, and management of patients with Barrett esophagus (BE), the precursor lesion for reflux-induced adenocarcinoma, are in a state of confusion and total disarray. It is an entity without consensus regarding definition, diagnosis, and management. This was shown very well at a recent American Gastroenterology Association workshop (2). This workshop consisted of 18 experts in the field (15 gastroenterologists, 2 surgeons, and 1 pathologist) from four countries. This group developed 42 statements on the subject of Barrett esophagus. They individually and collectively evaluated the available evidence for each statement, and after discussion they ascribed a numerical grade for the nature of the evidence (graded I to V, I being best) and level of subgroup support for the statement based on the presented evidence (A to E, A being best).

After further discussion, each expert voted his or her level of support for each statement. The grading was as follows: A = accept recommendation completely; B = accept with some reservation; C = accept with major reservation; D = reject with reservations; E = reject completely. Table 1.1 sets out the results for the 10 statements we selected as most important.

The lack of consensus among the world's most recognized experts for the most basic and crucial of statements regarding Barrett esophagus is astounding. There was a lack of consensus even about the most accepted of these statements such as the need for biopsy-proven intestinal metaplasia for a diagnosis of Barrett esophagus, which received a C; only 5 of 18 experts agreed completely with this statement, and 2 rejected it.

Even more amazing is the assessment of the experts regarding the nature of the evidence that exists in the literature. The experts looked at the best evidence relating to these statements and concluded that a grade IV or V was appropriate in 30 of 41 questions (grade V = insufficient evidence to form an opinion; grade IV = opinions of respected authorities based on clinical experience, descriptive studies, or reports of expert committees; grade III = evidence obtained from case series, case reports, or flawed clinical trials). The conclusion reflects an appalling state of affairs, given that these 42 statements represent the most critical that this group could come up with. Further, the wide disparity of opinion rather than consensus among the 18 experts on these important issues makes it fair to say that the understanding of this disease is in a state of total disarray and profound confusion.

The impression that we get when we read the summary of the workshop is one of great futility. Criteria for diagnosis are not agreed on, screening for the

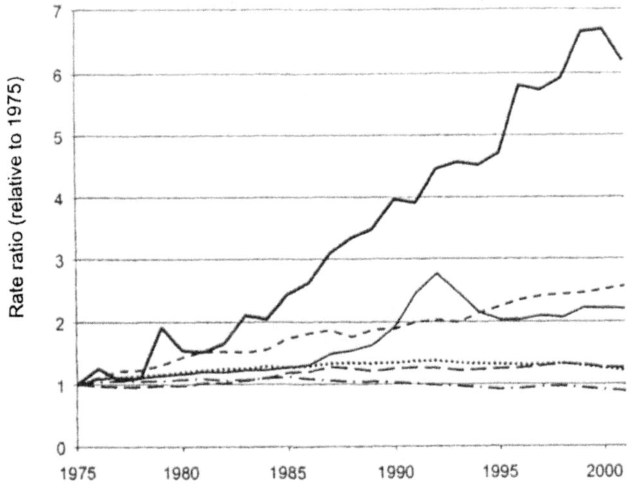

FIGURE 1.1 Relative change in incidence of esophageal adenocarcinoma and other malignancies in the period 1975 to 2000. Data from the NCI's SEER program. Solid black line = esophageal adenocarcinoma; dashed line = breast cancer; line = prostate cancer; other lines represent melanoma, lung cancer, and colorectal cancer. Reproduced with permission from Pohl H, Welch G. The role of overdiagnosis and reclassification in the marked increase of esophageal adenocarcinoma incidence. J Nat Cancer Inst 2005;97:143.

disease is generally felt to be worthless and cost-ineffective, the enthusiasm for the value of surveillance for Barrett esophagus is tepid at best (although this is a standard recommendation for patients diagnosed with Barrett esophagus), and all of today's available treatment methods (acid suppression, antireflux surgery, and ablation) are considered ineffective in preventing adenocarcinoma.

These experts seem to be telling us that we really do not clearly know what Barrett esophagus is; we do not want to do anything to detect it even though we know it is the precursor of adenocarcinoma; once we detect it we really do not want to even do what we are doing now (surveillance); and we clearly have nothing to offer the patient in terms of preventing cancer. We have a word of caution here: there was a huge predominance of gastroenterologists among the experts. We know it is possible to gather a group of surgeons who would agree with statement 39 that "antireflux surgery has not been proven to have a major effect against the development of esophageal adenocarcinoma," but they would immediately add that it has

TABLE 1.1 Results of Workshop Voting for Selected Statements out of the Total of 42[a]

Statement	Nature of evidence	Subgroup support	Group Grading				
			A	B	C	D	E
1. Esophageal intestinal metaplasia documented by histology is a prerequisite criterion for the diagnosis of BE.	IV	C	5	8	3	1	1
7. The proximal margin of the gastric rugal folds is a reliable endoscopic marker for the gastroesophageal junction.	IV	C	0	14	4	0	0
9. The normal appearing and normally located squamocolumnar junction should not be biopsied.	IV	C	14	4	0	0	0
12. Routine endoscopic screening for the detection of BE should be recommended for all adults >50 yrs of age.	V	D	0	0	0	8	10
16. Endoscopic screening for BE and dysplasia should be performed in all adults >50 yrs of age with heartburn.	IV	D	0	0	3	14	1
22. Endoscopic surveillance in patients with BE has been shown to prolong survival.	III	B	0	5	4	8	1
23. Endoscopic surveillance detects curable neoplasia in patients with BE.	III	B	7	8	3	0	0
35. Normalization of esophageal acid exposure by acid suppression reduces the risk for development of adenocarcinoma.	V	D	0	0	2	9	7
39. In patients with BE, antireflux surgery has not proven to have a major effect against the development of adenocarcinoma.	III	A	15	3	0	0	0
41. Mucosal ablation with acid suppression/antireflux surgery prevents adenocarcinoma in patients with BE without dysplasia.	IV	E	1	0	0	0	17

[a]The statements are selected as being the most important. The group grading represents the number of experts voting for each level of support for the statement (the number is expressed as a percentage in the publication and has been converted into an absolute number).

never been tested. If properly tested, it is likely that antireflux surgery would have a greater effect against the development of cancer than acid suppressive drug therapy.

The penalty for this confusion is paid by patients who develop and die from reflux-induced adenocarcinoma. This is the most rapidly increasing cancer type in the United States and Western Europe (Fig. 1). Expressed in another way, we are floundering in the midst of an epidemic of reflux-induced cancer without offering the population at large any hope that we can do anything to stop it.

In this book, we will show that the careful application of basic principles of cellular pathology to gastroesophageal reflux disease has the ability to provide understanding of the mechanism of the disease, develop accurate and reproducible criteria for definition and diagnosis of the disease, and lead to appropriate research to find logical methods of controlling the disease and potentially decreasing mortality. Many of these principles are derived from the study of cellular pathology, using the time-honored technique of microscopy.

PHYSIOLOGICAL VERSUS PATHOLOGICAL REFLUX

Gastroesophageal reflux is one of the most common human afflictions, ranking with atherosclerosis in being almost ubiquitous in the Western world. If a detection system (such as a pH electrode or an impedance catheter) is placed in the lower esophagus, virtually everyone will have sporadic evidence of a retrograde entry of gastric contents into the esophagus. The prevalence of gastroesophageal reflux is therefore close to 100% in the part of the human race living in the United States and Western Europe.

Gastroesophageal reflux can be considered to be a physiological event designed to vent a stomach that becomes distended with swallowed air. This is belching; the gas-distended stomach causes the lower esophageal sphincter to relax (transient relaxation), permitting the air to escape into the esophagus. The anatomy of the stomach is such that swallowed air collects immediately below the gastroesophageal junction, and belching therefore is limited to the reflux of air into the esophagus.

The occasional reflux of liquid gastric contents into the esophagus is probably a normal physiological occurrence. Consequently, we must draw strict guidelines as to the level reflux must reach before we consider it a disease. A physiological event does not produce a pathologic lesion. "Physiological reflux" must be carefully defined in a manner that is designed to exclude any pathologic abnormality that can be ascribed to reflux. The danger to this is that there may be pathologic changes in cells that we cannot see by present technology or do not recognize because of faulty understanding. We know that reflux causes dilated intercellular spaces that may be visible only by electron microscopy (3). The fact that our methods do not permit us to see these minimal pathologic changes does not make the reflux "physiological." The better term is "subclinical" because this is a moving definition; as our ability to detect abnormalities improves, the definition of what is pathologic will change. Perhaps the safest course is to regard all liquid reflux into the esophagus as being pathologic rather than physiological.

The concept of regarding a pathologic lesion as "physiological" or "normal" has been used commonly and with much detrimental effect in reflux disease. Long-segment columnar lined esophagus was ignored until 1953 because physicians defined the esophagus as ending at the squamocolumnar junction (4). Later, when properly described by Allison and Johnstone in 1953 (5) and confirmed by Barrett in 1957 (6), it was initially passed off as an insignificant congenital anomaly. When it was recognized as being caused by reflux, physicians devised an imaginary 2 cm of "normal" columnar lining in the esophagus to ignore short-segment Barrett esophagus for three decades (7).

Presently, there is a disturbing trend in gastroenterology to believe that things that are not visible to the endoscope do not exist. Thus, Barrett esophagus is presently defined as intestinal metaplasia occurring in a biopsy taken from *an endoscopically visible* columnar lined esophagus. By this reasoning, the presence of intestinal metaplasia in a biopsy taken from an endoscopically normal person is not Barrett esophagus.

A definition is a human device that is frequently wrong. We have seen the definition of Barrett esophagus change many times depending on the opinion of the moment. Surely, there must be a stage of Barrett esophagus that precedes the ability of the endoscope to detect it. The fact that we ignore patients who truly have Barrett esophagus just because they do not fit the definition of the moment does not mean that they are not at risk to develop carcinoma.

PREVALENCE OF GASTROESOPHAGEAL REFLUX DISEASE

The prevalence of gastroesophageal reflux disease is unknown because there is presently no definition of gastroesophageal reflux disease. This is an amazing statement of fact. Any definition of reflux disease must include all patients who have any pathologic abnormality resulting from gastroesophageal reflux and who are at risk of developing complications of reflux disease in the future.

The most commonly used definition of gastroesophageal reflux to assess prevalence is the presence of classical symptoms such as heartburn and regurgitation at some frequency defined by the whim of the investigator (9). Despite the fact that this is a hopelessly flawed definition, no better definition has ever been proposed. Population-based studies report that 4% to 9% of adults experience daily heartburn; an additional 10% to 15% have heartburn at least once a week; and an additional 10% to 15% have heartburn once a month (10, 11). In Nebel et al.'s study, the incidence of symptomatic gastroesophageal reflux was evaluated by a questionnaire in 446 hospitalized and 558 nonhospitalized subjects. Of 385 control subjects, 7% experienced heartburn daily, 14% noted heartburn weekly, and 15% experienced it once a month, meaning that a total of 36% of subjects had heartburn at least monthly. Daily heartburn occurred at a significantly greater rate for 246 medical inpatients (14%) and for 121 patients seen in an outpatient gastroenterological clinic (15%). Pregnant women seen in uncomplicated obstetrical clinics had symptoms of significantly greater incidence, as 25% reported daily heartburn and 52% experienced heartburn at least once monthly. Experts have never agreed on the exact frequency of heartburn and regurgitation that defines reflux disease.

Another way to assess the prevalence of reflux is to study trends in the use of acid suppressive drugs, though this assumes that acid suppression is used mainly for reflux disease, which is only partially true. Acid suppressive drugs are commonly prescribed. In a study of pharmacy billing data for two insurers within a large eastern Massachusetts provider network, 4684 of 168,727 patients (2%) were prescribed chronic (>90 day) acid suppressive drugs; 47% were taking H2 receptor antagonists and 57% were taking proton pump inhibitors (4% were taking both). Diagnostic testing was uncommon in these patients, as only 19% of the patients on acid suppressive drugs had undergone esophagogastroduodenoscopy within the prior 2 years (12). When one adds shorter term use of prescribed acid suppressive drugs and the massive unregulated use of over-the-counter antacids by patients who do not seek medical treatment, the clinical problem of gastroesophageal reflux disease is enormous. It is estimated that the use of acid suppressive drugs has a cost of more than $10 billion per year in the United States. Television abounds in direct marketing of acid suppressive agents to the consumer, and the shelves of drugstores are replete with a wide array of drugs that induce the population to cure themselves of heartburn using self-medication with the most powerful acid suppressive agents (Figs. 1.2 and 1.3).

FIGURE 1.2 The antacid section of a local pharmacy showing the shelf space devoted to over-the-counter acid suppressive drugs.

The use of classical symptoms to define reflux fails miserably when the end point of diagnosis is Barrett esophagus or adenocarcinoma of the esophagus because it is of inadequate sensitivity. In a screening study, Rex et al. (13) showed that Barrett esophagus was present in 5.6% of patients who had never had any heartburn; 0.36% had long-segment Barrett esophagus. Forty percent of patients who develop esophageal adenocarcinoma give no history of symptomatic reflux disease (9). The use of a definition of reflux disease based on symptoms will therefore fail to identify patients who will present for the first time with advanced adenocarcinoma.

A better definition of gastroesophageal reflux that has the ability to more precisely categorize asymptomatic and symptomatic patients in terms of their risk or lack of risk for adenocarcinoma is desperately needed.

HISTOLOGIC DEFINITION OF GASTROESOPHAGEAL REFLUX DISEASE

Reflux causes squamous epithelial damage (erosions, ulcers) as well as columnar transformation. Most patients will have pathologic changes occurring in both the squamous epithelium and the metaplastic columnar epithelium; these changes may not always be visible at endoscopy. Symptoms are most likely dependent on squamous epithelial damage; metaplastic columnar epithelia in the esophagus are less sensitive. A patient with predominant columnar metaplasia is more likely to be asymptomatic.

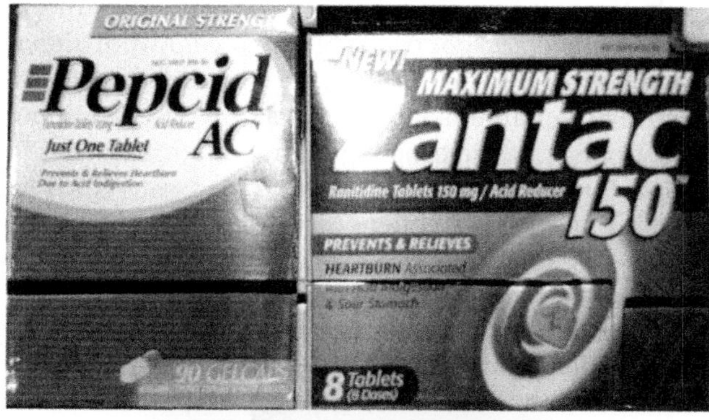

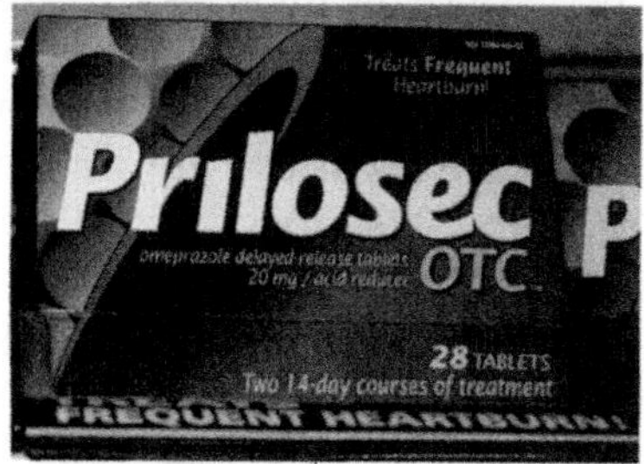

FIGURE 1.3 The labels on these drugs encourage the population at large to use over-the-counter acid suppression to control heartburn without seeking medical care.

Carcinogenesis occurs only in the metaplastic columnar epithelium; squamous carcinoma incidence is not increased by reflux. It is likely that the tendency to carcinogenesis is primarily dependent on luminally acting molecules in the gastroesophageal refluxate. Patients who have a highly carcinogenic refluxate will progress quickly to adenocarcinoma, but only when they develop metaplastic columnar epithelia that are susceptible to the carcinogens. If the development of cancer is more dependent on factors in the refluxate than on the actual damage caused by reflux, it is possible that cancer develops in patients with minimal features of reflux damage (no symptoms, no erosive esophagitis, and small amounts of metaplastic columnar epithelium in the esophagus).

Therefore, there is a possible disconnect but synergistic effect between reflux damage and carcinogenesis. In this setting, it is vitally important to understand the exact pathway by which reflux causes the esophageal squamous epithelium to transform into an adenocarcinoma (the reflux-adenocarcinoma sequence). It is only when this is done accurately that we can select the point in the sequence that best defines reflux disease to include all patients at risk for adenocarcinoma and exclude all patients not at risk.

At present, we have selected Barrett esophagus as this point because we do not recognize any earlier pathologic stage that may exist between squamous epithelium and Barrett esophagus. Barrett esophagus is defined as the presence of intestinal metaplasia in a biopsy taken from an endoscopically visualized columnar lined esophagus. Patients without an endoscopic abnormality are not subject to biopsy and, by the edict of the most powerful opinion of the moment, are defined as not having Barrett esophagus. This is indeed a powerful edict; the 18 experts in the Chicago workshop all agreed with statement 9: "The normal appearing and normally located squamocolumnar junction should not be biopsied" (2).

It is clear that our present methods of defining risk are failing; the incidence of adenocarcinoma of the esophagus is continuing to rise rapidly in the United States and Western Europe. We desperately need to detect these patients at an earlier stage and find effective ways of preventing this disease if we are to have any chance of reversing this trend. Attacking the progression of disease in its earliest stage will be key to preventing reflux-induced adenocarcinoma.

PREVALENCE OF BARRETT ESOPHAGUS

The prevalence of Barrett esophagus is likely to be greatly underestimated because its present definition limits the diagnosis to patients who have an endoscopically visible columnar lined esophagus. All patients with this disease are ignored from the time of onset to the time when the disease becomes visible on endoscopy. It is highly likely that our present definition of Barrett esophagus has the effect of underestimating the true prevalence of Barrett esophagus, because, in general, the earliest stage of a disease (i.e., when there are minimal changes) is the most prevalent.

Endoscopy performed in patients with symptomatic gastroesophageal reflux disease shows the presence of long-segment Barrett esophagus in 3% to 5% and short-segment Barrett esophagus in 10% to 15% (14). When screening endoscopy is performed (this is not routinely recommended), the prevalence of Barrett esophagus in patients who have never had any heartburn is around 5%; of these, the majority of patients have short-segment disease (13). The prevalence of long-segment Barrett esophagus in Rex et al.'s screening study was 0.36% in asymptomatic patients. This means that by the definition of Barrett esophagus that existed from the 1960s through the 1990s (which included only long-segment Barrett esophagus), only 0.36% of asymptomatic people and 3% to 5% of symptomatic people would have the condition. When we recognize the earliest stages of the disease that we now ignore by our present definition, the prevalence will increase. It is important to recognize that this true prevalence exists despite the fact that our definition underestimates it. Between 1960 and 1990, patients with short-segment Barrett esophagus who were considered "normal" by the definition of the time had the risks associated with short-segment Barrett esophagus despite the fact that we ignored them. Similarly, today every patient with Barrett esophagus that is too small to be seen by the endoscope is ignored while these patients continue to be at some undetermined risk of adenocarcinoma.

MANAGEMENT OF BARRETT ESOPHAGUS

Patients falling into the present definition of Barrett esophagus enter long-term surveillance with regular endoscopy and biopsy to detect dysplasia or early cancer for the purpose of either preventing cancer or treating an early cancer at a stage that decreases the likelihood of death. There is evidence that the survival rate in patients who develop cancer detected during surveillance is better than it is in those that present with symptomatic cancer (15). Surveillance is therefore successful in terms of decreasing the risk of death in these patients.

There is little evidence, however, that surveillance has any significant impact on the death rate from esophageal adenocarcinoma as a whole. Less than 5% of patients with esophageal adenocarcinoma were known to have had Barrett esophagus before they presented with symptomatic cancer (16).

By present definition, then, the occurrence of most esophageal adenocarcinoma appears inevitable, and these patients are doomed to die of the disease. The primary objective of this book is to show that this is not true. We will show that the present definition of Barrett esophagus is incorrect and represents part of the problem. Correction of the definition and recognition of earlier stages in the reflux-adenocarcinoma sequence will permit the development of manipulations that can prevent or significantly impact the incidence of esophageal adenocarcinoma. We will attempt to show that esophageal adenocarcinoma is a preventable disease.

PREVALENCE OF REFLUX-INDUCED ADENOCARCINOMA

In the half century spanning the years 1926 to 1976, only 0.8% to 3.7% of esophageal cancers were adenocarcinoma (17). In the early 1990s, adenocarcinoma overtook squamous carcinoma as the most common malignancy of the esophagus in white males (18, 19). This is largely as a result of a rapid increase in the incidence of esophageal adenocarcinoma, although squamous carcinoma has slowly declined. In an editorial comment, Haggitt (20) suggested in 1992 that adenocarcinoma in Barrett esophagus was a new epidemic.

The incidence rates for esophageal adenocarcinoma in the United States and Western Europe have shown a sharp increase since the mid-1970s. Devesa et al. (21), assessing data from the Surveillance, Epidemiology, and End-Results (SEER) program of the National Cancer Institute, reported that the annual incidence rate among white males rose from 0.7 per 100,000 population during the 1974–1976 period to 3.2 per 100,000 population during the 1992–1994 period. This is an increase of greater than 350%. There was a similar increase among black males and white females, although the incidence in these groups was much lower than it was among white males. A study by Pohl and Welch (1) showed that the overall esophageal adenocarcinoma incidence has continued to rise (Fig. 1.4). In this report, the incidence increased more than six-fold between 1975 and 2001 from 3.8 to 23.3 cases per million. The incidence of adenocarcinoma of the esophagus, which was one-sixth that of esophageal squamous carcinoma in 1975, overtook squamous carcinoma incidence in the mid-1990s (Fig. 1.4).

Adenocarcinoma of the gastric cardia (also called adenocarcinoma of the proximal stomach) has epidemiologic features that are similar to esophageal adenocarcinoma (17). Lagergren et al. (9) showed that adenocarcinoma of the gastric cardia has a causal relationship to symptomatic gastroesophageal reflux, albeit less than for esophageal adenocarcinoma. Adenocarcinoma of the gastric cardiac has shown a curve of increasing incidence that is similar to that of esophageal adenocarcinoma and contrasts with the declining incidence curve for distal gastric adenocarcinoma (Figs. 1.5 and 1.6). The rate of increase is less than it is for esophageal adenocarcinoma; in white males, the incidence increased from 2.1 to 3.3 per 100,000 from the 1974–1976 period to the 1992–1994 period (21). The curve of increasing rate for adenocarcinoma of the gastric cardia seems to be flattening (1). The similarity in the increasing trends and clinico-pathologic characteristics suggest that esophageal adenocarcinoma and gastric cardiac adenocarcinoma are best considered as variations of reflux-induced adenocarcinoma (22).

Similar increasing trends have been reported in Europe, Australia, and New Zealand. The prevalence of adenocarcinoma of the gastric cardia rose from 2.8 to 5.2 per 100,000 population in Oxfordshire, England, between the 1960–1964 period and the 1984–1988 period (23).

The increasing incidence of adenocarcinoma of the esophagus since the mid-1970s has been paralleled by a similar increase in mortality from this disease (Fig. 1.7). Esophageal adenocarcinoma is a devastating disease; when it occurs, it rapidly progresses to symptoms and death, usually within 1 to 2 years of diagnosis if not treated. Unlike prostate and breast cancer, which have also shown an increased incidence since the mid-1970s (see Fig. 1.1), esophageal adenocarcinoma is almost never found incidentally at autopsy. Its natural history does not make it likely that there are occult cases, the improved detection of which would be responsible for the increased incidence.

It is difficult to find data that show the actual number of adenocarcinomas of the esophagus and cardia. Estimates have been made that suggest the total number of esophageal adenocarcinomas to be approximately 8000 per year in the United States annually, with a similar or greater number of patients developing adenocarcinoma of the gastric cardia. This total number is small, and when spread among the large number of gastroenterologists in the United States, the

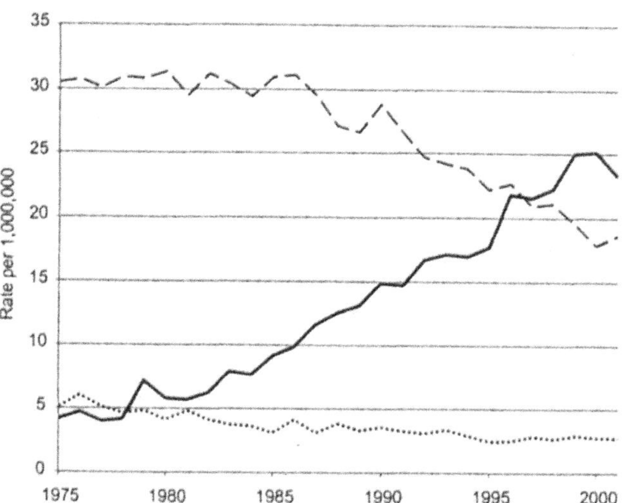

FIGURE 1.4 Changing incidence of esophageal adenocarcinoma (solid black line) compared with esophageal squamous carcinoma (dashed line) in the period 1975–2000. The dotted line represents cases that were identified as esophageal cancer, not otherwise specified. The absolute number of patients developing esophageal adenocarcinoma overtook squamous carcinoma around 1996. Reproduced with permission from Pohl H, Welch G. The role of overdiagnosis and reclassification in the marked increase of esophageal adenocarcinoma incidence. J Nat Cancer Inst 2005;97:143.

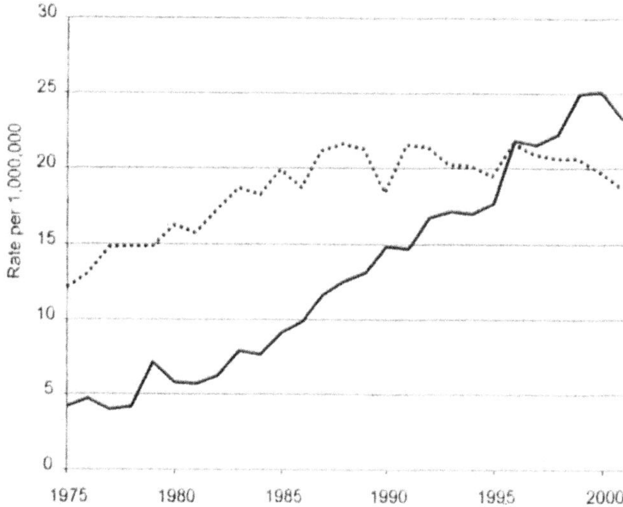

FIGURE 1.5 Trends in the incidence of adenocarcinoma of the esophagus (solid black line) compared with adenocarcinoma of the gastric cardia (dotted line). The lines are parallel with a flattening of the increase for gastric cardia cancer in the last decade of the 20th century. Reproduced with permission from Pohl H, Welch G. The role of overdiagnosis and reclassification in the marked increase of esophageal adenocarcinoma incidence. J Nat Cancer Inst 2005;97:144.

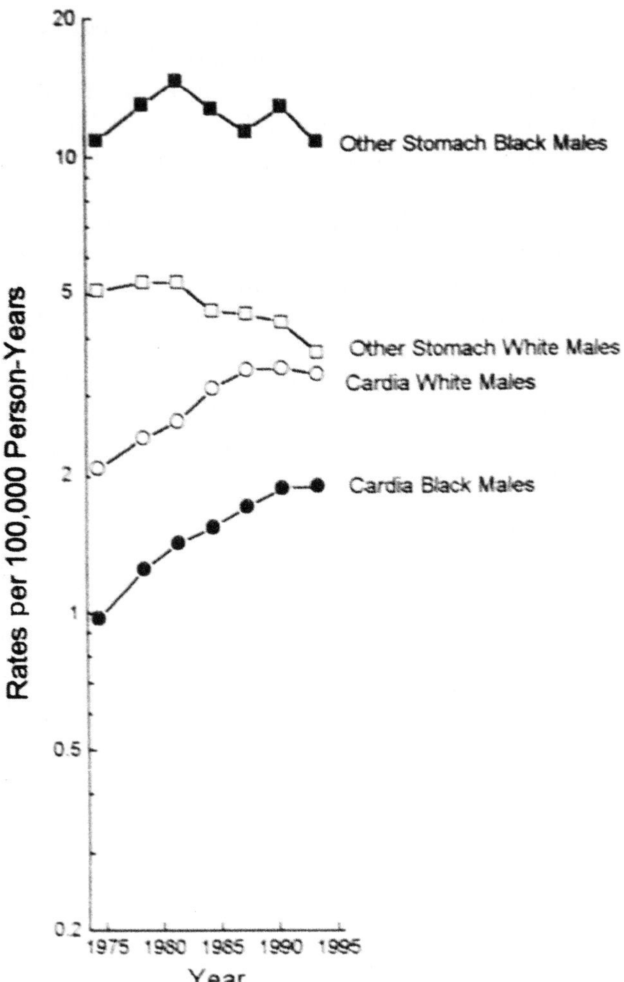

FIGURE 1.6 Trends in the incidence of adenocarcinoma of the stomach among males in the United States during the 1975–1995 period. The incidence of cardia cancers, which is more common in white than black males, has shown a sustained increase. This contrasts with adenocarcinoma of the distal stomach, which is more common in black than white males and which has shown a decline, particularly in white males. Reproduced with permission from Devesa SS, Blot WJ, Fraumeni JF Jr. Changing patterns in the incidence of esophageal and gastric carcinoma in the United States. Cancer 1998;83:2049–2053.

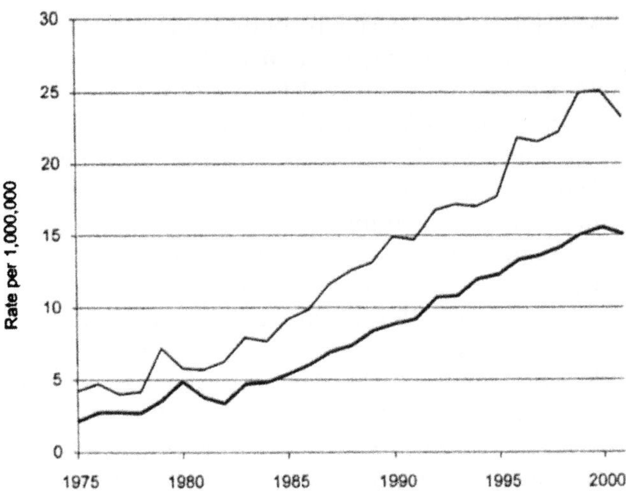

FIGURE 1.7 Disease-specific mortality and incidence of esophageal adenocarcinoma (1975–2001). The absolute number of deaths (solid black line) has paralleled the increasing incidence (line). Reproduced with permission from Pohl H, Welch G. The role of overdiagnosis and reclassification in the marked increase of esophageal adenocarcinoma incidence. J Nat Cancer Inst 2005;97:144.

impression is that it is a very rare tumor. This impression is partly responsible for the lack of urgency. However, the continuing upward trend of the incidence of reflux-induced adenocarcinoma is an ominous sign for a looming public health problem.

As a pathologist and surgeon, we have a very different perspective of the incidence of esophageal adenocarcinoma. When Dr. Chandrasoma started practicing surgical pathology at the Los Angeles County–University of Southern California Medical Center in 1978, he saw 10 squamous carcinomas of the esophagus and 10 distal gastric adenocarcinomas for every one adenocarcinoma of the esophagus or cardia; it was unusual to see more than one esophageal adenocarcinoma in a given year. This has changed dramatically. Part of the change is due to the fact that a surgeon, Dr. DeMeester, came to the University of Southern California in 1990, and due to his interest in esophageal disease he attracted a referral population. Dr. Chandrasoma became an increasing source for expert consultations. One of his most striking experiences has been the marked increase in the number of cases of high-grade dysplasia and Barrett adenocarcinoma that come across his desk. Dr. Chandrasoma now sees four to eight new patients per week. From a personal unscientific and anecdotal standpoint, the epidemic of adenocarcinoma is frightening.

References

1. Pohl H, Welch HG. The role of overdiagnosis and reclassification in the marked increase of esophageal adenocarcinoma. J Natl Cancer Inst 2005;19:142–146.
2. Sharma P, McQuaid K, Dent J, Fennerty B, Sampliner R, Spechler S, Cameron A, Corley D, Falk G, Goldblum J, Hunter J, Jankowski J, Lundell L, Reid B, Shaheen N, Sonnenberg A, Wang K, Weinstein W. A critical review of the diagnosis and management of Barrett's esophagus: The AGA Chicago Workshop. Gastroenterology 2004;127:310–330.
3. Tobey NA, Carson JL, Alkiek RA, et al. Dilated intercellular spaces: A morphological feature of acid reflux-damaged human esophageal epithelium. Gastroenterology 1996;111:1200–1205.
4. Barrett NR. Chronic peptic ulcer of the oesophagus and "oesophagitis." Br J Surg 1950;38:175–182.

5. Allison PR, Johnstone AS. The oesophagus lined with gastric mucous membrane. Thorax 1953;8:87–101.
6. Barrett NR. The lower esophagus lined by columnar epithelium. Surgery 1957;41:881–894.
7. Hayward J. The lower end of the oesophagus. Thorax 1961; 16:36–41.
8. Sampliner RE. Practice guidelines on the diagnosis, surveillance, and therapy of Barrett's esophagus. Am J Gastroenterol 1998;93:1028–1031.
9. Lagergren J, Bergstrom R, Lindgren A, Nyren O. Symptomatic gastroesophageal reflux as a risk factor for esophageal adenocarcinoma. N Engl J Med 1999;340:825–831.
10. Nebel OT, Fornes MF, Castell DO. Symptomatic gastroesophageal reflux: Incidence and precipitating factors. Dig Dis Sci 1976;21:953–956.
11. Talley NJ, Zinsmeister AR, Schlek CD, et al. Dyspepsia and dyspepsia subgroups: A population based study. Gastroenterology 1992;102:1259–1268.
12. Jacobson BC, Ferris TG, Shea TL, Mahlis EM, Lee TH, Wang TC. Who is using chronic acid suppression therapy and why? Am J Gastroenterol 2003;98:51–58.
13. Rex DK, Cummings OW, Shaw M, Cumings MD, Wong RK, Vasudeva RS, Dunne D, Rahmani EY, Helper DJ. Screening for Barrett's esophagus in colonoscopy patients with and without heartburn. Gastroenterology 2003;125:1670–1677.
14. Hirota WK, Loughney TM, Lazas DJ, Maydonovitch CL, Rholl V, Wong RKH. Specialized intestinal metaplasia, dysplasia, and cancer of the esophagus and esophagogastric junction: Prevalence and clinical data. Gastroenterology 1999;116:277–285.
15. Peters JH, Clark GWB, Ireland AP, Chandrasoma P, Smyrk TC, DeMeester TR. Outcome of adenocarcinoma arising in Barrett's esophagus in endoscopically surveyed and nonsurveyed patients. J Thorac Cardiovasc Surg 1994;108:813–821.
16. Dulai GS, Guha S, Kahn KL, Gornbein J, Weinstein WM. Preoperative prevlance of Barrett's esophagus in esophageal adenocarcinoma: A systematic review. Gastroenterology 2002; 122:26–33.
17. Cameron AJ. Epidemiology of columnar lined esophagus and adenocarcinoma. Gastroenterol Clin N Am 1997;26:487–494.
18. Blot WJ, Devesa SS, Kneller RW, Fraumeni JF Jr. Rising incidence of adenocarcinoma of the esophagus and gastric cardia. JAMA 1991;265:1287–1289.
19. Blot WJ, Devesa SS, Fraumeni JF. Continued climb in rates of esophageal adenocarcinoma: An update. JAMA 1993;270:1320.
20. Haggitt RC. Adenocarcinoma in Barrett's esophagus: A new epidemic? Hum Pathol 1992;23:475–476.
21. Devesa SS, Blot WJ, Fraumeni JF Jr. Changing patterns in the incidence of esophageal and gastric carcinoma in the United States. Cancer 1998;83:2049–2053.
22. Portale G, Peters JH, Hagen JA, DeMeeter SR, Gandamihardja T, Tharavej C, Hsieh CC, DeMeester TR. Comparison of the clinical and histological characteristics and survival of distal esophageal-gastroesohageal junction adenocarcinoma in patients with and without Barrett mucosa. Arch Surg 2005; 140(6):570–575.
23. Rios-Castellanos E, Sitas F, Shepherd NA, Jewell DP. Changing patterns of gastric cancer in Oxfordshire. Gut 1992;33: 1312–1317.

CHAPTER

2

The Past, Present, and Future of Columnar-Lined (Barrett) Esophagus

The columnar-lined esophagus that has Norman Barrett's name associated with it has probably confused physicians more than any other human disease for the past 50 years (Fig. 2.1). Before Allison's description in 1953 (1), its very existence was denied. Physicians of that era, including Barrett himself (2), regarded the esophagus as ending at the squamocolumnar junction. This led to the initial false belief that the columnar-lined esophagus was actually a tubular intrathoracic stomach resulting from a congenitally short esophagus (3).

Allison, who recognized the columnar-lined esophagus and provided the means to differentiate it from an intrathoracic stomach (sliding hiatal hernia), contributed to the confusion in 1953 by using the term "esophagus lined with gastric mucous membrane" (1). This introduced a unique and confusing concept that one organ (the esophagus) could be lined by the mucosa of another (the stomach). This confusion still lingers.

Barrett, in 1957, while agreeing with Allison on the existence of the entity, recognized that it was more accurate to regard this condition as an esophagus lined by columnar esophageal epithelium rather than gastric mucous membrane (4). Barrett pointed out the obvious that still eludes many people: *the esophagus is always lined by esophageal epithelium, whatever its type and the stomach is lined by gastric epithelium* (Fig. 2.2).

Barrett's and Allison's definitions and terminology illustrate well the confusion that has always existed in differentiating between the distal esophagus and proximal stomach. In 1950, Barrett was applying mucosal criteria to define the esophagus and ignoring the anatomic criteria. In 1953, Allison was applying nonmucosal anatomic criteria such as the peritoneal reflection, type of musculature, and the presence of submucosal glands to define the structure as an esophagus but was willing to call the mucosal lining gastric (Allison and Johnstone 1953).

Spechler, reviewing the columnar-lined esophagus in 1997, expressed the problem well (5):

> Conceptually, [columnar-lined esophagus] can be defined simply as the condition in which the stratified squamous epithelium of the esophagus is replaced by a metaplastic columnar epithelium. Practically, it can be difficult to apply this conceptual definition in clinical situations because of difficulties in delimiting the distal esophagus and the proximal stomach and in distinguishing normal from metaplastic columnar epithelium on the basis of endoscopic appearance.

Spechler is correct; however, he ignores the fact that endoscopy is not the only or best modality available to solve this problem. The best available "scope" is the microscope.

In this book, we will develop the thesis that these objectives are easily achieved by the correct use of histology. Much of the confusion that exists today is rooted in the fact that the most influential physicians on the subject simply do not look through the microscope and have no true understanding of histology. This is one price we pay for super-specialization in medicine. To be successful, the super-specialized clinical esophagologist must understand the significance of the microscopic pathology of the biopsies that are taken,

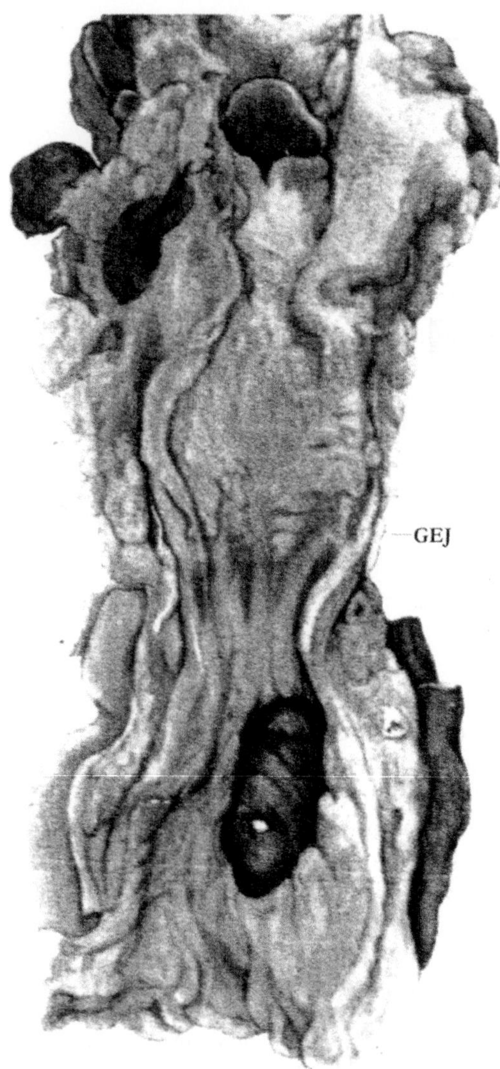

FIGURE 2.1 Barrett (1950): "The specimen shows a large chronic gastric ulcer which has perforated into one of the pulmonary veins just below the aortic arch. The mucous membrane in relation to this ulcer is everywhere gastric in type, and columnar epithelium extends up for a distance of 1.5 inches above the top edge of the ulcer in one continuous expanse" (p. 180). According to Barrett, the esophagus ends at the squamocolumnar junction (GEJ), and the entire columnar-lined structure distal to this is a tubular intrathoracic stomach (with a gastric ulcer). The columnar-lined esophagus has an inauspicious beginning; this is about as incorrect as it can possibly get. Reproduced with permission from Barrett NR. Chronic peptic ulcer of the oesophagus and "oesophagitis." Br J Surg 1950; 38:175–182.

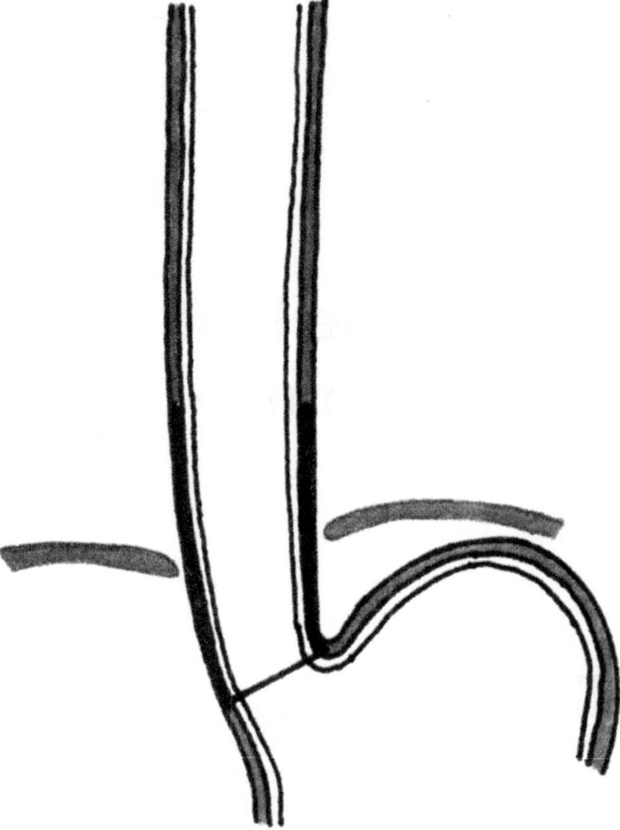

FIGURE 2.2 In this drawing of normal anatomy, an imaginary line is drawn between the end of the esophagus and the stomach. This is the gastroesophageal junction. Note that the dark part of the wall is the normal lower esophageal sphincter, which ends at the end of the esophagus. The mucosa, whatever its histologic type, above the line is *esophageal*; the mucosa, whatever its histologic type, below the line is *gastric*.

and the super-specialized esophageal pathologist must understand the clinical basis of columnar-lined esophagus. Histology has a proven record in solving clinical problems in many areas and it should be no surprise that it will solve this one as well.

THE HISTORY OF COLUMNAR-LINED ESOPHAGUS

The Pre-1900 Era: Clinical and Autopsy Studies

Norman Barrett detailed the early history of the entity in his classical 1950 study titled "Chronic Peptic Ulcer of the Oesophagus and 'Oesophagitis'" (2). He attributed the first description of esophagitis to John Peter Frank in the 18th century, with numerous early contributors including Boehm (1722); Bleuland (1785); Mondiere (1829); Billard (1828) who reported its occurrence in children; and Morell Mackenzie (1884)—Bleuland and Mondiere were both afflicted with the disease and recorded its clinical findings. In 1884, Mackenzie defined esophagitis as "acute idiopathic inflammation of the mucous membrane of the oesophagus giving rise to extreme odynophagia and often to aphagia." These early anecdotal reports appear to be describing the most severe symptoms of what we now call reflux esophagitis, including "acute pain which reached down even to the stomach and which was accompanied by hiccup and a constant flow of serum from the mouth" (Boehm 1722).

Barrett lamented that these early clinical reports of esophagitis were ignored in the first half of the 20th century because of the emergence of a pathologic entity of the lower esophagus defined by the presence of ulcers. Barrett stated, in 1950:

> The terms "oesophagitis" and "peptic ulcer of the oesophagus" connote one thing to some people and something quite different to others. Confusion has overtaken us partly because the rich legacy of clinical observations recorded by Morell Mackenzie and his contemporaries have not been sufficiently aligned with the recent advances in the pathology of the living oesophagus. (p. 175)

Barrett's words are prophetic. While the questions today are different, confusion has overtaken the present-day physician because of the profusion of terms now in use such as Barrett esophagus (long, short, and ultrashort segment), columnar-lined esophagus, cardia, gastric cardia, junctional region, proximal stomach, cardiac mucosa, carditis, intestinal metaplasia of the cardia, specialized intestinal metaplasia, gastric intestinal metaplasia, reflux, and reflux disease. Different researchers define these terms in different ways because there are no standard and universally accepted definitions. These terms truly "connote one thing to some people and something quite different to others" (2, p. 175). The field has become a medical Tower of Babel. Confusion will cease only when we define these terms precisely and reproducibly and start talking the same language. Today's issues may differ from those Barrett encountered in 1950, but the basis of the problem has eerie similarities.

The Pre- and Early Endoscopy Era (1900 to 1950): Peptic Ulcer of the Lower Esophagus, Ectopic Gastric Mucosa, Short Esophagus, and Reflux Esophagitis (Table 2.1)

In the late 19th century, physicians started reporting the rare autopsy finding of ulcers in the lower tubular esophagus that resembled peptic ulcers of the stomach. A 1906 review by Wilder Tileston, a consulting physician at the Massachusetts Charitable Eye and Ear Infirmary in Boston, examined the literature up to that time. Tileston (6) defined peptic ulcer of the esophagus by the resemblance of its pathologic features to acute and chronic peptic ulcer of the stomach and by excluding ulcers caused by carcinoma, corrosive ingestion, foreign bodies, acute infections (diphtheria, scarlatina, variola, typhoid fever, pneumonia), thrush, decubitus, aneurysm of the aorta, catarrhal, and those associated with traction diverticula, tuberculosis, syphilis, and varicose ulcers. To be defined as a peptic ulcer of the esophagus, pathologic confirmation was necessary. The inability to perform endoscopy at the time meant that autopsy confirmation was

TABLE 2.1 Historical Evolution of Reflux-Induced Damage of the Esophagus and Columnar-Lined Esophagus until 1961

Year	Author	Finding	Comment
Pre-1948	Multiple	Ulcers in lower esophagus	Based on autopsy studies
1948	Allison	Describes reflux esophagitis	Accurate
		Describes hiatal hernia	Accurate
1950	Barrett	Coins the term reflux esophagitis	Accurate
		Disputes existence of heterotopic gastric mucosa	Accurate
		Calls columnar-lined esophagus tubular stomach	Wrong
1953	Allison & Johnstone	Describes esophagus lined by columnar epithelium	Accurate
		Calls columnar lining of esophagus "gastric mucous membrane"	Wrong
		Believes columnar-lined esophagus is congenital	Wrong
1957	Barrett	Coins the term "columnar-lined esophagus"	Accurate
		Believes columnar-lined esophagus is congenital	Wrong
1961	Hayward	Proposes that columnar-lined esophagus is reflux-induced	Accurate
		Suggests that 2 cm of columnar lining is normal in distal esophagus	Wrong; leads to a definition of Barrett esophagus that excludes short segment BE
		Suggests cardiac mucosa is esophageal, not gastric	Accurate

required and the diagnosis could not be made during life.

Tileston attributed the first report of esophageal ulcers to Albers in 1839, and gave credit to Quincke for establishing esophageal peptic ulcer as an entity in 1879 by convincing microscopic examination. Tileston collected 41 cases from the literature that he believed satisfied strict diagnostic criteria and added 3 cases of his own. He suggested that the disease may be more common, quoting that 6 of 4496 autopsies at the Massachusetts General, Boston City, and Long Island hospitals noted the presence of peptic ulcers in the esophagus for a prevalence of 0.13%.

In his careful review, Tileston concluded that the cause of the ulcers was regurgitation of gastric juice through an insufficient cardia. The ulcers were mainly in the lower esophagus, below the arch of the aorta. There is no mention of histology in most of the cases in this review. Tileston mentioned finding mucous glands resembling gastric glands at the edge of the ulcer in one of his cases, and there is an illustration that shows submucosal glands at the ulcer base and what looks like columnar epithelium at the ulcer edge, but the concept of columnar-lined esophagus did not exist at the time.

Several other researchers between 1906 and 1948, including Stewart and Hartfall (7), Lyall (8), and Chamberlin (9), confirmed the occurrence of peptic ulcers in the lower esophagus. The later studies were in the endoscopy era and the diagnosis was made during life. Ulcers were considered to be an infrequent cause of esophageal symptoms. The etiology of these ulcers, which involved the tubular esophagus, was initially believed to be regurgitation of gastric juice through a patent cardia. In the early studies, there is hardly any mention of histology, and the tacit assumption is that these ulcers occurred in the squamous-lined esophagus.

In the late 1920s, some researchers suggested that esophageal ulcers were a complication of islets of heterotopic gastric mucosa in the lower esophagus (7, 9, 10). The concept of islets of heterotopic gastric mucosa within squamous lined esophagus appears to be derived from scattered mention of the finding of "heterotopic gastric mucosa" at the edge of ulcers of the lower esophagus. Friedenwald et al. (11) reported the mucosa adjacent to the ulcer as thickened and red in two of his cases, but no histologic data were given to confirm whether this was inflamed squamous epithelium or columnar epithelium. Many early papers refer to Stoerk as being the first to observe gastric-type glands at the edge of an esophageal ulcer in a case that appears in Tileston's review (6). Because of the severe autolysis that almost invariably accompanied these autopsy studies, the "gastric heterotopia" was never described as anything other than "islets within the squamous epithelium" usually adjacent to the ulcers. Such "islets of gastric mucosa" were known to be present in some people in the upper esophagus (these are what are called inlet patches today), and the conjecture was that similar heterotopic islets in the lower esophagus could secrete acid and cause ulceration in the lower esophagus.

In the 1931 summer meeting of the Royal Society of Medicine in London, Frindlay and Kelly reported on nine children with congenital esophageal stenosis in the mid-thorax. Though this type of patient had been reported previously, there had never been any suggestion prior to this paper that the tubular segment distal to the stenosis was anything other than the continuation of the esophagus. For the first time, Frindlay and Kelly reported, "by endoscopy the junction of the oesophagus and stomach [cardia] to be situated close to the lower end of the stenosis and that the parts beyond presented the usual aspect of gastric mucosa." They took this as evidence that the structure between the end of the stenosis and the diaphragmatic hiatus was an intrathoracic stomach associated with a congenitally short esophagus. Frindlay and Kelly were the first to suggest a tubular intrathoracic stomach. The universal opinion at the time was that the esophagus was always lined by squamous epithelium, except for the exceptionally rare presence of islets of heterotopic gastric mucosa within it. No one was willing to suggest that Frindlay and Kelly's patients had heterotopic islets of gastric mucosa involving the entire circumference of the distal esophagus. Rather, they reverted to the definition of the esophagus as a structure lined by squamous epithelium and declared the columnar-lined tubular structure distal to it as stomach. By definition, then, the esophagus had to be shortened, and the assumption was that the shortening was congenital. This fit well with the young age of Frindlay and Kelly's patients.

In 1939, Chamberlin, reported on seven adults from the Lahey Clinic in Massachusetts with similar shortening of the esophagus; this was associated with diaphragmatic hernia in four cases. The fact that these patients were adults was taken to suggest that the esophageal shortening was more likely acquired as a result of gastroesophageal reflux than a congenital anomaly.

Endoscopic examination was becoming commonplace at this time. In 1929, Jackson reported the finding of esophagitis at endoscopy at a much higher frequency than had been recognized by pathologic studies predating endoscopy (12). Allison, in a series of reports from 1943 to 1948, also reported increasing numbers of patients with reflux-induced pathology of the lower esophagus (13, 14). This work culminated in Allison's 1948 paper, which is the first accurate description of reflux esophagitis in the literature and established that most cases of esophageal shortening, inflammation, and ulceration were due to gastroesophageal reflux rather than secondary to a congenital shortening of the esophagus.

Literature Review

Allison PR. Peptic ulcer of the oesophagus. Thorax 1948;3:20–42.

Allison (15) refuted the existing opinion that ulcers in the lower esophagus were related to congenital malformation of the esophagus:

> A lack of detailed investigation in the past led to the easy assumption that the oesophagus suffered acid digestion as a result of congenital maldevelopment. Its failure to reach its full length left the fundus of the stomach in the posterior mediastinum so that the normal barriers to acid reflux were ineffective. Peptic ulcer of the oesophagus and congenitally short oesophagus became almost like the Beaver and the Butcher: "You could never meet either alone!" Although such congenital abnormality cannot be denied, the burden of this report is that *short esophagus is usually an acquired condition due to defects in the diaphragm which allow a sliding hernia of the stomach*. In the early stages the oesophagus may remain elastic so that the hernia can be reduced, but, after peptic ulceration and fibrosis have developed, permanent shortening occurs (Fig. 2.3). This hernia may be present in infancy or be acquired later in life. As will be seen, it is the predisposing cause of most peptic ulcers of the oesophagus. (p. 20)

Allison reported on 74 patients treated by him in the thoracic surgical unit at the General Infirmary at Leeds; 63 were adults (22 to 80 years old; 89% over 50 years old). He gave detailed and extremely accurate descriptions of the symptoms of heartburn, dysphagia, and regurgitation. His discussion on regurgitation is especially interesting: "One important symptom of obstruction was described by patients as 'vomiting of bile.' If this had been taken at face value it would have been difficult to explain, but more detailed questioning revealed that it was in fact regurgitation of mucus. The belief that mucus is 'clear bile' may be local to Yorkshire, but it is a point worthy of emphasis" (p. 21). In light of present knowledge about the frequency of duodeno-gastric reflux, it is likely that the patient's description of "vomiting of bile" was more accurate than Allison's interpretation.

Allison detailed radiologic findings, quoting from his earlier 1943 and 1946 studies, emphasizing the importance of screen examination over "snap pictures." In his description of the esophagoscopic appearance, he detailed the four stages in the natural history of the disease: esophagitis,

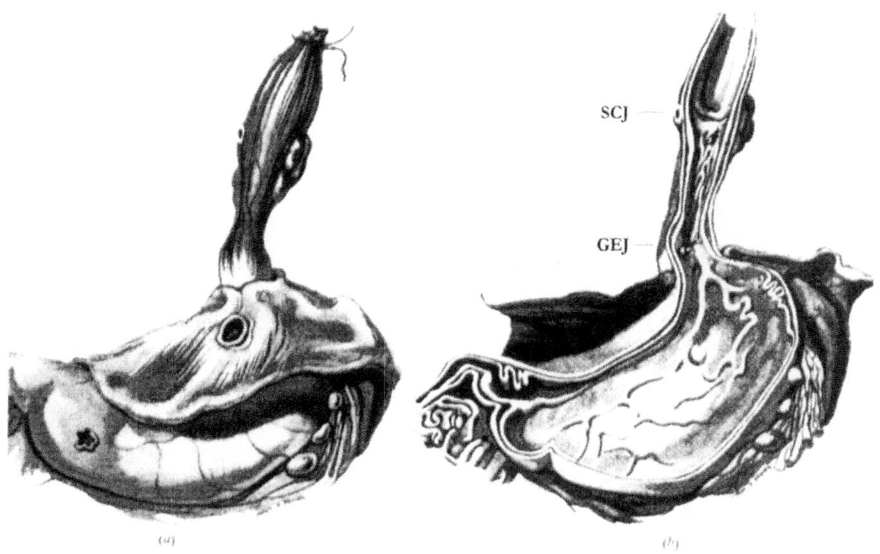

FIGURE 2.3 Allison (1948): "(a) Post-mortem specimen distended with formalin and removed with the diaphragm attached showing a long stricture of the oesophagus extending up to the level of the azygos vein. A small pouch of stomach is shown above the diaphragm. (b) The specimen cut open to show the junction of gastric and oesophageal mucosa above the diaphragm and the dense fibrous stricture extending up from this" (p. 34). Allison is placing the gastroesophageal junction at the top of the small pouch (GEJ); he does not say whether the lining of the dense fibrous stricture (which appears to have longitudinal rugal folds) is columnar or squamous. Reproduced with permission from Allison PR. Peptic ulcer of the esophagus. Thorax 1948;3:20–42.

esophagitis with acute ulceration, esophagitis with chronic ulcer, and healed fibrous stenosis. A discussion of biopsy findings concentrates only on the importance of distinguishing benign ulcers from carcinoma.

In his discussion, Allison detailed the normal function of the cardia, which he equated with the valve in the lower esophagus: "The only structure in the body which is known to have natural powers to resist peptic digestion is the mucous membrane of the stomach and duodenal cap. . . . The esophagus has no such resistance, but there is a mechanism at the cardia which normally prevents acid from reaching it. . . . A failure of this mechanism will allow acid to reach the oesophagus, and in time this leads inevitably to inflammation and ulceration."

Allison proceeded to accurately describe the different types of hiatal hernia as paraesophageal, sliding, sliding with a paraesophageal pouch and what he calls a "bulging hernia". His diagrams of these different hernia types are extremely lucid (Figs. 2.4 and 2.5).

Allison's diagrams depict squamous epithelium as a straight line and gastric mucosa as a serrated line. His rendering is very close to modern belief that squamous epithelium comes down to the end of the tubular esophagus where it flares out into the stomach. This is well shown in his drawing of the paraesophageal hernia (Fig. 2.3), where the squamous epithelium extends to the end of the tubular esophagus below the level of the diaphragm. In his Figure 5 (Fig. 2.6), which shows normal anatomy without a sliding hiatal hernia, Allison described the presence of "heterotopic gastric mucous membrane in the oesophagus, either in the form of islands or by a direct extension upwards from the stomach" (p. 40). The columnar epithelium is depicted as extending above the diaphragm in the tubular esophagus and is labeled "heterotopic gastric mucosa." This is the first diagram in the literature to show columnar-lined esophagus, an entity whose recognition was still 5 years away.

Few papers have achieved as much as Allison's 1948 paper. It established reflux disease as a much more common entity than previously believed. It accurately described reflux esophagitis clinically, radiologically, and endoscopically, establishing the natural history of the disease from the point of inflammation through acute and chronic ulceration to stricture formation. It detailed the importance of the sphincter mechanism of the cardia, describing normal and abnormal function of the sphincter. It described the various types of hiatal hernias and hinted at the existence of a

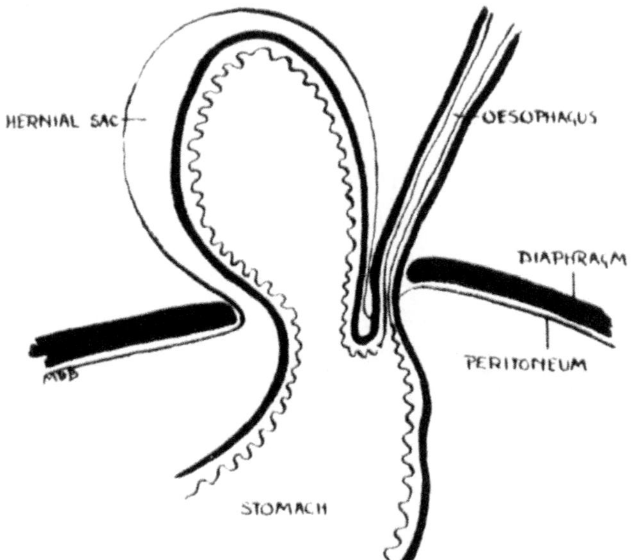

FIGURE 2.4 Allison (1948): "Para-oesophageal hiatal hernia." Note that, in this diagram, the esophageal squamous epithelium (straight line) is depicted very accurately as ending at the end of the tubular esophagus below the level of the diaphragm. The serrated gastric mucosa lines the saccular stomach. Reproduced with permission from Allison PR. Peptic ulcer of the esophagus. Thorax 1948;3:40.

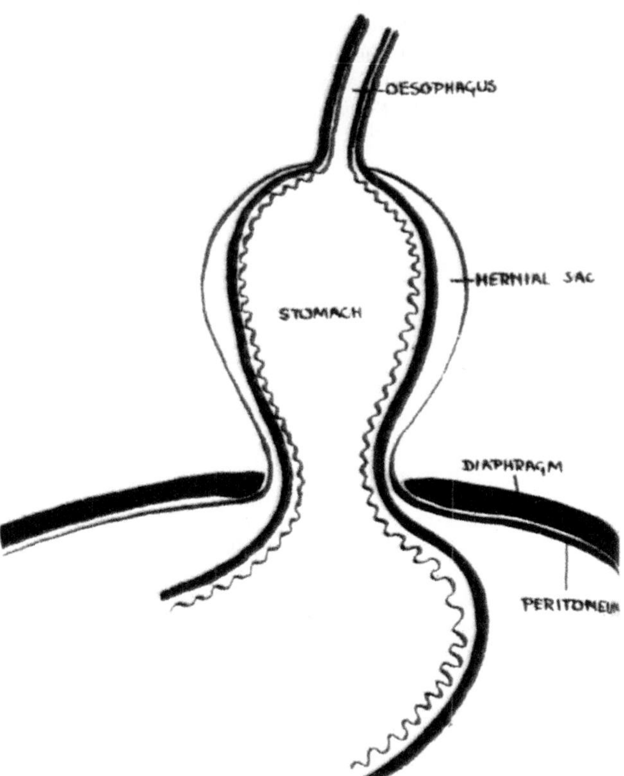

FIGURE 2.5 Allison (1948): "Sliding hiatal hernia with oesophageal shortening." Again, the anatomy is correctly shown; the sliding hiatal hernia is covered by a sac of peritoneum and the squamous epithelium ends at the end of the tubular esophagus at the point of the peritoneal reflection. There is no columnar-lined esophagus shown here. Reproduced with permission from Allison PR. Peptic ulcer of the esophagus. Thorax 1948;3:40.

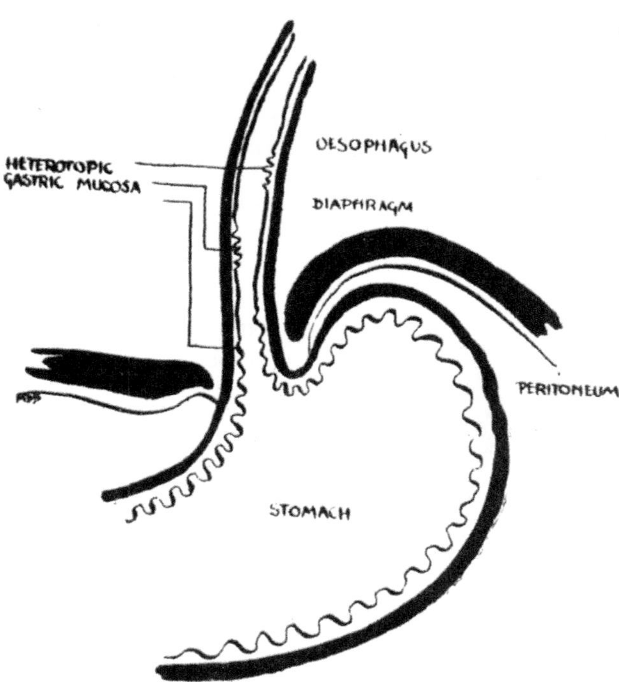

FIGURE 2.6 Allison (1948): "Heterotopic gastric mucous membrane without hernia." This diagram shows serrated mucosa extending into the esophagus, both continuous with the gastric mucosa and as islets in squamous epithelium. Allison called this "heterotopic gastric mucosa"; this is the first depiction in the literature of columnar-lined esophagus. Reproduced with permission from Allison PR. Peptic ulcer of the esophagus. Thorax 1948;3:41.

columnar-lined esophagus, which involved the esophagus circumferentially and continuously with gastric mucosa rather than the previously described columnar islets embedded in squamous epithelium.

Literature Review

Barrett N. Chronic peptic ulcer of the oesophagus and "oesophagitis." Br J Surg 1950;38:175–182.

Norman Barrett's 1950 study (2), which followed Allison's treatise, appears to be largely a vigorous support of Allison's study. His objective appears to be to separate Allison's acquired reflux esophagitis and what was at the time called chronic peptic ulcer of the esophagus, complicating congenitally short esophagus and gastric heterotopia of the lower esophagus. His own data were limited to a few beautifully illustrated cases he had encountered, and his report is really a review of the existing literature by an influential surgeon of the time in London. This is a truly classic article that abounds in conclusions that were ultimately proven incorrect!

Barrett reviewed the literature relating to chronic peptic ulcers of the esophagus that were defined at autopsy as reported by Tileston and Lyall and the emerging endoscopic literature of ulceration of the lower esophagus resulting from reflux esophagitis reported by Jackson and Allison:

> The confusion in which we find ourselves today dates from (Tileston's) time, because the pathologists, the clinicians, and the endoscopists assumed that they . . . were talking about the same entity when they referred to "peptic ulcer of the oesophagus" or to "oesophagitis." By 1929 it was clear all was not well, for Chevalier Jackson claimed to have seen 88 cases in 4000 consecutive endoscopies, whereas Stewart and Hartfall (7) had found but 1 example in 10,000 consecutive autopsies. (p. 175)

Allison had opined that these were simply two consecutive stages in the pathology of reflux esophagitis (15). Barrett suggested that these two entities are different: that the chronic peptic ulcers are rare and of little clinical significance, whereas reflux esophagitis reported by Jackson and Allison are common and important entities. Barrett noted, "I suggest that the pathology of the former has been tacked on to the symptomatology of the latter, and that chaos has resulted" (p. 176).

He described the sequence of reflux esophagitis "which I consider to be the best name for the lesion so ably described by Allison (1948)" (p. 176) (15), indicating that this paper was the official birth of the term "reflux esophagitis." Essentially reiterating what Allison had already stated, Barrett stated that in the normal person, the esophagus is completely separated from the stomach by a mechanism at, or immediately above, the esophagogastric junction, which allows food and liquids to pass down but strictly prevents reflux. When this mechanism becomes inefficient, Barrett described the changes that occur: (a) In some patients no harm appears to result from reflux of gastric secretions into the lower gullet; (b) in others, an inflammatory lesion ensues with edema, congestion, and erosion of the squamous epithelium. He likened the change to what happens in the skin surrounding a gastric fistula. The ulcers are at first shallow and associated with regeneration and can remain this way for a long time. If successfully treated, the esophagus returns to normal. (c) If the cause persists and digestion continues, the inflammation spreads beyond the submucosa, producing fibrous strictures that can ascend up the esophagus. The affected gullet becomes contracted and strictured both in its longitudinal and transverse directions. Above the stricture is normal squamous-lined esophagus. "Below the stricture is a stretch of gut, which, at first sight, is apt to be mistaken for oesophagus, but which is generally a pouch of stomach partially enveloped by peritoneum and drawn up by scar tissue into the mediastinum" (p. 177). Barrett was reiterating Allison's concept of a sliding hiatal hernia resulting from reflux-induced acquired shortening of the esophagus (Figs. 2.7 and 2.8). However, in his pathologic sequence of reflux esophagitis, he does not emphasize chronic ulcers.

Barrett was trying to convince the reader that all cases that have been reported previously as peptic ulcers of the lower esophagus are really gastric ulcers and that they are occurring in a tubular stomach rather than the lower esophagus. Barrett has recognized that, at least in some cases, the

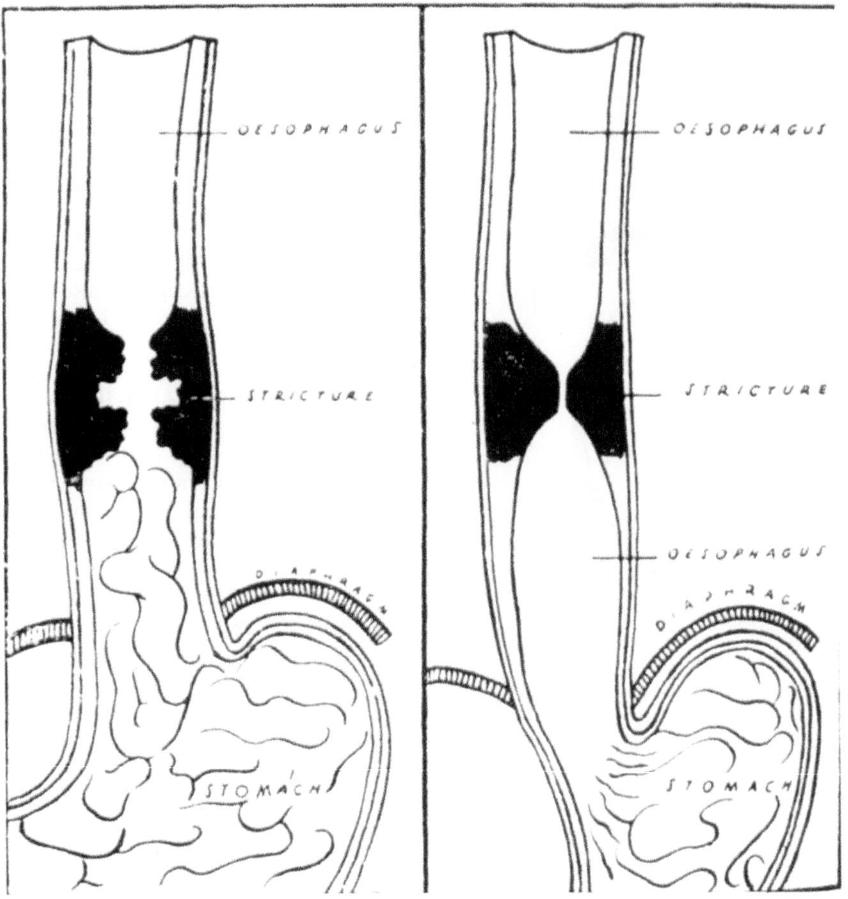

FIGURE 2.7 Barrett (1950): "Diagram devised to stress the anatomical differences between a stricture which has drawn a pouch of stomach up into the mediastinum and one which has not." In neither of these drawings is a columnar-lined esophagus shown. Note that Barrett showed the entire esophagus lined by squamous epithelium (white) and gastric mucosa beginning at the end of the tubular esophagus (with rugal folds). Reproduced with permission from Barrett NR. Chronic peptic ulcer of the oesophagus and "oesophagitis." Br J Surg 1950;38:175–182.

mucosa in which the ulcers occur are columnar and not squamous. It seems likely that Allison had also made this observation but had not mentioned it in his 1948 paper except for his Figure 5 (Fig. 2.6). In Barrett's Figures 258 (Fig. 2.1) and 260, which are clear today as being long segments of columnar-lined esophagus with large ulcers in the columnar-lined segment, Barrett places the end of the esophagus at the squamocolumnar junction and proclaims the ulcer as being a gastric ulcer in a tubular stomach because it is in an area lined by columnar epithelium.

To make his argument, Barrett asked the question: "First, what is the esophagus?" He answered his own question as follows:

> To the anatomist it is the part of the alimentary canal which extends from the pharynx to the stomach, and the implication is that it traverses the mediastinum, whereas the stomach lies in the abdomen.... But these criteria are not relevant to the physiologist, the surgeon, or the physician. To them the oesophagus is that part of the foregut, distal to the cricopharyngeal sphincter, which is lined by squamous epithelium; that is, the emphasis is on the nature of the secretions produced into the lumen, rather than upon the musculature or the serous covering of the walls. A piece of gullet lined by gastric mucosa, whatever the external or superficial appearances may be, functions as stomach—secreting hydrochloric acid and digestive juice—and is heir to the ailments which afflict the stomach. (p. 177)

Barrett went on to suggest that all previous cases reported as chronic peptic ulcers of the lower esophagus are not related to heterotopic gastric mucosa, rather they are all gastric ulcers occurring in patients with congenitally short esophagus with a tubular stomach. He extensively reviews the literature relating to heterotopic gastric mucosa in the esophagus, providing convincing data that such islets are

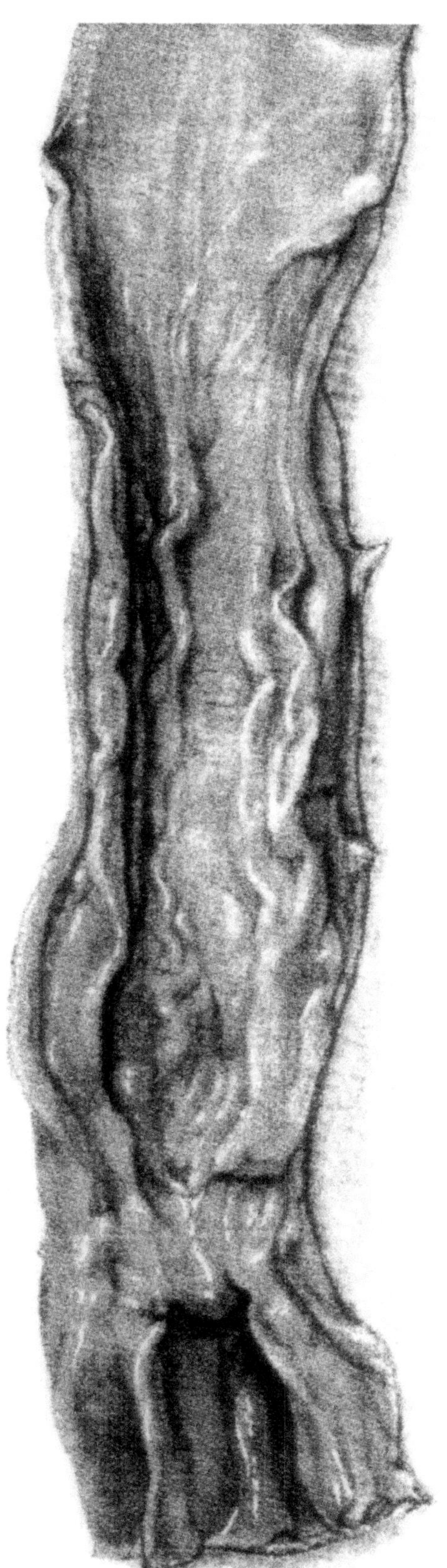

almost always found in the post-cricoid region and almost never in the distal esophagus. He dismissed the concept that acid from heterotopic gastric islets can cause peptic ulcers in the esophagus: "I cannot believe that the tiny volume of acid, diluted by pints of saliva, would be likely to harm the lower reaches of the gullet" (pp. 179–180).

Barrett carefully reviewed the reports of peptic ulcer of the esophagus by Tileston, Stewart and Hartfall, and Lyall and emphasized that none of these had descriptions of the mucosa around the ulcers. He quoted Lyall's description of one of his patients that showed columnar and not squamous epithelium around the ulcer to prove that this ulcer occurred in the stomach and not the esophagus. He also illustrated gross pictures of three cases in his experience where the squamocolumnar junction is in the midesophagus and a chronic ulcer is present in the area of what he called tubular stomach because it is lined by what he called gastric mucosa (Fig. 2.1). According to Barrett, this represents congenital short esophagus with a gastric ulcer in the tubular stomach. He claimed that the absence of inflammation and stricture makes this different from the esophageal ulcers that occur in reflux esophagitis, which involve squamous epithelium.

Barrett achieved the two main objectives of his paper. First, he correctly discredited the concept of ulcers arising in islets of heterotopic gastric mucosa in the lower esophagus. It is difficult to find any significant reference to "islets of gastric mucosa in the lower esophagus" in the literature following this paper. Second, he separated two entities: (a) The rare chronic peptic ulcer occurring in what Barrett called the tubular intrathoracic stomach below the esophagus, which, by his definition, ended at the squamocolumnar junction. According to Barrett, the stomach had been pulled into the thorax in these patients because of congenital short esophagus (in effect, a congenital sliding hiatal hernia). These ulcers become known as "Barrett ulcers." (b) The common acquired reflux esophagitis, which causes ulcers in the true squamous-lined esophagus, often associated with strictures,

FIGURE 2.8 Barrett (1950): Oesophagus which shows a chronic fibrous stricture in the lower part; this is circumferential, shows superficial surface ulceration and inflammation is transmural. "The gullet above the stricture is neither inflamed nor unduly dilated; it is lined by squamous epithelium. The gut below the stricture, which at first sight might be regarded as oesophagus, is a pouch of stomach, drawn up into the mediastinum by the contracting scar tissue of the stricture; it is lined by columnar epithelium. This pouch of stomach should not be described as 'ectopic.'" Barrett used the squamocolumnar junction to define the end of the esophagus, thereby misinterpreting columnar-lined esophagus as a tubular stomach. Reproduced with permission from Barrett NR. Chronic peptic ulcer of the oesophagus and "oesophagitis." Br J Surg 1950;38:177.

acquired shortening of the esophagus, which sometimes resulted in an acquired sliding hiatal hernia. The ulcers in these patients occurred in the squamous-lined esophagus and not in the gastric mucosa of the sliding hiatal hernia.

Barrett's paper appears to have had a tremendous impact and acceptance, probably a testament to his great influence. His definition of the esophagus by its squamous lining fit well with the existing concepts of congenital short esophagus put forward by Frindlay and Kelly and the newer ideas of acquired shortening secondary to reflux esophagitis in Allison's paper.

In his 1953 paper that challenged Barrett's concept of the tubular stomach (1), Allison praised Barrett for clearing the confusion between reflux esophagitis and peptic ulcer of the lower esophagus: "the clarification of this point has been so important . . . that it would be justifiable to refer to (peptic ulcers arising in what Barrett calls the tubular stomach complicating congenital short esophagus) as Barrett's ulcer" (p. 87). Allison's intent with the eponym was to apply Barrett's name to the chronic gastric ulcers that everyone at this point agreed were very rare and occurred in a tubular shaped intrathoracic stomach caused by congenital shortening of the esophagus.

1953 to 1957: "Columnar-Lined Esophagus" Is Born

For several decades preceding 1953, columnar-lined esophagus was misinterpreted as a tubular stomach distal to a shortened esophagus. This resulted entirely from the prevailing false definition that the esophagus ended at the squamocolumnar junction. Frindlay and Kelly's (3) series in children and Chamberlin's (9) series in adults are very likely examples of columnar-lined esophagus.

The first clear description of columnar-lined esophagus is to be found in Lyall's paper in 1937 (8) where he reported eight cases of peptic ulcer of the esophagus. He made no mention of epithelial type in seven cases but described his last case as follows:

> Close examination of the mucous membrane in the region of the ulcer showed the presence of a remarkable state of affairs. The intact mucosa separating the lateral edges of the ulcer was found to be heterotopic gastric mucosa which extended as a tongue-shaped process of well preserved tissue upwards from that of the fundus of the stomach. . . . The mucosa bore a resemblance to that normally found towards the pyloric end of the stomach. (p. 540)

Lyall was surely describing columnar-lined esophagus and cardiac mucosa, although he interpreted it as "heterotopic gastric mucosa in the esophagus."

In Barrett's paper of 1950 (2), he described an excellent example of columnar-lined esophagus:

> I was shown a specimen (Fig. 2.1) by Professor Barnard. His patient was a child, aged 13, who had several other congenital deformities, and who had died from perforation of a chronic ulcer into a large pulmonary vein. . . . There are two small islets of ectopic mucosa at the top of the gullet and below these the oesophageal mucosa stops short. Below the level of the arch of the aorta the mucosa is everywhere histologically gastric in type and the chronic ulcer which has perforated into one of the great vessels is a stomach ulcer. (p. 182)

This specimen would be interpreted today as an extremely long columnar-lined esophagus.

In 1951, Boscher and Taylor (10) described a similar patient. This 63-year-old patient had an ulcerated stricture in the squamous-lined esophagus at the level of the aortic arch. Below the stricture was a tubular structure that resembled the esophagus from the outside, had esophageal type submucosal glands in the wall, but was lined by "gastric mucous membrane without oxyntic cells." This is another clear example of a columnar-lined esophagus in the early literature.

It is obvious that Barrett's definition of the esophagus as that part of the foregut lined by squamous epithelium did not find universal acceptance. Many of the researchers of the time probably felt uncomfortable with designating the tubular-lined structure they were seeing below a squamocolumnar junction as a tubular stomach because it contradicted well-accepted criteria of anatomical definition of the esophagus. In their 1953 paper (1), Allison and Johnstone stated that their praise for Barrett's paper and suggestion of the eponym "Barrett's ulcer" does not imply agreement with Barrett's description of an esophagus lined with gastric mucous membrane as "stomach." Allison and Johnstone appear to be expressing a widely held disagreement in the medical community regarding Barrett's definition of the esophagus as "that part of the foregut lined by squamous epithelium."

Literature Review

Allison PR, Johnstone AS. The oesophagus lined with gastric mucous membrane. Thorax 1953;8:87–101.

This article detailed for the first time the existence of a structure between the squamocolumnar junction and the stomach in some patients (Figs. 2.9 and 2.10). Allison and Johnstone reported seeing 125 patients with esophageal ulcer and stenosis. Of these, 10 patients had the stenosis associated with Barrett's ulcer, and 115 had a reflux-induced stricture in the squamous-lined esophagus (in another 23

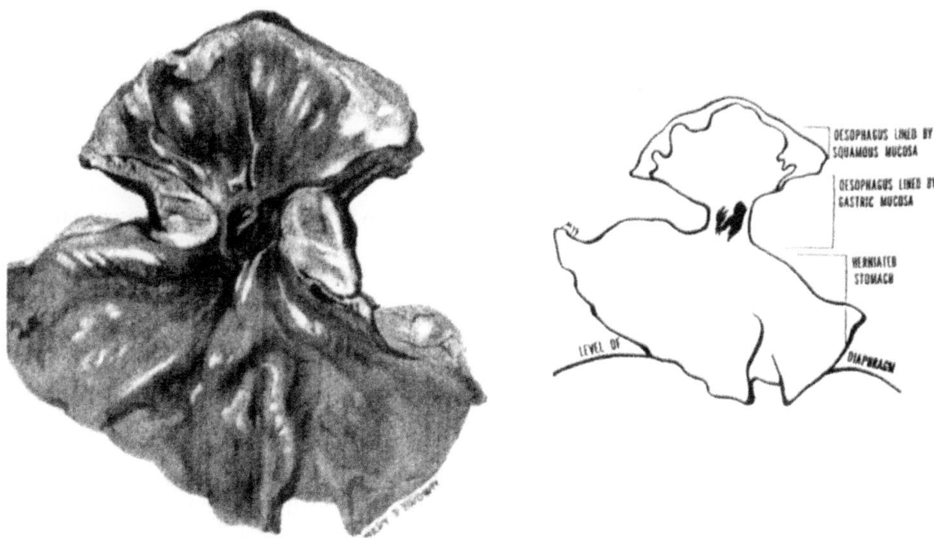

FIGURE 2.9 Allison and Johnstone (1953): Excised specimen from his case 4 showing a typical stenosing, penetrating Barrett's ulcer. "Below this is part of the herniated stomach; above the ulcer is oesophagus lined by gastric mucus membrane with a small strip of squamous epithelium above." The columnar-lined esophagus is now correctly identified as esophagus, but its mucosa is incorrectly called "gastric." Reproduced with permission from Allison PR, Johnstone AS. The oesophagus lined with gastric mucous membrane. Thorax 1953;8:88.

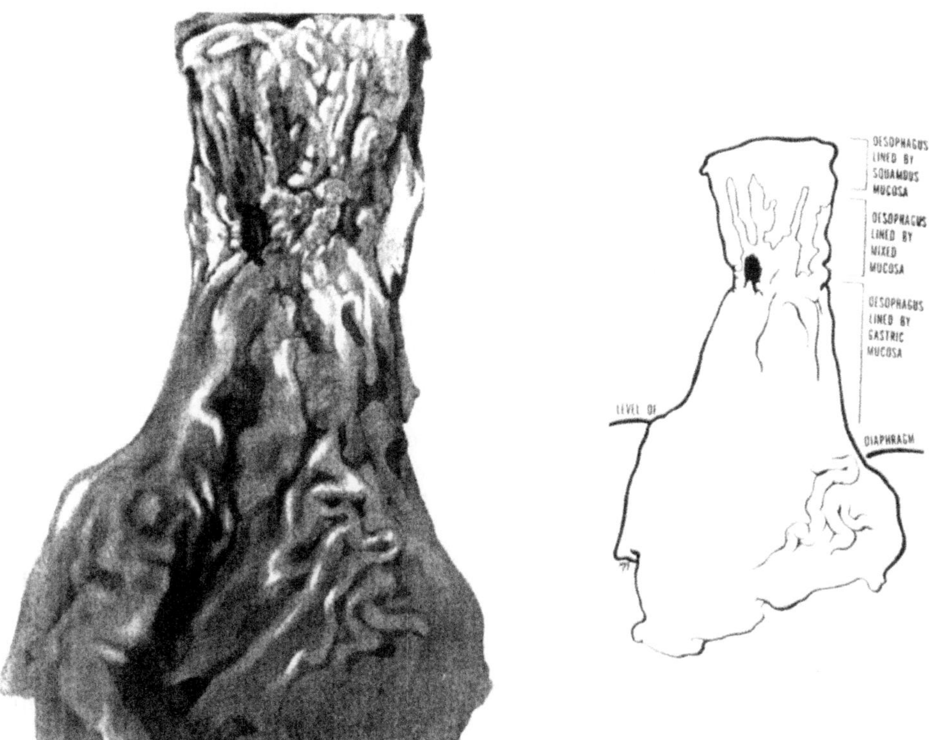

FIGURE 2.10 Allison and Johnstone (1953): Excised specimen from their case 5 showing an extremely irregular junction of squamous and gastric mucous membrane. The ulcer straddles the squamocolumnar junction. In the drawing, a long segment of "oesophagus lined by gastric mucosa" is shown below the ulcer and above the diaphragm. Note that there is no attempt to define the lower limit of the esophagus. To Allison and Johnstone, the entire distal esophagus and stomach are lined by gastric mucosa; the drawing shows rugal folds in the distal esophagus. Reproduced with permission from Allison PR, Johnstone AS. The oesophagus lined with gastric mucous membrane. Thorax 1953;8:91.

patients, the stricture had a doubtful anatomy). Eleven of the 115 patients with esophageal stricture had "indisputable evidence of a segment of oesophagus lined with gastric mucosa between the stenosis and the hernia" (p. 98). The other 104 patients with esophageal strictures and ulcers did not have any recognizable columnar lining in the esophagus. Allison and Johnstone stated: "These figures must not be taken to indicate either the frequency of gastric mucosa in the oesophagus, which was 21/125, or of Barrett's ulcer, which was 10 out of 125, but they do give a very rough idea of the relative chances of a peptic stenosis being oesophageal or gastric, about 10 to 1" (p. 98). It should be noted that Allison and Johnstone included the 10 cases of Barrett's ulcers in their total of patients with "gastric mucosa in the esophagus." In effect, they equated "Barrett's ulcers" as being ulcers occurring in columnar-lined esophagus rather than gastric ulcers as Barrett had suggested. Their analysis is correct although they continued to recognize "Barrett's ulcers" as a separate entity.

Allison and Johnstone provided a detailed description of the clinical, radiological, endoscopic, and pathologic (by biopsy and resection) features of seven of their patients with a segment of columnar-lined esophagus interposed between the squamocolumnar junction and the stomach. They described why they believed this tubular segment is esophagus and not stomach: "More careful examination of such a specimen shows that it has no peritoneal covering, that the musculature is that of the normal oesophagus, that there may be islands of squamous epithelium within it, that there are no oxyntic cells in the mucosa, and that in addition to gastric glands there are present typical esophageal mucous glands" (p. 87). It should be noted that Allison and Johnstone used the absence of parietal (oxyntic) cells in the columnar epithelium to denote that the mucosa is esophageal. They also indicated how this segment can be distinguished from a sliding hiatal hernia: "The dilated sac covered with peritoneum and lined with gastric mucous membrane is obviously stomach whether it lies in the abdomen or is herniated into the mediastinum" (p. 87).

Allison and Johnstone very accurately described three zones in these seven patients: (a) reflux induced stenosis, which marks the lower limit of squamous epithelium; (b) the intermediate abnormal tubular esophagus lined by "gastric mucosa" (which is neither ulcerated nor the site of stricture), followed by (c) the stomach (Figs. 2.9 and 2.10). "The position of the cardia [which Allison and Johnstone used synonymously with the gastroesophageal junction] can be identified where the lumen widens again to form the sac of the herniated stomach" (p. 99).

There is no doubt when reading this paper that Allison and Johnstone were convinced of two things: (a) the correct conclusion that the tubular structure distal to the squamocolumnar junction was esophagus and not a tubular stomach as defined by Barrett and (b) the incorrect conclusion that the lining was gastric mucosa. This was an obvious conflict that the authors could not reconcile. This is surprising because they clearly paid a lot of attention to the gross and microscopic features of biopsies and resection specimens. They described the columnar epithelium in the esophagus as having few oxyntic cells and that on occasion it had goblet cells and that it differed from the mucosa in the stomach. They were describing what we would now recognize as metaplastic columnar epithelia in the esophagus of cardiac (junctional) and intestinal (specialized) types.

In their conclusion, the authors stated:

> A variable amount of the oesophagus below the aortic arch may be lined by gastric mucosa of cardiac type. This is presumably, but not necessarily in every case, a congenital abnormality, and in the series presented has always been associated with herniation of the true stomach through the diaphragmatic hiatus, with the cardia [that is, the gastroesophageal junction] lying in the mediastinum. (p. 100)

The way Allison and Johnstone put these various observations together is a harbinger for much of the future confusion that resulted. Allison and Johnstone were masterful in their observations that have incredible clarity and accuracy. They define the cardiac mucosa in the esophagus above the gastroesophageal junction as being very different from gastric mucosa below the junction, and as an epithelium showing evidence of injury. In what must be the most precise description ever recorded, the pathologist Dr. D. H. Collins reported on the dissection of the pathologic specimen in their Case 1:

> The esophagus was separated . . . from the stomach along the line of the peritoneal reflection. . . . The stomach below the anatomical junction with the oesophagus is lined by gastric mucosa of fundal type. . . . Cardiac glands and cardiac gastric mucosa do not appear until 0.6 cm up the anatomical oesophagus. . . . The first islet of squamous epithelium appears 3 cm up the anatomical esophagus, but predominantly gastric mucosa continues to the upper limit of the specimen. (p. 94)

His histologic description of cardiac mucosa in this esophageal segment is incredibly accurate: "The rather villous type of cardiac mucosa, its lack of depth, and a diffuse fibrosis of the submucosa in the zone for 2 to 4 cm above the stomach orifice, suggest healing of previous shallow ulcerations" (p. 94). It is likely that the authors were merely reiterating in the title of the paper their pathologist's conclusion that this mucosa lining what they both believed was the anatomic esophagus was "gastric mucosa." Despite recognizing that the mucosa lined the anatomic esophagus, and despite recognizing that the mucosa was histologically different from the mucosa distal to the anatomic gastroesophageal junction, the authors called this columnar-lined esophagus "gastric mucosa." They had proof that this cardiac mucosa was entirely esophageal because it began 0.6 cm proximal to the gastroesophageal junction. *This is the proof that what is not accepted even today and what we are trying to establish in this book existed in 1953!*

A feature of note in this paper is that Case 7 was one of the first patients in the literature with an adenocarcinoma clearly recognized as originating from columnar-lined esophagus (Fig. 2.11). They described this as "gastric carcinoma of the oesophagus." While this seems oxymoronic, we should recognize that we still could be making the same mistake when we say that "adenocarcinoma of the gastric

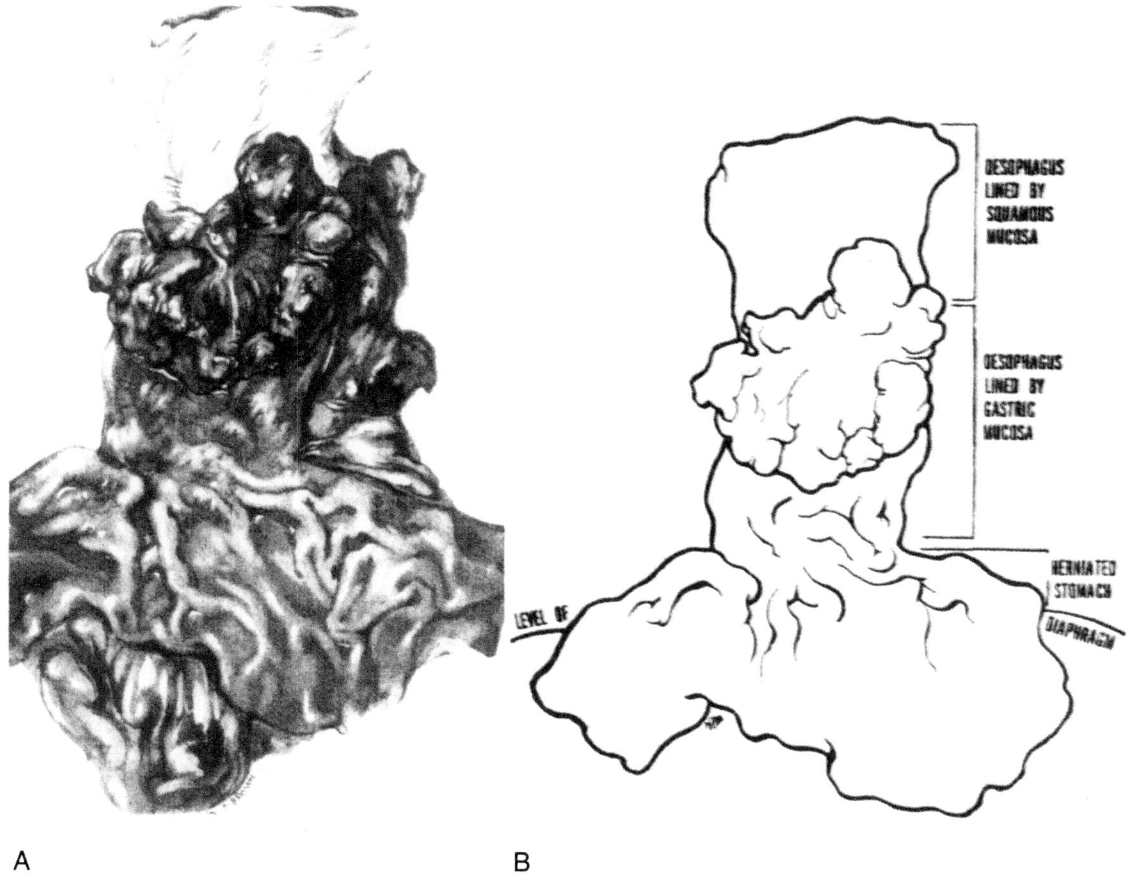

FIGURE 2.11 (a) and (b) Allison and Johnstone (1953): This is called "gastric carcinoma of the oesophagus" arising from an area of oesophagus lined by gastric mucous membrane. This is one of the earliest examples of adenocarcinoma arising in Barrett esophagus. Reproduced with permission from Allison PR, Johnstone AS. The oesophagus lined with gastric mucous membrane. Thorax 1953;8:93.

cardia" has an epidemiologic relationship with gastroesophageal reflux disease.

Allison and Johnstone's contention that Barrett's tubular stomach was actually esophagus must have created a significant argument at the time between these two influential surgeons in Leeds and London, respectively (Fig. 2.12). It is unusual for major controversy in medicine to lead to a resolution within 4 years, as was the case here. In 1957, Norman Barrett agreed with Allison and Johnstone's concept and reversed the opinions he had expressed in 1950.

Literature Review

Barrett NR. The lower esophagus lined by columnar epithelium. Surgery 1957;41:881–894.

Barrett's landmark paper of 1957 (4) is actually the text of a lecture given by him in many places including the Mayo

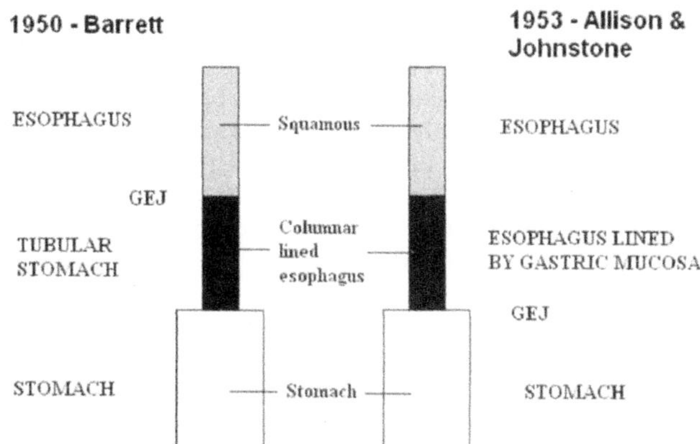

FIGURE 2.12 Diagrammatic representation of columnar-lined esophagus as interpreted by Barrett (1950) and Allison and Johnstone (1953). Barrett defined the gastroesophageal junction as the end of squamous epithelium (histologic). He then interpreted the columnar-lined esophagus as "a tubular intrathroacic stomach." Allison and Johnstone correctly recognized that what Barrett called a tubular stomach is indeed esophagus, but they called the mucosal lining "gastric." They did not define the gastroesophageal junction histologically because they did not see a difference between "gastric mucosa" lining the esophagus and stomach.

Clinic in Rochester, Minnesota, in November 1956. In his opening statement, Barrett noted, "The ideas discussed here are not based on statistics nor upon a large collection of specimens; they are the results of thinking about a few unusual cases of esophageal disease.... Some may be worried because I have changed my opinion relating to certain matters, but progress is not static and there is no subject which does not yield more knowledge as the depths are sounded" (p. 881). It is interesting that the original paper that describes Barrett's esophagus is not evidence based; it is based on "thinking about a few unusual cases."

> This paper concerns a condition whose existence is denied by some, misunderstood by others, and ignored by the majority of surgeons. It has been called a variety of names which have confused the story because they have suggested incorrect etiologic explanations; congenital short esophagus, ectopic gastric mucosa, short esophagus, and the lower esophagus lined by gastric epithelium are but a few. *At the present time, the most accurate description is that it is a state in which the lower end of the esophagus is lined by columnar epithelium.* This does not commit us to ideas which could be wrong, but it carries certain implications which must be clarified. (p. 881)

Barrett discussed the three potential causes for the finding of columnar epithelium in the gullet (a term he uses for the structure connecting the pharynx to the abdominal part of the stomach below the diaphragm; "gullet" includes the esophagus and any abnormal intrathoracic stomach): (a) a sliding hiatal hernia, (b) the condition described in this paper where the lower part of the esophagus is lined by columnar epithelium extending upward for a long or a short distance above the esophagogastric junction, and (c) true islets of ectopic columnar epithelium. Barrett strongly reiterated his correct belief that ectopic islets of gastric mucosa are not found with any frequency in the lower esophagus in adults.

Barrett defined columnar-lined esophagus as follows: "When the lower esophagus is found to be lined by columnar cells, the abnormally placed mucous membrane extends upwards from the esophago-gastric junction in a continuous, unbroken sheet.... The extent of the anomaly varies from a few centimeters to the upper esophagus" (p. 883). Barrett affirmed that the anatomy of the esophagus and stomach are not altered, including the musculature, mediastinum, blood supply, the relationship to the diaphragm, and the peritoneal reflection.

It is important to understand why Barrett rejected Allison and Johnstone's term "esophagus lined by gastric mucous membrane." In describing the columnar epithelium he stated: "Although there is no doubt that in these cases the mucosa lining the lower esophagus looks, in every way, identical to the stomach (being velvety in texture, bluish in color, and raised in typical rugae), there are those who argue whether or not it is stomach" (p. 884). Careful review of Figure 260 in Barrett's 1950 paper shows that Barrett recognized that the mucosa in what he was calling the tubular stomach (in reality, the columnar-lined esophagus) did not contain rugal folds.

He described the histology of the columnar-lined esophagus with great accuracy and compared it to the normal gastric mucosa.

> In the upper part of the columnar-lined segment the columnar cells are flat and arranged in shallow, tubular glands amongst which lie mucus-secreting units. There are no oxyntic cells.... Lower, in the esophagus the simple tubular crypts give place to more typical gastric mucous membrane. Scattered oxyntic cells appear.... At any level in this abnormal segment, perfectly formed patches of squamous epithelium appear. These findings, which are similar to those described by Allison and Johnstone, suggest that the abnormal epithelium, despite its looks, does not function exactly as stomach. (p. 884)

Barrett, with great acumen, was resisting Allison and Johnstone's designation of this epithelium as gastric mucous membrane. It is clear that he regarded it as an abnormal columnar epithelium that is different from gastric mucosa.

Barrett did not know the etiology of this columnar-lined esophagus. He used the excellent embryologic data available at the time (16) to suggest that columnar-lined esophagus may result from an arrest in the normal embryonic transformation of foregut columnar epithelium to squamous, but he questioned why this would only involve the lower esophagus. In his analysis, he touched on the correct etiology:

> The explanation could be that if the cardiac valve of a normal person were to become incompetent and if the lower esophagus were, as a result, to be bathed for a long time by digestive gastric juice, the squamous epithelium could be eaten away and totally replaced by more quickly growing columnar cells. This concept might explain the site of the deformity, the fact that many cases occur in patients who have an incompetent cardia, due to sliding hiatal hernia, and the fact that many patients are elderly and have a history of heartburn dating back many years. (p. 885)

Barrett was close to the correct answer but he was confused by experimental animal data: "But defects, produced experimentally in the squamous epithelium of the esophagus of the dogs, heal by squamous regeneration." Neither Allison nor Barrett mentioned epithelial metaplasia, where the squamous epithelium could transform into columnar epithelium without first being denuded.

With agreement between Allison and Barrett and apparent universal acceptance by the entire medical community, the correct conclusion that some patients had a columnar-lined lower esophagus, always beginning at the gastroesophageal junction and varying in length, was permanently established as fact in 1957.

What is most interesting is that there was no confusion about the anatomy and histology of columnar-lined esophagus at this time. Both Allison and Johnstone's and Barrett's papers described the normal state with the normal esophagus lined by squamous epithelium to the cardia (which they both equated with the gastroesophageal junction and the peritoneal reflection) at which point the stomach begins and the mucosa changes to gastric fundal (oxyntic) type with parietal cells. They both recognized that columnar-lined esophagus is an uncommon anomaly that begins at the distal end and varies in length from a few centimeters to a very long segment that could reach the proximal esophagus. They both described the three zones in this abnormal state: (a) the proximal squamous-lined esophagus, often with strictures and reflux damage; (b) the intermediate columnar-lined esophagus, which always "extends upward from the esophago-gastric junction in a continuous unbroken sheet" (p. 885); and (c) the true stomach distal to the gastroesophageal junction, which is normally found in the abdomen but could be found in the thorax as a reflux-induced sliding hiatal hernia in patients with columnar-lined esophagus. They both showed that the columnar-lined esophagus is lined by cardiac mucosa with evidence of injury that is different in appearance than the fundic (oxyntic) mucosa of the stomach below the gastroesophageal junction.

If we stop at this point in history, what we will suggest in this book is very similar to what was originally stated by both Barrett and Allison. Much of the confusion came later in the next 50 years. Both Allison and Barrett paid considerable attention to histology and recognized the stomach as being lined by normal fundal (oxyntic) mucosa and the abnormal columnar-lined esophagus being lined by a different columnar epithelium they called cardiac mucosa, often devoid of parietal cells. Neither ever placed cardiac mucosa in the true stomach. Allison and Johnstone (1) categorically stated: "A variable amount of the oesophagus below the aortic arch may be lined by gastric mucosa of cardiac type" (p. 883). They both described the occurrence of scattered oxyntic (parietal) cells in the lower segment of the columnar-lined esophagus. They were describing what we now call oxynto-cardiac mucosa in the esophagus. The only mistake that Allison and Johnstone made was to defer to their pathologist's designation of cardiac mucosa as "gastric." Barrett, in changing the name to "columnar-lined esophagus" corrected Allison and Johnstone's error, but we have yet to recognize the significance of the observation that *cardiac mucosa is the epithelium that lines the columnar-lined esophagus, it is not gastric.* Before continuing, we should stop a moment to explain how Barrett's name came to be used for an entity that Allison described.

The effect of Allison's 1948 (15) and Barrett's 1950 (2) treatises on the subject had the clear effect of incorrectly ascribing different etiologies to two entities in the lower esophagus: (a) the rare peptic ulcer of the esophagus occurring in columnar epithelium that Barrett incorrectly classified as a gastric ulcer in a tubular stomach resulting from congenital short esophagus and (b) the common entity of reflux esophagitis, which affected squamous epithelium with and without ulcers.

In their 1953 paper (1), Allison and Johnstone considered this distinction so important that they suggested applying the eponym "Barrett's ulcer" to gastric ulcers occurring in intrathoracic stomach. It appears that the medical community accepted Allison's suggestion, because in his 1957 paper Barrett freely used the term "Barrett's ulcer."

In the same 1953 paper that Allison and Johnstone suggested using the eponym "Barrett's ulcer" for a gastric ulcer occurring in what Barrett had called a tubular stomach in 1950, they described for the first time the lower esophagus lined by gastric mucous membrane. In 1957 Barrett agreed with the existence of this entity, but he disagreed with Allison and Johnstone's suggestion that the esophagus was lined by gastric mucous membrane and coined the term "columnar-lined esophagus," which immediately gained wide acceptance in the medical community.

Over the next few years, it was slowly recognized that most patients with "Barrett's ulcers" in reality had their ulcers in the columnar-lined segment of the lower esophagus rather than in a stomach drawn up into the mediastinum. In fact, even Allison and Johnstone in their 1953 paper (1) refer to 10 patients with Barrett's ulcer and suggest that the gastric epithelium surrounding the ulcers is really in the esophagus rather than the stomach. In their conclusion, Allison and Johnstone stated: "Patients with the oesophagus lined by gastric mucous membrane are subject to gastric ulcers . . . and it is suggested that if these become chronic, they might be known as Barrett's ulcers" (p. 100).

With the recognition that Barrett's ulcers occurred in columnar-lined esophagus rather than a tubular stomach, the term Barrett's ulcer lost its original meaning. It was, however, a natural extension to apply Barrett's name to the entire spectrum of changes associated with columnar-lined esophagus.

In reading the literature, one feels a sense of injustice that Allison's name is not applied to an entity that he so clearly recognized for the first time in 1948 and crystallized by detailed study of many patients in his 1953 paper. This is in stark contrast to Barrett's paper of 1957 (4) which was "not based upon statistics nor upon a large collection of specimens; they are the results of thinking" (p. 881). It is ironic that it was Allison and Johnstone who initially suggested the eponym "Barrett's ulcer." There may be justification for using Barrett's name, however, because it was he who resisted concluding that the columnar lining in the distal esophagus was gastric mucous membrane and suggested the more accurate term "columnar-lined esophagus." It would be wonderful if we could go back in history and call this Allison-Barrett esophagus and give credit to both surgeons who so beautifully developed rational criteria for a new entity.

1957 to 1970: Controversy as to the Cause of Columnar-Lined Esophagus: Congenital Heterotopic Gastric Mucosa or Acquired Columnar Mucosa Secondary to Gastroesophageal Reflux?

Into the mid-1950s, there seemed to be universal agreement that the finding of columnar mucosa in the lower esophagus was a congenital anomaly. Allison and Barrett both agreed on this. To Allison, the epithelium was an upward extension of gastric mucosa into the esophagus. To Barrett, it was more likely a failure of squamous transformation of fetal esophageal epithelium, which embryologists of the time had accurately described as being columnar. However, both Barrett and Allison also recognized the frequent coexistence of reflux esophagitis and sliding hiatal hernia in patients who harbored a columnar-lined esophagus. In Allison and Johnstone's original series of seven patients, reflux esophagitis and sliding hiatal hernia were present in all, and four had progressed to fibrous stricture.

Despite the fact that they recognized the strong association with reflux and the fact that most of their patients were elderly, Allison and Johnstone (13) believed that columnar-lined esophagus was a congenital anomaly. Their wording suggested unease rather than certainty about this: "This is presumably, but not necessarily in every case, a congenital abnormality" (p. 100). Allison and Johnstone argued that the congenital gastric mucosa in the lower esophagus was not symptomatic or harmful and that the reflux was actually the result of the coexisting hiatal hernia. They were suggesting that the columnar mucosa in the esophagus was an incidentally found congenital abnormality in patients whose esophagitis and peptic stenosis was caused by the coexisting reflux and hiatal hernia.

Barrett, in 1957 (4), expressed an even greater degree of uncertainty about the congenital nature of columnar-lined esophagus, although his ultimate conclusion was that it is most likely a congenital anomaly.

The recognition that columnar-lined esophagus was an abnormality caused by reflux appears to have occurred as a slow change of opinion in the late 1950s and cannot be traced to any single paper. By the time Hayward published his highly influential paper in 1961, it appears that the medical community had accepted the viewpoint that columnar-lined esophagus was acquired as a result of gastroesophageal reflux. Hayward (17) described the process of reflux-induced columnar transformation of the esophagus beautifully:

> When the normal sphincteric and valvular mechanism in the lower oesophagus and oesophago-gastric junction, i.e., what I call the cardia, fails . . . reflux from the stomach occurs and acid and pepsin reach the squamous epithelium and begin to digest it. . . . In quiet periods some healing occurs, and in these periods the destroyed squamous epithelium may re-form, often with leukoplakia, or junctional (= *cardiac*) epithelium, usually not very healthy-looking, may replace it. Where this occurs the area is given considerable protection from future reflux because normal junctional (= *cardiac*) epithelium resists acid-peptic digestion. Further reflux therefore attacks principally the squamous epithelium higher up. In the next remission it may be replaced by more junctional (= *cardiac*) epithelium as a further protective reaction. With repetition over a long period the metaplastic junctional (= *cardiac*) epithelium may creep higher and higher until it reaches the level of the arch of the aorta. It seldom extends higher than this. (p. 40)

Hayward's original description of the pathogenesis of reflux induced columnar esophagus in his 1961 paper (17) is well nigh perfect.

The final proof that columnar-lined esophagus was an acquired condition resulting from gastroesophageal reflux rather than a congenital anomaly should be credited to the experimental study by Bremner et al. (18). They divided 35 dogs into three groups: (a) in 14, the lower esophageal sphincter was destroyed by myotomy, the distal esophageal squamous epithelium removed, and a hiatal hernia created; (b) 11 others were prepared in a manner identical with group 1, but 2 weeks after operation, 15 mg histamine was given by intramuscular injection on alternate days for periods ranging from 2 to 12 weeks; (c) in 10 dogs, the esophageal mucosa was excised without myotomy or creation of a hiatal hernia.

The dogs in groups 1 and 2, who were exposed to gastroesophageal reflux, showed columnar regeneration of the denuded esophagus (Fig. 2.13), and dogs in group 3, who had their sphincter intact and therefore not exposed to reflux, showed largely squamous regeneration (Fig. 2.14). The dogs in group 2, who had the highest level of acid exposure because of histamine stimulation, had the most columnar epithelium in the esophagus. Bremner et al. (18) reported the histology as follows: "The regenerated columnar epithelium was distinguishable from gastric epithelium by the . . . lack of parietal cells. The new epithelium was cuboidal or low columnar and was clearly mucus producing" (pp. 212–213). In the dogs exposed to gastroesophageal reflux, a cardiac type columnar epithelium replaced the original squamous mucosa in the distal

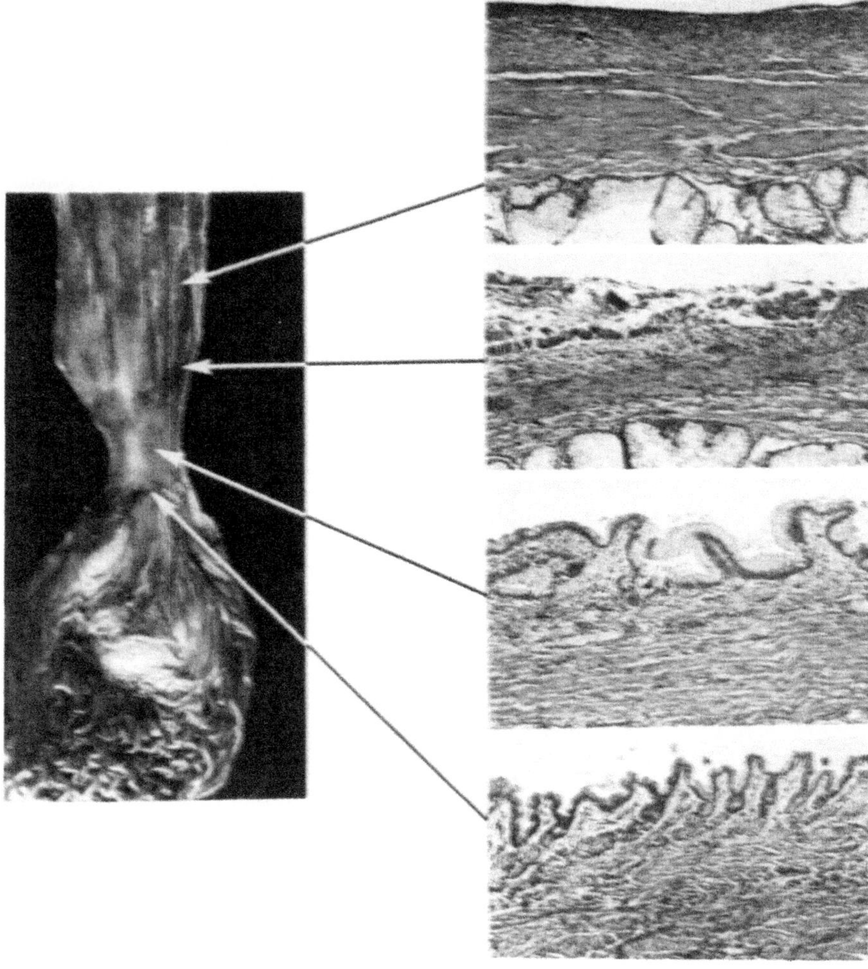

FIGURE 2.13 Bremner (1970): Specimen from dog in group 1 (reflux model) 109 days after operation, showing histologic appearance of mucosal lining at various levels of the regenerating esophagus. The denuded segment has been covered almost entirely by columnar epithelium.

esophagus. Bremner et al. had reproduced columnar-lined esophagus in an experimental animal and shown that gastroesophageal reflux was the factor that caused columnar rather than squamous regeneration of the denuded mucosa.

Dr. Cedric Bremner has been at the University of Southern California for the past 10 years as part of the foregut group, and we had the pleasure of seeing the slides from these classical experiments. The mucosa in the columnar epithelium is indeed a thin cardiac mucosa with no parietal cells. This experiment therefore also showed that the first type of columnar epithelium that develops from squamous epithelium in the esophagus is cardiac in type.

When columnar-lined esophagus was believed to be a congenital anomaly, it was not necessary to define a limit of what would be regarded as normal. However, when columnar-lined esophagus was recognized as an acquired disease, it became necessary to define the disease. The necessity to define the disease and the failure to define it correctly, beginning with Hayward's paper in 1961 (17), has caused all the confusion regarding columnar-lined esophagus in the ensuing five decades. We still cannot accurately define the limits of what is normal or what constitutes a disease state in a manner that is universally accepted or reproducible. Until such definition is achieved, confusion will remain.

THE REASONS FOR CONFUSION

A study of the historic evolution of Barrett esophagus suggests several reasons why confusion still exists in the understanding of this disease. By recognizing how these problems arose in the first place, we can

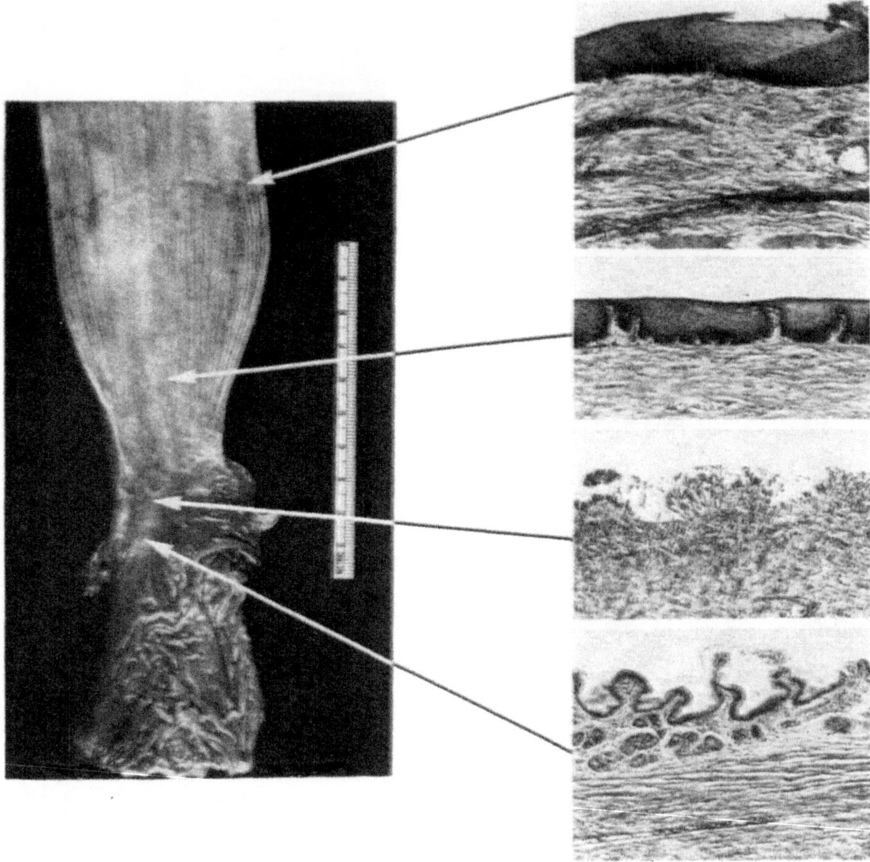

FIGURE 2.14 Bremner (1970): Specimen from dog in group 3 (no reflux model) 425 days after operation. The distal esophagus has a nearly normal appearance with most of the denuded area reepithelialized by squamous epithelium.

correct the misinterpretation and begin to understand the true evolution of the pathologic change.

Differences in Viewpoint

To the surgeon, the external landmarks used to define the esophagus and stomach are the peritoneal reflection, diaphragmatic hiatus, phrenoesophageal ligaments, and the point of flaring of the tubular esophagus into the saccular stomach. Of these, only the peritoneal reflection and possibly the point of attachment of the phrenoesophageal ligaments to the esophagus remain constant after the esophagus is damaged. The relationship between the diaphragmatic hiatus and gastroesophageal junction changes with the occurrence of esophageal shortening and hiatal hernia. The point of flaring of the tubular esophagus also becomes less defined in pathologic states and may disappear when a sliding hernia occurs. It must also be recognized that the surgeon is usually operating on a seriously diseased rather than a normal esophagus; this is not the best way to determine normal anatomy.

The endoscopist has a completely different viewpoint, seeing only the luminal aspect and the mucosal lining. The location of the diaphragm and the point of flaring of the esophagus can be discerned but not with precision in every case. There is no way to discern the peritoneal reflection or the phrenoesophageal ligament at endoscopy.

The landmark that was obvious from the outset was the squamocolumnar junction (also called the Z line). Until 1953, the squamocolumnar junction was universally equated with the gastroesophageal junction and therefore the distal end of the esophagus. Histopathologists, who had used the squamocolumnar junction to define the gastroesophageal junction, had already introduced the initial confusion by calling everything distal to the squamocolumnar junction "gastric mucosal." Allison, when he described the columnar-lined esophagus for the first time in 1953, called the lining "gastric mucous membrane" partly because his pathologist equated the columnar lining distal to the squamocolumnar junction as "gastric."

After Allison defined the columnar-lined esophagus in 1953, the endoscopic recognition of the true gastroesophageal junction became difficult because of the inability to accurately differentiate metaplastic columnar epithelium in the distal esophagus from normal gastric mucosa. Endoscopists have used different methods to define the gastroesophageal junction throughout history: though the present criterion for the endoscopic gastroesophageal junction (which is the proximal limit of the rugal folds) is widely accepted as true, it has never been proven to be the real gastroesophageal junction. Like all historic definitions that were widely accepted to be true in their time, this is also likely to be proven wrong. The reality is probably a more difficult pill than omeprazole for the gastroenterologist to swallow: *It is very likely impossible to precisely define the gastroesophageal junction by endoscopy, particularly in an esophagus whose lower end has been damaged by reflux.*

Pathologists use different criteria than both surgeons and endoscopists when they evaluate esophagogastrectomy specimens. Before 1953, pathologists used the squamocolumnar junction to define the gastroesophageal junction and reported everything distal to it as "gastric." This is the original basis by which "cardiac mucosa" was defined as lining the proximal part of the stomach. While the basis for the definition has been proved wrong, the misconception created by it has persisted in the literature, and most people still believe that cardiac mucosa is a normal lining of the proximal stomach.

Today, pathologists are directed to draw a horizontal line across the end of the tubular esophagus at the point where it flares into the saccular stomach and use this as the gastroesophageal junction, as noted by the Association of Directors of Anatomic and Surgical Pathology (ADASP) (19). This is necessary because when there are junctional tumors and ulcers, the rugal folds are frequently distorted. Where the esophagus tapers gradually rather than sharply into the saccular stomach, the exact point of flaring can be quite subjective.

It should be recognized that the different methods used by pathologists and endoscopists result in differences in the way tumors of this region are classified. Data based on pathologic reports differ from those that use clinical methods to define the gastroesophageal junction, leading to potential differences in epidemiologic studies based on pathologic versus clinical data (19, 20).

These varied viewpoints lead us to the danger of different definitions of the gastroesophageal junction: the surgeon's peritoneal reflection, the endoscopist's proximal limit of the rugal folds, and the pathologist's point of flaring of the tubular esophagus. It must be recognized that *there is only one true gastroesophageal junction, and this remains constant in normal and diseased states.* For all these definitions to be correct, they must all be coincidental. Do we know that the proximal limit of rugal folds is always coincident with the peritoneal reflection and the point of flaring of the tubular esophagus? We must; if we do not, we may be contributing to the confusion by using incorrect and variable definitions.

Inaccessibility and the Way Normalcy Is Defined

The thoracic esophagus is one of the least accessible organs of the gastrointestinal tract. As Allison so elegantly stated in his 1948 paper (15):

> The patient with peptic ulceration of the lower oesophagus has for a long time suffered his ulcer in a sort of "no-man's land" to which the abdominal surgeon could not reach and to which the physician had no access. The radiologist passed it by as quickly as the force of swallowing would take him. The endoscopist saw it as through a glass darkly on the distant boundary of his territory, but was glad to withdraw to more familiar fields. (p. 20)

In addition to its lack of access to the clinician, the distal esophagus and stomach was also not conducive to autopsy study. The columnar epithelia of this region undergo such rapid autolysis that detailed histologic study is difficult at postmortem examination. The first detailed autopsy study was reported in 2000 (21). We had to examine thousands of autopsy cases in our files to find 71 cases where a reasonably preserved histologic section of the gastroesophageal junctional region was available for study.

The difficulties associated with autopsy study of the esophagus have been a major reason for the confusion that exists. The criteria of normalcy for almost all human organs are established by examining autopsy tissues that provide material from patients who never suffered from disease during life. This was not done in the esophagus until the dawn of the new millennium (21). The early autopsy studies on esophageal ulcers reported by Tileston (6), Lyall (8), Stewart and Hartfall (7), and Chamberlin (9) have hardly any mention of the mucosal features because of the autolysis factor.

In the absence of autopsy study of normal patients (defined as a person who never contacted a physician with any complaint remotely related to the esophagus, let alone someone who had esophageal surgery or subjected himself or herself to endoscopic examination), the definition of normalcy with regard to the lining of the esophagus and stomach has progressed in a reverse direction.

The first histologic studies were done on esophagectomy specimens. These were often in the elderly and those with the most severe abnormalities in the esophagus such as cancer and strictures. Torek, a New York surgeon, reported the first successful removal of the thoracic esophagus in 1913. Torek (22) resected the thoracic esophagus, bringing out the transected cervical end as a fistula in the base of the neck. He placed a long rubber gastrostomy tube and instructed his patient to insert the proximal end of the tube into the cervical esophageal fistula during meals. He reported that his patient was extremely happy with his external rubber esophagus! Thoracic esophagectomy remained a rare procedure until the 1940s; Santy (23) reported that by 1943 there had been fewer than 100 successful thoracic esophagectomies.

The use of endoscopy began in the 1920s but was limited in application until the 1950s because the endoscopes were rigid and the expertise was limited. In this setting, only patients with the most severe esophageal diseases were subject to endoscopy; the initial focus was essentially the diagnosis of esophageal cancer. As endoscopic techniques became easier, the expertise became more widespread and both gastroenterologists and endoscopies increased in number. However, even today, endoscopy is restricted to patients who present to gastroenterologists and agree to endoscopy; this is far from a "normal" population.

Hayward (17) is given credit for the initial statement of normalcy: "The lower centimetre or two of this tube is normally lined by columnar epithelium . . . the view that the oesophagus is a tube lined only by squamous epithelium is rejected" (p. 41). Hayward, when he established these "normal" criteria, based his opinion on a highly selected abnormal patient population that he saw clinically; this was a highly abnormal method of defining normalcy.

The Desire to Limit the Number of Patients with Columnar-Lined Esophagus

The desire to limit the number of patients with a columnar-lined esophagus is one of the most important reasons why a normal esophagus is not defined correctly. Gastroesophageal reflux is an extremely common occurrence in humans, and it is logically feasible that minimal amounts of reflux-induced columnar-lined esophagus may exist in a large percentage of the population. This is similar to atherosclerosis of the aorta, which is almost universal in people over 40 years in the Western world. Though universal, no one will argue that aortic atherosclerosis is normal; in fact, recognition of the high prevalence of atherosclerosis is what has led to its control and the decline in the rates of atherosclerosis-induced diseases. Similarly, if many patients have columnar-lined esophagus, we should not hesitate to call it abnormal simply because of a desire to avoid an enormous problem. Ignoring an enormous problem will result in failure of disease recognition and control. We believe this is one reason why the incidence of reflux-induced adenocarcinoma is the most rapidly increasing cancer in the Western world.

Hayward, in 1961 (17), probably encountered many patients with small amounts of columnar lining in the distal esophagus admixed with others who had squamous epithelium lining the entire tubular esophagus. His choice was either to state that everyone who had any columnar lining in the esophagus was abnormal (the truth as we now know it) or to define that an arbitrary length of columnar lining was normal. Sadly, he chose the latter, largely because this definition would establish a "physiologic columnar-lined esophagus" and thereby restrict the number of patients who suffered from the pathologic state of columnar-lined esophagus. Based on Hayward's quite arbitrary edict that had no scientific basis, it became accepted that columnar epithelium normally lined the distal tubular esophagus to a length of less than 2 cm (Fig. 2.15). The absence of this "normal" columnar epithelium did not constitute a problem; patients with squamous epithelium lining their entire esophagus fell within this "normal" definition because zero columnar lining was less than 2 cm.

For the next 30 years, and even today in the minds of some gastroenterologists, up to 2 cm (or 3 cm for

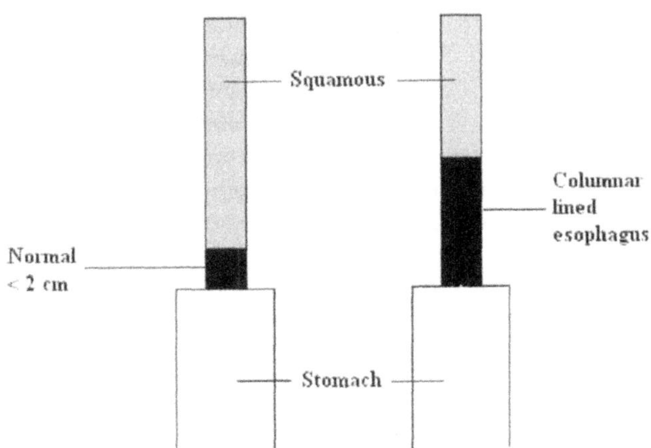

FIGURE 2.15 Diagrammatic representation of Hayward's definition of the normal state. Hayward rejected Barrett and Allison's correct concept that the entire esophagus was lined by squamous epithelium, stating (incorrectly) that the distal 2 cm of the esophagus could normally be lined by columnar (junctional/cardiac) mucosa (left). Columnar-lined esophagus could therefore be diagnosed only when the length exceeded 2 cm (right).

some) of columnar lining in the lower esophagus was regarded as normal. These normal patients did not need to be evaluated by biopsy; they were ignored because they fell within the definition of what was defined as normal. As we retrospectively look at these patients, we must recognize that the medical community ignored all patients with short-segment Barrett esophagus because of this incorrect definition from the 1960s to the mid-1990s.

As less severely symptomatic patients were subject to endoscopy, we have come to recognize that it is actually quite common for patients to have squamous epithelium extending all the way down to the end of the tubular esophagus. The present definition of endoscopic normalcy has the squamous epithelium reaching the end of the tubular esophagus where it transitions to the saccular stomach with its normal gastric rugal folds without intervening columnar-lined esophagus. This is a correct definition, but it must lead to the recognition of the converse, which is that any columnar lining in the distal esophagus is abnormal.

The correct question to be asked is not how much columnar-lined mucosa is normal; it is *how little columnar-lined esophagus is abnormal?* In Allison's drawing from 1948, it is clear that the squamous epithelium came all the way down to the end of the esophagus and transitioned to gastric oxyntic (fundal) mucosa at the gastroesophageal junction (Fig. 2.6). Allison regarded the complete absence of columnar epithelium in the esophagus to be normal and any columnar epithelium in the esophagus to be a congenital abnormality. It took more than 40 years for endoscopists to overcome the confusion created by Hayward's definition of normal and recognize that any columnar-lined esophagus is abnormal, at least from an endoscopic standpoint. The present practice guidelines of the American Gastroenterology Association recommend that all patients who have any visible columnar epithelium in their tubular esophagus above the gastroesophageal junction should undergo biopsy (24).

The same practice guidelines of the American Gastroenterology Association recommend that endoscopically normal patients should not be biopsied. Those making this recommendation have data that show that if they take biopsies of the junction in endoscopically normal people, a significant number (5% to 15% of the population) will have intestinal metaplasia. This number is so large for a disease that at least theoretically has a malignant potential that it is considered better not to find it because the resources needed to keep these patients under surveillance are inadequate. Ignorance may be bliss in terms of patient management, but being an ostrich that hides his head in the sand is not an attitude that leads to the understanding of the scientific truth. Until the medical community recognizes the enormity of the problem of columnar-lined esophagus by routinely taking biopsies of endoscopically normal patients, this disease will not be adequately controlled.

HISTORICAL EVOLUTION OF COLUMNAR-LINED (BARRETT) ESOPHAGUS (Table 2.2)

Though this heading suggests that columnar-lined esophagus is synonymous with Barrett esophagus, this is not true today. When Barrett defined columnar-

TABLE 2.2 Evolution of the Definition of Columnar-Lined (Barrett) Esophagus from 1961 to the End of the Millennium

Year	Author	Definition/change	Comment
1961	Hayward	Barrett esophagus is the presence of columnar epithelium in the distal esophagus exceeding 2 cm	>3 cm used by many authors Wrong
1976	Paull et al.	Classified BE into three histologic types No change in definition	Accurate
1980s	Many authors	Recognized that BE predisposed to adenocarcinoma	Accurate
1980s	Reid; Haggitt	Recognized that only the intestinal type of BE predisposed to cancer	Accurate
1980s	Many authors	Definition of BE required intestinal metaplasia in addition to the 2/3 cm rule	Accurate
		Other histologic types of BE were ignored	Wrong
1990s	Spechler	Recognized short segment BE (intestinal metaplasia in columnar epithelium <2 cm long)	Accurate
1990s	Many authors	Definition of short segment BE as intestinal metaplasia in any visible segment of columnar epithelium	Accurate

lined esophagus in 1957, it simply referred to the presence of columnar epithelium in the tubular esophagus. Because it was considered to be a congenital anomaly rather than a disease at the time, there was no need to define any limits. In the early 1960s, when it was recognized that columnar-lined esophagus was an acquired condition caused by reflux, there was a need to define criteria by which it could be diagnosed. When Hayward established the acceptable normal limit of columnar lining in the tubular esophagus as 2 cm, Barrett esophagus was defined by the presence of a columnar-lined esophagus >2 cm (or, more commonly, 3 cm).

In 1976, Paull et al. in Goyal's laboratory at Harvard Medical School recognized the histologic variation that existed in the columnar-lined esophagus and developed a classification of columnar-lined esophagus.

Literature Review

Paull A, Trier JS, Dalton MD, Camp RC, Loeb P, Goyal RK. The histologic spectrum of Barrett's esophagus. N Engl J Med 1976;295:476–480.

This study consisted of 11 patients with biopsy-proven columnar-lined esophagus. Strictures or ulcers were present in 8 patients; the other 3 had symptoms of gastroesophageal reflux. All patients had a small hiatal hernia on barium swallow examination.

Biopsies were taken under manometric control from the lower esophageal sphincter zone and at intervals of 1 to 3 cm above the upper border of the sphincter in the esophageal body. These biopsies were therefore from the esophagus proximal to the manometrically defined gastroesophageal junction.

Paull et al. (25) classified the epithelia in the columnar-lined esophagus into three types by the cell types that were present (Fig. 2.16): (a) gastric-fundic type epithelium characterized by the presence of glands that contained mucous, parietal, and chief cells; goblet cells were absent; (b) junctional-type epithelium characterized by only mucous cells without goblet, parietal, or chief cells; (c) specialized columnar-type epithelium characterized by goblet cells and mucous cells (Table 2.3).

In a detailed description of the distribution of these epithelial types in the columnar-lined esophagus, Paull et al. (25) stated:

> In nine [patients] this segment included an area of specialized columnar epithelium. . . . That extended over at least a length of 4.0 to 14.5 cm in seven patients but was present in only one biopsy in the other two. Parietal cells were never seen in areas lined by specialized columnar epithelium. In five of the patients in whom this specialized columnar epithelium was present either gastric-fundic-type epithelium . . . or junctional-type epithelium . . . was interposed between the specialized columnar-type epithelium and the lower esophageal sphincter zone. The interposed gastric-fundic-type epithelium extended over at least 2, 6, and 10 cm above the lower esophageal sphincter in three of these patients. (p. 477)

The only criticism of this brilliant paper is the use of the term "gastric-fundic-type" for the esophageal epithelium that contained parietal cells. It is significant that Paull et al. (25) used the term "gastric-fundic-type" rather than "gastric fundic." This suggests that they recognized that this esophageal epithelium was different to true gastric fundic epithelium where the glands do not contain any mucous cells. Unfortunately, the medical community shortened Paull et al.'s accurate term to simply "fundic epithelium," which became the standard term; this tended to perpetuate the incorrect belief that this was gastric mucosa in the esophagus. We have suggested the term "oxynto-cardiac mucosa" for metaplastic columnar epithelium of the esophagus that contains parietal cells (26). Paull et al.'s "specialized epithelium" is equivalent to the epithelium that defines Barrett esophagus today; synonymous terms are "intestinal metaplasia" and "specialized intestinal metaplasia." Paull's junctional epithelium is today called cardiac mucosa by most authorities.

After Paull et al.'s paper (25), the definition of Barrett esophagus remained as before (i.e., greater than 3 cm of columnar-lined esophagus as determined at endoscopy), but the entity was classified into junctional (cardiac), fundic (oxyntocardiac), and specialized (intestinal) types based on the histology of biopsies that were taken. Patients with less than 2 to 3 cm of columnar-lined esophagus did not undergo biopsy because this amount was considered normal.

Over the next decade, Barrett esophagus became recognized as the precursor premalignant lesion for adenocarcinoma of the esophagus. When Haggitt and Reid at the University of Washington provided evidence that adenocarcinoma occurred only in patients who had specialized columnar epithelium with goblet cells (27, 28), the definition of Barrett esophagus became more restrictive. It was no longer sufficient to see 3 cm of columnar lining in the tubular esophagus; biopsies were required to document the presence of goblet cells.

Two things happened here: (a) Barrett esophagus became defined by a combination of endoscopic (more than 3 cm of columnar lining, which led to biopsy) and histologic (presence of goblet cells) criteria rather than by endoscopic criteria alone. (b) Endoscopic columnar-lined esophagus of more than 3 cm without goblet cells in the biopsy sample was ignored completely, signaling the belief that histology was more important than

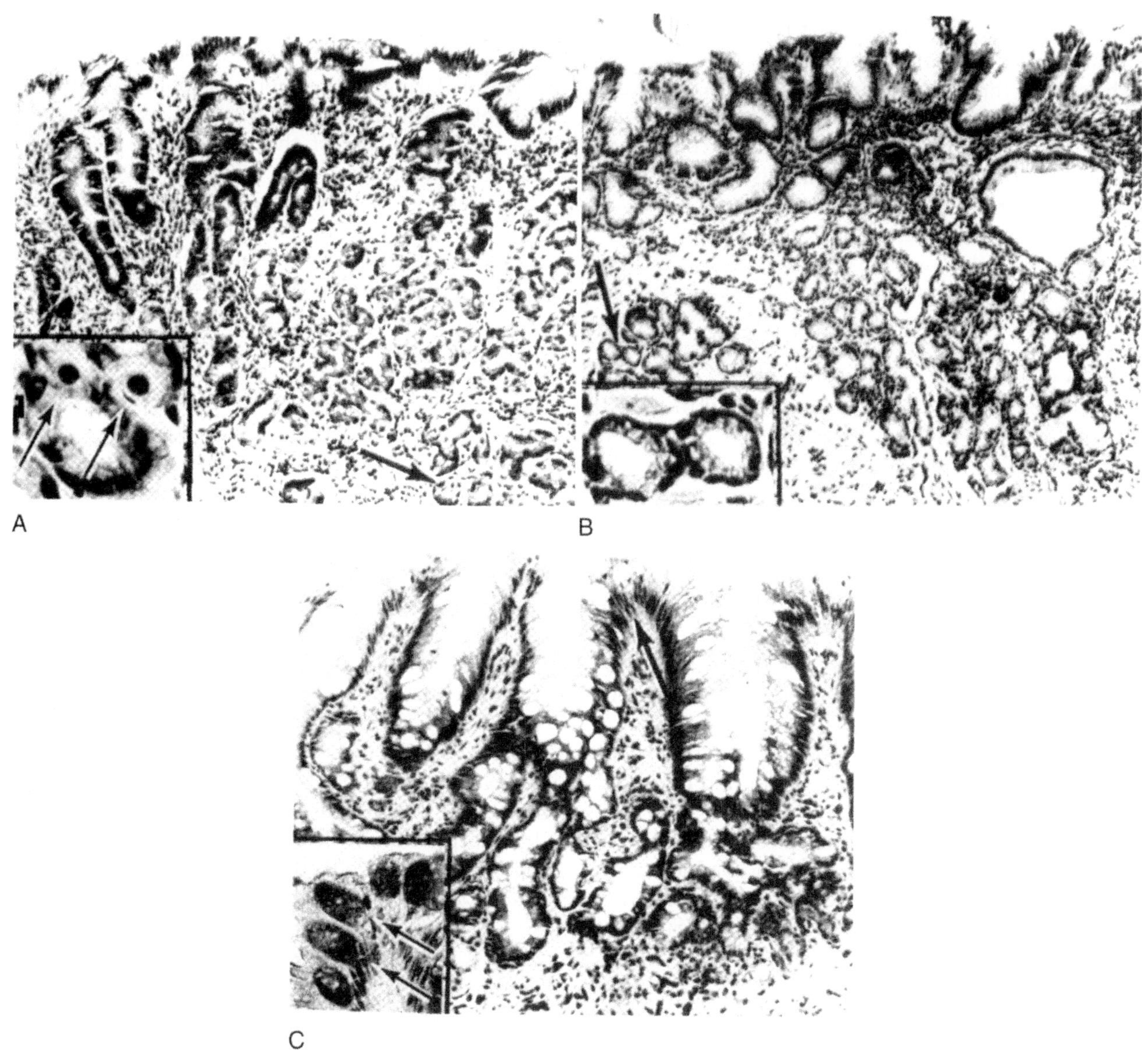

FIGURE 2.16 Paull (1976): Histologic types of epithelia in columnar-lined esophagus. The definitions are perfect. (a) Fundic-type epithelium where the glands contain a mixture of parietal cells (shown in inset) and mucous cells. (b) Junctional-type epithelium where the glands are composed only of mucous cells (inset). (c) Specialized-type characterized by the presence of goblet cells.

TABLE 2.3 Criteria for Classifying Esophageal Columnar Epithelium (25)

	Features of surface layer			Features of glandular layer		
Classification	**Villiform folds**	**Foveoli**	**Goblet cells**	**Mucous cells**	**Parietal cells**	**Chief cells**
Gastric-fundic type	–	+	–	+	+	+
Junctional-type	–	+	–	+	–	–
Specialized-type	+	–	+	+	–	–

endoscopy in the diagnosis of true Barrett esophagus in terms of its malignant potential.

The change that occurred in the definition of Barrett esophagus was illogical. Gastroesophageal reflux was known to be the cause of Barrett's original columnar-lined esophagus, irrespective of its histologic epithelial type. The reason for restricting the definition of Barrett esophagus to intestinal metaplasia was the recognition that this was the only epithelial type that had malignant potential. It had nothing to do with the etiology of columnar-lined esophagus being gastroesophageal reflux. The logical outcome of the new evidence that showed intestinal metaplasia to be premalignant should have been to separate out the intestinal metaplastic type as a premalignant epithelium within the context of Barrett esophagus, which should have retained its definition as columnar-lined esophagus. A logical solution would have been to call the intestinal type Haggitt's variant of Barrett's esophagus. This would have led to the retention of the recognition that columnar-lined esophagus of all types was caused by reflux and that the histologically defined intestinal subtype was the only one that was premalignant. Instead, columnar-lined esophagus without intestinal metaplasia simply disappeared (it became "not-Barrett esophagus" and therefore ignored) when the definition of Barrett esophagus required intestinal metaplasia.

In 1994, Spechler et al. at Harvard broke with the tradition of not biopsying patients who had less than 2 cm of columnar-lined esophagus and therefore were regarded as being endoscopically normal (29). Spechler et al. (29) clearly defined what they called the gastroesophageal junction in their patients. They used the point of flaring of the tubular esophagus unless the patient had a sliding hiatal hernia, when they used the proximal limit of the rugal folds to define the gastroesophageal junction. They reported that 18% of these "endoscopically normal" patients had specialized epithelium with goblet cells in biopsies taken immediately distal to the squamocolumnar junction. After a period of controversy as to the significance of this finding, it was recognized that this entity was also at risk for adenocarcinoma. This resulted in the definition of "short-segment Barrett esophagus," which was the occurrence of specialized intestinal epithelium in patients considered to be endoscopically normal at the time—that is, less than 2 cm of columnar-lined esophagus.

By the mid-1990s, short-segment Barrett esophagus was regarded as a premalignant condition and it became necessary to redefine normal endoscopic appearance to provide the endoscopist guidance as to who should be biopsied. For this reason, normal endoscopy was redefined as the presence of squamous epithelium all the way down to the end of the tubular esophagus—that is, to the point of the proximal limit of the gastric rugal folds (24) (Table 2.4). A normal appearance at endoscopy became the coincidence of the Z line and the gastroesophageal junction defined by the proximal limit of the rugal folds. Abnormal endoscopy was defined as the presence of any abnormal columnar epithelium between the Z line and the proximal limit of the rugal folds. This included

TABLE 2.4 Presently Accepted Endoscopic and Biopsy Definitions of Normal and Pathologic States

	Definition
Gastroesophageal junction	The proximal limit of the gastric rugal folds
Normal endoscopy	The coincidence of the squamo-columnar junction (Z line) and the gastroesophageal junction.
Abnormal endoscopy	Presence of columnar lining, either as tongues of mucosa extending into the squamous epithelium or a circumferential columnar-lined region in the tubular esophagus between the squamocolumnar junction and the proximal limit of the rugal folds
Reflux esophagitis	No standard definition; classical symptoms; erosions; histologic changes
Barrett esophagus	Presence of intestinal metaplasia in a biopsy taken from an endoscopically visualized columnar-lined esophagus
Long segment Barrett esophagus	Presence of intestinal metaplasia in a biopsy taken from an endoscopically visualized columnar-lined esophagus >3 cm in length
Short segment Barrett esophagus	Presence of intestinal metaplasia in a biopsy taken from an endoscopically visualized columnar-lined esophagus <3 cm in length
Carditis	Presence of inflammation in a mucosal biopsy taken distal to the endoscopic gastroesophageal junction
Intestinal metaplasia of the gastric cardia	Presence of intestinal metaplasia in a mucosal biopsy taken distal to the endoscopic gastroesophageal junction

circumferential columnar lining or tongues of columnar mucosa extending into the squamous epithelium causing an irregular Z line. Endoscopists were directed to biopsy these patients to find the intestinal metaplasia that would establish the diagnosis of short-segment Barrett esophagus. This led to the presently accepted definition of Barrett esophagus as the presence of specialized epithelium with goblet cells in a biopsy taken from an endoscopically visualized abnormal columnar epithelium of any length between the proximal limit of the rugal folds and the Z line (24).

One situation that is worthy of discussion is what happens when the endoscopist sees a columnar-lined esophagus and the biopsies do not show intestinal metaplasia. We are not sure what conclusion endoscopists reach or whether they simply ignore this situation without reaching a conclusion. The absence of intestinal metaplasia is not uncommon with short lengths of columnar-lined mucosa (30). The majority of endoscopists will classify these patients as "not-Barrett esophagus" or "normal" despite the fact that the taking of the biopsy was precipitated by the recognition of an endoscopic abnormality. These patients cannot be called normal if there was an endoscopic abnormality, and "not-Barrett esophagus" does not define the abnormality.

The mistake made here is identical to Allison's original mistake in 1953. When Allison saw an abnormality that he recognized was a columnar-lined esophagus and biopsied it, he saw cardiac mucosa and described it in abnormal terms with a villiform reactive surface and inflammation. Without coming to the obvious conclusion that this was abnormal epithelium, Allison concluded that it was normal because histologists of the time had proclaimed that cardiac mucosa lined the proximal stomach. The histologists were in error because they were using the incorrect criterion of the squamocolumnar junction to define the gastroesophageal junction. Today's endoscopists see a columnar-lined esophagus and correctly recognize it as being abnormal, appropriately biopsy it, but then call it "normal" when the biopsy does not show intestinal metaplasia. Not only do they call it normal but they often conclude that the biopsy is gastric because their pathologist reports that their *esophageal* biopsies are lined with cardiac mucosa, which, according to the dogma in the pathology textbooks, is the epithelium that normally lines the proximal stomach. History repeats itself; the incorrect histologic dogma is as true in 2005 as it was in 1953 and leads to the same incorrect conclusion.

It is dangerous to believe pathologists who are among the most super-specialized physicians and usually far removed from patients. We should recognize the obvious: (a) a biopsy taken from above the gastroesophageal junction is esophageal; (b) if this biopsy shows columnar epithelium, it is diagnostic of columnar-lined esophagus, which is caused by reflux; (c) three types of columnar epithelium occur in columnar-lined esophagus: junctional (= cardiac), fundic (= oxynto-cardiac), and specialized (= intestinal); (d) if the biopsy shows intestinal metaplasia, the patient has Barrett esophagus by today's definition, which has malignant potential. *Recognize the obvious: the esophagus is lined by esophageal epithelium. If you biopsy the esophagus and see cardiac mucosa, it is columnar-lined esophagus, not "normal gastric mucosa."* Be like Barrett (columnar-lined esophagus, 1957), not like Allison (esophagus lined by gastric mucous membrane, 1953).

THE STATE OF THE ART AND TODAY'S PROBLEMS

At present, two entities are recognized in the minds of most physicians treating these esophageal diseases: gastroesophageal reflux disease and Barrett esophagus. Gastroesophageal reflux disease is a medical problem causing symptoms and erosive disease that is easily treatable with acid suppression in most patients. Barrett esophagus is a premalignant disease that needs to be identified and patients suffering from it placed under surveillance for early detection of cancer. Despite the fact that it is clearly recognized that Barrett esophagus is caused by gastroesophageal reflux, most physicians think of these as separate entities and do not really understand or contemplate the nature of the link that must exist between the two.

Most physicians look at gastroesophageal reflux disease as a squamous epithelial disease. There is no accepted standard definition for reflux disease. It is recognized by classical symptoms of heartburn and regurgitation (Table 2.4). At endoscopy, the diagnosis of reflux disease focuses on the recognition of squamous epithelial injury, classifying reflux esophagitis as erosive and nonerosive esophagitis. Endoscopists assess the severity of reflux disease by endoscopic abnormalities in the squamous epithelium in the Savary-Miller and Los Angeles grading systems. They assess the effect of treatment of reflux disease by reversal of the changes in the squamous epithelium, mainly by the healing of erosions.

Universally accepted histopathologic criteria for the diagnosis of reflux disease are limited to changes in the squamous epithelium such as basal cell hyperplasia, papillary elongation, and the presence of intraepithelial eosinophils. All of these criteria can be seen in esophageal diseases other than reflux; the criteria are not specific. Biopsies taken from patients with clinical reflux disease often do not show any squamous epithelial abnormalities; the sensitivity of pathologic diagnosis by present criteria in the squamous epithelium is so low that biopsy is not considered clinically useful or necessary for the diagnosis of reflux disease. Reflux disease is managed without any help from microscopic pathology; many patients do not have any biopsies.

When endoscopists see a columnar-lined esophagus, they are only intent on making the diagnosis of Barrett esophagus, dysplasia, and carcinoma. They obtain biopsies from the columnar-lined esophagus and make the diagnosis of Barrett esophagus when the pathologist reports the presence of intestinal metaplasia (goblet cells). These patients are placed under surveillance with random biopsies at defined time intervals to detect dysplasia and early adenocarcinoma. Treatment of these patients is not really directed at the Barrett esophagus; it is largely directed at the associated reflux disease and aimed at controlling symptoms and erosions in the squamous epithelium if these are present. There is little or no evidence that acid suppression with drugs impacts the natural history of Barrett esophagus and many gastroenterologists become incomprehensible in their logic as they treat patients with asymptomatic Barrett esophagus with acid suppressive drugs.

Endoscopists are presently instructed not to biopsy endoscopically normal patients. However, when endoscopists defy this instruction and biopsy patients who are endoscopically normal, it is not uncommon to see specialized epithelium with goblet cells identical to that seen in Barrett esophagus immediately distal to the squamous epithelium, with the incidence in some series being as high as 15%. There is much controversy regarding this situation, which is not dissimilar to the controversy that arose when short-segment Barrett esophagus was first reported (Table 2.4). Some authorities consider this to be "ultrashort segment Barrett esophagus," while others call this "intestinal metaplasia of the cardia." The former term would suggest that the patient is at some risk for cancer and needs to be kept under surveillance; the latter would be less ominous because intestinal metaplasia of the cardia is not an indication for surveillance at the present time. Most gastroenterologists that we encounter are sufficiently uncomfortable about this entity that they place the patient under some kind of surveillance because it is the safer option.

On the surface, it would seem that some things at least are consistent and universal. Most authorities would agree that the definition of Barrett esophagus today absolutely requires the presence of intestinal metaplasia (= specialized epithelium) demonstrated histologically (31). It is therefore surprising to find the following statement regarding the criteria for diagnosis of Barrett esophagus in the "Methods" section of the highly influential paper in the *New England Journal of Medicine* by Lagergren et al. in 1999 (32): "Barrett's esophagus was defined as columnar-cell metaplasia of the specialized type, with goblet cells and villiform surface configuration of the mucosa resembling the features of the intestines. *Five patients with junctional or fundus metaplasia recorded more than 3 cm proximal to the gastroesophageal junction were also classified as having Barrett's esophagus*" (p. 826). By definition, junctional and fundus metaplasia do not have goblet cells, and Lagergren et al. were including these within the definition of Barrett esophagus. If such an influential paper in such an influential journal is permitted to classify Barrett esophagus without the presence of intestinal metaplasia, how can we fault the average physician at large for being confused about this entity.

SOLUTIONS TO THE PROBLEM AND WHAT WE HOPE TO SHOW

The confusion that exists in the understanding of reflux disease is a problem relating to the failure of precise definition. The solution to the confusion is simple. Two organs are involved, the esophagus and the stomach. Intestinal metaplasia occurs in both organs but it is caused by completely different mechanisms and etiologies. Barrett esophagus is intestinal metaplasia of the esophagus and is caused by gastroesophageal reflux. Gastric intestinal metaplasia occurs primarily in chronic atrophic gastritis and is caused mainly by *Helicobacter pylori* infection and autoimmune gastritis. For enlightenment to happen, researchers must be able to recognize the esophagus and stomach accurately, simply, and reproducibly. *We will show that histologic study is the only method of accurately defining the esophagus and stomach, and that this method is simple, accurate, and reproducible.*

This book is really a description of more than a decade's experience with examination of between 600 and 1000 patients per year. This has evolved into a classical scientific experiment in the true sense of the word. It is characterized by the following features.

Availability of Appropriate Material

Patients treated by the University of Southern California (USC) Foregut Surgery group routinely have biopsies performed in a standardized manner that permits complete evaluation of the esophagus, junctional region, and stomach (Fig. 2.17). Endoscopy is performed by the members of the surgical team. Biopsies are taken whether or not the patient has an endoscopic abnormality. This type of material is rare in the United States because in most centers it is the gastroenterologist who performs the biopsies and because of the existing American Gastroenterological Association recommendation to only biopsy endoscopically abnormal patients. Many patients also have a complete clinical examination, video-esophagography, 24-hour pH studies, and manometry, providing an optimal method of clinicopathologic correlation.

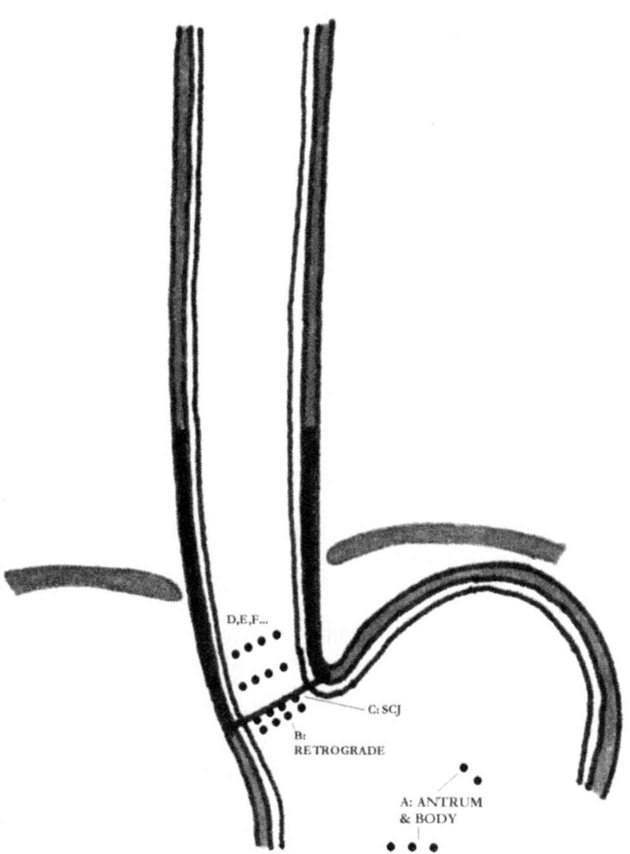

FIGURE 2.17 The standard biopsy protocol. Five biopsy specimens are taken from the distal stomach (A: antrum and body); 2 to 4 biopsies from the proximal limit of the rugal folds with the endoscope in the retrograde position (B: retrograde); 3 to 4 biopsies are taken from the squamocolumnar junction (C: SCJ), attempting to straddle the junction. In a patient who is endoscopically normal, the squamocolumnar junction (biopsy C) is immediately adjacent to the retrograde biopsy (B), and these three biopsies complete the examination. In a patient with a visible columnar-lined esophagus, measured biopsies are taken from this segment at 1 to 2 cm intervals (D, E, F, . . .) between the retrograde biopsy and the biopsy from the proximally displaced SCJ whose designation changes. Measurement is made from the incisor teeth.

Observation

Examination of thousands of these patients sampled in a repetitive manner led to observations regarding the frequent absence of cardiac mucosa at the junctional region. This unexpected observation refuted the existing literature. We also observed that cardiac mucosa, when it occurred, was found mainly above the gastroesophageal junction but was also found in areas that were thought to be hiatal hernias and proximal stomach. In some biopsies with cardiac mucosa, esophageal location was apparent because of the presence of esophageal gland ducts.

Frequent meetings between pathologist, the clinical surgical team, and research members ensure that there is complete understanding and correlation of clinical, pathological, and basic research findings. These meetings are a source of discussion and critical evaluation of new ideas and research.

Hypothesis

The observations relating to cardiac mucosa led to a radical new hypothesis that was very different from accepted dogma. According to our hypothesis, the esophagus was lined entirely by squamous epithelium and the stomach entirely by oxyntic gastric mucosa, which contained parietal cells. Cardiac mucosa was not normal; it was an abnormal epithelium produced from squamous epithelium as a result of reflux-induced damage. Based on this hypothesis, new histologic definitions were developed for esophagus, stomach, gastroesophageal junction, and reflux disease.

Experiments

We tested this hypothesis by numerous experiments that will be described. Every experiment that we have performed using strict methodology has supported this hypothesis. More important, the new definitions provided us with a practical method of using histology to manage patients by predicting accurately the severity of reflux-induced cellular damage and risk of adenocarcinoma. We have used this evidence-based system successfully in patient care since the 1990s with considerable success.

Duplication by Other Observers

Our hypothesis and experimental data were initially received with skepticism by the gastroenterology world in general but generated sufficient interest for other researchers to attempt to duplicate or refute the hypothesis. This is the highest compliment that can be paid to a new hypothesis. Over the years, despite strong resistance to it, the concept has continued to slowly grow. Every attempt at duplicating our findings has shown that there is little doubt that our hypothesis is true. Every attempt at refuting our concept has led to conclusions that are invalid either because the data of the study have been misinterpreted or the methodology of the study has been flawed. We will present an analysis of the more significant literature demonstrating this fact. At present, the hypothesis is rejected largely by people with influence in the field that use their reputation rather than fact and evidence to disprove it. As time passes and new voices emerge, the hypothesis will gain in acceptance. We predict that this hypothesis will form the basis whereby reflux disease will be understood and managed in the future.

The concept of the normal amount of cardiac mucosa that is present in humans has shrunk during this period, always tending toward our claim that it is zero. From the 2 to 3 cm or more of cardiac mucosa that was accepted as being normally present in the gastroesophageal junctional region before the hypothesis was put forward, it is now believed that the maximum extent of "native" cardiac mucosa is less than 4 mm (33). "Less than 4 mm" without a definition for a low size limit would mean that this normal organ can be absent, because zero is less than 4 mm; the situation replicates Hayward's flawed device in 1961 (17) when he stated that the esophagus was normally lined by less than 2 cm of cardiac mucosa. The dogma of normal cardiac mucosa is so strong that many authorities cling to this less than 4 mm of "normal" cardiac mucosa almost like Custer's last stand. It now takes only the reversal of this "less than 4 mm belief" for the hypothesis to be accepted completely. Custer's army still has many men standing, but their ammunition is running out. Zero to 4 mm equals about 0 to 100 epithelial cells and 0 to 30 foveolar-gland complexes. Can we reasonably hold the belief that this defines a normal structure?

The accurate and reproducible definitions of the esophagus, gastroesophageal junction, and stomach are prerequisites to understanding and treating patients with reflux disease. Most gastroenterologists and pathologists still adhere to the unproven endoscopic definition of the gastroesophageal junction to designate a given biopsy as esophagus or stomach; we will show that significant error is associated with this tendency. We will show that an accurate and reproducible definition is only possible when precise histologic criteria are used. *Histology will solve this problem, just as it has resolved the controversies of gastritis, colitis, hepatitis, and glomerulonephritis, all of which progressed from confusing clinical entities to precisely defined histologic entities where etiology and treatment became intelligible only when histologic criteria were developed and followed.*

References

1. Allison PR, Johnstone AS. The oesophagus lined with gastric mucous membrane. Thorax 1953;8:87–101.
2. Barrett NR. Chronic peptic ulcer of the oesophagus and "oesophagitis." Br J Surg 1950;38:175–182.
3. Frindlay L, Kelly AB. Congenital shortening of the esophagus and the thoracic stomach resulting therefrom. Proc R Soc Med 1931;24:1561–1578.
4. Barrett NR. The lower esophagus lined by columnar epithelium. Surgery 1957;41:881–894.
5. Spechler SJ. The columnar-lined esophagus. History, terminology, and clinical issues. Gastroenterol Clin North Am 1997;26:455–466.
6. Tileston W. Peptic ulcer of the esophagus. Am J Med Sci 1906;132:240–265.
7. Stewart MJ, Hartfall SJ. Ulcer of the esophagus. J Path Bact 1929;32:9–14.
8. Lyall A. Chronic peptic ulcer of the esophagus: A report of eight cases. Br J Surg 1937;24:534–547.
9. Chamberlin DT. Peptic ulcer of the esophagus. Am J Dig Dis 1939;5:725–730.
10. Bosher LH, Taylor FH. Heterotopic gastric mucosa in the esophagus with ulceration and stricture formation. J Thorac Surg 1951;21:306–312.
11. Friedenwald J, Feldman M, Zinn WF. Peptic ulcer of the esophagus. Am J Med Sci 1929;177:1–14.
12. Jackson C. Peptic ulcer of the esophagus. JAMA 1929;92: 369–372.
13. Allison PR, Johnstone AS, Royce GB. Short esophagus with simple peptic ulceration. J Thorac Surg 1943;12:432–457.
14. Allison PR. Peptic ulcer of the esophagus. J Thorac Surg 1946;15:308–317.
15. Allison PR. Peptic ulcer of the esophagus. Thorax 1948;3: 20–42.
16. Johns BAE. Developmental changes in the oesophageal epithelium in man. J Anat 1952;86:431–442.
17. Hayward J. The lower end of the oesophagus. Thorax 1961;16:36–41.
18. Bremner CG, Lynch VP, Ellis FH. Barrett's esophagus: congenital or acquired? An experimental study of esophageal mucosal regeneration in the dog. Surgery 1970;68:209–216.
19. Association of Directors of Anatomic and Surgical Pathology (ADASP). Recommendations for reporting of resected esophageal adenocarcinomas. Am J Surg Pathology 2000;31: 1188–1190.
20. Siewert JR, Stein HJ. Classification of adenocarcinoma of the oesophagogastric junction. Br J Surg 1998;85:1457–1459.
21. Chandrasoma PT, Der R, Ma Y, et al. Histology of the gastroesophageal junction: An autopsy study. Am J Surg Pathol 2000;24:402–409.
22. Torek F. The first successful case of resection of the esophagus for carcinoma. Surg Gynecol Obstet 1913;16:614–617.
23. Santy P, Ballivert MD, Berard M. Thoracic esophagectomy for cancer. J Thorac Surg 1943;12:397–431.
24. Sampliner RE. Practice guidelines on the diagnosis, surveillance, and therapy of Barrett's esophagus. Am J Gastroenterol 1998;93:1028–1031.
25. Paull A, Trier JS, Dalton MD, Camp RC, Loeb P, Goyal RK. The histologic spectrum of Barrett's esophagus. N Engl J Med 1976;295:476–480.

26. Chandrasoma PT, Lokuhetty DM, DeMeester, TR, et al. Definition of histopathologic changes in gastroesophageal reflux disease. Am J Surg Pathol 2000;24:344–351.
27. Haggitt RC, Tryzelaar J, Ellis FH, et al. Adenocarcinoma complicating columnar epithelium-lined (Barrett's) esophagus. Am J Clin Pathol 1978;70:1–5.
28. Reid BJ, Weinstein WM. Barrett's and adenocarcinoma. Annu Rev Med 1987;38:477–492.
29. Spechler SJ, Zeroogian JM, Antonioli DA, Wang HH, Goyal RK. Prevalence of metaplasia at the gastroesophageal junction. Lancet 1994;344:1533–1536.
30. Chandrasoma PT, Der R, Ma Y, et al. Histologic classification of patients based on mapping biopsies of the gastroesophageal junction. Am J Surg Pathol 2003;27:929–936.
31. Weinstein WM, Ippoliti AF. The diagnosis of Barrett's esophagus: Goblets, goblets, goblets. Gastrointest Endosc 1996;44:91–95.
32. Lagergren J, Bergstrom R, Lindgren A, Nyren O. Symptomatic gastroesophageal reflux as a risk factor for esophageal adenocarcinoma. N Engl J Med 1999;340:825–831.
33. Odze RD. Unraveling the mystery of the gastroesophageal junction: A pathologist's perspective. Am J Gastroenterol 2005; 100:1853–1867.

CHAPTER

3

Fetal Development of the Esophagus and Stomach

The esophageal and gastric epithelia develop from endodermal progenitor cells in the fetal foregut. These endodermal epithelial progenitor cells proliferate and differentiate in a well-regulated manner during fetal life, where external influences are limited. In association with mesodermal progenitor cells, they form the entire gastrointestinal tract. Development of the gastric epithelium is completed very early in fetal life and that in much of the esophagus is completed in the third trimester. The gastroesophageal junctional region is the last to develop with completion at or soon after birth. The junction matures in step with the lower esophageal sphincter, which often completes its development in early neonatal life.

The early endodermal progenitor cell of the embryonic disk must have the potential to ultimately give rise to all the epithelial types seen in the adult gastrointestinal tract. Fetal development at any given point in the gastrointestinal tract can therefore be defined genetically as the programmatic transformation at a specific site of the fetal progenitor cell to the adult progenitor cell (Fig. 3.1). The transformation of the fetal epithelium to adult epithelium in the stomach and intestines is simple with the adult type epithelia developing in early fetal life without intervening steps. In contrast, esophageal development is highly complex with expression of several different types of fetal columnar epithelia in the fetal esophagus before the adult squamous epithelium forms (Fig. 3.1).

The genetic signals that direct epithelial differentiation in different regions are in the process of being elucidated. In gastric epithelium, for example, the adult progenitor cell is directed by a gastric-type genetic signal to produce surface columnar mucous cells typical of the stomach as well as glandular cells that differentiate into parietal and chief cells. The adult progenitor cell in the esophagus in its fully developed state is directed by a different esophageal-type genetic signal to produce squamous cells. In the adult gastric progenitor cell, the genetic signal that causes the endodermal cell to differentiate as squamous cells is suppressed; in the esophageal progenitor cell, the signal to produce columnar cells and parietal cells is suppressed. Similarly in the intestine and colon, other genetic signals are activated to drive differentiation in the appropriate manner.

While many of the early genetic signaling mechanisms remain unknown, evidence for expression of genes in normal adult epithelia provides evidence for activation of specific genes in different sites. In the adult stomach, for example, parietal cell differentiation is associated with the expression of the Sonic Hedgehog gene (1). This gene, which is expressed in the stomach, is minimally expressed in other sites in the gastrointestinal tract except Meckel's diverticula, which contain parietal cells in ectopic gastric mucosa. In the intestine and colon, the homeobox genes Cdx-1 and Cdx-2 are expressed and believed to drive differentiation into intestinal and colonic epithelial types (2). These genes are normally not expressed in the esophagus and stomach but become activated when intestinal metaplasia occurs as a pathologic event in these locations (3–6).

It is likely that suppression of genes is not permanent in any cell of the body. The likelihood of reactivation of a suppressed gene appears to be related to the point in time during fetal development that genetic suppression occurs. For example, it is far more likely

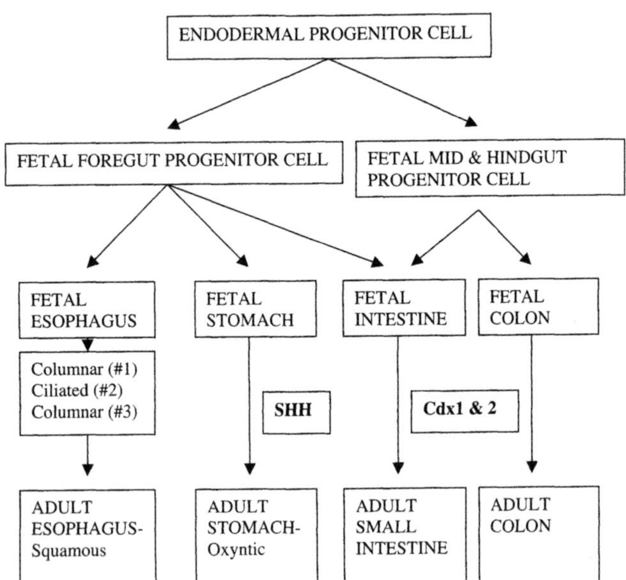

FIGURE 3.1 The development of the GI tract from the endodermal plate. The cells of the endodermal plate differentiate into foregut, midgut, and hindgut under the influence of early genetic signals. The foregut differentiates further into esophageal and gastric fetal stem cells. The fetal esophagus passes through several different types of columnar epithelia before the adult squamous genetic signal becomes established, usually in early infancy. The adult gastric signal (possibly activation of the Sonic Hedgehog gene) is established in the 17th week of gestation and results in oxyntic mucosa with parietal cells. The differentiation of adult intestinal and colonic epithelia are believed to be directed by the Cdx-1 and Cdx-2 gene complex.

for the progenitor cell in the adult esophageal epithelium to reactivate a suppressed endodermal genetic signal than an ectodermal signal. This means that the esophagus is far more likely to activate a signal that will cause it to differentiate into other gastrointestinal epithelial types than it is to activate a signal that will cause it to produce neurons, glial cells, and skin adnexal structures such as hairs. This principle probably applies to the likelihood of activating suppressed endodermal signals as well. If this is true, when the esophagus undergoes columnar metaplasia, it is more likely to reactivate late fetal esophageal columnar epithelial signals than either early esophageal columnar signals or gastric or intestinal signals.

The adult steady state that is recognized as histologic normalcy actually represents a well-ordered pattern of proliferation and differentiation of the adult esophageal and gastric progenitor cells throughout life that is designed to continually renew these epithelia. In pathologic states, they are capable of shifting their differentiation, by reactivation of suppressed genes, to produce a variety of metaplastic epithelia that display the full gamut of the fetal endodermal progenitor cells. While many of these exact genetic changes have not been elucidated, comparing the observed morphologic changes that occur during metaplasia to changes that occur during fetal development can help to explain the metaplastic process.

THE STUDY OF EMBRYOLOGY OF THE FOREGUT

It is intuitive to believe that the question of whether a structure is normal after birth is best answered by whether that structure is present in fetal life. According to this argument, if the structure is present in the fetus, then it must be a normal structure that is present in the adult. In 1961, Hayward (7), discussing Allison's and Barrett's (8, 9) contention that columnar-lined esophagus was a congenital anomaly, stated:

> I do not believe that any convincing evidence has been advanced to show that any of the type of case described with more junctional epithelium than normal in the oesophagus is congenital. *The only evidence that would convince me that the condition may sometimes be congenital would be the occasional finding of junctional epithelium extending an unusual distance up the oesophagus in stillborn babies.* So far as I am aware, this observation has not been made. . . . It is a wise principle in pathology never to regard anything as congenital until there is definite evidence that it was present at birth. (pp. 39–40)

When we initially suggested that cardiac mucosa was an acquired and not a "normal," "congenital," or "native" epithelial type in the gastroesophageal junctional region, one of the most common questions we were asked was whether cardiac mucosa was present during fetal life and at birth. We therefore struggled to find fetal specimens where we could test this question. It was quickly apparent that in early fetal life, the entire esophagus was lined by columnar epithelium. This fetal columnar epithelium was composed only of mucous cells and satisfied the definitional criteria of cardiac mucosa. Was this, then, proof that cardiac mucosa was a congenital structure?

The fetus begins as a zygote that passes through various fetal tissues during embryologic development before finally differentiating into adult epithelial types. Embryologic development in many organs is not complete until after birth. Tissues found in the fetus and early infant life are of two types: (a) developing fetal type tissues that are transforming into adult tissues (these disappear when fetal development is complete) and (b) differentiated adult-type tissues that arise from fetal tissues before birth. The main characteristic that distinguishes between fetal and adult types of tissue is that fetal-type tissues progressively decrease as the fetus matures, whereas adult-type tissues continue to increase as the fetus grows. The critical test to decide whether a given tissue in the fetus is fetal or adult in type is that when fetuses of increasing age are examined, fetal tissue decreases in quantity and adult tissue increases. Care has to be taken here; there is an individual variation in the rate at which fetuses develop, which makes it dangerous to come to conclusions regarding sizes with only a small number of fetuses.

The fetal kidney illustrates well the difference between fetal and adult types of tissue. The kidney has a layer of undifferentiated nephroblastic cells in the cortical region during fetal life and, in many babies, in the first year after birth. Nephroblast is a fetal-type tissue; it progressively differentiates into adult renal tubules during fetal life and infancy and is not seen in the adult kidney. Though congenital in terms of being present at birth, nephroblast is not a normal adult-type tissue; it is a fetal tissue that disappears during infancy. In contrast, renal tubules, which appear in fetal life, are an adult-type tissue. As the fetus and then the baby grows, the renal tubules grow exponentially to form the adult kidney. Depending on what stage in fetal life the kidney is examined, one will see varying amounts of fetal nephroblast and adult renal tubules. Sequential examination is necessary to demonstrate the progressive disappearance of nephroblast and the progressive increase in adult renal tubules. If one uses the same reasoning for columnar epithelium in the esophagus, the fact that a given epithelial type is present during fetal life or at birth does not prove that it is an adult type of epithelium. If sequential examinations show a decrease in the amount of that columnar epithelium in the esophagus as the fetus grows, it proves it is a fetal-type epithelium.

Hayward was correct when he said that junctional (= cardiac) epithelium had never been described in the midesophagus in stillborn babies (7). But I am not certain whether he was correct because he was unaware of the embryologic data that existed or whether he showed great acumen; evidence suggests the latter. In 1952, Johns published an exhaustive study of esophageal epithelium in embryos ranging in size from 3 mm to 230 mm crown-rump length and showed that the esophagus in early and mid fetal life is lined by columnar epithelium composed of ciliated and nonciliated mucous cells (10). Johns's findings are quoted extensively in Barrett's 1957 landmark paper, which is referenced by Hayward (7, 9). It therefore suggests that Hayward knew that the fetal esophagus was lined by columnar epithelium in much of its extent but correctly regarded this as a fetal columnar epithelium that was different to the junctional (= cardiac) epithelium he saw in his adult patients.

Embryologic study of foregut epithelium is very difficult. In the United States, ethical considerations prevent research on aborted embryos. Stillborn babies are subjected to autopsy, but they almost always show extreme maceration with gastroesophageal epithelia being unrecognizable in most cases. Except for a few stillborn autopsies that showed poor preservation, I have not yet been able to collect adequate material for study of fetal esophagi. A few autopsy studies have been published based on the evaluation of stillborn fetuses and premature infants dying at a variable time after birth. Some of these show excellent preservation of the epithelium and provide valuable information. We have had the benefit of reviewing the histologic material of one of these studies by Derdoy et al. (11) at the Children's Hospital of Los Angeles and some of the excellent material of Dr. Liebermann-Meffert of Freiberg, Germany.

EARLY DEVELOPMENT OF THE GASTROINTESTINAL TRACT

The gastrointestinal tract is derived from two germ layers: the endoderm and mesoderm. The endoderm is first recognizable on about the 8th day after conception as an epithelial cell layer lining the yolk sac. The mesoderm first appears on the 15th day as a layer of cells that forms between the ectodermal and endodermal cell layers in the embryonic disk. During the 3rd week, the endodermal lined yolk sac is an ovoid sac on the ventral aspect of the flat embryonic disk; the roof of the yolk sac consists of the endodermal component of the embryonic disk (Fig. 3.2).

When the embryonic disk develops its head and lateral folds, the rostral part of the yolk sac becomes included within the folded embryo. The compressed rostral part of the yolk sac now forms an endoderm-lined tube in the anterior part of the embryo located dorsal to the pericardial sac and in the midline as the embryo folds over it on all sides (Fig. 3.2). This is the future foregut, which includes the oral cavity, pharynx, esophagus, stomach, and proximal small intestine. The foregut connects with the stomodeum invaginating from the ectoderm to form the opening of the oral cavity when the buccopharyngeal membrane that separates the endodermal foregut from the stomodeum breaks down.

A similar folding takes place in the caudal part of the embryo with the tail and lateral folds compressing the caudal region of the yolk sac into a tube, which forms the hindgut from which the distal colorectum

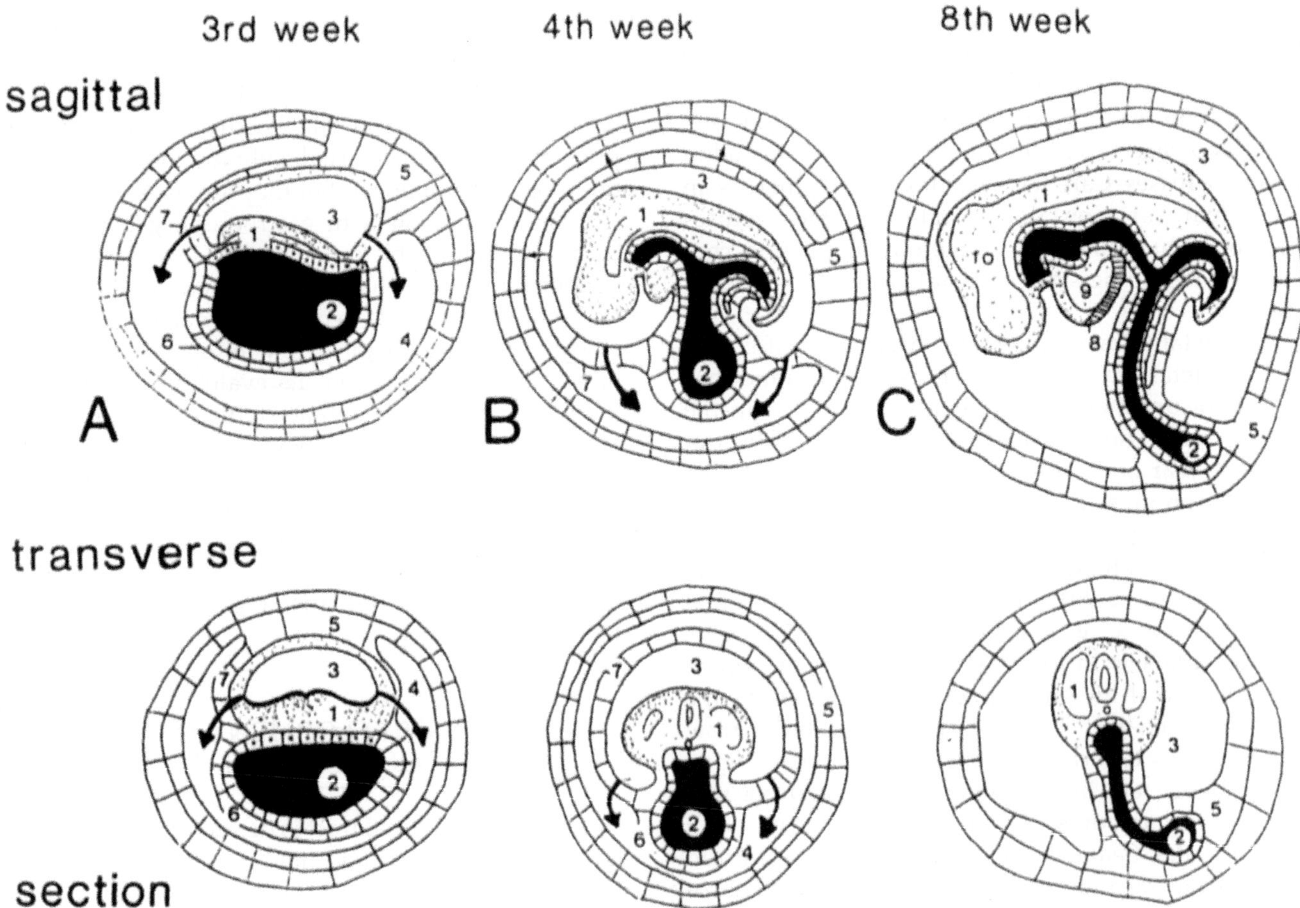

FIGURE 3.2 The early changes in the primitive endodermal tube. In the third week of gestation (a), before the formation of the head fold, the yolk sac is an ovoid cavity. Its roof is the endoderm, which is the underlayer of the embryonic disk. In the 4th week (b), with the formation of the head fold, the rostral part of the yolk sac becomes included within the embryo as an endoderm-lined tube, which extends from the stomodeum (future mouth) to a point distal to the septum transversum (future diaphragm). This is in the midline, dorsal to the developing heart and bounded laterally by the bronchial mesoderm. In the eighth week (c), this tube has elongated substantially with the growth of the heart and flexion of the head. 1 = embryo; 2 = yolk sac; 3 = amniotic cavity; 4 = extraembryonic coelom; 5 = future placenta; 8 = septum transversum; 9 = cardiac tube. Reproduced with permission from Liebermann-Meffert D, Duranceau A, Stein HJ. Anatomy and Embryology, Chapter 1. *In*: Orringer MB, Heitmiller R (eds), The Esophagus, Vol 1, Zudeima GD, Yeo CJ. Shackelford's Surgery of the Alimentary Tract, 5th ed. Saunders, Philadelphia, 2002, pp 3–39.

develops. The continued folding of the embryo progressively narrows the communication between the yolk sac and the developing midgut. In the 4th week, the yolk sac is seen as a long narrow vitelline tube that passes from the midgut into the umbilicus. Ultimately the yolk sac (now the vitelline tube) separates completely from the embryonic gut. By the end of the 4th week, the primitive gut has completely formed from that part of the yolk sac that was enclosed within the embryo. This ultimately forms the entire gastrointestinal tract, including the accessory glands such as pancreas and liver, and the respiratory system, which develops as a lung bud from the ventral aspect of the developing foregut. The extraembryonic yolk sac that becomes separated from the developing gut usually disappears by about the 12th week. In about 2% of the population, the part of the vitelline duct closest to the midgut remains as a Meckel's diverticulum; this is situated 2 to 3 feet proximal to the ileocecal junction. In rare cases, the vitelline duct becomes obliterated and, instead of disappearing, remains as a fibrous cord that connects the apex of a Meckel diverticulum with the umbilicus. It is interesting that the epithelium of a Meckel's diverticulum not infrequently shows differentiation into gastric oxyntic type epithelium and pancreatic epithelium, representing the retained ability of the endodermal progenitor cells in this region to differentiate into typically foregut elements.

EARLY DEVELOPMENT OF THE FOREGUT

In the 3rd to 4th week gestational period, extremely rapid development of the primitive foregut occurs. Initially a tube lined by endoderm and surrounded by embryonic mesoderm, the rostral part of the foregut develops multiple sacculations that give rise to the pharynx and the thyroid and tracheal buds that develop from the ventral aspect of the pharyngeal sacculation. The tracheal bud is visible on the 21st day.

In the 3rd week, the esophagus is very short, extending from the developing tracheal bud to the point where the stomach is starting to develop in the region of the septum transversum, which will form the diaphragm. In the 4th week, the stomach begins to develop a fusiform dilatation on its left side distal to the septum transversum. The left side (greater curvature) of the stomach has a higher rate of cell proliferation than the right side (lesser curvature), causing the left side of the stomach to bulge outward and form the greater curvature (12). The maximal cell proliferation is in the region of the future fundus, which bulges outward and upward toward the septum transversum. In this way, the developing fundus forms the acute angle of His with the esophagus at the gastroesophageal junction.

After the 4th week, the esophagus rapidly elongates. This is stimulated mainly by the rapid development of the head and the unbending of the head region by straightening of the cervical spine, which causes increased separation of the heart and head. This elongation mainly involves the distal half of the esophagus. By the 7th week (20-mm embryo), the esophagus is a muscular tube that extends from the tracheal bud, passing in the midline dorsal to the developing trachea and heart, through the developing septum transversum to the stomach, which can be recognized as a saccular structure bulging to the left and upward from the original tube. At this time, the basic external structure of the esophagus is developed; further development of the organ itself will consist of rapid growth and differentiation keeping pace with that of the fetus as a whole.

The esophageal muscle develops from the mesoderm, becoming visible as a ringlike condensation of cells with elongated nuclei around the endodermal tube in the 10-mm embryo. This becomes the circular muscle layer; the longitudinal muscle develops later. There is no development of a well-defined circular muscle layer in the stomach at this early stage. At the 90-mm stage, the muscular layers of the esophagus and stomach resemble that of the adult. The embryonic mesoderm around the distal esophagus also contributes to the formation of the diaphragmatic crura and the attachments of the esophagus to the diaphragm such as the phrenoesophageal ligament.

EPITHELIAL DEVELOPMENT IN THE FETAL ESOPHAGUS

Embryologic study of the epithelial development in the foregut reached a high level of understanding very early because of the availability of aborted embryos. Though there were many contributors such as Schridde (13), Johnson (14), and Lewis (15), credit for the definitive study of esophageal epithelial development must be given to Johns's well nigh perfect study published in 1952.

Literature Review

Johns BAE. Developmental changes in the oesophageal epithelium in man. J Anat (Lond) 1952;86:431–442.

Johns studied serial transverse sections of the esophagus in human embryos ranging in crown-rump length from 3 mm to 230 mm with embryos of all sizes between these being included. In 14 embryos under 42 mm in crown-rump length, the whole embryo was sectioned transversely at 10 μm; in those of greater length, the esophagus was dissected complete with its attachments to the pharynx above and stomach below and then serially sectioned at 10 μm. He also examined serial transverse sections in four patients ages 2, 11, 47, and 58 years and every 10th section of a transverse series from the esophagus in 20 other subjects whose ages ranged from 20 to 60 years. The study is based on hematoxylin and eosin stained sections. It must be recognized that transverse sections used by Johns limit the recognition of changes in a vertical orientation, which can be important at the gastroesophageal junction. Johns's study does not permit clear understanding of the changes in the gastroesophageal junctional region.

As the appearance of the epithelium is described from a microscopic standpoint, it is important to keep in mind that the embryo is in an extremely rapid growth phase. An epithelium that is of equal thickness at two gestational ages means that it has kept pace with the growth of the fetus. However, an epithelium that is of equal length at two ages means that the epithelium has decreased relative to another epithelial type because it has not kept pace with the elongation of the organ. After development is complete, the epithelium will always increase in absolute length in a manner that is coordinated with the growth of the baby. After growth is complete; the epithelial length remains constant. Extreme care is necessary in interpreting the significance of differences in length and thickness of epithelia in fetal tissues.

Johns outlines the epithelial appearance with increasing size of these sequential embryos as follows:

Between 3 mm (25 days old) and 13 mm (6 weeks). The epithelium lining the esophagus is stratified columnar in type, being in general three cells deep, of uniform thickness, and the oval nuclei are scattered through the entire thickness of the epithelium. This is a nonciliated epithelium that is presumably composed of the undifferentiated fetal endodermal progenitor cells (Fig. 3.3).

13- and 16-mm embryo (6 to 7 weeks old). The process of *vacuolation* begins (Fig. 3.4). These vacuoles appear between the epithelial cells and progressively become larger, sometimes exceeding the size of the lumen (Fig. 3.4). They compress the lumen. The vacuoles attain their largest size in the 25-mm embryo. The vacuoles are always separated from the lumen by epithelial cell membranes. The epithelium apart from the vacuoles is similar to that in the younger embryos. The vacuoles decrease in size after the 25-mm stage but can remain until the 72-mm stage. No explanation or meaning has been ascribed to this vacuolation.

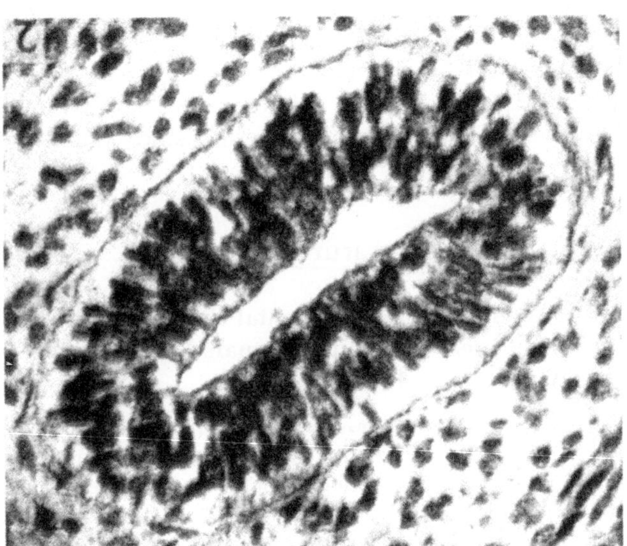

FIGURE 3.3 Transverse section of the middle of the esophagus in a 7-mm embryo showing the primitive stratified columnar epithelium. This epithelium (fetal columnar epithelium type 1) lines the entire esophagus in the first trimester. Reproduced with permission from Johns BAE. Developmental changes in the oesophageal epithelium in man. J Anat 1952;86:431–442.

23- to 62-mm embryo (late first to early second trimester). In the 40-mm embryo, the superficial layer of cells *becomes ciliated.* This process begins in the midesophagus and extends throughout the esophagus (Fig. 3.5). By the 62-mm stage, the stratified ciliated columnar epithelium has replaced the primitive stratified columnar epithelium in the entire esophagus except for a small area at both ends. At the upper and lower end, there are small areas of nonciliated epithelium lined by a single layer of tall columnar cells.

72- and 78-mm embryo (second trimester). Simple columnar epithelium is present in a small area at the upper and lower end of the esophagus. These cells have increased in height and their nuclei now occupy a basal position. In the 72-mm embryo these *tall columnar cell* areas are separated from the ciliated epithelium, which lines the rest of the esophagus, by a narrow zone of nonciliated stratified columnar epithelium (Fig. 3.6). With the disappearance of the vacuoles, the lumen of the esophagus has widened at this stage. At the lower end of the esophagus, the simple tall columnar epithelium is continuous with the simple columnar epithelium of the stomach with no line of demarcation between the two. All the cells of the simple columnar epithelium contain mucin, which stain positively with periodic acid Schiff reagent.

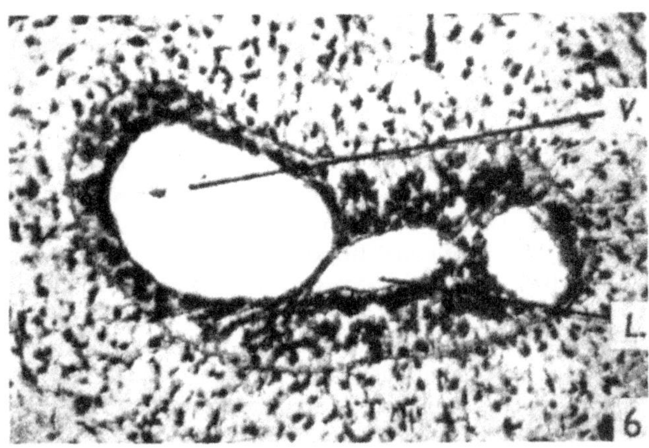

FIGURE 3.4 Transverse section of the upper esophagus in a 25 mm embryo. Two vacuoles (V) are seen with the lumen of the esophagus (L) between them. The epithelial cells are compressed with crowded nuclei. Reproduced with permission from Johns BAE. Developmental changes in the oesophageal epithelium in man. J Anat 1952;86:431–442.

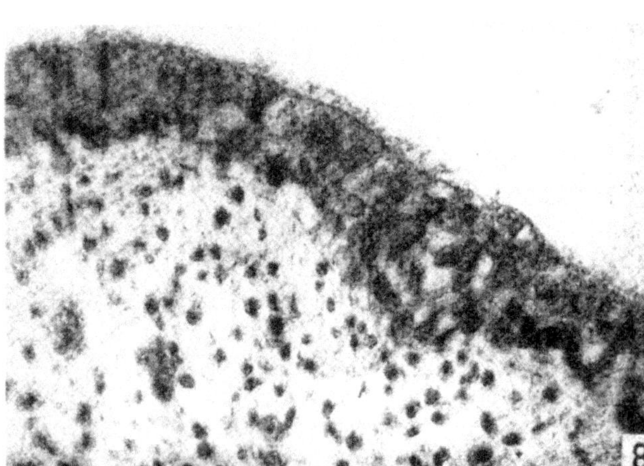

FIGURE 3.5 Middle third of the esophagus of a 42-mm embryo showing the epithelium to be composed of stratified ciliated epithelium. This epithelium (fetal columnar epithelium type 2) lines the entire esophagus in the second trimester. Reproduced with permission from Johns BAE. Developmental changes in the oesophageal epithelium in man. J Anat 1952;86:431–442.

90- to 130-mm embryo. Most of the esophagus is lined with stratified columnar ciliated epithelium. The total area occupied by the nonciliated columnar cells at both ends is smaller than in the 72-mm embryo. At the lower end, the longitudinal columns of epithelium have in part lost their continuity with the epithelium of the stomach, so that small discrete patches remain, and extend for about 1 mm above the cardioesophageal junction. The 130-mm embryo also shows the first appearance of an esophageal gland as a tubular downgrowth of tall columnar epithelium that reaches the muscularis mucosae (Fig. 3.7). In the midesophagus, *stratified squamous epithelium* appears in discrete patches in the ciliated epithelium.

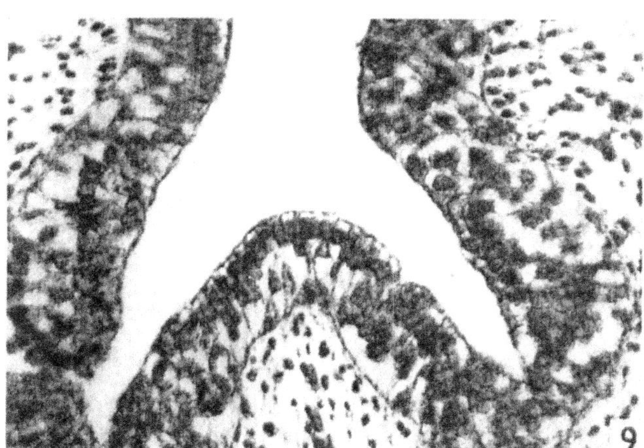

FIGURE 3.6 Epithelium in a 62 mm embryo. To the left and below the epithelium is tall columnar, while to the right it is stratified with occasional ciliated cells. Reproduced with permission from Johns BAE. Developmental changes in the oesophageal epithelium in man. J Anat 1952;86:431–442.

130- to 172-mm embryos (early third trimester). The middle third of the esophagus is increasingly lined by stratified squamous epithelium, which has replaced the stratified columnar ciliated epithelium in this area (Fig. 3.8). At the upper and lower end are areas of simple tall columnar epithelium, which are of lesser extent than in the previous embryo (Fig. 3.8). These cells stain positively for mucin. More glands have formed throughout the esophagus; these are still superficial without penetrating the muscularis mucosae.

210- to 230-mm embryo (third trimester). Stratified squamous epithelium has extended to the lower and upper esophagus, progressively replacing the stratified ciliated columnar epithelium (Fig. 3.9). The ciliated epithelium has now largely disappeared except for small residual areas, particularly in the upper esophagus; these disappear by full term. Simple nonciliated columnar epithelium is present at the lower end, but its extent is not given. Glands composed of mucous cells are more numerous and open into the surface covered by squamous epithelium, but they are still confined to the mucosa (Fig. 3.9). These persist into adult life.

Four full-term and 20 adult specimens. The emphasis of the description in these specimens shifts dramatically to a discussion of the glands of the esophagus, describing how these glands form as tubular outpouchings from squamous epithelium (Fig. 3.10) and extend through the muscularis mucosae during postnatal life to form secretory glands in the submucosa that contain mucous and serous cells (Fig. 3.11). Unfortunately, there is no further description of the esophageal epithelium in these postnatal

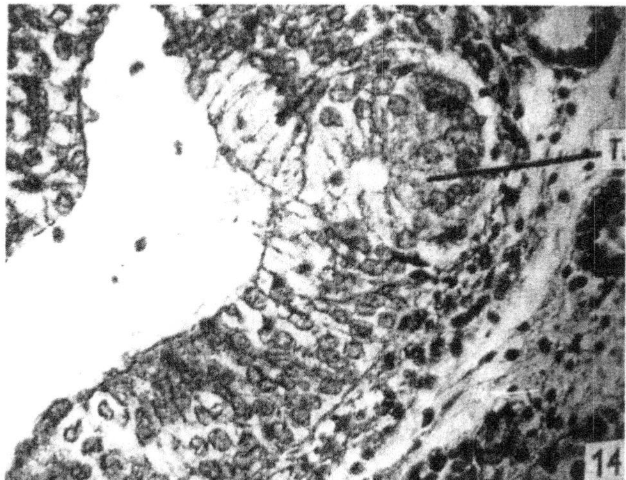

FIGURE 3.7 Esophagus in a 130 mm embryo. The epithelium is largely stratified ciliated with a small area of tall columnar epithelium from which is seen a small tubule cut transversely (T); this is the first stage in the development of superficial glands. Reproduced with permission from Johns BAE. Developmental changes in the oesophageal epithelium in man. J Anat 1952;86:431–442.

FIGURE 3.8 Esophagus in 210-mm embryo. This shows an area of tall columnar cells with a mucous gland in a more advanced stage of development. The gland is in the mucosa above the muscularis mucosae. Elsewhere the epithelium is stratified squamous (SS) with small patches of ciliated epithelium (C). Reproduced with permission from Johns BAE. Developmental changes in the oesophageal epithelium in man. J Anat 1952;86:431–442.

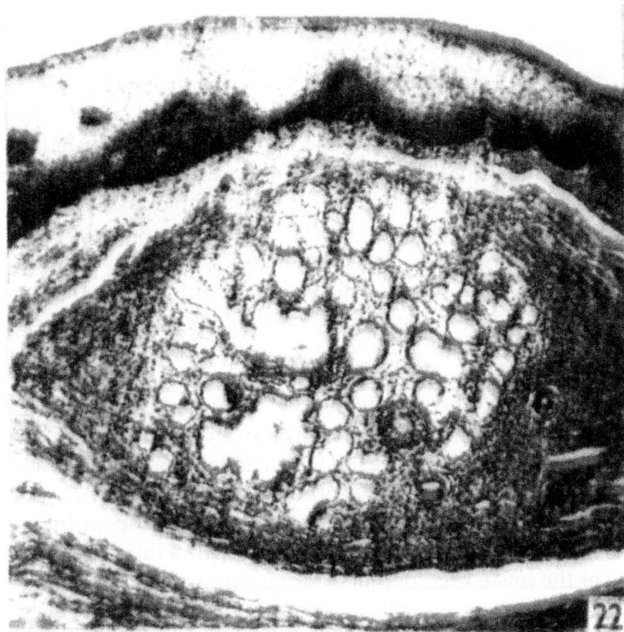

FIGURE 3.9 Transverse section of upper third of the esophagus in an adult to show superficial mucous gland in the mucosa with overlying squamous epithelium. Reproduced with permission from Johns BAE. Developmental changes in the oesophageal epithelium in man. J Anat 1952;86:431–442.

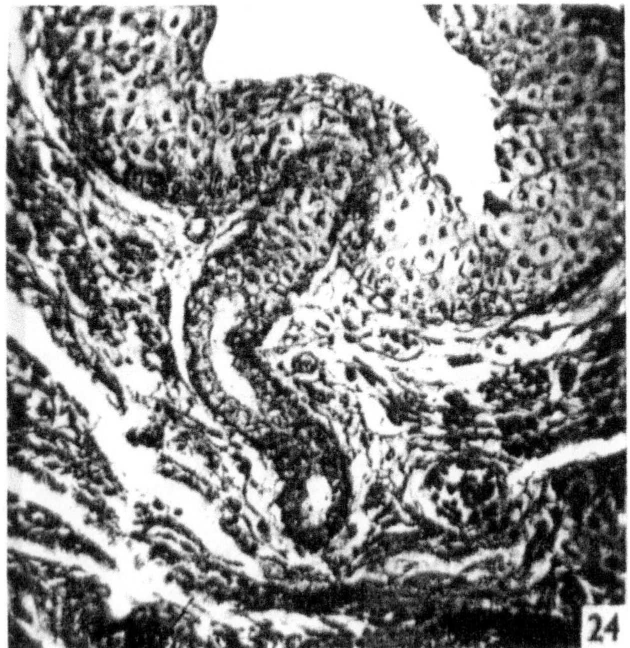

FIGURE 3.10 Middle third of the esophagus in a full term fetus, showing downgrowth of a duct from the stratified squamous epithelium. This penterates the muscularis mucosae and will later give rise to a submucosal gland. No secretory units have yet formed. Reproduced with permission from Johns BAE. Developmental changes in the oesophageal epithelium in man. J Anat 1952;86:431–442.

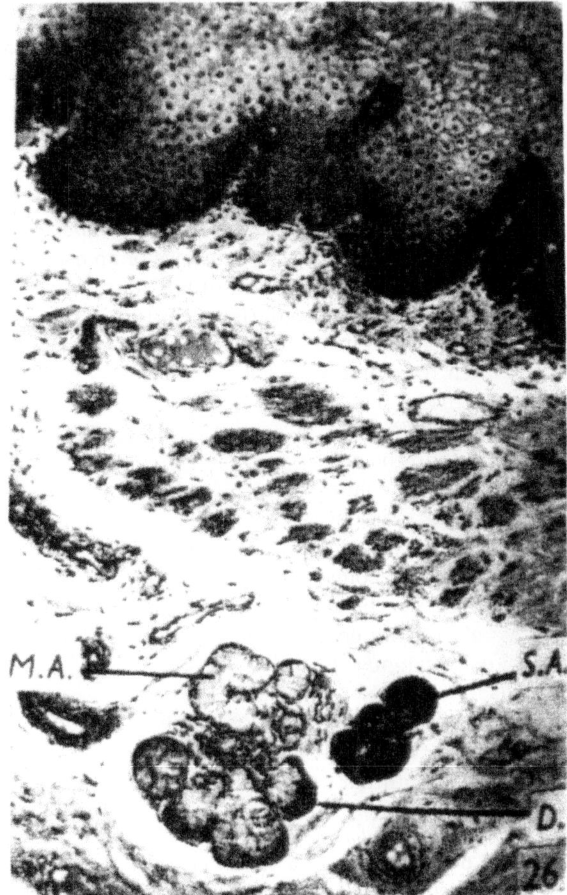

FIGURE 3.11 Esophagus in an adult showing a deep (submucosal) gland composed of mucous and serous cells. Reproduced with permission from Johns BAE. Developmental changes in the oesophageal epithelium in man. J Anat 1952;86:431–442.

specimens, and the fate of the columnar epithelium seen at the lower end of the esophagus in the 230-mm embryo is not addressed.

Johns's discussion is largely a review of the preceding literature, which essentially is in agreement with his findings. He made the following conclusions about his observations:

Tall columnar epithelium. "In the present series the tall columnar cells make their first appearance at the upper and lower ends of the esophagus in the 62 mm embryo. Although the areas occupied by this simple columnar epithelium are of about the same extent in the 62, 72 and 78 mm embryos, they are reduced in the 90, 100, 130 and 160 mm embryos, but in the 162 and 172 mm embryos the areas are as large as those in the 62 mm embryo. It cannot, therefore, be stated that there is in these stages a progressive reduction in the extent of the areas of simple columnar epithelium and the findings may be attributable to variation in the single specimens" (p. 437). Johns' does not take into consideration that the overall esophageal length increases greatly as the crown-rump length increases from 62 mm to 172 mm. The fact that the absolute measured extent remains the same means that a much smaller percentage of the esophagus is covered by these epithelia, a characteristic of a developing fetal-type epithelium. This is in contrast to squamous epithelium, which acts like an adult-type tissue, increasing in total length from the time of its first appearance until development has been completed.

Superficial esophageal glands. "The superficial oesophageal glands in both the upper and lower ends of the oesophagus have their origin in the tall columnar epithelium, from which they first differentiate and grow as tubules which extend as far as the muscularis mucosae at the 130 mm stage [Figs. 3.7, 3.8, and 3.9]. The localization of these glands in the adult corresponds to the localization of the simple tall columnar epithelium in the embryo. As this columnar epithelium is so constant in the embryo, and yet may be absent in the newborn, it appears that on occasions it may completely disappear without giving origin to any superficial glands" (pp. 437–438). Superficial mucous glands are encountered in the adult esophagus under the squamous epithelium (Fig. 3.9). They are well-circumscribed lobulated mucous glands and not associated with inflammation. It is probable that they remain as mucosal glands after the surface epithelium changes from the fetal tall columnar epithelium to the adult squamous epithelium.

Deep (submucosal) esophageal glands (Figs. 3.10 and 3.11). In a discussion of the deep glands of the esophagus, Johns noted that they are not seen in any of the embryos. "The present work suggests that the deep glands do not appear until the epithelium has become stratified squamous in character, and that they develop as outgrowths from this epithelium. . . . Goetsch [16] estimated that in the adult there are anything from 60–741 deep oesophageal glands. Since the deep glands are so commonly present and widely distributed in the adult, the present work indicates that their development must be mainly postnatal" (p. 438). These deep submucosal esophageal glands are clearly an adult type of tissue, which develops after birth from squamous-lined esophagus and persist and grow as the esophagus grows into its adult state. While their numbers vary in different individuals, deep esophageal glands are always present in adults, draining to the squamous surface via gland ducts that pass through the muscularis mucosae and the mucosa (Fig. 3.12). These submucosal glands and their gland ducts in the mucosa are a useful marker for esophageal location in esophagectomy and biopsy specimens because their embryology dictates that they are developed as outpouchings of stratified squamous epithelium, which is restricted to the esophagus in humans.

Johns's description of the development of the esophagus during early embryonic life is unparalleled in its thoroughness and accuracy. Except for slight differences in the timing of events, even the most modern descriptions of esophageal epithelial development parallel those described by Johns. The only epithelial type that has been described that is not evident in Johns's study is the presence of intestinal type goblet cells. This is recorded by Ellison et al. (17) as a rare occurrence in a 23-week fetus. We have also seen goblet cells in fetal esophagus in the material shared with me by Dr. Liebermann-Meffert.

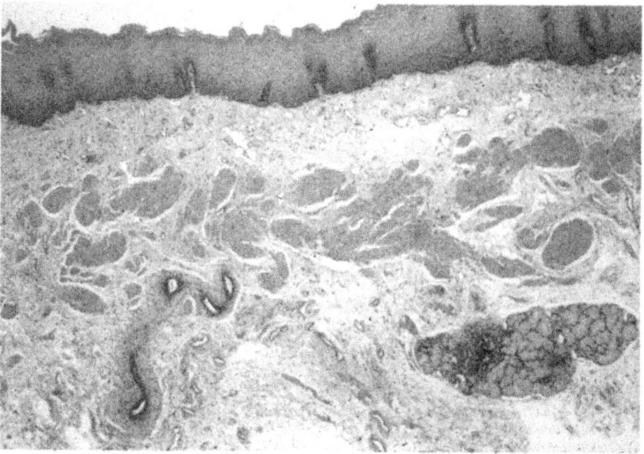

FIGURE 3.12 Esophagectomy specimen in an adult, showing submucous gland and a gland duct. In the completely developed adult state, the esophageal submucosal glands are composed almost entirely of mucous cells, resembling minor salivary glands. (Cross reference: Color Plate 1.3)

In particular, Johns is very clear about the fetal (developmental) nature of the esophageal columnar epithelia, describin g how they all disappear at different stages of fetal life (Fig. 3.13). The early undifferentiated stratified columnar epithelium is replaced by stratified ciliated epithelium except at the ends of the esophagus where it is replaced by tall columnar epithelium. The ciliated epithelium is progressively replaced by stratified squamous epithelium and disappears in the third trimester. The tall columnar epithelium is the last to disappear; according to Johns, it may be absent in the newborn.

EPITHELIAL DEVELOPMENT IN THE FETAL STOMACH

The development of gastric epithelium is much less complicated than esophageal epithelium. Numerous investigators have described the developmental sequence in identical terms. The most detailed and accurate study of gastric epithelium development is that presented by Salenius in 1962 (18). A 1990 study by Menard and Arsenault (19) from the University of Sherbrooke in Quebec, Canada, is chosen for review because it provides excellent information regarding cell kinetics.

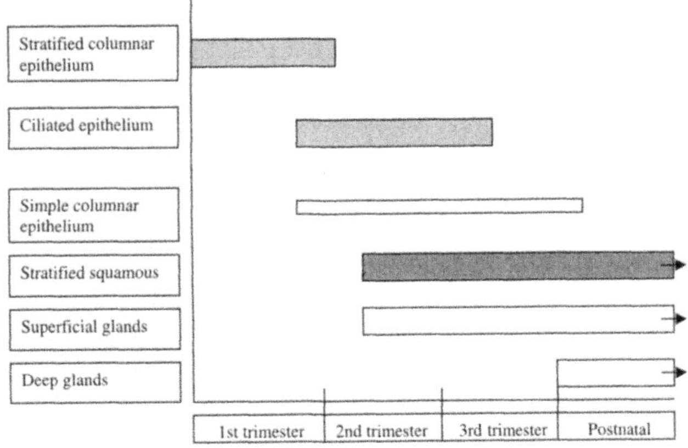

FIGURE 3.13 The various types of epithelia that occur in the esophagus at various stages of development. The stratified columnar epithelium (type 1) of early fetal life is replaced by ciliated epithelium (type 2) in the second trimester, which in turn is replaced by the adult squamous epithelium beginning in the late second trimester. The fetal simple columnar epithelium (type 3) is present as small areas at both ends of the esophagus in the second and third trimester, being replaced by adult squamous epithelium in early infancy. The arrows indicate that these are permanent adult epithelia and glands that persist throughout life. The gray shading shows epithelia that line the entire esophagus at different stages. The type 3 simple epithelium is largely restricted to the two ends of the esophagus in the second and third trimester.

Literature Review

Menard D, Arsenault P. Cell proliferation in developing human stomach. Anat Embryol 1990;182:509–516.

This study used tritiated thymidine labeling of gastric explants; this detects DNA uptake by the cells during the mitotic cycle and thereby identifies proliferative cells. Tissues from 22 fetuses from legal abortions were used. The body of the stomach was dissected out with the pyloric and cardiac ends removed. Longitudinal strips of gastric wall were cultured *in vitro*. Radioautography was done to detect radioactive labeling in the cells after the explant in culture was exposed for 2 hours to tritiated thymidine. The labeling index of cells was defined as the number of labeled cells (five or more grains overlying the nucleus) expressed as a percentage of the total number of nuclei counted.

The following observations were made:

8 to 10 weeks of gestation. The gastric epithelium is a stratified columnar epithelium that is essentially identical to that seen in the early fetal esophagus (Fig. 3.14a). On electron microscopy, these stratified columnar cells were undifferentiated with large amounts of glycogen and very few apical microvilli. Its thickness averaged 0.029 mm. Radiolabeled cells showing uptake of tritiated thymidine were abundant and scattered throughout the entire thickness of the epithelium in random fashion. Approximately 9% of the epithelial cells showed positive labeling.

11 to 17 weeks of gestation. The stratified epithelium showed the first evidence of *invagination into foveolar pits*. The first differentiated glandular epithelial cells (parietal or oxyntic cells) appeared at the base of the rudimentary foveolar pit (Fig. 3.14b). Even at this early stage, the parietal cells showed immunoreactivity for intrinsic factor. The stratified columnar epithelium at the surface was replaced by a simple mucous columnar epithelium with basal nuclei that lines the surface at 17 weeks. The supranuclear cytoplasm was now filled with mucous droplets, and the apical membrane showed more microvilli (Fig. 3.14b). The total thickness of the surface epithelium did not change during this period; the stratification disappeared and the single layer of mucous columnar cells became taller. Between the first appearance of parietal cells at around 13 weeks, gastric glands continued to elongate with increasing numbers of parietal cells. The length of the pit gland structure invaginating from the surface epithelium increased from 0.050 mm at 11 weeks to 0.097 mm by 17 weeks (Fig. 3.14c). This near doubling of gland length was occurring at the time the stomach was also increasing in size, which makes the thickness increase extremely impressive.

At 11 weeks when the surface epithelial differentiation began and the foveolar pit-gland complex started developing, the radioactive labeling moved away from the surface to the developing foveolar pit-gland complex (Fig. 3.14d). Between 14 and 17 weeks, the labeled epithelial cells were mainly confined to the pit/neck region of the gastric glands, and there was a decreasing labeling gradient from the bottom of the gland to the surface epithelium. At 17 weeks, labeled cells were concentrated in the deep foveolar pit region and the neck of the gland with both the deeper part of the gland and the surface showing significantly lower labeling. Uptake of tritiated thymidine was never seen in a differentiated parietal cell. In their discussion, Menard and Arsenault commented that *the developing stomach acquires the typical adult proliferation pattern at the beginning of the second trimester of gestation*.

The further development of the stomach in fetal life is represented by growth, not by any change in the differentiation characteristics of the cells. In the neonate, the gastric mucosa is thin with substantial increase in thickness of the mucosa occurring after birth (Fig. 3.15). The adult gastric oxyntic mucosa shows a marked elongation of the straight tubular gland, representing postnatal growth (Fig. 3.16a). The proliferating pattern, however, remains the same with the proliferative zone being in the deep foveolar region and gland neck after the 17th week of gestation (Figs. 3.14d and 3.16b). The human stomach is much less variable in its epithelial development than is the esophagus.

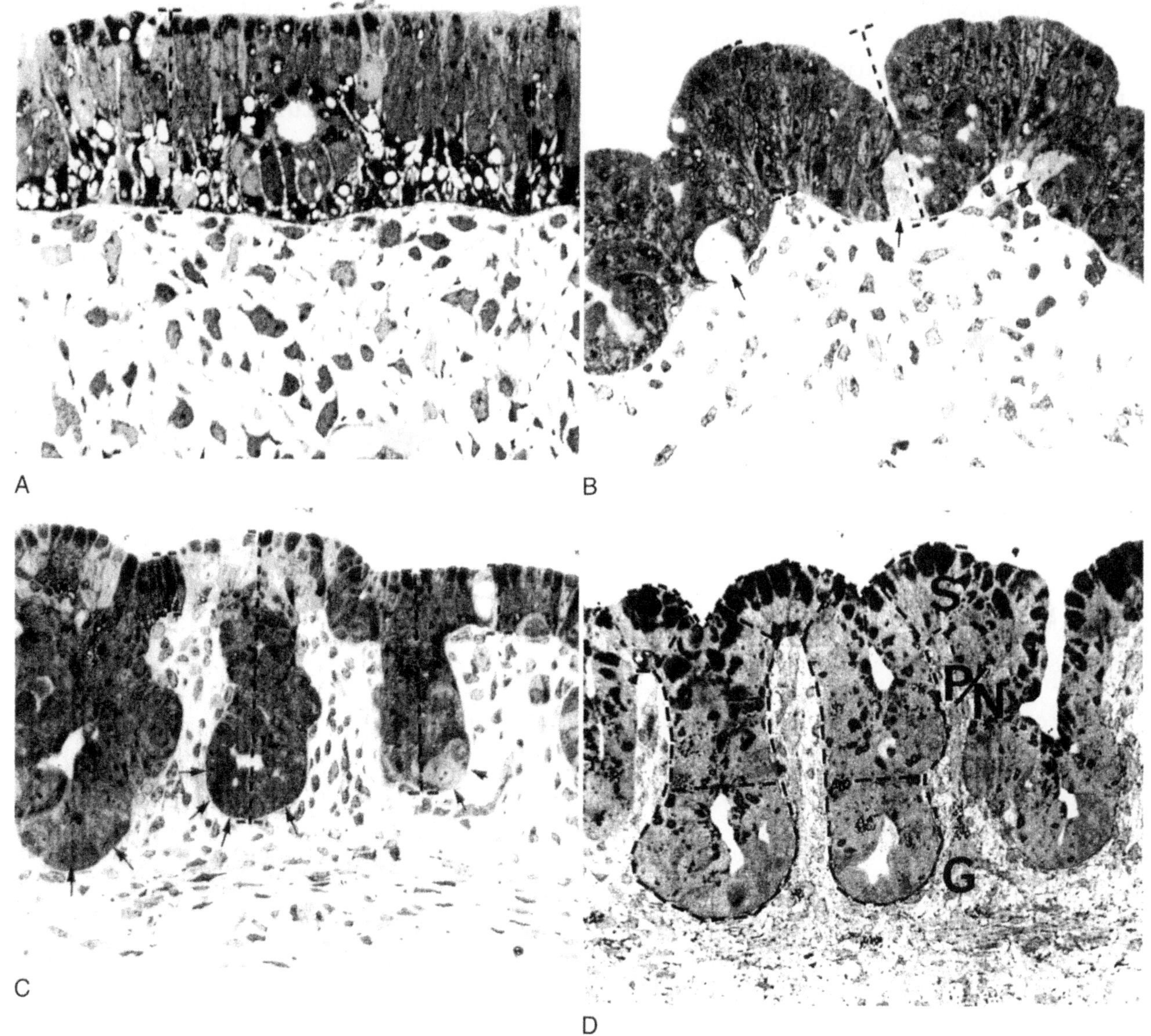

FIGURE 3.14 Human fetal gastric epithelium. (a) 8 to 11 weeks showing undifferentiated stratified columnar epithelium with proliferative activity at all levels. (b) 13 weeks, showing the earliest appearance of parietal cells (arrow) at the bottom of rudimentary pit-gland structures. (c) 17 weeks, showing adult-type architecture. The surface layer of columnar epithelium shows basal nuclei and apical accumulation of mucin while numerous parietal cells (arrows) are observed at the bottom of the developing gastric glands. (d) 17 weeks, showing labeled cells in the different compartments of developing gastric mucosa (S: surface epithelium; P/N: pit/neck of gland; G: base of gland). Reproduced with permission from Menard D, Arsenault P. Cell proliferation in developing human stomach. Anat Embryol 1990;182:509–516.

EPITHELIAL DEVELOPMENT IN THE FETAL GASTROESOPHAGEAL JUNCTION

In the early studies of esophageal and gastric development, a vacuum existed regarding the development of the gastroesophageal junctional region. In Johns's study, for example, the esophagus and stomach was examined in continuity only in embryos less than 42 mm, and the sections were transverse, making it difficult to evaluate vertical relationships of epithelial types. Older embryos were too large for complete serial sectioning. In these, the examination consisted of transverse sections from the end of the tubular esophagus. No significant data are available in Johns's study regarding the gastroesophageal junctional region in older embryos. In most studies of gastric epithelial development, the gastric body receives the most attention and descriptions of the most proximal stomach and the gastroesophageal junction are difficult to interpret. In Menard and Arsenault's study, the proximal stomach is removed and not included in the study material.

In the first three months of 2003, three studies of the epithelium in the gastroesophageal junctional region in late fetal and early neonatal life were published (11, 20, 21). Of these, de Hertogh et al.'s paper has the most careful methods and provides the best information (20). In particular, de Hertogh et al.'s precise definitions and measurements provide wonderful clarity regarding the exact location of the epithelial types that are described.

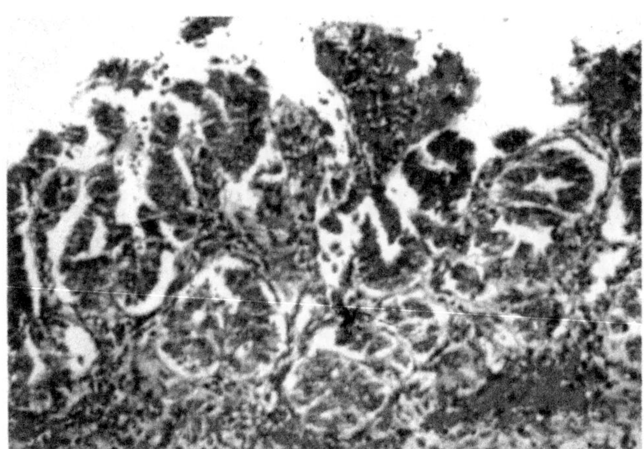

FIGURE 3.15 Gastric mucosa in a neonate, showing short glands containing parietal and chief cells below the surface epithelium and foveolar pit that is composed of mucous cells with basal nuclei and apical mucin.

Literature Review

De Hertogh G, Van Eyken P, Ectors N, Tack J, Geboes K. On the existence and location of cardiac mucosa: An autopsy study in embryos, fetuses, and infants. Gut 2003;52:791–796.

This study is based on tissue samples obtained from 31 embryonic, fetal, perinatal, and infant autopsies in the Department of Pathology, Leuven University Hospitals in Belgium. Subjects ranged in age from 13 weeks' gestational to seven months after term delivery. In reality, this study is limited to the 13- to 24-week period because there are only two cases outside this range, one a 41-week fetus and the other a 7-month baby. Conclusions that are based on the examination of one patient should be considered with caution.

Autopsies were performed between 6 and 24 hours after death. The distal esophagus and stomach were excised in one piece. In 24 cases (age range 13 to 23 weeks), the whole specimen was sectioned longitudinally. In six cases (age range 24 to 41 weeks), the specimen was so large that it had to be embedded top down, after which it was sectioned transversally. The number of sections selected varied between 10 (small specimens) and 40 (large specimens).

The authors used excellent anatomic definitions as follows: (a) The gastroesophageal junction is defined as the angle of His (Fig. 3.17). (b) They precisely define what they mean by the "cardia": "Anatomists have applied the term 'cardia' to that part of the stomach which lies around the orifice of the tubular oesophagus. There is no anatomical landmark for the distal margin of the so-defined cardia. Its proximal margin

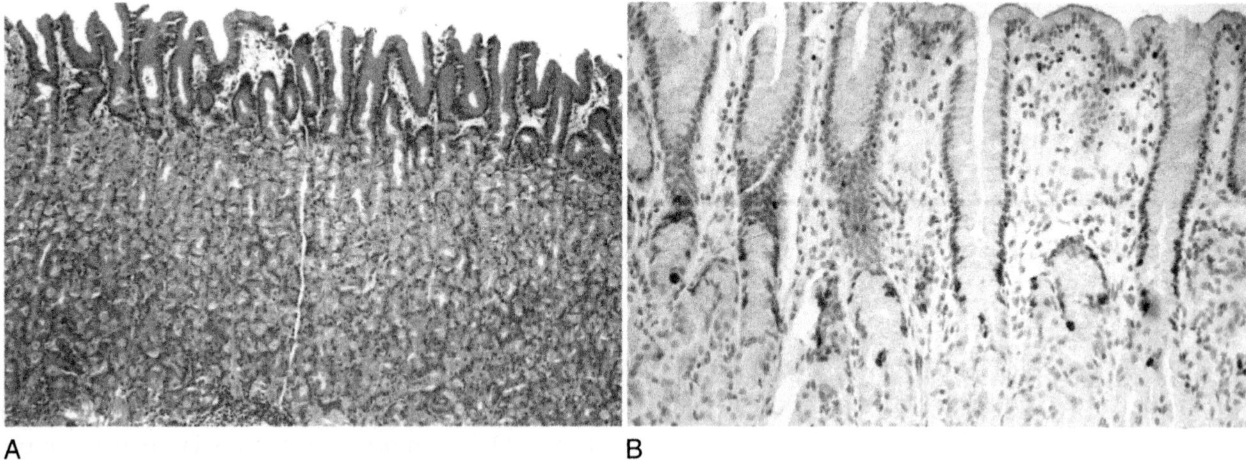

FIGURE 3.16 Adult gastric oxyntic epithelium, showing the fully developed structure. (a) H&E. (b) Immunoperoxidase stain for Ki67. The elongated straight tubular glands are composed entirely of parietal and chief cells and drain into the foveolar pit. The foveolar and surface epithelia are mucous cells with apical mucin. Proliferative activity, seen as Ki67 positive cells, is restricted to the deep foveolar region. (Cross reference: Color Plate 1.7)

is the gastro-oesophageal junction which, according to anatomists, is localised at the level of the angle of His" (p. 571). (c) The length of the abdominal segment of the esophagus is defined as the distance from the lower rim of the diaphragm to the angle of His; (d) The length of columnar-lined esophagus is defined as the distance from the squamocolumnar junction to the angle of His. Anything above the angle of His is the esophagus and anything below is the stomach. This study is valuable because it clearly defines the gastroesophageal junction as a point and the anatomic gastric cardia as being distal to this point (Fig. 3.17).

The histologic definitions used in this paper are critical to understanding what the authors are describing. It is unfortunate that they do not cite Johns's classical paper or use his terminology and definitions, making it somewhat difficult to compare the two papers. However, the two authors are seeing the same thing though they use different terms.

The authors classify mucosae in the gastroesophageal junctional region into the following types:

1. Primitive esophageal mucosa (PEM), consisting of a stratified columnar epithelium covered by ciliated cells (this is clearly the same as Johns's stratified columnar ciliated epithelium).
2. Squamous epithelium, which is unambiguous.
3. Primitive stomach mucosa (PSM), defined as a layer of columnar epithelial cells with no glandular structures present. This epithelium was limited to two fetuses of the lowest gestational age (13 and 14 weeks) and seems to be the same as the primitive stratified nonciliated columnar epithelium described by Johns as lining the entire esophagus before the ciliated epithelium develops and described by Menard and Arsenault as lining the stomach before parietal cells appear in the 13- to 17-week period. It is unfortunate that de Hertogh et al. use the term "primitive stomach mucosa" to describe an epithelium that is clearly in the esophagus by their own definitions.
4. Cardiac mucosa, which is defined as a lining composed of foveolar and surface epithelium overlying glandular structures containing no parietal cells. They define the distal end of the cardiac mucosa by the first parietal cell that is encountered.

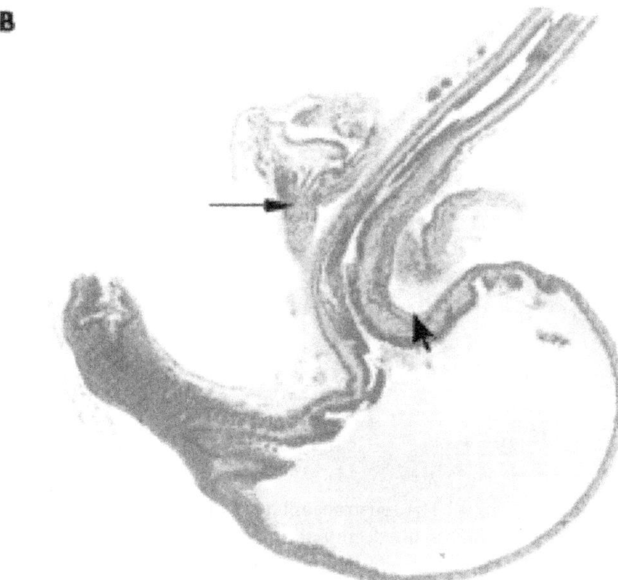

FIGURE 3.17 Fetal gross specimen of esophagus and stomach showing the diaphragm (long arrow) and the angle of His (short arrow). The part between the two is the abdominal portion of the esophagus. Reproduced with permission from De Hertogh G, Van Eyken P, Ectors N, Tack J, Geboes K. On the existence and location of cardiac mucosa: An autopsy study in embryos, fetuses, and infants. Gut 2003;52:791–796.

The illustration of cardiac mucosa in a 13-week-old fetus (Fig. 3.18) points to what appears to be an epithelium lined

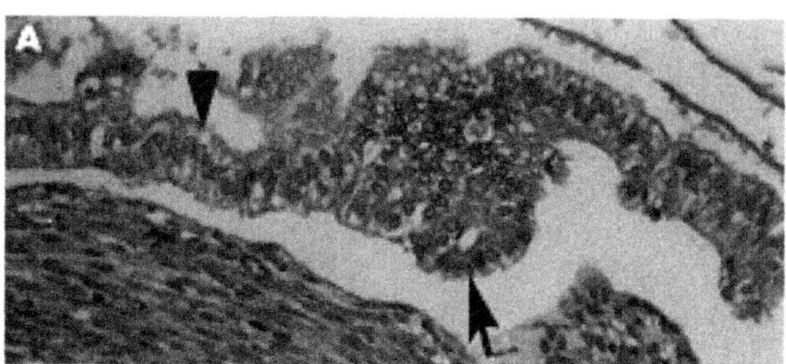

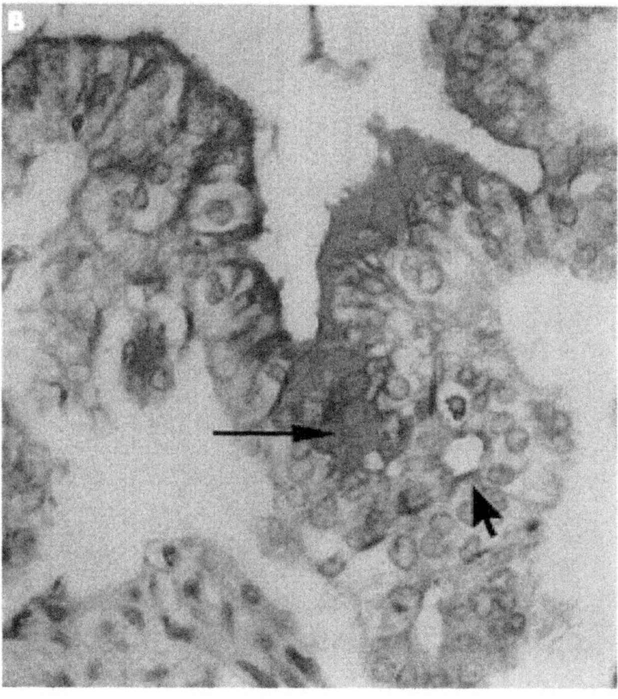

FIGURE 3.18 What the authors designate as "cardiac mucosa" in a 13-week embryo. This shows a columnar nonciliated epithelium with a rudimentary outpouching, which is termed a "gland." This epithelium closely resembles Figure 3.7 (Johns's 130-mm embryo). Reproduced with permission from De Hertogh G, Van Eyken P, Ectors N, Tack J, Geboes K. On the existence and location of cardiac mucosa: An autopsy study in embryos, fetuses, and infants. Gut 2003;52:791–796.

entirely by a stratified columnar epithelium with what is exactly like a tubular outpouching of a superficial gland described in the 130-mm embryo (somewhat older) in Johns's paper (Fig. 3.7). In the second illustration of "cardiac mucosa" in a 24-week fetus (Fig. 3.19), there is a parietal cell at the extreme edge of the picture; this must represent stomach; in Johns's study of fetal esophageal epithelium, there was not a single case with parietal cells. In both fetuses, the "cardiac mucosa" has what the authors call a gland coming down from the columnar epithelium (Figs. 3.18 and 3.19). These are not true glands because they have no secretory elements; they resemble the origin of tubular outpouchings that Johns described as being the precursors of superficial glands. The columnar epithelium shows features identical to Johns's tall columnar epithelium consisting of a single undulating layer of tall columnar cells with apical mucin and no glands under the foveolar region.

In the illustration of "cardiac mucosa" in the neonate (1 week, Fig. 3.20), there is a well-developed superficial mucous

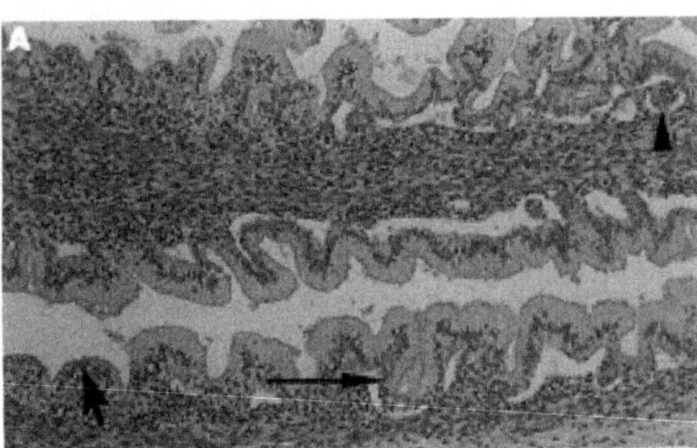

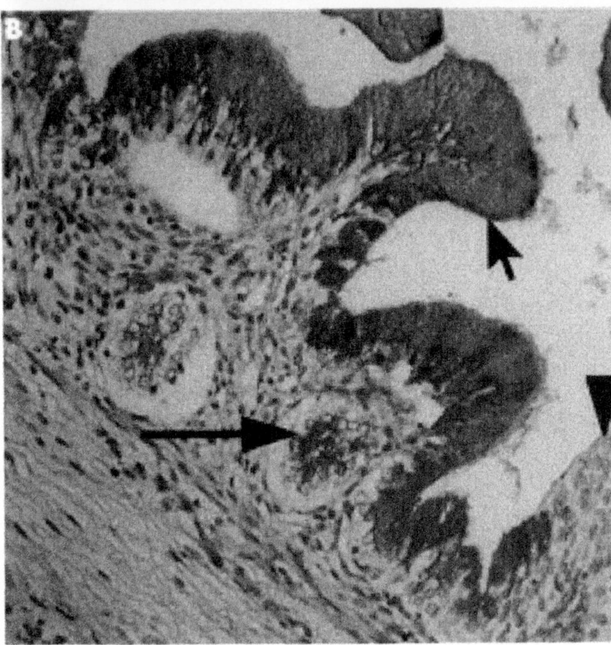

FIGURE 3.19 What the authors designate as "cardiac mucosa" in a 24-week fetus. This consists of squamous epithelium with residual ciliated cells (short arrow) and "cardiac mucosa" containing no parietal cells (long arrow). An arrowhead points to the first gland containing a parietal cell. The image at the bottom depicts two mucosal "glands." Reproduced with permission from De Hertogh G, Van Eyken P, Ectors N, Tack J, Geboes K. On the existence and location of cardiac mucosa: An autopsy study in embryos, fetuses, and infants. Gut 2003;52:791–796.

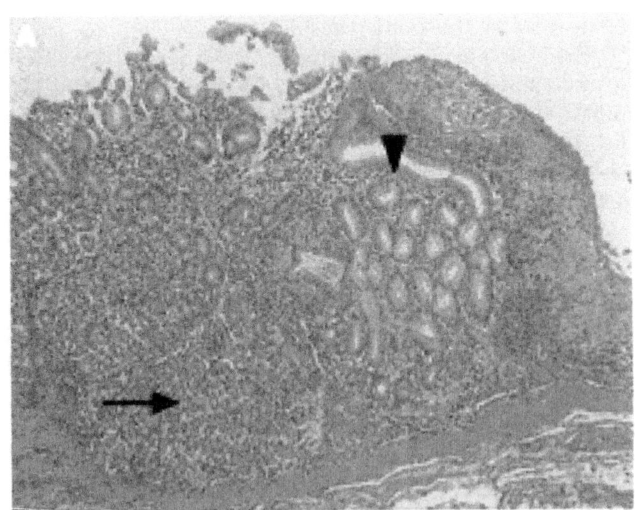

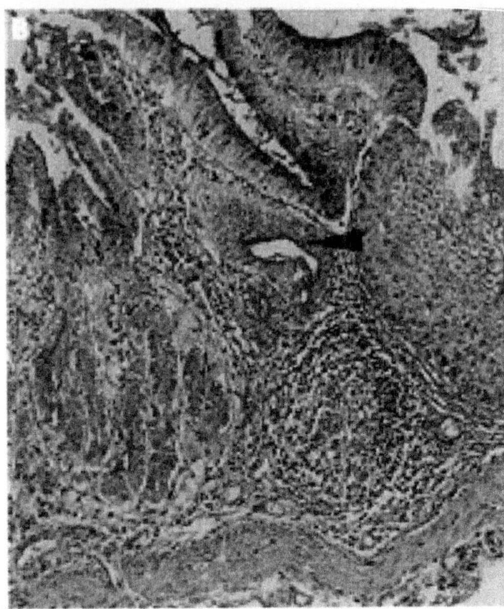

FIGURE 3.20 The gastroesophageal junction in a neonate showing squamous lined epithelium, a superficial mucous gland under the squamous epithelium that the authors call "cardiac mucosa" (arrowhead) and parietal cell containing oxyntic mucosa (arrow). The image at the bottom shows squamous and oxyntic mucosa with a "cardiac mucosal gland" between the two. The "gland" has the appearance of an esophageal gland duct without any secretory gland complex. Reproduced with permission from De Hertogh G, Van Eyken P, Ectors N, Tack J, Geboes K. On the existence and location of cardiac mucosa: An autopsy study in embryos, fetuses, and infants. Gut 2003;52:791–796.

gland underneath the squamous epithelium. In this picture, it appears that gastric oxyntic mucosa with parietal cells is immediately distal to the squamous surface and the superficial mucous gland; it is difficult to determine whether there is any cardiac mucosa between the superficial gland and the parietal cell containing gastric mucosa. What is crucial is that what the authors describe as "cardiac mucosa" in their 13- and 24-week fetuses (Figs. 3.18 and 3.19) are no longer seen in the neonate.

It should be noted that Johns did not use the term "cardiac mucosa" for any epithelium he described. It is unfortunate that de Hertogh et al. (20) apply the term "cardiac mucosa" to describe this fetal esophageal columnar epithelium. This fetal "cardiac mucosa" bears no resemblance to adult cardiac mucosa, and we have never seen anything like it in an adult esophagus. Also, the authors clearly placed their "cardiac mucosa" anatomically proximal to the angle of His in all their cases (their Table 2); why they decided to use the highly controversial term "cardiac mucosa" for this fetal columnar epithelium—which is in the esophagus and not in their defined "cardia," which is distal to the angle of His—is incomprehensible. The fact that what they called "cardiac mucosa" in the 13- and 24-week fetuses is not present in their neonatal specimen proves that this is a fetal type of epithelium and not the equivalent of adult cardiac mucosa (22).

The study confirms the presence of various columnar epithelial types that Johns described in the esophagus in the first two trimesters. The most interesting data in this study are that their fetal "cardiac mucosa" was in the esophagus, not in the stomach. "When one accepts the angle of His as a landmark for the gastro-oesophageal junction, the cardiac mucosa was located proximal to or at the gastro-oesphageal junction in all of these cases. Therefore, according to anatomical definitions, cardiac mucosa was located in the abdominal segment of the tubular oesophagus at birth. The distance from the cardiac mucosa to the angle of His was small however (0.0 mm at 41 weeks and 0.3 mm at seven months)." *By finding parietal cells at or slightly above the angle of His in all their cases, this study proves that cardiac mucosa does not line the proximal stomach.* Parietal cells are terminally differentiated gastric cells whose numbers continue to increase after birth, and therefore finding them at the gastroesophageal junction indicates that the junction is lined by parietal cell containing adult gastric epithelial type in late fetal life. It is unlikely that adult type parietal cells disappear; it is far more likely that they continue to increase with growth of the fetus.

This study provides data to determine whether what the authors defined as cardiac mucosa is a fetal epithelium or an adult epithelium. The researchers stated: "The mean length of the cardiac mucosa . . . reached a maximum at 16 weeks (1.2 mm). Its length at and after birth was very short (0.3 to 0.6 mm)." During the 13- to 24-week gestational period, the esophagus elongates dramatically with the growth of the fetus. The decreasing length of "cardiac mucosa" with increasing fetal age provides strong evidence that this is a fetal-type epithelium that is similar to and probably the same as that described as "tall columnar epithelium" by Johns. De Hertogh et al. were describing the late fetal columnar epithelium that is seen in the distal esophagus in late fetal life and early infancy; this is not the cardiac mucosa seen in adults (22).

De Hertogh et al. stated: "If cardiac mucosa develops during pregnancy, it is a normal structure at birth." This is an obviously incorrect conclusion. The entire esophagus is lined by stratified columnar epithelium until the second trimester; this is not a normal adult epithelium. The entire esophagus is lined by a stratified ciliated epithelium until the 22nd week of gestation; this is not a normal adult epithelium. Intestinal type epithelium with goblet cells has been shown to be present in fetal esophagus (17); Barrett esophagus is certainly not a normal adult epithelium. So how can the authors conclude that their "cardiac mucosa" is a normal adult epithelium simply because they found it in fetal life?

Their extremely accurate anatomic findings should have prevented them from using the terms "primitive stomach mucosa" and "cardiac mucosa" for these fetal epithelia that they so clearly localized to the esophagus above the gastroesophageal junction. Surely "stomach mucosa" should line the stomach and "cardiac mucosa" should line the proximal stomach or cardia by their definition. Much confusion would have been prevented if the authors had used the term "fetal esophageal columnar epithelia, types I, II, and III." If they had done this, they would have reached the correct conclusion that these were all developing fetal epithelia in the esophagus at different gestational ages that was not seen in the adult esophagus (22).

Coincidentally, April 2003 also brought two other reports from South Korea and Los Angeles on the same subject. In the three studies, including de Hertogh et al., two sets of authors conclude that cardiac mucosa is always present during fetal life and at birth and one concludes that cardiac mucosa is never found as a normal developmental structure. Such diametrically opposing viewpoints on the same subject are always interesting, and careful analysis of the data in the reports is required to determine the reasons for the different conclusions.

The study by Derdoy et al. (11) from Children's Hospital of Los Angeles retrospectively examined one longitudinal section taken across the gastroesophageal junction in autopsies performed in babies and children from gestational age 26 weeks to 18 years. This study therefore focused on babies that were almost all older than de Hertogh's series. Unfortunately, the authors determined age by the point of birth without correcting for premature birth. This precludes specific recognition of late fetal epithelial changes and comparison with de Hertogh's and Johns's findings. Also, many of these patients lived for a variable time after birth, making it possible that they had unrecorded trauma such as by nasogastric tubes. The only interpretable item of data in this paper is one very nice figure, #2 (in the youngest patient, gestational age 26 weeks), that shows a fetal epithelium that appears to be identical to what de Hertogh et al. called cardiac mucosa with only surface and foveolar regions without any definite glands. Derdoy et al. also called this "cardiac mucosa."

As noted, Derdoy et al. are from Children's Hospital of Los Angeles, which is an affiliated hospital of the University of Southern California, where I am based. Before the study was published, the authors kindly gave me the opportunity to review their histologic slides. Review of that material made me reach conclusions that were completely different from what they published. I saw fetal columnar esophageal epithelium, largely of the tall columnar type by Johns's definition, with the youngest patient showing foci of residual ciliated columnar epithelium. There was a decrease in the amount of fetal columnar epithelium as the age increased; and in some of the older patients, there was a direct transition from squamous epithelium to parietal cell containing gastric epithelium without intervening cardiac mucosa. Dr. Chandrasoma freely shared his observations and opinions with the authors, but these were not included in the paper.

The illustrations of "cardiac mucosa" in de Hertogh et al. are similar to the "cardiac mucosa" reported by Derdoy et al., which we have examined (Fig. 3.21). This mucosa is different from adult cardiac mucosa in the following respects:

1. It is very thin and uniform, consisting of mucous cells forming a surface and very short foveolar region without glands. The mucous cells are columnar and very uniform in appearance. The thin

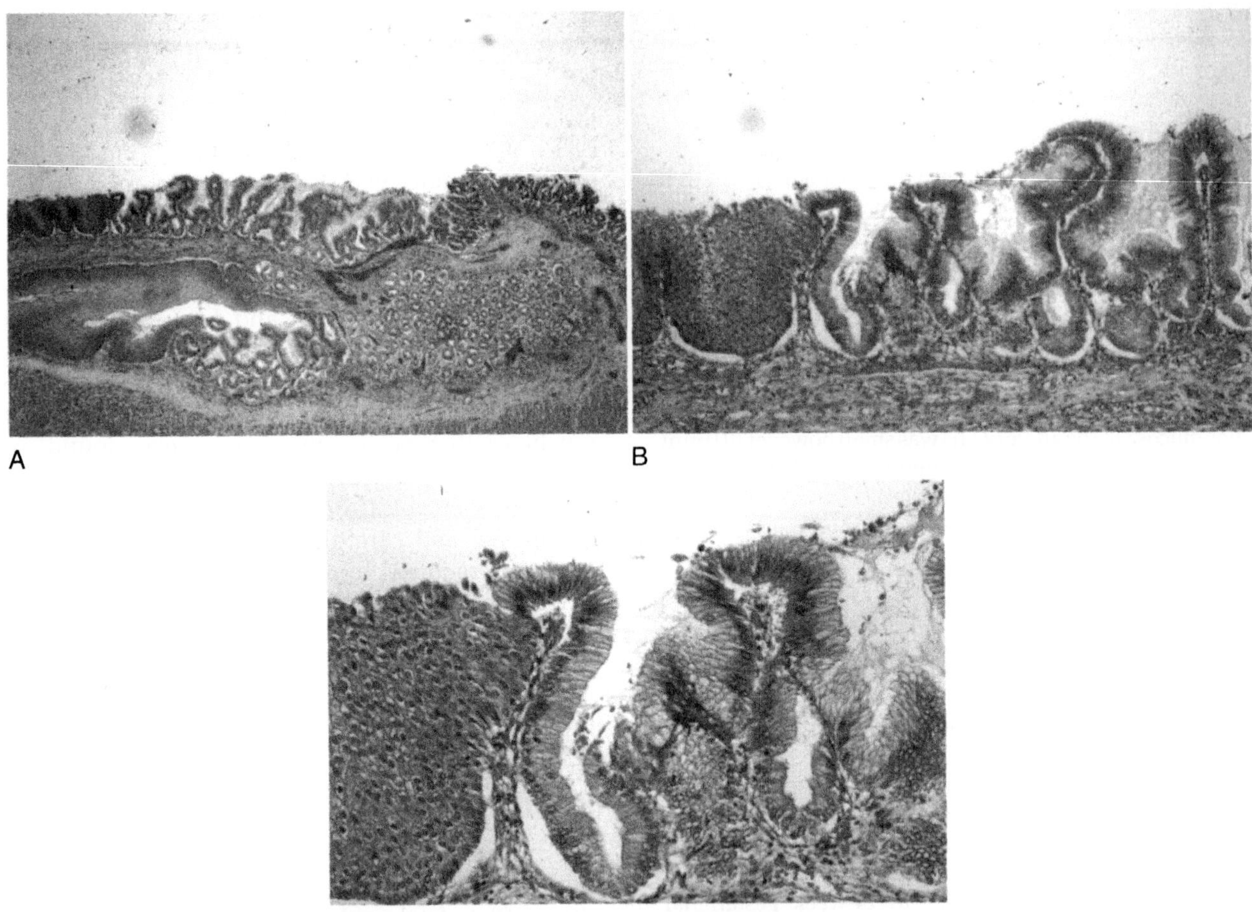

FIGURE 3.21 (a) Fetal columnar epithelium across the gastroesophageal junction in a third trimester fetus showing squamous epithelium in the distal esophagus to the left, parietal cell containing gastric mucosa to the left and simple fetal columnar epithelium (type 3) between the two as a single layer of columnar epithelium thrown into folds. (b) and (c) show the tall columnar epithelium as an undulating single layer of cells devoid of glands and any inflammatory cells lying on the thin muscularis mucosae. (Cross reference: Color Plate 1.2)

epithelium has a flat muscularis mucosae immediately beneath it.

2. It does not contain a single inflammatory cell; the lamina propria is very sparse and consists of uniform nondescript mesenchymal cells. Adult cardiac mucosa is very different; it commonly has a reactive foveolar zone, mucous glands that are frequently lobulated and distorted, a lamina propria that invariably contains lymphocytes, plasma cells and eosinophils, as well as frequent fibrosis and smooth muscle proliferation.

The third study, from Seoul National University Children's Hospital by Park et al. (21), is a prospective study of 23 autopsies of 15 prenatal cases ranging in gestational age from 18 to 34 weeks and 8 postnatal cases ranging from 9 days to 15 years. This is a mixture of completely examined prospective autopsies and retrospective material. There is no attempt to identify the gastroesophageal junction anatomically. The authors measured the distance from the squamocolumnar junction to the most proximal parietal cell identified. They reported that 5 cases (3 prenatal cases 19, 23, and 27 weeks old, and 2 postnatal cases 1 month and 5 years old) had a direct transition from squamous epithelium to a mucosa with fundic glands. Their Figure 2A (from a 5-year-old child) shows a convincing illustration of this phenomenon (Fig. 3.22). This figure shows squamous epithelium adjacent to mature oxyntic mucosa without any rudimentary fetal epithelia intervening. The glandular epithelium shows the typical straight tubular glands of gastric oxyntic mucosa with numerous parietal cells in the glands. This figure is identical to our illustration of a child with a normal squamous to gastric oxyntic transition in our autopsy study (Fig. 3.23) (23).

Park et al. reported that 18 of the 23 cases had a "transitional zone" between the squamous epithelium and mucosa with parietal cell containing gastric oxyntic mucosa. Figure 2B in this report illustrates what the authors called the transitional zone, showing only surface and foveolar mucous cells in the 400 μm segment between squamous epithelium and gland-containing epithelium (Fig. 3.24); this is similar to the "cardiac mucosa" of Derdoy et al.'s Figure 2 and de Hertogh's Figure 4. Even at this magnification, this figure clearly shows a parietal cell in the first rudimen-

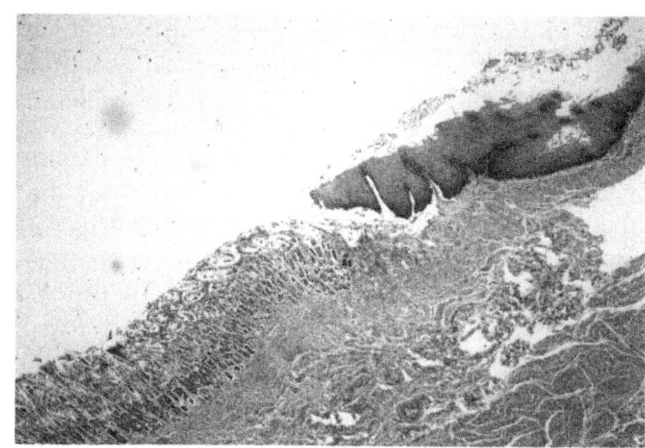

FIGURE 3.23 Gastroesophageal junction in an autopsy specimen from a 6-year-old child showing transition from squamous epithelium to gastric oxyntic mucosa with straight tubular glands containing parietal cells without any intervening cardiac mucosa. Despite the autolysis, the straight glands and the darker parietal cells are clearly seen. (Cross reference: Color Plate 1.5)

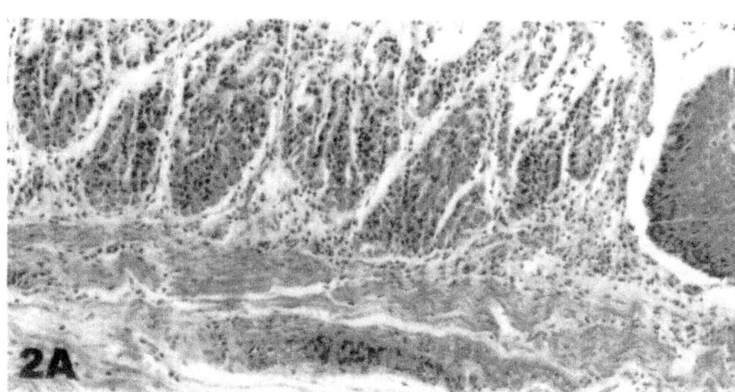

FIGURE 3.22 Park—Fig 2A. Transition from squamous to gastric oxyntic mucosa in a child without any intervening cardiac mucosa. Note that the glandular epithelium contains straight tubular glands with parietal cells typical of gastric oxyntic mucosa. There is neither cardiac mucosa nor fetal columnar epithelium between the esophageal squamous epithelium and gastric oxyntic mucosa.

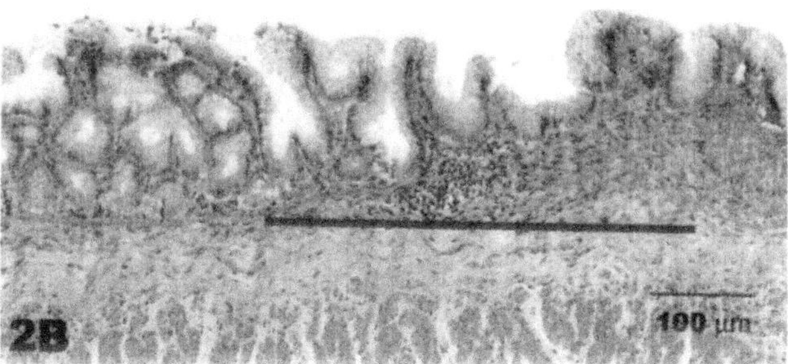

FIGURE 3.24 A less developed fetal gastroesophageal junction in Park et al. The esophageal squamous epithelium on the right is followed by an undulating tall columnar epithelium composed of mucous cells and devoid of glands (fetal tall columnar epithelium; designated transitional zone by Park et al.). This is followed by an epithelium containing glands on the left; the glands contain parietal cells and resemble rudimentary gastric body epithelium seen in the neonate (compare with Fig. 3.15).

tary gland beneath the foveolar region at the distal end of their "transitional zone." Park et al. correctly recognized that this "transitional zone" is different from cardiac mucosa. Their transitional zone consists of the same tall columnar fetal epithelium that Johns noted, which represents the last fetal epithelial columnar epithelium to survive, frequently into the postnatal period.

Figures 2A, 3, and 4 of Park et al.'s report all show the presence of parietal cells immediately adjacent to squamous epithelium. In these cases, the mucous cell's only transitional zone has disappeared completely. These figures contradict the conclusion of de Hertogh et al. and Derdoy et al. that there is *always* a mucosa distal to the squamous epithelium that contains glands devoid of parietal cells and composed only of mucous cells. de Hertogh et al. was accurate in their report; their embryos were predominantly of an age when some fetal columnar epithelium separated squamous from parietal cell containing gastric mucosa. What de Hertogh et al. showed clearly was that this fetal columnar epithelium was *esophageal*. If one recognizes this, then it is clear that as the fetus develops further, de Hertogh et al.'s fetuses evolve into Park et al.'s older patients who have completely lost this "transitional zone" of fetal columnar epithelium. The histology of the gastroesophageal junction has reached its fully developed adult stage with the entire fetal columnar epithelium of the esophagus transformed into stratified squamous epithelium and the entire fetal columnar epithelium of the stomach transformed into adult parietal cell containing gastric mucosa. By de Hertogh et al.'s data, this squamogastric junction is almost exactly at the gastroesophageal junction defined as the angle of His.

In our autopsy study (23), we also illustrated the transition of squamous epithelium to parietal cell-containing mucosa similar to Park et al.'s (Fig. 3.23). It should be recognized that photographs can prove the absence of a structure but never its universal presence. The absence of cardiac mucosa in illustrations of Chandrasoma et al. and Park et al. *proves* that cardiac mucosa is *not universally* present distal to the squamous epithelium.

The only other study of the fetal gastroesophageal junctional region is that by Zhou et al. (24) from the New York University School of Medicine. This paper is stimulated by our autopsy study of 2000 and tests our finding that cardiac mucosa is not normally present against the traditional view that cardiac mucosa normally separates the esophageal squamous epithelium from gastric oxyntic mucosa.

In their introduction, the authors misquote us: "Recently, [the traditional] concept has been challenged by those who advocate that the cardiac region does not exist early in life and that initially there is direct continuity between esophageal squamous and oxyntic mucosa" (p. 358). This is not a suggestion that we have ever made because it has always been obvious to us that the fetal esophagus is lined by columnar epithelium. The youngest patient in our prospective autopsy study was 3 years of age. It is also interesting to note that "the traditional concept" refers only to citations in textbooks; this reflects the incredible fact that no primary papers have ever put forward this traditional concept.

In the first part of the study, the authors examined 32 fetuses ranging in gestational age between 15 and 39 weeks with an equal representation throughout fetal life. They sectioned the entire squamocolumnar junction vertically, permitting evaluation of the glandular mucosa immediately distal to the squamous epithelium. They did not attempt to identify the esophagus and stomach anatomically. Any columnar epithelium between squamous epithelium and oxyntic mucosa was termed "transitional zone." A transitional zone was absent if the squamous epithelium was in direct continuity with oxyntic mucosa. There was no attempt at quantitation of the length of the transitional zone by measurement. The transitional zone was classified as mucous type if the glandular component was formed by only mucous cells and mixed type if glands contained parietal cells admixed with the mucous cells.

The authors found that a transitional zone always separated the squamous epithelium from oxyntic mucosa. Before 20 weeks gestational age, the transitional zone was composed of surface epithelium only. This most likely represents the fetal tall columnar epithelium noted by Johns. It is incredible that the authors did not find ciliated epithelium and that the squamous epithelium is so well developed in 15- to 20-week-old fetuses. The authors reported that mixed glands were evident in the transitional zone by 15 weeks and mucous glands by 22 weeks. It is difficult to interpret this information without details of location. Certainly, parietal cells can have appeared in the proximal stomach by the 15th week, and columnar epithelium is normally present in the esophagus distal to the squamous epithelium. The authors stated that none of fetuses had a transitional zone composed exclusively of oxyntic glands. This is not surprising; it is a normal and expected finding in fetuses who normally have developing fetal columnar esophageal epithelia distal to the squamous epithelium, separating it from parietal cell containing gastric oxyntic mucosa. The authors made the important observation that no inflammation was seen in the transitional zone in these fetal cases.

This was in contrast to the high incidence of inflammation in cardiac mucosa seen in the postnatal cases.

In their discussion, the authors stated:

> We postulated that the presence of a transitional zone between squamous esophageal and gastric oxyntic mucosa early in gestational development was indicative of a gastric origin, whereas its absence, characterized by direct continuity between squamous esophageal and gastric oxyntic mucosa, would support a metaplastic esophageal origin later in life.

This postulate is so incorrect as to be incomprehensible; if true, the entire esophagus, which is lined by columnar epithelium early in fetal life before squamous epithelium appears, would be gastric in origin.

Zhou et al.'s results in their postnatal autopsies will be discussed later in the consideration of whether cardiac mucosa is present universally in all people after development has been completed.

SUMMARY OF EPITHELIAL DEVELOPMENT OF THE ESOPHAGUS

Why the esophagus passes through several different types of columnar epithelia at different stages of fetal development is unknown (Fig. 3.13). The entire esophagus in the first and early second trimester is lined first by an undifferentiated stratified columnar epithelium (Fig. 3.3) and then by a stratified ciliated columnar epithelium with a small area of nonciliated columnar epithelium at both ends of the esophagus (Figs. 3.5 and 3.6). Stratified squamous epithelium replaces the columnar ciliated epithelium, beginning around 22 weeks at the midesophagus and rapidly extending proximally and distally (Fig. 3.8). Ciliated epithelium disappears from the esophageal surface epithelial lining in the third trimester, never to appear in adult life under any circumstance. There remains, in the third trimester, a small area of undifferentiated nonciliated columnar epithelium in the most distal 1 to 2 mm of the esophagus between the squamous epithelium and parietal cell containing gastric oxyntic mucosa (Figs. 3.19, 3.21, and 3.24). This is the epithelium that de Hertogh et al. and Derdoy et al. called "cardiac mucosa" and Park et al. called "transitional zone"; this is Johns's nonciliated columnar epithelium (tall and short). *I have never observed this epithelium after infancy.*

As shown in de Hertogh et al. and Derdoy et al., this tall columnar epithelium in the distal esophagus progressively decreases in length as the fetus matures, a sign of a differentiating fetal epithelium rather than a differentiated adult epithelium. As shown beautifully by Park et al. as well in our autopsy study (23) (Figs. 3.22 and 3.23), this epithelium disappears completely in the neonate, where the squamous epithelium transitions directly to a columnar epithelium that contains parietal cells. "Cardiac mucosa" of de Hertogh et al. and Derdoy et al., "the non-ciliated columnar epithelium (short and tall)" of Johns, and the "transitional zone" of Park et al. have finally developed completely into adult epithelia, which are squamous and parietal cell containing gastric oxyntic mucosa. If we use the term "fetal esophageal columnar epithelium type III" for this epithelium, we will recognize it as the last remnant of fetal columnar epithelium to disappear from the esophagus as it develops into a squamous lined tube.

This makes total sense if one recognizes that the lower esophageal sphincter completes development during early infancy, and epithelial differentiation into squamous and gastric must remain flexible until the full development of the sphincter establishes the true physiologic gastroesophageal junction. It is when esophagus and stomach have been defined that the fetal columnar epithelium at the junction ("nonciliated columnar epithelium" of Johns, "cardiac mucosa" of De Hertogh et al. and Derdoy et al., and "transitional zone" of Park et al. and Zhou et al.) finally differentiates into squamous epithelium in the esophagus and gastric mucosa in the stomach.

One important fact to note is that in all of their variety during fetal life, the esophageal columnar epithelia are composed entirely of mucous cells. Parietal cells never occur in the fetal esophagus; not a single case in Johns's exhaustive study showed parietal cells. If true, this would mean that the presence of parietal cells either establishes the epithelium as gastric in location or, if found in the esophagus, represents aberrant differentiation.

Goblet cells have been described as a transient feature in the fetal esophagus; an excellent illustration appears in a 23-week fetus (17). Like all other columnar epithelia that occur in the esophagus during fetal life, goblet cells disappear; their presence in fetal esophagus should not be taken as evidence that Barrett's esophagus is a congenital abnormality.

CONTROL OF FOREGUT EPITHELIAL DEVELOPMENT

Because the fetal environment is largely devoid of external environmental influences, the initial control of fetal epithelial development must be largely controlled by inherent genetic programming that gives the cells appropriate genetic signals to proliferate and differentiate in a predetermined manner. External influences on the developing epithelium become possible when the fetus begins swallowing amniotic fluid at 4 to 5 months. There is controversy regarding the exact time that developing parietal cells in fetal gastric mucosa (which first appear around the 13th to 17th week of gestation) begin to secrete significant amounts of acid into the gastric lumen. There is a possibility, therefore, of external influences affecting fetal epithelium development during the last half of fetal life, but this is likely to be negligible.

If genetic programming is the main driving force of fetal epithelial development, there must be a predetermined activation and suppression of specific genetic signals. A signal must be present that induces the mesodermal cells in the left side of the gastric wall to proliferate more rapidly than the right side, resulting in the bulging of the fundus and greater curvature and producing the normal gastric contour. Similarly, programmed genetic signals must direct the endodermal progenitor cells in the esophagus to differentiate sequentially into ciliated, tall columnar, and squamous cells and those in the stomach to form glands containing parietal cells and chief cells. These genetic signals are largely unknown, but their phenotypic results can be recognized by morphologic study.

Menard and Arsenault (25), in 1987, provided experimental support for the existence of genetic programs that direct differentiation of foregut epithelial cells. They took explants of esophagus from early stage human fetuses and grew and maintained them in organ culture. When followed ultrastructurally, the esophageal epithelium was shown to develop islands of ciliated cells exactly like in the developing embryo.

I would like to suggest a uniform and reasonable terminology for the esophageal epithelia seen during fetal life that should clear the confusion (Table 3.1). This is essentially identical to Johns's terminology except for the use of the term "fetal esophageal columnar epithelium" to describe all epithelia and designating different types based on the gestational age that they appear (Table 3.1). Thus, the undifferentiated 2- to 3-layer endodermal progenitor cell epithelium that covers the entire esophagus in the first trimester is designated "fetal esophageal columnar epithelium, type I" (Fig. 3.3). This undergoes vacuolation and changes in thickness but remain as a stratified nonciliated columnar epithelium till the 40-mm embryo stage (Fig. 3.4). It is replaced in the second trimester by a stratified ciliated epithelium ("fetal esophageal columnar epithelium, type II"), beginning in the 40-mm embryo and completed by the 62-mm embryo (Figs. 3.5 and 3.6). "Fetal esophageal columnar epithelium, type I," will never be expressed in the esophagus after the second trimester. The ciliated epithelium covers the entire esophagus except a small area at the upper and lower ends until the 22nd week of gestation. At the two ends, the esophagus has 1 to 2 mm of a short or tall, nonstratified, nonciliated columnar epithelium and may be associated with superficial mucosal glands ("fetal esophageal columnar epithelium, type III" Figs. 3.6, 3.7, 3.8, 3.19, 3.21, and 3.24). Beginning in the 22nd week of gestation in the midesophagus, the esophageal mucosal surface becomes progressively lined by stratified squamous epithelium (Figs. 3.8 and 3.9). By the mid-third trimester, the entire area covered by ciliated epithelium has been replaced by stratified

TABLE 3.1 Epithelial Differentiation in the Fetal Esophagus

Type	Description	Location	Gestational age
Fetal esophageal columnar type I	Stratified columnar epithelium 2– 3 layers; undifferentiated	Entire length	First trimester
Fetal esophageal columnar type II	Stratified ciliated epithelium	Entire length	Second and early third trimester
Fetal esophageal columnar type III	Nonstratified tall or short columnar epithelium with short foveolar pit	Upper and lower ends	Third trimester; rarely after birth
Fetal esophageal columnar type IV	Intestinal type with goblet cells	Lower end—rare	Rare; transient
Superficial glands (fetal and adult)	Initially associated with nonstratified columnar epithelium; later under squamous epithelium	Entire length	Third trimester; persist into adult life
Squamous (fetal and adult)	Stratified squamous epithelium	Begins in middle third	22nd week to adult
Deep glands (adult)	Submucosal glands with gland ducts draining into surface	Entire length	Postnatal to adult life

squamous epithelium. Fetal esophageal columnar epithelium, type II, will never be expressed in the esophagus after the third trimester.

Fetal esophageal columnar epithelium, type III, remains at the lower end of the esophagus, above the gastroesophageal junction in late fetal life and often in the early postnatal period as a 1- to 2-mm zone between the squamous epithelium and parietal cell containing gastric mucosa, which begins at the true gastroesophageal junction (angle of His in de Hertogh et al.'s study). Fetal esophageal columnar epithelium, type III, in the most distal 1 to 2 mm of the esophagus is replaced by squamous epithelium in the first year of life, probably related to completion of development of the lower esophageal sphincter. If this epithelium contained superficial mucous glands, these may persist into adult life as normal mucous glands in the lamina propria under the esophageal squamous epithelium (Figs. 3.7–3.9).

The only other columnar epithelial type that has been reported is intestinal type epithelium with goblet cells. This was reported as a single focus in a 23-week fetus. This is beautifully and convincingly illustrated and reported to contain acid mucin by positive staining with Alcian blue at pH 2.5 (17). This can be designated "fetal esophageal columnar epithelium, type IV," with the recognition that it is a rare and transient manifestation in the fetal esophagus.

When development has been completed, the entirety of the fetal columnar epithelium of all types has been replaced by squamous epithelium and the esophagus is lined completely by squamous epithelium, transitioning to parietal cell containing gastric mucosa at the gastroesophageal junction (angle of His). Deep glands likely develop in the postnatal period as outpouchings of the squamous epithelium that pass through the muscularis mucosae into the submucosa (Figs. 3.10–3.12). These glands drain to the surface via ducts that are lined by either squamous epithelium, a multilayered columnar epithelium, or mixed squamous and columnar epithelium.

Fetal development of the gastric epithelium is much simpler (Table 3.2). The early undifferentiated endodermal progenitor cell epithelium (fetal gastric columnar epithelium, type I) is identical in appearance to fetal esophageal columnar epithelium, type I (Fig. 3.14a and b). Beginning in the 11th week of gestation, this develops into the fetal gastric columnar epithelium, type II, by the formation of foveolar-gland complexes that differentiate into surface mucous cells and glandular chief and parietal cells (Fig. 3.14c and d). This is the adult gastric epithelium; after the 17th week of gestation, gastric epithelial development is basically growth by increase in mucosal thickness and cell numbers to reach the adult size (Figs. 3.15 and 3.16). Parietal cells, which represent a specific morphologic marker of gastric epithelial differentiation, are seen in the epithelium at the angle of His in the second trimester.

Based on these observations of fetal epithelial development, the presumed genetic signaling mechanism can be constructed (Fig. 3.1, Table 3.3). The fetal endodermal genetic signal ("foregut columnar type I") directs the endodermal cell to form an undifferentiated stratified columnar epithelium, which lines both esophagus and stomach in the first trimester. In the esophagus, this primitive signal is suppressed in the second trimester, with simultaneous expression of the esophageal columnar type II signal that results in a stratified ciliated epithelium that covers the entire esophagus except at the two ends where there is expression of the esophageal columnar type III which results in a nonstratified columnar epithelium (type III). The esophageal columnar type II signal is suppressed progressively and replaced by the adult squamous signal, beginning in the midesophagus in the 22nd week and extending in both directions. The esophageal columnar type II signal is not expressed after the third trimester. The esophageal columnar type III signal is maintained until the lower esophageal sphincter completes development in the first year of life, permitting fine-tuning of the mucosal type to the exact physiologic gastroesophageal junction while the lower esophageal sphincter is developing. It is then replaced by the squamous signal in the esophagus. Distal to the angle of His in the stomach, the primitive

TABLE 3.2 Epithelial Differentiation in the Fetal Gastric Body

Type	Description	Location	Gestational age
Fetal gastric columnar type I	Stratified columnar epithelium 2–3 layers; undifferentiated	Entire	First trimester
Fetal gastric columnar type II (fetal and adult)	Foveolar-gland complexes with parietal cells	Entire	From 13th week to adult life

TABLE 3.3 Postulated Genetic Signals That Operate in Fetal Life during Development of the Esophagus and Stomach

Genetic signal	Manifestation	Age expressed	Age suppressed
Foregut columnar type I	Fetal esophageal and gastric epithelia type I	First trimester	Second trimester
Esophageal columnar type II	Fetal esophageal columnar epithelium, type II (ciliated)	Second trimester	Third trimester
Esophageal columnar type III	Fetal esophageal columnar epithelium, type III (columnar)	Second trimester	Third trimester or infancy
Esophageal columnar type IV	Fetal esophageal columnar epithelium, type IV (intestinal)	Second trimester (rare)	Second trimester (rare)
Esophageal squamous	Squamous epithelium	22nd week	Never without disease state
Gastric columnar epithelium, type II	Adult gastric epithelium	17th week	Never without disease state

foregut signal is replaced by the expression of the gastric columnar type II signal, which results in the development of the adult gastric mucosa. It is possible that this fine-tuning in signals is dependent on postnatal acid exposure of the epithelium, the presence of acid inducing gastric type signals and the absence of acid above the lower esophageal sphincter inducing the squamous signal.

Once development has been completed, the esophagus is under the squamous genetic signal and the stomach is under the gastric genetic signal throughout adult life. Changes of these signals occur only in pathologic states.

Little is known about the exact nature of the genetic signals that are operative in the development of the esophageal epithelium. In the stomach, however, evidence suggests that a gene known as the Sonic Hedgehog (SHH) gene is involved in directing the epithelium to form parietal cells (1). The Sonic Hedgehog gene is selectively expressed in gastric mucosa in all regions that show parietal (oxyntic/fundic) cells. It is lost in gastric atrophy where the fundic glands disappear and intestinal metaplasia occurs. Significant expression of the Sonic Hedgehog gene in other areas of the gastrointestinal tract has been reported in two other locations: (a) in Meckel's diverticulum, which shows heterotopic gastric mucosa with parietal cell containing fundic type glands, and (b) in columnar-lined esophagus, where its expression is restricted to epithelia that contain parietal cells (fundic or oxyntocardiac type). The Sonic Hedgehog gene is not present in metaplastic esophageal epithelia devoid of parietal cells such as cardiac and intestinal types. It is also not expressed in squamous epithelium.

It appears then that the sonic Hedgehog gene is a genetic signal that determines parietal cell differentiation in endodermal epithelia of the gastrointestinal tract and is expressed wherever parietal cells are seen. Similar genes must be present that dictate squamous epithelium in the esophagus as well as the various early fetal columnar epithelial types; these have yet to be identified.

Expression of Cdx-1 and Cdx-2 homeobox transcription factors are believed to play an important role in directing cell differentiation in the gut (2). In the mouse, these genes are expressed throughout the endoderm in early fetal development but become suppressed in the foregut. Their expression in the adult small intestine and colon are believed to be important in driving differentiation in those sites. It has been shown that mice carrying constitutional mutations in one Cdx2 allele develop polyps in the colon that contain stratified squamous epithelium and gastric epithelium (26, 27). These findings suggest that the absence of normal Cdx2 expression in the colon results in differentiation along lines normally seen in the esophagus and stomach.

The genetic signals involved in normal differentiation can be regarded as temporary fetal signals that produce fetal epithelia and adult signals that produce adult epithelia. Most of the fetal genetic signals in the foregut are unknown. The adult esophageal signal that dictates progenitor cell differentiation into squamous cells arises in the midesophagus in the second trimester, spreading to involve the entire esophagus in the first year of life, after which it remains throughout adult life in the physiologic state. The nature of this "squamous" genetic signal has not been elucidated; absence of Cdx1 and Cdx2 in the esophagus may contribute to squamous differentiation. In the stomach, the adult Sonic Hedgehog signal is presumably activated in the first trimester, coinciding with the development of the parietal cells in gastric mucosa. This also remains throughout adult life in the physiologic state. Changes in the normal adult-type genetic signals occur only in pathologic states of the esophagus and stomach, causing metaplasia.

References

1. van den Brink GR, Hardwick JC, Nielsen C, Xu C, ten Kate FJ, Glickman J, van Deventer SJ, Roberts DJ, Peppelenbosch MP. Sonic Hedgehog expression correlates with fundic gland differentiation in the adult gastrointestinal tract. Gut 2002; 51(5):628–633.
2. Silberg DG, Swain GP, Suh ER, Traber PG. Cdx1 and Cdx2 during intestinal development. Gastroenterology 2000;119: 961–971.
3. Silberg DG, Furth EE, Taylor JK, Schuck T, Chiou T, Traber PG. CDX1 protein expression in normal, metaplastic, and neoplastic human alimentary tract epithelium. Gastroenterology 1997; 113:478–486.
4. Eda A, Osawa H, Satoh K, Yanaka I, Kihira K, Ishino Y, Mutoh H, Sugano K. Aberrant expression of CDX-2 in Barrett's epithelium and inflammatory esophageal mucosa. J Gastroenterol 2003;38:14–22.
5. Phillips RW, Frierson HF, Moskaluk CA. Cdx2 as a marker of epithelial differentiation in the esophagus. Am J Surg Pathol 2003;27:1442–1447.
6. Vallbohmer D, DeMeester SR, Peters JH, et al. Cdx-2 expression in squamous and metaplastic columnar epithelia of the esophagus. Dis Esophagus 2005; (in press).
7. Hayward J. The lower end of the oesophagus. Thorax 1961;16:36–41; The phreno-esophageal ligament in hiatal hernia repair. Thorax 1961;16:41–45.
8. Allison PR, Johnstone AS. The oesophagus lined with gastric mucous membrane. Thorax 1953;8:87–101.
9. Barrett NR. The lower esophagus lined by columnar epithelium. Surgery 1957;41:881–894.
10. Johns BAE. Developmental changes in the oesophageal epithelium in man. J Anat 1952;86:431–442.
11. Derdoy JJ, Bergwerk A, Cohen H, Kline M, Monforte HL, Thomas DW. The gastric cardia: To be or not to be? Am J Surg Pathol 2003;27:499–504.
12. Liebermann-Meffert D, Duranceau A, Stein HJ. Anatomy and Embryology, Chapter 1. *In*: Orringer MB, Heitmiller R (eds). The Esophagus, Vol 1, Zudeima GD, Yeo CJ. Shackelford's Surgery of the Alimentary Tract, 5th ed. Saunders, Philadelphia, 2002, pp 3–39.
13. Schridde H. Uber magenschleimhautinseln vom Ban der Cardialdrusenzone und Fundusdrusenregion und den unteren oesophagealen Cardialdrusen gleichende Drusen im obersten Oesophagusabschnitte. Virchows Arch 1904;175:1–16.
14. Johnson FP. The development of the mucous membrane of the oesophagus, stomach and small intestine in the human embryo. Amer J Anat 1910;10:521–559.
15. Lewis FT. The development of the oesophagus. *In*: Keibel & Mall's Manual of Human Embryology, pp 2355–2368. Lippincott, Philadelphia and London.
16. Goetsch E. The structure of the mammalian oesophagus. Amer J Anat 1910;10:1–40.
17. Ellison E, Hassall E, Dimmick JE. Mucin histochemistry of the developing gastroesophageal junction. Pediat Path Lab Med 1996;16:195–206.
18. Salenius P. On the ontogenesis of the human gastric epithelial cells: A histologic and histochemical study. Acta Anat (Basel) 1962;(Suppl 46)50:S1–S70.
19. Menard D, Arsenault P. Cell proliferation in developing human stomach. Anat Embryol 1990;182:509–516.
20. De Hertogh G, Van Eyken P, Ectors N, Tack J, Geboes K. On the existence and location of cardiac mucosa: An autopsy study in embryos, fetuses, and infants. Gut 2003;52:791–796.
21. Park YS, Park HJ, Kang GH, Kim CJ, Chi JG. Histology of gastroesophageal junction in fetal and pediatric autopsy. Arch Pathol Lab Med 2003;127:451–455.
22. Chandrasoma PT. Fetal "cardiac mucosa" is not adult cardiac mucosa. Gut 2003;52:1798.
23. Chandrasoma PT, Der R, Ma Y, et al. Histology of the gastroesophageal junction: An autopsy study. Am J Surg Pathol 2000;24:402–409.
24. Zhou H, Greco MA, Daum F, et al. Origin of cardiac mucosa: Ontogenic considerations. Pediatr Dev Pathol 2001;4:358–363.
25. Menard D, Arsenault P. Maturation of human fetal esophagus maintained in organ culture. Anat Rec 1987;133:219–223.
26. Chawengsaksophak K, James R, Hammond VE, Kontgen F, Beck F. Homeosis and intestinal tumours in Cdx2 mutant mice. Nature 1997;386:84–87.
27. Beck F, Chawengsaksophak K, Waring P, Playford RJ, Furness JB. Reprogramming of intestinal differentiation and intercalary regeneration in CDx2 mutant mice. Proc Natl Acad Sci USA 1999;96:7318–7323.

CHAPTER

4

Normal Anatomy; Present Definition of the Gastroesophageal Junction

ANATOMY AND PHYSIOLOGY

The esophagus is a tube that is delimited by upper and lower sphincters. Its function is to transmit a bolus of food from the pharynx to the stomach. This function is performed by a coordinated muscle pump, which develops a peristaltic wave that propels the bolus downward. The sphincter at either end prevents the regurgitation of gastric contents into the esophagus and esophageal contents into the pharynx.

Deglutition is initiated voluntarily by the action of the tongue that pushes and compresses a bolus of food from the oral cavity into the pharynx. The entry of the food bolus into the pharynx stimulates the involuntary process of swallowing. The initial or pharyngeal phase of swallowing is mediated by the swallowing center via the cranial nerves. Contraction of the pharyngeal muscles increases intrapharyngeal pressure. All the orifices of the pharynx except the esophageal opening are shut off; the larynx elevates, the epiglottis closes, the palate moves up to close the nasopharynx, and the faucial pillars approximate to close off the oral cavity. Coördinated exactly with the increase in the pressure in the oropharynx, the cricopharyngeal (upper esophageal) sphincter relaxes, permitting the food bolus to enter into the esophagus. The movement of the bolus is assisted by the pressure gradient between the high pharyngeal pressure and the low intrathoracic/esophageal pressure. As the bolus passes into the esophagus the cricopharyngeal sphincter closes (i.e., reverts to its resting high pressure), but the contraction wave initiated in the pharynx continues down the esophagus pushing the bolus towards the stomach.

The muscle wall of the proximal esophagus consists of striated muscle that progressively and at a variable point in the upper esophagus gradually becomes admixed with smooth muscle (1). By the end of the upper third of the esophagus and sometimes more proximally, the esophageal wall consists entirely of smooth muscle, arranged as an inner circular layer and an outer longitudinal layer. Between the two is the myenteric plexus, which is the nerve center of the esophagus that communicates with the submucosal plexus and cranial nerves. The neuromuscular function of the esophagus is largely a local phenomenon and is not significantly altered by complete denervation. The exact mechanism by which the sensory input and motor activity is coordinated by the nerves and ganglion cells in the submucosal and myenteric plexuses is not well understood. The roles played by sensory nerve endings in the squamous epithelium, the neurotransmitters involved, and the mechanisms involved in generating the contractile wave and controlling sphincter function are largely unknown.

The esophageal phase of swallowing is controlled by the intrinsic function of the muscular wall of the esophagus, which generates a peristaltic wave. Propulsive peristalsis in the esophagus pushes the bolus very effectively down the esophagus. Waves that arise in the proximal esophagus as a result of entry of a bolus of food are known as primary waves; those that begin lower down in the esophagus are called secondary waves. The lower esophageal sphincter relaxes on the initiation of a swallow to accept the oncoming bolus and permits its passage into the stomach. The lower esophageal sphincter maintains a resting pressure of

15 to 20 mm Hg, which is restored after the food bolus passes into the stomach. This ensures that increased intragastric pressure, caused by transmitted abdominal pressure and gastric distention is not vented with gastric content into the esophagus.

Anatomically, the normal esophagus begins as an extension of the pharynx in the neck, passes through the posterior mediastinum, and enters the abdomen through the diaphragmatic hiatus (Fig. 4.1). The length of the esophagus is approximately 25 cm in the adult but varies with the height of the individual. The upper end is approximately 15 cm from the incisor teeth. The incisor teeth is a common reference point used for endoscopic measurement. The esophagus extends in an average person from 15 to 40 cm from the incisor teeth. The cricopharyngeal sphincter is 2 to 3 cm, and the lower esophageal sphincter is 3 to 4 cm long.

The esophagus ends distally in the lower esophageal sphincter. At the caudal end of the lower esophageal sphincter is the gastroesophageal junction. The esophagus has a muscle wall, a submucosal layer, and a mucosa; it is surrounded by adventitia, which merges with the mediastinal connective tissue above the diaphragmatic hiatus and with the connective tissue around the phrenoesophageal ligaments in its abdominal part. The esophagus is not covered by peritoneum, which becomes reflected off the diaphragm to the stomach at the gastroesophageal junction (Fig. 4.2).

In the early embryo, the mesoderm around the esophagus at the level of the septum transversum participates in the development of the crura of the diaphragm, which are intimately related to the wall of the esophagus (2). Approximately 2 cm of the esophagus lie below the diaphragm in the abdomen (Fig. 4.2). Within this portion of the esophagus the abdominal part of the lower esophageal sphincter is located; another 1 to 2 cm of the lower esophageal sphincter lie above the diaphragm in the mediastinum, and is called the thoracic portion of the sphincter giving a total sphincter length of 3 to 4 cm. There is no part of the esophagus in the abdomen that is devoid of a sphincter in normal people. The total sphincter length, the length

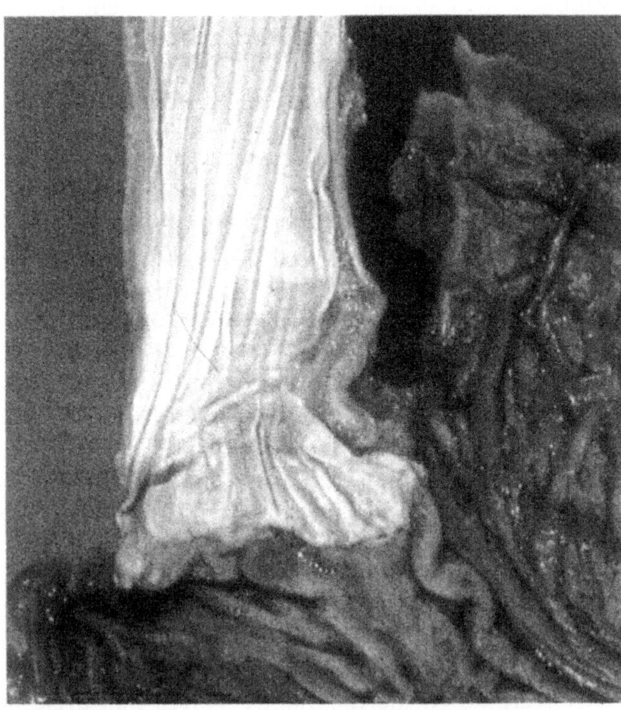

FIGURE 4.1 The normal gross appearance of the esophagus and stomach. The esophagus is a tube that is entirely lined by squamous epithelium. The squamocolumnar or Z line is horizontal and ends abruptly at the end of the tubular esophagus. The stomach is saccular at this point with an angle of His that is created by the fundus passing upward and to the left. The rugal folds of the stomach extend all the way to the Z line. (Cross reference: Color Plate 1.1)

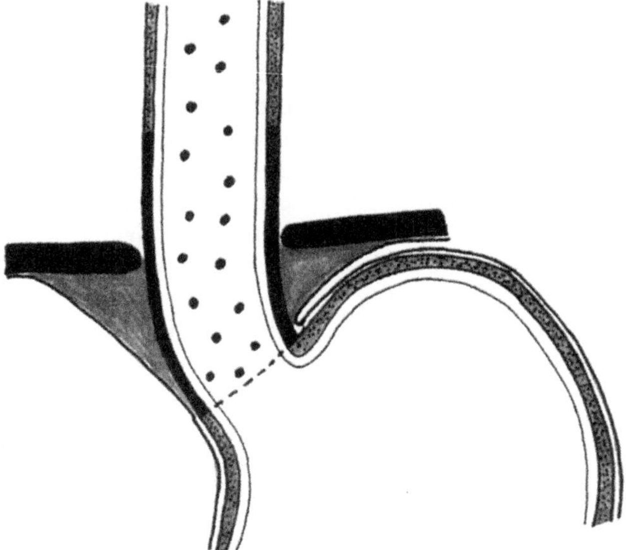

FIGURE 4.2 The anatomy of the distal esophagus and proximal stomach. The esophagus passes from thorax to the abdominal cavity through the diaphragmatic hiatus. The lower esophageal sphincter zone (solid black muscle wall) is the distal 3 to 4 cm of the esophagus, including the entire intraabdominal part of the esophagus. The gastroesophageal junction (end of the esophagus indicated by interrupted line) is the end of the lower esophageal sphincter at which point the peritoneum reflects off the gut to the undersurface of the diaphragm, the esophageal musculature becomes gastric and the esophageal mucosa (irrespective of histologic type) transitions to gastric mucosa. The abdominal esophagus is surrounded by connective tissue in which condensations (the phrenoesophageal ligaments) are present. The black dots are the submucosal glands, which are an accurate marker of the esophagus; they are absent in the stomach.

of its abdominal component, and the resting pressure of the lower esophageal sphincter are all critical in the normal functioning of the sphincter to prevent reflux of gastric contents into the esophagus (3, 4).

Physiologically, the abdominal part of the esophagus behaves like an abdominal organ. During manometry, the respiratory pressure changes in the abdominal part of the lower esophageal sphincter are similar to those in the stomach, increasing with inspiration and decreasing with expiration. Above the diaphragm, the thoracic portion of the lower esophageal sphincter and the esophageal body is physiologically in the thorax. At manometry, the point at which respiratory pressure excursions change from positive to negative with inspiration is known as the respiratory inversion point. In the normal person, this point is at or close to the diaphragmatic hiatus and the point of attachment of the superior leaf of the phrenoesophageal ligament (Fig. 4.3).

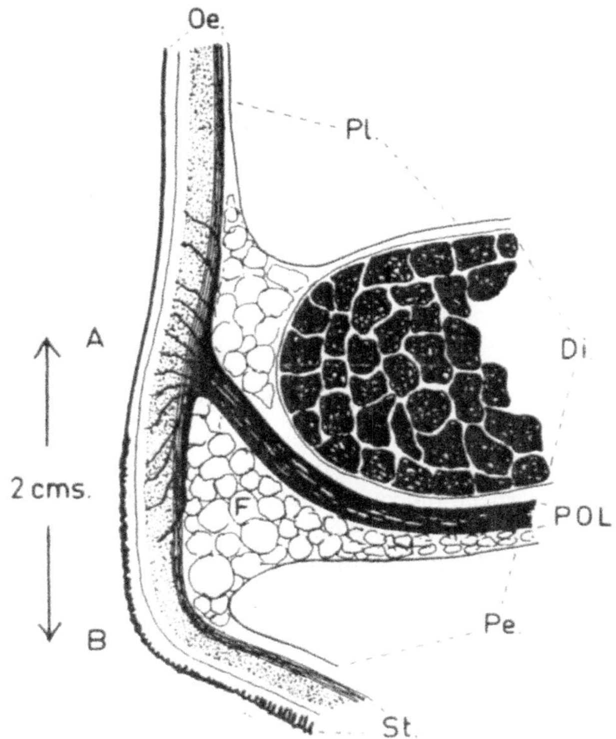

FIGURE 4.3 Anatomy of the distal esophagus (Oe), accurately shown by Hayward (5). The peritoneum (Pe) reflects off the undersurface of the diaphragm (Di) to the gastroesophageal junction located approximately 2 cm distal to the diaphragmatic hiatus. The abdominal segment of the esophagus is surrounded by extraperitoneal fat. The superior phrenoesophageal ligament (POL) is shown attaching to the esophagus at the diaphragmatic hiatus. Reproduced with permission from Hayward J. The phreno-esophageal ligament in hiatal hernia repair. Thorax 1961;16:41–45.

The anatomy around the abdominal esophagus is complicated (Figs. 4.2 and 4.3). The peritoneum on the undersurface of the diaphragm reflects off the diaphragm to begin covering the gastrointestinal tract. The connective tissue between the peritoneum and the undersurface of the diaphragm condenses near the diaphragmatic hiatus to form two ligaments, the superior and inferior leaf of the phrenoesophageal ligaments, that are variably developed in different people and sometimes not clearly recognized. Some authorities even question their existence. The inferior leaf of the phrenoesophageal ligament passes close to the peritoneal reflection and attaches to the abdominal esophagus close to the gastroesophageal junction; it is really the connective tissue under the peritoneum. The superior phrenoesophageal leaf, which is usually better defined, passes upward as a conelike structure to attach to the esophagus close to the diaphragmatic hiatus about 2 to 3 cm above the angle of His and the peritoneal reflection (5; Fig. 4.3). The two ligaments enclose fibro-adipose connective tissue that surrounds the abdominal 2 to 3 cm of the esophagus.

The stomach normally begins at the end of the tubular esophagus at the point where the peritoneum reflects onto the diaphragm. On the right side, the lesser curvature passes straight down and curves to the right toward the pylorus. On the left side, the embryonic overgrowth of the upper part of the greater curvature has produced an acute angle of His because the fundus of the stomach moves upward and to the left from the end of the esophagus to abut the undersurface of the diaphragm (Figs. 4.1 and 4.2).

It is interesting to consider the relationship of the intraluminal pressures of the esophagus and stomach and their shape (Fig. 4.4). In the mediastinum, the thoracic esophagus has a negative intraluminal pressure like all thoracic structures. As a result, the esophagus tends to retain its tubular shape and remain collapsed. The stomach, on the other hand, is an intraabdominal organ, which has a positive intraluminal pressure. As it fills, the positive intraluminal pressure slightly increases as the stomach distends, serving as its main function of meal storage. The stomach acts as a true reservoir, accepting the volume of a meal by filling to its capacity without much rise in intraluminal pressure due to the process of adaptive relaxation of its muscular walls. Under normal conditions, when the abdominal portion of the lower esophageal sphincter is exposed to the same intraabdominal pressure as the stomach a sphincter pressure of 15 to 20 mm Hg above abdominal pressure very effectively prevents reflux of gastric contents into the esophagus even after a meal. Overeating with the associated distension of the stomach beyond its normal capacity is necessary for a

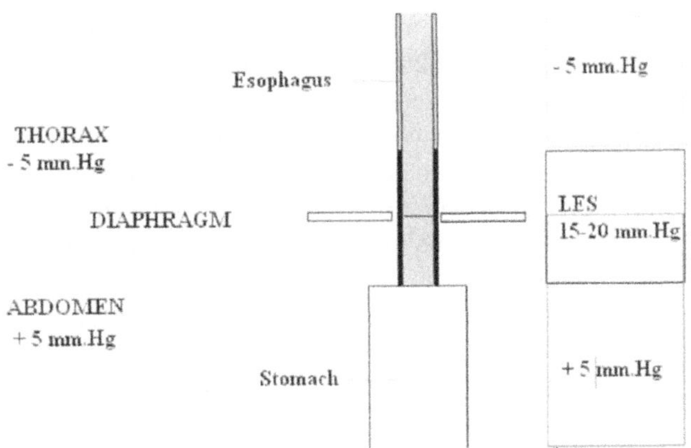

FIGURE 4.4 Diagrammatic representation of the distal esophagus and proximal stomach to show the pressures. The intrathoracic esophagus (above the line, which represents the demarcation of the thorax and abdomen; this is the respiratory inversion point at manometry) has an intraluminal pressure of −5 mmHg. The lower esophageal sphincter zone, which straddles the diaphragmatic hiatus, is a high-pressure zone with a resting pressure of 15 to 20 mmHg. The intraluminal pressure of the intraabdominal esophagus and stomach is +5 mmHg. However, the existence of the sphincter in the distal abdominal esophagus normally masks the positive intraluminal pressure therein.

significant increase in intragastric pressure. Overeating does result in a loss of length of the abdominal portion of the lower esophageal sphincter as it is taken up by the distending stomach much like the neck of a balloon shortens when it is inflated. If too much of the abdominal portion of the sphincter is taken up the sphincter gives way and reflux of gastric contents into the esophagus will occur.

An effective lower esophageal sphincter, by preventing gastric contents refluxing into the esophagus, also maintains a sharp pH gradient at the gastroesophageal junction. Distal to this point, in the stomach, the pH is highly acid (in the 1–3 range). Proximal to this point, in the esophagus, the pH is in the neutral (approximately 7) range. One effect of gastroesophageal reflux is to increase the length of this pH gradient; as the volume of reflux increases, increasing lengths of the distal esophagus become subject to pH levels that range between the acidity of the stomach and neutrality of the esophagus.

Anatomy of the Lower Esophageal Sphincter

There is controversy as to whether the lower esophageal sphincter can be recognized anatomically. Most authorities believe that the sphincter is purely physiologic and there is no recognizable difference in the muscle wall between the sphincter and the more proximal esophagus. Cursory examination of this area in esophagectomy specimens leads one to agree with this conclusion. Very careful examination shows subtle differences in the musculature in the region of the sphincter that are claimed to be related to the sphincter action. These are so difficult to characterize that this remains an academic issue only. In practice, any reasonable examination of the lower esophagus does not permit the anatomic definition of the lower sphincter or develop sufficiently precise criteria to recognize associated pathologic lesions. The definition of pathologic abnormalities of the lower esophageal sphincter is based entirely on manometric studies *in vivo*. There has never been any correlation with manometric abnormalities and specific morphologic changes in the musculature in conditions such as hypertensive lower esophageal sphincter, achalasia or the hypotensive sphincter associated with severe reflux disease. The only correlations that exist are the loss of ganglion cells in the entire distal esophagus in achalasia of the cardia and the muscular atrophy with fibrosis that accompanies esophageal motility abnormalities seen in scleroderma.

At the distal end of the esophagus, the two-layered structure of the muscle wall changes to that of the stomach where the circular and longitudinal layers are less well defined and a third oblique layer is present. The transition between the two muscle walls theoretically provides a method of identifying the gastroesophageal junction, but the transition is so gradual that its precise definition within millimeters is not practicable.

Esophageal Glands

The normal esophagus contains two types of glands in the wall.

1. The first of these are probably remnants of the fetal superficial mucous glands (6). These appear in the second trimester as outpouchings from the nonciliated columnar epithelium of the fetal esophagus that is commonly present at the two ends of the esophagus. These outpouchings do not penetrate the muscularis mucosae. When fully developed, they are seen as lobulated glands in the lamina propria of the mucosa immediately under the epithelium. In fetal life, the epithelium is a nonciliated columnar epithelium, but this is replaced by squamous epithelium at or soon after birth. In the adult, these mucous glands are seen in the lamina propria under the squamous epithelium in both resected specimens and biopsies (see Fig. 3.9). They resemble minor salivary glands and are relatively few in number.

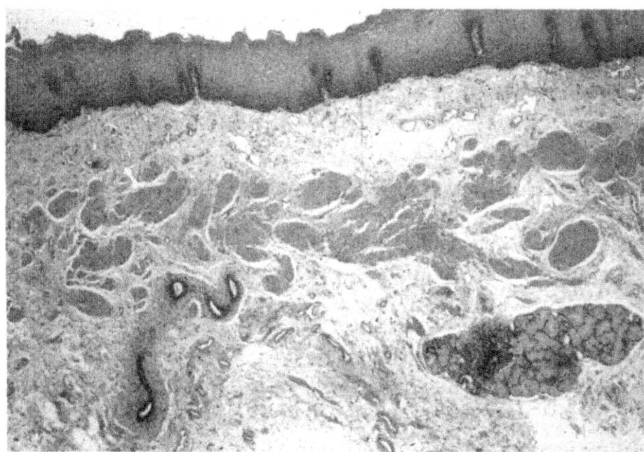

FIGURE 4.5 Submucosal gland in the esophagus. The gland is composed of mucous cells and has a duct. It is located under the muscularis mucosae. To the left is another duct without a secretory glandular unit. (Cross reference: Color Plate 1.3)

FIGURE 4.7 Gland duct as it traverses the mucosa and opens at the surface of cardiac mucosa in the esophagus. The section of the duct in the deeper mucosa shows a stratified columnar epithelial lining; at the point of opening at the surface, the epithelium is squamous.

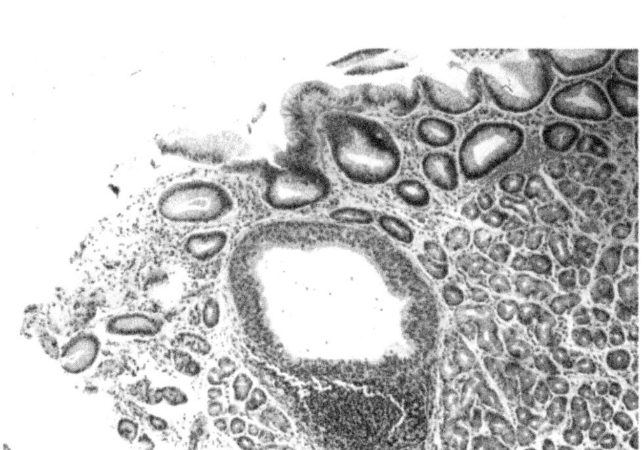

FIGURE 4.6 Duct of a submucosal gland as it traverses oxyntocardiac mucosa in the esophagus. The duct is dilated and lined by a stratified columnar epithelium.

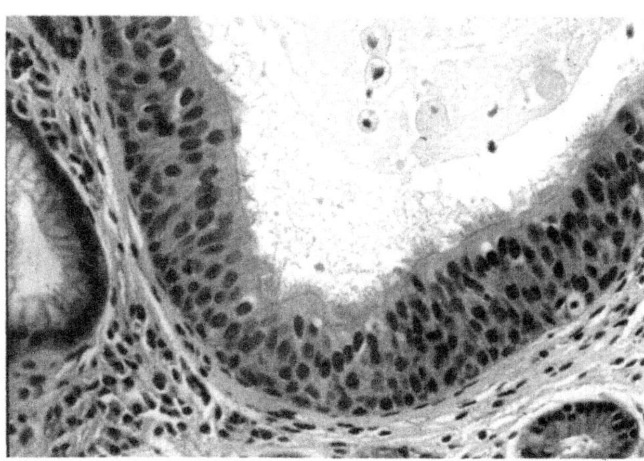

FIGURE 4.8 Ciliated cells in the stratified columnar epithelium of a gland duct in a mucosal biopsy of cardiac type metaplastic esophageal epithelium.

2. The deep (submucosal) glands are characteristic of the esophagus (7). Submucosal glands are invariably found in the esophagus and are uniformly distributed. Goetsch (8) estimated that in the adult there are 60 to 741 deep esophageal glands. These glands are seen as large, lobulated glands composed of mucous cells in the submucosa (Fig. 4.5). They have well-developed ducts that are lined by stratified columnar, squamous, or a mixed columnar and squamous epithelium (Figs. 4.6 and 4.7). Rarely, ciliated epithelium is present in the ducts (Fig. 4.8). These ducts pass up through the muscularis mucosae and lamina propria to penetrate the overlying squamous epithelium; when the ducts penetrate metaplastic esophageal columnar epithelium, their surface region commonly forms a residual squamous island (Fig. 4.7).

Johns (6) reported that these deep esophageal glands are not present in embryos and suggested that they form after birth:

> The present work suggests that the deep glands do not appear until the epithelium has become stratified squamous in character, and that they develop as outgrowths from this epithelium. . . . Since the deep glands are so commonly present and widely distributed in the adult, the present work indicates that their development must be mainly postnatal. (p. 438)

Lobulated and well-defined mucous glands are not present in gastric mucosa either in the lamina propria or in the submucosa. As such, these glands are a marker for the esophagus. Normally, these esophageal glands are covered by squamous epithelium. However, they remain in pathologic states where the squamous epithelium undergoes columnar metaplasia. In such cases, the presence of gland ducts and glands can be used to define a given point that contains these glands as esophageal and not gastric (9, 10). It is an esophageal characteristic that Allison and Johnstone (11) originally used in 1953 to prove to Barrett (12) that his tubular intrathoracic stomach was actually esophagus.

The Anatomic Gastroesophageal Junction

The normal end of the esophagus (or gastroesophageal junction) can be defined anatomically by (Figs. 4.1 and 4.2):

1. The peritoneal reflection.
2. The end of the tubular esophagus, recognized by the acute angle of His on the left side of the gastroesophageal junction. On the right side, the esophagus passes without significant flaring or angulation to become continuous with the lesser curvature of the stomach.
3. The change in the muscle wall structure from having two distinct layers (circular and longitudinal) in the esophagus to a less defined layering of the muscle of the gastric wall. This is not easy to determine by routine examination of specimens.
4. Manometrically, the normal esophagus ends at the distal end of the lower esophageal sphincter (Fig. 4.4).

It is not easy to correlate this external definition of the end of the esophagus with a point on the mucosal surface. The thickness of the wall of the esophagus is significant at this point and cuts can be made from the externally defined point of the peritoneal reflection to the mucosa in varying directions away from the vertical. This can result in a variation of several millimeters on either side of the vertical. This possible error must be kept in mind when studies attempting to define the mucosal gastroesophageal junction are evaluated. For example, in a study of fetuses by de Hertogh et al. (13), the reference point for the gastroesophageal junction was the angle of His (see Fig. 3.17). When the mucosal point that corresponded to this external gastroesophageal junction was examined, it was found that the most proximal parietal cell was either at or a maximum of 0.3 mm proximal to the angle of His. It must be recognized that 0.3 mm is within the error range of the exact point of the external gastroesophageal junction because of the possible differences in the direction of the vertical section that must be taken.

In resected specimens, the mucosa moves longitudinally in relation to the underlying muscle wall. The differential contraction of the divided muscle wall and the mucosa and the ability of the mucosa to slide relative to the muscle wall alter the relationship between the mucosa and muscle wall in a manner that is unpredictable. This makes correlation of the peritoneal reflection with a point on the mucosa fraught with considerable technical error.

The difficulty in finding the exact point of the mucosal gastroesophageal junction does not mean that there is no true gastroesophageal junction on the mucosal aspect. It is important, however, that any criterion used to define the mucosal junction must correlate with the external anatomically defined gastroesophageal junction to an extent that is within the range of technical error that may exist. *There is only one gastroesophageal junction; all definitions of it—anatomic, manometric, and endoscopic—must be concordant.*

Changes in Anatomy in Pathologic States

There is no requirement that normal anatomic characteristics are retained in pathologic states; in fact it is highly likely that they are not retained. Reflux disease of the esophagus causes maximal damage at the distal end of the esophagus and very likely distorts the normal anatomy, possibly in ways that have not yet been recognized. In patients with reflux, therefore, it can be a serious error to accept normal landmarks to define various structures.

For example, let us hypothesize that the lowermost 5 mm of the esophagus becomes dilated as a result of reflux-induced damage. If this happens, the point of flaring of the tubular esophagus and the angle of His will move 5 mm proximally, creating the illusion that the esophagus ends 5 mm more proximally than its true end if the point of flaring or angle of His is used as the criterion to define the gastroesophageal junction. In that event, the 5-mm segment of dilated reflux-damaged esophagus will be misinterpreted as being distal to the gastroesophageal junction and therefore part of the stomach. We will show in a later chapter that such dilatation of the reflux-damaged distal esophagus does occur (14). Possibilities such as this must be first conceptualized and then evaluated using meticulous criteria to provide accurate answers. This has not happened in esophageal anatomy; assumptions have been made too easily and dogmas established without adequate study let alone proof. This area is home to more unproved assertions of fact than most other areas of medicine. It behooves us to critically examine even our most cherished beliefs for the underlying proof.

The most severe anatomic abnormality of the distal esophagus is a sliding hiatal hernia (Fig. 4.9). This is associated with displacement of the abdominal esophagus above the diaphragm; but within the hernia sac. The portion of peritoneum that bulges up into the thorax functions as an extension of the abdominal cavity into the thorax. The angle of His is also lost and the stomach is above the diaphragm, also within the hernia sac. The stomach is recognizable as a dilatation but often not clearly separable from the end of the esophagus because there is no sharp distinction between the tubular esophagus and saccular stomach (Fig. 4.9a). The sliding hiatal hernia was described with near absolute accuracy by Allison (15) (Fig. 4.9b).

The only anatomic criteria used to define the normal gastroesophageal junction that do not alter in this pathologic state are the following: (a) One criterion is the peritoneal reflection; however, the location of the peritoneal reflection is impossible to determine accurately by endoscopy. (b) Another criterion is the point of attachment of the phrenoesophageal ligaments, which remains constant. The point of attachment of the superior phrenoesophageal ligament marks the original proximal limit of the abdominal part of the sphincter, being approximately 2 cm above the peritoneal reflection. Unfortunately, these ligaments become attenuated and difficult to define in disease states and lose their value as a reliable marker. Like the peritoneal reflection, the ligamentous attachment sites are not seen at endoscopy. (c) The third criterion is the submucosal esophageal mucous glands, which are usually unaffected by pathologic changes and remain as permanent markers of the esophagus in pathologic states. Because these glands are in the submucosa and drain via ducts into the mucosa (Figs. 4.5, 4.6, and 4.7), they retain a constant relationship to the mucosa of the

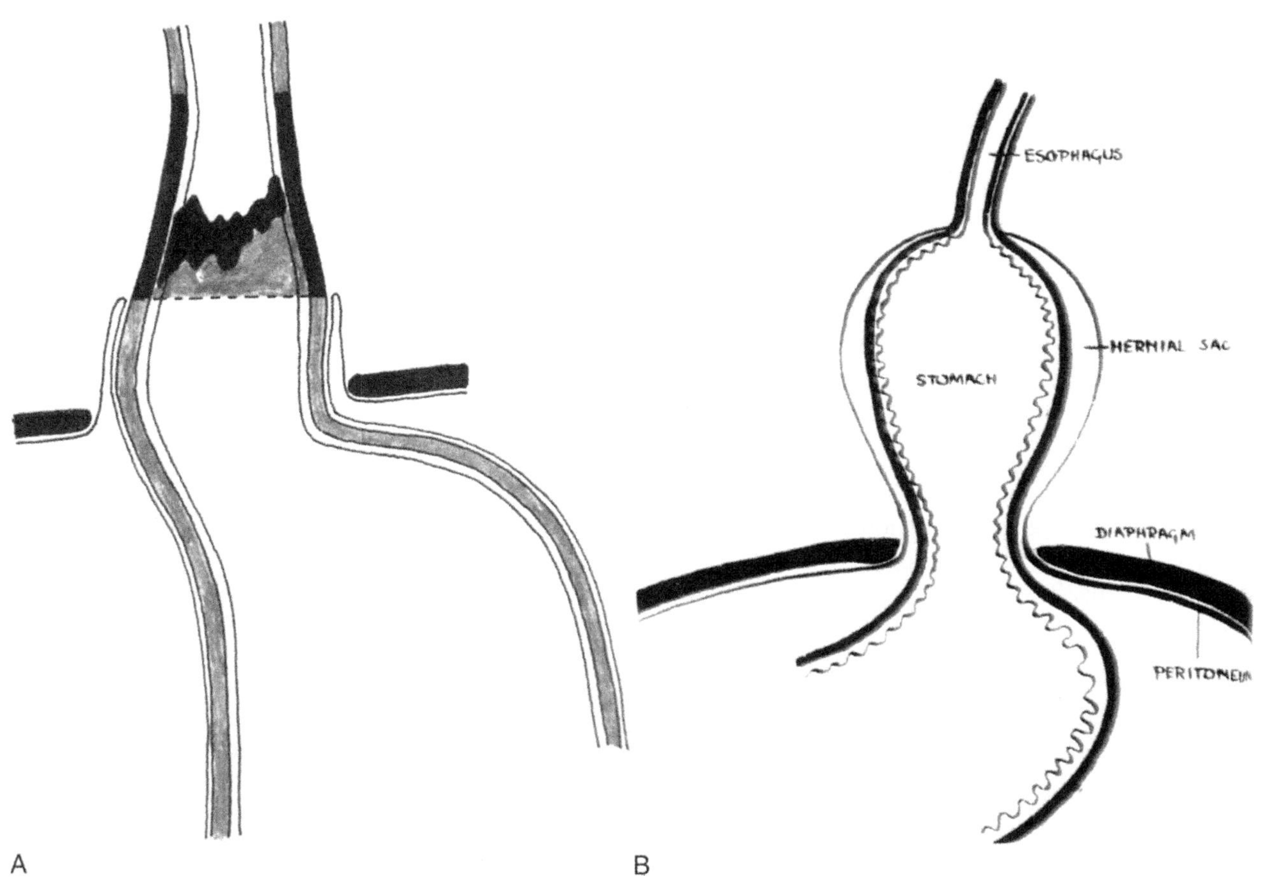

FIGURE 4.9 (a) Sliding hiatal hernia showing a shortened esophagus ending in the thorax and drawing up a peritoneum-lined hernia sac containing the proximal stomach. The gastroesophageal junction is at the peritoneal reflection; the entire sphincter is within the thorax. (b) From Allison (1948), showing a near-identical representation of the hiatal hernia. The only difference is that in Allison's drawing, there is a clearer demarcation between the end of the esophagus and the hiatal hernia. Reproduced with permission from Allison PR. Peptic ulcer of the oesophagus. Thorax 1948;3:20–42.

esophagus in histologic sections. Their presence is the best means of defining a given point anatomically as the esophagus. It should be recognized, however, that the glands are distributed sporadically in the esophagus and vary in number in different people. Not every section taken will show glands and the number of glands varies considerably in different esophagi. Therefore, while the presence of submucosal glands defines a point as the esophagus, the absence of such glands does not mean that the location is not esophageal. Because the most distal esophageal gland may be proximal to the true gastroesophageal junction, the most distal gland is not a precise marker for the junction. It may be located proximal to the true gastroesophageal junction. However, it is correct that *the true anatomic gastroesophageal junction must be either at or distal to the most distal esophageal submucosal gland in both normal and pathologic specimens.*

ENDOSCOPIC/GROSS LANDMARKS

At endoscopy and gross examination of esophagectomy and autopsy specimens, the following landmarks can be recognized with varying degrees of precision.

The Point of Flaring of the Tubular Esophagus

The point of flaring of the tubular esophagus generally corresponds to the angle of His. In normal people, this is sharply defined (Fig. 4.1); in patients with reflux-induced damage to the lower esophagus and particularly in patients with a hiatal hernia, the acute angle of His is progressively obliterated and the point of flaring becomes less defined (Figs. 4.9a and 4.10). The angle of His, therefore, is not a constant criterion of the anatomic gastroesophageal junction because it can be altered in patients with pathologic conditions of the distal esophagus. It therefore does not have sufficient precision to be a criterion that defines the gastroesophageal junction.

The Diaphragmatic Impression

The location of the diaphragmatic hiatus can often be recognized as an external constriction (Allison's pinchcock) of the esophagus at endoscopy, normally approximately 2 centimeters proximal to the point at which the tubular esophagus flares. This has never been regarded as sufficiently constant or precise as a landmark to permit any precise conclusion regarding anatomy, except to determine the presence of a hiatal hernia.

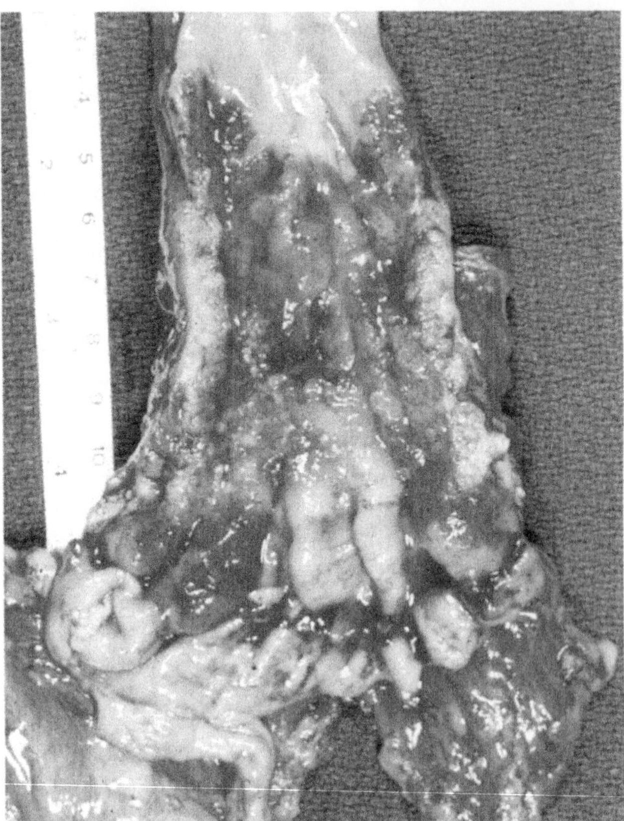

FIGURE 4.10 Esophagectomy specimen showing gradual flaring of the distal esophagus as it merges with the stomach. The serrated squamocolumnar junction is displaced proximally. The proximal limit of the rugal folds appears to be at multiple levels, making assessment of the gastroesophageal junction very difficult. (Cross reference: Color Plate 3.2)

The Squamocolumnar Junction (the Z Line)

The squamocolumnar junction or Z line can be accurately recognized in most cases. In some cases where there is inflammation and erosion of the squamous epithelium, the exact point of transition to columnar epithelium can be difficult to determine. Rarely, we receive biopsies labeled "rule out Barrett's" that show regenerative, inflamed squamous epithelium.

The squamocolumnar junction varies from horizontal (Fig. 4.1) to increasing degrees of irregularity and serration in different individuals (Figs. 4.11, 4.12, and 4.13).

Columnar Epithelium in Which There Are Longitudinal Rugal Folds

Longitudinal rugal folds are a normal feature of gastric mucosa. Rugated columnar epithelium is

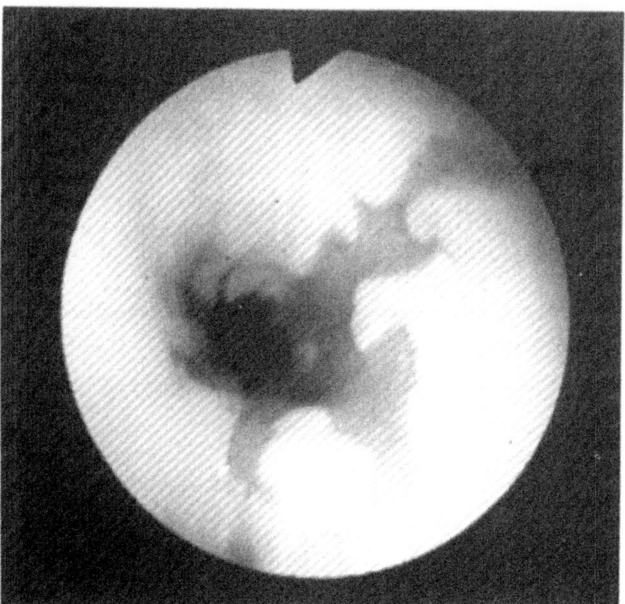

FIGURE 4.11 Endoscopic view of short-segment columnar-lined esophagus showing a serrated squamocolumnar junction with irregular tongues of columnar epithelium extending up into the squamous-lined esophagus from the gastroesophageal junction. Reproduced with permission from Tytgat GNJ. Endoscopic features of the columnar-lined esophagus. Gastroenterol Clin N Amer 1997;26:507–517.

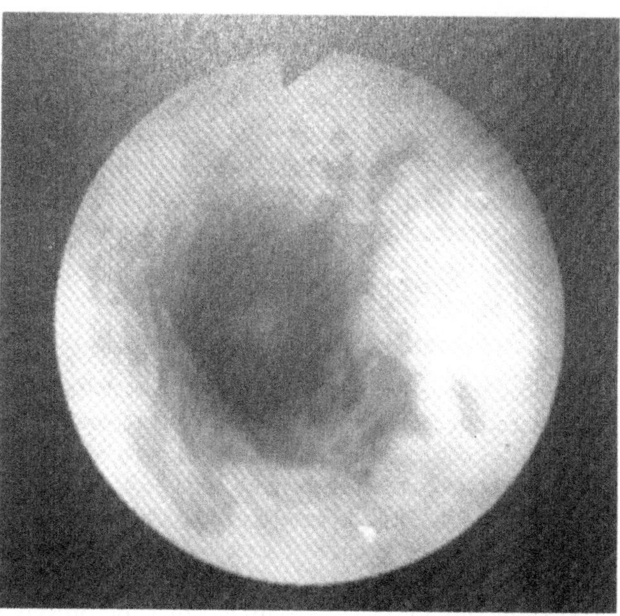

FIGURE 4.12 Endoscopic view of long-segment Barrett esophagus showing extension of the columnar lining in a circumferential manner into the midesophagus. The squamocolumnar junction is very irregular. Reproduced with permission from Tytgat GNJ. Endoscopic features of the columnar-lined esophagus. Gastroenterol Clin N Amer 1997;26:507–517.

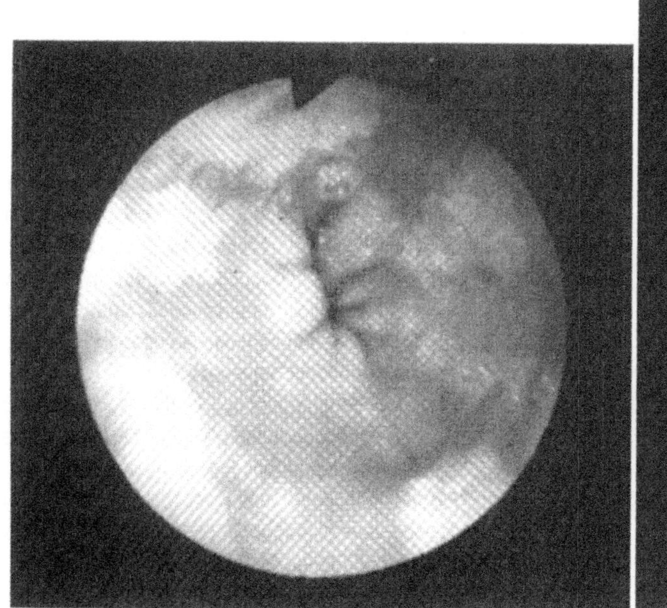

A

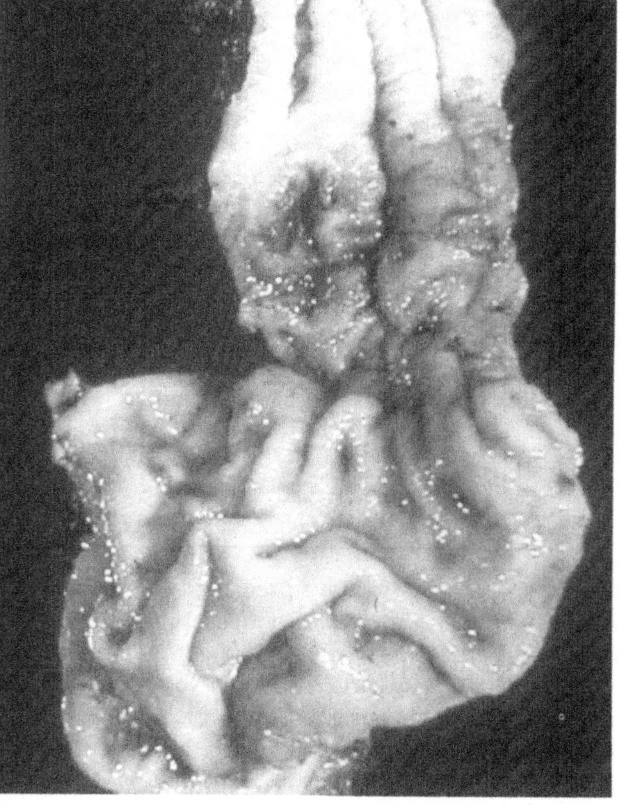

B

FIGURE 4.13 The proximal limit of the rugal folds. (a) Endoscopic view showing the proximal extent of rugal folds with a columnar-lined esophagus separting this from the serrated squamocolumnar junction. Note that the proximal limit of the rugal folds is not a sharply defined line. Reproduced with permission from Tytgat GNJ. Endoscopic features of the columnar-lined esophagus. Gastroenterol Clin N Amer 1997;26:507–517. (b) Gross specimen showing the proximal limit of the rugal folds at the end of the tubular esophagus. The rugal folds are well defined and merge with the columnar-lined esophagus. The squamocolumnar junction has been displaced proximally. (Cross reference: Color Plate 3.5)

usually seen in the saccular part at or beyond the point of flaring of the tubular esophagus (Figs. 4.13 and 4.14). In what is presently regarded as the normal state, the proximal limit of the rugal folds is coincidental with the horizontal Z line at the end of the tubular esophagus (Fig. 4.1).

Flat Columnar Epithelium

Flat columnar epithelium is usually seen distal to the proximally displaced squamocolumnar junction in the tubular esophagus and separates the often serrated Z line from the proximal limit of the rugal folds (Figs. 4.11, 4.12, and 4.13). The length of this flat epithelium that separates the proximal limit of the rugal folds from the Z line is the length of the columnar-lined esophagus. It should be recognized that the presence of a columnar-lined esophagus is not equivalent to Barrett esophagus; a biopsy from this flat epithelium must show intestinal metaplasia before the criteria for a diagnosis of Barrett esophagus are satisfied. Flat epithelium may sometimes be seen in the saccular part between the end of the tubular esophagus and the proximal limit of mucosa containing rugal folds (Fig. 4.14).

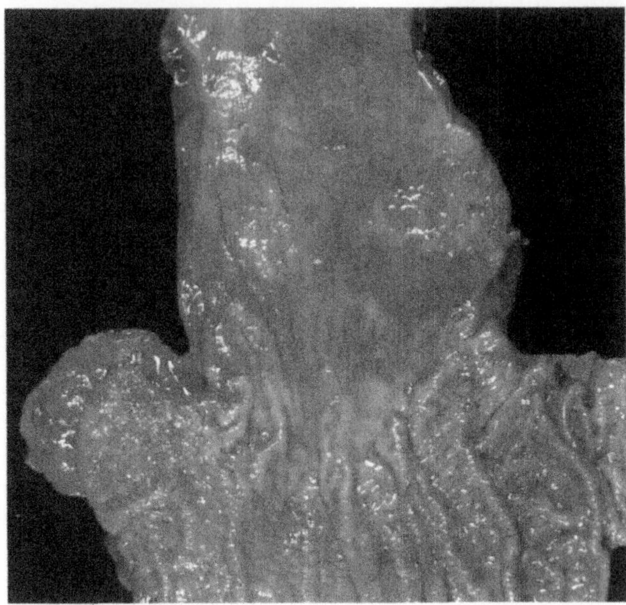

FIGURE 4.14 Gross specimen with columnar-lined tubular esophagus. The flat columnar epithelium lines the distal tubular esophagus and extends into the saccular organ distal to the end of the tubular esophagus, forming a concave border. Note the contrast with rugated epithelium distal to the flat epithelium. (Cross reference: Color Plate 3.3)

Squamous Islands in Columnar Mucosa

Another endoscopic/gross criterion that can be easily seen is the presence of squamous islands in columnar mucosa. They frequently occur in columnar-lined esophagus, usually related to the points of opening of the esophageal submucosal glands (Fig. 4.7). They can be seen endoscopically as pearly white islands in the flat columnar epithelium (Fig. 4.15). The presence of a squamous island in columnar mucosa is reliable evidence of esophageal location because they do not occur in gastric mucosa in the human.

THE MEANING OF ENDOSCOPIC/GROSS LANDMARKS

It is important not to have preconceived notions as to the meaning of any of these endoscopic/gross landmarks until these meanings have been subjected to testing and there is evidence that the meanings ascribed to them are correct. This is particularly true for flat and rugated columnar epithelium. While normal gastric mucosa typically has rugal folds, these commonly disappear in areas of gastritis. The absence of rugal folds, therefore, does not mean that the mucosa is not gastric. Similarly, metaplastic columnar epithelium in the esophagus is usually flat but can show rugal folds. When this occurs, "the proximal limit of the rugal folds" is not a well-defined horizontal line. Also, though it is universally accepted that the proximal limit of the rugal folds is the endoscopic gastroesophageal junction, this has never been proved to coincide with the true anatomic gastroesophageal junction. Flat and rugated columnar epithelia are simply endoscopically determinable landmarks; their significance in defining anatomy must be proved.

We must learn from the lesson Allison and Johnstone taught us in 1953 (11). They used persuasive scientific evidence to convince the medical community including Barrett that the accepted definition of the esophagus and gastroesophageal junction was incorrect. They showed that what Barrett was calling a tubular stomach (12) was a columnar-lined esophagus because "careful examination of such a specimen shows that it has no peritoneal covering, that the musculature is that of the normal oesophagus, that there may be islands of squamous epithelium within it, that there are no oxyntic cells in the mucosa, and that . . . there are present typical oesophageal mucous glands" (p. 87).

Of the many criteria that Allison and Johnstone used, the esophageal submucosal glands are the most constant and reliable marker to differentiate between esophagus and stomach. These glands retain a constant relationship to the mucosa in resected specimens

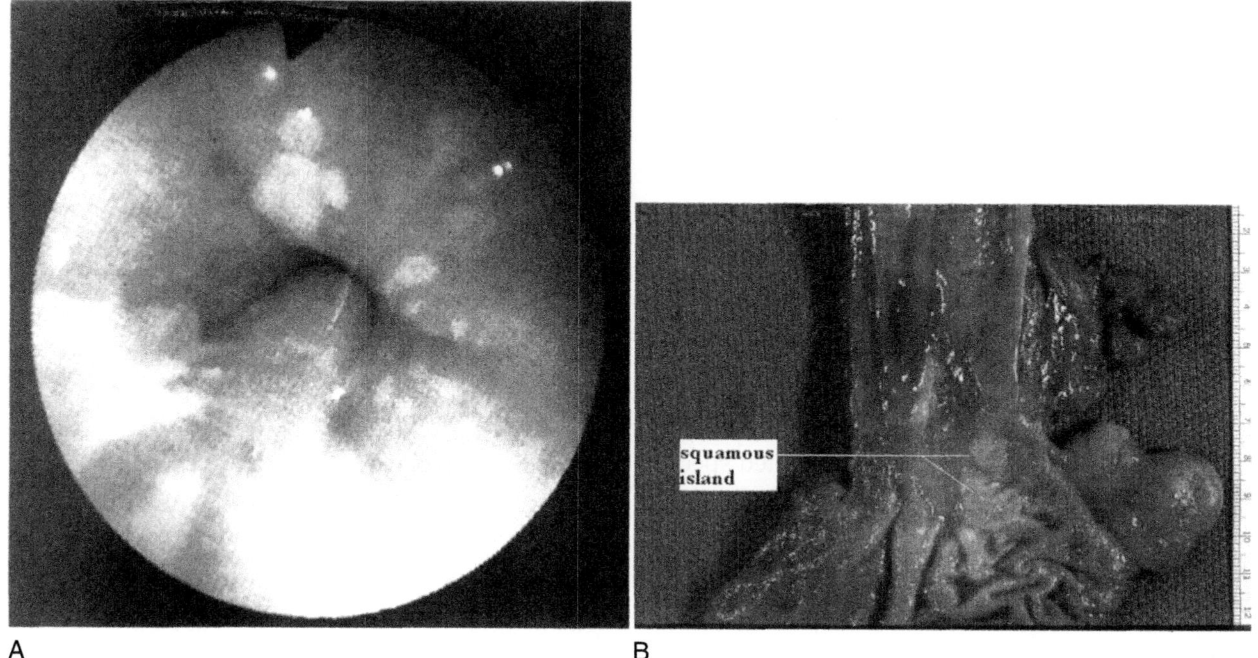

FIGURE 4.15 Squamous islands in the columnar-lined esophageal segment. (a) Endoscopic view showing the pearly white islands in the erythematous mucosa. Reproduced with permission from Tytgat GNJ. Endoscopic features of the columnar-lined esophagus. Gastroenterol Clin N Amer 1997;26:507–517. (b) Gross specimen showing a short segment of columnar-lined esophagus with two squamous islands. (Cross reference: Color Plate 3.1)

because of the fact that their ducts pass through the mucosa; they follow the mucosa with any longitudinal relative movement between the mucosa and muscle wall. Their presence characterizes any mucosal location as esophageal because these glands are not found in the stomach. Any criterion that is used to define the gastroesophageal junction, esophagus, and stomach should be tested by evaluating its relationship to submucosal glands to prove whether that defining criterion is true. Allison and Johnstone (11) showed that what Barrett was calling tubular stomach had submucosal glands under the mucosa, and therefore it was esophagus and not stomach. *The rule: If submucosal glands are present distal to the defined gastroesophageal junction, the definition is incorrect because these glands are not found in the stomach.*

PRESENT DEFINITION OF THE GASTROESOPHAGEAL JUNCTION

The endoscopic definition of the gastroesophageal junction has undergone periodic changes since the 1950s. Criteria used for definition have come and gone, and it is useful to examine the historical reasons for this. It is also important to question whether the gastroesophageal junction can be accurately and reproducibly defined endoscopically in every patient. Attempting to do things that are beyond the capability of a technique is a common source of serious error. For example, in the 1980s, endoscopists defined gastritis by endoscopic criteria leading to enormous confusion because these criteria were not reliable. It is only when it was recognized that endoscopy was unreliable in assessing gastritis and the disease was defined and classified by histologic criteria that clarity resulted. *We suggest that the present confusion in understanding distal esophageal pathology is the result of overemphasis of unreliable and unproven endoscopic definition of the gastroesophageal junction.*

Historically, endoscopists and anatomists have used the following landmarks at different times to define the gastroesophageal junction.

The Point of Flaring of the Tubular Esophagus

By tacit agreement, the point at which the tubular esophagus flares into the stomach has always been used to define the gastroesophageal junction. However, as Allison (15) showed, when a sliding hiatal hernia is present, this point is not well defined and the end of the esophagus tends to gradually flare like a bottle of chardonnay without the sharp angle of His on the left side that is typical in the normal patient. Spechler (16), recognizing this in his landmark study of 1994, defined the gastroesophageal junction as the end of the tubular esophagus, except when a hiatal hernia is present, when he used the criterion of the proximal limit of the longitudinal rugal folds.

Pathologists are directed by the Association of Directors of Anatomic and Surgical Pathology to use the point of flaring of the tubular esophagus to define the gastroesophageal junction in esophagogastrectomy specimens (17). This criterion is selected over the proximal limit of the rugal folds because many specimens have pathologic lesions in the distal esophagus and junctional region and the rugal folds are frequently distorted and difficult to recognize in resected specimens. The pathologic definition of tumors into esophageal, junctional, and gastric is based on the relationship of the epicenter of the tumor to the line drawn across the point of flaring of the tubular esophagus. When the esophagus flares gradually, the exact point selected by the pathologist probably has significant interobserver variation.

There has been no study that has verified whether the point of flaring of the esophagus satisfies Allison's test criteria. We do not know whether all points distal to the point of flaring are devoid of deep esophageal mucous glands. If all points distal to the flaring are devoid of deep esophageal glands in a reasonable number of specimens from normal and abnormal patients, we can correctly conclude that the point of flaring can be used to define the gastroesophageal junction. If, however, submucosal glands are present distal to the point of flaring in any patient, this is not a reliable definition of the true gastroesophageal junction. *The point of flaring of the esophagus is untested for veracity as a defining criterion for the gastroesophageal junction.*

In a study aimed at seeing whether the point of flaring of the tubular esophagus coincides with the anatomic gastroesophageal junction, we took esophagogastrectomy specimens and separated them at the point of flaring of the tubular esophagus. We then cut in the entire circumference of the specimen distal to the line of separation by full thickness vertical sections that were approximately 3 cm in length. We used the presence of submucosal mucous glands to differentiate between the esophagus and stomach. We found that submucosal glands were consistently present in the proximal part of the saccular organ that was classified as stomach by the definition of the gastroesophageal junction as the point of flaring to a distance up to 2.05 cm in one case. In this case, therefore, use of the point of flaring of the tubular esophagus would place the gastroesophageal junction at least 2.05 cm proximal to the true anatomic gastroesophageal junction. We will review this study more fully in a later chapter (14).

The Distal End of the Lower Esophageal Sphincter

Some endoscopists used the distal limit of the lower esophageal sphincter to define the true gastroesophageal junction. This has largely been abandoned because it was shown that the endoscopic determination of the lower esophageal sphincter was inaccurate by several centimeters when compared with manometric study (18).

Even if the distal limit of the lower esophageal sphincter can be precisely recognized at endoscopy, this is not a viable criterion for defining the true gastroesophageal junction. The sphincter is frequently severely damaged and shortened in reflux disease (19). While it is accurate in defining the end of the esophagus in a normal patient, it ceases to be accurate when the sphincter is shortened distally. In that event, the distal end of the esophagus will have lost its sphincter while still being the esophagus. *The distal end of the lower esophageal sphincter has been correctly abandoned as a criterion for defining the gastroesophageal junction.*

The Squamocolumnar Junction

Before 1953, there was general agreement in the medical community that the esophagus ended at the squamocolumnar junction, which was therefore the gastroesophageal junction (12). Anything distal to the squamocolumnar junction was regarded as the stomach. As Barrett stated vehemently in 1950 (12) in reply to his own question "What is the esophagus?": *"To the anatomist, it is that part of the alimentary canal which extends from the pharynx to the stomach, and the implication is that it traverses the mediastinum. . . . To (the physiologist, the surgeon, or the physician) the esophagus is that part of the foregut, distal to the cricopharyngeal sphincter, which is lined by squamous epithelium"* (p. 177).

Allison's 1948 article (15) has the first clear description of normal endoscopy:

> During the examination of an anatomically normal oesophagus the pinchcock constriction of the lower end by the diaphragm is noted, after which the patient's head must be depressed and moved to the right to allow the oesophagoscope to pass forwards and to the left into the stomach. At a precise moment the lumen of the viscus gapes, gastric mucosa prolapses into the field and fluid wells up. (p. 36)

Allison appears to be describing the squamocolumnar junction to be exactly at the point where "the lumen of the viscus gapes" (the end of the tubular esophagus) and "gastric mucosa prolapses into the field." Allison's line drawings show the squamous epithelium coming all the way down to the end of the tubular esophagus. This description is in line with modern thinking where the normal gastroesophageal junction is recognized as the coincidence of squamous epithelium with the rugal folds of gastric mucosa at the end of the tubular esophagus.

Allison went on to describe the abnormal situation associated with a sliding hiatal hernia (15):

> In the patients with short oesophagus the picture is quite different, for in the absence of stenosis the instrument is passed down from the oesophagus to the stomach without impediment or deflection. The viscus is lined by oesophageal mucosa at one level and by gastric the next; the level is higher than would be anticipated in the normal, and no pinchcock action of the diaphragm is observed. The level of the change varies with the degree of herniation, but it has been seen at 24 cm from the alveolar margin in a patient in whom the normal level would have been about 39 to 40 cm. (p. 36)

At this point in history, columnar-lined esophagus had not been described and everything distal to the squamocolumnar junction was regarded as stomach. The entire intrathoracic structure distal to the squamocolumnar junction was interpreted as stomach (a sliding hiatal hernia) simply because the gastroesophageal junction was erroneously defined as the squamocolumnar junction.

Five years later, in 1953, Allison and Johnstone (11) recognized that part of what they were calling a sliding hiatal hernia at endoscopy was actually esophagus. They did this by pathologic examination of *resected* specimens, not by endoscopy, by showing that part of the structure distal to the squamocolumnar junction had esophageal anatomic characteristics (no peritoneal covering, esophageal musculature, squamous mucosal islands, absence of parietal cells, and esophageal mucous glands). After Allison and Johnstone's report, which Barrett accepted in 1957 (20), *the squamocolumnar junction correctly ceased to be the definition of the endoscopic gastroesophageal junction because it was recognized that in some patients a columnar-lined esophagus is interposed between the squamocolumnar junction and the stomach.*

It is clear that the squamocolumnar junction moves cephalad as a result of injury caused by gastroesophageal reflux (Fig. 4.16). Csendes et al. (21) showed that the amount of cephalad migration of the squamocolumnar junction was directly proportional to the severity of reflux. This cephalad migration of the squamocolumnar junction occurs as a result of columnar metaplasia of the reflux-damaged esophagus.

The present state of the art is to assume that the squamocolumnar junction is the true gastroesophageal junction when no columnar lining is seen in tubular esophagus distal to the squamocolumnar junction. If rugal folds come all the way to this point, the patient is regarded as being endoscopically normal. No heed is paid to the possibility that the squamocolumnar junction may have moved cephalad to an extent that is not detectable at endoscopy. No heed is paid to the possibility that columnar-lined esophagus may be part of the saccular structure distal to the tubular esophagus. No heed is paid to the possibility that columnar-lined esophagus may develop rugal folds. If any of these possibilities is true, then the assumption that the squamocolumnar junction is the true gastroesophageal junction in endoscopically normal patients is incorrect. The error will result in the placement of the endoscopic gastroesophageal junction at a point that is proximal to the true gastroesophageal junction. Part of the esophagus will then be called stomach (proximal stomach or gastric cardia).

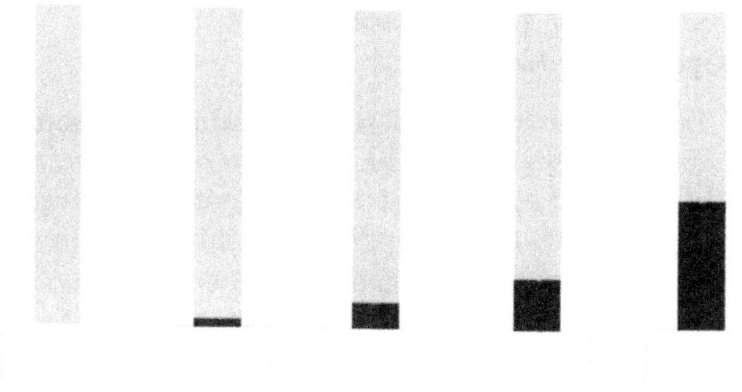

FIGURE 4.16 The squamocolumnar junction moves cephalad as columnar metaplasia of the esophagus occurs. At the left is the normal state with the entire tubular esophagus lined by squamous epithelium with increasing lengths of columnar-lined esophagus in the succeeding figures. The logical and correct view is that if the entire esophagus is lined by squamous epithelium, any cephalad migration of the squamocolumnar junction is abnormal. What is accepted today is that any migration of the squamocolumnar junction is abnormal *if it is visualized at endoscopy and contains intestinal metaplasia in a biopsy taken from it.*

The Proximal Limit of the "Gastric" Rugal Folds

When Allison and Johnstone described columnar-lined esophagus in 1953, it became obvious that in patients with a columnar-lined esophagus, *the true gastroesophageal junction is the junction between the columnar-lined esophagus and gastric mucosa.* Recognition of this fact created two problems.

How Much, If Any, Columnar-Lined Esophagus Is Normal?

The answer to this question has almost come full circle. It is clear from the writings of Allison and Barrett that they believed that the entire esophagus was lined by squamous epithelium and any columnar lining was abnormal ("heterotopic gastric mucosa") in the tubular esophagus (Fig. 4.17). However, because they believed this was a congenital anomaly and a normal anatomic variant, there was no need to define any limits. This changed with the recognition that columnar-lined esophagus was a pathologic entity resulting from gastroesophageal reflux.

The influential study on this subject is that of Hayward (5). Hayward did not produce data or evidence for his conclusions, which were probably based on endoscopic examination of patients and gross examination of resected specimens. Because endoscopy in this era was likely biased heavily toward patients with severe reflux disease, Hayward was seeing those patients who had the greatest likelihood of having columnar-lined esophagus. Hayward argued against the concept that was accepted at the time: "We have all been taught that the oesophagus is a tube lined by squamous epithelium" (p. 36). Hayward defined the anatomic esophagogastric junction as the point where the tube ends and wrote: "From the inside the only line of demarcation apparent is the wavy line of junction of stratified squamous with columnar epithelium; but it is not at the level of the oesophago-gastric junction as defined above. It is 1 or 2 cm higher" (Fig. 4.18). He concluded his treatise as follows:

> The oesophagus is defined as the tube conducting food from the throat to the stomach and all of this tube is regarded as oesophagus, irrespective of its lining. The lower centimeter or two of this tube is normally lined by columnar epithelium. . . . Thus the view that the oesophagus is a tube lined only by squamous epithelium is rejected. (p. 40)

Hayward was establishing a definition for normalcy that was likely based on examinations limited to

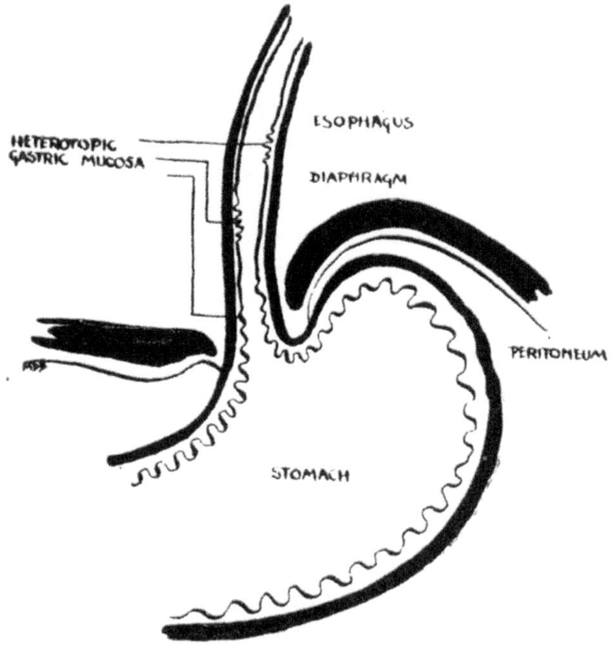

FIGURE 4.17 Diagrammatic representation by Allison in 1948 showing the presence of columnar ("gastric") epithelium in the distal esophagus. Allison referred to this as "heterotopic gastric mucosa" recognizing that it is a deviation from the normal state where the entire tubular esophagus is normally lined by squamous epithelium. Reproduced with permission from Allison PR. Peptic ulcer of the oesophagus. Thorax 1948;3:20–42.

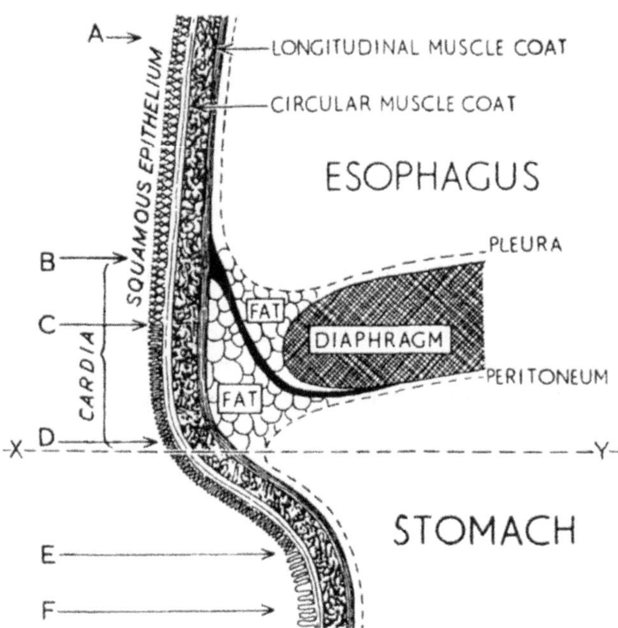

FIGURE 4.18 Hayward's flawed conception of the gastroesophageal junction, which caused three decades of confusion and failure to recognize short-segment Barrett esophagus. "Line XY crosses the oesophago-gastric junction. Everything above this line is oesophagus and everything below is stomach. A to C is lined by squamous epithelium. B is the level of the insertion of the phreno-oesophageal ligament. B to D is the cardia, i.e., the functioning muscular sphincter. C to E is the junctional (= cardiac) epithelium. F is lined by fundal epithelium. Thus the cardia is partially lined by squamous and partly by columnar epithelium and the junctional epithelium extends across the oesophago-gastric junction." To Hayward, the cardia is the sphincter and is not lined solely by cardiac mucosa and cardiac mucosa lines the stomach as well as the cardia. The confusing disconnect between cardia and cardiac mucosa is established. Reproduced with permission from Hayward J. The lower end of the oesophagus. Thorax 1961;16:36–41.

patients with the most severe reflux damage of the esophagus. This would be like examining the coronary arteries of patients who present with myocardial infarction to define the normal state of the coronary arteries.

Hayward's article had great influence. The medical community actually expanded Hayward's recommended limit of 2cm of columnar epithelium being normal in the tubular esophagus to 3cm. For more than three decades, gastroenterologists ignored up to 3cm of columnar-lined esophagus because of this erroneous definition of the normal state. Columnar-lined esophagus was diagnosed only when the length of columnar-lined esophagus exceeded 2 or 3cm (Fig. 4.19).

In the early 1990s, based on Spechler et al.'s studies that defined short-segment Barrett esophagus (16), the definition of endoscopic normalcy changed back to the pre-Hayward era. Any visible columnar lining in the tubular esophagus is now defined as sufficiently abnormal to require biopsy. The definition of normalcy has reverted to squamous epithelium lining the entire tubular esophagus. We have correctly rejected Hayward's rejection of this truth but have suffered its error in ignorance for more than three decades. There are still those who believe that 3cm of columnar epithelium normally line the distal tubular esophagus, but we can now classify these individuals as being not in line with modern thinking.

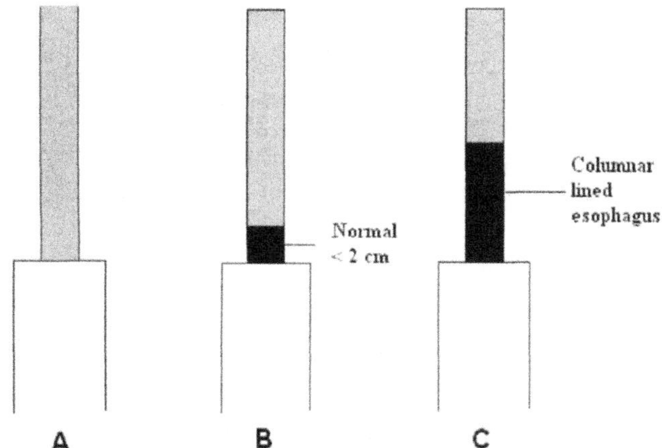

FIGURE 4.19 Until 1961, it was accepted that the esophagus was a tube lined entirely by squamous epithelium (a). We now know this is correct. Hayward rejected this viewpoint, stating that the presence of up to 2cm of "junctional" (= cardiac) mucosa was normal in the distal esophagus (b). According to this, columnar-lined esophagus was pathological only when there was greater than 2cm in the distal esophagus (c). Hayward's influence lasted for more than three decades.

Can the Columnar-Lined Esophagus Be Reliably Seen by Endoscopy Whenever It Is Present?

The answer to this question appears to be a vehement yes in the practice guidelines of the American Gastroenterological Association (22). The presently accepted definition of Barrett esophagus is based on this assumption: "Barrett esophagus is defined as the presence of intestinal metaplasia in a biopsy taken from an endoscopically visible segment of columnar-lined esophagus."

This answer that is universally accepted defies logic. It assumes that the endoscope is capable of seeing the first cell of the squamous epithelium that undergoes columnar metaplasia. Clearly, this is not true. Given the physics of the optics of the endoscopic instrument, a microscopic columnar-lined esophagus must exist before its amount becomes visible to the standard endoscope. As endoscopes become more sophisticated with the use of dyes and magnified images, the ability to recognize shorter segments of columnar-lined esophagus improves (23–25). Does this mean that the standard endoscope defines Barrett esophagus incorrectly? Of course it does. Does it mean that these new endoscopic modalities define Barrett esophagus correctly? Of course it does not. No dyes and no amount of magnification at endoscopy will ever achieve the magnification available by microscopy. The need for endoscopic visibility of columnar-lined esophagus to define Barrett esophagus must give way to microscopic visibility of columnar-lined esophagus if the definition is to be accurate. A definition cannot depend on a blind person seeing something; endoscopy is blind below its level of maximum resolution.

The truth with the ability of the endoscope to "visualize" columnar-lined esophagus depends on the sophistication of the endoscope and the experience of the observer. It has been shown that there is significant interobserver variation among experienced endoscopists in recognizing segments of columnar-lined esophagus that are less than 1cm long. *It seems reasonable that the present definition of Barrett esophagus has the potential to miss the diagnosis in patients with less than 1cm of columnar-lined esophagus because of endoscopic error.*

Can the Columnar-Lined Esophagus Be Reliably Distinguished from Gastric Mucosa by Endoscopy?

In 1987, McClave et al. (26) described a new endoscopic landmark, the proximal limit of the "gastric" rugal folds, and suggested that this marked the true gastroesophageal junction. *The proximal limit of the rugal folds has become accepted universally by gastroenterologists as the defining endoscopic criterion for the gastroesophageal junction.*

However, several problems exist with the use of the proximal limit of the rugal folds as the gastroesophageal junction: (a) It assumes that the metaplastic columnar epithelium of the distal esophagus will never form rugal folds and that gastric mucosa always has rugal folds. (b) The proximal limit of the rugal folds is a line that is not always definable with precision; the rugal folds sometimes merge indistinctly with the columnar-lined esophagus, particularly if there is coexisting gastritis. (c) The rugal folds are frequently distorted and altered when pathologic lesions such as ulcers and tumors are present in the area (Figs. 4.20 and 4.21). (d) Their proximal limit is never a point that is definable with millimetric precision because the top of the rugal fold merges with the flat nonrugated normal squamous or metaplastic columnar epithelium in a manner similar to an undulating hill rather than the sharp drop of a canyon edge (Fig. 4.22). There is no definition as to whether the proximal limit of the rugal fold is the point at which the fold begins its descent or reaches the point of the flat epithelium. Because of the gradual undulation, these points can be up to 0.5 cm apart, making this the limit of precision for this criterion (5). The tops of every rugal fold are commonly not at the same horizontal level, and the troughs between the folds have no proximal limit because they are continuous with the flat epithelium. This makes the line incomplete

FIGURE 4.20 Fresh esophagectomy specimen with a polypoid adenocarcinoma at the squamocolumnar junction. There are two to four rugal folds at the distal edge of the tumor. These seem to fade off before more prominent rugal folds become apparent. The prominent rugal folds are incomplete. Where is the proximal limit of the rugal folds? (Cross reference: Color Plate 5.1)

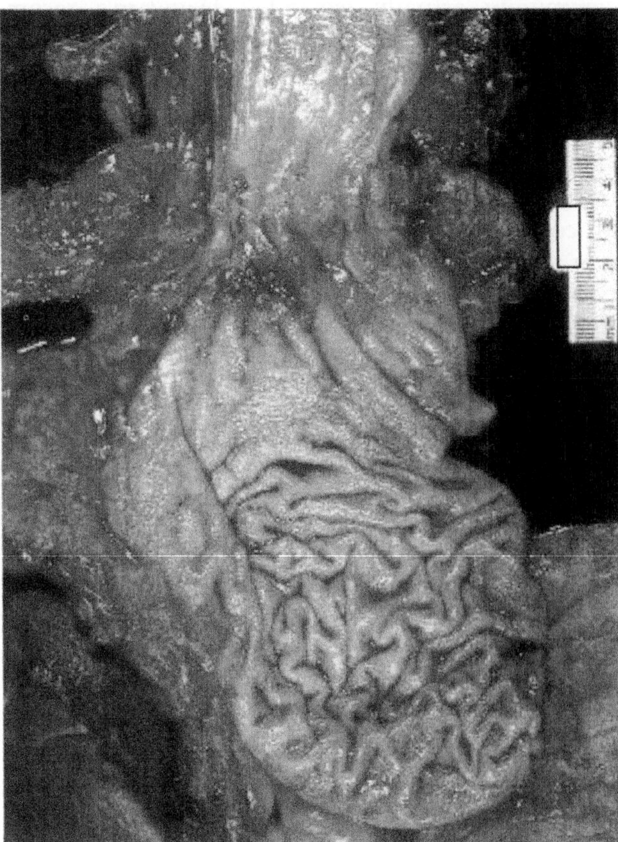

FIGURE 4.21 Adenocarcinoma of the esophagus. The ulcerated tumor arises at the point of flaring of the esophagus immediately distal to the squamocolumnar junction. However, the columnar epithelium in a large part of the proximal saccular organ distal to the tumor is flat. Rugated epithelium begins more distally. Where is the gastroesophageal junction? (Cross reference: Color Plate 5.2)

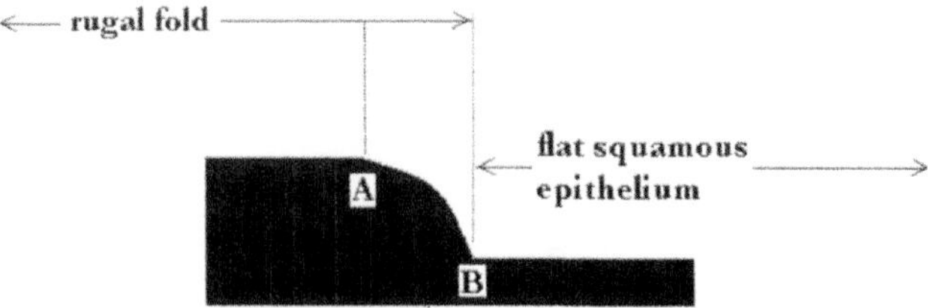

FIGURE 4.22 Diagrammatic representation of the transition between a rugal fold and flat proximal epithelium. The transition is gradual, and it is uncertain whether the proximal limit of the rugal fold is at point A where the fold starts ending or at point B where it joins the flat surface. Because this is gradual, there may be a 0.5 cm gap between points A and B, resulting in significant interobserver variation.

and its definition, at least partly, determined by imagination (Fig. 4.23; see also Fig. 4.10). While the separation of the tops of the rugal folds from the squamocolumnar junction can be recognized when it is of a significant length, it is unreasonable to believe that endoscopy can recognize microscopic separation in the 1- to 5-mm range. In fact, consensus meetings show that there is significant variation among experienced endoscopists in accurately ascribing a length to columnar-lined esophagus when it is less than 1 cm.

In addition to the lack of precision, there is also absolutely no evidence that the proximal limit of the rugal folds is in fact the true gastroesophageal junction. It is simply an endoscopic landmark that was described by an astute endoscopist and then accepted universally to be the true endoscopic gastroesophageal junction. No one has done any study to see if there is a correlation between the point defined by this criterion and the anatomic gastroesophageal junction. No one has shown that the areas of columnar mucosa covered by rugal folds are always devoid of submucosal esophageal glands. *The proximal limit of the rugal folds has never been proven as a reliable criterion for defining the true gastroesophageal junction.* In an era where so much emphasis is paid to "evidence-based medicine," the definition of something as basic as the gastroesophageal junction has been universally accepted to be fact without any supporting evidence.

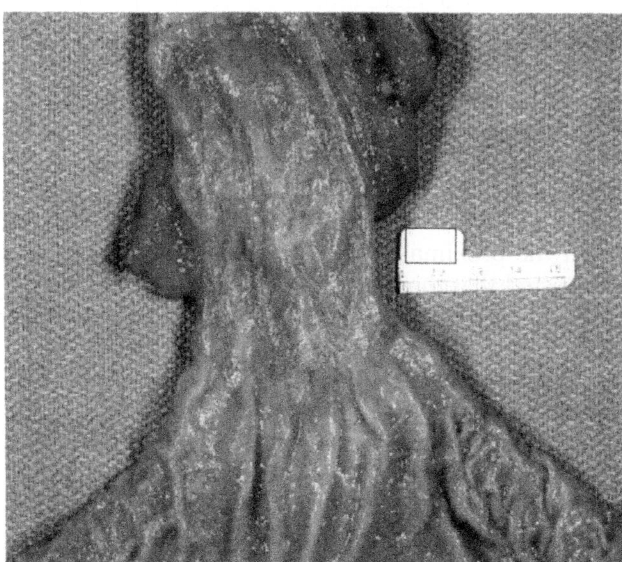

FIGURE 4.23 Long-segment columnar-lined esophagus with a fairly clear line representing the proximal limit of the rugal folds. There are significant flat areas between the proximal parts of the rugal folds. Where exactly is the gastroesophageal junction? Can this be determined with a precision of 0.5 cm?

Spechler, in his 1997 review of the columnar-lined esophagus (27), suggested that the columnar-lined esophagus cannot be reliably distinguished from gastric mucosa (27). In a later review, however, he iterated the widely held view that there is a clear distinction between the rugal folds and the flat metaplastic mucosa of the columnar-lined esophagus (28). In these descriptions, the rugal folds extend to the end of the tubular esophagus and the columnar-lined esophagus, when it is seen, involves the tubular esophagus.

In our experience, the columnar-lined tubular esophagus is usually flat, inflamed, and devoid of regular coarse folds and can be distinguished from the point where the folds begin. In most cases, the folds begin at or near the point at which the tubular esophagus flares. However, in many cases, the demarcation is not absolutely clear (Figs. 4.10, 4.20, and 4.21). This is particularly true when ulceration or tumors are present in the junctional region (Figs. 4.20 and 4.21), but it can also be true when there are no lesions (Fig. 4.10).

In the same experiment on esophagogastrectomy specimens described previously (14), we selected cases where the proximal limit of the rugal folds extended up to the point at which the tubular esophagus flares. Histologic assessment of the vertical sections from areas of mucosa covered by well-formed rugal folds therefore permitted the determination of whether there were submucosal mucous glands in this area of rugated mucosa. We found that submucosal glands were consistently present under mucosa that showed rugal folds. The maximum length of saccular organ covered by rugal folds that contained submucosal glands was 2.05 cm. In this case, therefore, use of the proximal limit of the rugal folds would place the gastroesophageal junction 2.05 cm proximal to the true anatomic gastroesophageal junction. *We concluded that the proximal limit of the rugal folds is not a reliable criterion to accurately define the gastroesophageal junction.* This experiment will be reviewed more fully in a later chapter.

What are rugal folds? From a logical standpoint, mucosal rugal folds most likely represent a marker for any organ that distends (e.g., the stomach, urinary bladder, gallbladder, and small intestine all have mucosal folds). In the normal patient, the abdominal esophagus is tubular and lined with squamous epithelium; except for transient distension as the bolus passes, the esophagus retains its tubular shape. The stomach is saccular, lined by gastric mucosa, and dilates with meals. In the empty condition, the stomach is collapsed and the mucosa shows rugal folds. When it dilates, the rugal folds become obliterated to accommodate the increased volume. *In the normal person, the proximal limit of the rugal folds accurately defines the gastroesophageal junction.*

The problem is that the anatomy of the reflux-damaged esophagus changes. If the distal end of the esophagus becomes damaged by reflux and loses the sphincter action in the wall, it can dilate. When this happens, the distal end of the esophagus can become a distensible organ and when it undergoes columnar metaplasia, that mucosa within the damaged distensible part of the esophagus can theoretically develop rugal folds. The presence of *rugal folds in the mucosa is not a criterion that defines the stomach; it is a criterion that defines a distensible organ.*

The answer to the question "Can the columnar-lined esophagus be reliably distinguished by endoscopy from gastric mucosa?" is *no.* The presence of submucosal esophageal glands under mucosa covered by rugal folds indicates that metaplastic columnar-lined esophagus can have rugal folds. This means that the accurate definition of the gastroesophageal junction is not possible at endoscopy. This fact must be recognized and acknowledged; otherwise endoscopists will erroneously declare that there is no columnar-lined esophagus, not because it is absent but because they cannot see it. If they fail to see a columnar-lined esophagus that is present, they will place the gastroesophageal junction at the squamocolumnar junction, proximal to the true gastroesophageal junction. *This is a common error in practice today.* The delusion of absolute endoscopic accuracy in defining the gastroesophageal junction must end. Exactly as it was with the understanding of gastritis, it is only when we recognize the natural limitation of the endoscope's capability that we will even attempt to develop more reliable criteria for the recognition of the true gastroesophageal junction.

Conclusion: The true gastroesophageal junction cannot be accurately defined by endoscopy.

Histologic Definition of the Gastroesophageal Junction

A little-recognized fact is that when Norman Barrett reversed his views in 1957 (20), the only histologic definition of the esophagus that has ever existed disappeared. Although the definition (the end of the squamous epithelium) was wrong, Barrett's idea was correct: without a histologic definition of the esophagus, gastroesophageal junction, and stomach, confusion would surely reign. Indeed, confusion has certainly reigned supreme in the years after Allison. To date, pathologists have no accepted method of accurately defining the gastroesophageal junction by histology.

When examining slides of this region, the only reference point that pathologists have is the squamocolumnar junction. There is no method whereby the proximal limit of the rugal folds, the point of flaring of the tubular esophagus, the peritoneal reflection, or the angle of His can be accurately translated to a microscopic slide. It has always been the tendency for pathologists to believe that the squamocolumnar junction marks the gastroesophageal junction unless there is evidence to the contrary. The only evidence to the contrary is if endoscopic or gross examination indicates that a columnar-lined esophagus is present. This is an error.

When we reported our autopsy study (29), we identified squamous epithelium, cardiac mucosa, oxyntocardiac mucosa, and gastric oxyntic mucosa and then measured the extent of cardiac and oxyntocardiac mucosa that was present between the squamocolumnar junction and gastric oxyntic mucosa. When we found that 56% of patients did not have cardiac mucosa and the overall length of cardiac and oxyntocardiac mucosa was less than 1 cm in the majority of patients, we concluded that it was highly likely that these epithelia were abnormal. Armed with data that showed that the presence of these epithelial types correlated with abnormal gastroesophageal reflux (30), we concluded that these epithelia represented metaplastic columnar-lined esophagus rather than normal gastric epithelia.

In a similar study by Kilgore et al. (31), the same measurement of the length of cardiac mucosa separating the squamocolumnar junction from gastric oxyntic mucosa was made. Their data led to a similar conclusion that only a small amount of cardiac mucosa was present (0.1 to 0.4 mm). However, Kilgore et al. concluded that this represented "normal" or "native" gastric cardia. They reached this conclusion by taking the gross observation that the squamocolumnar junction was coincidental with the proximal limit of the rugal folds. This made their measurement from the squamocolumnar junction equivalent to measurement from the proximal limit of the rugal folds. We have suggested that the proximal limit of the rugal folds is not a precise point (Figs. 4.22 and 4.23) and its gross determination is imprecise. In addition, Kilgore et al.'s conclusion is impossible by principles of the physics of optics. The resolution of the human eye when examining a specimen grossly is always less than the microscope. This must mean that columnar-lined esophagus must first be microscopic before it becomes visible to the naked eye. It is therefore an error to assume that the coincidence of the squamocolumnar junction and proximal limit of rugal folds determined by gross examination is proof that there is no microscopic columnar-lined esophagus. The researchers' conclusion that the cardiac mucosa distal to the squamous epithelium represents proximal stomach is based on this self-evident error in scientific method; as such, the conclusion must be rejected.

We will show in later chapters that the true gastroesophageal junction is the proximal limit of gastric oxyntic mucosa. The squamous, cardiac, and oxyntocardiac epithelia that are found proximal to this point represent esophagus.

WHAT IS THE CARDIA? LET'S REMOVE THIS TERM FROM OUR VOCABULARY

"Cardia" is an incredibly ambiguous medical term (Table 4.1, Fig. 4.24). Its usage has evolved in an insidious manner over the years, beginning as a term that was synonymous with the physiological mechanism at the gastroesophageal junction that prevented reflux, to a term that was synonymous with the lower esophageal sphincter, to a term that is now applied to the proximal part of the stomach.

Historical Definitions of the Cardia

In the classic articles by Allison and Barrett in the 1950s, the cardia is clearly synonymous with the mechanism at the gastroesophageal junction that prevented reflux. Allison, in 1948 (15) stated: "The change from the oesophageal mucosal pattern to that of the coarse gastric folds marks the position of the cardia" (p. 22). Barrett, in 1957 (20), stated: "The cardia describes the mechanism (flap valve or whatever it be) which prevents reflux from the stomach into the esophagus. The word does not refer to the site at which the gullet widens out into the stomach nor to a hypothetical muscle sphincter which some believe separates the two" (p. 881). To Allison and Barrett, the cardia was a physiological valve, not an anatomic structure.

In the late 1950s, the lower esophageal sphincter was becoming recognized not as a valvular mechanism at the gastroesophageal junction, as believed by Allison and Barrett, but rather as a longer segment of the lower esophagus. Hayward, in 1961 (5), asked the question "What is the cardia?" and answered his own question in a manner that indicates his frustration with the term:

> I have never seen or heard the cardia clearly defined. Is it just the gastro-esophageal junction? . . . Is it the continuously varying portions of the lower esophagus and upper stomach lined by junctional (= cardiac) epithelium? . . . Is it the part which is narrow in the disease called achalasia of the cardia or cardiospasm? . . . is it merely gastric and composed only of that portion of the stomach immediately surrounding the gastro-esophageal junction? The literature does not answer these questions. The cardia is often referred to as if it were part of the stomach, and a malignant growth

TABLE 4.1 Definitions of the Term "Cardia" as Appearing in Selected Research in the Literature

Reference	Definition
Allison (15)	Synonymous with gastroesophageal junction
Barrett (12, 20)	Synonymous with the lower esophageal sphincter mechanism
Hayward (5)	The lower esophageal sphincter equal to the distal 2 cm of the esophagus between the gastroesophageal junction and the attachment of the phrenoesophageal ligament
Lagergren et al. (33)	An area bounded by being 2 cm proximal or 3 cm distal to the gastroesophageal junction
Present general usage	An undefined extent of the proximal stomach distal to the gastroesophageal junction
Odze (32)	<0.4 cm between the proximal limit of the "gastric" folds and the beginning of gastric oxyntic mucosa

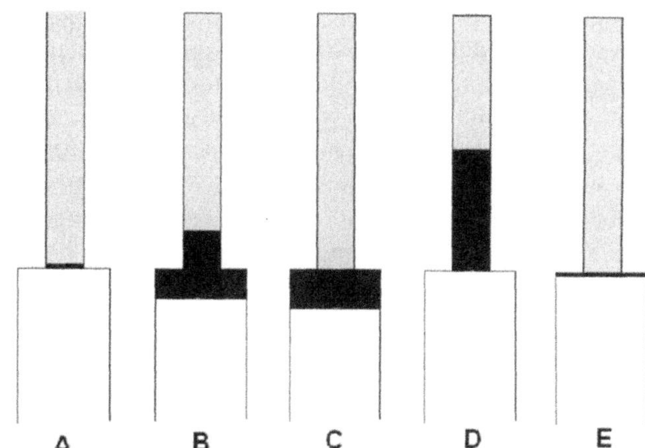

FIGURE 4.24 Various interpretations of the term "cardia." (a) Allison's and Barrett's concept of the cardia as the physiological valvular mechanism at the gastroesophageal junction that prevents reflux. (b) Hayward's (1961) view of the normal extent of junctional (cardiac) mucosa, which he said normally lined the 2 cm of distal esophagus and variable amount of proximal stomach. (c) When short segment Barrett esophagus was defined, the entire tubular esophagus was believed (correctly) to be normally lined by squamous epithelium. The cardiac mucosa and "cardia" then shifted to that part lining a variable part (2 cm or so) of the proximal stomach. This is the prevalent definition of the cardia. (d) The term cardia is also used at this time synonymously with the lower esophageal sphincter, as in achalasia of the cardia. (e) Odze's (2005) "true gastric cardia" recently defined as the area between the proximal limit of gastric rugal folds (by endoscopy) and the beginning of gastric oxyntic mucosa (by histology). This area is said by Odze to have a maximum normal extent of 0.4 cm.

> arising in the junctional epithelium is called a carcinoma of the cardia and always regarded as a gastric neoplasm. At the same time the narrow zone in achalasia of the cardia is accepted as lower esophagus and the inconsistency is conveniently ignored. I do not know exactly what the cardia is. (p. 37)

It is incredible that we have not progressed much from the state of confusion that Hayward described in 1961 (5); even today, "achalasia of the cardia" and "adenocarcinoma of the cardia" do not refer to pathology in the same anatomic region. What is absolutely the same is how physicians continue to conveniently ignore their own double-speak just as our predecessors did. Hayward, in 1961, tried to provide an answer to the confusion as follows: "Since it is wise not to continue to use a word which cannot be clearly defined, two possible courses are open: either to drop the word altogether, or to find some precise meaning for it. I propose the second" (p. 38). Hayward went on to precisely define the cardia as a term that should be applied to the lower esophageal sphincter: "It is the sphincteric lower part of the oesophagus between the insertion of the phreno-oesophageal ligament and the gastro-oesophageal junction" (p. 38).

Hayward made a mistake; he should have opted for the first of is two choices and dropped the term "cardia" altogether and used the term "lower esophageal sphincter" to define that structure. Physicians listened to Hayward's suggestion to retain the term "cardia" but completely ignored his precise definition and opted to retain the confusing panoply of definitions for the cardia. This has the advantage of convenience; when one is uncertain as to whether something is distal esophageal or proximal gastric, one can fall back on the wonderfully ambiguous "cardia."

Present Definition of the Cardia

At present, most people use the term "cardia" as synonymous with the proximal stomach ("gastric cardia"). Anatomy texts will describe the stomach as being composed of four areas: the cardia, body, fundus, and antrum. In their 2003 article (13), de Hertogh et al. accurately defined the cardia in terms that are generally accepted today:

> Anatomists have applied the term "cardia" to that part of the stomach which lies around the orifice of the tubular oesophagus. There is no anatomical landmark for the distal margin of the so-defined cardia. Its proximal margin is the gastro-oesophageal junction which, according to anatomists, is localized at the level of the angle of His. (pgs. 791–796)

Most people today follow this definition of the cardia as being part of the stomach that is limited proximally by the gastroesophageal junction and extends into the stomach for a "variable" distance that has never been defined (Fig. 4.24c). In general, studies that attempt to define the cardia use an arbitrary distance ranging from 0.5 to 3 cm from the gastroesophageal junction.

Odze (32), in a review article, defined the "true" gastric cardia as the area of mucosa located distal to the anatomic gastroesophageal junction (defined as the proximal limit of the gastric folds) and proximal to the portion of stomach (corpus) that is composed entirely of oxyntic glands. According to Ozde, this true gastric cardia normally measures less than 0.4 cm (their abstract mistakenly states 0.4 mm) and "is typically composed of pure mucous glands, or mixed mucous/oxyntic glands that are histologically indistinguishable from metaplastic mucinous columnar epithelium of the distal esophagus" (p. 1853). It should be noted that Ozde does not give a minimum extent to this true gastric cardia, only a maximum; one is left to assume then that this normal organ can be zero or absent in some people (i.e., the true gastric cardia has a normal extent of 0 to 4 mm). Odze's article is the first to use a histologic criterion, the gastric oxyntic mucosa, to define the distal limit of the cardia.

Histology of the Cardia

Histology textbooks will suggest that the gastric cardia is lined by cardiac mucosa, the body and fundus by oxyntic mucosa, and the pyloric antrum by antral mucosa. However, there has never been any attempt to use the extent of cardiac mucosa to define the extent of the anatomic gastric cardia. Almost by tacit agreement, pathologists will accept that the cardia is lined by cardiac mucosa, and when they see cardiac mucosa in a biopsy, they assume that it represents the stomach.

We will consider cardiac mucosa and histology in great detail in the next chapter. However, it is important to state that cardiac mucosa is not defined as the mucosa that lines the cardia; it is defined by histologic criteria. This is clearly seen in Hayward's 1961 article (Fig. 4.18), where the anatomically defined cardia is entirely in the distal esophagus and lined proximally by squamous epithelium and distally by junctional (= cardiac) mucosa. Hayward also showed junctional (= cardiac) mucosa extending distal to his anatomic cardia and into the proximal stomach (Figs. 4.18 and 4.24b). Hayward established the wonderfully convenient disconnect between "cardia" and "cardiac mucosa."

This disconnect continues to the present. In Odze's 2005 article (32), the "true cardia" has shrunk dramatically from Hayward's time to "a maximum of 0.4 cm," and moved from being entirely in the distal esophagus to now being entirely in the proximal stomach (Fig. 4.24e). Odze's "true" gastric cardia can be lined by a mucous-only (cardiac) mucosa as well as a mixed mucous/oxyntic (oxyntocardiac) mucosa that contains parietal cells. According to Odze, this same cardiac and oxyntocardiac epithelia are seen in the distal esophagus in patients with reflux damage.

The entire definition of Odze's cardiac mucosa is based on the precise definition of the endoscopic gastroesophageal junction. If there is a 0.4-cm error in the endoscopic definition of the proximal limit of the rugal folds, Ozde's "true" cardiac mucosa disappears. We have suggested that the proximal limit of the rugal folds is not a sharp point and that its endoscopic determination is imprecise in the 0.4-cm range that Odze claimed is "true" gastric cardia. In fact, in a later section of the article, in discussing the differential diagnosis of inflammation of cardiac mucosa, Odze stated as much:

> As it is difficult, if not impossible, to know the precise anatomic location (i.e., esophagus or stomach) of a particular biopsy obtained from the GEJ [gastroesophageal junction] at the time of endoscopy,... clinicians have relied on pathologists to help establish the correct location and etiology of the inflammatory condition in mucosal biopsies from the GEJ region." (p. 1862)

This is not true; Dr. Chandrasoma is the only pathologist who has ever advocated that we disregard the endoscopist's impression and use only histology to determine whether a biopsy is located in the esophagus and stomach. All other pathologists strongly believe that correlation with endoscopic findings is essential in making this determination. The only reason we can recognize esophagus from stomach in a biopsy is because *we know that cardiac and oxyntocardiac mucosae are always esophageal and never gastric.* If Odze believes that cardiac and oxyntocardiac mucosa exists both in the esophagus and his "true" gastric cardia, *he cannot help establish the correct location* of cardiac mucosa; he is always dependent on his clinician's endoscopic impression of the gastroesophageal junction to differentiate between esophageal and true gastric cardiac mucosa. It is interesting that he depends on a finding that, in his own opinion, is *"impossible, to know the precise anatomic location (i.e., esophagus or stomach) of a particular biopsy obtained from the GEJ at the time of endoscopy."* This same "impossible" endoscopic finding is the one he used to base his definition of "true" gastric cardia.

Our Definition of the Cardia

We would like to re-ask Hayward's question: "What is the cardia?" However, we propose at this time that we abandon the use of this term in scientific literature. It is not a necessary term. There is no need to have a synonym for the lower esophageal sphincter or for the proximal stomach. The term "cardia" only gives the uncommitted mind a crutch to hold on to. By saying "cardia," the uncommitted mind can sit on the fence between the esophagus and stomach without committing to either; he or she can use it to widen the point that is the gastroesophageal junction both proximally into the esophagus and distally into the stomach to any desired extent that fits the moment. An excellent example of this tendency is Lagergren et al.'s influential article in the *New England Journal of Medicine* in 1999 (33), which described the risk of adenocarcinoma of the esophagus and "gastric cardia" in patients with symptomatic gastroesophageal reflux. Citing from their methods section as it relates to the definition of the "gastric cardia": "For a case to be classified as a cancer of the gastric cardia, the tumor had to have its center within 2 cm proximal or 3 cm distal, to the gastroesophageal junction" (p. 826). The "gastric cardia" to Lagergren et al. is a 5-cm zone on either side of the gastroesophageal junction; more incredibly, the "gastric cardia" includes 2 cm of the distal esophagus! This is similar to Hayward's (1961) view of the extent of "normal" junctional mucosa (Fig. 4.24b).

The first sentence of Spechler's abstract in his 1999 review of the subject (34) is similar: "Adenocarcinomas at the gastroesophageal junction appear to arise from foci of intestinal metaplasia that develop either in the distal esophagus or the proximal stomach (the gastric cardia)" (p. 218). This is an incorrect statement. The gastroesophageal junction is an imaginary line that is drawn across the point between the esophagus and the stomach. It is neither an organ nor a zone with a measurable length. As Spechler stated, there are only two organs here, and adenocarcinomas either arise in the stomach or in the esophagus. Spechler also accurately described how the term "gastric cardia" or "gastroesophageal junction" is used to include both the distal esophagus and the proximal stomach on either side of the gastroesophageal junction to dodge the issue of the point of origin because the user does not know whether the tumor originates in the stomach or esophagus. The same is true of the term "intestinal metaplasia of the cardia." The terms "adenocarcinoma" and "intestinal metaplasia of the cardia," by giving a false suggestion that these are real pathologic entities, remove the need to try to accurately ascribe an esophageal or gastric origin to these entities. We can hide in the ambiguity of the term "cardia" and avoid looking foolish; if we were honest, we would state: "this is an adenocarcinoma (or intestinal metaplasia) that is in an area that I have no clue is esophagus or stomach."

TABLE 4.2 Consequences (or Lack Thereof) of Deleting the Term "Cardia" from the Medical Literature

Uses of the word "cardia"	New term
Achalasia of the cardia	Achalasia (or achalasia of the lower esophageal sphincter)
Adenocarcinoma of the cardia	Adenocarcinoma of the proximal stomach
Intestinal metaplasia of the cardia	Intestinal metaplasia of the proximal stomach
Gastric cardia	Proximal stomach
Cardia	Never to be used

Our most precise definition of the cardia is exactly this: The cardia is a term used to describe a location when one is uncertain as to whether it is esophagus or stomach. To people who object to this definition and say that when they use the term "cardia," they mean a point in the proximal stomach distal to the gastroesophageal junction, we say, "call it proximal stomach, not cardia." If you then want to be precise, you can say the proximal stomach within 5mm or 3cm (or any number you choose because there is no defined distal limit unless you believe Kilgore's and Odze's flawed 4-mm limit) from the gastroesophageal junction. If this is done, we will be forced to explain how adenocarcinoma and intestinal metaplasia of the proximal stomach have an epidemiologic association with gastroesophageal reflux. How can gastroesophageal reflux cause gastric disease? This is almost oxymoronic but, however incredible it may be, this is the present belief of the entire medical community and most experts in this field of study.

The term "cardia" introduces only ambiguity to any discussion on the topic. This is a strong reason to never use this term in the scientific literature. There is no penalty for not using the term (Table 4.2).

THE LOGICAL CONCLUSION THAT SHOULD BE TESTED

If it is true that adenocarcinoma of the proximal stomach and intestinal metaplasia of the proximal stomach are etiologically associated with gastroesophageal reflux, there should be considerable alarm. Surely, the reasoning must go, there is something radically wrong with this premise. The proximal stomach is not damaged by gastroesophageal reflux; so how could reflux cause pathologic changes in the proximal stomach? Gastroesophageal reflux damages the esophagus. *Is it possible that what we are now calling the proximal stomach is in reality the reflux-damaged esophagus?*

In general, the reliability of the definition of "proximal stomach" depends on the reliability of the definition of the gastroesophageal junction. *If the definitions are true, then proximal gastric lesions must be unrelated to gastroesophageal reflux.* If the definition of the gastroesophageal junction is incorrect and the defined junction is more proximal than the true junction, any study that defines a lesion in the proximal stomach by reference to the incorrectly defined gastroesophageal junction will have an admixture of lower esophageal and proximal gastric pathology. This admixture of esophageal pathology will result in an association between gastroesophageal reflux disease and disease of the erroneously defined "proximal stomach."

The degree of admixture of lower esophageal diseases within the definition of "proximal stomach" is dependent on the length from the incorrect gastroesophageal junction that is used to define the proximal stomach. A series where proximal gastric cancer is defined as a tumor centered within 1cm of the gastroesophageal junction will have a greater percentage of esophageal tumors than a series where the definition uses 5cm. If there is a greater association between gastroesophageal reflux disease and "proximal gastric cancer" defined as a cancer with its center less than 1cm from the defined gastroesophageal junction than less than 5cm from the gastroesophageal junction, it strongly suggests that the definition of the gastroesophageal junction is inaccurate.

We will show in subsequent chapters that answers to these questions all strongly point to the fact that the gastroesophageal junction as defined endoscopically at the present time by the proximal limit of the rugal folds is not correct. Also, the gastroesophageal junction as defined by gross examination by a line drawn across the end of the tubular esophagus is not correct.

When faced with logical dilemmas, it is important to evaluate the criteria that are presently used to define these structures. What is the esophagus? What is the stomach? What is the gastroesophageal junction? What criteria are used to define these structures? What proof exists that these criteria are correct? We will examine these questions in the later discussion of the pathologic changes associated with reflux disease of the esophagus and develop accurate and reproducible definitions using histology.

References

1. Meyer GW, Austin RM, Brady CE, Castell DO. Muscle anatomy of the human esophagus. J Clin Gastroenterol 1986;8:131–134.
2. Liebermann-Meffert D, Duranceau A, Stein HJ. Anatomy and Embryology, Chapter 1. *In:* Orringer MB, Heitmiller R (eds),

The Esophagus, Vol 1, Zudeima GD, Yeo CJ. Shackelford's Surgery of the Alimentary Tract, 5th ed. Saunders, Philadelphia, 2002, pp 3–39.

3. Zaninotto G, DeMeester TR, Schwizer W, Johansson K-E, Cheng SC. The lower esophageal sphincter in health and disease. Am J Surg 1988;155:104–111.
4. DeMeester TR, Wernly JA, Bryant GH, Little AG, Skinner DB. Clinical and in vitro analysis of gastroesophageal competence: A study of the principles of antireflux surgery. Am J Surg 1979;137:39–46.
5. Hayward J. The phreno-esophageal ligament in hiatal hernia repair. Thorax 1961;16:41–45.
6. Johns BAE. Developmental changes in the oesophageal epithelium in man. J Anat 1952;86:431–442.
7. De Nardi FG, Riddell RH. Esophagus. *In:* Sternberg SS (ed), Histology for pathologists. Raven Press, New York, 1992, pp 524–526.
8. Goetsch E. The structure of the mammalian oesophagus. Amer J Anat 1910;10:1–40.
9. Sarbia M, Donner A, Gabbert HE. Histopathology of the gastroesophageal junction: A study on 36 operation specimens. Am J Surg Pathol 2002;26:1207–1212.
10. Shi L, Der R, Ma Y, Peters J, DeMeester T, Chandrasoma P. Gland ducts and multilayered epithelium in mucosal biopsies from gastroesophageal junction region are useful in characterizing esophageal location. Dis Esophagus 2005;18:87–92.
11. Allison PR, Johnstone AS. The oesophagus lined with gastric mucous membrane. Thorax 1953;8:87–101.
12. Barrett NR. Chronic peptic ulcer of the oesophagus and "oesophagitis." Br J Surg 1950;38:175–182.
13. De Hertogh G, Van Eyken P, Ectors N, Tack J, Geboes K. On the existence and location of cardiac mucosa; an autopsy study in embryos, fetuses, and infants. Gut 2003;52:791–796.
14. Chandrasoma P, Makarewicz K, Wickramasinghe K, Ma Y, DeMeester T. A proposal for a new validated histologic definition of the gastroesophageal junction. Hum Pathol 2006;37: 40–47.
15. Allison PR. Peptic ulcer of the oesophagus. Thorax 1948;3: 20–42.
16. Spechler SJ, Zeroogian JM, Antonioli DA, Wang HH, Goyal RK. Prevalence of metaplasia at the gastroesophageal junction. Lancet 1994;344:1533–1536.
17. Association of Directors of Anatomic and Surgical Pathology. Recommendations for reporting of resected esophageal adenocarcinomas. Am J Surg Pathology 2000;31:1188–1190.
18. Kim SL, Waring PJ, Spechler SJ, et al, and the Department of Veterans Affairs Gastroesophageal Reflux Study Group. Diagnostic inconsistencies in Barrett's esophagus. Gastroenterology 1994;107:945–949.
19. DeMeester TR, Peters JH, Bremner CG, Chandrasoma P. Biology of gastroesophageal reflux disease: Pathophysiology relating to medical and surgical treatment. Ann Rev Med 1999;50: 469–506.
20. Barrett NR. The lower esophagus lined by columnar epithelium. Surgery 1957;41:881–894.
21. Csendes A, Maluenda F, Braghetto I, et al. Location of the lower esophageal sphincter and the squamocolumnar mucosal junction in 109 healthy controls and 778 patients with different degrees of endoscopic esophagitis. Gut 1993;34:21–27.
22. Sampliner RE. Practice guidelines on the diagnosis, surveillance, and therapy of Barrett's esophagus. Am J Gastroenterol 1998;93:1028–1031.
23. Sharma P, Topalovski M, Mayo MS, et al. Methylene blue chromoendoscopy for detection of short-segment Barrett's esophagus. Gastrointest Endosc 2001;54:289–293.
24. Yagi K, Nakamura A, Sekine A. Accuracy of magnifying endoscopy with methylene blue in the diagnosis of specialized intestinal metaplasia and short-segment Barrett's esophagus in Japanese patients without *Helicobacter pylori* infection. Gastrointest Endosc 2003;58:189–195.
25. Sharma P, Weston AP, Topalovski M, et al. Magnification chromoendoscopy for detection of intestinal metaplasia and dysplasia in Barrett's esophagus. Gut 2003;52:24–27.
26. McClave SA, Boyce HW Jr, Gottfried MR. Early diagnosis of columnar lined esophagus: A new endoscopic diagnostic criterion. Gastrointest Endosc 1987;33:413–416.
27. Spechler SJ. Esophageal columnar metaplasia (Barrett's esophagus). Gastrointest Endosc Clin North Am 1997;7:1–18.
28. Spechler SJ. Barrett's esophagus. N Engl J Med 2002;346: 836–842.
29. Chandrasoma PT, Der R, Ma Y, et al. Histology of the gastroesophageal junction. An autopsy study. Am J Surg Pathol 2000; 24:402–409.
30. Oberg S, Peters JH, DeMeester TR, et al. Inflammation and specialized intestinal metaplasia of cardiac mucosa is a manifestation of gastroesophageal reflux disease. Ann Surg 1997; 226:522–532.
31. Kilgore SP, Ormsby AH, Gramlich TL, et al. The gastric cardia: Fact of fiction? Am J Gastroenterol 2000;95:921–924.
32. Odze RD. Unraveling the mystery of the gastroesophageal junction: A pathologist's perspective. Am J Gastroenterol 2005;100: 1853–1867.
33. Lagergren J, Bergstrom R, Lindgren A, Nyren O. Symptomatic gastroesophageal reflux as a risk factor for esophageal adenocarcinoma. N Engl J Med 1999;340:825–831.
34. Spechler SJ. The role of gastric carditis in metaplasia and neoplasia at the gastroesophageal junction. Gastroenterology 1999;117:218–228.

CHAPTER

5

Histologic Definitions and Diagnosis of Epithelial Types

DEFINITIONS

There is so much controversy about the histology of the esophagus and stomach that it is important to begin by establishing accurate and reproducible definitions. While there are differences in terminology even today, all pathologists will agree that the histologic types as defined here are easily and reproducibly recognized. *The starting point of understanding is the use of these definitions precisely and without any modification.*

At this point, it is important to avoid ascribing any correlation between the epithelium defined histologically and the anatomy as we recognize it. Thus, cardiac mucosa is simply as it is defined histologically, not a mucosa that lines any anatomically defined region.

If one begins in the proximal esophagus and ends at the point where the gastric body mucosa changes to pyloric antral mucosa, a limited number of different epithelial types can be defined after development is completed (Table 5.1). In fetuses and infants under 1 year, developing fetal columnar epithelia may be seen as discussed in Chapter 2.

Stratified Squamous Epithelium

Stratified squamous epithelium of the esophagus is of the nonkeratinizing type (Fig. 5.1). This consists of two to three layers of rounded basal cells in the basal region that are small with a high nuclear : cytoplasmic ratio and hyperchromatic nuclei. Normally, the basal cell region occupies less than 20% of the thickness of the epithelium. Above the basal layer are intermediate cells (keratinocytes), which extend to the surface. The keratinocytes progressively show increasing amounts of eosinophilic cytoplasm and tend to flatten at the top where the nuclei also become flattened. Normally, these nucleated keratinocytes are lost from the surface without losing their nuclei (i.e., there is no stratum corneum or keratin layer at the surface). Mitotic activity is normally limited to the basal region; the keratinocytes are terminally differentiated nonproliferative cells.

The subepithelial lamina propria extends into the epithelium as papillae, giving the epithelium a wavy appearance at the base with rete pegs alternating with the papillae. The papillae contain fibrovascular tissue and are usually devoid of inflammatory cells. Normally, the esophageal papillae are short and extend less than 60% of the thickness of the epithelium (Fig. 5.1). This structure of the squamous epithelium is well suited to the transient dilatation required during swallowing to allow the bolus of food to pass. The squamous epithelium flattens during this process and loses its papillae.

Stratified squamous epithelium is also well suited to withstand the abrasion that accompanies the repeated passage of solid food, which is forced down the tube by peristalsis. The loss of surface cells during this physiologic process is rapidly and continuously replaced from below in an orderly manner. The squamous epithelium also provides an excellent barrier that prevents access to luminal molecules. The keratinocytes have tight cell junctions that make the epithelium impervious to even small molecules in the normal state (1).

TABLE 5.1 Different Epithelial Types That Are Seen between the Proximal Esophagus and the Distal Stomach (Antrum)

Epithelial type	Definition	Location
Stratified squamous	Squamous; may have mucosal and submucosal glands and gland ducts	Only esophagus
Oxyntic mucosa	Columnar epithelium with straight tubular glands composed only of parietal and chief cells	Only stomach
Atrophic oxyntic mucosa	Oxyntic mucosa with complete or partial loss of parietal cells	Only stomach
Cardiac mucosa	Columnar epithelium composed only of mucous cells	Esophagus and within 3 cm distal to the presently defined gastroesophageal junction
Oxyntocardiac mucosa	Columnar epithelium with glands composed of a mixture of mucous and parietal cells	Esophagus and within 3 cm distal to the presently defined gastroesophageal junction
Esophageal intestinal metaplasia	Goblet cells in cardiac mucosa	Esophagus and within 3 cm distal to the presently defined gastroesophageal junction
Gastric intestinal metaplasia	Goblet cells in oxyntic (usually atrophic) mucosa	Only stomach

Note: Notice that the presence of chief cells, Paneth cells, serous (pancreatic) cells, and neuroendocrine cells does not affect the definition of these epithelial types.

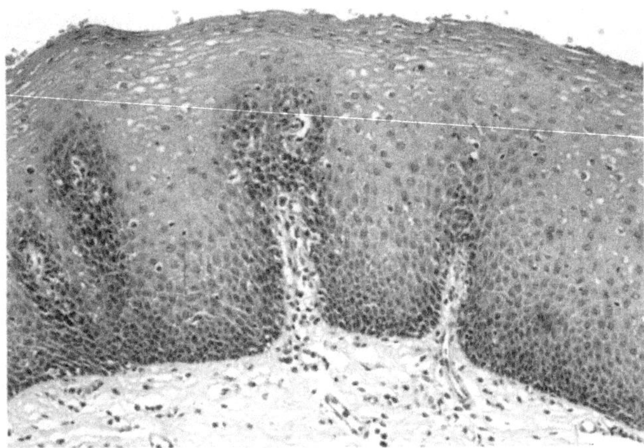

FIGURE 5.1 Normal esophageal stratified squamous epithelium hematoxylin and eosin. The basal cells are cuboidal and have scanty cytoplasm. The cells enlarge and flatten as they differentiate into mature squamous cells at the surface. The only direction of maturation is from the base to the surface. (Cross reference: Color Plate 1.4)

In the human, stratified squamous epithelium is never seen distal to the esophagus. This is not true of all species. The most lucid description of comparative anatomy is found in Barrett's original 1957 article (2) where he established columnar-lined esophagus as an entity. He wrote:

> The mucous membrane of the esophagus and of the stomach varies in its distribution in different species of mammals. It is principally in the carnivore that the epithelium changes abruptly from squamous to columnar precisely at the level of the "cardiac valve." And in this group the esophagus widens into the stomach at the same place as that at which the epithelium changes. The anatomy of the mucous membrane of the foregut depends upon the food habits of the animal concerned. Those which have teeth in the mouth and which eat meat are arranged as we are; but in ruminants and species which are edentulous in the mouth, the stomach not only masticates the food but stores it before digestion. Thus it comes about that in most vertebrates there is no "cardiac valve" mechanism and an important part of the stomach is lined by horny squamous epithelium. The extreme example is ornithorynchus (the duck-billed platypus) in which the squamous epithelium of the esophagus continues onward as far as the pylorus. But an important area of the stomach of animals as widely different as the horse, the cow, the rat, and the rabbit is covered by squamous cells. (pp. 883–884)

Barrett was trying to use comparative anatomy to explain the occurrence of columnar lining in the esophagus. He concluded: "This anomaly of the lower esophagus cannot be explained as a reversion to a more primitive zoologic type. . . . The tendency in the mammalian world is for squamous epithelium to continue onward toward the pylorus, not for columnar epithelium to grow up the esophagus" (p. 884).

Stratified squamous epithelium is always present in the adult esophagus. In the fetus, squamous epithelium is not seen until about the 22nd week of gestation when it begins to replace fetal columnar epithelium (3) (see Chapter 3). The ciliated epithelium (type II fetal columnar epithelium) disappears in the third trimester. Vestiges of nonciliated fetal columnar epithelium (type III fetal columnar epithelium) frequently remain at the lower end of the esophagus after birth but are replaced by squamous epithelium in the first year of life.

Opinion as to the extent of the esophagus that is lined by squamous epithelium has come full circle. Allison in 1948 (4) and Barrett in 1957 (2) held the opinion that the entire tubular esophagus in the adult is lined by stratified squamous epithelium and that presence of columnar epithelium in the tubular esophagus was not normal. As Barrett defined columnar-lined esophagus in 1957 (2): "The abnormally placed mucous membrane extends upward from the esophago-gastric junction in a continuous, unbroken sheet.... The extent of (columnar lined esophagus) varies from a few centimeters to the upper esophagus" (p. 884). In 1961, Hayward (5) opined that the idea that the esophagus was lined entirely by squamous epithelium should be rejected. Without any supporting data, he suggested that the distal 1 to 2 cm of the tubular esophagus was normally lined by columnar epithelium. This unsubstantiated dogma was widely accepted until the recognition of short-segment Barrett esophagus, which is defined as the occurrence of intestinal metaplasia in any endoscopically visible columnar lining in the esophagus. The present endoscopic definition of normalcy is that the entire tubular esophagus is normally lined by squamous epithelium (6–8).

The belief that squamous epithelium is limited to the esophagus in humans is not universal. In 2000, Fass and Sampliner reported the presence of squamous epithelium distal to the endoscopic gastroesophageal junction in 16 of 547 (3%) of patients. The squamous epithelium extended to a maximum of 3 cm. The authors' implicit faith in the reliability of the endoscopic gastroesophageal junction makes them conclude that the squamous epithelium is in the stomach. We will show in Chapter 11, that it is more likely that this squamous epithelium is actually in esophagus; the misinterpretation results from using an incorrect definition of the gastroesophageal junction at endoscopy.

Stratified squamous epithelium is a dynamic tissue with a high rate of turnover. Squamous epithelium develops in the second trimester of fetal life when programmed activation of unknown genetic signals causes the fetal endodermal progenitor cells to direct their daughter cells to differentiate into squamous cells. At this point, the previous fetal genetic signals that resulted in fetal columnar epithelia become suppressed. The fully developed adult stratified squamous epithelium is normally under the direction of the squamous genetic signal. The adult progenitor cells in the basal layer continuously multiply to renew the epithelium. The daughters of these basal cells move upward in the epithelium, becoming increasingly keratinized and establishing the typical intercellular bridges between adjacent cells that are typical of stratified squamous epithelium. The superficial keratinized squamous cells that are lost at the surface are continuously replaced by cells from below. The average time for a new daughter cell arising from the adult stem cell at the base to move up the epithelium and be lost at the surface is 7.5 days (9).

Staining with Ki67, which is an antigen expressed in the nuclei of proliferative cells, by immunoperoxidase shows positivity restricted to the suprabasal cell region (Fig. 5.2). The primitive progenitor cells in the basal layer usually do not show Ki67 positivity. By the time the midepithelium is reached, the keratinocytes have lost their ability to undergo mitotic division.

Gastric Oxyntic Mucosa

This is the normal lining of the body and fundus of the stomach and frequently extends into the pyloric antrum where it transitions with antral type mucosa. Oxyntic gastric mucosa is always present in the normal fundus and body of the stomach (Figs. 5.3 and 5.4). Oxyntic gastric mucosa is characterized by (a) a surface layer of mucous cells with small, basally located nuclei and uniform apical cytoplasm that contains neutral mucin (Fig. 5.4); (b) a foveolar region that extends down vertically from the surface as a pit, which is usually very short and lined by mucous cells (Fig. 5.4); and (c) a straight tubular gland that is lined by parietal and chief cells and scattered neuroendocrine cells. Parietal cell is a term that is synonymous with oxyntic cell; it is the acid-secreting cell of gastric mucosa that is recognized as a large round cell with abundant eosinophilic cytoplasm (Fig. 5.4). The chief cells secrete pepsinogen and have more basophilic cytoplasm. There are a few mucous cells in the neck region of the gland ("mucous neck cells") but the deeper part of the

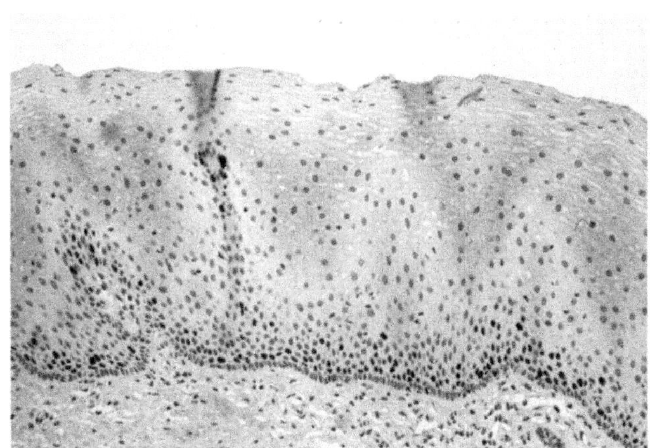

FIGURE 5.2 Stratified squamous epithelium, Ki67 stain. Note the proliferative Ki67-positive pool of cells in the suprabasal region. These are derived from the division of the stem cells in the basal layer. The cells have a signal to differentiate into mature squamous cells toward the surface.

gland is normally devoid of mucous cells. The glands are tightly packed so that there is minimal lamina propria. No inflammatory cells are normally present in oxyntic mucosa. The presence of inflammatory cells signifies gastritis, a pathologic state.

Atrophy of gastric oxyntic mucosa is defined by a loss of parietal cells and may occur as a result of *Helicobacter pylori* infection and autoimmune gastritis. When atrophy is severe, there can be complete loss of parietal cells and the gastric mucosa is lined by mucous cells, often with intestinal metaplasia (Figs. 5.5 and 5.6). In the pathologic state of atrophy, therefore, gastric oxyntic mucosa may not have any parietal cells. Therefore, the absence of parietal cells cannot be used as a criterion to exclude a mucosa as being gastric oxyntic mucosa.

Normal gastric oxyntic mucosa is a dynamic, continually renewing epithelium. In fetal life, the adult genetic signals that dictate gastric development are developed early, in the 11th to 17th week of gestation

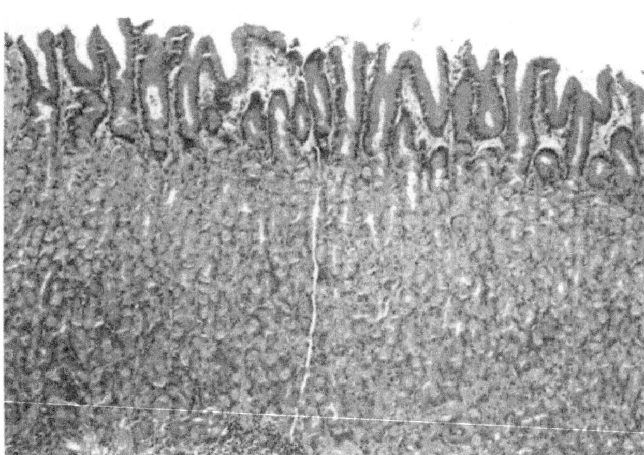

FIGURE 5.3 Oxyntic mucosa, H&E, low power. The surface and short foveolar region are lined by mucous cells with basal nuclei and apical pale cytoplasm. The glands arise at the bottom of the foveolar pit and consist of large secretory parietal cells and chief cells. (Cross reference: Color Plate 1.7)

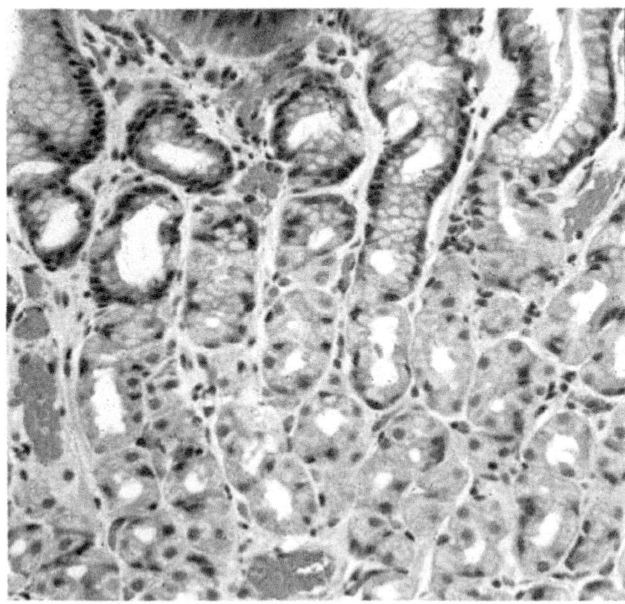

FIGURE 5.4 Oxyntic mucosa, H&E, high power, showing the mucous cells in the surface and foveolar region and parietal and chief cells in the glands. (Cross reference: Color Plate 1.8)

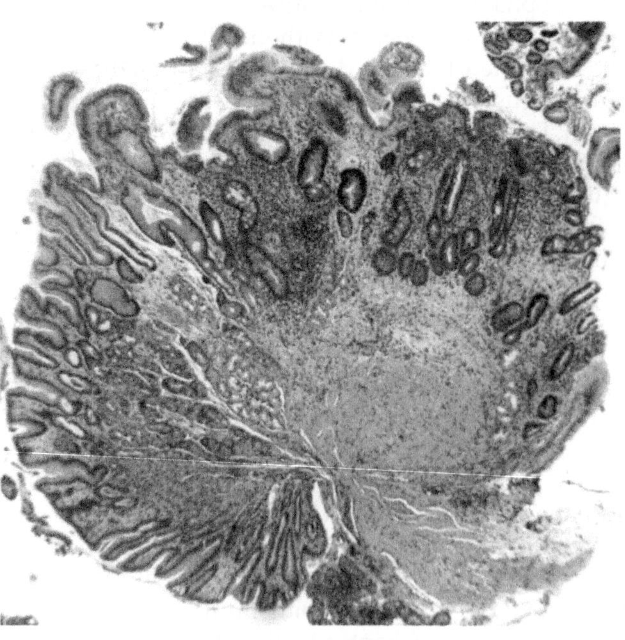

FIGURE 5.5 Atrophic gastritis showing normal oxyntic mucosa on lower left and an area of inflammation with loss of glands on upper right. (Cross reference: Color Plate 4.6)

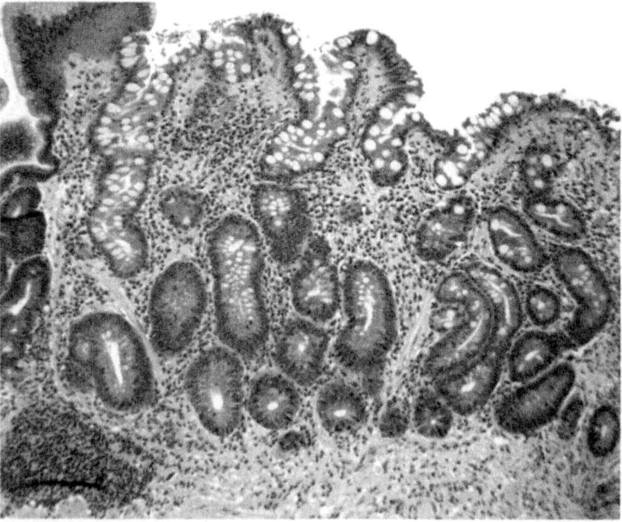

FIGURE 5.6 Atrophic gastritis showing loss of parietal cell–containing glands, chronic inflammation, and gastric intestinal metaplasia.

(10) (see Chapter 3). At this point, the fetal progenitor cells complete the programmed genetic activation and suppression to become the adult progenitor cell that remains in the epithelium throughout life to renew the epithelium. In the normal stomach, the progenitor cells are found in the deep region of the foveolar pit and the isthmus of the gastric gland just below the foveolar pit, where their actively proliferating daughter cells can be recognized by their positive staining with Ki67 (Fig. 5.7). Adult progenitor cell division results in two lineages of cells based on whether they move superficially toward the lumen or deep into the gland (Fig. 5.8) (9). The cells that move superficially become terminally differentiated surface-type cells by the midpoint of the foveolar pit above which there is normally no proliferative activity (11). These cells ultimately replace the mucous cells at the surface that are being continuously lost. They have a rapid turnover of around 3 to 6 days. The cells that move downward into the glands differentiate into mucous neck cells, parietal cells, and zymogenic cells. Parietal cells and chief cells do not proliferate; they are terminally differentiated secretory cells with a programmed life span; they undergo apoptosis at the end of their life span. The glandular parietal and chief cells have a slow turnover, measured at 54 and 194 days, respectively.

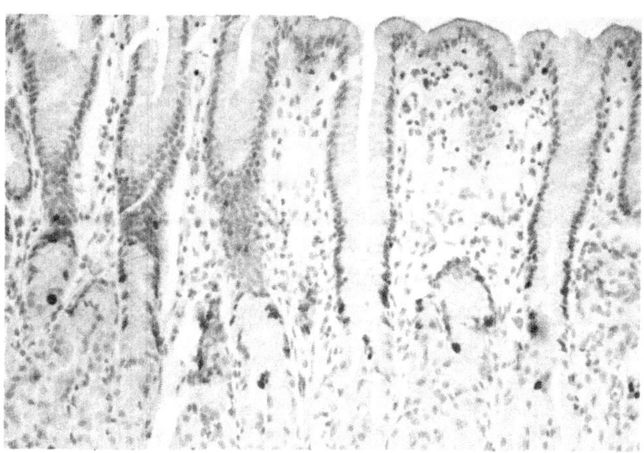

FIGURE 5.7 Oxyntic mucosa, Ki67. The proliferative cell zone is a very short area of the deep part of the foveolar pit. The superficial half of the foveolar pit contains terminally differentiated nonmitotic cells, which are continually lost at the surface to be replenished from below. The glands are long-lived stable cells that are not mitotically active. They undergo programmed death (apoptosis) and are replaced by stem cells in the deep foveolar pit.

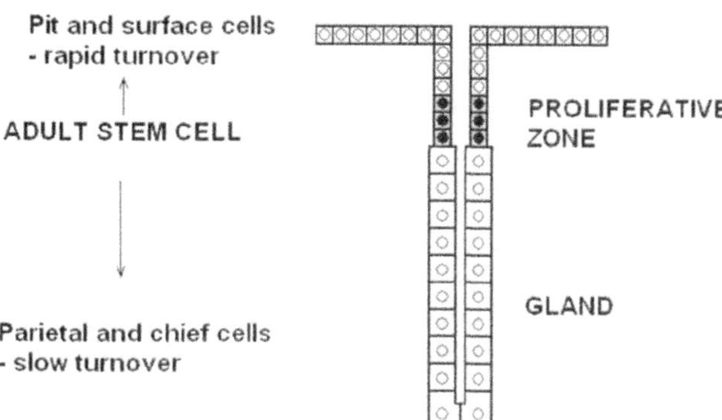

FIGURE 5.8 Diagram of proliferation and differentiation in normal gastric mucosa. The adult stem cells reside in the deep part of the foveolar pit and gland neck region immediately below the foveolar pit. Stem cell divisions produce two types of cells: those moving up into the pit to replenish surface cells and those moving down to replenish glandular (parietal and chief) cells undergoing apoptosis.

Cardiac Mucosa

Cardiac mucosa is defined as a mucosa in the region of the foregut proximal to the pyloric antrum that is composed entirely of mucous cells (Figs. 5.9 and 5.10). The definition is one of exclusion; cardiac mucosa does not have any parietal or goblet cells. Cardiac mucosa is permitted by definition to have Paneth cells (Fig. 5.11), serous (or pancreatic) cells (Fig. 5.12), and neuroendocrine cells, but never parietal cells or goblet cells. Some publications refer to pancreatic metaplasia as an entity but this is not a defined epithelial type; it is merely a metaplasia occurring in another epithelial type (usually cardiac or oxyntocardiac).

The presence of pancreatic metaplasia has been studied and shown to be a finding that is of little clinical significance in a study of 76 esophagogastrectomy specimens with a normally located gastroesophageal junction (12). Pancreatic metaplasia, defined as the presence of acinar cells filled with large red granules

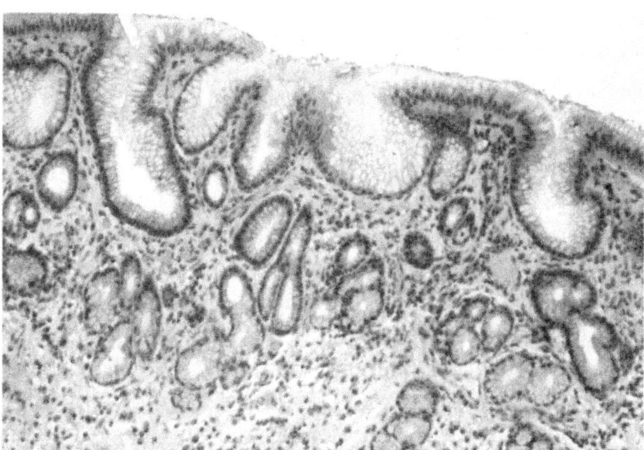

FIGURE 5.9 Cardiac mucosa showing a flat surface with chronic inflammation. Note the slightly elongated foveolar region and lobulated glands, all composed of mucous cells. There are no parietal cells or goblet cells.

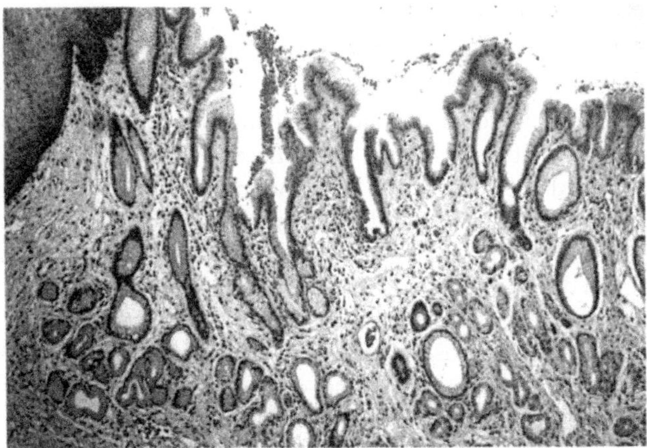

FIGURE 5.10 Cardiac mucosa immediately distal to squamous epithelium showing a villiform surface resulting from foveolar hyperplasia, chronic inflammation, and disorganized mucous glands in the deep part of the mucosa.

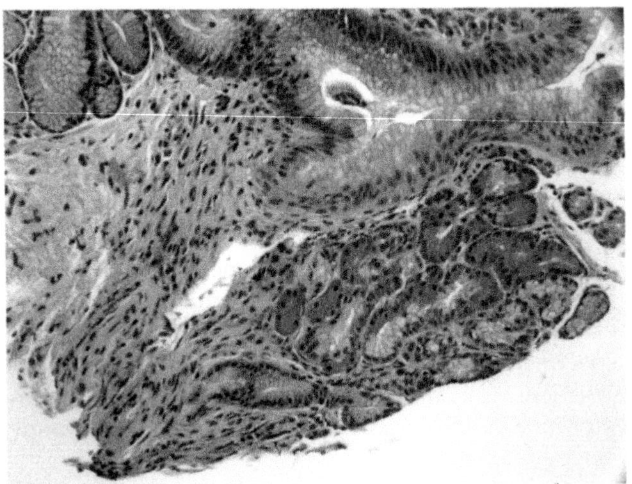

FIGURE 5.11 Paneth cells in gland of cardiac mucosa.

in the cytoplasm and confirmed by antilipase immunostaining, was present within 2 cm of the squamocolumnar junction and the end of the tubular esophagus in 11 patients (14%). This may be an underestimation of the prevalence because the study used archival tissue with a mean of 1.4 sections per patient, indicating less than circumferential sampling of the junction. The presence of pancreatic metaplasia was associated with mucosal inflammation. There was no expression of c-*erb*B-2 or K-*ras* mutations in pancreatic metaplasia. The authors concluded that there is no indication that pancreatic metaplasia has neoplastic potential. In my experience, pancreatic metaplasia is most commonly seen in oxyntocardiac mucosa (Fig. 5.12).

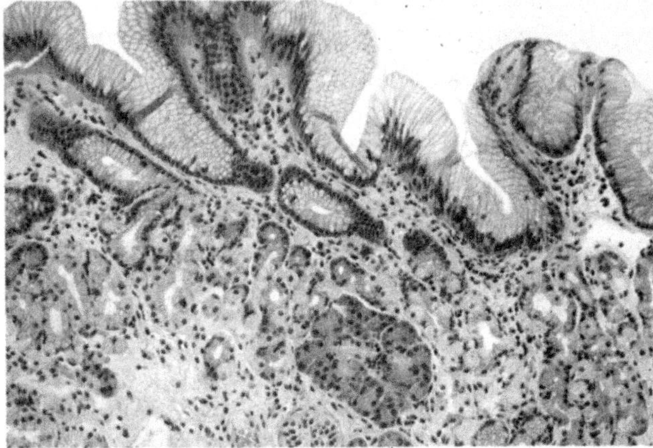

FIGURE 5.12 Pancreatic metaplasia, seen as a lobule composed of serous cells within oxyntocardiac mucosa.

Similarly, it is present belief that the presence of neuroendocrine cells and Paneth cells are not of any great significance. There are no studies that have verified or refuted this belief.

Cardiac mucosa must be distinguished from gastric oxyntic mucosa that has undergone severe atrophy with complete loss of parietal cells. Being an epithelium that is devoid of parietal cells, it can fall within the strict definition of cardiac mucosa. The mucous-cell-only epithelium of atrophic gastritis differs from true cardiac mucosa in the following respects: (a) it is an atrophic epithelium, which has a flat surface compared to the hyperplastic true cardiac mucosa (which frequently has a villiform surface); and (b) it involves large areas of the stomach, unlike true cardiac mucosa, which is found only in the distal esophagus and 3 cm distal to the presently defined gastroesophageal junction in what is called proximal stomach. In cases where true cardiac mucosa coexists with atrophic gastritis, the distinction between the two can be difficult, but this does not alter the fact that they are different epithelial types.

The following synonyms are used for true cardiac mucosa:

1. *Junctional mucosa*. This term was coined by Hayward and used by Paull et al. in their landmark study (13) that defined histologic epithelial types in columnar-lined esophagus. The gastroesophageal junction must be regarded as an imaginary line drawn at the junction of the esophagus and stomach. The term "junctional mucosa" creates the illusion that the gastroesophageal junction is a zone with a longitudinal extent. The term "junctional mucosa" has largely been abandoned in general usage though it sometimes rears its head in publications.

2. *Mucous cell only mucosa.* This term has been used by the Harvard group (14). This is an acceptable term because it is descriptive of the definition and has no connotation to location. However, it has a greater likelihood of confusion with other gastrointestinal tract epithelia composed of mucous cells only (atrophic gastritis, pyloric antral mucosa). This term has not been accepted into general usage and "cardiac mucosa" still remains the most common term for this epithelium.

In reality, *the best term for cardiac mucosa is "metaplastic esophageal columnar epithelium, type 1,"* but we fear this is too cumbersome for general consumption and at this time too controversial.

Cardiac mucosa may exist as a single-layered flat epithelium that lines the surface, most frequently in the healing phase of an erosion (Fig. 5.13). It may be multilayered with an appearance that resembles nonciliated pseudo-stratified columnar epithelium (Figs. 5.14 and 5.15) (15, 16). There may be admixtures of squamous cells or basaloid "reserve cells" with the mucous cells producing a mixed squamous-cardiac epithelium, which will fall under the definition of cardiac mucosa. It may exist as a thin foveolar epithelium with short foveolar pits in addition to the surface layer but without well-formed glands (Fig. 5.16). It may have glands under the foveolar region (Fig. 5.10). In general, when glands are present, they are disorganized lobulated glands that are, by definition, composed only of mucous cells without parietal or goblet cells. We have not seen ciliated cells in adult cardiac mucosa except in the gland ducts of submucosal glands (see Chapter 4).

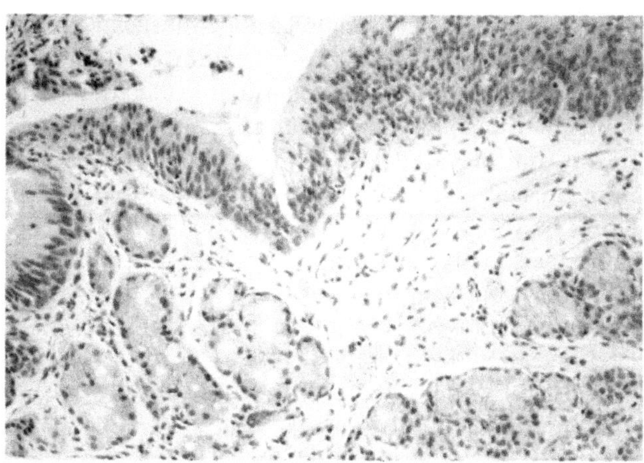

FIGURE 5.14 Multilayered epithelium with basal cell hyperplasia in cardiac mucosa.

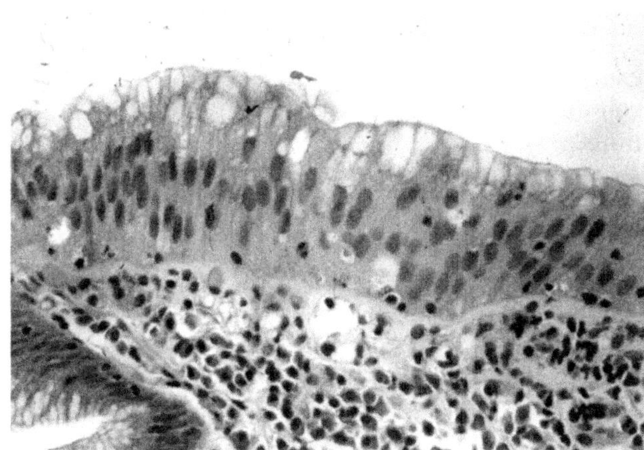

FIGURE 5.15 Multilayered epithelium with nuclear pseudostratification.

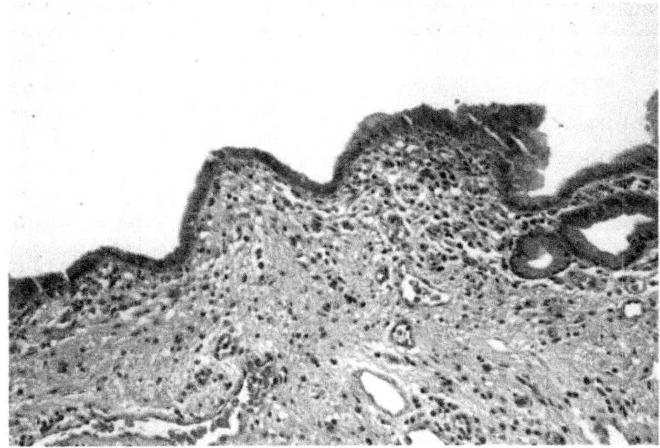

FIGURE 5.13 Flat cardiac mucosa. A section from a metaplastic columnar esophageal lining composed of a single layer of columnar cells. These are mucous cells. At the right extreme, the surface epithelium invaginates to form a foveolar pit.

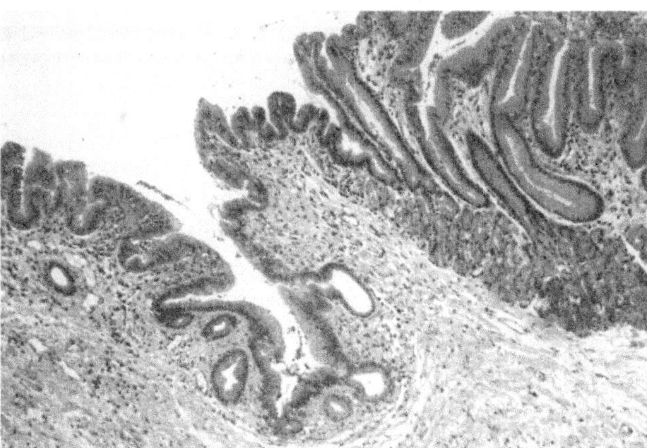

FIGURE 5.16 Cardiac mucosa, foveolar type. Sections across metaplastic columnar esophagus shows an epithelium composed of foveolar pits only on the left and the presence of glands on the right. (Cross reference: Color Plate 3.12)

The surface epithelium of cardiac mucosa is generally composed of a single layer of columnar epithelium with apical neutral mucin that is identical to that in the stomach. In some cases, the surface epithelium is focally multilayered (Fig. 5.15). Multilayered epithelium consists of four to seven layers of cells that appear either as pseudo-stratified nonciliated columnar epithelium as a mixture of basaloid squamous cells in the lower part and columnar cells in the superficial layer (15, 16). The presence of multilayered epithelium has been associated with reflux.

Confusion may arise with terminology in fetuses and neonates. Because fetal esophagus is lined by columnar epithelium of different types at varying stages of embryonic development, and because these epithelia consist entirely of mucous cells of varying morphologic types, they can be considered to fall within the strict definition of "cardiac mucosa." In fact, two articles designate fetal columnar epithelia as cardiac mucosa (17, 18). Other studies refer to these same epithelial types seen in fetal esophagi as "tall columnar epithelium" (3) and "transitional epithelium" (19). It is important to recognize that fetal columnar epithelia are different to what is described here as cardiac mucosa (see Chapter 3) (20). When evaluating biopsies from neonates, it is important to consider the possibility that a mucosa composed only of mucous cells found in the gastroesophageal junctional region could be a developing fetal columnar esophageal epithelial rather than adult cardiac mucosa. This is not a consideration after infancy, when development of the esophageal epithelium is completed and fetal columnar epithelium has disappeared.

Oxyntocardiac Mucosa

Oxyntocardiac mucosa is defined by the presence of glands under the foveolar region that are composed of a mixture of mucous cells and parietal (= oxyntic) cells. The number of each cell type varies. When parietal cells are few, oxyntocardiac mucosa resembles cardiac mucosa and a careful search for the presence of parietal cells is required to differentiate the two (Fig. 5.17). The presence of even one parietal cell in a single glandular unit (defined as a unit that drains into one foveolar pit) excludes cardiac mucosa and satisfies the criteria for oxyntocardiac mucosa. When parietal cells are numerous and mucous cells are few, oxyntocardiac mucosa resembles gastric oxyntic mucosa and a careful search for admixed mucous cells within the glands is necessary to establish the mucosa as oxyntocardiac (Fig. 5.18). Paneth cells, serous (pancreatic) cells, and neuroendocrine cells may be present within oxyntocardiac mucosa without altering this definition.

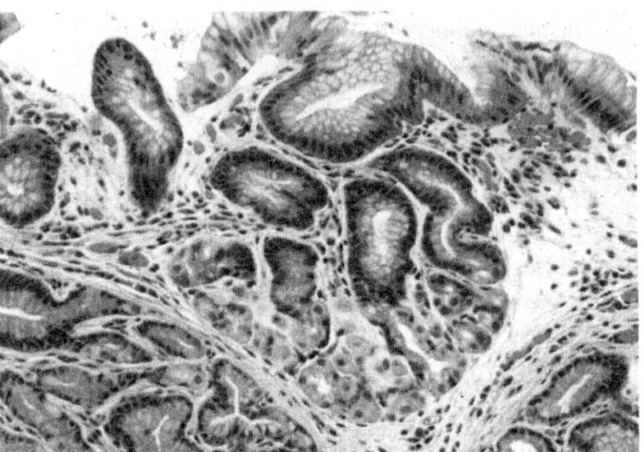

FIGURE 5.17 Oxyntocardiac mucosa with few parietal cells. The glands are lobulated and contain parietal cells admixed with mucous cells. There is moderate inflammation. This epithelium differs from cardiac mucosa only in that it contains parietal cells. (Cross reference: Color Plate 3.14)

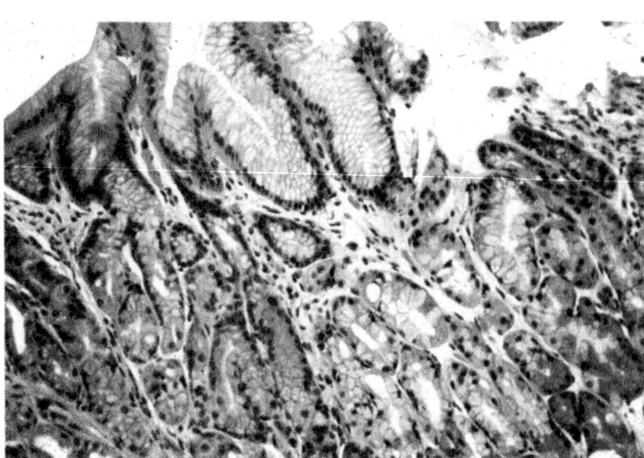

FIGURE 5.18 Oxyntocardiac mucosa with many parietal cells. The glands are straighter and the inflammation is less than that shown in Figure 5.17. This epithelium differs from oxyntocardiac mucosa only in the presence of mucous cells deep in the glands.

"Oxyntocardiac mucosa" is a term that Dr. Chandrasoma invented, and it has gained acceptance in publications originating in Europe but not in the United States. The common synonym used for oxyntocardiac mucosa is "fundic epithelium" after Paull et al.'s original description of this as "gastric-fundic-type epithelium." It is strongly recommended that "fundic epithelium" not be used for this epithelium; it is never located in the fundus of the stomach, which is lined by gastric oxyntic mucosa. The only resemblance between oxyntocardiac mucosa and fundic mucosa is the presence of parietal cells. Paull et al. used the term "fundic-type" for an epithelium that they clearly defined in their study as being in the esophagus. The term "fundic-type" was accurate because they were merely stating that it was an esophageal columnar epithelium that resembled fundic mucosa in that it contained parietal cells. Unfortunately, the medical community shortened this to "fundic" with resulting confusion. The other synonyms that are used for oxyntocardiac mucosa are "mixed glandular mucosa or mixed mucous/oxyntic mucosa" as distinct from "mucous-cell-only mucosa" (14) and "transitional epithelium/mucosa" (21). Mixed glandular mucosa means nothing because it does not indicate the mixture that is present; "mixed mucous/oxyntic" is better because the "mixture" is strictly defined by the presence of mucous and parietal cells. The term "oxyntocardiac" seems appropriate (*oxynto* denoting parietal cell presence, and *cardiac* denoting the mucous cells) and is presently gaining acceptance, albeit slowly. The term "transitional mucosa" is rarely used and there is no reason why it should be introduced. *Our preferred term for oxyntocardiac mucosa is "metaplastic esophageal columnar epithelium, type 2,"* but we fear this is too cumbersome for general consumption and at this time too controversial.

Intestinal Metaplastic Epithelium

Intestinal metaplasia is characterized by the presence of well-formed goblet cells, which maybe present in the surface epithelium or the foveolar region (Fig. 5.19). Goblet cells are large round vacuoles filled with acid mucin (Fig. 5.20). The diagnosis of intestinal metaplasia depends on the identification of at least one well-defined goblet cell in a routine hematoxylin and eosin-stained section (Fig. 5.21).

Intestinal metaplasia may occur in both the esophagus and the stomach. The mechanism and etiology of intestinal metaplasia are very different in these two organs, although they are both defined morphologically by the same criterion, the goblet cell. To distinguish between the two, we use the following terminology: For intestinal metaplasia of the esophagus, we describe the fact that the goblet cells occur in cardiac mucosa and use the term "intestinal (Barrett-type) metaplasia." An equivalent and probably better term in the future would be "Esophageal intestinal metaplasia."

"Intestinal metaplasia of the esophagus" is defined by the presence of goblet cells in cardiac mucosa.

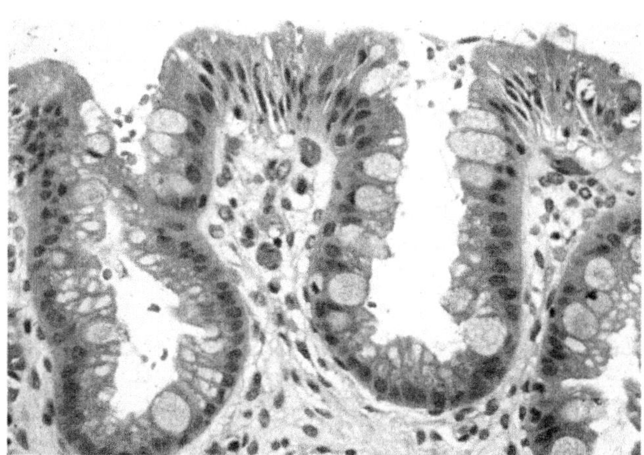

FIGURE 5.20 Intestinal metaplasia with blue goblet cells. The goblet cells have a blue tinge to the cytoplasm because the mucin has not been dissolved out during fixation and has taken the hematoxylin stain. (Cross reference: Color Plate 4.3)

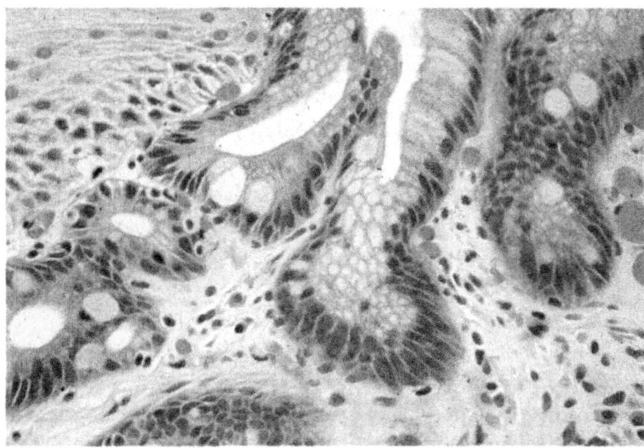

FIGURE 5.19 Intestinal metaplasia, showing goblet cells in cardiac mucosa. The goblet cells are interspersed and contain a large, round, empty vacuole that distends the lateral borders of the cells.

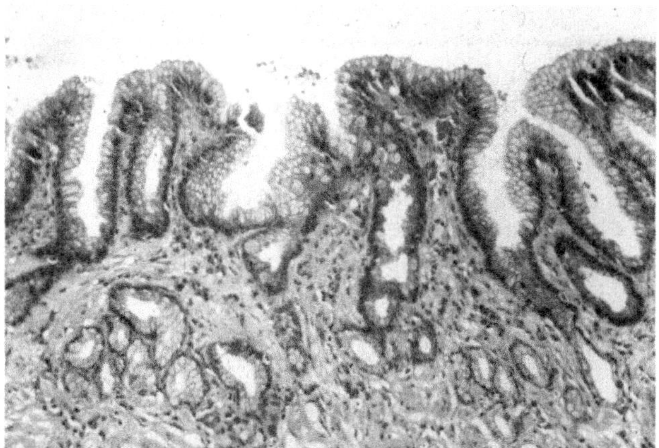

FIGURE 5.21 Cardiac mucosa with a few goblet cells in the central part of the field. Note that intestinal (Barrett-type) metaplasia is diagnosed when the first definite goblet cell is recognized in cardiac mucosa.

Intestinal metaplasia of the esophagus never contains parietal cells (13); it only rarely contains Paneth cells and pancreatic cells. Paull et al. used the term "specialized epithelium" to describe this epithelium with a mixture of goblet cells and nongoblet mucous cells. Other synonymous terms that have been used are "specialized columnar epithelium (SCE)" and "specialized intestinal metaplasia (SIM)." The word "specialized" means nothing in the absence of "unspecialized" intestinal metaplasia or "unspecialized" columnar epithelium. These synonyms are commonly used in the gastroenterology literature today; they should be dropped in favor of "esophageal intestinal metaplasia" or "intestinal (Barrett-type) metaplasia." Esophageal intestinal metaplasia is equivalent to Barrett esophagus by its present-day definition and it is the result of the effect of gastroesophageal reflux on the esophagus. It is generally regarded as a premalignant epithelium.

"Intestinal metaplasia of the stomach (gastric intestinal metaplasia or GIM)" is defined by the occurrence of goblet cells in gastric mucosa, both gastric oxyntic mucosa of the body and fundus and pyloric antral mucosa (Figs. 5.5 and 5.6). The mechanism of gastric intestinal metaplasia is mainly through atrophic gastritis caused by *H. pylori* infection and autoimmune gastritis, and in the healing reactive phase of erosive lesions. In *H. pylori* infection, the pathology is maximal in the pyloric antrum but frequently involves the body to cause a pan-gastritis. In autoimmune gastritis, which is an immune destruction of parietal cells, the atrophy and intestinal metaplasia is body predominant and spares the antrum where parietal cells are not found. Gastric intestinal metaplasia is believed to be associated with a slightly increased risk of malignancy with the potential for carcinoma being too low to justify surveillance. It is therefore different from esophageal intestinal metaplasia, which carries a sufficiently high malignant potential to justify surveillance.

"Intestinal metaplasia of the cardia" is a term applied when an endoscopically normal patient is biopsied (against present practice guidelines of the American Gastroenterology Association) and intestinal metaplasia is found. Some gastroenterologists ignore the guidelines and biopsy endoscopically normal people when they have symptoms of reflux; when they do this, 5% to 15% of endoscopically normal people have intestinal metaplasia at their squamocolumnar junction (22). The prevalence is much less in asymptomatic screened populations (23). There is controversy as to what the finding of intestinal metaplasia at an endoscopically normal junction means as well as how the patient should be managed.

The term "intestinal metaplasia of the cardia (CIM)" is used as if there is a third organ between the esophagus and stomach. If our definition of "cardia" is applied (see Chapter 4), this becomes "intestinal metaplasia in a place that I am uncertain is esophagus or stomach" and it is clear that uncertainty exists between whether this is intestinal metaplasia of the esophagus (Barrett esophagus) or intestinal metaplasia of the stomach.

Intestinal metaplasia of the cardia is a phantom entity created by the failure of accurate definition of the gastroesophageal junction. If the junction can be defined accurately, there is no "cardia" interposed between the only two organs that exist in this region: the esophagus and stomach. "Intestinal metaplasia of the cardia" is a term that tacitly accepts the failure of accurately defining the gastroesophageal junction. As such, the entity will always be an admixture of intestinal metaplasia of the esophagus (Barrett esophagus) and gastric intestinal metaplasia (atrophic gastritis). When we develop the ability to define the gastroesophageal junction with accuracy, precision, and reproducibility, intestinal metaplasia of the cardia will resolve itself into esophageal intestinal metaplasia (Barrett esophagus) and gastric intestinal metaplasia. When this happens, we will recognize that the present belief that intestinal metaplasia of the cardia is caused by both gastroesophageal reflux and *H. pylori* (24) will resolve into patients with esophageal intestinal metaplasia where the etiology is gastroesophageal reflux and patients with gastric intestinal metaplasia where the etiology are the causes of atrophic gastritis.

PROBLEMS WITH THE DEFINITIONS

Stratified squamous epithelium, gastric oxyntic mucosa (including the pathologic state of total atrophy resulting in a mucous-cell-only epithelium that resembles cardiac mucosa), cardiac mucosa, oxyntocardiac mucosa, and intestinal metaplastic epithelium (subdivided into esophageal and gastric) are the only epithelial types encountered in the esophagus and stomach proximal to the beginning of pyloric antral mucosa. Adherence to these strict definitions is easy and highly reproducible. The only problems arise in the following situations.

Failing to Apply Definitions to Single Foveolar-Gland Units

A single biopsy piece frequently has multiple of these epithelial types in any combination (Figs. 5.22, 5.23, and 5.24). When only one of the epithelial types is represented, there is no problem with definition.

When there are mixtures of oxyntic, oxyntocardiac, cardiac, and intestinal epithelia, it is important to apply the definitions to each foveolar-gland unit. One unit may have glands devoid of parietal cells, while the next unit has a mixture of parietal and mucous cells; this is an epithelium that has both cardiac and oxyntocardiac mucosa (Fig. 5.23). One area may have goblet cells and the next may be composed of mucous cells without goblet cells and parietal cells; this is an epithelium that is classified as intestinal plus cardiac (Fig. 5.24).

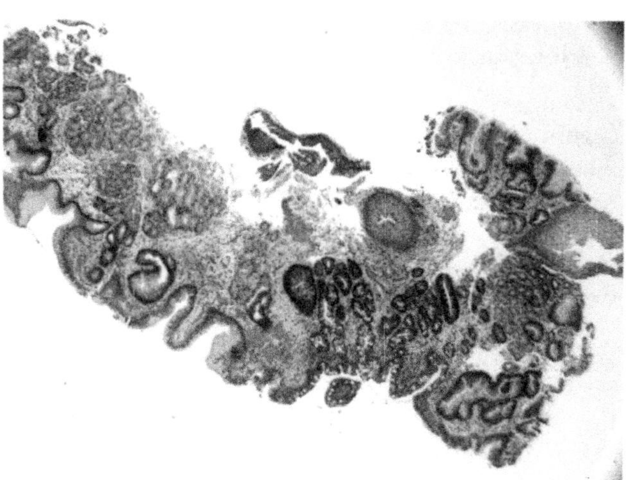

FIGURE 5.22 Single biopsy containing cardiac mucosa, intestinal metaplasia with goblet cells, and oxyntocardiac mucosa. This biopsy also shows a squamous-lined duct, proving an esophageal location. (Cross reference: Color Plate 3.4)

In patient care related biopsies, we usually report the most significant epithelial types present at any level for purposes of brevity. Intestinal (Barrett) metaplasia and cardiac mucosa are always reported when present, and when intestinal metaplasia is absent, a statement to that effect is always specifically included in the microscopic description. Oxyntocardiac mucosa is only reported when intestinal (Barrett) and cardiac mucosae are absent. Gastric oxyntic mucosa is reported when it is the only type of columnar epithelium present. For research, we note the presence of all epithelial types. This is probably more accurate although it is somewhat tedious and probably unnecessary in pathology reports for clinical usage.

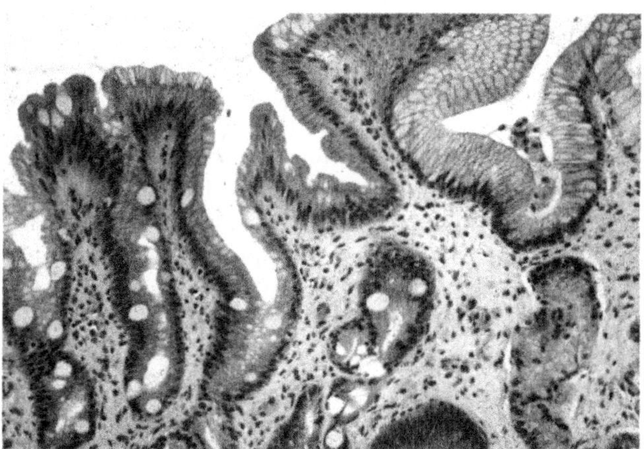

FIGURE 5.24 Biopsy showing intestinal metaplasia on the left and cardiac mucosa without goblet cells on the right.

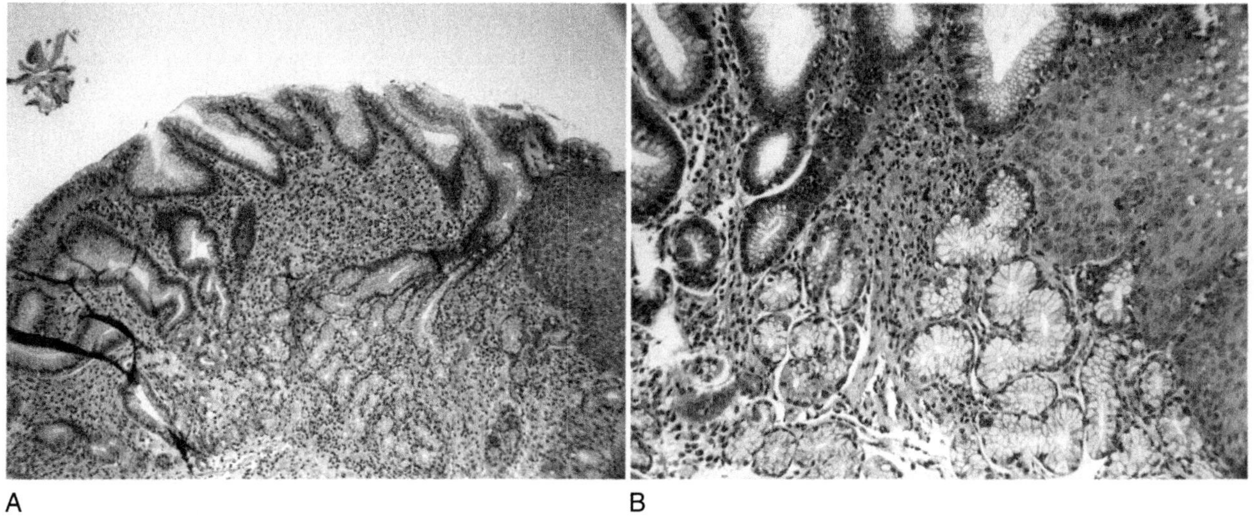

FIGURE 5.23 Cardiac mucosa + oxyntocardiac mucosa. Cardiac mucosa is represented by a single foveolar-gland unit between squamous epithelium and oxyntocardiac mucosa in (a) low magnification and (b) higher magnification. (Cross reference: Color Plate 3.8)

Overinterpreting Inadequate Biopsy Specimens

The definition of cardiac mucosa can only be applied accurately when the entire thickness of the epithelium is present for evaluation because it depends on the *absence* of both parietal cells and goblet cells. Superficial biopsies that contain only the surface and foveolar region will contain only mucous cells in cardiac, oxyntocardiac, and oxyntic mucosae. The definitions of oxyntocardiac and oxyntic mucosae depend on the composition of the cells in the glands under the foveolar region and therefore glands must be present to distinguish these from each other and from cardiac mucosa. Superficial specimens are adequate for definition if goblet cells are present because this defines intestinal metaplasia irrespective of any other criterion.

There is a danger of erroneously calling an inadequate superficial mucosal biopsy "cardiac" because only mucous cells are seen (Fig. 5.25); this is not an epithelium composed only of mucous cells, it is an epithelium where the deeper glandular tissue is not represented in the biopsy. Theoretically, such specimens should be reported as "superficial epithelial elements only, inadequate for classification of epithelial type." This is our practice for research diagnosis. In clinical practice, because biopsy is a precious specimen obtained from a significant procedure, we try to avoid using the term "inadequate" unless absolutely necessary. When a superficial biopsy is composed of mucous cells only, and when there is significant chronic inflammation and foveolar hyperplasia, we frequently report this as "cardiac mucosa" on the basis that this is likely rather than with certainty. The clinical impact of an error that results in the overdiagnosis of cardiac mucosa is zero.

Care must also be taken with tangentially sectioned biopsies (Fig. 5.26). These may consist of a horizontal section through the superficial foveolar region and appear to be composed of mucous cells only. Again, there is the danger of an erroneous diagnosis of the epithelium as "cardiac." In these cases, deeper serial sections are necessary to see the composition of the glands and ensure that parietal cells are absent before classifying the epithelium as cardiac.

Failing to Strictly Adhere to the Definitions

See the discussion of cardiac mucosa in Chapter 6.

Failing to Differentiate Intestinal Metaplasia of the Esophagus and Stomach

This is not a problem when the intestinal metaplasia occurs in an abnormal columnar-lined esophagus when it defines Barrett esophagus and when it occurs in a location clearly (>3 cm) distal to the endoscopic gastroesophageal junction and involves a large area of the stomach when it defines gastric intestinal metaplasia.

The problem arises when there is no endoscopic abnormality and a biopsy taken from a point within 3 cm distal to the endoscopic gastroesophageal junction shows intestinal metaplasia in cardiac mucosa. When the remainder of the stomach is normal, including the immediately adjacent gastric oxyntic mucosa, one is faced with the dilemma of a very small area of localized intestinal metaplasia in what has been defined as the gastric cardia by endoscopy. This problem will be addressed later in the discussion of the differential diagnosis of esophageal and gastric intestinal metaplasia.

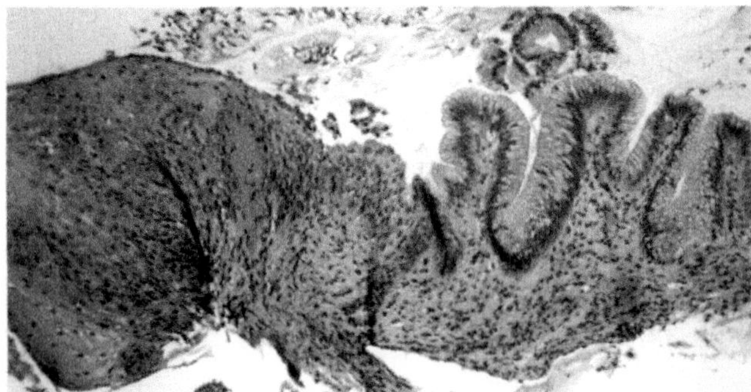

FIGURE 5.25 Superficial biopsy mistaken for cardiac mucosa. In deeper sections, glands containing parietal cells were present, making this oxyntocardiac mucosa.

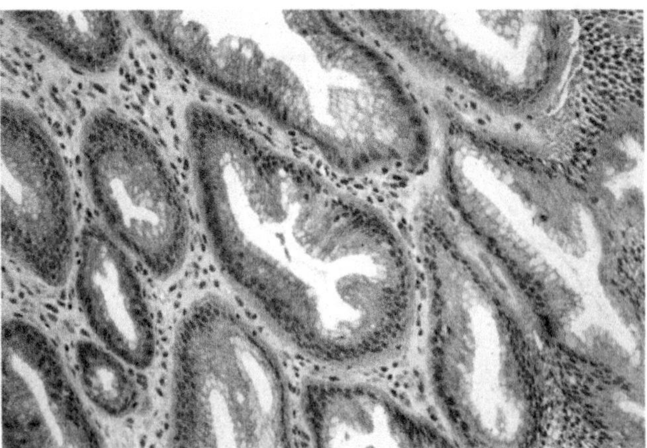

FIGURE 5.26 Tangential biopsy showing a cross section of the hyperplastic foveolar region only. Theoretically, this cannot be diagnosed as cardiac mucosa because the glands below the foveolar region are not seen. Deeper sections are needed.

DIAGNOSIS OF DIFFERENT EPITHELIAL TYPES

Apart from the technical problems relating to inadequate specimens and tangentially sectioned biopsies that must be recognized, the diagnosis of all epithelial types is based on definitions that are easily assessed in routine hematoxylin and eosin-stained sections. Mucous cells, parietal (oxyntic) cells, chief cells, serous (pancreatic) cells, Paneth cells, goblet cells, and squamous cells have well-defined and firmly established histologic features that are easily recognized by pathologists in routinely stained sections.

Use Only Hematoxylin and Eosin Stain for Diagnosis; Be Careful about Interpreting Mucin Stains

These epithelial cell types are *not* defined by their mucin content.

The use of mucin stains only complicates these definitions, and they should not be used in routine practice because they are a source of potential error. Mucin stains are useful only in the academic study of these epithelia. Several mucin types are present in fetal and adult columnar epithelia in the esophagus and stomach (Table 5.2). These include neutral mucin (which stains a bright magenta color with periodic acid-Schiff (PAS) stain) and acid mucins (which stain blue with Alcian blue stain at pH 2.5). PAS stain is a relatively nonspecific stain that stains many cellular components other than mucin; these include glycogen and mucopolysaccharide organelle components. It not infrequently stains mitochondrion-rich parietal cells positively. In Alcian blue stain at pH 2.5, cells will show varying shades of blue depending on the concentration of acid mucin in the cytoplasm. A combined Alcian blue–PAS stain (AB-PAS) is frequently used in practice; this shows goblet cells with acid mucin staining deep blue and mucous cells with neutral mucin staining magenta. Cells that contain an admixture of mucin types stain an intermediate color between magenta and blue.

There are various classifications of intestinal metaplasia. The first is into complete and incomplete based on whether or not there are absorptive cells (evidenced by the presence of a brush border) associated with the goblet cells. The second is into categories based on the type of acid mucin present, divided into intestinal metaplasia with sulfated mucin (staining positively with high iron-diamine stain) and intestinal metaplasia without sulfated mucin but containing only sialomucin (negative with high iron-diamine and positive with Alcian blue pH 2.5).

Among the different types of intestinal metaplasia in the esophagus, the type that is considered to be of greatest concern was the one where there was a predominance of sulfated mucin in the columnar cells (but not the goblet cells) of Barrett's specialized epithelium. The significance of this specific type was carefully evaluated by Haggitt et al. (26).

Literature Review

Haggitt RC, Reid BJ, Rabinovitch PS, Rubin CE. Barrett's esophagus: Correlation between mucin histochemistry, flow cytometry, and histologic diagnosis for predicting increased cancer risk. Am J Pathol 1988;131:53–61.

A total of 271 biopsies from 42 patients with Barrett esophagus, defined by the presence of specialized metaplastic epithelium in the tubular esophagus, were studied. These biopsies were classified into (a) the type of mucosa present (fundal gland, cardiac gland, cardiofundal, or specialized metaplastic epithelium; in our terminology this would be equivalent to oxyntic, cardiac, oxyntocardiac, and intestinal); (b) negative, indefinite, and positive for low- or high-grade dysplasia and carcinoma; and (c) special mucin histochemistry using two stain combinations. The first was the Alcian blue pH 2.5-periodic acid Schiff method (AB-PAS). This stains neutral mucins magenta and acidic mucins blue; if both neutral and acidic mucins are present in the same cell, the resultant staining varies between magenta and blue depending on the relative proportion of each mucin. The second stain was the high iron-diamine-Alcian blue pH 2.5 stain (HID-AB). This stains sulfated acid mucins dark brown to black and nonsulfated acid mucins (sialomucins) blue.

TABLE 5.2 Mucin Types and Staining Characteristics

Mucin type	Found in	PAS	Alcian blue pH 2.5	High iron diamine
Neutral mucin	All columnar epithelial cells, esophageal glands	+	−	−
Acid mucin	Goblet cells, nongoblet mucous cells, gland ducts	+/−	+	+/−
Sulfated acid mucin	Intestinal metaplasia; rarely in cardiac mucosa	+/−	+	+
Sialomucin (acid mucin)	Intestinal metaplasia; rarely cardiac mucosa	+/−	+	−

The staining in each component part of the specialized metaplastic epithelium was evaluated separately. The component parts included the goblet cells and the intervening columnar cells in the surface, foveolar pits, and glands. The HID-AB–stained biopsies were also scored as containing a predominance of sulfated mucin when more than half the nongoblet columnar cells were positive, as being positive for sulfated mucin when fewer than half the cells stained, and negative (containing only sialomucins) when the stain was negative for sulfated mucin.

Of the 271 biopsies, 66 contained only fundal (oxyntic) mucosa, 8 only cardiofundal (oxyntocardiac) mucosa, and 45 only cardiac mucosa. Of the 119 biopsies without intestinal metaplasia, small amounts of sulfated mucin were occasionally seen in cardiac mucosa but not in fundal (oxyntic) or cardiofundal (oxyntocardiac) mucosa. The 152 biopsies with intestinal metaplasia included 22 patients (58 biopsies) with no dysplasia, 9 patients (39 biopsies) who were indefinite for dysplasia, and 11 patients (55 biopsies) who showed dysplasia or carcinoma. Sulfated mucin was positive in 8 of 11 patients (73%) with histologic dysplasia or carcinoma, 7 of 9 patients (78%) who were indefinite for dysplasia, and 12 of 22 patients (53%) whose biopsies were negative for dysplasia ($p = 0.37$). Sulfated mucins predominated in 9%, 22%, and 9% of these patient groups, respectively ($p = 0.56$).

The authors concluded that neither the presence nor the predominance of sulfated mucins in the specialized metaplastic epithelium of Barrett esophagus has sufficiently high sensitivity or specificity for dysplasia or carcinoma to be of value in managing patients.

This study led to the virtual abandonment of the use of high iron-diamine–Alcian blue pH 2.5 (HID-AB) in routine evaluation of esophageal biopsies. This stain still appears in learned articles, but there has been no study that clearly shows it has any practical value. As such its use is not recommended. We do not use it either in research or in practice.

Diagnosis of Intestinal Metaplasia

In most cases of intestinal metaplasia, numerous goblet cells are present, often with blue mucin in the vacuoles, and diagnosis is easy, accurate, and highly reproducible (Figs. 5.19 and 5.20). On infrequent occasion, the diagnosis of intestinal metaplasia is not easy, and there is significant interobserver error in a few borderline cases. The main reason for error is that the mucous cells in cardiac mucosa can become distended with vacuoles of mucin, which can resemble goblet cells. These are sometimes called "pseuodgoblet cells" and are not easy to differentiate from true goblet cells (Figs. 5.27 and 5.28). Pseudogloblet cells are extremely common in cardiac mucosa and most likely represent a reactive phenomenon that is completely unrelated to intestinal metaplasia. Failure to distinguish pseudogoblet cells from true goblet cells results in a high frequency of false positive diagnoses of intestinal metaplasia. Pseudogoblet cells are most prevalent in reactive epithelia such as that of cardiac mucosa and reactive gastropathy; they are less prevalent in atrophic gastritis.

For most pathologists, the recognition of goblet cells in biopsies falls into three categories: (a) definite and unquestionable goblet cells are present, either alone or admixed with pseudogoblet cells (Figs. 5.19 and 5.20); (b) there is doubt as to whether or not goblet cells are present because of the difficulty in differentiating them from pseudogoblet cells (Fig. 5.28); and (c) there are definitely no goblet cells (Fig. 5.10). We make the diagnosis of intestinal metaplasia only in the first category. If there is any doubt that we are looking at

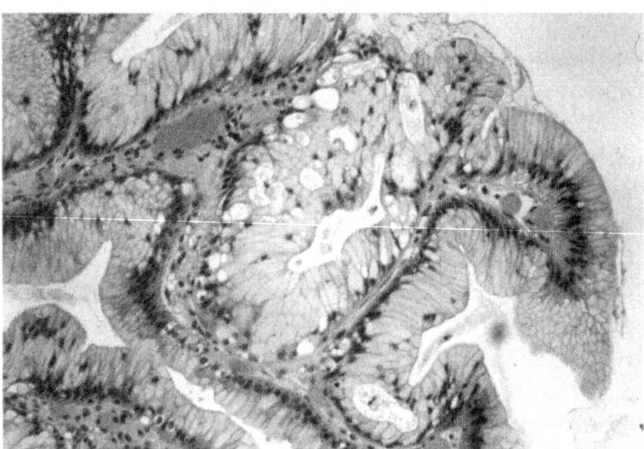

FIGURE 5.27 Pseudogoblet cells in cardiac mucosa showing marked vacuolation of cells due to mucin distension. The vacuoles are empty and of irregular shapes and sizes and are present in numerous adjacent cells.

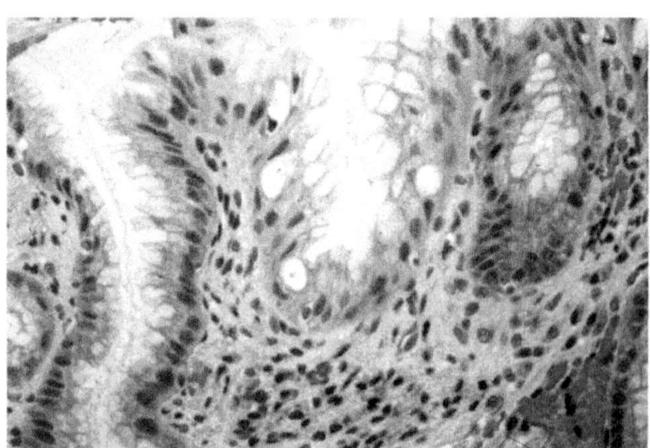

FIGURE 5.28 Pseudogoblet cells in cardiac mucosa showing scattered vacuolated cells that resemble goblet cells but lack the definition of true goblet cells.

goblet cells, we do not make the diagnosis of intestinal metaplasia. From a practical standpoint, this high degree of specificity for the diagnosis of intestinal metaplasia is appropriate. Specificity of diagnosis is more important than sensitivity because the presence of intestinal metaplasia is, at least in the esophagus, an indication for lifelong surveillance, which is a significant expense and discomfort to the patient and should be restricted to those patients where the diagnosis is certain.

However, though the diagnosis of intestinal metaplasia is usually made at a gut level, it is necessary to attempt to define specific features that can be used to differentiate pseudogoblet from true goblet cells (27): (a) True goblet cells have a single large round vacuole between the basal nucleus and surface membrane that distends the lateral borders of the cell uniformly (Figs. 5.19 and 5.20); pseudogoblet cells may have multiple vacuoles or less defined and imperfectly circular single vacuoles. (b) The vacuoles of true goblet cells frequently show basophilic mucinous material (Fig. 5.20), whereas pseudogoblet cells are empty and clear (Fig. 5.28). This is a reliable criterion only when positive for basophilia; true goblet cells sometimes have clear vacuoles (Fig. 5.19). Basophilia varies with different hematoxylin types used in routine staining. (c) True goblet cells occur sporadically and are separated by nonvacuolated cells, giving the appearance of a deeper staining epithelium punctuated by vacuolated cells (Fig. 20). Pseudogoblet cells tend to involve many adjacent cells and have a lighter staining quality (Fig. 5.19).

It is important to recognize that pseudogoblet cells stain with varying degrees of positivity for acid mucin with Alcian blue stain at pH 2.5 (Fig. 5.29, "columnar blue cells"). The use of Alcian blue stain therefore increases rather than decreases the likelihood of interpreting a pseudogoblet cell as a true goblet cell, increasing the rate of false positive diagnosis of intestinal metaplasia. If Alcian blue stain is used, it should be considered positive only when there is deep blue staining of histologically obvious goblet cells. Because there is a danger in depending on Alcian blue positivity in uncertain cases, its use will usually result in an increased false positive diagnosis of Barrett esophagus. For this reason, routine Alcian blue stains are not recommended.

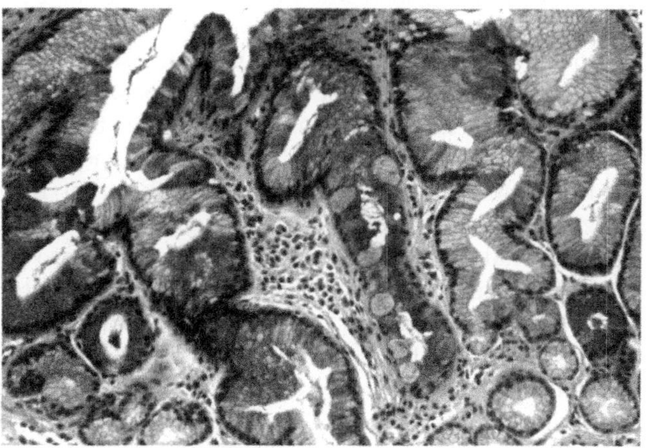

FIGURE 5.29 Alcian blue showing goblet cells (perfectly round vacuoles staining positively with Alcian blue) and pseudogoblet cells (columnar cells staining deeply with Alcian blue). (Cross reference: Color Plate 4.4)

Some authorities suggest that failure to use Alcian blue results in an overdiagnosis of intestinal metaplasia; this is the reverse of what was stated earlier. This may be a true statement if one is dealing with an outdated pathologist to whom any columnar-lined esophagus represents Barrett esophagus. For these pathologists, the requirement of a positive Alcian blue stain will decrease their extreme overdiagnosis of Barrett esophagus. However, they will still continue to have a high rate of overdiagnosis because many Alcian blue positive cases will not have goblet cells. The answer is to educate misinformed pathologists that Barrett esophagus requires the presence of definite goblet cells, not to encourage them to use a less accurate Alcian blue stain.

Despite these difficulties, if the pathologist accepts the principle that intestinal metaplasia should be diagnosed only when definite goblet cells are present, the accuracy and reproducibility of the diagnosis is high. Overdiagnosis of intestinal metaplasia results from calling all vacuolated cells true goblet cells and the error of defining any cell that is positive with the Alcian blue stain a true goblet cell.

Differentiation of Esophageal (Barrett-Type) Intestinal Metaplasia from Gastric Intestinal Metaplasia

This difficulty arises when endoscopy is normal and a biopsy immediately distal to the squamocolumnar junction reveals intestinal metaplasia. Intestinal metaplasia at the junction will be considered pathologic by all authorities because intestinal metaplasia is not normal in either the esophagus or the stomach. However, it is crucial to distinguish those cases of esophageal intestinal metaplasia from cases of gastric intestinal metaplasia because the former is more worrisome in terms of risk of future malignant transformation. Although the premalignant risk of small amounts of esophageal (Barrett-type) intestinal metaplasia is not proven, most gastroenterologists worry sufficiently about this possibility as to place the patient on surveillance. Reasons for this concern are (a) that data show that short-segment Barrett esophagus has a risk of malignancy that is similar to that for long-segment Barrett esophagus, suggesting that cancer risk is not related to the amount of intestinal metaplasia (28), and (b) the incidence of adenocarcinoma of the "cardia" or "proximal stomach" has increased at a similar alarming rate as adenocarcinoma of the esophagus (29).

The finding of intestinal metaplasia in a biopsy from a patient with normal endoscopy should result in a diagnosis of esophageal (Barrett-type) intestinal metaplasia when the following features are present: (a) when there is clinical evidence of reflux or evidence of reflux damage to the squamous epithelium, (b) when the intestinal metaplasia is restricted to the region within 3 cm distal to the squamocolumnar junction and there is an absence of any pathology in the gastric body and antrum (it is very useful to take biopsies of the distal stomach whenever the gastroesophageal junction is biopsied), and (c) when the intestinal metaplasia is typical of that seen in the esophagus (i.e., it has the features of what Paull et al. described as specialized epithelium, which is simply the occurrence of goblet cells within a reactive cardiac mucosa).

The only reason not to go along with this reasoning is to have such absolute and illogical faith in the accuracy of the endoscopic definition of the gastroesophageal junction that there is a complete unwillingness to accept that the biopsy may be from the esophagus and not the stomach. This is a common occurrence, and when faced with the situation, some of the best gastroenterologists we know will vehemently state: "I am certain of the location of the gastroesophageal junction." This is a true statement, but their certainty is based on a criterion (the proximal limit of the rugal folds) that has never been established as the true gastroesophageal junction by any study. The fact that the criterion is accepted by everyone does not mean that is a proven criterion; it is yet another of those things that we accept that will ultimately be proven wrong.

This dogmatic belief in the accuracy of the endoscopic gastroesophageal junction results in a series of false conclusions. Because the intestinal metaplasia is perceived as being distal to the gastroesophageal junction, a diagnosis of intestinal metaplasia of the cardia (i.e., the proximal stomach) is made. If the patient has symptoms of gastroesophageal reflux, this is ascribed as a cause of this entity. The obvious is ignored; gastroesophageal reflux causes esophageal disease only, it cannot cause gastric disease.

Intestinal metaplasia may occur in the proximal stomach in patients with *H. pylori* gastritis and autoimmune gastritis. In these patients, there are usually no symptoms of gastroesophageal reflux, there are no reflux changes in the squamous epithelium, and the intestinal metaplasia occurs in a chronic gastritis that involves the entire gastric body. This is simply chronic atrophic gastritis involving the most proximal part of the stomach.

In a few patients in whom reflux changes coexist with chronic atrophic gastritis, the interpretation of intestinal metaplasia at the squamocolumnar junction in a patient with an endoscopically normal gastroesophageal junction can be difficult. The two possibilities are intestinal metaplasia occurring in atrophic gastritis in the most proximal part of the stomach and esophageal (Barrett) intestinal metaplasia. The following methods are available for this differentiation:

1. Careful morphologic examination usually permits the distinction. Chronic atrophic gastritis occurs in oxyntic mucosa, and the intestinal metaplasia may occur in a mucosa that contains scattered parietal cells. The inflammation in atrophic gastritis is usually a diffuse lymphoplasmacytic infiltrate, and the mucosal surface is flat with atrophic features. In contrast, esophageal (Barrett) intestinal metaplasia occurs in cardiac mucosa devoid of parietal cells and showing reactive features including a villiform surface, foveolar hyperplasia, inflammation with numerous eosinophils, and frequently fibrosis and smooth muscle proliferation. In these cases, when cardiac and gastric oxyntic mucosae coexist, the histologic difference in appearance between the two is often striking.
2. Special immunohistochemical studies are sometimes helpful in this situation. We use these rarely because in virtually all cases, the differential diagnosis is easily made purely by morphologic examination. The monoclonal antibody $7E_{12}H_{12}$ (IgM isotype), also known as m-DAS-1, has been shown to stain esophageal (Barrett) but not gastric intestinal metaplasia (30). This is an antibody that was developed against colonic epitopes, and positivity in Barrett esophagus has been taken to suggest that esophageal (Barrett) intestinal metaplasia resembled colonic mucosa and is different from gastric intestinal metaplasia, which has more small intestinal characteristics. The staining patterns with cytokeratins 7 and 20 of esophageal (Barrett) intestinal metaplasia and gastric intestinal metaplasia are reported to be different (31). Esophageal (Barrett) intestinal metaplasia showed superficial CK20 staining and strong CK7 staining of both superficial and deep glands (Figs. 5.30). This pattern was not seen in gastric intestinal metaplasia. In Dr. Chandrasoma's experience, CK7/CK20 patterns are unreliable in differentiating between the two types of

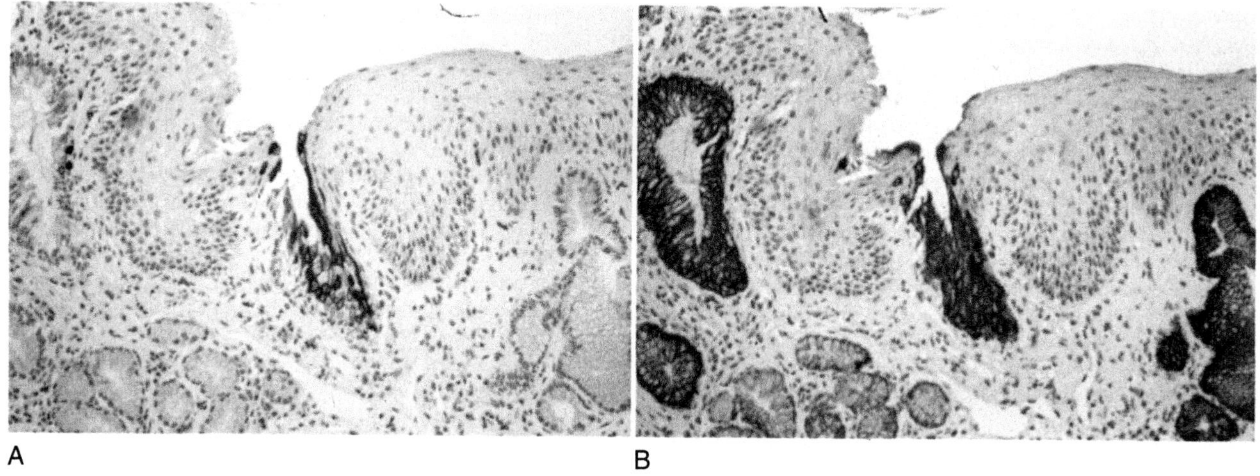

FIGURE 5.30 Cytokeratin staining pattern in Barrett esophagus. (a) Cytokeratin 20 showing superficial staining only. (b) Cytokeratin 7 showing superficial and deep staining.

intestinal metaplasia; He has long since given up using them routinely in his practice.

In summary, the vast majority of cases where intestinal metaplasia is found in a biopsy taken at the endoscopically normal squamocolumnar junction can be classified as esophageal (Barrett) or gastric by purely morphologic criteria. A small number of cases remain where this is not possible. In these cases of doubt, it is our practice to not make a diagnosis of Barrett esophagus and subject the patient to lifelong surveillance. The risk of cancer in small foci of Barrett-type intestinal metaplasia and the benefit of long-term surveillance in these patients are unproven, and such a course is clearly justified by present evidence in the few cases where the diagnosis is in doubt.

References

1. Tobey NA, Hosseini SS, Argore CM, Dobrucali AM, Awayda MS, Orlando RC. Dilated intercellular spaces and shunt permeability in non-erosive acid-damaged esophageal epithelium. Am J Gastroenterol 2004;99:13–22.
2. Barrett NR. The lower esophagus lined by columnar epithelium. Surgery 1957;41:881–894.
3. Johns BAE. Developmental changes in the oesophageal epithelium in man. J Anat 1952;86:431–442.
4. Allison PR. Peptic ulcer of the oesophagus. Thorax 1948; 3:20–42.
5. Hayward J. The lower end of the oesophagus. Thorax 1961; 16:36–41.
6. Sampliner RE. Practice guidelines on the diagnosis, surveillance, and therapy of Barrett's esophagus. Am J Gastroenterol 1998;93:1028–1031.
7. Spechler SJ. Barrett's esophagus. N Engl J Med 2002;346:836–842.
8. Fass R, Sampliner RE. Extension of squamous epithelium into the proximal stomach: A newly recognized mucosal abnormality. Endoscopy 2000;32:27–32.
9. Karam SM. Lineage commitment and maturation of epithelial cells in the gut: Frontiers in bioscience 1999;4:286–298.
10. Menard D, Arsenault P. Cell proliferation in developing human stomach. Anat Embryol 1990;182:509–516.
11. Olvera M, Wickramasinghe K, Brynes R, Bu X, Ma Y, Chandrasoma P. Ki67 expression in different epithelial types in columnar lined oesophagus indicates varying levels of expanded and aberrant proliferative patterns. Histopathology 2005;47: 132–140.
12. Polkowski W, van Lanschot JJB, ten Kate FJW, Rolf TM, Polak M, Tytgat GNJ, Obertop H, Offerhaus GJA. Intestinal and pancreatic metaplasia at the esophagogastric junction in patients without Barrett's esophagus. Am J Gastroenterol 2000;95: 617–625.
13. Paull A, Trier JS, Dalton MD, Camp RC, Loeb P, Goyal RK. The histologic spectrum of Barrett's esophagus. N Engl J Med 1976;295:476–480.
14. Glickman JN, Fox V, Antonioli DA, Wang HH, Odze RD. Morphology of the cardia and significance of carditis in pediatric patients. Am J Surg Pathol 2002;26:1032–1039.
15. Shields HM, Rosenberg SJ, Zwas FR, et al. Prospective evaluation of multilayered epithelium in Barrett's esophagus. Am J Gastroenterol 2001;96:3268–3273.
16. Glickman JN, Chen Y-Y, Wang HH, Antonioli DA, Odze RD. Phenotypic characteristics of a distinctive multilayered epithelium suggests that it is a precursor in the development of Barrett's esophagus. Am J Surg Pathol 2001;25:569–578.
17. De Hertogh G, Van Eyken P, Ectors N, Tack J, Geboes K. On the existence and location of cardiac mucosa: An autopsy study in embryos, fetuses, and infants. Gut 2003;52:791–796.
18. Derdoy JJ, Bergwerk A, Cohen H, Kline M, Monforte HL, Thomas DW. The gastric cardia: To be or not to be? Am J Surg Pathol 2003;27:499–504.
19. Park YS, Park HJ, Kang GH, Kim CJ, Chi JG. Histology of gastroesophageal junction in fetal and pediatric autopsy. Arch Pathol Lab Med 2003;127:451–455.

20. Chandrasoma PT. Fetal "cardiac mucosa" is not adult cardiac mucosa. Gut 2003;52:1798.
21. Zhou H, Greco MA, Daum F, et al. Origin of cardiac mucosa: Ontogenic considerations. Pediatr Dev Pathol 2001;4:358–363.
22. Hirota WK, Loughney TM, Lazas DJ, Maydonovitch CL, Rholl V, Wong RKH. Specialized intestinal metaplasia, dysplasia, and cancer of the esophagus and esophagogastric junction: Prevalence and clinical data. Gastroenterology 1999;116:277–285.
23. Rex DK, Cummings OW, Shaw M, Cumings MD, Wong RKH, Vasudeva RS, Dunne D, Rahmani EY, Helper DJ. Screening for Barrett's esophagus in colonoscopy patients with and without heartburn. Gastroenterology 2003;125:1670–1677.
24. Goldblum JR, Vicari JJ, Falk GW, et al. Inflammation and intestinal metaplasia of the gastric cardia: The role of gastroesophageal reflux and *H. pylori* infection. Gasteroenterology 1998;114: 633–639.
25. Goldblum JR. Inflammation and intestinal metaplasia of the gastric cardia: *Helicobacter pylori*, gastroesophageal reflux disease, or both. Dig Dis 2000;18:14–19.
26. Haggitt RC, Reid BJ, Rabinovitch PS, Rubin CE. Barrett's esophagus: Correlation between mucin histochemistry, flow cytometry, and histologic diagnosis for predicting increased cancer risk. Am J Pathol 1988;131:53–61.
27. Chandrasoma PT, Der R, Dalton P, Kobayashi G, Ma Y, Peters J, DeMeester T. Distribution and significance of epithelial types in columnar lined esophagus. Am J Surg Pathol 2001;25: 1188–1193.
28. Rudolph RE, Vaughan TL, Storer BE, Haggitt RC, Rabinovitch PS, Levine DS, Reid BJ. Effect of segment length on risk for neoplastic progression in patients with Barrett esophagus. Ann Intern Med 2000;132:612–620.
29. Devesa SS, Blot WJ, Fraumeni JF, et al. Changing patterns in the incidence of esophageal and gastric carcinoma in the United States. Cancer 1998;83:2049–2053.
30. Das KM, Prasad I, Garla S, Amenta PS. Detection of a shared colon epithelial epitope on Barrett epithelium by a novel monoclonal antibody. Ann Intern Med 1994;120:753–756.
31. Glickman JN, Wang HH, Das KM, et al. Phenotype of Barrett's esophagus and intestinal metaplasia of the distal esophagus and gastroesophageal junction: An immunohistochemical study of cytokeratins 7 and 20, Das-1 and 45MI. Am J Surg Pathol 2001;25:87–94.

CHAPTER

6

Cardiac Mucosa

WHAT IS CARDIAC MUCOSA?

Two ways of defining cardiac mucosa have always existed in the literature; this ambiguity in definition results in massive confusion and the inability to compare different articles. The first and correct definition of cardiac mucosa is by its histologic characteristics irrespective of any perceived anatomic or endoscopic location. This definition will be described in this chapter. All reports that use the strict histologic definition produce data that are intelligible and worthy of study. The reason why the histologic definition is correct is because it does not depend on any unproved or subjective information; it is on a microscopic slide and therefore is verifiable and reproducible. Histologically defined cardiac mucosa is not present in everyone (1, 2, 4). When found, it is located in either the columnar-lined tubular esophagus or less than 3 cm distal to the endoscopically defined gastroesophageal junction (proximal limit of the rugal folds) (2, 3) (Fig. 6.1).

The second definition of cardiac mucosa is that it is the epithelium that lines the anatomic or endoscopic cardia (or gastric cardia). In this method of definition, the "cardia" or "gastric cardia" is defined endoscopically (see Chapter 4), a biopsy is taken, and any mucosal type found therein is called "cardiac mucosa." The clearest recent example of this is in a study by Gulmann et al. (5) that aims to "investigate the relationship between the type of cardiac mucosa and carditis with various histological and clinical parameters" (p. 69). To these authors, "cardiac mucosa" is any epithelium lining the "cardia" (defined endoscopically "as 0.5 cm distal to the squamocolumnar junction"). In classifying cardiac mucosal types, they state: "The type of mucosa from the cardia . . . was defined as mucinous . . . or oxyntic." The problem with this definition is that the definition of the "cardia" is not reproducible (see Chapter 4). The proximal limit of the cardia is the endoscopic gastroesophageal junction, which at present is the proximal limit of the rugal folds; this has never been proven to be the true gastroesophageal junction, and we will suggest that it is not the true gastroesophageal junction. The distal limit of the cardia has never been defined; the authors use 0.5 cm distal to the proximal limit of the rugal folds, but this is a figure taken from one reference, not a uniformly accepted definition. The interobserver variation between different endoscopists in defining a 0.5 segment of this area and the problems in accurately taking a biopsy from this area make such definitions extremely suspect. Lagergren et al. (6) accepted as "gastric cardia" any point up to 3 cm distal to the gastroesophageal junction.

Articles that use this anatomic/endoscopic definition of cardiac mucosa produce data that are confusing. They add the clutter to the literature that makes study very difficult. They also permit those reading these articles superficially at an abstract level to extract the erroneous conclusions in them and quote these as fact. This is particularly true when the literature is from highly reputable authorities who are not immune to this error.

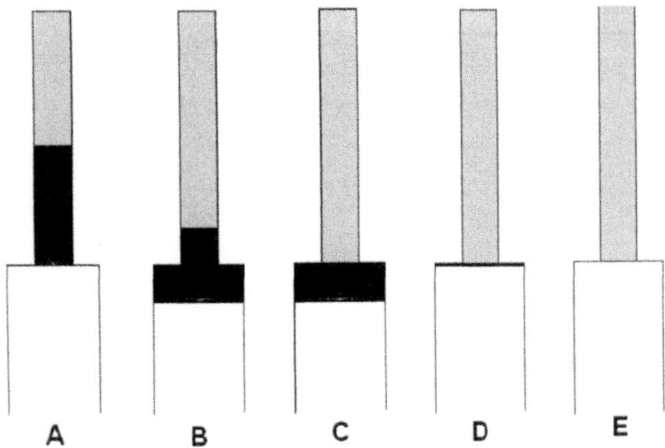

FIGURE 6.1 Location of "normal cardiac mucosa." In the 1950s, Allison and Barrett recognized that while the normal esophagus was entirely lined by squamous epithelium, some people had cardiac mucosa extending up to the arch of the aorta (a). They believed this was a congenital anomaly. In 1961, Hayward defined the extent of normal cardiac mucosa as "lining the distal 2cm of the tubular esophagus and a variable extent of the proximal stomach" (b). This view is still held by some (6). When the endoscopic definition of normalcy recognized that the entire tubular esophagus was lined by squamous epithelium, cardiac mucosa shifted to being a lining of the proximal stomach (c). Most textbooks state this view. When many autopsy studies showed that most people had less than 5mm of cardiac mucosa, the amount accepted as normal shrank; present belief is that 1 to 4mm of cardiac mucosa is normally present (d). We are inexorably but infuriatingly slowly moving toward the truth, which is the cardiac mucosa is normally absent (e).

The Definition of Cardiac Mucosa by Histologic Criteria

There has never been any ambiguity or question about the histologic definition of cardiac mucosa. Cardiac mucosa is defined as an epithelium in this region that is composed only of mucous cells. While this has been accepted from the inception of histologic study of this region, the definition was crystallized by Paull et al.'s (7) landmark study in 1976 of the histologic spectrum of columnar-lined esophagus (Fig. 6.2). It must be recognized at the outset that Paull et al. was defining esophageal epithelia, not gastric. Cardiac (= junctional) epithelium was one of "three types of columnar epithelium . . . found in esophageal biopsies taken from above the level identified by manometry as the lower–esophageal-sphincter zone" (p. 477).

Paull et al. assessed the following features in columnar epithelia for their classification: "surface architecture; surface cell types—whether consisting only of gastric surface cells or gastric surface cells combined with intestinal-type goblet cells; and composition of glandular layers—whether containing mucous cells only or also parietal and chief cells" (p. 477). Based on their findings, they classified columnar esophageal epithelia into three types (Fig. 6.2):

1. Gastric-fundic-type epithelium with chief and parietal cells in the glandular layer (Fig. 6.2a)
2. Junctional-type epithelium with cardiac type mucosal glands and no parietal cells (Fig. 6.2b)
3. Specialized columnar epithelium with a villiform surface, mucous glands, Alcian blue staining intestinal-type goblet cells, and no parietal and chief cells (Fig. 6.2c)

These definitions are precise and easy to follow from a histopathologist's standpoint. Paull et al.'s terminology is regrettable in that it perpetuates the misconception that columnar esophageal epithelia are in some way gastric. While proving their material was from the esophagus, they make Allison and Johnstone's (8) mistake of ascribing gastric qualities to them. The surface epithelial cells are called gastric surface cells; the epithelium with glands composed of mucous and parietal cells is called gastric-fundic type; the epithelium composed of only mucous cells is called junctional type even though it is from a biopsy proven to be proximal to the gastroesophageal junction. It would have been wonderful if they had followed Barrett and called these metaplastic columnar esophageal epithelia, types 1, 2, and 3, rather than reverting to Allison and Johnstone's original mistake of attributing a gastric nature to them. In Chapter 5, we have suggested that the best and most likely ultimately accepted terms for junctional (= cardiac), gastric-fundic-type (= oxyntocardiac) and specialized (= intestinal) *are metaplastic esophageal columnar epithelia, types 1, 2, and 3*. Much of today's confusion will disappear if we accept these terms universally.

Paull et al.'s terms are still used sporadically in the literature, and there is no problem with that as long as the definitions are adhered to. We use different terms (1). We use the term "cardiac mucosa" as synonymous with Paull et al.'s junctional mucosa; this has become the more common term in usage today. We use the term "oxyntocardiac mucosa" for Paull et al.'s gastric-fundic-type mucosa. "Oxyntocardiac" is purely descriptive without the false suggestion that the epithelium has something to do with the gastric fundus; the term "oxyntocardiac mucosa" is slowly gaining acceptance, particularly in research originating in Europe (4, 9). We use the term "intestinal (Barrett-type) metaplasia" as synonymous with specialized columnar epithelium.

It should be recognized that these three epithelial types that constitute columnar-lined esophagus can be regarded as being variants of cardiac mucosa. Cardiac mucosa represents the least differentiated epithelial type composed only of mucous cells without any

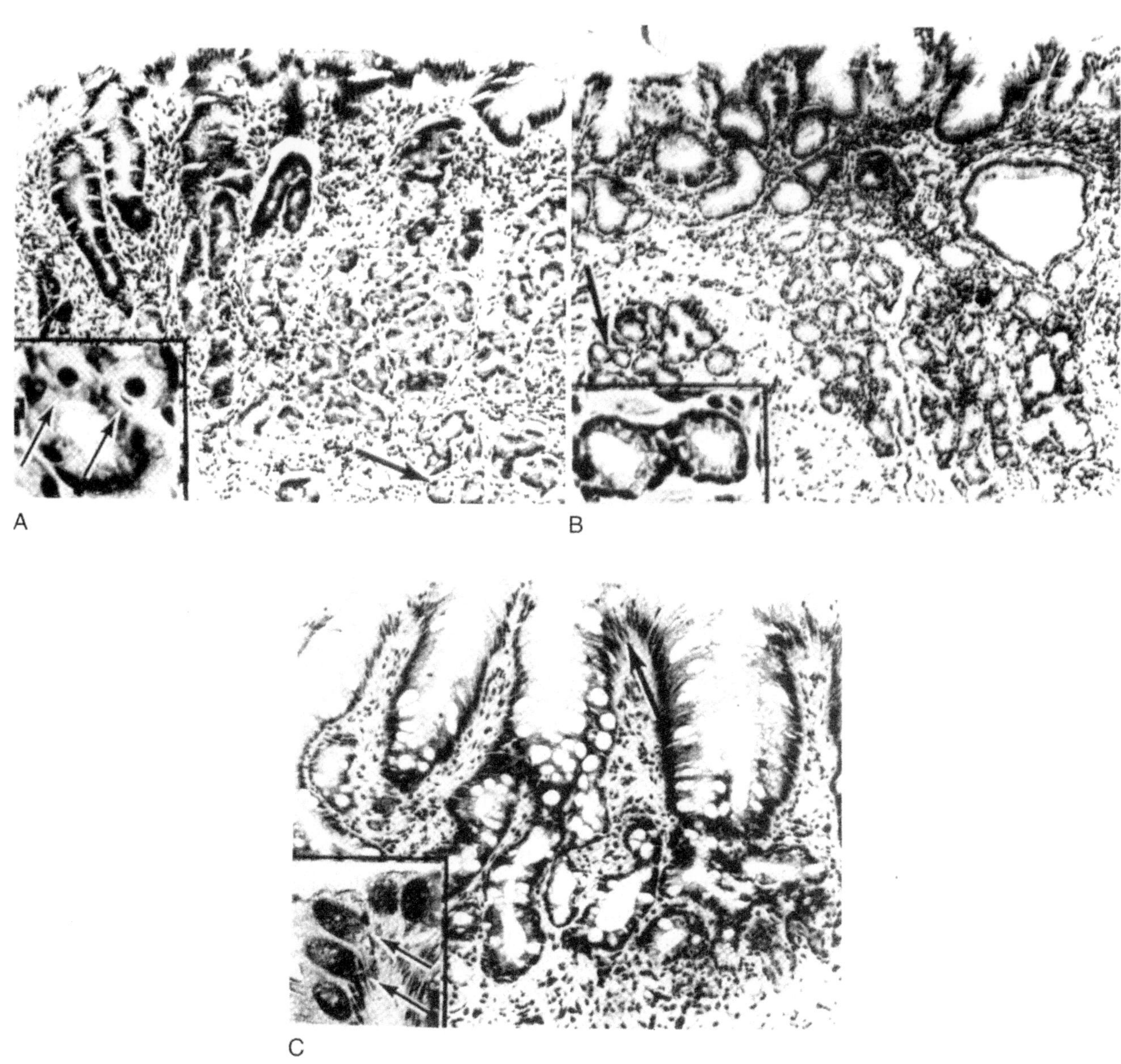

FIGURE 6.2 Paull et al.'s (7) histologic classification of epithelial types found in columnar-lined esophagus. (a) Gastric-fundic type (= oxyntocardiac) epithelium. The inset demonstrates parietal cells in glands. (b) Junctional type (= cardiac) epithelium. The inset shows mucous glands with no parietal cells. (c) Specialized columnar (intestinal, Barrett-type) epithelium with goblet cells (inset shows positive staining of goblet cells with Alcian blue stain).

specialized secretory cells. Oxyntocardiac mucosa is merely the presence of parietal cells in cardiac mucosa. Intestinal metaplasia is merely the presence of goblet cells in cardiac mucosa. We will show later that cardiac mucosa is the first epithelium that arises when squamous epithelium undergoes columnar metaplasia. The occurrence of specialized cells in cardiac mucosa, which lead to either oxyntocardiac or intestinal metaplasia, result from different damage environments associated with reflux (Fig. 6.3).

Failure to Adhere to the Histologic Definition of Cardiac Mucosa

"Allowing a Few Parietal Cells"

Despite the fact that definition of these three mucosal types is crystal clear, there is a tendency to confuse even this straightforward issue. The most disturbing thing that happens is that some pathologists allow a few parietal cells within their definition of cardiac mucosa.

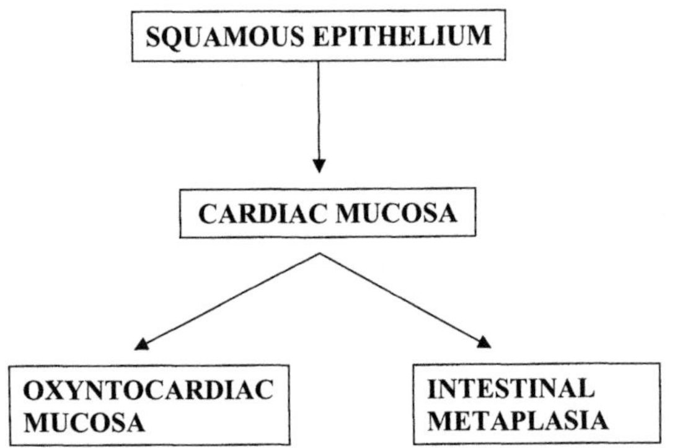

FIGURE 6.3 The interrelationship between different epithelial types in the esophagus. Columnar metaplasia of squamous epithelium caused by reflux results in a mucous cell–only epithelium that we call cardiac mucosa. Cardiac mucosa evolves into the other columnar epithelial types by the development of specialized cells. When parietal cells develop in cardiac mucosa, the mucosa is oxyntocardiac, and when goblet cells arise, it is intestinal (Barrett type).

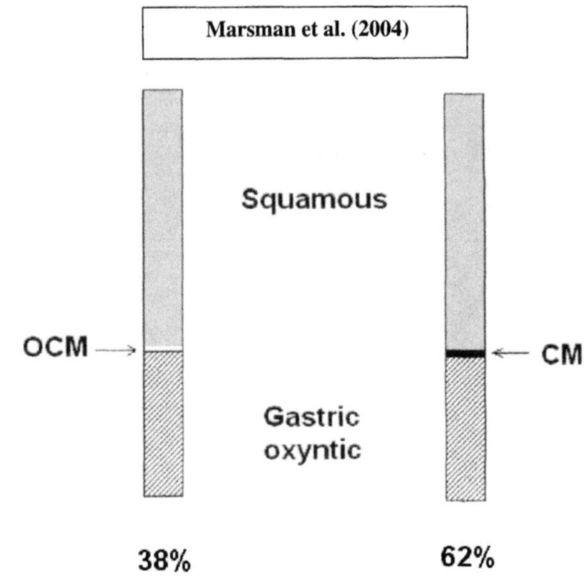

FIGURE 6.4 Findings at biopsy as reported by Marsman et al. (4). Biopsies straddling the squamocolumnar junction showed cardiac mucosa (solid black) adjacent to squamous epithelium in 62% of patients; the rest had oxyntocardiac mucosa (solid white) between squamous epithelium (solid gray) and gastric oxyntic mucosa (striped) without any cardiac mucosa being present.

In an article by Marsman et al. (4), the authors reported the biopsy findings at the gastroesophageal junction in 63 patients who had the histologic squamocolumnar junction in one biopsy sample. They accurately classified the glandular epithelium into three types: purely cardiac (composed of mucous cells only), oxyntocardiac (containing parietal cells admixed with the mucous cells), and oxyntic (containing parietal and chief cells). They reported that 39/63 (62%) of patients had purely cardiac mucosa, and the other 24 (38%) patients had oxyntocardiac mucosa immediately distal to the squamous epithelium (Fig. 6.4). After producing these wonderful data, the authors concluded that "cardiac mucosa was uniformly present adjacent to the squamous epithelium at the gastroesophageal junction" (p. 212). This incomprehensible conclusion is contrary to the data they reported. The obviously correct conclusion is that cardiac mucosa was found in 39/63 (62%) and absent in 24/63 (38%) of their patients. The only way their conclusion that cardiac mucosa is present in everyone can be reached is if they equate oxyntocardiac mucosa with cardiac mucosa, which is equivalent to "allowing a few parietal cells to be present in cardiac mucosa" (p. 212).

The most personally hurtful example of this is in the study by Gulmann et al. (5), quoted above. To these authors, "cardiac mucosa" is any epithelium lining the "cardia" (defined endoscopically "as 0.5 cm distal to the squamocolumnar junction"). In classifying cardiac mucosal types, they stated: "The type of mucosa from the cardia . . . was defined as mucinous if there was no or only very occasional oxyntic glands (as previously published by others (20)). If the mucosa consisted of packed oxyntic glands similar to those found in the corpus it was defined as oxyntic mucosa" (p. 69). The hurtful part of this is that the reference #20 they cite for this is Dr. Chandrasoma's. At no point, let alone in the cited article, has he ever used any definition that remotely resembles this one. In that paper, he wrote: "The epithelial types found were defined . . . into the following five types: . . . (3) Pure cardiac mucosa in which the glands were composed of mucous cells only, without parietal cells . . . (4) Oxyntocardiac mucosa in which the glands contained a mixture of mucous and parietal cells in varying numbers" (10, p. 345). The practice of misquoting papers is one that should be proscribed.

These histologic definitions are not new or devised by the authors. As clearly stated in all of our research, they are the classical definitions presented in 1976 in the landmark paper by Paull et al. from Dr. Goyal's laboratory at Harvard (7). In examining thousands of biopsies from this region, we have confirmed the accuracy and absolute reproducibility of Paull et al.'s definitions. Perfection has existed for 30 years; let's just stick with it. We have only suggested newer, less ambiguous terms, all of which have accepted usage in the present literature.

Many other authors similarly combine cardiac and oxyntocardiac mucosa under the definition of cardiac mucosa. At an international Barrett's esophagus conference in Hamburg, after Dr. Chandrasoma had presented his paper on the histology of epithelia in this region, Dr. Fred Weinstein, his esteemed gastroenterologist colleague from UCLA, proclaimed that he was not willing to be subject to the "histologic apartheid of USC" where not a single parietal cell is permitted within the definition of cardiac mucosa. According to him, parietal cells frequently extended into the pyloric region of the stomach and therefore they could extend into the cardiac region.

Why is this not true? It is not true for the following reasons: (a) There is no difficulty differentiating histologically between cardiac mucosa and oxyntocardiac mucosa; one has parietal cells admixed with mucous cells and the other does not. No author who has tried to follow these definitions has expressed any difficulty with it; the differentiation is highly reproducible as long as the first parietal cell causes the epithelium to be called oxyntocardiac rather than cardiac. (b) We will show that there is a difference in the clinical significance in the finding of cardiac and oxyntocardiac mucosa in the junction. To understand this difference and to use it in practice, one must adhere strictly to the definition. If there was no clinical significance to histologically differentiating between cardiac and oxyntocardiac mucosa, Dr. Weinstein would be correct. As far as is known, there is no clinical significance to the finding of parietal cells admixed with mucous cells in antral mucosa.

Use of Unique and Self-Devised Histologic Definitions

Rarely, articles use their own histologic criteria to define cardiac mucosa in preference to the well-established criterion that defines cardiac mucosa as an epithelium consisting of mucous cells only on a routine hematoxylin and eosin section.

The most egregious example of this is in Kilgore et al.'s (11) autopsy study in pediatric patients designed to "determine the presence and extent of cardiac mucosa" (p. 922). In the methods sections, the authors describe the histologic criteria: "In each case, 5 µm sections were stained with hematoxylin and eosin as well as Alcian blue/periodic acid-Schiff (PAS). . . . All slides were evaluated in a blinded fashion by two observers looking for the presence of cardiac-type mucosa characterized by unequivocal PAS-positive mucous glands arranged in lobular configuration" (p. 922). This is the first and only time in the literature that cardiac mucosa has been primarily defined by the PAS stain and the first time that the definition has been made by the *presence of mucous glands rather than the absence of parietal cells.* The authors suggested in the next sentence that parietal cells were absent in their cardiac mucosa: "The length of the cardiac-type mucosa, the distance between the most distal portion of the squamous mucosa and the identification of the most proximal parietal cell, was measured" (p. 922). However, because the cytoplasm of parietal cells frequently stain weakly with PAS, exclusion of parietal cells is not always clear. Careful examination of the illustrations is revealing. In their Figure 4, they show an admixture of PAS-positive glands and parietal-cell containing glands, indicative of oxyntocardiac mucosa (which they term "transitional mucosa"). When their Figure 2 is examined, it shows a low power view of the squamocolumnar junction where "squamous mucosa . . . is found adjacent to cardiac-type mucosa characterized by a lobular configuration of PAS-positive mucous-secreting glands." First, the glands appear more straight tubular than lobulated; second, not all the cells of the gland are strongly PAS-positive; third, there appear to be a population of smaller, PAS-negative cells between the surface and the deep glands that look identical to the parietal cells in their Figure 4.

The critical point here is that cardiac mucosa is not defined by the presence of PAS-positive mucous-secreting cells; it is defined by the fact that *all* the cells must be mucous cells *in an H&E-stained section.* PAS-stained sections have not been used in any other paper in the literature to define cardiac mucosa. If PAS-stained sections are used, all the cells must be strongly PAS-positive mucous-secreting cells, and the entire mucosa must be lobulated. Every columnar epithelium in the esophagus and stomach has a surface-lining epithelium composed of mucous cells containing neutral mucin and therefore strongly PAS positive. Kilgore et al.'s report is important because it is the only study in the literature that reports the universal presence of cardiac mucosa at the junction. The fact that it uses a nonstandard definition of cardiac mucosa makes the data suspect. I raised this question in a letter to the editor (12); the author's answer to these questions was less than satisfactory, at least to my mind.

WHERE IS CARDIAC MUCOSA LOCATED (FIGURE 6.1)?

Allison (13) describing reflux esophagitis in 1948, says very little in his descriptions about cardiac mucosa. At this time, the squamocolumnar junction was accepted as the end of the esophagus, and cardiac mucosa, by definition, lined the proximal stomach. In the illustrations that accompany this paper, Allison seems to depict squamous epithelium as a straight line and gastric mucosa as a serrated line. He showed the entire esophagus lined by squamous epithelium up to the peritoneal reflection in his Figures 1, 2, 3, and 4; the serrated gastric mucosa starts at the peritoneal reflection (Fig. 6.5). There is no description of the histologic type of this gastric mucosa. In his Figure 5, which is a depiction of "heterotopic gastric mucous membrane without hernia" (p. 41), Allison showed gastric mucosa in the tubular esophagus above the peritoneal reflection (Fig. 6.6). It seems reasonable to interpret this to mean that Allison believed that the finding of "gastric mucosa" in the tubular esophagus is not the norm.

Review of Allison and Johnstone's 1953 article (8), which describes columnar-lined esophagus for the first time, shows a dichotomy in what Allison has called the esophagus and stomach. He defined the esophagus by its external anatomic criteria but called the columnar epithelium lining that segment of tubular esophagus "gastric mucous membrane." There is a detailed description of the histology of Case 1 as follows:

> The esophagus was separated . . . from the stomach along the line of the peritoneal reflection. . . . The stomach below the anatomical junction with the esophagus is lined by gastric mucosa of fundal type. . . . *Cardiac glands and cardiac gastric mucosa do not appear until 0.6 cm up the anatomical esophagus* . . . The rather villous type of cardiac mucosa, its lack of depth, and a diffuse fibrosis of the submucosa in the zone from 2 to 4 cm. above the stomach orifice, suggest healing of previous shallow ulcerations.

Allison and Johnstone (8) clearly placed the cardiac mucosa in the esophagus above the peritoneal reflection (Fig. 6.1a). They also recognized that cardiac mucosa shows evidence of injury. The gastric mucosa at the peritoneal reflection is not cardiac, it is gastric fundic with parietal cells.

Norman Barrett, in his 1957 treatise that established Barrett esophagus as an entity (14), wrote:

> Histologically the upper part of the esophagus is normal and there is an obvious and abrupt transition between this and the unusual segment at the bottom. Just below this boundary. . . . the columnar cells are flat and arranged in shallow, tubular glands amongst which lie mucus secreting units. There are no oxyntic cells. . . . Lower, in the esophagus the simple tubular crypts give rise to more typical gastric mucous membrane. Scattered oxyntic cells appear. At any level in this abnormal segment, perfectly formed patches of squamous epithelium occur. These findings, which are similar to those described by Allison and Johnstone (1953), suggest that the abnormal epithelium, despite its looks, does not function exactly as stomach. (p. 884)

Barrett was describing the presence of cardiac mucosa in the proximal segment and oxyntocardiac

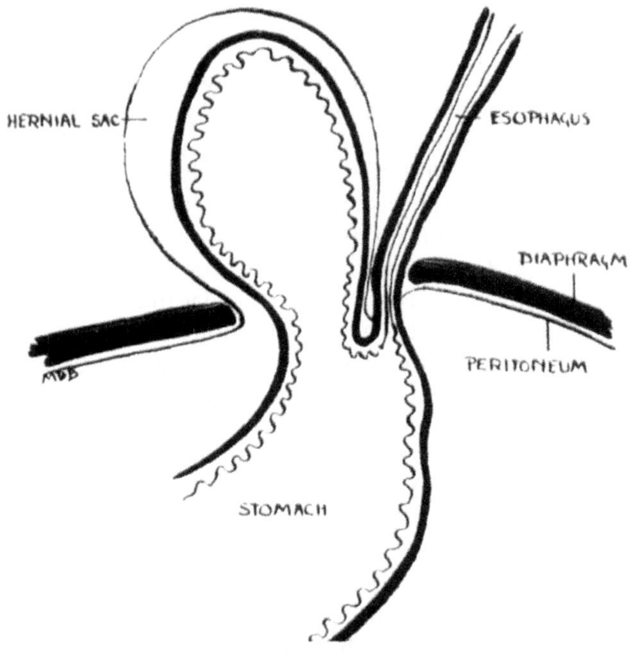

FIGURE 6.5 Allison's diagrammatic representation of a paraesophageal hernia. The tubular esophagus and its junction with the stomach are normal. Allison shows squamous epithelium (straight line) lining the entire tubular esophagus, transitioning to gastric mucosa (serrated line) at the end of the tubular esophagus. Reproduced with permission from Allison PR. Peptic ulcer of the oesophagus. Thorax 1948;3:20–42.

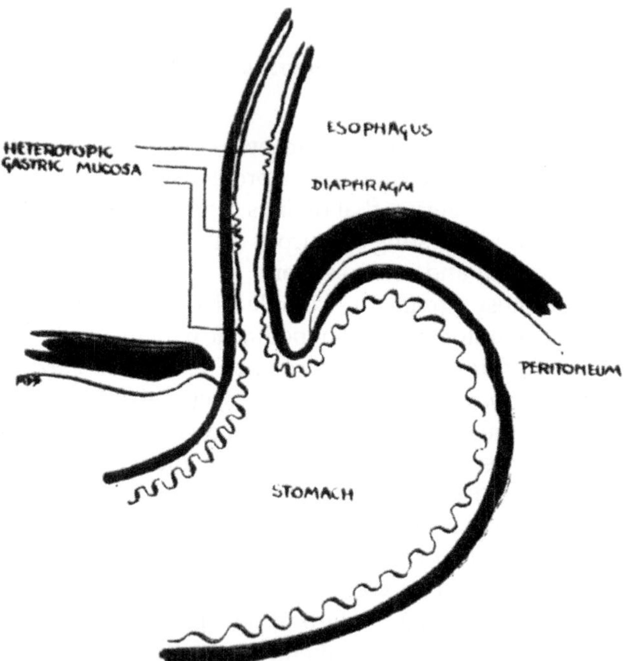

FIGURE 6.6 Allison's diagrammatic representation of "heterotopic gastric mucosa" in the most distal esophagus. The heterotopic gastric mucosa is shown as a serrated line with the squamous epithelium as a straight line. Reproduced with permission from Allison PR. Peptic ulcer of the oesophagus. Thorax 1948;3:20–42.

mucosa in the distal segment of columnar-lined esophagus. Barrett rejected Allison and Johnstone's designation that this is "gastric mucous membrane" and correctly calls this "columnar-lined esophagus."

From the very moment that columnar-lined esophagus was defined and accepted into the literature, cardiac mucosa was known to be the histologic type that predominated in the columnar-lined esophagus.

The next influential research regarding cardiac mucosa is that of Hayward in 1961 (15) who was attempting to define the "normal limit of cardiac mucosa" in the esophagus and to define the pathological entity of columnar-lined esophagus. Hayward described how the distal 1 to 2 cm of the tubular esophagus is normally lined with junctional (= cardiac) epithelium. He also indicated that this junctional (= cardiac) epithelium extends into the proximal stomach for a variable distance (Fig. 6.7 and Fig. 6.1b). Hayward wrote: "This columnar epithelium (in the distal 1 to 2 cm of the esophagus) has always been regarded as gastric in type and has been called cardiac epithelium, a rather absurd term which is saved from ambiguity because there happens to be no epithelium in the heart" (p. 36).

Hayward postulated that "if squamous epithelium joined gastric epithelium of fundal type directly, it

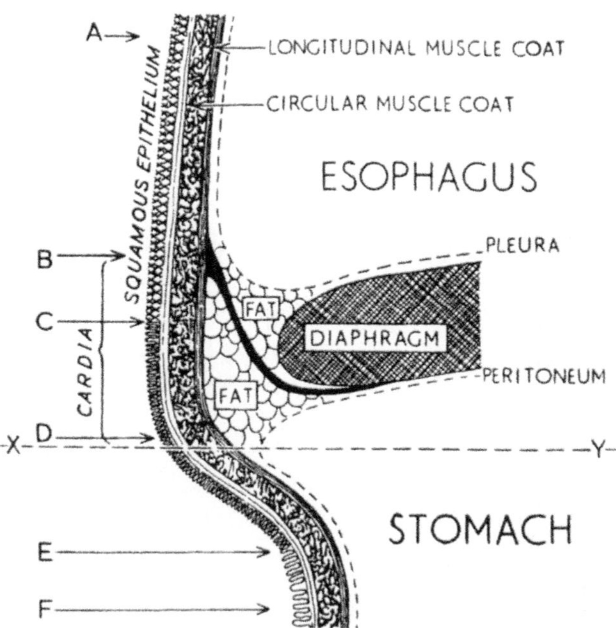

FIGURE 6.7 Hayward's representation of cardiac mucosa, which he shows lining the distal 2 cm of the tubular esophagus and extending into the proximal stomach for approximately 2 cm. Reproduced with permission from Hayward J. The lower end of the oesophagus. Thorax 1961;16:36–41.

would be liable to digestion at the junction. The buffer zone of junctional epithelium, which does not secrete acid or pepsin but is resistant to them, has to be interposed" (p. 37). This is an incorrect statement. The lower esophageal sphincter represents an effective barrier that prevents digestion of squamous epithelium by gastric acid. The function of protecting the squamous epithelium has never been ascribed to cardiac mucosa by anyone other than Hayward.

In different places in the article, Hayward made the following contradictory statements regarding the location of cardiac (or junctional) mucosa, which he clearly distinguished histologically from fundal and pyloric epithelium: (a) "[The] lower 1 to 2 cm (of the esophagus) is lined by columnar epithelium, which also extends a little way into the stomach" (p. 36); (b) "[P]art of it is always in the stomach" (p. 36); (c) "It must lie, as it does astride the gastroesophageal junction partly in the oesophagus and partly in the stomach" (p. 37); (d) "I suggest that junctional (= cardiac) epithelium should be regarded as oesophageal" (p. 37); (e) "Thus, the oesophageal lining should be described as having mainly stratified squamous epithelium . . . but partly columnar epithelium . . . round the whole circumference in the lower one or two centimeters. The stomach should be described as lined by two sorts of epithelium, fundal and pyloric, except for a small area around the oesophageal opening where the oesophageal junctional epithelium protrudes into it" (p. 37); (f) "In [cases of reflux esophagitis] junctional epithelium may be found extending for varying distances up the oesophagus, sometimes even as far as the level of the arch of the aorta. Many previous papers have referred to this state of affairs as 'gastric' epithelium in the oesophagus. . . . In my opinion it can be accepted that 'gastric' epithelium in the oesophagus is, in this type of case, almost always, if not always, of 'cardiac' type, identical with what is here termed junctional oesophageal epithelium" (p. 39).

Hayward appears to be arguing that cardiac (junctional) mucosa is esophageal in location, but he confused the issue by suggesting that it may also extend into the proximal stomach. His conclusion stated: "All the literature about gastric epithelium in the oesophagus in cases of reflux oesophagitis becomes invalid because with the new outlook it is simply oesophageal epithelium in the oesophagus. It is probably neither ectopic, nor congenital, nor permanent, nor in need of resection but metaplastic and reversible" (p. 41).

Unfortunately, Hayward illustrated the anatomy and histology of the region and showed cardiac mucosa extending a considerable distance into the proximal stomach below the level of the peritoneal reflection (Fig. 6.7). This is the first time in the medical literature that cardiac mucosa had been described distal to the peritoneal reflection in the proximal stomach. It should be recognized that Hayward produced no data to support this drawing, and it contradicts Allison's report that the mucosa at the peritoneal reflection is gastric fundal in type.

It is possible that Hayward was trying to reconcile existing pathologic opinion that the proximal stomach was lined with cardiac mucosa. In his introduction, he touched on the problem: "Under the microscope the oesophagogastric junction cannot be identified, so it is not surprising that histologists adopted the only line of demarcation available to them, namely the sudden change from squamous to columnar epithelium, and called the squamous epithelium oesophageal and the columnar epithelium gastric" (p. 36). Hayward is absolutely correct; this is exactly what happened. The normal histology of this region was defined by examination of esophagectomy specimens at a time when it was accepted that the squamocolumnar junction was the end of the esophagus. The frequent finding of cardiac type mucosa distal to the squamous epithelium in these very abnormal specimens led to the erroneous belief that cardiac mucosa was always present and in the proximal stomach. It is also possible that Hayward was trying to reconcile his clinical observations. By this time, Hayward had decided that the distal end of the tubular esophagus was the gastroesophageal junction. It would not have been surprising if his ideas were not generated by the fact that he was finding cardiac mucosa distal to this point.

After Hayward's article in 1961 (15), the dogma that cardiac mucosa normally lines both the distal 1 to 2 cm of the esophagus as well as the proximal stomach became well established as fact (Fig. 6.1b). Again, it is a fact for which there is no evidence. No significant test of this fact appeared in the literature until recently. This dogma appears in every textbook. In the 1997 edition of the widely used text *Histology for Pathologists*, DeNardi and Riddell (16), in describing the histology of the esophagus, wrote: "The mucosal gastroesophageal junction. . . . Normally lies within the LES (lower esophageal sphincter), found usually within 2 cm of the muscular junction as defined by the proximal edge of the gastric folds; thus, the distal 2 cm of the esophagus is lined by columnar epithelium that is identical to that found in the gastric cardia. The length of these extensions may be unusually exaggerated, sometimes extending 3 cm into the esophagus and thus endoscopically resembling early Barrett's esophagus" (p. 467). In the chapter on the histology of the stomach in the same book, Owen (17) wrote: "Extending distally from the gastroesophageal junction for approximately 1 to 2 cm is the cardiac mucosa, where the glands are mucus secreting. . . . At the lower end of the esophagus, there is a change from nonkeratinizing squamous epithelium to columnar epithelium, which is abrupt, both grossly and microscopically. The position of this squamocolumnar junction is variable and does not coincide with the strict anatomic esophagogastric junction, that is, the point where the tubular esophagus becomes the saccular stomach. The mucosal junction commonly is located 0.5 to 2.5 cm proximal to the anatomic junction and is often serrated, rather than being a regular circumferential line (Z line). The lower portion of esophagus, below the Z line, is therefore covered by cardiac-type gastric mucosa" (pp. 483–484). The combination of the two indicates almost complete acceptance of Hayward's illustration for the normal extent of cardiac mucosa (Fig 6.7). Note also the different definitions of the gastroesophageal junctions used by these two authorities.

The concept that columnar epithelium normally lines the distal 1 to 2 cm of the tubular esophagus has changed since the recognition of short-segment Barrett esophagus (18). To define short-segment Barrett esophagus, it has become necessary to define endoscopic normalcy as the absence of *any* visible columnar epithelium in the tubular esophagus. The entire esophagus is now considered to be lined by squamous epithelium (Fig. 6.8). The presence of any visible columnar-lined esophagus is considered abnormal enough to dictate that a biopsy be taken (19, 20). This has removed the concept that 1 to 2 cm of cardiac mucosa normally lines the distal tubular esophagus. Today, most people believe that cardiac mucosa is only normally found in the proximal stomach (Fig. 6.1c and d).

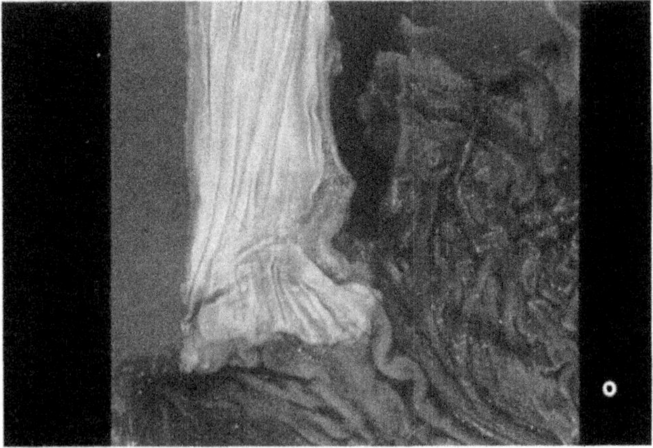

FIGURE 6.8 Grossly normal esophagus and stomach. The entire tubular esophagus is lined by squamous epithelium; the squamocolumnar junction is horizontal and the rugal folds come all the way to the squamocolumnar junction. (Cross reference: Color Plate 1.1)

The article looking at fetal esophagi by de Hertogh et al. (21) reflected the presently held view regarding the cardia: "Anatomists have applied the term 'cardia' to that part of the stomach which lies around the orifice of the tubular oesophagus. There is no anatomical landmark for the distal margin of the so-defined cardia. Its proximal margin is the GOJ (gastro-oesophageal junction), which, according to anatomists, is localised at the level of the angle of His. This is the point where the tubular oesophagus joins the saccular stomach" (p. 571). This is an extremely precise definition of the anatomic gastric cardia as it is presently regarded.

De Hertogh et al.'s (21) study precisely localized the relation of the mucosae to the gastroesophageal junction, which the authors defined as the angle of His. Their conclusion about the location of what they called "cardiac mucosa" is as follows:

> The second observation we made in reviewing our results relates to the position of the cardiac mucosa. *When one accepts the angle of His as a landmark for the GOJ, the cardiac mucosa was located proximal to or at the GOJ in all of our cases.* Therefore, according to anatomical definitions, cardiac mucosa was located in the abdominal segment of the tubular oesophagus at birth. (p. 575)

In all their cases, the mucosa at the point of the angle of His contained parietal cells, an observation that is identical to that made by Allison and Johnstone (8). It is quite clear that de Hertogh et al.'s "cardiac mucosa" is in the distal esophagus, not in their defined gastric cardia. In Chapter 3 we stressed that de Hertogh's "cardiac mucosa" is not the equivalent of adult cardiac mucosa; rather it represents a fetal type of epithelium that undergoes future transformation to squamous epithelium (22). The most important item of data in de Hertogh's paper is that *there is no cardiac mucosa at or distal to the gastroesophageal junction; even in fetuses, parietal cells are present in the epithelium at the junction.*

De Hertogh et al.'s data essentially solve the controversy of where cardiac mucosa is located. While their intent is to show that fetal esophagus contains cardiac mucosa, what their study clearly shows is that *cardiac mucosa is not found in the fetal stomach; in their fetuses, the mucosa at and distal to the gastroesophageal junction (the angle of His) contains parietal cells.* This is identical to the finding of parietal cells at the gastroesophageal junction defined as the point of the peritoneal reflection in Allison and Johnstone's article. Parietal cell differentiation in the stomach in the fetus is the end point of gastric development; parietal cells are adult-type cells that remain throughout the life of the individual. It is unlikely that the adult-type parietal cells will disappear after they have developed. Allison and Johnstone's (8) and de Hertogh's (21) finding that the mucosa at the gastroesophageal junction in the adult and fetus contains parietal cells proves that cardiac mucosa is not present in the stomach.

The controversy that exists as to the location of cardiac mucosa is dependent on how researchers define the gastroesophageal junction. It is only by using an incorrect and unproved definition of the gastroesophageal junction can cardiac mucosa ever be placed in the stomach. Let us review two articles in terms of how they localize cardiac mucosa and evaluate the effect of the definition of the gastroesophageal junction (Fig. 6.9).

In the first of these articles, Kilgore et al. (11) measured the extent of cardiac mucosa (as shown previously in this chapter, they used a definition of cardiac mucosa that is not the accepted definition) from the squamocolumnar junction, found 1 to 4 mm (mean 1.8 mm) of cardiac mucosa distal to this, and declared that cardiac mucosa is present in the proximal stomach. The correct conclusion is that 1 to 4 mm (mean 1.8 mm) of what they called cardiac mucosa is present distal to the squamocolumnar junction. To reach the conclusion that cardiac mucosa is in the stomach, the authors declared that in all their cases, the squamocolumnar junction, proximal limit of the rugal folds, and peritoneal reflection were exactly coincidental (Fig. 6.9). They

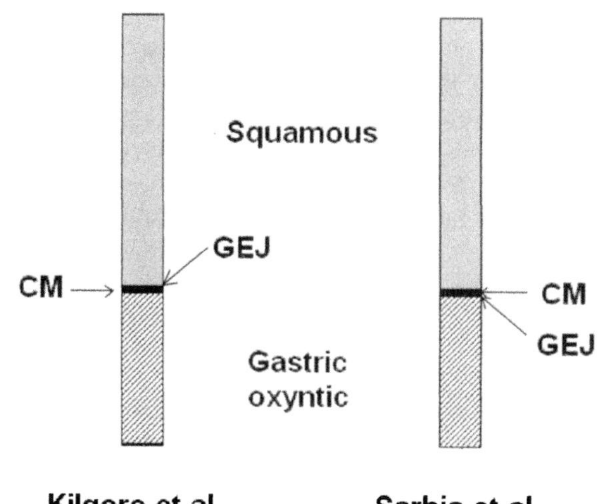

FIGURE 6.9 Similar findings in two papers looking at a vertical section across the squamocolumnar junction in which a small amount of cardiac mucosa was found between the squamous epithelium and gastric oxyntic mucosa. Kilgore et al. (10) (*left*) interpreted the cardiac mucosa as gastric by assuming the squamocolumnar junction and gastroesophageal junction (proximal limit of rugal folds) are coincidental. Sarbia et al. (9) (*right*) interpreted the cardiac mucosa as esophageal because they found esophageal submucosal glands under cardiac mucosa in 25% of patients.

must make this assumption because they have no point of reference on a microscopic slide to identify the proximal limit of the rugal folds. They can measure only from the squamocolumnar junction (1).

This absolute certainty of the coincidence of the squamocolumnar junction, peritoneal reflection and proximal limit of the rugal folds within a 1 to 4mm error range borders on the ridiculous. A slightly non-vertical cut across the wall at the peritoneal reflection, or contraction of the muscle wall of the esophagus can create a technical error of 1 to 4mm between the point of the peritoneal reflection and the mucosal surface. The rugal folds never end with a degree of abruptness that permits accuracy of determining its proximal limit within 1 to 4mm. Finally, the resolution of the human eye cannot detect a microscopic cephalad displacement of the squamocolumnar junction. Kilgore et al.'s certainty that the 1 to 4mm of cardiac mucosa they found was distal to the proximal limit of the rugal folds is not credible

Even if Kilgore et al.'s conclusion that they can be certain that there is no microscopic gap between the proximal limit of the rugal folds and squamocolumnar junction is correct, they use the proximal limit of the rugal folds to define the junction. This is unproven as the true junction and if it is not the true junction, their conclusion will be wrong. Conclusions based on unproven assumptions are not valid. Kilgore et al'.s reasoning should be compared with the study by Sarbia et al.

The article by Sarbia et al. (9) is an excellent study that looks at 36 esophagogastrectomy specimens removed in patients with squamous carcinoma where the distal edge of the tumor was greater than 5cm above the gastroesophageal junction. The authors correctly stated that this represents a good control group to study the gastroesophageal junction because there is no known association between squamous carcinoma of the esophagus and gastroesophageal reflux. Despite the good logic, the control group is problematic because evidence of esophagitis was present in the squamous epithelium in over 50% of the patients.

In all the cases, the entire tubular esophagus was lined by squamous epithelium and transitioned into "the brown-red gastric mucosa." (The authors really meant brown-red columnar epithelium.) They made no attempt to identify the proximal limit of the rugal folds; they had no predetermined bias as to the location of the gastroesophageal junction. This is not a bad thing because there is no way to identify the top of the rugal folds in the histologic sections; the point of measurement must be from the distal limit of squamous epithelium. They sectioned the entire junction by longitudinal sections and measured the minimal and maximal lengths of cardiac mucosa and oxyntocardiac mucosa found distal to the squamous epithelium. Their definitions of cardiac and oxyntocardiac mucosa are perfect.

Sarbia et al. (9) reported that cardiac mucosa was present in the entire circumference of the squamocolumnar junction in 20 patients, was present in part of the circumference in 15 patients, and was completely absent in one patient. In the one patient without cardiac mucosa, oxyntocardiac mucosa was present distal to the squamous epithelium. The minimum length of cardiac mucosa ranged between 0 and 3mm (median 1mm); the maximum length ranged between 1 and 15mm (median 5mm). The combined length of cardiac plus oxyntocardiac mucosa was a minimum of 1 to 12mm (median 4mm), and the maximum length was between 5 and 28mm (median 11mm).

Sarbia et al. (9) next did something that is indicative of true scientists. They did not jump to the conclusion that this cardiac and oxyntocardiac mucosa that is found distal to the squamous epithelium *and distal to the tubular esophagus* is gastric. Instead, they looked for objective evidence to prove the location of these mucosae by evaluating their relationship to submucosal glands which define the esophagus. Their conclusion is startling: "Intraesophageal location of cardiac mucosa and/or oxyntocardiac mucosa could be verified histologically in nine of 36 cases. Thus, in eight cases, CM/OCM was situated over submucosal mucous glands. In the ninth case, CM/OCM was situated over squamous epithelium-lined ducts, a finding that was also present in three of the eight aforementioned cases." Thus, Sarbia et al. proved that in 25% of their patients, the cardiac and oxyntocardiac mucosa was esophageal in location by virtue of the fact that they were associated with the esophageal submucosal gland-duct complexes (Fig. 6.9). Because these glands are distributed sporadically in the esophagus, the absence of submucosal glands does not prove that the location is not esophageal, particularly when the length of tissue is in the 5 to 10mm range. Because of the fact that cardiac and oxyntocardiac mucosae were histologically identical whether or not they were associated with submucosal glands, it is reasonable to assume that the cardiac and oxyntocardiac mucosae in the other 75% of patients were also esophageal, but there is no proof of that.

The correct conclusion to be drawn from Kilgore et al. (11) and Sarbia et al. (2000) is as follows: cardiac mucosa was present distal to the squamocolumnar junction in the entire circumference in 20/36 and in at least part of the circumference in all but one patient in Sarbia et al. and in all patients in one point of the circumference in Kilgore et al. (11) (if one believes Kilgore

TABLE 6.1 The Tom DeMeester Biopsy Protocol

In addition to sampling any visible lesions, the following samples are taken routinely in every patient:

1. Five biopsies from the antrum and body of the stomach. This samples the stomach well away from the gsatroesophageal junctional region. In some patients, an additional sample is taken from the fundus. This sample is routinely stained by Giemsa stain in addition to H&E for evaluation of *Helicobacter pylori.*
2. A retrograde sampling of the mucosa at and 1 to 2 cm distal to the proximal limit of the rugal folds. This is an unmeasured biopsy because the endoscope is in the retroflexed position.
3. Antegrade biopsies (four quadrants) of the mucosa at the squamocolumnar junction. This is a measured biopsy that aims to sample the columnar epithelium immediately distal to the squamocolumnar junction. Optimally, this biopsy tries to sample the junction itself with one specimen containing both squamous and glandular epithelium.
4. Antegrade measured biopsies (four quadrants at each circumferential level) at 1 to 2 cm intervals when there is any gap between the squamocolumnar junction and the proximal limit of the rugal folds. The number of levels sampled varies from 0 to 10 depending on the degree of cephalad migration of the squamocolumnar junction.
5. Biopsy of the squamous epithelium immediately above the squamocolumnar junction.

et al.'s definition of cardiac mucosa, which we do not). This was located distal to the end of the tubular esophagus and, in Kilgore et al.'s article, under rugated mucosa. When Kilgore et al. assumed that the squamocolumnar junction was the true gastroesophageal junction because no separation could be seen with the naked eye between it and the proximal limit of the rugal folds, the conclusion reached was that cardiac mucosa was gastric in location. When Sarbia et al. (9) used an objective criterion of esophageal location (submucosal glands), the cardiac mucosa was localized to the esophagus (Fig. 6.9). We have performed studies that are similar to that of Sarbia et al. and confirm their findings that submucosal gland-duct complexes are often present under cardiac and oxyntocardiac mucosa (3). We believe this is extremely convincing evidence for an esophageal location of cardiac mucosa. This is true even when the cardiac and oxyntocardiac mucosa is found distal to the end of the tubular esophagus and in mucosa that shows rugal folds.

Conclusion: Cardiac mucosa and its variants, oxyntocardiac and intestinal (Barrett-type) epithelia, are esophageal in location, irrespective of any contrary endoscopic or gross pathologic opinion.

IS CARDIAC MUCOSA PRESENT IN EVERYONE?

When Dr. Tom DeMeester came to the University of Southern California as chairman of surgery in 1991, Dr. Chandrasoma was assigned to work with him. He educated himself on the literature relating to reflux disease and normal histology. He had no reason to believe that the prevailing universal opinion that cardiac mucosa normally lined a 2- to 3-cm region between the squamocolumnar junction and gastric

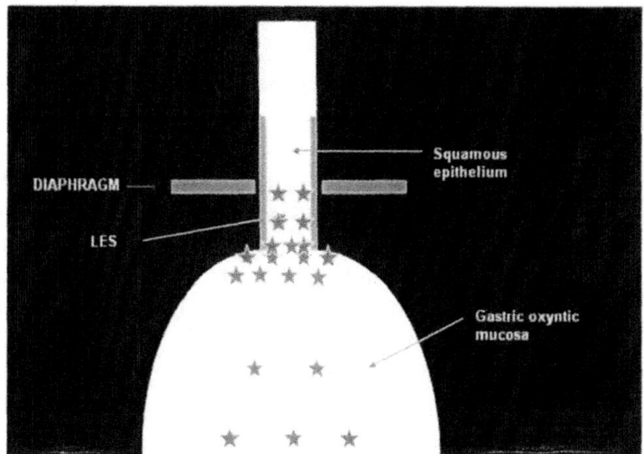

FIGURE 6.10 The DeMeester biopsy protocol. Every patient has biopsies of the antrum and body of the stomach. Some have additional sampling of the fundus. Every patient has a biopsy from the proximal limit of the rugal folds and 1 to 2 cm distal to it with the endoscope in the retrograde position. In patients who are endoscopically normal, a biopsy is taken from the squamocolumnar junction immediately above the retrograde biopsies. This biopsy attempts to straddle the squamocolumnar junction. In patients who have a visible columnar-lined segment of esophagus, measured biopsies are taken at 1 to 2 cm intervals until the squamocolumnar junction is reached.

oxyntic mucosa was incorrect. At the time this 2 to 3 cm of cardiac mucosa was thought to straddle the gastroesophageal junction, being normally present in both the distal esophagus (1 to 2 cm) and the proximal stomach (Fig. 6.1b).

Dr. DeMeester brought with him a standardized biopsy protocol that he had developed (Table 6.1, Fig. 6.10). Every patient, normal and abnormal, was biopsied according to the protocol. This protocol provides a minimum of three specimens in endoscopically normal patients: the distal gastric sampling, the retroflex biopsy of the tops of the rugal folds that are immediately (practically within 0 to 10 mm) distal to the squamocolumnar junction, and the squamocolumnar junction itself, which attempts to include the actual junction and always has squamous epithelium. When a visible segment of columnar-lined esophagus is seen, four-quadrant measured biopsies are taken from this zone at 1- to 2-cm intervals. Each of these level biopsies is evaluated for mucosal type permitting histologic mapping of the area between the gastric oxyntic mucosa usually present in the retroflex biopsy and the squamocolumnar junction (23). The histologic mapping is based on the precise criteria of diagnosis for the various epithelial types as described in Chapter 5. This is done irrespective of any preconceived endoscopic determination of the gastroesophageal junction.

The only use of the proximal limit of the rugal folds is to define a place at endoscopy that most likely represents the proximal limit of gastric oxyntic mucosa. An adequate biopsy set requires that the most proximal biopsy must contain squamous epithelium and the most distal biopsy must contain gastric oxyntic mucosa. There is never a problem with squamous epithelium; the Z line is easily visible. There is sometimes a problem with adequacy at the distal end in that gastric oxyntic mucosa is not present. In some patients, gastric oxyntic mucosa begins up to 3 cm distal to the proximal limit of the rugal folds (2, 3), and the most distal biopsy is too proximal to sample gastric oxyntic mucosa. With increasing understanding, the point of the distal biopsies has been extended distally, and we now have fewer patients without gastric oxyntic mucosa than in the 1990s.

By this mapping biopsy technique, the length of columnar lining composed of cardiac, oxyntocardiac, and intestinal metaplasia between the squamous epithelium and gastric oxyntic mucosa can be defined. When intestinal metaplasia is present, its length can also be defined independently of the length of columnar-lined esophagus. These histologically defined measurements are used to define length in all of our patients; they supersede any endoscopic measurement that is made. These are the most precise measurements within the context of the technical capability of the procedure based on well-defined histologic criteria. This is the starting point for determining the meaning of these different epithelial types.

From the outset, Dr. Chandrasoma received biopsies from two to five patients every day from Dr. DeMeester's service. Since the early 1990s, he has seen biopsies from more than 5000 endoscopies performed according to this unwavering standard biopsy protocol. This is a unique experience not in the number of patients seen but in the standardized nature of the biopsies. Many of these patients have 24-hour pH studies, manometry, complete clinical evaluation, and video-esophagography in addition to the endoscopy, permitting exact correlation between observed changes and each histologic parameter that is identified.

Within the first few months, Dr. Chandrasoma was struck by the fact that there were several patients in whom biopsies did not contain any cardiac mucosa. His initial belief that this must be a sampling error was quickly removed when he realized that these patients had six to eight biopsy samples from the squamocolumnar junction within a 0- to 10-mm distance. It was obvious that the dogma that 2 to 3 cm of cardiac mucosa was normally present was incorrect.

This led to the hypothesis that cardiac mucosa is not a normal structure but rather an abnormal metaplastic epithelium that results from the effect of gastroesophageal reflux on esophageal squamous epithelium. When this hypothesis was proposed in 1994 (24), the belief was that 2 to 3 cm of cardiac mucosa was normally present between esophageal squamous epithelium and gastric oxyntic mucosa and that this "straddled" the gastroesophageal junction (Fig. 6.1b). When, in the mid-1990s, the recognition of short-segment Barrett esophagus caused the normal endoscopic appearance to be defined as a tubular esophagus lined entirely by squamous epithelium, cardiac mucosa was deemed to line a 2- to 3-cm length of the proximal stomach only (Fig. 6.1c).

Because the definition of normal endoscopic appearance was that the gastric rugal folds came up to the squamocolumnar junction, it followed that cardiac mucosa was present in mucosa containing rugal folds in the proximal stomach. When an abnormal columnar epithelium was recognized between the squamocolumnar junction and the proximal limit of the rugal folds by endoscopy, columnar-lined esophagus was diagnosed at endoscopy; when this abnormal columnar epithelium was subject to biopsy, it was composed of cardiac mucosa, oxyntocardiac mucosa, and intestinal metaplastic epithelium (Barrett-type epithelium). Therefore, cardiac mucosa was considered to be present both normally in the area under the rugal folds and as an abnormal metaplastic epithelium in the columnar-lined esophagus. No morphologic difference has ever been reported between "normal gastric" and "abnormal esophageal" cardiac mucosa (32).

The beginning of trying to establish normalcy must have a basis in autopsy studies; this is the only situation where "normal" people who have never consulted a physician with a complaint related to reflux disease can be found. When a search of the literature showed the almost complete absence of any detailed histologic study of this region at autopsy, we embarked on such a study.

Literature Review

Chandrasoma PT, Der R, Ma Y, Dalton P, Taira M. Histology of the gastroesophageal junction: An autopsy study. Am J Surg Pathol 2000;24:402–409.

We completed this autopsy study in 1996 submitting it for publication at around this time. Being the first-ever autopsy study of the histology of the gastroesophageal junction, the scientific community should have enthusiastically embraced it. This did not happen. In what is an all-too-common occurrence in medicine, this report was rejected by several pathology and gastroenterology journals for 3 years. In essence, the reason for rejection was that the idea that cardiac mucosa was not present was too revolutionary to be believed. No question was ever raised about the data.

This article has two parts. We initially reviewed retrospective autopsy material in the archives of the Los Angeles County—University of Southern California Medical Center. The autopsy department had been directed for many years by Dr. Dorothy Tatter, who was extremely compulsive about collecting material from every autopsy. Dr. Chandrasoma trained under Dr. Tatter and learned that an autopsy was not complete unless there were four parathyroid glands and a longitudinal section across the junction of the esophagus and stomach. We reviewed hundreds of autopsies and found that the inevitable autolysis precluded accurate evaluation of the histology in most cases. We were able to find 72 cases with sufficient preservation to permit recognition of histologic types. Of these 72 patients, 21 (29%) had only squamous and oxyntic mucosa, 32 (44%) had only oxyntocardiac mucosa between squamous and oxyntic mucosa, and 19 (26%) patients had cardiac mucosa; 53 of 72 (74%) did not have any cardiac mucosa at the junction. In 61 of these patients, orientation was adequate to measure the length of cardiac and oxyntocardiac mucosa that was found; the combined cardiac and oxyntocardiac mucosal length was zero in 21 patients, less than 0.5 cm in 25 patients, 0.5 to 1 cm in 12 patients, and greater than 1 cm in 3 patients (1.1 cm, 1.1 cm, and 1.5 cm lengths in the 3 patients).

Recognizing that a single longitudinal section across the junction was not an adequate study, we prospectively evaluated 18 patients; we tried to select younger patients, and there were 12 patients under the age of 25 years. In these 18 patients, the entire junctional circumference was pinned out and examined by vertical, well-oriented sections that permitted measurement. Cardiac mucosa was completely absent in 10 (56%) patients; in the other 8 patients it was present in only a part of the circumference. Oxyntocardiac mucosa was present in some part of the circumference in all patients but was present in the entire circumference only in 9 (50%) cases. In 9 cases, there was at least one point of the circumference where the squamous epithelium transitioned directly to gastric oxyntic mucosa without any cardiac or oxyntocardiac mucosa (Fig. 6.11). The maximum cardiac mucosal length ranged between 0 and 2.75 mm; the maximum combined cardiac and oxyntocardiac mucosal length ranged from 0.475 to 8.05 mm; 14 (76%) of the patients had a maximum combined cardiac and oxyntocardiac mucosal length of less than 5 mm separating squamous from gastric oxyntic mucosa (Fig. 6.12).

In four patients we found the presence of submucosal mucous glands under cardiac and oxyntocardiac mucosa. At least in these four cases, this finding established cardiac and oxyntocardiac mucosa being located in the esophagus rather than the stomach.

If the data in this study are believed, it clearly establishes that cardiac mucosa is not universally present in humans (Fig. 6.11). When present, the combined

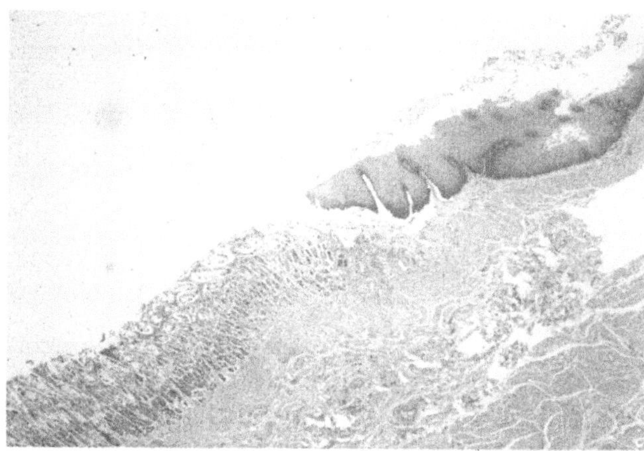

FIGURE 6.11 Longitudinal section across the squamocolumnar junction at autopsy in a 4-year-old child showing squamous epithelium directly transitioning into gastric oxyntic mucosa characterized by straight tubular glands containing parietal cells under the foveolar region. (Cross reference: Color Plate 1.5)

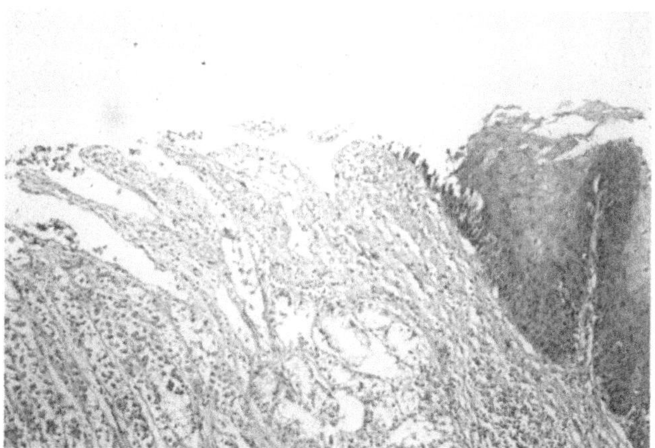

FIGURE 6.12 Longitudinal section across the squamocolumnar junction in a 14-year-old child showing a lobulated gland containing mucous cells between the squamous epithelium and parietal cell containing gastric oxyntic mucosa. This represents cardiac mucosa of less than 1 mm in length consisting of one foveolar-gland complex. The true gastroesophageal junction is the point at which gastric oxyntic mucosa (straight glands without mucous cells) begins distal to the cardiac mucosa.

cardiac and oxyntocardiac mucosal length was less than 1 cm in the vast majority of patients, and there was a tendency for cardiac and oxyntocardiac mucosal lengths to be greater in the older patients.

As with all scientific data, independent substantiation of the data in this study must be reported before these data can be accepted as fact. It is therefore valuable to review the subsequent literature on the subject (Table 6.2). The studies must be examined carefully in the following respects:

1. What is the population being studied? Is it normal or likely to have a significant number of patients with reflux? If our hypothesis that cardiac mucosa is an abnormal structure induced by gastroesophageal reflux is correct, the prevalence of cardiac mucosa will be higher as the likelihood of patients with reflux in the study population increases.
2. What are the definitions used for histologic diagnosis of cardiac and oxyntocardiac mucosa? These must be precise and according to the standard, well-established definitions for the study to be valid.
3. What potential impact is there from sampling error? The optimal sampling method is a complete histologic study of the full circumference of the area between squamous and oxyntic mucosa in autopsy and resection specimens. Studies with a single vertical section across the junction are incomplete, as are endoscopic biopsies. In studies based on endoscopic biopsy, the most reliable are those where a single biopsy specimen shows the squamocolumnar junction, permitting measurement of cardiac and oxyntocardiac mucosa, if they are found. In other endoscopic studies, the likelihood of sampling error decreases as the number of biopsies increases.
4. What is the age distribution? The prevalence of cardiac mucosa will tend to increase with age as the damage caused by cumulative reflux increases.

Three excellent studies evaluate the fetal esophagus and junctional region for the presence of cardiac mucosa. These have been carefully reviewed in Chapter 3, and the reader is referred there. It is important to recognize that the entire esophagus in early fetal life is lined by columnar cells, which, by virtue of absence of goblet and parietal cells, can be classified as "cardiac mucosa." Careful study of these reports suggests that what the authors call "cardiac mucosa" represents fetal esophageal columnar epithelium (22). Fetal columnar epithelium is progressively replaced by squamous epithelium, the process completing in the first year of life. To answer the question about the universal presence of cardiac mucosa, it is important to limit studies to children over the age of 1 year and adults because of the potential confusion resulting from the presence of fetal columnar epithelia in fetuses and infants. Park et al.'s (28) study shows what is probably the best illustration of squamous epithelium transitioning directly into gastric oxyntic mucosa, proving the absence of cardiac epithelium in this patient (Fig. 6.13).

TABLE 6.2 Literature Relating to the Prevalence of Cardiac Mucosa at the Squamocolumnar Junction

Reference	Type of study	Sampling	Percentage without cardiac mucosa	Length of cardiac mucosa
Chandrasoma et al. (1)	Autopsy (prospective)	Circumferential	56	0–2.75 mm
Chandrasoma et al. (1)	Autopsy (retrospective)	One vertical section	74	<10 mm in 58/61
Kilgore et al. (11)	Autopsy	One vertical section	0	1–4 mm
Jain et al. (2)	Clinical; 25/31 normal endoscopy	Extensive (4-quadrant) biopsies	65	N/A
Oberg et al. (25)	Clinical; normal endoscopy	Extensive biopsies	26[a]	N/A
Marsman et al. (4)	Clinical; normal endoscopy	Biopsies straddling squamocolumnar junction	38	N/A
Chandrasoma et al. (23)	Clinical; all patients	DeMeester protocol	39	N/A
Glickman et al. (26)	Pediatric clinical	Biopsies	19	N/A
Sarbia et al. (9)	Elderly; squamous carcinoma	Circumferential	3	0–15 mm
Zhou et al. (27)	Autopsy; 18 over 1 yr old	Circumferential	46	Not given

[a]In Oberg et al. (27), cardiac and oxyntocardiac mucosa were combined; the 26% represents the percentage of patients without either of these two epithelial types.

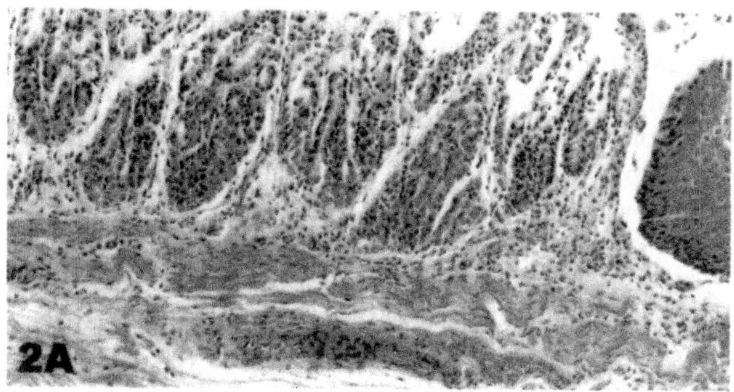

FIGURE 6.13 Squamocolumnar junction at autopsy in a child showing transition from squamous epithelium to parietal cell containing gastric oxyntic mucosa without any intervening cardiac mucosa. Reproduced with permission from Park YS, Park HJ, Kang GH, Kim CJ, Chi JG. Histology of gastroesophageal junction in fetal and pediatric autopsy. Arch Pathol Lab Med 2003;127:451–455.

Studies in children and adults are now numerous:

Jain R, Aquino D, Harford WV, Lee E, Spechler SJ. Cardiac epithelium is found infrequently in the gastric cardia. Gastroenterology 1998;114:A160 (Abstract)

This was a biopsy study of a group of 31 patients; 25 had a normal endoscopic appearance. Extensive biopsies were taken from the squamocolumnar junction and esophagogastric junction. Cardiac mucosa was reported to be absent in 65% of patients at the esophagogastric junction. The authors concluded: "Even at the esophagogastric junction, cardiac epithelium is found in only a minority of patients. These findings challenge the traditional view of a gastric cardia lined predominantly by cardiac epithelium." *Surely, this finding provides evidence that there is a high likelihood that cardiac mucosa is not universally present in all people.*

Jain et al.'s (2) study presents the usual problem of the possibility of inadequate sampling. Though described as "extensive (four quadrant biopsies at the Z-line, EGJ, 1 cm below the EGJ and 2 cm below the EGJ)" and the senior author is Dr. Spechler, one of the most authoritative gastroenterologists in the field, endoscopic biopsies do not sample the entire junction and may miss cardiac mucosa that is present. Although the article does not permit the statement that cardiac mucosa is definitely absent from 65% of patients, the fact that it was not found means that it is at most present in very small amounts and very likely absent in at least some patients. It is interesting that Jain et al.'s number of 65% for absence of cardiac mucosa falls between the numbers in our retrospective sample (74%) and prospective sample (56%). This suggests that the level of sampling in Jain et al.'s "extensive endoscopic biopsy protocol" is intermediate between full circumference examination and examination of a single vertical section from the junction at autopsy.

Oberg S, Peters JH, DeMeester TR, et al. Inflammation and specialized intestinal metaplasia of cardiac mucosa is a manifestation of gastroesophageal reflux disease. Ann Surg 1997;226:522–532

In a similar study from our institution, Oberg et al. (25) reported on the histologic findings in biopsies from the squamocolumnar junctional region in 334 patients with normal endoscopic appearance. These patients were biopsied according to what has been described as the DeMeester protocol (Fig. 6.10). The results showed that 246 (74%) of patients had cardiac and/or oxyntocardiac mucosa; the other 88 (26%) had neither, with biopsies containing only squamous and gastric oxyntic mucosa. The patients in this study were patients being evaluated for foregut disease in a clinical setting with a high likelihood of containing patients with reflux. The prevalence percentage in this study is not comparable to that in Jain et al.'s study because Oberg et al. combined cardiac and oxyntocardiac epithelia, unlike Jain et al. who were reporting on the absence of cardiac mucosa only. *Surely, the failure to find either cardiac or oxyntocardiac mucosa in 88 of 334 patients provides evidence that cardiac mucosa is not universally present even in a population with a likelihood of having reflux.*

Marsman WA, van Sandyck JW, Tytgat GNJ, ten Kate FJW, van Lanschot JJB. The presence and mucin histochemistry of cardiac type mucosa at the esophagogastric junction. Am J Gastroenterol 2004;99:212–217

In this study, the authors selected patients undergoing endoscopy for any reason who had a normal endoscopic gastroesophageal junction. In a clinical gastroenterology service, this population would, like Oberg et al.'s study (25), include many patients with significant reflux. The authors selected patients in whom biopsies showed the actual transition of squamous to columnar epithelium, making this study more reliable than that of Oberg et al. in terms of not missing small areas of cardiac mucosa because of sampling error. The results showed that 24 of 63 patients (38%) had a direct transition from squamous to oxyntocardiac mucosa without any intervening pure cardiac mucosa. This shows that cardiac mucosa was absent at least in this point of the circumference of the squamocolumnar junction. This study is comparable to that of

Jain et al.; the percentage of patients without cardiac mucosa in this study (38%) is lower than that in Jain et al. (65%). While some of this difference can be explained by the sampling factor, it is likely that the higher prevalence of cardiac mucosa in this study (62%) compared to Jain et al. (35%) is due to the fact that the authors limited their evaluation only to those cases that had the actual squamocolumnar junction in one biopsy piece. In fact, in patients who did not have the actual junction in one biopsy piece, cardiac mucosa was absent in 98%. *Surely, the data in this study provide evidence that cardiac mucosa is not universally present in all people, even when they have reflux.*

Chandrasoma PT, Der R, Ma Y, Peters JH, DeMeester TR. Histologic classification of patients based on mapping biopsies of the gastroesophageal junction. Am J Surg Pathol 2003;27:929–936 (23)

We reported the findings in 959 patients who were biopsied under the DeMeester biopsy protocol described previously. This study was in a population that was biased toward patients suffering from clinical reflux. We found that 811 (85%) patients had a less than 1-cm separation of squamous epithelium from gastric oxyntic mucosa. Of these, 161 (20%) patients had no cardiac or oxyntocardiac mucosa, and 158 (19%) had oxyntocardiac mucosa without any cardiac mucosa. The total number of patients without cardiac mucosa in this study (39%) is remarkably similar to the study of Marsman et al. (4) (38%). These numbers are lower than in our autopsy study (56%), strongly suggesting that the prevalence of cardiac mucosa in a population increases as the number of patients with reflux increases. *Surely, this provides evidence that cardiac mucosa is not universally present in all people.*

Glickman JN, Fox V, Antonioli DA, Wang HH, Odze RD. Morphology of the cardia and significance of carditis in pediatric patients. Am J Surg Pathol 2002;26:1032–1039 (26)

This is an endoscopic study of a clinical pediatric population with a bias toward patients with clinical reflux. For this study, 74 patients were selected by the following criteria: (a) age under 19 years, (b) absence of endoscopically apparent columnar-lined esophagus, (c) availability of clinical and endoscopic data for review, and (d) the presence of squamocolumnar junctional epithelium in the biopsy specimen.

Histologic examination included the classification of the columnar epithelia into pure mucous glands (mucous cell only), pure oxyntic-type glands (mixture of parietal cells and/or chief cells), and mixed mucous/oxyntic glands (mixture of mucous cells and parietal and/or chief cells). The definitions are adequately descriptive to permit interpretation of these unorthodox terms for the epithelia as being equivalent to cardiac, oxyntic, and oxyntocardiac, respectively.

The authors reported: Sixty patients (81%) had mucous-type glands within 1.0 mm of the squamocolumnar junction. . . . The remaining 14 patients in the study group (19%) had mixed mucous/oxyntic type glands present within 1.0 mm of the squamocolumnar junction. . . . None of the patients (0%) had pure oxyntic glands (without mucous glands) in the columnar mucosa within 1.0 mm of the squamocolumnar junction. According to the terminology used in this chapter, this means that cardiac mucosa was absent at the squamocolumnar junction in 19% of patients. *Surely, this provides evidence that cardiac mucosa is not universally present in all patients, including a population of children with a significant prevalence of reflux.*

Sarbia M, Donner A, Gabbert HE. Histopathology of the gastroesophageal junction: A study on 36 operation specimens. Am J Surg Pathol 2002;26:1207–1212

This study looked at the histology of the junction in esophagogastrectomy specimens in a population with esophageal squamous carcinoma, a disease with no known association with reflux. The entire circumference of the junction was examined with impeccable histologic definitions that are identical to those we have suggested. Of 36 patients, cardiac mucosa was present in the entire circumference of the squamocolumnar junction in 20 patients, in part of the circumference of the junction in 15 patients, and entirely absent in one case.

The authors discuss the difference between their paper and our autopsy study in the number of patients that did not have cardiac mucosa: "the higher mean age of our study group (55 years) than that of Chandrasoma et al. (20 years) may at least partially explain why in our study cardiac mucosa was observed in a substantially higher percentage" (9, p. 1211). This explanation is reasonable. The other explanation is that the authors reported that over 50% of their patients showed evidence of reflux esophagitis in the squamous epithelium. This is likely to be a much higher percentage than in a normal autopsy population. It actually suggests a very high prevalence of reflux in this population given that the sensitivity of histologic evidence of reflux in the squamous epithelium is around 50% in patients with proven reflux.

The difference in the two studies is, however, in the percentage of patients without cardiac mucosa. Even with complete circumferential sampling, this study showed that cardiac mucosa is not universally present. *Surely, this shows that cardiac mucosa is not universally present even in an elderly population with esophageal disease including a high percentage of patients with reflux esophagitis.*

Zhou H, Greco MA, Daum F, Kahn E. Origin of cardiac mucosa: Ontogenic consideration. Pediatr Dev Pathol 2001;4:358–363 (27)

This study focused on 45 children in whom the entire squamocolumnar junction was examined by means of vertical sections. Of these, 15 were neonates under 1 month old, 13 were between 1 month and 1 year, and 17 were between 1 and 17 years. The authors defined a transitional zone between the esophageal squamous epithelium and gastric oxyntic mucosa. This transitional zone is absent if the squamous epithelium is in direct continuity with gastric oxyntic mucosa. When present, the transitional zone is divided into a mucous type if the glands contain only mucous cells (= cardiac mucosa) and a mixed type if the glands contained parietal cells in addition to mucous cells (= oxyntocardiac mucosa). Unfortunately, the authors gave no details regarding the length of these epithelia or their presence in the entire of part of the circumference of the squamocolumnar junction.

In the neonates only, there were two patients who did not have any glands. This was similar to the fetal cases and not seen after 1 month of age, making it highly likely that this was fetal columnar epithelium persisting after birth. One patient (5.5%) in this group had no transitional zone, with gastric oxyntic mucosa being present immediately distal to squamous epithelium. Cardiac mucosa was present in the transitional zone in 22 of the 45 patients (49%); cardiac mucosa was therefore absent in 24 of the 45 (53%), including the one patient who had only oxyntic mucosa. The other 21 patients had oxyntocardiac mucosa between the squamous and gastric oxyntic mucosa.

If one excludes patients under 1 year old, where there is a question regarding fetal and adult type tissues, there were 18 patients (the number in the table is one more than in the text, a minor error). Of these, 1 (5.5%) had no cardiac or oxyntocardiac mucosa, 10 (55%) had cardiac mucosa, and 7 (39%) had oxyntocardiac mucosa immediately distal to the squamous epithelium.

If one looks at the data in this article, they are remarkably similar to the findings in our autopsy study, in which there were no children under 1 year old. In fact, the data provide a more persuasive argument for cardiac mucosa being not universally present than did our study, because one of the children had no cardiac or oxyntocardiac mucosa at all. The authors conclusion, however, is completely the reverse: "Our study revealed that a transitional zone with the microscopic characteristics of cardiac mucosa was universally present at the squamo-columnar junction" (27, p. 358). This conclusion is clearly not supported by the data they provide; cardiac mucosa was present in only 22 of the 45 patients (49%). The way the authors get around this data is to include oxyntocardiac mucosa within their definition of cardiac mucosa as they stated at the end: "Cardiac mucosa contains both mucous and mixed glands" (27, p. 362). No it does not; it is the authors' definition of the transitional zone that they classify as having these pure mucous and mixed types of glands. They never define cardiac mucosa in their article, and in the end they equate cardiac mucosa with the transitional zone rather than use the standard definition of cardiac mucosa that is restricted to the mucous only epithelium devoid of parietal cells. The authors cannot change the histologic definition of cardiac mucosa at their whim to suit their conclusion. The incredible thing about this is that even when the authors include oxyntocardiac mucosa within the definition of cardiac mucosa, their data still do not support their conclusion that "cardiac mucosa was universally present at the squamocolumnar junction" because one of their patients had neither cardiac or oxyntocardiac mucosa. "Universal" cannot be tweaked to 94.5%; it means 100%. *Surely, the data in this paper provide evidence that cardiac mucosa is not present in everyone.*

In all the studies quoted here, there is evidence in the data presented in the articles that cardiac mucosa is not universally present at the squamocolumnar junction. The percentage of patients without cardiac mucosa tends to be greatest in normal populations and progressively decreases in populations with a high likelihood of having patients with significant reflux. A second factor is age; the prevalence of cardiac mucosa increases with increasing age of the population.

Of all the studies to date, only Kilgore et al.'s autopsy study reported that cardiac mucosa was universally present in all patients studied. The length of cardiac mucosa varied from 1 to 4 mm with a mean of 1.8 mm. This study is extremely important to review critically to see if there is a reason for this contrary finding.

Kilgore SP, Ormsby AH, Gramlich TL, et al. The gastric cardia: Fact of fiction? Am J Gastroenterol 2000;95:921–924

This is an autopsy study of pediatric patients with no propensity to gastroesophageal reflux disease during life; it is therefore an excellent control population, similar to that in our autopsy study. Unfortunately, as described in detail in the section relating to definitions, the authors used a flawed definition of cardiac mucosa: "All slides were evaluated in a blinded fashion by two observers looking for the presence of cardiac-type mucosa characterized by unequivocal PAS-positive mucous glands arranged in lobular configuration" (11, p. 922). This is not an acceptable definition of cardiac mucosa because "PAS-positive mucous glands arranged in lobular configuration" will be found in both cardiac and oxyntocardiac mucosa. In fact, when the only illustration that the authors provided of the junction is carefully examined, PAS-positive glands are restricted to only the deeper part of the epithelium and are more straight and not really lobulated immediately adjacent to the squamous epithelium. There appears to be a smaller cell population that is PAS negative in the superficial part of the gland that seems to be exactly like the parietal cells the authors illustrated in their figure of transitional (oxyntocardiac) mucosa. In fact, the only figure provided appears to show absence of cardiac mucosa at the junction, contrary to that reported (11).

Because this is the only study to report that cardiac mucosa is universally present, it cannot be accepted unless there is independent corroboration by other workers. This review includes no study that shows the complete absence of cardiac mucosa in all study patients. Kilgore et al.'s (11) study is therefore uncorroborated. There is good reason to believe that the cause for lack of corroboration is that the methods and data in the study are fatally flawed.

One would think that review of all the available data would lead to the inevitable conclusion that cardiac mucosa is not universally present in the human squamocolumnar junction. Nothing is further from the truth. Jain et al. concluded that the absence of cardiac mucosa at the esophagogastric junction in a majority of patients "challenge the traditional view," but this is in an abstract that never reached publication. Most other authors of the articles that are reviewed here express the viewpoint that cardiac mucosa is present as a normal structure at the junction. (a) Marsman et al., after reporting that 38% of patients did not have pure cardiac mucosa at the junction, concluded that cardiac mucosa is universally present at the junction. They did this by including oxyntocardiac mucosa under the definition of cardiac mucosa in their conclusion, although they defined cardiac mucosa correctly in their methods. (b) Glickman et al. (26), after reporting that cardiac mucosa was absent in 19% of their patients, concluded: In summary, a small amount of pure mucous-type glands is present in the cardia in most pediatric patients, a finding that supports the congenital origin for this type of epithelium. Most (81%) is certainly not universal. (c) Zhou et al. (27) reported that cardiac mucosa was universally present at the squamocolumnar junction despite the fact that they report data that showed cardiac mucosa in only 22 of 45 postnatal patients. (d) Kilgore et al. (11) concluded their flawed report: A short segment of cardiac mucosa was consistently present on the gastric side of the esophago-gastric junction, independent of gender and age. These results support the concept that the gastric cardia is present from birth as a normal structure. Kilgore et al.'s conclusion is supported by their findings, but the fact that their findings were based on flawed methods suggests that their conclusion is not scientifically valid.

The conclusions in these influential papers have led to the following analysis by de Hertogh et al. (21):

> Recently, the existence of cardiac mucosa as a normal structure has been questioned by Chandrasoma and colleagues. They suggest that cardiac mucosa is abnormal and results from metaplasia of the oesophageal squamous epithelium as a consequence of reflux damage. In contrast, Kilgore and colleagues, Zhou and colleagues, and Glickman and colleagues support the concept of the gastric cardia as a normal structure which is present from birth. (p. 791)

Chandrasoma et al. are still perceived as being a lone voice with an uncorroborated hypothesis with much contrary data. Nothing is further from the truth. It is the opinions of these authors that are different; careful review of the data in their studies provides excellent corroboration for the absence of cardiac mucosa in a significant number of normal people. It is a confounding fact that researchers will find cardiac mucosa is absent in 19% (Glickman et al. (26)), 51% (Zhou et al. (27)), and 38% (Marsman et al. (4)) of patients in their studies and conclude that cardiac mucosa is present as a normal structure in all patients. *Reading conclusions of articles in abstracts is not sufficient; it is critically important to see if the opinions expressed in the conclusions are supported by the facts presented in the report.*

Conclusion: Cardiac mucosa is frequently absent in humans; its prevalence increases with age and in study populations with a high likelihood of reflux.

WHAT IS OXYNTOCARDIAC MUCOSA, AND WHERE IS IT?

Oxyntocardiac mucosa is defined histologically as an epithelium in this region that contains glands composed of a mixture of mucous and parietal cells (Fig. 6.14). In his 1957 paper, Barrett (14) recognized that

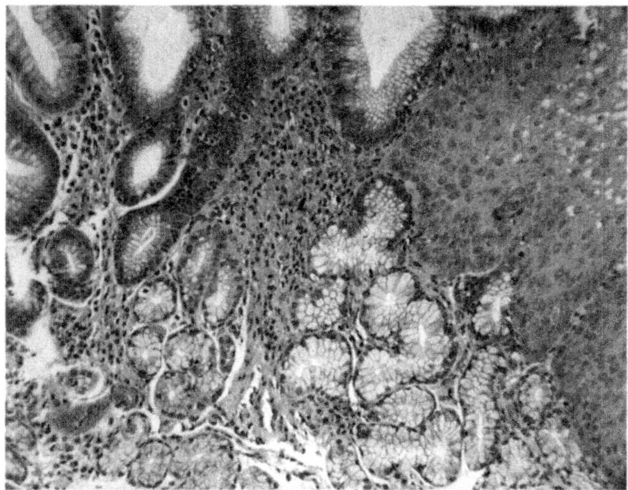

FIGURE 6.14 The squamocolumnar junction in a biopsy from an adult patient with heartburn showing squamous epithelium on the right, less than 1 mm of cardiac mucosa in the center, and oxyntocardiac mucosa with parietal cells on the left. Note that the true gastroesophageal junction is not shown; it is distal to the left edge of the picture where oxyntocardiac mucosa becomes gastric oxyntic mucosa. (Cross reference: Color Plate 3.8)

oxyntocardiac mucosa was present in the lower region of the columnar-lined esophageal segment. This was clearly established by Paull et al. (7) in 1976 when he showed that an epithelium with a mixture of parietal and mucous cells (Paull et al.'s "gastric fundal mucosa," which is equivalent to oxyntocardiac mucosa) was present above the location of the manometrically defined lower esophageal sphincter (Fig. 6.15). Many workers have shown that oxyntocardiac mucosa is one of the histologic elements in the metaplastic columnar esophageal epithelium (1).

The distinction between cardiac and oxyntocardiac mucosa is of fundamental importance. During embryologic development, parietal cells never occur in the esophagus (29). While cardiac mucosa resembles some normal fetal esophageal epithelia composed of only mucous cells, oxyntocardiac mucosa represents an aberrant epithelium for the esophagus. In the stomach, parietal cells develop, probably under the genetic influence of the Sonic Hedgehog gene whose expression is constant in gastric oxyntic mucosa (30). The presence of parietal cells in oxyntocardiac mucosa in the esophagus indicates that the epithelium is very likely under the Sonic Hedgehog (SHH) genetic signal. Studies have shown this to be true (31). Absence of parietal cells in cardiac mucosa indicates that this signal is not operative; the Sonic Hedgehog gene is not expressed in cardiac type mucosa in the esophagus.

The importance of recognizing oxyntocardiac mucosa and differentiating it from cardiac mucosa is

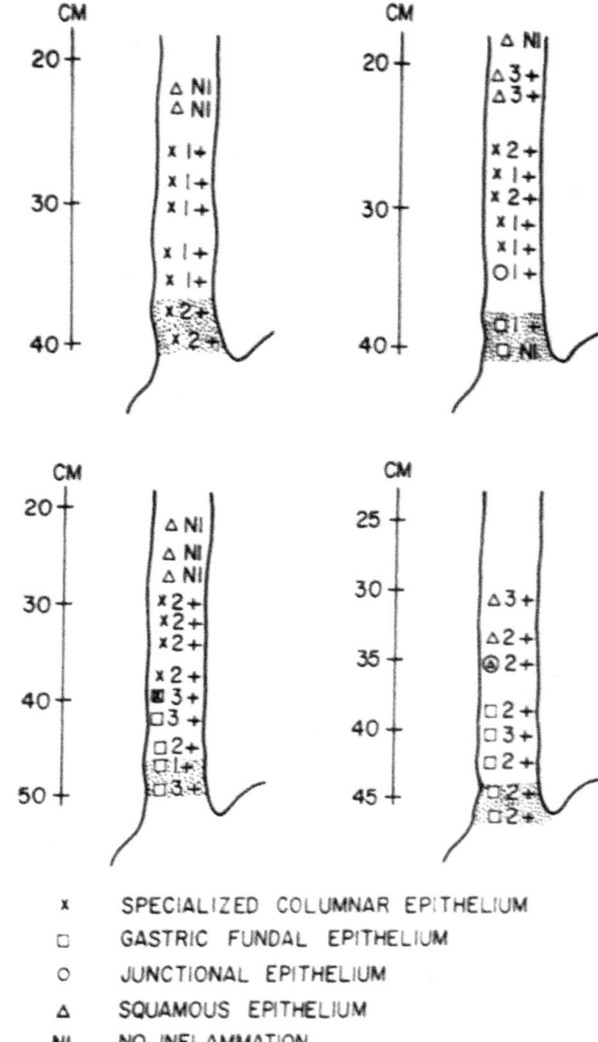

FIGURE 6.15 Mapping of epithelial types in columnar-lined esophagus. Junctional (= cardiac) and "gastric fundal" epithelia are shown above the upper limit of the lower esophageal sphincter, which was determined by manometry in this study. Reproduced with permission from Paull A, Trier JS, Dalton MD, Camp RC, Loeb P, Goyal RK. The histologic spectrum of Barrett's esophagus. N Engl J Med 1976;295:476–480.

that the expression of the genetic signal that dictates parietal cell development in esophageal metaplastic columnar epithelium appears to preclude the genetic changes that direct intestinal metaplasia. Intestinal metaplasia of the esophagus only occurs in cardiac mucosa and never in oxyntocardiac mucosa (7); this is of incredible significance and therefore an excellent reason to vigorously defend the histologic distinction between cardiac and oxyntocardiac mucosa.

Oxyntocardiac mucosa is anatomically located distal to cardiac mucosa in the metaplastic esophageal columnar-lined segment (Figs. 6.14 and 6.15). This was first indicated by Barrett and established in 1976 by Paull et al. (7) The three epithelial types in columnar-lined esophagus are distributed in a fairly consistent manner with intestinal metaplasia (when present) being mainly in the proximal segment and oxyntocardiac mucosa (which is almost always present) being in the most distal segment.

In 2001 (31), we mapped the three different epithelial types (cardiac, oxyntocardiac, and intestinal) in 32 patients with Barrett esophagus (1). In these patients who had a columnar-lined esophagus measuring between 2 to 16 cm, oxyntocardiac mucosa was not seen at the most proximal level immediately distal to the squamocolumnar junction in any patient. It was present at the most distal level in 16 patients and in the retrograde biopsy in another 3 patients. It was seen in more proximal biopsies when there was a long segment of columnar-lined esophagus where it was admixed with cardiac mucosa. It never showed intestinal metaplasia (i.e., a single foveolar-gland complex never had both parietal cells and goblet cells).

This study showed that oxyntocardiac mucosa is esophageal in location, found in the most distal part of the columnar-lined esophagus, usually interposed between cardiac mucosa proximally and gastric oxyntic mucosa distally. It transitions distally into gastric oxyntic mucosa at the true gastroesophageal junction (Fig. 6.14).

Oxyntocardiac mucosa is almost universally present at the junction in adults. Only rare cases have been reported where it has not been found at all if adequate sampling has been achieved. In our autopsy study, 21 of 72 patients had neither oxyntocardiac nor cardiac mucosa in the retrospective series of patients where only one vertical section was examined. In the prospective part of the study where the entire junction was examined by multiple sections, all patients had oxyntocardiac mucosa. However, in this group, oxyntocardiac mucosa was not present in the entire circumference of the junction; it was absent in some part of the circumference in 9 of the 18 patients (50%). In Zhou et al.'s study, which also examined the entire gastroesophageal junction, one patient was reported as having neither cardiac nor oxyntocardiac mucosa.

Oxyntocardiac mucosa may occur without cardiac mucosa. In our autopsy study, the 10 of 18 patients who did not have cardiac mucosa had only oxyntocardiac mucosa interposed between squamous epithelium and gastric oxyntic mucosa. In these patients, the mean length of oxyntocardiac mucosa varied between 0.25 and 1.92 mm. In Zhou et al.'s study, 7 of 18 patients over the age of 1 year had only oxyntocardiac mucosa in the transitional zone between squamous and gastric oxyntic epithelia. In endoscopic studies where sampling is less complete, oxyntocardiac mucosa is seen less than universally.

Our biopsy protocol permits the classification of patients into different lengths of histologically defined columnar-lined esophagus and permits the mapping of each epithelial type within this segment (23). In 148 of 959 patients who had a columnar-lined esophagus measuring 1 cm or greater, oxyntocardiac mucosa was associated with cardiac mucosa in all but one patient. In the group of 811 patients whose columnar-lined esophagus measured less than 1 cm, oxyntocardiac mucosa was present as the only columnar epithelium between squamous and gastric oxyntic in 158 (19.4%) of patients. This study showed that oxyntocardiac mucosa exists as the only epithelial type between squamous epithelium and gastric oxyntic mucosa only in patients who have very short (less than 1 cm) cephalad migration of the squamocolumnar junction. It is likely that the majority of these patients are perceived as being endoscopically normal.

It appears, then, that oxyntocardiac mucosa is present in nearly everyone. In patients with only oxyntocardiac mucosa without cardiac mucosa, oxyntocardiac mucosa is commonly seen in only a part of the circumference of the gastroesophageal junction. In these patients, the remainder of the junction shows squamous epithelium transitioning directly into gastric oxyntic mucosa. Oxyntocardiac mucosa occurs without cardiac mucosa only in patients with a columnar-lined esophagus that measures less than 1 cm (23). When the length of columnar-lined esophagus exceeds 1 cm, oxyntocardiac mucosa occupies the most distal part of the columnar-lined segment with cardiac mucosa (with and without intestinal metaplasia) being present more proximally.

We must look upon oxyntocardiac mucosa as an extremely common marker for reflux-induced damage of the distal esophagus. The fact that it is present in nearly everyone is a testament to the ubiquity of gastroesophageal reflux in humans. The situation is analogous to the finding of aortic atherosclerosis in the Western world; nearly everyone over the age of 40 years will have some evidence of it as a testament to the frequency of factors that produce it. We will show later that oxyntocardiac mucosa, though abnormal and caused by reflux, is a benign epithelium that has little clinical significance. We classify a patient with oxyntocardiac mucosa as the only metaplastic epithelium in the esophagus as having "compensated reflux."

HOW MUCH CARDIAC AND OXYNTOCARDIAC MUCOSA ARE PRESENT?

Barrett (14), in describing columnar-lined esophagus in 1957, wrote: "When the lower esophagus is found to be lined by columnar cells, the abnormally placed mucous membrane extends upwards from the esophago-gastric junction in a continuous, unbroken sheet. . . . The extent of the anomaly varies from a few centimeters to the upper esophagus" (p. 883). Both Barrett and Allison before him reported that the mucosa lining this segment of the esophagus was devoid of parietal cells (i.e., cardiac mucosa) in the upper part and one that had a few parietal cells (oxyntocardiac mucosa) in its more distal extent. They were seeing the most severe cases with long segments of columnar epithelium frequently reaching up to aortic arch. Hayward, also seeing a very abnormal population, sought to limit the number of patients with "pathological" columnar-lined esophagus by defining (incorrectly) that the lower 2 cm of the esophagus was lined normally by junctional (= cardiac) mucosa. Hayward felt no need to define the minimum amount of cardiac mucosa that was found; he was only interested in defining the maximum amount to define the abnormal state of columnar-lined esophagus.

Because it was universally accepted that cardiac mucosa was a normal epithelium between the squamous epithelium and gastric oxyntic mucosa (first straddling the gastroesophageal junction [Fig. 6.1b] and then in the proximal stomach [Fig. 6.1c]), there was never any need to define the lower limit of cardiac mucosa that was normal. It was therefore a dramatic change when we suggested that cardiac mucosa was never normal (i.e., that the lower limit of normal cardiac mucosa was zero) (Fig. 6.1e). This was the first time anyone had attempted to define the normal lower limit for cardiac mucosa; it was also such a radical concept that there was universal disbelief that it could be true.

This initial disbelief that cardiac mucosa could be completely absent from the gastroesophageal junction has never disappeared despite the fact that every study but the flawed autopsy study of Kilgore et al. has shown that cardiac mucosa is absent at least in some patients if one excludes fetuses (where transient developing fetal columnar epithelia fall within the strict definition of "cardiac mucosa" and are mistakenly called as such by some authorities).

The present belief among pathologists is confusing and difficult to define. Kilgore et al.'s (10) autopsy study, which indicated that they found 1 to 4 mm (mean 1.8 mm) of cardiac mucosa in all people, appears to be the one that most pathologists believe. They give no reason why the data in Kilgore et al. should be believed over the opposing data in our autopsy study (1). Kilgore et al.'s (11) study appeared in the gastroenterology literature where the possibility of reviewers with less than complete knowledge of histology is greater. While using Kilgore et al.'s study to define the extent of what he calls "true" gastric cardia, Dr. Odze from the Harvard group (32) stated that this maximum length of "true" gastric cardia can be lined by both cardiac and oxyntocardiac mucosa. He even mentioned that it is acceptable in rare cases that this "true" cardia can be lined by gastric oxyntic mucosa without any cardiac mucosa. So it appears that we are reaching the inevitable conclusion that the only acceptable length of the lower limit of cardiac mucosa is zero. The truth will finally prevail, but not yet. The interpretation of pathologic material from this region will change from its highly complicated present state to a very simple system only when we recognize that cardiac mucosa is always abnormal and always represents metaplastic esophageal columnar epithelium (33) (Fig. 6.16).

The problem is not a small one that befits an epithelium that even the naysayers agree has a 0- to 4-mm extent. If cardiac mucosa, however small in extent, is regarded as a normal structure at the squamocolumnar junction, it cannot be regarded as an abnormal epithelium when seen in a biopsy unless quantitated. Thus, today's pathologists will say that 1 to 4 mm of cardiac mucosa is normal as a justification for calling cardiac mucosa in a biopsy normal. One should develop a vision of 1 mm: it is approximately 30 epithelial cells and probably four to five foveolar-gland complexes. If a person has only 1 mm of cardiac mucosa at the junction, it will not be seen unless the exact squamocolumnar junction is present in the biopsy, as Marsman et al. illustrated when they showed that while 62% of patients whose biopsy had the actual junction had cardiac mucosa, only 2% of patients who had biopsies that showed only glandular mucosa had cardiac mucosa. This shows that if the amount of cardiac mucosa is 1 mm, the biopsy fails to sample cardiac mucosa if it misses the squamocolumnar junction by 30 cells or five foveolar pits.

In many biopsies that show cardiac mucosa, there is more than 1 to 4 mm of cardiac mucosa; at least these should be called quantitatively abnormal, *but they are not* (Fig. 6.16). It is still almost universal among pathologists to regard all cardiac mucosa in a biopsy as being normal. With all the evidence in the literature that shows that the amount of cardiac mucosa they see is abnormal, they still regard this as a normal finding. By calling it "normal," they are not required to explain how it came to be there and they can hide behind the

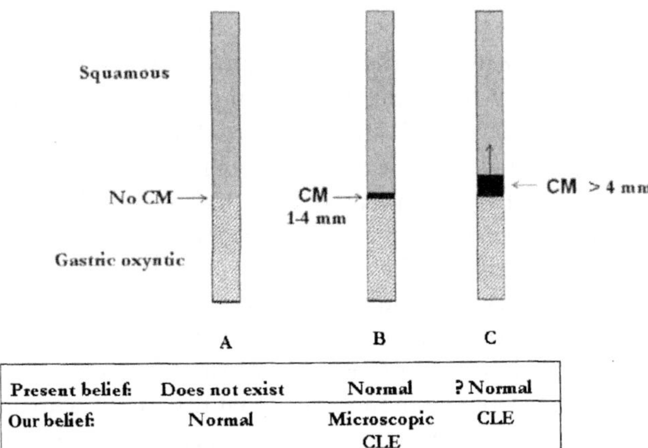

Present belief:	Does not exist	Normal	? Normal
Our belief:	Normal	Microscopic CLE	CLE

FIGURE 6.16 Diagrammatic representation of presently accepted and our viewpoint regarding the finding of various amounts of cardiac mucosa between squamous epithelium and gastric oxyntic mucosa. (a) It is presently believed that everyone has some cardiac mucosa; we do not believe this. (b) It is presently believed that 1 to 4mm of cardiac mucosa is normal; we believe this is the earliest microscopic stage of reflux-induced columnar-lined esophagus. (c) More than 4mm of cardiac mucosa is ignored unless intestinal metaplasia is present; we believe this is increasingly severe reflux-induced columnar-lined esophagus.

normalcy. This point of view is well stated by Glickman et al. (26): "Our finding of an association between length of mucosa occupied by pure mucous glands and active esophagitis suggests that injury and repair related to gastroesophageal reflux disease may contribute to expansion of the zone occupied by cardia-type mucous glands" (p. 1038).

The present state of the art as accepted by the medical community is as follows: About 1 to 4mm of cardiac mucosa exists normally at the squamocolumnar junction (Figs. 6.1b and 16.1d). With reflux-induced damage, this zone of cardiac mucosa expands, presumably by metaplasia of squamous epithelium, and can involve a large part of the 25-cm length of the esophagus in some cases. However, we shall never call cardiac mucosa abnormal because the most distal 1 to 4mm of a long segment of cardiac mucosa is normal. This borders on the ridiculous; surely, it is obvious that the most distal 1 to 4mm of cardiac mucosa that everyone calls "normal" is the first microscopic stage of the process that results in long segments of columnar metaplasia of the esophagus (Fig. 6.16). The degree of absurdity worsens because everyone agrees (32) that the "normal" cardiac mucosa that occurs in the proximal stomach *is histologically indistinguishable from the cardiac mucosa that is found in columnar-lined esophagus.*

In fact, Glickman et al.'s article from the Harvard group (26) proves that the viewpoint that 1mm of cardiac mucosa is normally present is incorrect. In their study of pediatric patients with a significant number suffering from clinical reflux, they reported that 19% had no cardiac mucosa. In the remainder, those with less than 1mm of cardiac mucosa had less evidence of reflux than those with greater than 1mm of cardiac mucosa. In Glickman et al.'s conclusion, the statement regarding the "normal" zone of cardiac mucosa from which expansion occurs is as follows: "a small amount of pure mucous-type glands is present in the cardia in *most* pediatric patients" (p. 1039). *Most is not all; in fact in their own data, 19% of patients had zero cardiac mucosa. The expansion that occurs with reflux disease starts from zero, not from "a small amount."*

WHAT DOES THE PRESENCE/ABSENCE AND AMOUNT OF CARDIAC MUCOSA MEAN?

The presence of histologically defined cardiac mucosa, even in microscopic amounts, has always been shown to correlate with the presence of clinical evidence of reflux disease (26), abnormal gastroesophageal reflux, and lower esophageal sphincter abnormalities (25). The absence of cardiac mucosa at the junction has always been shown to correlate with a more normal physiologic state. If this is true, cardiac mucosa cannot be a normal structure. Surely, the absence of a normal structure must be associated with a state that is more pathologic and less physiologic than its presence. We always take the corollary of the human head. The reason why the human head is a normal structure is not because it is present in everyone; it is not. The reason is that the absence of the human head in anencephaly is a less physiologic and more pathologic state than the presence of the human head. If cardiac mucosa were normal, its absence should be associated with a more pathologic state when compared to a person who has cardiac mucosa. If the reverse is true, it is proof beyond doubt that cardiac mucosa is an abnormal structure.

Let us review some of the same studies to see whether there are any data within them that answer this important question.

Oberg S, Peters JH, DeMeester TR, et al. Inflammation and specialized intestinal metaplasia of cardiac mucosa is a manifestation of gastroesophageal reflux disease. Ann Surg 1997;226:522–532

This article is from our group, and Dr. Chandrasoma was the pathologist who looked at all the slides. We looked at 334 patients who had a normal endoscopic appearance (i.e., no visible columnar-lined esophagus). These patients had biopsies according to the DeMeester protocol (Fig. 6.10); a minimum of five biopsies were taken from the region of the proximal limit of the rugal folds and squamocolumnar junction, which were endoscopically coincidental. Of the 334 patients studied, 246 (74%) had either cardiac or oxyntocardiac mucosa in their biopsies; the other 88 patients (26%) had only squamous and gastric oxyntic mucosa. The patients with cardiac and oxyntocardiac mucosa had significantly higher probability of an abnormal pH score, a hiatal hernia, erosive esophagitis, and a structurally defective lower esophageal sphincter compared with the patients who did not have these mucosal types (Fig. 6.17). The absence of cardiac or oxyntocardiac mucosa was associated with a more physiologic state than its presence. *This is not compatible with cardiac mucosa being a normal structure and is proof that cardiac mucosa is abnormal.*

Glickman JN, Fox V, Antonioli DA, Wang HH, Odze RD. Morphology of the cardia and significance of carditis in pediatric patients. Am J Surg Pathol 2002;26:1032–1039

This study correlated the histologic findings at the squamocolumnar junction with the clinical and pathological features. The authors reported that patients who had pure mucous glands located both within and >1 mm from the squamocolumnar junction showed . . . an increased prevalence of active esophagitis (55% versus 21%, p = 0.04) compared with patients who had a combination of pure mucous glands within 1 mm of the squamocolumnar junction and either mixed mucous/oxyntic glands or pure oxyntic glands >1 mm from the squamocolumnar junction. This is a convoluted way of saying that patients with more than 1 mm of cardiac mucosa had a significantly greater association with reflux than those with less than 1 mm of cardiac mucosa (Fig. 6.18). This association was present despite the fact that the authors used active esophagitis, defined as the presence of histologic changes of reflux in the squamous epithelium, to assess prevalence of reflux. Active esophagitis in the squamous epithelium is a notoriously low sensitivity indicator for reflux. Calculating from the data given, 55% of the 32 patients (18 patients) with cardiac mucosa of more than 1 mm had active esophagitis. Of the total of 74 patients in the entire study group, the prevalence of active esophagitis was 28 (38%). This means that only 10 patients without cardiac mucosa greater than 1 mm had active esophagitis. Unfortunately, the authors did not tell us whether these patients had less than 1 mm of cardiac mucosa or noncardiac mucosa.

According to the authors' data, the presence of cardiac mucosa correlated at a 1 mm length with an increased prevalence of reflux. It is important to note that 21% of these pediatric patients with cardiac mucosa less than 1 mm had evidence of active esophagitis in the squamous epithelium (this assumes that the prevalence of active esophagitis was equal in the less than 1 mm cardiac mucosa group and those without cardiac mucosa at all, because this group was

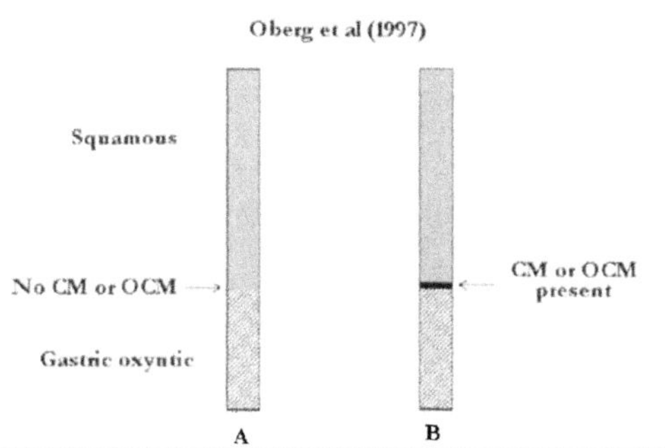

Number of cases:	88/334 (26%)	246/334 (74%)
% time pH < 4:	1.1 +/- 4.6	6.0 +/- 7.4 (p = < 0.01)
LES pressure (mm.Hg):	13.2 +/- 12.8	8.0 +/- 8.0 (p = < 0.01)
Abd length of LES (cm):	1.6 +/- 1.1	1.0 +/- 1.2 (p = < 0.01)
% Defective LES	27.2	62.3 (p = < 0.01)

FIGURE 6.17 Oberg et al. (25) study showing that patients who have cardiac or oxyntocardiac mucosa (solid black) at the squamocolumnar junction have significantly more acid exposure defined by an abnormal 24-hour pH test and a greater likelihood of having esophageal sphincter abnormalities than a patient who does not have cardiac or oxyntocardiac mucosa between squamous epithelium (solid gray) and gastric oxyntic mucosa (striped).

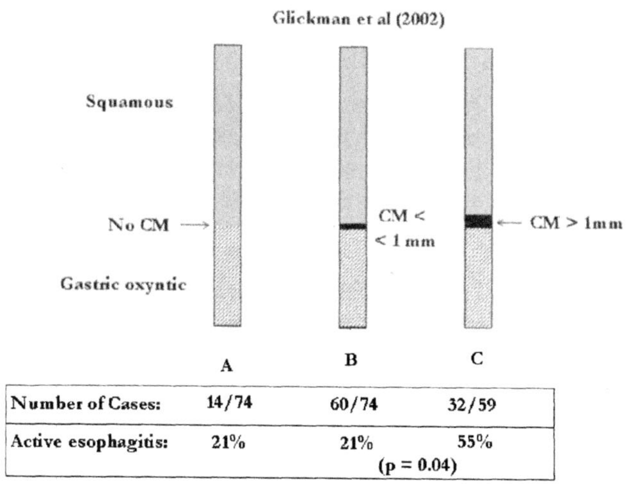

Number of Cases:	14/74	60/74	32/59
Active esophagitis:	21%	21%	55% (p = 0.04)

FIGURE 6.18 Glickman et al. (26) study showing that children who have more than 1 mm of cardiac mucosa at the junction (c) have more evidence of active esophagitis than those with less than 1 mm of cardiac mucosa (b) and no cardiac mucosa (a).

considered as a whole). The fact that the group with more than 1 mm of cardiac mucosa had significantly *more* reflux than those with less than 1 mm of cardiac mucosa does not mean that the latter group did not have reflux; in fact, they did. *The presence of cardiac mucosa correlates with reflux severity at lengths as short as 1 mm.*

The only other study in the literature that correlated the presence of small amounts of cardiac mucosa in the junction with reflux was that by Sarbia et al. (9). This study correlated the length of cardiac mucosa, oxyntocardiac mucosa, and the combined length of the two with histologic evidence of GERD in the squamous epithelium. They reported that the minimal length of cardiac mucosa, . . . oxyntocardiac mucosa, and junctional mucosa [which they define as CM + OCM] tended to be larger in the presence of GERD. This did not reach statistical significance because of the small number of patients, but the trend is similar to the statistically significant finding of Glickman et al. that *very short lengths of cardiac mucosa are associated with reflux.*

These studies collectively show the following:

1. Patients who are normal endoscopically and who do not have cardiac or oxyntocardiac mucosa at their junction in their biopsies will have a lower likelihood of having abnormal acid exposure by a 24-hour pH study, a lower likelihood of associated reflux changes in the squamous epithelium, and a lower likelihood of having lower esophageal sphincter abnormalities by manometry than patients who have cardiac or oxyntocardiac mucosa (Fig. 6.17).
2. Patients with cardiac mucosa that measures greater than 1 mm in length have a greater association with reflux symptoms and evidence of reflux in the squamous epithelium than patients with less than 1 mm of cardiac mucosa (Fig. 6.18).

Conclusion: The proof that cardiac mucosa is associated with reflux at the smallest lengths should be obvious to any scientifically trained individual with an objective mind.

WHAT DOES INCREASING LENGTH OF CARDIAC MUCOSA MEAN?

Cardiac mucosa and its companion histologic variants of oxyntocardiac mucosa (= cardiac mucosa with parietal cells) and intestinal metaplasia (= cardiac mucosa with goblet cells) are the components of columnar-lined esophagus. There is ample evidence in the literature without any dissent that there is a correlation between the length of columnar-lined esophagus and the severity of reflux (Fig. 6.19). Csendes (34) clearly showed this in 1993. In this study, the location of the lower esophageal sphincter measured by manometry and the location of the squamous columnar junction measured by endoscopy were determined in 109 healthy controls and 778 patients with different degrees of endoscopic esophagitis.

The total length of the esophagus, measured by the location of the distal end of the lower esophageal sphincter, was similar in all patients; however, the location of the squamous columnar junction extended more proximally and was related to the increasing severity of endoscopic esophagitis. The manometric defects at the cardia were more frequent in severe esophagitis ($p < 0.001$). The authors concluded that their results suggest that, during the course of reflux esophagitis, the squamocolumnar junction is displaced proximally.

The best study showing the relationship of cardiac mucosal length to likelihood and severity of gastroesophageal reflux is that of Glickman et al. (26), reviewed earlier. In this study, the presence of more than 1 mm of cardiac mucosa had a significantly greater association with reflux than those with less than 1 mm of cardiac mucosa. It is important to note that 21% of patients with cardiac mucosa that is less than 1 mm had evidence of active esophagitis in the squamous epithelium. This study showed that the

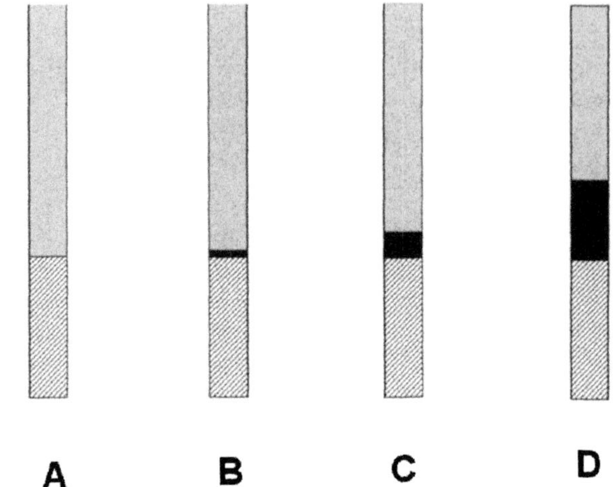

FIGURE 6.19 Diagram showing the relationship of increasing lengths of columnar metaplastic epithelium (solid black) interposed between the squamocolumnar junction (solid gray) and gastric oxyntic mucosa to severity of acid exposure by 24-h pH test. The severity of acid exposure increases as one passes from (a) normal without cardiac or oxyntocardiac mucosa to (d) long (>2 cm) segment of columnar lined esophagus. (b) represents patients who are normal on endoscopy who have CM/OCM in biopsies, and (c) patients with a columnar lined esophagus measuring less than 2 cm.

relationship between reflux and cardiac mucosal presence began at less than 1mm of cardiac mucosal length.

This relationship continues as the length of columnar-lined esophagus increases. Our study (17) showed that patients with a columnar-lined esophageal length of more than 2cm have a highly significantly greater degree of reflux as shown by a 24-hour pH study than patients with less than 2cm of columnar-lined esophagus. This was a study of 71 patients who had multiple measured biopsies according to our standard protocol. Ambulatory 24-hour pH studies were available in 53 patients. The biopsies were used to map the region by epithelial type, permitting assessment of the length of the various epithelial types present. Cardiac mucosa was present in 68 of the 71 patients, whereas the other 3 had oxyntocardiac mucosa, 22 patients had a CM + OCM length of greater than 2cm, whereas the other 49 patients had a CM + OCM length less than 2cm. Intestinal metaplasia was present in all 22 patients who had a CM + OCM length of greater than 2cm and 20 of 49 patients with a CM + OCM length less than 2cm.

Correlation of the CM + OCM length with acid exposure as measured in the 24-hour pH study showed that 15 patients with CM + OCM length of greater than 2cm had a markedly increased acid exposure measured as percent time pH < 4 (median 30.6; range: 3.6 to 54.3; interquartile range: 16.1, 41.1). The levels were elevated but at a significantly lower level in the 38 patients with CM + OCM lengths less than 2cm (median 7.1; range: 1.1 to 39.7; interquartile range: 5.2, 12.1). It should be noted that the level of significance is extremely high; there is no overlap between the interquartile range values. When one compares the abnormal 24-hour pH values observed in the patients of this study, who all had an endoscopically visible columnar-lined esophagus, with those of Oberg et al.'s study where there was no visible columnar-lined esophagus at endoscopy, there is a progressive increase in acid exposure as measured by the 24-hour pH test as the length of cardiac + oxyntocardiac mucosa increases (Fig. 6.19).

Another important finding in this study was the fact that cardiac mucosa was present in 68 of 71 patients (96%). These patients all had multiple measured level biopsies; these are only done in our biopsy protocol when there is a visible columnar-lined segment between the proximal limit of the rugal folds and squamous epithelium. This shows that the vast majority of patients with a visible segment of columnar-lined esophagus will have cardiac mucosa in their biopsies (23). It is easy to understand how the dogma relating to the normal presence of cardiac mucosa lingers. The American Gastroenterology Association practice guidelines recommend that biopsies only be taken when an abnormal columnar-lined segment of esophagus is seen endoscopically. If this recommendation is followed, the universal presence of cardiac mucosa becomes a self-fulfilling prophecy; nearly all patients undergoing biopsy will have cardiac mucosa. It is only when an aggressive biopsy protocol that biopsies endoscopically normal individuals is followed that patients without cardiac mucosa will be encountered.

A HUMAN EXPERIMENT

Patients who undergo esophagogastrectomy with anastomosis of the gastric body to the proximal esophagus represent a human experiment to evaluate mucosal changes occurring in reflux. Because the lower esophageal sphincter has been removed, these patients almost invariably have gastroesophageal reflux. Because the entire gastroesophageal junctional region has been removed surgically, there is no possibility of the existence of any "normal" cardiac mucosa after surgery. The anastomosis is between gastric oxyntic mucosa and esophageal squamous epithelium. Follow-up sequential biopsy in these patients will show how the esophageal squamous and gastric oxyntic epithelia are altered by reflux.

Literature Review

Dresner SM, Griffin SM, Wayman J, Bennett MK, Hayes N, Raimes SA. Human model of duodenogastro-oesophageal reflux in the development of Barrett's metaplasia. Br J Surg 2003;90:1120–1128 (35).

This excellent study from the Northern Oesophago-gastric Cancer Unit at the Royal Victoria Infirmary in Newcastle upon Tyne looked at what happened to 40 patients treated for Barrett's adenocarcinoma and high-grade dysplasia (26 patients) and squamous neoplasia (14 patients) with an intrathoracic oesophagogastrostomy (i.e., gastric pullup) after subtotal oesophagectomy. The authors reported that the patients experienced profound duodenogastro-oesophageal reflux, verified in 30 patients by combined 24-hour ambulatory pH and bilirubin monitoring. The authors pointed out that these patients, whose operation abolishes the normal antireflux mechanism, represent "an ideal model in which to study the early events in the pathologenesis of Barrett's metaplasia from normal human squamous epithelium subjected to duodenogastroesophageal reflux" (p. 1121).

Serial endoscopic assessment and systematic biopsy at the oesophagogastric anastomosis was done over a 36-month period. These patients had a total of 130 (median 3, range 1 to 8) examinations. The authors used an unconventional definition of Barrett esophagus, but their clear description permits clear interpretation of the data: "The definition of Barrett's mucosa included both specialized and non-specialized columnar epithelium from the oesophageal remnant. The former epithelium was identified by the presence of goblet cells and is referred to as 'intestinal metaplasia-type' Barrett's mucosa. Non-specialized epithelium, which has no goblet cells and has features similar to those of the mucosa of the gastric cardia or fundus, is referred to as 'cardiac-type' or 'fundic-type' Barrett's mucosa respectively" (pp. 1121–1122).

At the end of the study, 7 patients had normal squamous epithelium, 14 had reflux esophagitis of varying grades, and 19 patients had esophageal columnar epithelium (Fig. 6.20). Biopsies from columnar regeneration revealed cardiac-type epithelium in 10 patients and intestinal metaplasia in 9. Endoscopically, the columnar epithelium was seen as single tongues (6), multiple tongues (2), noncircumferential patches (5), and circumferential segments (6). The measured extent of columnar epithelium ranged from 0.5 to 3.5 cm (median 1.9 cm). Coexistent reflux esophagitis in the squamous epithelium was present in 14 of 19 patients with columnar metaplasia.

At the index endoscopy in this group, only 9 patients showed columnar metaplasia. In the other 10 patients, metaplastic progression was observed over 6 to 36 months. In these patients there was evidence of squamous esophagitis before progression to columnar metaplasia.

The initial detection of columnar epithelium in the esophagus was made a median of 14 (range 3 to 118) months after surgery. Of the 9 patients with columnar mucosa at the index endoscopy, 7 had cardiac-type mucosa and 2 had intestinal metaplasia. All 10 patients who were found to have developed columnar epithelium during follow-up initially had cardiac-type epithelium; intestinal metaplasia was identified only later in 7. The median time to the development of cardiac-type mucosa was 14 (range 3 to 22) months, whereas intestinal metaplasia was first detected significantly later at a median of 27 (range 11 to 118) months (Fig. 6.20).

The gastric mucosa below the anastomosis demonstrated "gastric body epithelium in all cases, with various degrees of quiescent and active gastritis" (pp. 1123–1124).

The incidence of Barrett's metaplasia was similar irrespective of the histological subtype of the resected tumour. Patients with esophageal columnar epithelium had significantly higher acid ($p = 0.015$) and bilirubin ($p = 0.011$) reflux compared with patients who had only squamous epithelium at the end of the study period.

The authors concluded that severe duodeno-gastro-esophageal reflux occurring after subtotal oesophagectomy causes a sequence of changes in the esophageal squamous epithelium, which shows a sequential temporal progression from reflux esophagitis to columnar metaplasia of cardiac type without goblet cells and eventually to intestinal metaplasia.

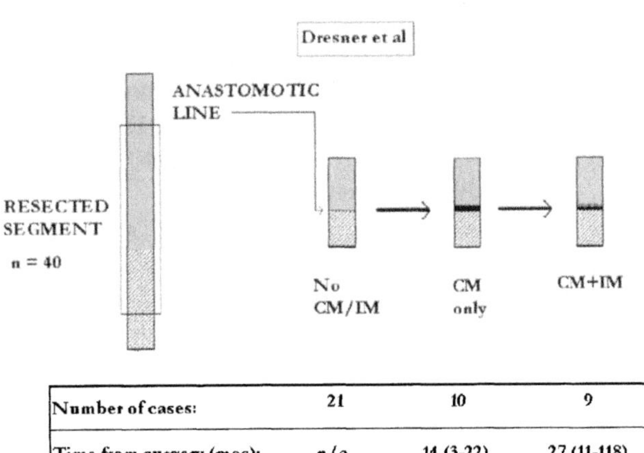

Number of cases:	21	10	9
Time from surgery (mos):	n/a	14 (3-22)	27 (11-118)

FIGURE 6.20 Dresner et al. (35) showing changes above the squamo-oxyntic anastomotic line after resection and esophagogastrostomy. Columnar-lined esophagus developed above the anastomotic line during the study period in 19 of 40 patients; 10 had cardiac mucosa, and 9 had intestinal metaplasia. Seven of 9 patients who developed intestinal metaplasia showed a sequential progression from cardiac mucosa in serial biopsies.

This study is, in essence, a human experiment that proves that cardiac mucosa results from squamous metaplasia as a consequence of reflux. These patients had their entire gastroesophageal junctional region removed at surgery with an anastomosis that was made between squamous lined-esophagus and gastric body mucosa. They had no possibility of having any cardiac mucosa after the surgery. Reflux affected only the squamous epithelium above the anastomotic line. This showed a sequential change characterized by inflammation (reflux esophagitis) columnar metaplasia into cardiac mucosa and, finally, intestinal metaplasia. Cardiac mucosa did not develop in the gastric body mucosa below the anastomosis. The columnar metaplasia of the squamous epithelium was shown to be the result of reflux, whose presence was objectively assessed by 24-hour pH and bilirubin monitoring and shown to be significantly higher in those patients with columnar metaplasia compared with those that did not have columnar metaplasia at the end of the study.

SUMMARY STATEMENT REGARDING CARDIAC MUCOSA

In summary, based on the preceding review of the literature and our experience since the early 1990s, which are concordant:

1. Cardiac mucosa is not universally present at the squamocolumnar junction in fully developed humans (over 1 year old).
2. When present, cardiac mucosal length is less than 5 mm in the majority of patients, and often less than 1 mm.
3. The prevalence and length of cardiac mucosa tend to increase with age.
4. The prevalence of cardiac mucosa is lowest in normal autopsy populations (Chandrasoma et al.—44%) and biopsies obtained from a population of patients with a bias toward having a normal gastroesophageal junction at endoscopy (Jain et al.—35%).
5. The prevalence of cardiac mucosa is higher in populations where there are patients with clinical evidence of reflux. When endoscopy is normal, cardiac mucosa is present in 60% to 62% (Chandrasoma et al., Marsman et al.). When endoscopically visible columnar epithelium is present, the prevalence of cardiac mucosa approaches 100% (23).
6. The presence of cardiac mucosa at the junction, even when endoscopy is normal and the length of cardiac mucosa is in the 1-mm range, has a strong association with gastroesophageal reflux.
7. The length of cardiac mucosa has a direct correlation with severity of reflux.
8. When evaluated by objective criteria for defining the gastroesophageal junction such as the peritoneal reflection, the angle of His, or the association with esophageal submucosal glands, cardiac mucosa is esophageal in location, not gastric.
9. In patients who have had esophagogastrectomy, cardiac mucosa develops by metaplasia of squamous epithelium above the anastomotic line. This columnar metaplasia correlates with the occurrence of duodeno-gastro-esophageal reflux.

If one combines these into a conclusive statement: cardiac mucosa is an abnormal epithelium that occurs in the esophagus due to reflux-induced columnar metaplasia of the stratified squamous epithelium. *The hypothesis is proved, as demonstrated here.*

References

1. Chandrasoma PT, Der R, Ma Y, et al. Histology of the gastroesophageal junction: An autopsy study. Am J Surg Pathol 2000; 24:402–409.
2. Jain R. Aquino D, Harford WV, Lee E, Spechler SJ. Cardiac epithelium is found infrequently in the gastric cardia. Gastroenterology 1998;114:A160 (Abstract).
3. Chandrasoma P, Makarewicz K, Wickramasinghe K, Ma YL, DeMeester TR. A proposal for a new validated histologic definition of the gastroesophageal junction. Human Pathol 2006;37: 40–47.
4. Marsman WA, van Sandyck JW, Tytgat GNJ, ten Kate FJW, van Lanschot JJB. The presence and mucin histochemistry of cardiac type mucosa at the esophagogastric junction. Am J Gastroenterol 2004;99:212–217.
5. Gulmann C, Rathore O, Grace A, Hegarty H, O'Grady A, Leader M, Patchett S, Kay E. "Cardiac-type" mucosa and carditis are both associated with *Helicobacter pylori*-related gastritis. Eur J Gastroenterol Hepatol 2004;16(1):69–74.
6. Lagergren J, Bergstrom R, Lindgren A, Nyren O. Symptomatic gastroesophageal reflux as a risk factor for esophageal adenocarcinoma. N Engl J Med 1999;340:825–831.
7. Paull A, Trier JS, Dalton MD, Camp RC, Loeb P, Goyal RK. The histologic spectrum of Barrett's esophagus. N Engl J Med 1976;295:476–480.
8. Allison PR, Johnstone AS. The oesophagus lined with gastric mucous membrane. Thorax 1953;8:87–101.
9. Sarbia M, Donner A, Gabbert HE. Histopathology of the gastroesophageal junction. A study on 36 operation specimens. Am J Surg Pathol 2002;26:1207–1212.
10. Chandrasoma PT, Lokuhetty DM, DeMeester, TR, et al. Definition of histopathologic changes in gastroesophageal reflux disease. Am J Surg Pathol 2000;24:344–351.
11. Kilgore SP, Ormsby AH, Gramlich TL, et al. The gastric cardia: Fact or fiction? Am J Gastroenterol 2000;95:921–924.
12. Chandrasoma P. Letter to the editor: Re: Kilgore et al., The gastric cardia: Fact of fiction? Am J Gastroenterol 2000;95: 2384–2385.
13. Allison PR. Peptic ulcer of the esophagus. Thorax 1948;3: 20–42.
14. Barrett NR. The lower esophagus lined by columnar epithelium. Surgery 1957;41:881–894.
15. Hayward J. The lower end of the oesophagus. Thorax 1961;16: 36–41.
16. De Nardi FG, Riddell RH. Esophagus. *In* Sternberg SS (ed) Histology for pathologists, second edition. Lippincott-Raven Publishers, Philadelphia, 1997. pp 461–480.
17. Owen DA. Stomach. *In* Sternberg SS (ed) Histology for pathologists, second edition. Lippincott-Raven Publishers, Philadelphia, 1997. pp 481–493.
18. Spechler SJ, Zeroogian JM, Antonioli DA, Wang HH, Goyal RK. Prevalence of metaplasia at the gastroesophageal junction. Lancet 1994;344:1533–1536.
19. Sampliner RE. Practice guidelines on the diagnosis, surveillance, and therapy of Barrett's esophagus. Am J Gastroenterol 1998;93:1028–1031.
20. Spechler SJ. Barrett's esophagus. N Engl J Med 2002;346: 836–842.
21. De Hertogh G, Van Eyken P, Ectors N, Tack J, Geboes K. On the existence and location of cardiac mucosa; an autopsy study in embryos, fetuses, and infants. Gut 2003;52:791–796.
22. Chandrasoma PT. Fetal "cardiac mucosa" is not adult cardiac mucosa. Gut 2003;52:1798.
23. Chandrasoma PT, Der R, Ma Y, Peters J, DeMeester T. Histologic classification of patients based on mapping biopsies of the gastroesophageal junction. Am J Surg Pathol 2003;27:929–936.
24. Clark GWB, Ireland AP, Chandrasoma P, DeMeester TR, Peters JH, Bremner CG. Inflammation and metaplasia in the transitional mucosa of the epithelium of the gastroesophageal junction: A new marker for gastroesophageal reflux disease. Gastroenterology 1994;106:A63.
25. Oberg S, Peters JH, DeMeester TR, et al. Inflammation and specialized intestinal metaplasia of cardiac mucosa is a manifestation of gastroesophageal reflux disease. Ann Surg 1997;226: 522–532.

26. Glickman JN, Fox V, Antonioli DA, Wang HH, Odze RD. Morphology of the cardia and significance of carditis in pediatric patients. Am J Surg Pathol 2002;26:1032–1039.
27. Zhou H, Greco MA, Daum F, et al. Origin of cardiac mucosa: Ontogenic considerations. Pediatr Dev Pathol 2001;4:358–363.
28. Park YS, Park HJ, Kang GH, Kim CJ, Chi JG. Histology of gastroesophageal junction in fetal and pediatric autopsy. Arch Pathol Lab Med 2003;127:451–455.
29. Johns BAE. Developmental changes in the oesophageal epithelium in man. J Anat 1952;86:431–442.
30. van den Brink GR, Hardwick JC, Nielsen C, Xu C, ten Kate FJ, Glickman J, van Deventer SJ, Roberts DJ, Peppelenbosch MP. Sonic hedgehog expression correlates with fundic gland differentiation in the adult gastrointestinal tract. Gut 2002;51(5): 628–633.
31. Chandrasoma PT, Der R, Dalton P, Kobayashi G, Ma Y, Peters JH, DeMeester TR. Distribution and significance of epithelial types in columnar-lined esophagus. Am J Surg Pathol 2001;25: 1188–1193.
32. Odze RD. Unraveling the mystery of the gastroesophageal junction: A pathologist's perspective. Am J Gastroenterol 2005;100: 1853–1867.
33. Chandrasoma P. Controversies of the cardiac mucosa and Barrett's esophagus. Histopathol 2005;46:361–373.
34. Csendes A, Maluenda F, Braghetto I, et al. Location of the lower esophageal sphincter and the squamocolumnar mucosal junction in 109 healthy controls and 778 patients with different degrees of endoscopic esophagitis. Gut 1993;34:21–27.
35. Dresner SM, Griffin SM, Wayman J, Bennett MK, Hayes N, Raimes SA. Human model of duodenogastro-oesophageal reflux in the development of Barrett's metaplasia. Br J Surg 2003;90:1120–1128.

CHAPTER

7

New Histologic Definitions of Esophagus, Stomach, and Gastroesophageal Junction

The gastroesophageal junction must be defined as an imaginary line that is drawn at the point where the esophagus ends and the stomach begins. In the normal person, this junction is easily recognized (Figs. 7.1, 7.2, and 7.3): it is the junction between the tubular esophagus and the saccular stomach, it is the angle of His, it is the peritoneal reflection, it is the end of the esophageal squamous epithelium, it is the beginning of gastric oxyntic mucosa, it is the proximal limit of the rugal folds, it is the end of the lower esophageal sphincter, it is a point beyond which no submucosal esophageal glands are found. Any one of these criteria defines the true gastroesophageal junction in the normal person.

When the distal esophagus becomes damaged by reflux, it changes in appearance, but the gastroesophageal junction remains the same; the damaged esophagus remains above this line. *The main source of controversy is the failure to accurately recognize the nature of this damaged distal esophagus.* There is a tendency to believe that normal anatomic landmarks are retained in the damaged esophagus. We will show that this is not true; the reflux-damaged distal esophagus is so devastatingly altered that it ceases to look anything like the original organ that it was (Fig. 7.4).

The failure to accurately recognize the damaged esophagus results in a failure to recognize the original gastroesophageal junction because most of the criteria just listed that define the normal gastroesophageal junction become altered. The squamous lining of the esophagus is transformed to columnar and the squamocolumnar junction migrates cephalad (1) (Fig. 7.5). The distal limit of the residual lower esophageal sphincter moves cephalad as the sphincter is destroyed from below upward (2). We will show later that the destruction of the distal abdominal sphincter results in a loss of the tubular shape of the distal esophagus and the angle of His. As the distal esophagus dilates, rugal folds begin to line the columnar-lined esophagus in its dilated lower part. The present endoscopic and gross criteria of recognizing the true gastroesophageal junction produce a significant (up to a maximum of 3 cm) error.

The only criteria that remain to mark the gastroesophageal junction are the point of the peritoneal reflection, the association of esophageal submucosal glands with esophageal epithelium, and the proximal limit of gastric oxyntic mucosa (Fig. 7.5). The peritoneal reflection is not endoscopically detectable. The esophageal submucosal glands drain by ducts that open at the lumen and are a useful histologic marker for esophagus (3). However, these glands and ducts are sporadically present in the esophagus; they are not found in every biopsy of the esophagus and their absence does not mean that the biopsy is not esophageal. Gastric oxyntic mucosa, which is not damaged by reflux, remains in the same position but this cannot be differentiated reliably from the lowest part of the columnar-lined esophagus by any means other than histology.

LET US ESTABLISH COMMON GROUND IN HISTOLOGY

In trying to understand the truth, let us first find common ground.

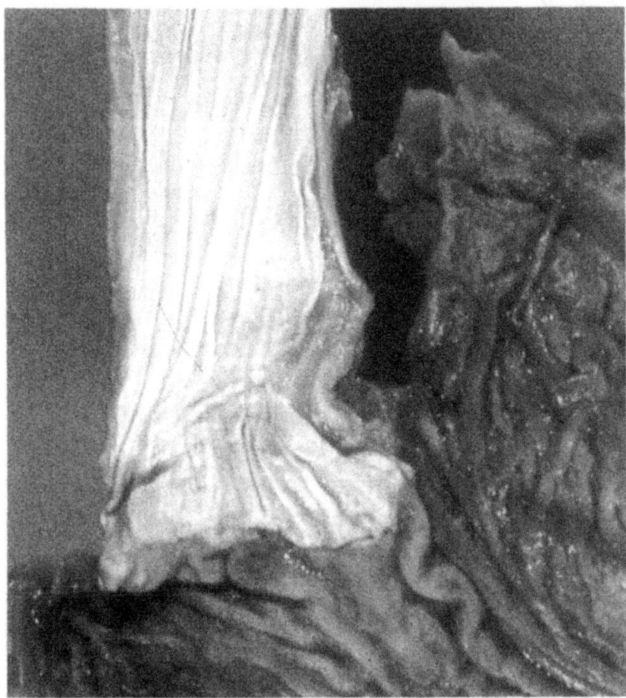

FIGURE 7.1 Normal appearance of distal esophagus. Note that this is only "normal" within the resolution capability of the naked eye. The presence of microscopic amounts of metaplastic columnar esophageal epithelium between gastric oxyntic mucosa and squamous epithelium cannot be excluded without microscopy. Also, the determination of the proximal limit of gastric oxyntic mucosa is most reliably assessed by microscopic examination, not by the gross position of the proximal limit of the rugal folds. (Cross reference: Color Plate 1.1)

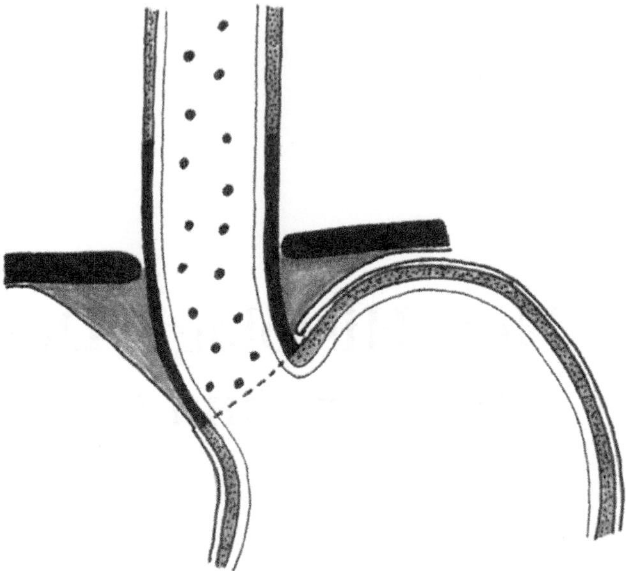

FIGURE 7.2 Drawing of the normal anatomy. The true gastroesophageal junction (dotted line) is marked by the point of the peritoneal reflection at which point the left side passes upward and to the left to create the angle of His. The diaphragm is approximately 2 to 3cm proximal to this point, and the abdominal part of the esophagus is surrounded by connective tissue (solid gray). The lower esophageal sphincter (solid black wall) envelops the entire abdominal esophagus and 2 to 3cm of the distal thoracic esophagus. Esophageal submucosal glands are shown as black dots in the esophagus.

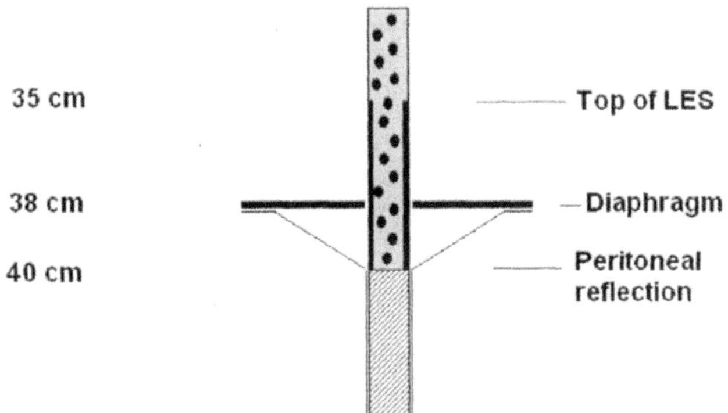

FIGURE 7.3 Diagrammatic representation of normal anatomy with histology interposed, shown as a vertical section across the gastroesophageal junction (solid line). The esophagus is lined by squamous epithelium (solid gray) and the stomach by gastric oxyntic mucosa (diagonal stripe). There is normally no metaplastic columnar epithelium between the two. In an average person, the gastroesophageal junction is at around 40cm from the incisors. The diaphragmatic hiatus is 2cm proximal and the lower esophageal sphincter (black wall) lines the distal 3 to 5cm of the esophagus. The esophageal submucosal glands (black circles) are randomly distributed in the esophagus.

Histologic Epithelial Types Are Reproducibly Definable

Ever since Paull et al. (4) defined the histologic types of epithelia in columnar-lined esophagus, *there has been no controversy* that there are a limited number of epithelial types if one examines the mucosa of the entire esophagus and proximal stomach (up to the beginning of antral type mucosa in the distal stomach) by histology. These epithelial types are accurately and reproducibly definable by histology (Table 7.1).

Unfortunately, as we have shown in Chapters 5 and 6, not all pathologists adhere to these definitions; studies that do not adhere to these strict definitions lose their value and should not have been accepted into the literature.

Location of These Epithelial Types

There is no controversy that squamous epithelium never lines the stomach in humans. *There is no controversy* that gastric oxyntic mucosa, either normal or showing atrophy and intestinal metaplasia, never lines the esophagus.

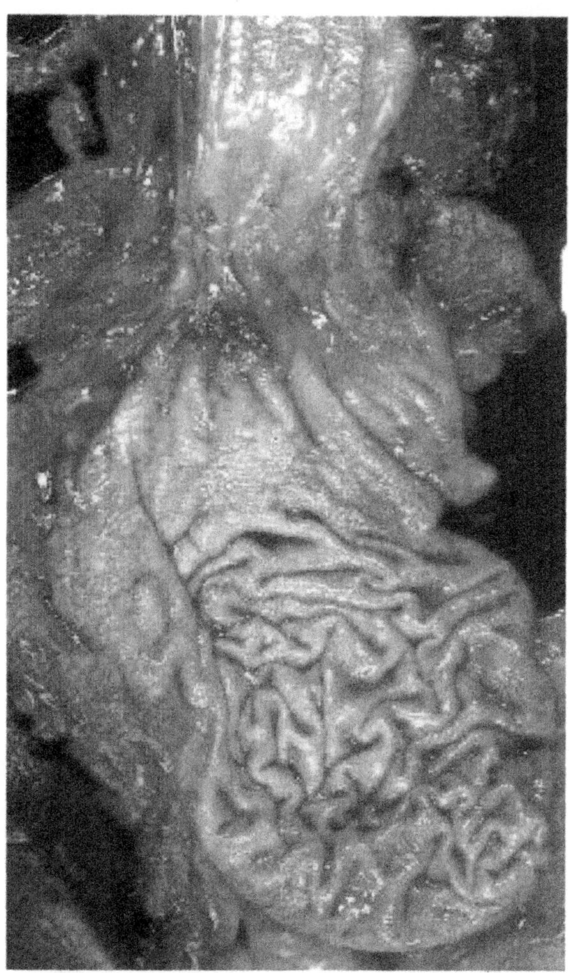

FIGURE 7.4 Severe reflux damage to the distal esophagus with an adenocarcinoma at the squamocolumnar junction. The end of the tubular esophagus is widely separated from the proximal limit of the rugal folds. The area between the two is saccular and lined by flat columnar epithelium. By present definitions of the gastroesophageal junction, it is not a hiatal hernia (stomach) because it does not contain rugal folds; it is not esophagus because it has a saccular shape and is distal to the end of the tubular esophagus. It is not possible to accurately define the true location of the gastroesophageal junction by gross (or endoscopic) examination when such severely distorted anatomy is present. (Cross reference: Color Plate 5.2)

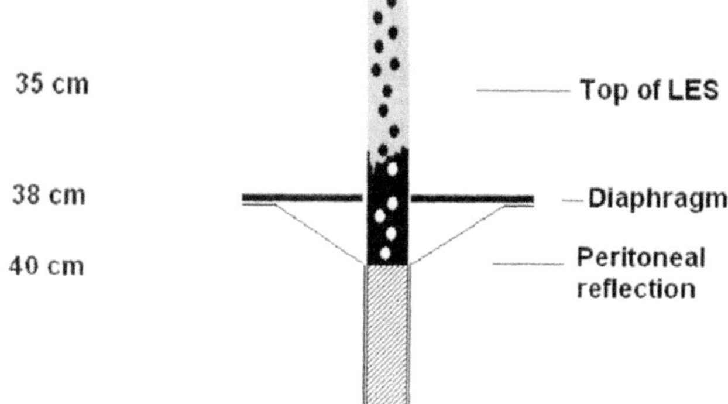

FIGURE 7.5 Reflux-damaged esophagus, shown as a vertical section. The squamocolumnar junction moves proximally to a length that is related to the severity of reflux (shown as a dark area above the original gastroesophageal junction). The point of the original gastroesophageal junction, the peritoneal reflection, and the proximal limit of gastric oxyntic mucosa (diagnonal stripe) remain constant. The submucosal glands remain, now opening in the columnar surface (white dots). A columnar metaplastic segment (black) has been interposed between the original gastroesophageal junction and the proximally migrated squamocolumnar junction.

On occasion, it may be difficult to differentiate atrophic gastritis and gastric intestinal metaplasia from cardiac mucosa and esophageal intestinal metaplasia, respectively. Similarly, there may be difficulty differentiating gastric oxyntic mucosa from oxyntocardiac mucosa. It is important to recognize that difficulty in distinguishing between two entities does not mean that they are the same. For example, changes in the gastric body of chronic atrophic gastritis caused by *Helicobacter pylori* and autoimmune gastritis are often indistinguishable on a biopsy; this does not mean they are not different entities. In Chapter 5 we showed that the use of careful histologic criteria can separate these with almost complete accuracy.

There is no controversy about the fact that cardiac mucosa, oxyntocardiac mucosa, and esophageal intestinal metaplasia (specialized columnar epithelium) are the three epithelial types found in columnar-lined esophagus (4). By present definitions, these epithelial types are recognized as representing an endoscopic abnormality when they are found in the tubular esophagus above the level of the proximal limit of the gastric rugal folds (5). However, the medical community recognizes this abnormality as pathological only when intestinal metaplasia is present (this defines Barrett esophagus). They tend to ignore cardiac and oxyntocardiac mucosa above the endoscopic gastroesophageal junction, although they recognize that this is an area of endoscopic abnormality. It is now accepted that the normal tubular esophagus is entirely lined by squamous epithelium.

There is no controversy that cardiac mucosa, oxyntocardiac mucosa, and intestinal metaplasia occurring in cardiac mucosa can be found in specimens taken distal to the gastroesophageal junction as defined by endoscopy (the proximal limit of the rugal folds) or gross examination (a line drawn across the end of the tubular esophagus (6)). Jain et al. (7) demonstrated that biopsies 1 and 2 cm distal to the endoscopic gastroesophageal junction (the proximal limit of rugal folds) contained cardiac mucosa. Chandrasoma et al. (8) showed that oxyntocardiac and cardiac mucosa with and without intestinal metaplasia was present in a

TABLE 7.1 Different Epithelial Types That Are Seen between the Proximal Esophagus and the Distal Stomach (Antrum)

Epithelial type	Definition	Location
Stratified squamous	Squamous; may have mucosal and submucosal glands and gland ducts	Only esophagus
Oxyntic mucosa	Columnar epithelium with straight tubular glands composed only of parietal and chief cells	Only stomach
Atrophic oxyntic mucosa	Oxyntic mucosa with loss of parietal cells	Only stomach
Cardiac mucosa	Columnar epithelium composed only of mucous cells	Esophagus and within 3 cm distal to presently defined gastroesophageal junction
Oxyntocardiac mucosa	Columnar epithelium with glands composed only of mucous and parietal cells	Esophagus and within 3 cm distal to presently defined gastroesophageal junction
Esophageal intestinal metaplasia	Goblet cells in cardiac mucosa	Esophagus and within 3 cm distal to presently defined gastroesophageal junction
Gastric intestinal metaplasia	Goblet cells in oxyntic (usually atrophic) or antral mucosa	Only stomach

Note: Notice that the presence of chief cells, Paneth cells, serous (pancreatic) cells, and neuroendocrine cells does not affect the definition of these epithelial types.

length of 0.31 to 2.05 cm distal to the end of the tubular esophagus and proximal limit of the rugal folds in 10 esophagectomy specimens.

The Meaning of Cardiac Mucosa When It Is Found

There is controversy as to whether or not cardiac mucosa is present in everyone. A review of Chapter 6 should convince you that the literature overwhelmingly indicates that cardiac mucosa is absent in a significant number of people (Figs. 7.6 and 7.7; see also Figs. 3.22 and 3.23). Despite this evidence, the presently accepted general belief is there is a "very small amount," usually quantitated as "approximately 1 to 4 mm" of "normal" cardiac mucosa, in humans. This is based almost entirely on Kilgore et al.'s autopsy study (9); this study is flawed and has never been corroborated (see Chapter 6). Every other study in the literature has shown that a variable percentage of people do not have cardiac mucosa between the squamous epithelium and parietal cell containing mucosa. This controversy exists at present not because of the existing evidence but because of the strength of an old dogma that refuses to die. The medical community appears unable to believe that a cherished fact such as the normalcy of a gastric cardia lined by cardiac mucosa is a figment of the human imagination. This will change; *the evidence indicates that cardiac mucosa is not a normal epithelium in humans.*

There is no controversy that cardiac mucosa occurs as a result of reflux-induced columnar transformation of the esophagus. In Glickman et al.'s study (10), the presence of 1 mm of cardiac mucosa correlated with the presence of reflux. It is also agreed that increasing length of columnar-lined esophagus (the three types of epithelium described by (4) correlate with increasing reflux. The present belief among pathologists considered experts in the field is that the "normal" 1 mm of cardiac mucosa in the stomach "expands" into the esophagus as a result of reflux-induced damage (11).

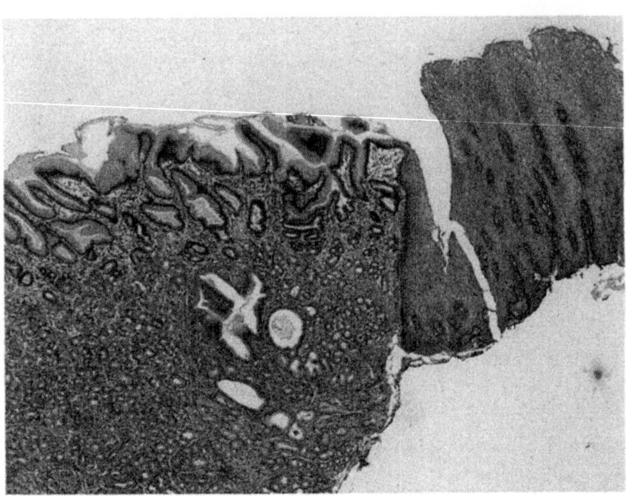

FIGURE 7.6 Section taken across the squamocolumnar junction in a 67-year-old person who had a resection for a large squamous papilloma of the esophagus. The squamous epithelium transitions directly to gastric oxyntic mucosa without intervening columnar-lined esophagus. This is the normal gastroesophageal junction at a microscopic level.

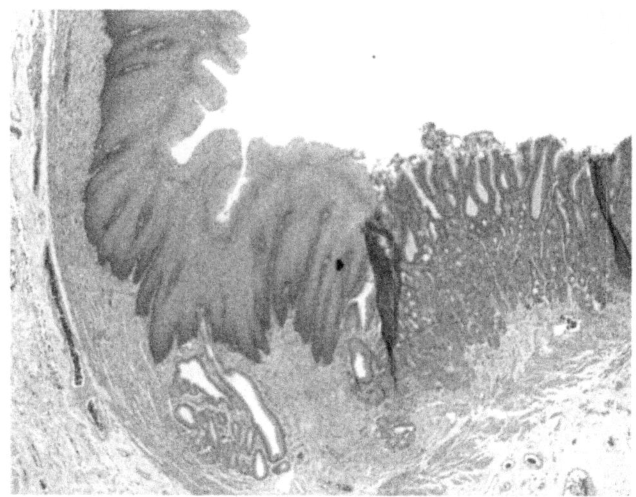

FIGURE 7.7 A section taken across the squamocolumnar junction in a 72-year-old person who had a resection for a squamous carcinoma of the esophagus. The squamous epithelium transitions directly to gastric oxyntic mucosa without intervening columnar-lined esophagus. This is the normal gastroesophageal junction at microscopic level. Note the presence of a normal mucous gland in the lamina propria under the squamous epithelium. (Cross reference: Color Plate 1.6)

When one combines the statements in the preceding two paragraphs, the logical conclusion should be that when cardiac mucosa is seen in a biopsy, it is very likely an abnormal finding; the maximum error associated with this belief is 1 mm. Nothing could be further from the truth. *Most pathologists and clinicians will accept virtually any amount of cardiac mucosa as a normal finding wherever it is located.* Thus, cardiac mucosa found anywhere distal to the endoscopic gastroesophageal junction is normal; cardiac mucosa found 3 cm up the tubular esophagus is normal as long as there is no intestinal metaplasia. A 1-mm "normal" structure has incredibly expanded to a 5-cm "normal" structure. Despite the fact that "expansion" of cardiac mucosa is known to be associated with reflux, no one will attach any significance to any amount of cardiac mucosa.

To us, cardiac mucosa always represents reflux-induced metaplastic esophageal columnar epithelium, irrespective of its location. If our belief that cardiac mucosa is always abnormal is true, there is no error in this viewpoint. *This is considered to be a highly controversial statement.* The only error possible with this belief is that we will call abnormal the 1 mm of cardiac mucosa that most people believe normally exists. An error that is a maximum 1-mm error is much better than the potential 5-cm (×50) error of the people holding the opposite viewpoint of the existence of cardiac mucosa as a normal structure in the large area that they will accept it as a "normal" finding. In fact, no pathologic entity accepted today is based on the presence of cardiac mucosa, despite the fact that its presence is known to be associated with reflux.

The Distal Limit of Cardiac Mucosa

This has never been defined. However, there is no controversy that cardiac mucosa is limited to the "proximal stomach." The only attempt of defining a distal limit was the rather feeble statement by Hayward (12) that cardiac mucosa "extends to a variable distance into the proximal stomach." We suggest that we establish a distal limit for cardiac mucosa as 3 cm distal to the presently defined endoscopic gastroesophageal junction (proximal limit of rugal folds).

This suggestion is based on following excellent study by Jain et al. (7), which looks at the mucosal types at the endoscopic gastroesophageal junction and at 1 and 2 cm distal to this point. This is an abstract, which unfortunately never reached formal publication.

Literature Review

Jain R, Aquino D, Harford WV, Lee E, Spechler SJ. Cardiac epithelium is found infrequently in the gastric cardia. Gastroenterology 1998;114:A160.

The methods in this excellent study were as follows:

> At endoscopy, both the squamocolumnar junction (the Z-line) and the anatomic esophagogastric junction (EGJ) were identified and biopsy specimens were taken from the columnar epithelium at each of 4 quadrants in the following locations: 1) the Z-line, 2) the EGJ, 3) the gastric cardia 1 cm below the EGJ, and 4) the gastric cardia 2 cm below the EGJ. Cardiac epithelium was identified in specimens with mucus-secreting glands devoid of parietal and chief cells, fundic epithelium in specimens with abundant parietal and chief cells, and intestinal metaplasia in specimens with goblet cells.

The methods are beautifully precise except for one thing: they pass from cardiac epithelium, which is "devoid of parietal and chief cells," to fundic epithelium "with abundant parietal and chief cells." There is ambiguity here because the researchers do not tell us where an epithelium with a few parietal cells would fall in their classification because it is neither devoid nor abundant. They evade recognizing oxyntocardiac mucosa, which is one of the epithelia that we recognize. This is, however, a minor deficiency for the purposes of our discussion and is only mentioned to demonstrate imprecision in definition, even in excellent studies.

The researchers studied 31 patients. Their endoscopic findings are as follows: "The Z-line and EGJ were located at the same level in 25 patients, whereas in 6 the Z-line was located >1 cm proximal to the EGJ." This is as honest a statement of endoscopic capability as can be found in the literature. The authors seem to be admitting that when the Z-line is within 1 cm from the EGJ, it is impossible to distinguish from the state where the Z line and EGJ are coincident. This is a reasonable limit of the power of optical resolution of the endoscope that will be accurate with most observers; some may be better but most will be able to be accurate within this standard.

Their results for the frequency of finding the different epithelial types in at least one of the four biopsy specimens obtained at the various locations are tabulated as follows:

Biopsy site	Numbers of patients	Cardiac epith	Fundic epith	Intestinal met
EGJ	31	11 (35%)	21 (68%)	10 (32%)
1 cm below EGJ	29	4 (14%)	23 (79%)	7 (24%)
2 cm below EGJ	30	1 (3%)	30 (100%)	2 (7%)

These results are fascinating because the data are produced without any specific agenda. Cardiac epithelium was found in only 1 patient at 2 cm distal to the endoscopic gastroesophageal junction; at this point all patients had oxyntic (fundic) epithelium. The authors concluded:

> Both cardiac epithelium and intestinal metaplasia are found with decreasing frequency as the biopsy sites move distally down the gastric cardia. Even at the EGJ, cardiac epithelium is found in only a minority of patients. These findings challenge the traditional view of a gastric cardia lined predominantly by cardiac epithelium.

There are two ways to rationally explain the decreasing frequency of cardiac mucosa and intestinal metaplasia as the location of the biopsy moves distally away from the EGJ (Fig. 7.8). The first is that the esophagogastric junction is defined correctly and the proximal stomach (gastric cardia) is normally lined to a variable length by cardiac epithelium. The second is that the esophagogastric junction is defined incorrectly and that the correct esophagogastric junction is the junction between the cardiac (and oxyntocardiac) epithelium and fundic epithelium. Note that the frequency of fundic epithelium increases as the biopsies move distally and 100% of patients have fundic epithelium 2 cm distal to their defined esophagogastric junction.

The fact that cardiac epithelium was absent in 65% of patients at the presently defined junction raises doubt that this is a normal epithelium. The correct answer to this important question cannot, however, be determined by the data in this study. It must be answered by studies that ask the following four questions: (a) Is the cardiac mucosa histologically normal? (b) Is the absence of cardiac mucosa associated with a pathologic state, as would be expected if cardiac mucosa was a normal structure? (c) Is the presence of cardiac mucosa associated with a pathologic state, as would be expected if cardiac mucosa was an abnormal structure? (d) Is the area lined by cardiac mucosa anatomically stomach or esophagus, as shown by the absence of peritoneum on the outside or the presence of submucosal glands, both markers of esophagus rather than stomach?

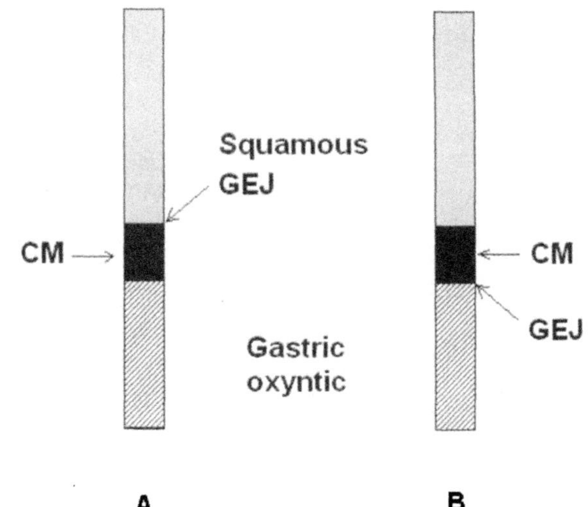

FIGURE 7.8 Jain et al. (7) found cardiac mucosa (solid black) to be present 2 cm distal to the endoscopic gastroesophageal junction in one patient. By present definition, this means that cardiac mucosa lined the proximal 2 cm of the stomach (a). However, if the definition of the true gastroesophageal junction was incorrect and the true junction is defined by the proximal limit of gastric oxyntic mucosa (b), the interpretation would be different. It would then be that this person has 2 cm of abnormal metaplastic esophageal columnar epithelium (mistaken as proximal stomach because of an incorrect definition of the gastroesophageal junction at endoscopy).

Chapter 6 shows that there are excellent data in the literature to indicate that cardiac mucosa is always histologically abnormal (Figs. 7.9 and 7.10) (13, 14), its absence is not associated with any abnormality, its presence always indicates the presence of reflux (10, 15), and its location is the anatomic esophagus (16–18).

Therefore, when cardiac mucosa is found in what is called the proximal stomach by the present endoscopic definition of the gastroesophageal junction (proximal limit of the rugal folds), it actually lines the esophagus. The endoscopic definition of the gastroesophageal junction must be incorrect. The distal reflux-damaged esophagus, when present, is mistaken as the proximal stomach.

In Chapter 11 we show that the extent of cardiac mucosa distal to the tubular esophagus and the proxi-

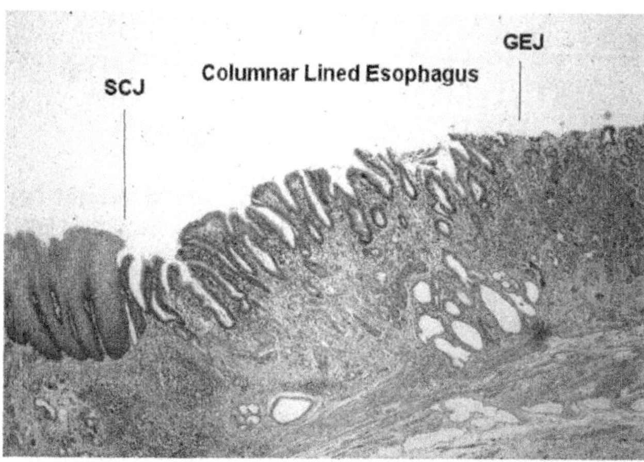

FIGURE 7.9 A vertical section taken from the grossly normal squamocolumnar junction at esophagectomy for squamous carcinoma in a 78-year-old male. The section shows an area of metaplastic columnar esophagus between squamous epithelium (left) and gastric oxyntic mucosa (right). This is cardiac mucosa and shows foveolar hyperplasia and chronic inflammation. It is very likely that this patient will be "grossly and endoscopically normal" and the cardiac mucosa between the squamocolumnar junction and oxyntic mucosa will be called "normal gastric mucosa." The correct interpretation, which is based on histologic separation of the squamocolumnar junction from the oxyntic mucosa (the true gastroesophageal junction), is that this cardiac mucosa is columnar-lined esophagus. (Cross reference: Color Plate 3.9)

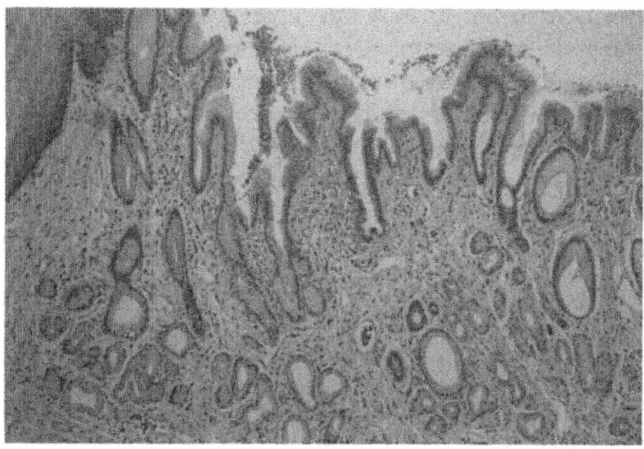

FIGURE 7.10 Higher power of metaplastic columnar epithelium showing cardiac mucosa with foveolar hyperplasia and marked chronic inflammation. (Cross reference: Color Plate 3.10)

mal limit of the rugal folds varies between zero (normal) and a maximum of 3 cm (severe reflux-induced damage) (8). This is the error that is created by the present incorrect definition of the gastroesophageal junction. Pathology in this part of the esophagus is presently localized incorrectly to the stomach.

LET US UNDERSTAND THE PROBLEM

If it is believed that cardiac mucosa is normally present, there is no way to distinguish histologically between this "normal" cardiac mucosa and metaplastic cardiac mucosa that is known to occur in the esophagus as a result of gastroesophageal reflux. It becomes impossible to use the presence of cardiac mucosa as a marker of a pathologic state as long as it is believed to be a normal epithelium. We are in the ridiculous position as pathologists of inventing a phantom normal 1 to 4 mm of cardiac mucosa at the squamocolumnar junction that has been proven to be absent in many people to prevent us from regarding cardiac mucosa as being a pathologic epithelium.

Let us imagine 1 mm; it is 1000 µm; it is the size of a large pinhead; it is approximately 30 epithelial cells; it is approximately six to eight foveolar pits with their normal separation; it is 150 erythrocytes; it is 70 plasma cells. If this is the extent of normal cardiac mucosa, it is likely to be absent in a biopsy of the columnar epithelium distal to the squamocolumnar junction if you miss the actual junction. This was beautifully shown in the study by Marsman et al. (19). In this excellent study, the authors precisely defined endoscopic landmarks and used precise histologic definitions. They reported on the findings of biopsies taken "from the gastric cardia just below the squamocolumnar junction" in patients without visible columnar-lined esophagus (either tongues or circumferential). They described the normal squamocolumnar junction as a straight line between the white (squamous) and red (columnar) mucosa, coinciding with the gastroesophageal junction (the proximal margin of the gastric folds). Of the 198 patients studied, 63 patients had the actual squamocolumnar junction in one biopsy sample; in the other 135, the actual junction was not present.

Their results are astounding. In the 63 patients in whom the actual squamocolumnar junction was present in one biopsy specimen, 39 (62%) showed cardiac mucosa adjacent to the squamous epithelium; in the other 24 patients (38%), oxyntocardiac mucosa was seen adjacent to the squamous epithelium. Of the 135 patients who did not have the actual junction in one biopsy specimen, the epithelial types in the "gastric cardia just below the squamocolumnar junction" were as follows: 90 (67%) had pure oxyntic mucosa; 43 (32%) had oxyntocardiac mucosa; only 2 (1%) had cardiac mucosa.

This study shows that the finding of cardiac mucosa by endoscopic biopsy in a patient with a normal endoscopic appearance is very difficult and requires a biopsy that actually straddles the squamocolumnar junction. The important lesson to be learned from this study is that if there is a biopsy distal to the squamocolumnar junction and it contains any cardiac mucosa, it is likely that *the volume of cardiac mucosa exceeds the normal limit that is accepted at present, which is 1 to 4 mm.* This is not very different from saying that the presence of cardiac mucosa in a biopsy is always abnormal, which is based on what we believe is the true situation that cardiac mucosa is never a normal epithelium.

The correct thinking relating to the problem is this: (a) If we accept that only 1 mm of cardiac mucosa is normally present at the squamocolumnar junction, we would not see it in a biopsy unless the actual squamocolumnar junction is present in the biopsy. (b) If the actual squamocolumnar junction is present, we can measure the length of cardiac mucosa present; if this is greater than 1 mm, it is abnormal. (c) If there is a biopsy that does not contain the actual junction and it contains cardiac mucosa, *it is quantitatively abnormal* and must be considered a pathological structure whose existence must be explained.

The incorrect thinking relating to the problem, and one that is almost universally followed by pathologists, is this: (a) I accept that only 1 mm of cardiac mucosa is normally present at the junction. (b) When I see cardiac mucosa in a biopsy, I classify it as a normal finding irrespective of its quantity because cardiac mucosa is normally present at the junction.

The truth differs from both of these scenarios. The truth is that cardiac mucosa is normally absent; its presence indicates a pathological abnormality caused by gastroesophageal reflux. Ockham's razor works: the simplest answer is most likely to be the truth.

NORMAL HISTOLOGY OF THE ESOPHAGUS AND STOMACH: A STATEMENT OF FACT AND NEW HISTOLOGIC DEFINITIONS

This most confusing and controversial subject has a simple statement of truth: *The entire esophagus is lined by squamous epithelium. The entire stomach distal to the true gastroesophageal junction and proximal to the beginning of pyloric mucosa in the distal stomach is lined by gastric oxyntic mucosa. Cardiac mucosa does not exist normally.*

Recognition of these simple truths is the key to understanding and defining the normal state and pathologic states (Table 7.2; Figs. 7.3 and 7.5). Histologic normalcy in the region is defined as the presence of two organs (the esophagus and the stomach) and two epithelial types (esophageal squamous epithelium and gastric oxyntic mucosa) (Fig. 7.3). Histologic abnormality is the interposition of a columnar-lined esophagus between the squamous epithelium and gastric oxyntic mucosa (Fig. 7.5). Columnar-lined esophagus results from gastroesophageal reflux, which injures the squamous epithelium of the esophagus and causes it to undergo columnar metaplasia. Columnar metaplasia of squamous epithelium is induced only by reflux and results in cardiac mucosa. The gastric oxyntic mucosa is not altered by gastroesophageal reflux; its proximal limit continues to be an accurate marker of the true mucosal gastroesophageal junction even when the squamocolumnar junction has migrated proximally as a result of reflux-induced columnar metaplasia. *The proximal limit of gastric oxyntic mucosa is the true gastroesophageal junction both in normal people and patients with columnar-lined esophagus.*

TABLE 7.2 Histologic Definitions of Anatomy and Histology

Normal columnar epithelium: The only columnar epithelium present normally in the esophagus and proximal stomach is gastric oxyntic mucosa. This is subject to pathologic changes that occur in it like gastritis, gastric intestinal metaplasia, and gastric adenocarcinoma.

Metaplastic columnar esophageal epithelium: Cardiac mucosa (with and without intestinal metaplasia) and oxyntocardiac mucosa. This is always caused by gastroesophageal reflux. It is always histologically abnormal.

The esophagus: The esophagus is that part of the foregut that is lined by squamous epithelium and metaplastic columnar esophageal epithelia.

The stomach: The stomach is that part of the foregut that is lined by gastric oxyntic and antral mucosa. It is unaffected by reflux and its proximal limit remains constant.

The true gastroesophageal junction: This is the proximal limit of gastric oxyntic mucosa, defined histologically. In normal patients, the squamous epithelium is coincidental with this point. In patients with reflux damage, the squamous epithelium moves cephalad due to columnar metaplasia of the esophagus. The gastroesophageal junction is the junction between metaplastic columnar esophageal epithelium and gastric oxyntic mucosa; it can only be defined by histology.

Columnar-lined esophagus: The segment of distal esophagus lined by metaplastic columnar epithelium that is interposed between the gastroesophageal junction (proximal limit of gastric oxyntic mucosa) and squamocolumnar junction. This is first microscopic and then becomes visible at gross examination and endoscopy.

The endoscopic gastroesophageal junction: Endoscopy has no ability to localize the true gastroesophageal junction (proximal limit of gastric oxyntic mucosa) because it cannot differentiate metaplastic columnar esophageal epithelia from gastric oxyntic mucosa.

Recognition of the proximal limit of gastric oxyntic mucosa is not possible by any method other than histology. The proximal limit of gastric oxyntic mucosa does not always coincide with the proximal limit of the rugal folds. Mucosa with endoscopic and gross rugal folds may be lined by cardiac and oxyntocardiac mucosa; the fact that this is esophageal is proved by the fact that there are submucosal esophageal glands and mucosal gland ducts in cardiac and oxyntocardiac epithelia lining rugated mucosa (see Chapter 11) (8).

These histologic definitions are true irrespective of the anatomic appearance of the structures. To really understand them, one should look at this lesion as a vertical section without anatomic form (neither tubular nor saccular) or gross mucosal characteristics (flat or containing rugal folds). When this is done, the correct definitions are easy to understand (Figs. 7.3 and 7.5).

We will show in a later chapter that reflux damage to the distal esophagus causes (a) destruction of the lower esophageal sphincter, beginning at its distal end, (b) dilatation of the damaged distal esophagus, and (c) columnar metaplasia of the damaged distal esophagus with rugal folds developing in the dilated segment of damaged esophagus. This causes changes in the gross anatomy of the reflux-damaged distal esophagus. Damage by reflux is not a mild change; it is a devastating pathologic state where the damaged esophagus loses all resemblance to normal and creates a panoply of abnormalities that have confused the medical community for half a century (see Chapter 11).

APPLICATION OF THESE DEFINITIONS TO PRACTICE

In Resection Specimens and at Autopsy

In resection specimens and at autopsy, the true gastroesophageal junction can be identified by taking a vertical section to include gastric oxyntic mucosa in its distal part. If a vertical section begins at the end of the tubular esophagus and measures 3 cm, gastric oxyntic mucosa will always be encountered at the distal end because the limit of cardiac mucosa distal to the tubular esophagus is 3 cm (Fig. 7.11).

The point at which gastric oxyntic mucosa changes to squamous epithelium (in normal) or metaplastic columnar esophageal epithelium (usually oxyntocardiac because this is the most distal epithelial type in patients with columnar-lined esophagus) is the true gastroesophageal junction (Figs. 7.6, 7.7, and 7.9; see also Figs. 3.22, 3.23, 5.23, and 6.12). This transition can be difficult to determine with certainty because oxyntocardiac mucosa with many parietal cells may closely resemble gastric oxyntic mucosa; the presence of mucous cells admixed with parietal cells in the glands is the defining feature. In such cases, the localization of the most distal submucosal gland provides absolute proof of esophageal location (Fig. 7.12).

When one traverses the vertical section from the proximal limit of gastric oxyntic mucosa (the true histologic gastroesophageal junction) upward into the esophagus, there is usually a variable length of

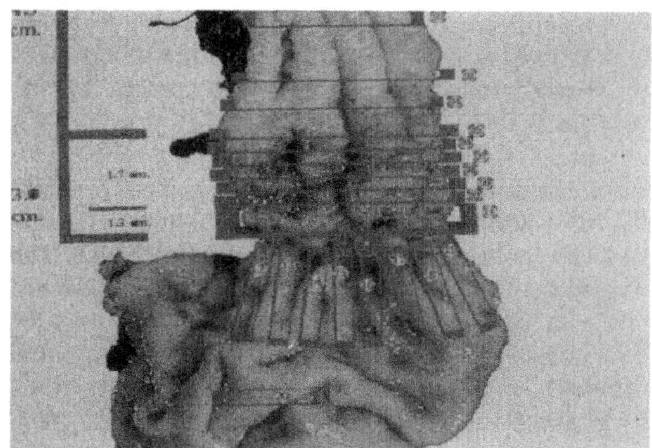

FIGURE 7.11 Method of determining the true gastroesophageal junction in a resection specimen (similar for autopsy specimen). The esophagus is cut horizontally at the end of its tubular part. Vertical sections are taken from this point distally, permitting histologic examination of the entire region. The point of the proximal limit of gastric oxyntic mucosa can be measured on the slide and represents the true gastroesophageal junction irrespective of the shape of the structure or the presence of rugal folds.

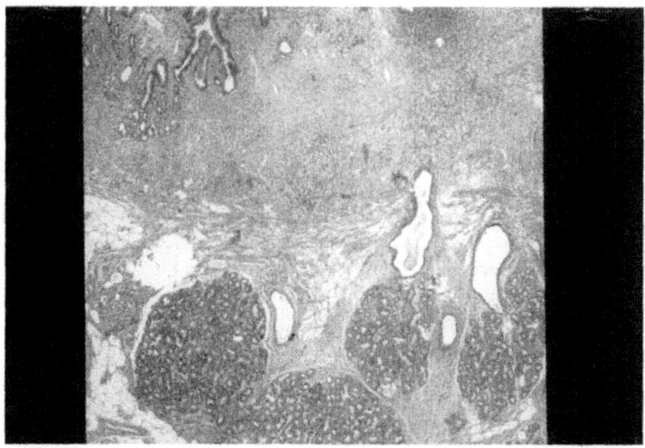

FIGURE 7.12 Section of cardiac mucosa under rugal folds and distal to the end of the tubular esophagus in specimen 7.11, showing the presence of submucosal glands, confirming that this area represents esophagus.

oxyntocardiac mucosa before cardiac mucosa is encountered. The distinction between oxyntocardiac mucosa and cardiac mucosa is easy; cardiac mucosa is characterized by a foveolar-gland complex devoid of any parietal cells.

Resection specimens and autopsy permit examination of the entire circumference of this region by multiple vertical sections; usually 10 to 15 are required to sample the entire circumference (Fig. 7.12). The maximum endoscopic sample is a four-quadrant biopsy, which samples only about 25% of the full circumference. When the entire circumference is examined histologically, it is usual for the length of cardiac and oxyntocardiac mucosa separating squamous from oxyntic mucosa to be very different in different parts of the circumference (20). This reflects the effect of irregular metaplastic columnar epithelial transformation within the circumference of the reflux-damaged esophagus. The only element of this that is grossly visible is the wavy line of the squamocolumnar junction (Z line) that is a hallmark of columnar-lined esophagus. The irregular distribution of various types of columnar epithelia within the metaplastic segment cannot be seen grossly because it is impossible to differentiate the different histologic types of columnar metaplastic esophageal epithelia from themselves and from normal gastric oxyntic mucosa. Microscopic mapping is necessary to show this immense variation of epithelial composition within the columnar-lined esophagus. We will show in later chapters that understanding the exact histologic composition of the columnar-lined esophagus provides insights into the mechanism of progression of pathologic events in reflux disease.

At Endoscopy

At endoscopy, the gastroesophageal junction can only be defined by mapping biopsies (Fig. 7.13). If the patient is endoscopically normal (as described by Marsman et al., the normal squamocolumnar junction is a straight line between the white squamous and red columnar mucosa, coinciding with the proximal margin of the rugal folds), an attempt should be made to take a biopsy that straddles the squamocolumnar junction. If this is not possible, the biopsy should attempt to sample the columnar mucosa within 3 mm from the distal limit of squamous epithelium. Otherwise, very small lengths of metaplastic columnar esophageal epithelium can be missed.

Normalcy will be found within a subpopulation of endoscopically normal patients who can be divided into those who are histologically normal and those who have a microscopic columnar-lined esophagus (Table 7.3).

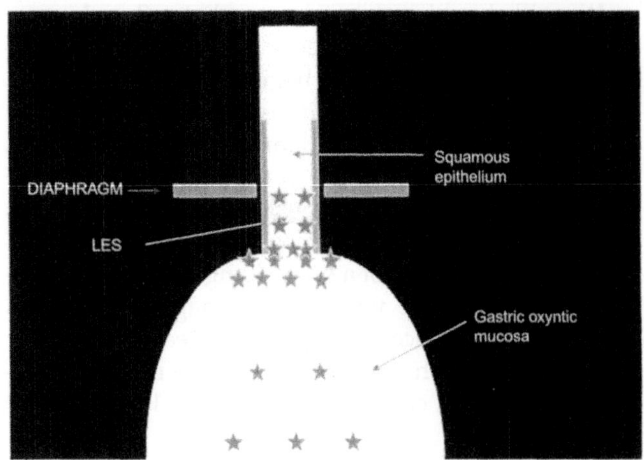

FIGURE 7.13 Biopsy protocol designed to histologically map the area immediately proximal to the true gastroesophageal junction. The true junction (the proximal limit of the gastric oxyntic mucosa) cannot be seen endoscopically. It is present in most persons 0 to 2.5 cm of the end of the tubular esophagus, or where this is not a clear landmark, within 2.5 cm of the proximal limit of the rugal folds.

TABLE 7.3 Interpretation of Histologic Findings in Vertical Sections of Resected Specimens (or Serial Biopsies[a]) Taken across the Squamocolumnar Junction in Patients Who Have No Visible Abnormality on Endoscopy or Gross Examination

Microscopic finding	Interpretation
Squamous epithelium directly transitioning to gastric oxyntic mucosa; no other epithelium found	Normal
Cardiac mucosa with or without intestinal metaplasia, or oxyntocardiac mucosa interposed between squamous and oxyntic mucosa	Reflux-induced columnar-lined esophagus (microscopic stage)

[a]In serial biopsies, there will always be a sampling error. Squamous and gastric oxyntic mucosae are easy to find because they are present in large amounts. The failure to find oxyntic mucosae means that all biopsies are above the gastroesophageal junction. Cardiac and oxyntocardiac mucosae are often found in such small amounts (1 mm range) that adequate sampling is necessary to confidently exclude them. "Adequate biopsy sampling" will always be less than complete sectioning of the entire circumference in resected specimens.

If the patient has a visible separation of the proximal limit of the rugal folds and squamocolumnar junction, this defines a later stage of columnar-lined esophagus than the microscopic stage where the endoscopy is normal. These patients invariably have cardiac mucosa with and without intestinal metaplasia or oxyntocardiac mucosa. In these patients with more severe abnormality, the most distal biopsy should be taken 1 to 2 cm distal to the proximal limit of the rugal folds. Additional biopsies (four-quadrant biopsies at measured levels) are taken above this level at 1- to 2-cm intervals until the squamocolumnar junction is reached. In general, the true gastroesophageal junction will tend to be most distal to the end of the tubular esophagus and the proximal limit of the rugal folds in patients who do not have clear-cut landmarks and whose endoscopy is perceived as very abnormal (8), (see Chapter 11). In these rare patients, the most distal biopsy should be taken 2 to 3 cm distal to the perceived gastroesophageal junction.

An adequate series of biopsies must show oxyntic mucosa in the most distal sample and squamous epithelium in the most proximal sample; otherwise the entire extent of the columnar-lined esophagus has not been sampled. A common and significant error in sampling is that the most distal biopsies do not contain oxyntic mucosa; this is most common in patients with severe reflux disease with long segments of columnar-lined esophagus. Absence of oxyntic mucosa in the most distal sample means that the true gastroesophageal junction and the total length of the columnar-lined esophagus remain unknown.

It is vitally important to recognize that *histology is the only means of defining the true gastroesophageal junction and the true extent of the columnar-lined esophagus.* The proximal limit of the rugal folds is only accurate enough to act as an endoscopic guide to direct biopsies to identify the proximal limit of gastric oxyntic mucosa; it is not the true junction. The true gastroesophageal junction can be as much as 3 cm distal to this point.

With such a biopsy protocol, normal and abnormal can be defined accurately. Normal patients without any reflux damage will have only squamous epithelium and oxyntic mucosa (Figs. 7.3, 7.6, and 7.7; see also Figs. 3.22 and 3.23). Patients with reflux-induced damage will have an interposed columnar metaplastic epithelium between squamous epithelium and oxyntic mucosa. This may consist of cardiac mucosa without intestinal metaplasia (Figs. 7.5 and 7.9; see also Figs. 5.23 and 6.12), cardiac mucosa with intestinal metaplasia, or oxyntocardiac mucosa (see Figs. 5.19 to 5.24).

Up to this point, we have considered the histologic changes only from the standpoint of a vertical section taken across the distal esophagus and proximal stomach. We have not considered the gross anantomic changes that are associated with this. This subject will be discussed in Chapter 11 once we have considered the mechanism of these changes. In that chapter, the histologic findings described in this chapter will be correlated with the gross anatomic abnormalities that will provide a comprehensive solution to the understanding of the pathology of gastroesophageal reflux disease.

References

1. Csendes A, Maluenda F, Braghetto I, et al. Location of the lower esophageal sphincter and the squamocolumnar mucosal junction in 109 healthy controls and 778 patients with different degrees of endoscopic esophagitis. Gut 1993;34:21–27.
2. DeMeester TR, Peters JH, Bremner CG, Chandrasoma P. Biology of gastroesophageal reflux disease: Pathophysiology relating to medical and surgical treatment. Annu Rev Med 1999;50: 469–506.
3. Shi L, Der R, Ma Y, Peters J, DeMeester T, Chandrasoma P. Gland ducts and multilayered epithelium in mucosal biopsies from gastroesophageal junction region are useful in characterizing esophageal location. Dis Esophagus 2005;18:87–92.
4. Paull A, Trier JS, Dalton MD, Camp RC, Loeb P, Goyal RK. The histologic spectrum of Barrett's esophagus. N Engl J Med 1976;295:476–480.
5. Sampliner RE. Practice guidelines on the diagnosis, surveillance, and therapy of Barrett's esophagus. Am J Gastroenterol 1998;93:1028–1031.
6. Association of Directors of Anatomic and Surgical Pathology. Recommendations for reporting of resected esophageal adenocarcinomas. Am J Surg Pathology 2000;31:1188–1190.
7. Jain R, Aquino D, Harford WV, Lee E, Spechler SJ. Cardiac epithelium is found infrequently in the gastric cardia. Gastroenterology 1998;114:A160 (Abstract).
8. Chandrasoma P. Controversies of the cardiac mucosa and Barrett's esophagus. Histopathol 2005;46:361–373.
9. Kilgore SP, Ormsby AH, Gramlich TL, et al. The gastric cardia: Fact of fiction? Am J Gastroenterol 2000;95:921–924.
10. Glickman JN, Fox V, Antonioli DA, Wang HH, Odze RD. Morphology of the cardia and significance of carditis in pediatric patients. Am J Surg Pathol 2002;26:1032–1039.
11. Odze RD. Unraveling the mystery of the gastroesophageal junction: A pathologist's perspective. Am J Gastroenterol 2005;100: 1853–1867.
12. Hayward J. The lower end of the oesophagus. Thorax 1961;16: 36–41.
13. Der R, Tsao-Wei DD, DeMeester T, et al. Carditis: A manifestation of gastroesophageal reflux disease. Am J Surg Pathol 2001;25:245–252.
14. Riddell RH. The biopsy diagnosis of gastroesophageal reflux disease, "carditis," and Barrett's esophagus, and sequelae of therapy. Am J Surg Pathol 1996;20(Suppl 1):S31–S50.
15. Oberg S, Peters JH, DeMeester TR, et al. Inflammation and specialized intestinal metaplasia of cardiac mucosa is a manifestation of gastroesophageal reflux disease. Ann Surg 1997; 226:522–532.

16. Allison PR, Johnstone AS. The oesophagus lined with gastric mucous membrane. Thorax 1953;8:87–101.
17. Sarbia M, Donner A, Gabbert HE. Histopathology of the gastroesophageal junction: A study on 36 operation specimens. Am J Surg Pathol 2002;26:1207–1212.
18. De Hertogh G, Van Eyken P, Ectors N, Tack J, Geboes K. On the existence and location of cardiac mucosa: An autopsy study in embryos, fetuses, and infants. Gut 2003;52:791–796
19. Marsman WA, van Sandyck JW, Tytgat GNJ, ten Kate FJW, van Lanschot JJB. The presence and mucin histochemistry of cardiac type mucosa at the esophagogastric junction. Am J Gastroenterol 2004;99:212–217.
20. Chandrasoma PT, Der R, Ma Y, et al. Histology of the gastroesophageal junction: An autopsy study. Am J Surg Pathol 2000;24:402–409.

CHAPTER

8

Pathology of Reflux Disease at a Cellular Level: Part 1—Damage to Squamous Epithelium and Transformation into Cardiac Mucosa

The best way to study cellular changes of reflux is by examining histologic changes across the distal esophagus and proximal stomach without considering either anatomy or the gross appearance of the mucosa. We will use the device of a diagrammatic vertical section for this purpose (Fig. 8.1). When normalcy and the pathologic cellular changes of reflux have been identified, we can superimpose the anatomic and gross mucosal structure on the observed histologic changes (Chapter 11). In this way, we use the most reliable modality (histology) to define the anatomic and gross mucosal changes of reflux. In the next three chapters, we will describe the cellular events of reflux disease. In subsequent chapters, we will correlate these events with anatomic and gross mucosal changes to develop a complete picture of the pathology of reflux disease.

The study of pathology must begin with the understanding of the normal. In the previous chapters, we presented evidence of normalcy in this region as defined by histology, anatomy, physiology, and gross mucosal architecture. This normal state is easy to understand (Fig. 8.1). It is an esophagus that is entirely tubular and entirely lined by squamous epithelium, which transitions to the stomach at the imaginary line of the gastroesophageal junction. This line is marked by the end of the lower esophageal sphincter, the angle of His, and the peritoneal reflection. The stomach, which begins at this line, is entirely saccular, is entirely lined by gastric oxyntic mucosa, and shows rugal folds in the mucosa. We hope the previous chapters have convinced you that there is normally no metaplastic columnar-lined esophagus that is defined as cardiac mucosa, with and without intestinal metaplasia, or oxyntocardiac mucosa. The present belief that still recognizes "a very small amount" or "less than 4 mm" of cardiac mucosa should be recognized as incorrect.

Taking our definition of normal anatomy and histology, it follows that the pathology of reflux must begin with reflux-induced damage of the squamous epithelium of the esophagus. The only other epithelial structure present normally, which is gastric oxyntic mucosa, is designed to withstand gastric juice. However, reflux disease ends with the occurrence of adenocarcinoma, not squamous carcinoma. Therefore, somewhere in the pathogenesis of cellular injury, the squamous epithelium must transform into glandular epithelium. It is well known that this happens in reflux; this is what we call columnar-lined esophagus. Columnar lined esophagus is defined histologically by cardiac mucosa (with and without intestinal metaplasia) and oxyntocardiac mucosa. It is first detectable by microscopy until its extent becomes sufficient to become visible to the naked eye and endoscope.

There has been an unusual dichotomy in the minds of physicians treating patients with reflux, and this persists to date (Fig. 8.2). Reflux disease is seen as a disease with changes in the squamous epithelium, classified by symptoms as typical and atypical, classified by endoscopy as erosive and nonerosive, graded by systems such as the Savary-Miller and Los Angeles systems by endoscopic changes in the squamous epithelium, complicated by the occurrence of ulcers and

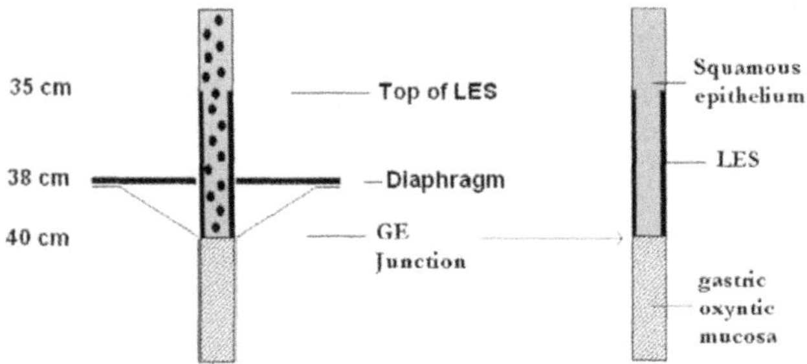

FIGURE 8.1 Vertical section showing the esophagus lined by squamous epithelium (solid gray) and stomach lined by gastric oxyntic mucosa (diagonal stripe). The submucosal glands of the esophagus (black circles), diaphragm, and peritoneal reflection are shown in the figure on the left. On the right is shown the endoscopic appearance where the submucosal glands and peritoneal reflection are not seen and the diaphragm not as clearly definable. The gastroesophageal junction can be recognized histologically as the junction between squamous epithelium and gastric oxyntic mucosa.

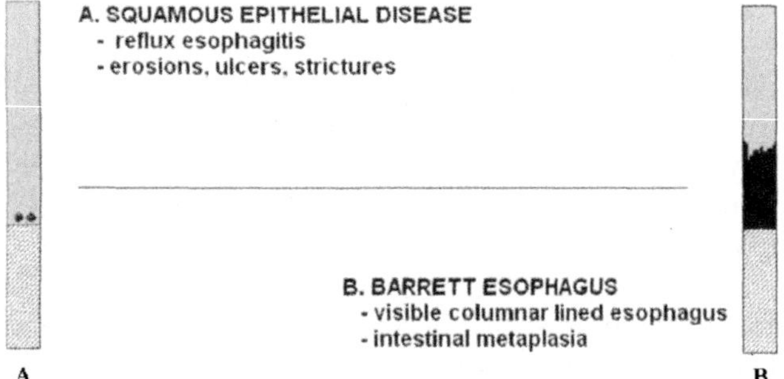

FIGURE 8.2 Gastroesophageal reflux disease affects the squamous epithelium. There is a tendency to artificially separate the disease into two distinct entities: reflux disease (a) manifested by squamous epithelial abnormalities only, and Barrett esophagus (b) manifested by the presence of intestinal metaplasia in a visible columnar-lined esophagus. In reality, both changes almost always occur simultaneously and columnar-lined esophagus without intestinal metaplasia precedes Barrett esophagus.

strictures, and diagnosed by histologic changes in the squamous epithelium. Patients with reflux disease are treated by acid suppression and the success of treatment is defined by the relief of symptoms and healing of erosions. Unless columnar-lined esophagus is seen at endoscopy, these patients are deemed to have only squamous epithelial disease; this is almost always incorrect.

Although everyone agrees that columnar-lined esophagus is caused by reflux, this entity is recognized only when the endoscopist sees a columnar-lined esophagus *and* a biopsy reveals intestinal metaplasia (Fig. 8.2). This is Barrett esophagus; a patient with Barrett esophagus enters an endoscopic surveillance program to detect the occurrence of dysplasia and early adenocarcinoma. When a patient is diagnosed as having Barrett esophagus, scant attention is paid to the fact that he or she commonly also has reflux damage of the residual squamous epithelium. Scant attention is paid to the fact that the columnar-lined esophagus has intestinal metaplasia admixed with cardiac and oxyntocardiac mucosa in various proportions. It is often incorrectly assumed that the entire columnar-lined segment is lined by intestinal metaplastic epithelium in the patient with Barrett esophagus.

Though it has been known for half a century that columnar-lined esophagus is a manifestation of reflux disease, there has never been any recognition of the fact that columnar-lined esophagus can be used to diagnose and classify reflux disease. Nonintestinalized columnar-lined esophagus is completely ignored. Present criteria for the diagnosis of reflux disease are restricted to squamous epithelial changes, both at endoscopy and histology.

The study of the pathology of reflux must begin with a study of the earliest changes in the least affected people. These are patients who do not present to physicians because they have no significant symptoms. They can be studied only at autopsy. In our autopsy study, the only constant abnormalities (by the definitions we proposed in the previous chapter) were (a) the presence of very small amounts of oxyntocardiac mucosa in all patients (in 50%, this was present only in part of the circumference of the squamocolumnar junction), (b) the absence of any cardiac mucosa in 56% of patients, and (c) the very short length of cardiac mucosa in the 44% of patients in whom this epithelium was found (the mean length of cardiac mucosa was 0.036 to 1.038 mm in these patients) (1). It is important to understand that 0.036 mm is equal to 36 μm, which is the equivalent of one foveolar-gland complex. In Marsman et al.'s study (2) of an endoscopically normal clinical population, oxyntocardiac or cardiac mucosa was present in all patients who had the actual squamocolumnar junction in the biopsy specimen. When the actual junction was not present, 90 of 135 patients had neither cardiac nor oxyntocardiac mucosa, indicating that the amount of oxyntocardiac and cardiac mucosa even in this endoscopically normal clinical population was very small.

If oxyntocardiac and cardiac mucosa are reflux-induced columnar metaplastic epithelium in the esophagus, as we suggest, cellular changes of reflux are present in almost all adult humans. It should not be surprising that mild reflux damage is as ubiquitous as carbon dust in the lungs and atherosclerosis in the abdominal aorta. What is surprising is that the evidence of reflux damage in these minimally affected people is the presence of metaplastic columnar epithelium, not changes in the squamous epithelium. Columnar metaplasia is a much more sensitive indicator of reflux damage than the manifestation of reflux esophagitis in squamous epithelium. The medical community has ignored this manifestation of reflux disease by the false device of finding a way to call this normal.

A crucial fact is that the patient with reflux-damaged esophageal mucosa shows a complexity of changes in both squamous and columnar metaplastic epithelium. Every patient very likely has both components, even when the squamous epithelium is normal at endoscopy and histology and even when the endoscopist does not see any columnar-lined esophagus. We cannot stress enough that our lack of ability to see things does not mean that there is no pathologic change. Our methods of visualization have limitations that create error, but only if we depend on them beyond their technical capability of resolution.

The cause of the present epidemic of esophageal adenocarcinoma is unknown and, as such, we are helpless in preventing it. However, we must recognize that ignorance and error may contribute to the epidemic. For over three decades until the 1990s, we declared by edict without proof that what we now recognize as short-segment Barrett esophagus was a normal finding. It is uncertain what effect surveillance of that group would have had on the increasing incidence of adenocarcinoma if it was instituted in 1961 rather than in the 1990s. Today, we declare by edict without proof that the presence of cardiac mucosa and small areas of intestinal metaplasia distal to the squamous epithelium are either normal or of uncertain significance if there is no endoscopic abnormality. If this assumption is wrong, it represents ignorance that can contribute to the continuation of the epidemic. Accurate definition of the problem is essential before the medical community can address it. By ignoring that the problem may exist and recommending that patients with normal endoscopy need not have biopsy, we are not furthering our understanding.

REFLUX-INDUCED DAMAGE OF THE SQUAMOUS EPITHELIUM

Normal Resistance

The structure of the stratified squamous epithelium is designed to withstand injury; squamous cells have tight cell junctions that form an effective barrier that resists penetration by luminal molecules. It has an unknown but very likely small amount of resistance when exposed to gastroesophageal reflux.

The normal resistance of the squamous epithelium to a significant amount of reflux is not great. Esophagogastrectomy represents a model where severe reflux occurs postoperatively because the sphincter mechanism is removed. Dresner et al. (3), in a study of 40 such patients, showed that these patients, rapidly pass through a sequence of changes in the esophageal squamous epithelium, which progress from reflux esophagitis in the squamous epithelium, to columnar metaplasia of cardiac type without goblet cells, and eventually to intestinal metaplasia. We reviewed this study completely in Chapter 7. It is interesting that this model of severe reflux is different from the normal population with mild reflux in the type of reflux-induced damage that is seen. Unlike in the normal autopsy population (1), where mild chronic reflux most commonly produces oxyntocardiac mucosa with much less evidence of reflux esophagitis or cardiac or intestinal epithelia, this postsurgical population with severe acute reflux manifests the exact reverse—the absence of oxyntocardiac (the authors' "fundic-type") epithelium and the presence of severe reflux esophagitis, as well as cardiac and intestinal epithelia. These authors showed that the occurrence of columnar metaplasia is directly related to the presence of greater duodenogastro-esophageal reflux. The severity of reflux determines not only the amount of columnar metaplasia (4–6), but its histologic type. Mild chronic reflux promotes oxyntocardiac mucosa; increasing severity of reflux promotes cardiac mucosa, first without and then complicated by intestinal metaplasia.

There is very likely a difference between the two sexes in terms of squamous epithelial resistance to acid damage. The evidence is that within the same demographic groups, males and females show different manifestations of reflux disease. In particular, long-segment Barrett esophagus and lower esophageal adenocarcinoma are predominantly male diseases. The male predominance decreases as the length of esophagus involved in the metaplastic process decreases. Apart from the epidemiologic evidence, there have been few data to explain this difference between the sexes. We will attempt to explain this gender difference later. There are also ethnic differences in the prevalence of complications of reflux disease; adenocarcinoma of the esophagus is much more common in Caucasians than African Americans in the United States (7). While these differences may also suggest differences in epithelial resistance, other factors such as diet and socioeconomic status are more likely to explain them.

Conclusion: Esophageal squamous epithelium has only slight resistance to gastroesophageal reflux. Reflux-induced changes vary with the severity of reflux.

Nonerosive Reflux Disease (NERD)

The normal esophagus is a dynamic epithelium that continually divides to replace the cells normally lost at the surface. The dividing cells are the adult esophageal progenitor cells, which are found in the basal layer (Fig. 8.3). These cells have a normal proliferative rate; in sections stained by Ki67, a marker of proliferation, cells in the mitotic cycle are normally limited to the suprabasal 2 to 3 cell layers (Fig. 8.4). This is the normal proliferation necessary to maintain the squamous epithelium. After division, the cells move up in the epithelium, becoming nonmitotic, terminally differentiated squamous cells by the midpoint of the epithelium; these cells then reach the surface to be desquamated after about 7 to 8 days from the time they were generated from the adult stem cell (8).

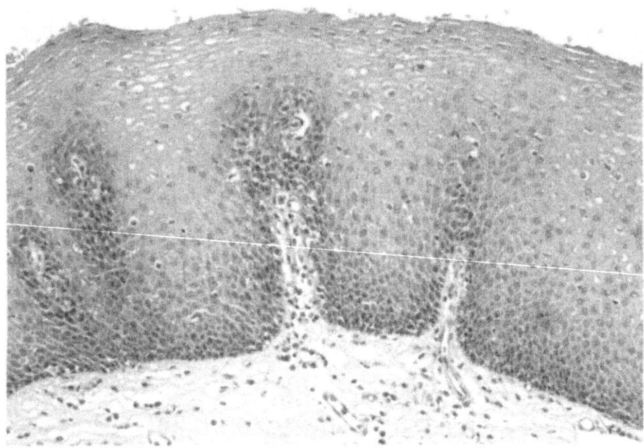

FIGURE 8.3 Normal squamous epithelium showing short papillary extensions and a 2- to 3-layer basal layer zone. (Cross reference: Color Plate 1.4)

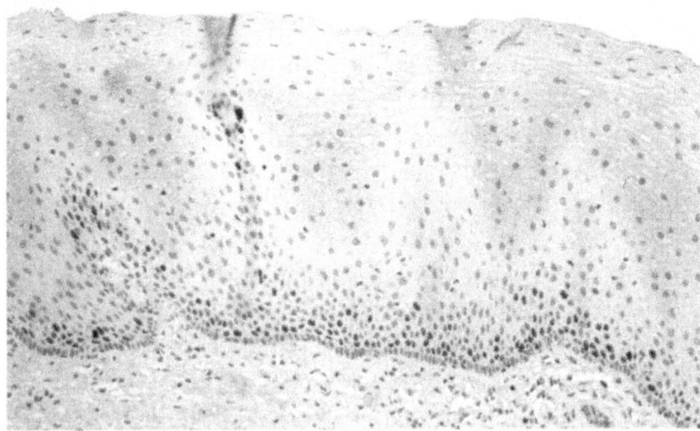

FIGURE 8.4 Normal squamous epithelium stained with Ki67 showing the proliferative zone above the adult progenitor cell zone in the basal layer. The proliferative zone is normally 2 to 3 cell layers only.

As the amount of acid exposure increases, the ability of the squamous epithelium to resist injury is progressively overwhelmed. The pathologic changes of reflux cause (Table 8.1): (a) symptoms of reflux, typically heartburn and regurgitation; (b) changes that are visible at endoscopy, the earliest of which is erosion (erosive esophagitis); (c) changes that are invisible by standard endoscopy but visible with magnification endoscopy; (d) changes that are visible by histology; and (e) changes that are invisible by all modalities available at the present time. Nonerosive reflux disease (NERD) is defined as being present when classical symptoms of reflux are not associated with any visible endoscopic abnormality, assuming that only standard endoscopy is used. This is not a logical definition; the condition these patients have should be termed "symptomatic reflux without visible endoscopic change." Defining NERD by the presence of symptoms is also illogical because we know that asymptomatic people develop complications of reflux such as adenocarcinoma. NERD should therefore be considered a spectrum of early disease that includes symptomatic and asymptomatic patients defined by any cellular change caused by reflux.

The first change in reflux disease is most likely an increased rate of the loss of surface cells resulting from acid-induced damage. This stimulates a compensatory increase in the rate of proliferation of the basal esophageal progenitor cells to maintain the epithelium. This increased rate of cell turnover in the squamous epithelium is manifested by an expansion of the proliferative basal cell zone. This is seen histologically as basal cell hyperplasia (Fig. 8.5). The expression of proliferative markers such as Ki67 is increased with more than the usual two to three basal layers containing mitotically active cells (Fig. 8.6). These are the first recognizable histologic criteria of reflux. No studies have evaluated Ki67 expression rates in reflux disease; we have observed this to be a common change.

Basal cell hyperplasia is usually accompanied by elongation of the lamina propria papillae, imparting a serration of the base of the squamous epithelium (Fig. 8.5). The elongated papillae reach near the surface, and their blood vessels can be seen at endoscopy if newer high-magnification endoscopic instruments are used (9). Studies evaluating the criteria of basal cell hyperplasia and papillary elongation have shown a

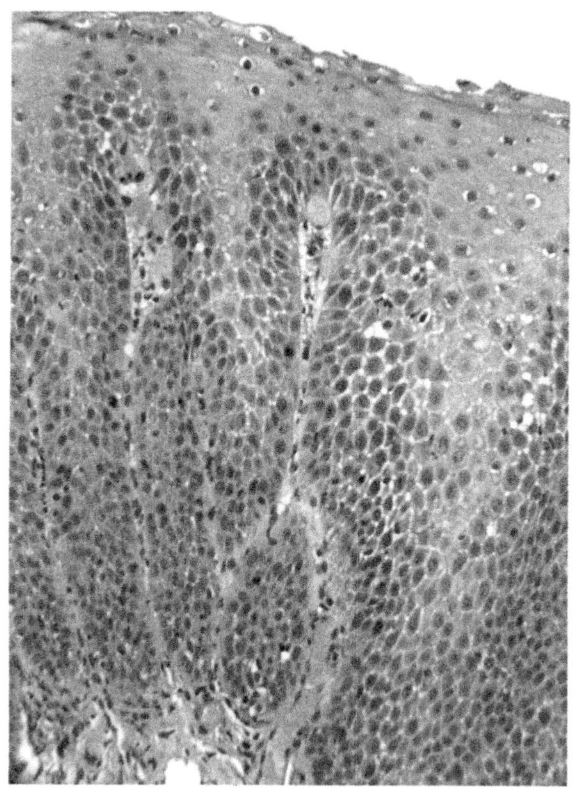

FIGURE 8.5 Squamous epithelium showing basal cell hyperplasia and papillary elongation. The basal layer region is expanded from normal and now occupies more than 20% of the epithelial thickness. The lamina propria papillae extend into the epithelium to a level that is more than 60% of its total thickness. Intraepithelial eosinophils are present. There are spaces between the epithelial cells (dilated intercellular spaces). This is the typical histologic appearance of untreated reflux esophagitis. (Cross reference: Color Plate 2.2)

significant correlation between clinical reflux when the basal cell layer has a thickness that is greater than 20% of total epithelial thickness and the papillae are greater than 60% of the total thickness of the epithelium (10). These are accepted criteria for the histologic diagnosis of reflux disease in biopsy specimens at the present time; they are of low specificity and sensitivity and not very useful for diagnosis.

It must be recognized that basal cell hyperplasia and papillary elongation are nonspecific manifestations of squamous epithelium to any injury resulting in an increased rate of cell loss. In the skin, these changes are typical of psoriasis and numerous agents that induce psoriasiform dermatitis; the latter is a common complication associated with a wide range of therapeutic agents. The possibility that drugs may produce changes in esophageal squamous epithelium that mimic the changes of reflux has never been systematically investigated. Papillary hyperplasia may occur in achalasia due to chronic irritation resulting from chronically retained food in the esophagus (Fig. 8.7).

More than half the patients with classical symptoms of gastroesophageal reflux disease show no endoscopic abnormality (NERD). Newer, high-magnification endoscopy, whose usage is still largely experimental, shows earlier changes in vascular patterns. The study by Kiesslich et al. (9) assessed the value of endoscopic and histological markers for the prediction of NERD before and after treatment with 20 mg of esomeprazole. Patients presenting for upper endoscopy were strati-

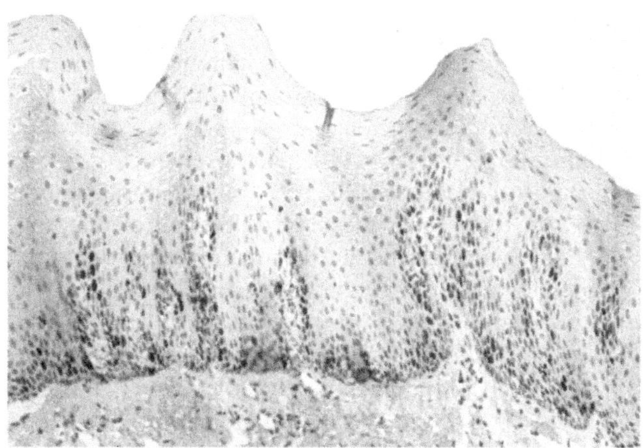

FIGURE 8.6 Squamous epithelium basal cell hyperplasia and papillary elongation stained by immunoperoxidase technique to show Ki67 staining. There is an increase in the number of Ki67-positive cells in the proliferative zone, representing compensation to increased surface cell loss.

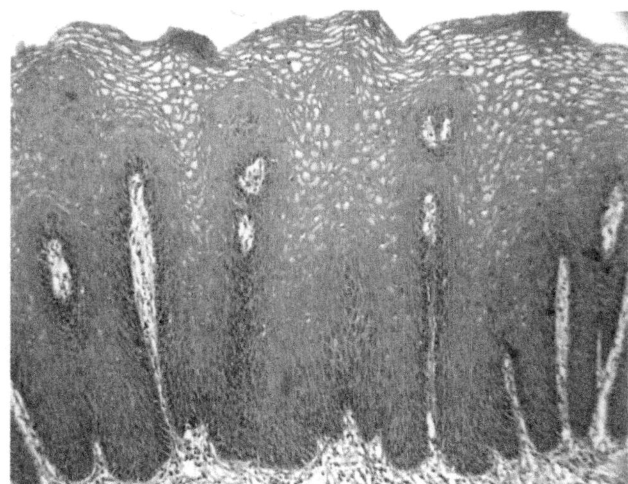

FIGURE 8.7 Esophageal epithelium in a patient with achalasia of the cardia. There is papillary elongation showing the lamina propria papillae extending into the upper part of the squamous epithelium. There are a few intraepithelial lymphocytes, and basal cell hyperplasia is not prominent.

fied into reflux and nonreflux patients (control group) based on the presence and absence, respectively, of typical reflux symptoms. Using magnifying endoscopes, minimal change esophagitis was defined by the presence of vascular injection or vascular spots above the Z line, villous mucosal surface, and islands of squamous cell epithelium below the Z line. Targeted and random biopsies were taken below and above the Z line. Patients with endoscopically visible classical signs of esophagitis (using the Los Angeles classification A–D) or histologically proven Barrett's esophagus were not further investigated in the study (i.e., they dropped out). The esophageal specimens were histologically evaluated for erosions, infiltration with leukocytes, hyperplasia of basal cells, and length of papillae. Patients with NERD were treated with 20 mg of esomeprazole per day for 4 weeks and reevaluated by endoscopy as described before.

The study focused on 39 patients with heartburn and 39 patients without reflux symptoms (controls). Patients with NERD significantly ($p = 0.005$) more often showed endoscopic signs of minimal change esophagitis at histology (27 of 39) than the control group (8 of 39). An increased length of papillae (14 of 39 versus 2 of 39; $p = 0.005$) and basal cell hyperplasia (17 of 39 versus 4 of 39; $p = 0.009$) was significantly more common in the heartburn group. After treatment with esomeprazole, no significant endoscopic or histologic differences between the NERD and control group could be observed.

The authors concluded that minimal change esophagitis can be seen with high-resolution magnifying endoscopy. By combining endoscopic and histological markers, NERD can be predicted with a sensitivity of 62% and a specificity of 74%. Treatment with esomeprazole for 4 weeks reverses the slight alterations to normal.

This study, in reality, shows that both magnifying endoscopy and histologic changes, while useful in identifying early changes in the squamous epithelium of patients with reflux, are worthless in the diagnosis of NERD in an individual patient because they are present in the control patients. The use of these criteria will therefore result in a significant false positive diagnosis of NERD. It is interesting that magnifying endoscopy appears to be more sensitive in detecting early changes than the presently used histologic markers.

If one removes the classification of patients as reflux or not based on the presence or absence of symptoms, the results of this study become very interesting. In all, 35 of 78 patients showed minimal change esophagitis by magnifying endoscopy (27 in patients with and 8 in patients without symptoms). It is possible, and highly likely, that these patients all have reflux disease based on the principle that they have a morphologically recognizable abnormality and the known fact that cellular changes of reflux can occur in asymptomatic patients. We have to learn that reflux cannot be defined by symptoms; the use of cellular changes is needed to identify asymptomatic reflux. Unfortunately, neither abnormalities detected at magnification endoscopy nor histologic changes are sufficiently specific features of reflux that can be used to define the disease (Table 8.1).

What this study shows is that magnifying endoscopy (and to a lesser extent, histologic changes) detects early morphologic changes in 27 of 39 patients with symptomatic reflux and 8 of 39 patients with asymptomatic reflux. The technology is better than standard endoscopy, which fails to detect early changes in any of these patients. However, the 12 of 39 patients with symptoms who remain negative by magnifying endoscopy are still highly likely to have invisible changes of reflux (unless their symptoms are incorrectly attributed to reflux). One must consider the obvious possibility that these patients may have cellular changes that are not being looked for in this study. One such change is dilated intercellular spaces (discussed later); another is columnar metaplasia that is endoscopically not visualized (discussed later). Also, it is unknown whether the 31 of 39 asymptomatic patients have evidence of reflux-induced changes because it is obvious that even the most sensitive endoscopic methods (and histologic markers) fail to detect the earliest changes because of their technical limitations.

The fact that the postomeprazole treatment biopsies of the patients in this study caused a reversal of the changes toward the controls shows that acid is the likely cause of basal cell hyperplasia, papillary elongation, and vascular changes in patients with reflux disease.

Though caused by acid in patients with reflux disease, these are relatively nonspecific changes that are seen with many causes of esophageal surface epithelial injury such as reflux, caustic chemicals, drugs, pills, allergy, achalasia (Fig. 8.7), or infectious agents. None of these "early markers for reflux disease" whether they are detected by magnifying endoscopy or histology, is likely to be specific for reflux. This was well shown in a study from Tokyo by Takubo et al. (11), which evaluated the sensitivity and specificity of histologic changes in the esophageal squamous mucosa in patients with gastroesophageal reflux disease, including dilated intercellular spaces, balloon cells, intrapapillary vessel dilation, elongated papillae, basal cell hyperplasia, acanthosis, intraepithelial eosinophils, Langerhans cells, and p53 protein overexpression. It should be noted that this study looked at

TABLE 8.1 Changes That Occur in the Squamous Epithelium as a Result of Gastroesophageal Reflux

Change	Stage	Cause	Sensitivity	Specificity
Increased Ki67 expression	Very early	Reflux; other causes of increased cell loss	Very high	Low
Basal cell hyperplasia	Early	Reflux; other causes of increased cell loss	Intermediate	Low
Papillary elongation	Early	Reflux; other causes of increased cell loss	Intermediate	Low
Abnormal vascular pattern in mucosa	Early	Reflux; other causes of increased cell loss	High	Low
Intraepithelial neutrophils	Intermediate	Reflux; infections, erosion	Intermediate	Low
Intraepithelial eosinophils	Intermediate	Reflux; allergy	Intermediate	Intermediate
Dilated intercellular spaces	Early	Reflux; other causes of cell injury	High	Low
Increased permeability	Early	Reflux; other causes of cell injury	High	Low
Chemokines	Early	Reflux; other causes of cell injury	High	Unknown
Esophageal pain	Early	Reflux (when typical)	Intermediate	High
Erosion	Late	Reflux; other causes of cell injury	Low	Intermediate
Ulceration	Late	Reflux; other causes of cell injury	Low	Intermediate
Strictures	Late	Reflux; other causes of cell injury	Low	Intermediate
Genetic changes	Early	Reflux; other causes of cell injury	Intermediate	Unknown
Columnar metaplasia	Very early	Reflux only	Very high	100%
Intestinal (Barrett-type) metaplasia	Late	Reflux only	Low	100%
Adenocarcinoma	Late	Reflux only	Very low	100%

histologic changes that are more numerous than that studied by Kiesselich et al. (9). To define a set of histologic changes that are invariably reflux associated, the researchers examined the histologic changes in esophageal specimens from normal controls, patients with GERD, patients without GERD but with a suspicion of other pathology, and patients with esophageal carcinoma. A definitive set of reflux-associated histologic changes could not be defined. The authors concluded that histologic changes indicative of GERD are likely to be found somewhere in the esophagus in all patients with GERD, but these changes are nonspecific.

Conclusion: Early, nonerosive reflux disease (NERD) can be detected by magnifying endoscopy and histologic criteria. However, these techniques are relatively insensitive and nonspecific. Their diagnostic value in reflux disease is limited; they certainly cannot be used to define reflux disease.

Erosion and Ulceration (Erosive Esophagitis)

When the rate of surface cell loss exceeds the compensatory capability of the squamous epithelium, erosion occurs. An erosion results in denudation of the epithelium, exposing the underlying lamina propria. This most likely first occurs in the thinned suprapapillary epithelium. Endoscopically, erosions are seen as small yellow areas, often linear, in the white squamous epithelium; they are surrounded by erythema (Fig. 8.8). They define "erosive esophagitis." The severity of erosions in terms of size and number are used to grade erosive esophagitis endoscopically by the Los Angeles and Savary-Miller systems (12).

An ulcer differs from an erosion in that it involves the full thickness of the mucosa, including the muscularis mucosae which is the deepest layer of the mucosa (Fig. 8.9). Erosions and ulcers are not specific for reflux disease. Any cause of epithelial surface injury—corrosives, pills, infectious agents like herpes simplex, and cytomegalovirus—can cause erosions and ulcers in the esophagus. The use of erosions to grade reflux is clinically feasible because over 99% of erosions in immunocompetent patients are caused by reflux and their distribution is typically in the lower esophagus and different than pill esophagitis, which occurs more proximally. Despite this, there is possibility of diagnostic error of erosive reflux disease because of the lack of complete specificity of the finding of erosions. This risk of error is probably much higher in immunocompromised patients in whom infectious causes of erosion are relatively common.

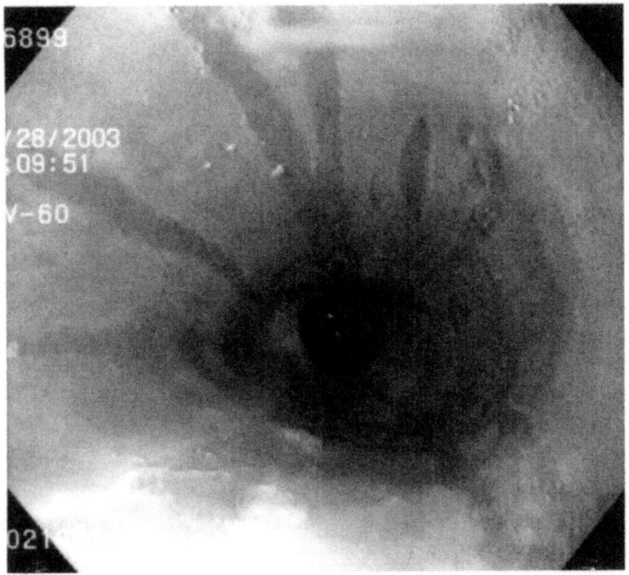

FIGURE 8.8 Erosive esophagitis, endoscopic appearance, showing linear erosions extending into the squamous epithelium from the gastroesophageal junction. The erosions are multiple and characterized by erythema. (Cross reference: Color Plate 2.1)

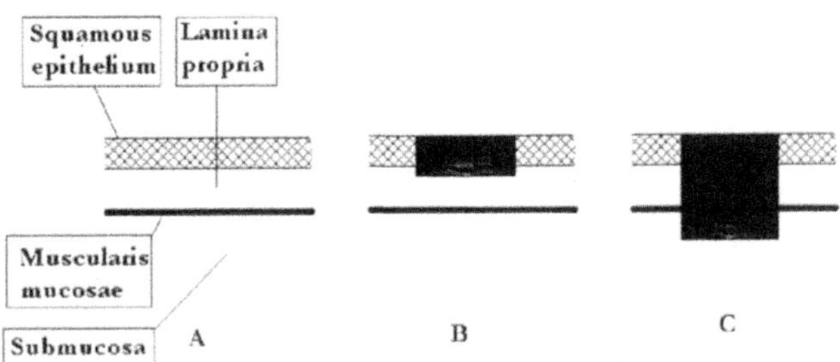

FIGURE 8.9 The difference between an erosion and ulcer. An erosion involves partial thickness of the mucosa extending only into the lamina propria under the squamous epithelium (b). An ulcer involves the muscularis mucosae (the deepest point of the mucosa) and extends below it (c); The muscle wall of the esophagus is under the submucosa and is not shown (a).

With ulceration, fibrosis of the submucosa and muscle wall may follow. Contraction of the collagen may occur both longitudinally, leading to shortening of the esophagus, and circumferentially, leading to strictures. In the early days, the main clinical problem associated with reflux disease was the occurrence of ulcers and strictures (13). With the increasing efficacy and use of acid suppressing agents in the treatment of reflux disease, complicated and nonhealing ulcers and complex strictures have become relatively uncommon, suggesting that acid is primarily responsible for the erosions and ulcerations that occur in squamous epithelium.

Conclusion: Erosion and ulceration are evidence of severe surface injury, often caused by reflux but also caused by other agents.

Separation of Squamous Cells—"Dilated Intercellular Spaces"—and Increased Epithelial Permeability

With increasing and probably more sustained damage, the cell junctions between squamous cells separate, resulting in dilated intercellular spaces. In a beautiful experimental study from Tulane University, Tobey et al. (14) showed by electron microscopic measurement that an increase in the intercellular spaces resulted from acid exposure of esophageal squamous epithelium. In this study, endoscopic biopsy specimens from 11 patients with recurrent heartburn (6 had erosive esophagitis, 5 had a normal-appearing squamous epithelium at endoscopy) and 13 control patients (no symptoms or endoscopic evidence of reflux esophagitis) were examined using transmission electron microscopy. Computer-assisted measurement of the intercellular space diameter in the electron photomicrographs was performed in each specimen. The intercellular space diameter was significantly greater in specimens from patients with heartburn (irrespective of whether or not they had endoscopic erosions) than in the control patients. Space diameters of 2.4 microns or greater were present in 8 of 11 patients with heartburn and in none of the controls.

The authors concluded that dilated intercellular spaces are a feature of reflux damage to the esophageal squamous epithelium, represent a morphologic marker of increased paracellular permeability, and may help explain the occurrence of heartburn in patients without endoscopic abnormalities.

In a study of patients with NERD compared with controls (no symptoms, pH study negative), Caviglia et al. (15) reported that patients with NERD (irrespective of whether their 24-hour pH study was normal or abnormal) had a significantly greater dilatation of intercellular spaces than controls. They, like Tobey et al., used transmission electron micrographs for computer-assisted measurement. The authors concluded that the criterion of dilated intercellular spaces is an objective structural marker of GERD symptoms.

Villanacci et al. (16) also showed that the severity of dilatation of intercellular spaces correlates with the severity of reflux and that this can be assessed by a semiquantitative morphometric study of routine histologic sections, bringing this criterion into the realm of routine diagnostic histopathology. In Villanacci et al.'s study, the squamous epithelium in biopsies from 21 patients with reflux symptoms was assessed by light microscopy for the degree of separation of the squamous cells. In a beautiful study, the authors developed criteria for grading dilated intercellular spaces on a dilated intercellular spaces score of 0 to 3 and confirmed its validity against a computer-assisted morphometric analysis. The grading was as follows: DIS-0 = absence of visible intercellular spaces; DIS-1 = focal presence of small-sized intercellular spaces; DIS-2 = moderate presence of large-sized intercellular spaces; DIS-3 = diffuse presence of very large intercellular spaces.

The authors showed that the DIS score is strongly associated with (a) the score of esophageal symptoms and (b) the histological score of esophagitis. Based on this study, the authors suggested that DIS is a useful light microscopic criterion for the diagnosis of reflux disease. In particular, because it occurs in noneroded squamous epithelium, it can be used to confirm nonerosive esophagitis where the endoscopy is normal.

In our study of esophageal biopsies, we noticed the frequent presence of dilated intercellular spaces as well as the variation in severity in biopsies from patients with reflux (Figs. 8.10–8.12). We have not correlated this with clinical reflux symptoms or 24-hour pH, largely because we believe the results of the studies by Tobey et al. and Villanacci et al.'s and do not feel the need to confirm them. We have recently started reporting dilated intercellular spaces as suggesting reflux disease, largely because dilated intercellular spaces has recently become more increasingly recognized in the United States as a change that may be useful in the diagnosis of reflux disease.

Squamous epithelial separation in patients with reflux has been shown to be associated with an increase in the permeability of the epithelium, permitting entry and passage of small luminal molecules. Tobey et al. (17) tested the changes in permeability to molecules of dextran of varying sizes and epidermal growth factor in an *in vitro* model of rabbit esophageal epithelium when exposed to acid and acid-pepsin damage. They showed that luminal HCl (pH 1.1), or HCl (pH 2.0) + pepsin 1 mg/ml, for 30 minutes caused a linear increase in permeability to 4 kD dextrans as well as an increase in permeability to 6 kD epidermal growth factor and dextrans as large as 20 kD but produced no gross erosions or histologic evidence of cell necrosis. Transmission electron microscopy, however, documented the presence of dilated intercellular spaces.

The authors concluded that in nonerosive acid-damaged esophageal epithelium, dilated intercellular spaces develop in association with and as a marker of increased permeability. This change in permeability upon acid or acid-pepsin exposure is substantial, permitting dextran molecules as large as 20 kD (33 A) and luminal EGF at 6 kD to diffuse across the acid-damaged epithelium and by so doing enabling it to access receptors on epithelial basal cells (Fig. 8.13). The authors hypothesized that the increased permeability to EGF may in part account for the development of a reparative phenomenon on esophageal biopsy in patients with nonerosive reflux disease known as basal cell hyperplasia.

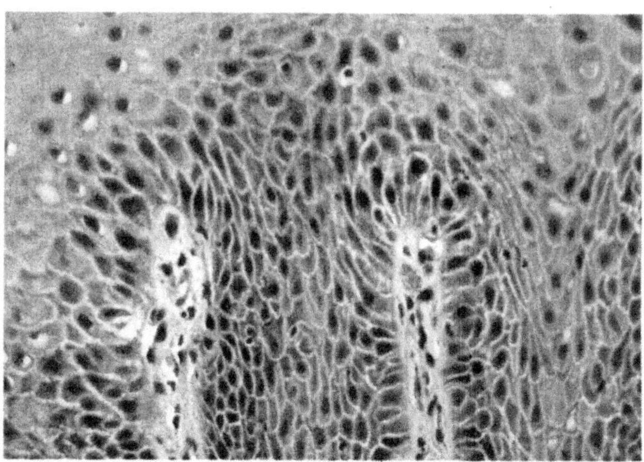

FIGURE 8.10 Dilated intercellular spaces in reflux-damaged squamous epithelium showing the separation of the squamous cells. As the cells separate, the spaces between them become visible and the intercellular bridges between cells become prominent. This is a case of mild reflux; there is no eosinophil infiltration, and basal cell hyperplasia is marginal. (Cross reference: Color Plate 2.3)

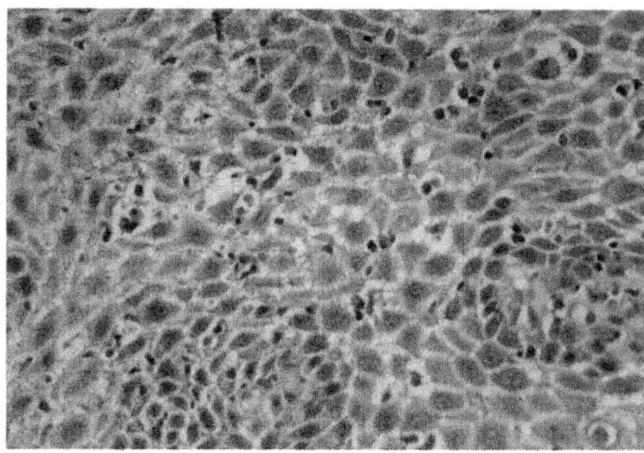

FIGURE 8.11 More pronounced separation of intercellular spaces in a patient with severe reflux disease showing also the presence of numerous intraepithelial eosinophils. This was an adult with a markedly abnormal pH score and no evidence of allergy.

The specificity of the squamous epithelial cell separation is unknown—that is, it is not known whether injurious agents in the refluxate other than acid can produce the same change. "Dilated intercellular spaces" is a new term applied to a well-recognized change in squamous epithelium. In the skin, this change is recognized as intercellular edema or "spongiosis." Spongiosis in squamous epithelia represents a common and relatively nonspecific manifestation of cellular injury, and there is no reason to believe otherwise in the esophageal squamous epithelium. Dilated intercellular spaces are likely to be a common change associated with any esophageal injury and therefore not a specific feature for reflux disease. For example, the most severe examples of dilated intercellular spaces occur in allergic (eosinophilic) esophagitis, a disease produced by a mechanism that has nothing to do with reflux (Fig. 8.12). As such, its specificity in the diagnosis of reflux remains unproved, although the presence of dilated intercellular spaces in patients with reflux damage is likely to be caused by acid rather than by another component of the refluxate.

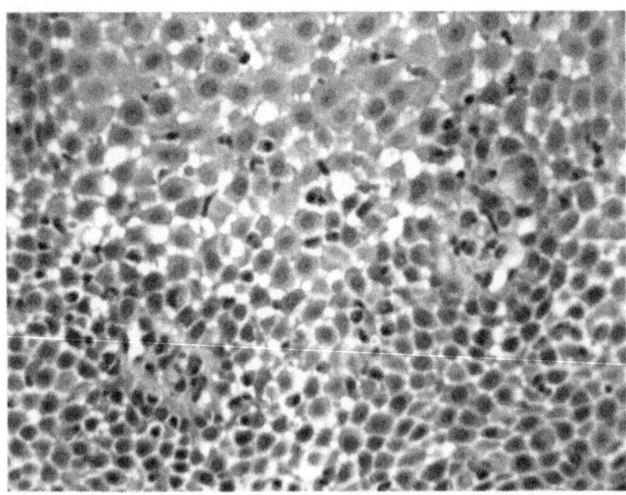

FIGURE 8.12 Eosinophilic esophagitis showing basal cell hyperplasia, numerous (>20/high power field) eosinophils in the lamina propria and markedly dilated intercellular spaces. This was from a child with allergic esophagitis without evidence of reflux. The distinction between eosinophilic esophagitis in the adult and reflux esophagitis with unusual intraepithelial eosinophilia is not possible by histology alone.

Conclusion: Squamous cell separation (dilated intercellular spaces) is an early but nonspecific change seen in reflux disease. In patients with reflux, it is caused by acid and makes the epithelium increasingly permeable, allowing size-dependent entry of molecules in the refluxate into the epithelium.

Esophageal Pain

Tobey et al.'s study shows that acid-induced separation of squamous cells makes the epithelium more permeable, permitting small molecules to enter the epithelium. It is important to recognize that the degree of separation probably acts like a filter; the greater the separation, the larger the size of molecule permitted to enter the epithelium and the greater the distance to which the molecules penetrate within the epithelium (Fig. 8.13).

Esophageal pain (heartburn) is almost certainly the result of this increased permeability and explains how pain can occur without erosion. Sensory nerve endings are present in the epithelium, and their stimulation by molecules entering the epithelium most likely causes heartburn (Fig. 8.13). The fact that this symptom is often effectively and immediately controlled by antac-

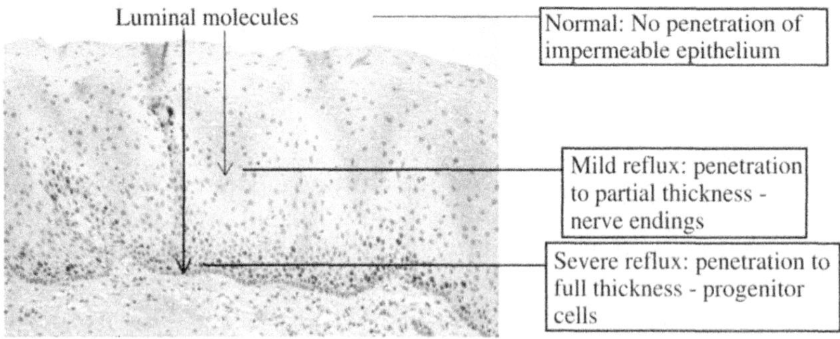

FIGURE 8.13 Potential effects of increased permeability of squamous epithelium resulting from reflux-induced damage. The normal epithelium is impermeable and forms an effective barrier to the entry of luminal molecules. With increasing permeability, molecules of increasing size enter the epithelium to varying levels. This can stimulate sensory nerve endings present in the epithelium and reach the proliferative pool of basal cells where they can induce genetic changes.

ids provides evidence that stimulation of nerve endings is largely an acid-dependent phenomenon. However, it is quite possible that molecules other than hydrogen ions can stimulate nerve endings in a permeable epithelium. This would explain how some patients continue to have heartburn while on effective acid suppressive medication; these episodes of heartburn have been shown by impedance studies to be caused by nonacid reflux. In these cases, pain is likely the result of stimulation of nerve endings in damaged epithelium by nonacid molecules; relief of pain will result only when the squamous epithelial permeability has been reversed.

The probability is that there are qualitative and quantitative differences in patients with regard to stimulation of nerve endings by luminal molecules. Patients with high resistance to stimulation will tend to be asymptomatic even with significant reflux. Patients with sensitivity restricted to acid will respond immediately to acid suppression. Those with sensitivity of nerve endings to nonacid components of the refluxate may continue to experience pain after acid suppression and will only be painless when the epithelial permeability increase has been controlled by effective acid suppression.

Conclusion: Esophageal pain in reflux is likely caused by stimulation of intraepithelial nerve endings by molecules entering the epithelium made permeable by acid.

Leukocytic Infiltration of Squamous Epithelium

One of the recognized criteria for a histologic diagnosis of reflux esophagitis is the presence of intraepithelial granulocytes (Fig. 8.11). Both neutrophils and eosinophils can be seen; eosinophils are considered more specific for reflux disease because neutrophils are more common in nonreflux causes of esophagitis (10). Neutrophils can be seen in any cause of infection or erosion of the esophagus. Even eosinophils are not specific for reflux; they can be seen in large numbers in patients with eosinophilic (allergic) esophagitis, a disease that is common in children but increasingly recognized in adults (Fig. 8.12). The presence of intraepithelial leukocytes has a high diagnostic specificity but a low sensitivity in patients with symptomatic reflux (18). In a study by Bowrey et al. (19), the presence of neutrophils in the epithelium was the only histologic criterion that correlated with reflux as assessed by a 24-hour pH test.

The literature suggests that a small number of intraepithelial eosinophils can be seen "normally" in the distal esophageal squamous epithelium (20). This is an incorrect statement to make in the absence of a reliable (100% sensitive) diagnostic criterion for gastroesophageal reflux. It is our experience that the presence of even rare intraepithelial eosinophils is very likely a fairly specific diagnostic criterion (assuming that eosinophilic esophagitis has been excluded) of reflux. The sensitivity of eosinophils is probably not high, and the absence of intraepithelial eosinophils is a common occurrence in patients with proven reflux.

The reason for granulocyte extravasation into intact squamous epithelium is not known. It can be hypothesized that injurious molecules in the refluxate entering the reflux-damaged permeable squamous epithelium cause cell damage with the release of cytokines (chemokines) that are chemotactic for leukocytes.

Chemokines (or cytokines) have been reported to be present in the squamous epithelium in patients with reflux disease. In a study by Isomoto et al. (21) from Nagasaki, Japan, 32 outpatients with reflux esophagitis, graded according to the Los Angeles classification, and 13 normal controls were studied. Paired biopsy specimens were taken from the esophagus 3 cm above the gastroesophageal junction; one biopsy was snap-frozen for measurement of mucosal levels of interleukin 8 (IL-8), monocyte chemoattractant protein 1 (MCP-1), regulated on activation normal T-cell expressed and presumably secreted (RANTES), and IL-1 beta by enzyme linked immunosorbent assays, while the other was formalin-fixed for histopathological evaluation. IL-8, MCP-1, and RANTES levels were significantly higher in esophageal mucosa of reflux esophagitis patients than those of the controls. IL-8 levels correlated significantly with the endoscopic severity of reflux esophagitis. Basal zone hyperplasia and papillary elongation, histopathological hallmarks of reflux esophagitis, were both associated with higher levels of IL-8 and MCP-1. The presence of intraepithelial neutrophils and eosinophils, which also indicate reflux esophagitis, was associated with high levels of IL-8 and RANTES, respectively. There were no significant differences in IL-1 beta levels between the reflux esophagitis and control groups, but IL-1 beta levels correlated significantly with the IL-8 production. The authors concluded that chemokines produced locally in the esophageal mucosa may be involved in the development and progression of reflux esophagitis.

In a follow-up study, these authors examined patients with endoscopy-negative reflux disease (NERD), taking biopsies of normal squamous epithelium 3 cm above the gastroesophageal junction (22). They measured interleukin-8 messenger ribonucleic acid (mRNA) expression levels assessed by real-time quantitative polymerase chain reaction. The relative IL-8 mRNA expression levels were significantly higher in esophageal mucosa of patients with endoscopy-negative GERD than those of the controls. The presence of basal zone hyperplasia and intraepithelial neutrophils, histopathological hallmarks of GERD, were associated with higher levels of IL-8 mRNA.

Interleukin-8 is involved in the chemoattraction of granulocytes to sites of inflammation and injury and its presence in squamous epithelium at a stage before the occurrence of erosions strongly supports the presence of cell injury in the epithelium in patients with reflux-induced damage.

Conclusion: Molecules entering the squamous epithelium cause cell damage and release of cytokines (chemokines) that are chemoattractive to leukocytes. Intraepithelial neutrophils and eosinophils are not specific for gastroesophageal reflux.

Activation of Cyclo-Oxygenase-2

There is evidence of the presence of genetic alterations in reflux-damaged squamous epithelium. One of the most significant genetic changes to be reported in esophageal squamous epithelium in reflux disease is activation of the cyclo-oxygenase-2 gene. There is evidence that this gene may be involved in multiple steps in the pathway that leads to adenocarcinoma through Barrett esophagus.

Kuramochi et al. (23) reported a study that was aimed at elucidating the role of cyclo-oxygenase 2 (Cox-2) and the isoform Cox-1 in the development of esophageal adenocarcinoma. The gene expressions (mRNA levels) of Cox-2 and Cox-1 were measured by real-time quantitative polymerase chain reaction in tissues from normal esophagus with and without erosive gastroesophageal reflux disease, Barrett esophagus, dysplasia, adenocarcinoma, and in healthy gastric antrum. All tissues were purified by laser capture microdissection from endoscopic or surgical resection specimens.

Median Cox-2 gene expression did not differ significantly among the esophageal control groups but was elevated 5-fold in Barrett's esophagus, 8-fold in dysplasia, and 16-fold in esophageal adenocarcinoma compared to normal esophageal controls with no erosive gastroesophageal reflux disease. Erosive gastroesophageal reflux disease tissue had higher median Cox-2 expression than that in a normal esophagus.

In a follow-up study, the same group studied Cox-2 gene expression by the same methodology in squamous epithelia obtained by biopsies taken 3 cm above the squamocolumnar junction in (a) normal control subjects (24-hour pH normal; endoscopy negative for erosive esophagitis), (b) patients with confirmed gastroesophageal reflux disease (positive 24-hour pH study and histologic evidence of reflux esophagitis or carditis or nondysplastic Barrett esophagus) before Nissen fundoplication, and (c) the same patients at least 6 months after Nissen fundoplication. Before the post-Nissen endoscopy, this group was confirmed at endoscopy to have an intact fundoplication.

This study showed that Cox-2 gene expression was significantly higher in the preoperative biopsies of the patients with gastroesophageal reflux disease compared to the normal control. In addition, Cox-2 expression was significantly reduced in the postfundoplication biopsies compared with the matched preoperative biopsies in these patients. The postoperative biopsies showed that the Cox-2 expression had reached levels that were similar to the normal controls.

This study suggests that some molecule in the refluxate interacts with the squamous epithelium of the esophagus in a manner that causes activation of the Cox-2 gene. It also shows that removal of this molecule causes this genetic change to reverse. Because fundoplication prevents reflux completely, it is not possible to determine which specific molecule in the refluxate is involved in the Cox-2 gene activation.

This is a finding of profound importance because it indicates that molecules that have the ability to produce genetic changes have gained access to the proliferating stem cell pool in the basal region of the squamous epithelium. Genetic transformations only occur in cells that undergo mitoses, and the presence of any genetic change is indirect evidence for a markedly increased permeability of the squamous epithelium.

Conclusion: Genetic changes occur as a result of action of refluxate molecular interactions with the squamous epithelial stem cell pool.

Loss of Sphincter Function

Lower esophageal sphincter abnormalities are demonstrable very early in reflux disease. The earliest changes relate to sphincter length with shortening of the abdominal component of the sphincter often being the first recognizable change (24, 25). This occurs long before there is erosion, ulceration, and extension of the process into the muscle wall where the sphincter action resides. The mechanism whereby this happens is unknown, and there is little speculation in the literature as to its pathogenesis. This is understandable, because the normal mechanism of action of the lower esophageal sphincter is also little understood.

This is the place for speculation in an attempt to provide at least a logical explanation that can stimulate research (Fig. 8.14). It is well known that sphincter function is not dependent on external innervation of the esophageal muscle because denervation does not cause loss of sphincter function. This must mean that maintenance of tonic muscle contraction in the sphincter is dependent on local phenomena. Let us imagine that this is some type of intramural reflex arc, the effector element of which is the motor nerve supplying the sphincter muscle and the center is a ganglion cell in the myenteric plexus. Where is the sensory input to this reflex arc? It is likely that this is related to the lumen, which is the first point of contact of the food bolus being propelled down the esophagus. If the sensory component of the reflex arc is represented by the nerve endings in the squamous epithelium, it is possible that damage, inhibition, or stimulation of these nerve endings can interfere with the local reflex arc and cause the sphincter muscle to lose its tonically contracted state on which the resting pressure of the sphincter depends. This will be seen manometrically as a shortening of the lower esophageal sphincter, beginning at its distal end. Similarly, stimulation and inhibition of intraepithelial sensory nerves by refluxate molecules can explain the occurrence of motility abnormalities in the esophageal body that may be associated with reflux disease.

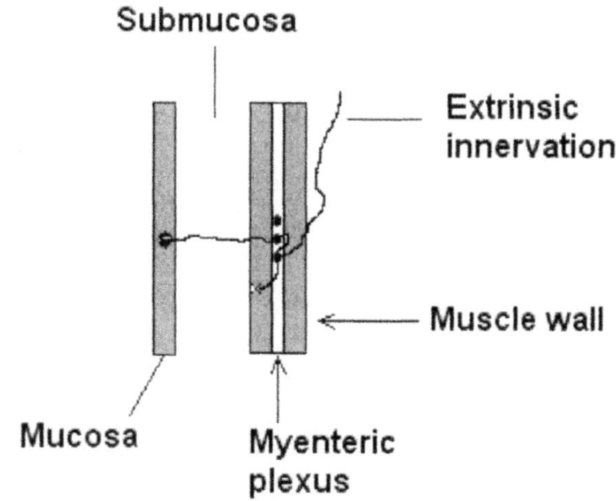

FIGURE 8.14 Innervation in the esophagus. Afferent nerve endings in the squamous epithelium transmit sensory input to the nerve center in the myenteric plexus where the neuronal bodies are present. The nerve cell collects these impulses, combines external impulses reaching the esophagus via vagal and glossopharyngeal nerves, and fires efferent impulses that control muscle tone, contraction, and relaxation of esophageal muscle, including sphincters.

Destruction, stimulation, and inhibition of the nerve endings in the squamous epithelium can occur due to interaction of luminal molecule entering the reflux-damaged epithelium, due to the action of cytokines (chemokines) liberated by damaged cells, or due to actual physical destruction. Columnar transformation of squamous epithelium will be inevitably associated with a complete loss of these nerve endings; columnar epithelia in the gastrointestinal tract do not have intraepithelial nerve endings.

While totally uncorroborated and somewhat fanciful, this mechanism can explain the neuromuscular abnormalities that occur in reflux disease, often at an early stage without deep wall injury. A fanciful hypothesis is better than no explanation at all! The fact of sphincter loss in early reflux disease is not fanciful; it is well documented by manometric studies.

Conclusion: Early reflux damage to the squamous epithelium is associated with a permanent loss of sphincter function by a mechanism that is unknown.

COLUMNAR METAPLASIA OF THE SQUAMOUS EPITHELIUM

When it was first described by Allison and Johnstone in 1953 and confirmed by Barrett in 1957, columnar-lined esophagus was believed to be a congenital anomaly characterized by the presence of heterotopic gastric mucosa in the squamous-lined esophagus. Over the next decade, evidence emerged that columnar-lined esophagus was a result of columnar metaplasia of the esophagus caused by gastroesophageal reflux. Despite this knowledge being available since that time, columnar metaplasia has never been used as a diagnostic criterion for reflux disease. The only point at which columnar-lined esophagus has ever been regarded as a clinically significant finding is in patients with Barrett esophagus. Though recognized as being caused by reflux, the significance of Barrett esophagus is not as a diagnostic criterion for reflux disease but as a premalignant lesion that is an indication for surveillance. Every diagnostic criterion for gastroesophageal reflux that has ever existed in the literature has been a change in the squamous epithelium, at both the endoscopic and histologic level. The fact that none of these squamous epithelial criteria is specific or sensitive enough has relegated histology to irrelevance in the diagnosis of gastresophageal reflux disease.

The failure to use columnar-lined metaplasia as a diagnostic criterion for reflux disease defies understanding. We will consider the pathogenesis of columnar metaplasia in the next three chapters and show in Chapters 13 and 14 that columnar metaplasia of the esophagus is an extremely specific as well as a highly sensitive diagnostic test for gastroesophageal reflux disease at a cellular level (Table 8.1).

The Reflux-Primed Squamous Epithelium

Whereas the normal esophageal squamous epithelium had the capacity to resist the entry of luminal molecules, the reflux-damaged squamous epithelium permits the entry of luminal molecules. Acid in the refluxate is very likely the key that opens the lock of the squamous epithelial barrier. Its action allows all the other molecular intruders in gastric refluxate to enter the squamous epithelium (Fig. 8.13).

With the maximum acid-induced damage, the squamous epithelium displays the characteristics of Tobey et al.'s experimental model where molecules up to a size of 20 kD penetrate the full thickness to reach the basal region of the squamous epithelium where the proliferative pool of cells resides (19). It is important to understand that no genetic changes can occur without injurious agents having access to proliferating cells. Without the acid-induced permeability, the basal region of the squamous epithelium is immune to luminal molecules.

The normal esophageal squamous epithelium is continually undergoing renewal directed by continuous proliferation of adult esophageal progenitor cells located in the basal layer of the epithelium. These adult stem cells are derived from the fetal foregut progenitor cells from which the esophagus developed. During the development process, the fetal endodermal progenitor cells progress through programmed genetic activation and suppression until the final squamous genetic signal was established in the adult stem cell. The nature of this genetic signal in the adult esophagus is unknown, but it first appears in the midesophagus in the 20th week of gestation. The squamous signal is established in the entire esophagus during the first year of life, when fetal columnar epithelium finally disappears in the esophagus (see Chapter 3).

The fetal esophageal columnar signals are progressively suppressed as development is completed. It should be recognized that the adult esophageal progenitor cell retains its multipotent capability to differentiate in any direction seen in the endoderm. The fact that it produces a squamous epithelium means only that it expresses the genetic signal that directs squamous differentiation while genes that direct other types of differentiation are suppressed.

The first division of the progenitor cell in the basal zone of the squamous epithelium produces a daughter cell that has been given the genetic signal to become squamous. By the time it reaches the midregion of the stratified squamous epithelium, the cell has become terminally differentiated and incapable of mitotic division.

In the epithelium that has been primed by reflux-damage, luminal molecules permeate through the epithelium and can reach the normally sequestered progenitor cells at the base of the epithelium. Interactions between these luminal molecules and cellular receptors on the actively proliferating adult stem cells now become possible. These interactions must form the basis for genetic changes that occur in these cells that drive the further evolution of this pathologic process.

Conclusion: The reflux-primed squamous epithelium is one where refluxate molecules can interact with cellular receptors in the proliferating progenitor cell pool.

The Process of Columnar Transformation

Columnar transformation has been long recognized as a complication of gastroesophageal reflux. In 1961, Hayward (26) described the process well:

> When the normal sphincteric and valvular mechanism in the lower oesophagus and oesophago-gastric junction . . . fails, . . . reflux from the stomach occurs and acid and pepsin reach the squamous epithelium and begin to digest it. . . . In quiet periods some healing occurs, and in these periods the destroyed squamous epithelium may re-form, often with . . . junctional epithelium, usually not very healthy-looking. . . . Further reflux therefore attacks principally the squamous epithelium higher up. In the next remission it may be replaced by more junctional epithelium. . . . With repetition over a long period the metaplastic junctional epithelium may creep higher and higher. (p. 40)

Hayward, as did Barrett before him, believed that this process required erosion of the squamous mucosa (26, 27). It is now apparent that columnar transformation occurs without ulceration. Noneroded, reflux-primed squamous epithelium permits small molecules in the refluxate to reach the basal region and interact with receptors on the surface of the proliferating adult progenitor cells. Such interactions can have many potential effects, but the one that is significant is the genetic switch that causes the columnar transformation of the squamous epithelium (Figs. 8.15 and 8.16). While it is certain that such a mechanism must exist to explain columnar metaplasia, little research has been directed at elucidating the mechanism of this step. The main reason for this is the false dogma that cardiac mucosa is normal; there is no need to study the causation of a normal epithelium.

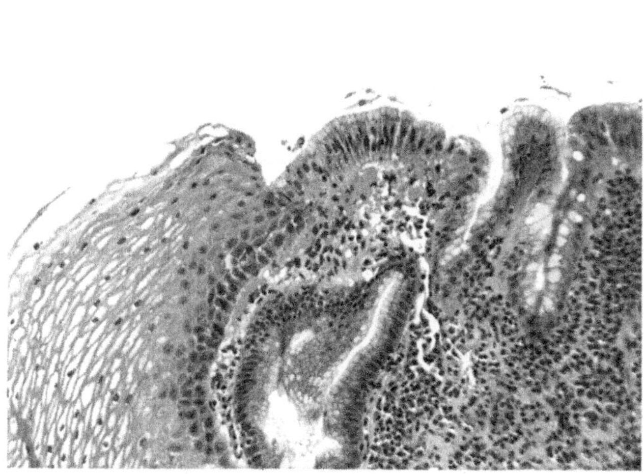

FIGURE 8.15 The metaplastic process at the squamocolumnar junction showing squamous epithelium, a sudden sharp transition to a multilayered cardiac mucosa, followed by single layered cardiac mucosa on the right. There is a suggestion of a plane of cleavage involving the 6 to 10 cells of the squamous epithelium nearest the junction. (Cross reference: Color Plate 3.7)

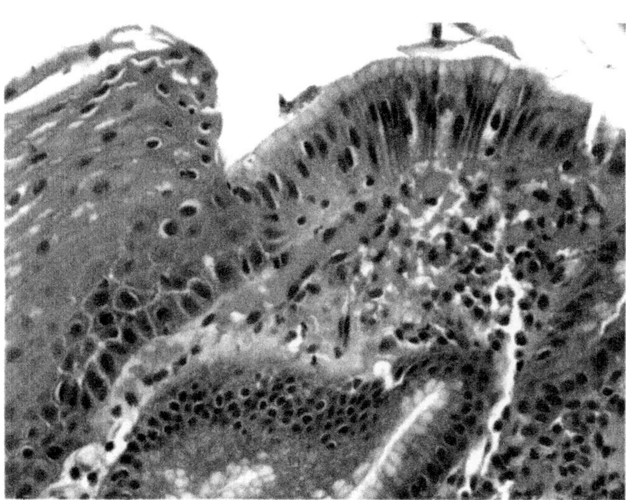

FIGURE 8.16 Higher power of Figure 8.15 showing the plane of cleavage between the basal layer of squamous epithelium and cells above it. This plane would result if the differentiating signal had changed from squamous to columnar in this region and cell attachment characteristics had changed. If separation were to occur in this plane, single or multilayered cardiac mucosa (right) would result.

Conclusion: Columnar metaplasia of the squamous epithelium must be the result of an unknown interaction between refluxate molecules and progenitor cells that results in a genetic switch that causes columnar metaplasia.

Cardiac Mucosa Is the Result of Columnar Transformation

There is a great misconception in the minds of many people that Barrett esophagus consists of intestinalized epithelium with goblet cells that covers the entire extent of columnar transformed esophagus. Some believe that patients go on directly from reflux esophagitis to intestinal metaplasia. The only explanation for such beliefs is that these people have never looked at columnar-lined esophagus under the microscope and have never comprehended the literature on the histology of the columnar-lined esophagus.

There has never been any question in the literature that the ubiquitous epithelial types in columnar-lined esophagus are cardiac mucosa and oxyntocardiac mucosa. Allison and Johnstone in the first description in 1953, and Barrett (27), clearly described cardiac mucosa. Paull et al. (28), in his classical histologic mapping of columnar-lined esophagus, showed that three epithelial types constitute columnar-lined esophagus. In our study of mapping biopsies, cardiac mucosa is ubiquitous in columnar lined esophagus except in patients with a very short segment consisting only of oxyntocardiac mucosa (29).

The three epithelial types found in columnar-lined esophagus can be classified into (a) those without specialized cells (cardiac mucosa, which has only mucous cells) and (b) those with specialized cells (goblet cells in intestinal metaplasia and parietal cells in oxyntocardiac mucosa). The specialized epithelia also contain mucous cells. It is much more likely that the epithelium without specialized cells is the first epithelium to result from the columnar transformation of squamous epithelium, and the other two evolve by the appearance of specialized cells at a later point in time. This is exactly what happens (Fig 8.17).

Dresner et al., in their study of the changes in squamous epithelium above the anastomotic line in postesophagectomy patients, showed that this is in fact what occurs as a result of reflux. The squamous epithelium shows evidence of damage and then undergoes metaplasia to cardiac mucosa. It is after a significant time interval, even in these patients with severe duodeno-gastroesophageal reflux, that intestinal metaplasia appears within the metaplastic cardiac mucosa.

Conclusion: The first result of columnar transformation of esophageal squamous epithelium is cardiac mucosa.

Why Does Cardiac Mucosa Result from Columnar Metaplasia?

In 1957, when Barrett was faced with the task of explaining the columnar-lined esophagus, he demonstrated great skill. He first investigated comparative anatomy to see whether the columnar-lined esophagus was a return to a lower evolutionary state.

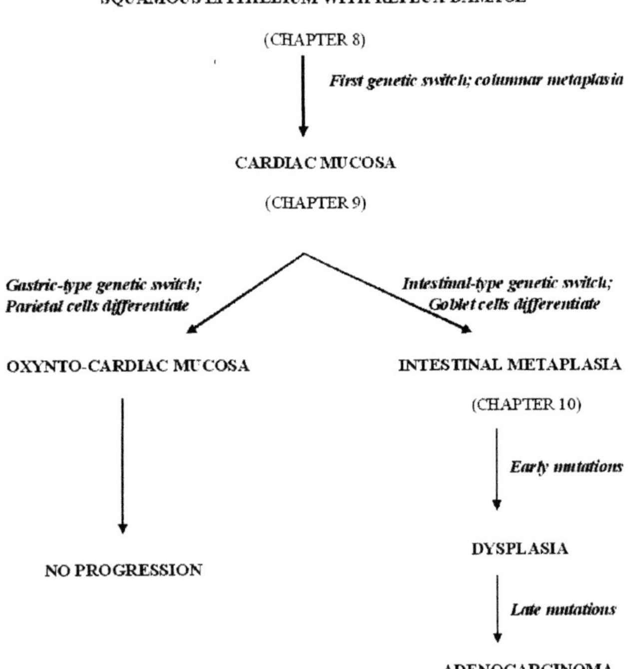

FIGURE 8.17 The sequence of changes in gastroesophageal reflux disease. This is the reflux-adenocarcinoma sequence and will be described in detail in this and the next two chapters.

He concluded: "This anomaly of the lower esophagus cannot be explained as a reversion to a more primitive zoologic state.... The tendency in the mammalian world is for squamous epithelium to continue onward toward the pylorus, not for columnar epithelium to grow up the esophagus" (p. 884).

He then considered the possibility that columnar-lined esophagus is a reversion to a more embryologic state. He carefully summarized the findings of Johns' 1952 article (30), commenting on the fact that fetal esophagus is lined entirely by columnar epithelium before squamous differentiation occurs. He concluded: "During the development of a human embryo the sequence of changes in the epithelia which line the foregut might be arrested or deviated from normal. Thus it should not surprise us to find columnar, ciliated, transitional, or squamous epithelium anywhere between the mouth and the duodenum" (p. 885). Barrett was trying to suggest that columnar-lined esophagus was a congenital anomaly resulting from a deviation or arrest of squamous transformation of the fetal columnar esophagus. It is obvious that he was uncomfortable with this theory of congenital origin. He suggested that the pattern of involvement is more in keeping with an acquired defect associated with reflux. He concluded his search for the true explanation of columnar-lined esophagus: "None of these suggestions smooth away all difficulties and the etiology [of columnar-lined esophagus] remains open to speculation" (p. 885).

We now know that columnar-lined esophagus does not occur as a failure of development; it is not a congenital anomaly. When development is complete, there is no columnar epithelium lining the esophagus. Columnar-lined esophagus is acquired and results from the effect of reflux on squamous epithelium. The exact mechanism by which this happens must involve a change in the genetic signal of the adult progenitor cells. This change can be either the suppression of the normal squamous genetic signal, permitting the epithelium to revert to its fetal columnar state, or the activation of a new genetic signal that directs columnar differentiation. The genetic basis for columnar metaplasia is not yet known, but its existence can be assumed by recognizing the phenotypic expression of the change, which is columnar transformation of the squamous epithelium.

It is reasonable to look at fetal esophagi to see if cardiac mucosa resembles any fetal columnar seen in the esophagus. When this is done, it is clear that cardiac mucosa is similar to the nonciliated columnar epithelium (tall and short; what we have designated "fetal esophageal columnar epithelium, type III," see Chapter 3) that occurs in late fetal life in the distal esophagus (Fig. 8.18). This epithelium is also the one that is associated with the superficial esophageal glands composed of mucous cells (30). It is relevant that this is the last columnar genetic signal to be suppressed during fetal development. It seems reasonable that the first genetic signal to be reactivated in columnar metaplasia is the last one to become suppressed during fetal development. Earlier fetal columnar epithelia, notably the ciliated epithelium that lines the entire esophagus in the second trimester, are never produced in reflux-induced metaplasia; ciliated epithelium has never been described in the adult esophageal epithelium (it is sometimes seen in the gland ducts that drain the submucosal esophageal glands). The nature of the genetic signals that causes the adult esophageal progenitor cells to shift from squamous to cardiac type columnar differentiation are not yet known, largely because they have not been looked for.

The daughter cell of the adult esophageal stem cell division that has been given the genetic signal to differentiate into a columnar cell develops features of columnar cells, not squamous cells. One inevitable effect is the loss of normal cell attachments, resulting in a loss of adhesion between the new columnar cells and its squamous neighbors above, leading to sloughing of the squamous cells and resulting in a columnar

FIGURE 8.18 Fetal columnar epithelium (simple columnar type, type 3) seen in the distal esophagus in the third trimester of pregnancy. This consists of simple columnar cells with a villiform appearance that resembles the acquired cardiac mucosa of adult life (compare to Figure 8.15) except for a total lack of inflammation. (Cross reference: Color Plate 1.2)

epithelium (Fig. 8.16). The area at the outset is short, possibly one cell, but as Hayward described, the change is cumulative, resulting in a progressive increase in the amount of squamous epithelium that undergoes columnar transformation. There is evidence to suggest that the first columnar epithelium to result from this metaplasia is a multilayered epithelium (Figs. 8.15 and 8.16; see also Figs. 5.14 and 5.15) (31). This is very likely a transient and unstable columnar epithelium that quickly becomes a single-layered columnar epithelium (Fig. 8.16).

The histologic product of the initial columnar metaplasia of esophageal epithelium is a columnar epithelium devoid of specialized cells; these cells are recognized histologically as mucous cells and the mucosa falls within the definition of cardiac mucosa. The newly formed cardiac mucosa is interposed between the squamocolumnar junction and gastric oxyntic mucosa, reflecting the fact that the maximum effect of gastroesophageal reflux is at the most distal part of the squamous-lined esophagus.

Conclusion: Cardiac metaplasia represents a reversion of differentiation from adult squamous to the last type of fetal columnar differentiation that occurred during development.

Cardiac Metaplasia Is Highly Specific for Reflux

Columnar transformation of the esophageal squamous epithelium to form cardiac mucosa is the result of a complex series of changes that are highly specific for reflux. It requires reflux-induced damage of the squamous epithelium to cause increased permeability followed by a genetic switch resulting from an interaction between an unknown and probably specific molecule in the refluxate and the esophageal progenitor cell. While the squamous epithelial damage is not specific for reflux and can occur in other diseases such as allergic esophagitis, the actual genetic switch that causes columnar metaplasia in the squamous epithelium appears to be highly specific for reflux. It is not reproduced by any other known agent producing injury in the esophagus—not by chemicals, pills, corrosives, or infectious agents.

Conclusion: Cardiac metaplasia is highly specific for reflux.

What Molecule in the Refluxate Causes Cardiac Metaplasia

While there is strong evidence that acid acts as the key that permits access to the squamous epithelium, it is very possible that a molecule other than acid is responsible for the actual genetic switch that leads to columnar transformation in the esophageal progenitor cell. The exact molecule in the gastric refluxate that attaches to the unknown progenitor cell receptor to induce the genetic switch that results in cardiac metaplasia is unknown. Acid is a possible agent, but any other molecule that gains access to the stem cells as a result of the increased permeability can be responsible.

A study of epidemiology and prevalence changes since the 1970s gives insight to etiology. If a given esophageal abnormality has declined in prevalence since the 1970s, it suggests that the abnormality is produced by acid because of the increasing use and efficacy of acid suppressive medications during this period. This is true for severe ulceration and strictures of the esophagus.

The change in prevalence and severity of columnar-lined esophagus (without intestinal metaplasia) over since the 1950s is difficult to elucidate from the literature. This is because after Barrett esophagus became defined by the presence of intestinal metaplasia, all research was directed at the intestinalized type of columnar-lined esophagus, and nonintestinalized columnar-lined esophagus (i.e., cardiac metaplasia) was essentially ignored. It is known from Allison and Johnstone's study in 1953 that columnar-lined esophagus of significant length was not uncommon at the time. Hayward's suggestion that 2 cm of cardiac mucosa was normal in the distal esophagus also suggests that cardiac metaplasia was common at the time.

This is in contrast to the prevalence of intestinal metaplasia (Barrett esophagus) and esophageal adenocarcinoma where evidence suggests a marked increase in prevalence since the 1970s. Without any knowledge of the changes in prevalence and extent of cardiac mucosa before and after acid suppression became common, it is not possible to say whether acid or some other molecule is responsible for the genetic switch from squamous to cardiac mucosa.

Conclusion: The exact molecule that causes the squamous-to-cardiac switch is unknown.

Does Cardiac Metaplasia Increase in Amount/Length?

Cardiac transformation of squamous esophageal epithelium is a cumulative change, as Hayward described so beautifully. As the number of reflux episodes increases, the amount of squamous epithelium that transforms into cardiac mucosa increases. Given the fact that some gastroesophageal reflux occurs in virtually all people, this explains the increase in the prevalence and length of cardiac mucosa that is seen with increasing age. The length of columnar transformation that is found in the esophagus in any patient is a function of the amount of exposure to reflux, the duration of exposure, and the susceptibility of that individual to reflux-induced damage. In any given individual, columnar transformation will tend to increase with increasing age and increasing exposure to reflux.

Conclusion: In patients with reflux, cardiac mucosa increases throughout life as the cumulative exposure of the squamous epithelium to gastroesophageal reflux increases.

What Happens to Cardiac Mucosa in the Esophagus?

In 1961, Hayward (26) described cardiac mucosa as being an epithelium that was resistant to reflux. However, there is much evidence that reflux damages cardiac mucosa. Cardiac mucosa invariably shows inflammation with the number of eosinophils and plasma cells correlating with the severity of reflux, and it almost invariably shows reactive hyperplasia of the foveolar region with elongation, serration, mucin distension of the cells, and smooth muscle proliferation (Fig. 8.19) (10, 32).

Studies with immunoperoxidase staining for the proliferation-associated antigen Ki67 show an increased rate of cell turnover in cardiac mucosa compared with normal gastric oxyntic mucosa, suggesting that cardiac mucosa is the site of rapid cell loss from the surface (33). The exact agents responsible for damage in cardiac mucosa are unknown.

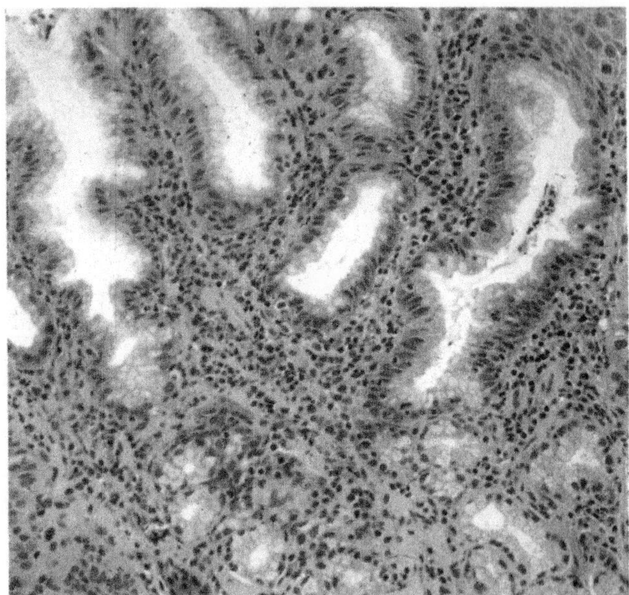

FIGURE 8.19 Cardiac mucosa with evidence of damage characterized by marked foveolar elongation, luminal serration, and severe chronic inflammation. (Cross reference: Color Plate 3.11)

The location of the adult esophageal progenitor cell in cardiac mucosa varies with the type of cardiac mucosa. In its earliest form at the point of columnar transformation of the squamous epithelium, cardiac mucosa is a single or multilayered flat epithelium covering the esophagus. This epithelium has strong Ki67 positivity, indicating that the proliferative progenitor cells are present in the surface (Fig. 8.20). The gastrointestinal tract abhors having the precious progenitor cells at the surface; as such, this purely surface type of cardiac mucosa is rarely seen and usually only in the region of transformation or covering healing erosions. Very quickly, the surface epithelium dips down to form a foveolar pit (Fig. 8.21). When this happens, the morphologically identical mucous cells become heterogeneous; the proliferative, Ki67-positive, pool of cells that includes the adult progenitor cells becomes located in the deepest part of the foveolar pit (Fig. 8.21). The superficial part of the foveolar pit and the surface epithelium are terminally differentiated cells that are nonmitotic (Ki67 negative). The cardiac mucosa has reorganized itself to sequester the stem cells in the deep foveolar region in a manner that is similar to all columnar epithelia in the gastrointestinal tract. This is the purely foveolar type of cardiac mucosa. In some cases, cardiac mucosa retains this purely foveolar appearance (Fig. 8.22). In others, glands develop

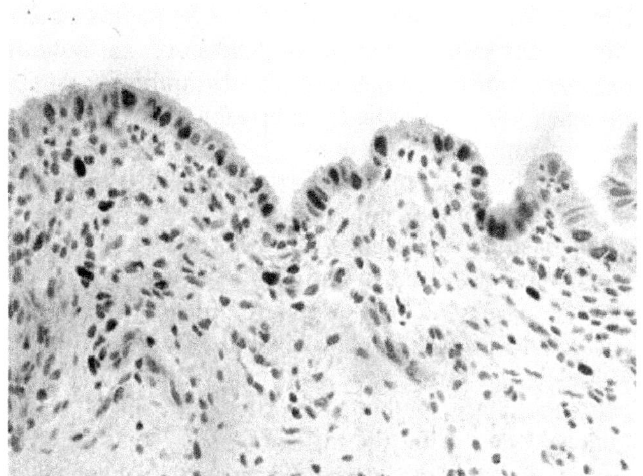

FIGURE 8.20 Newly formed cardiac mucosa consisting of a single layer of cuboidal epithelial cells lining the flat surface. This is stained for Ki67 and shows that the surface layer is positive, indicating proliferative activity.

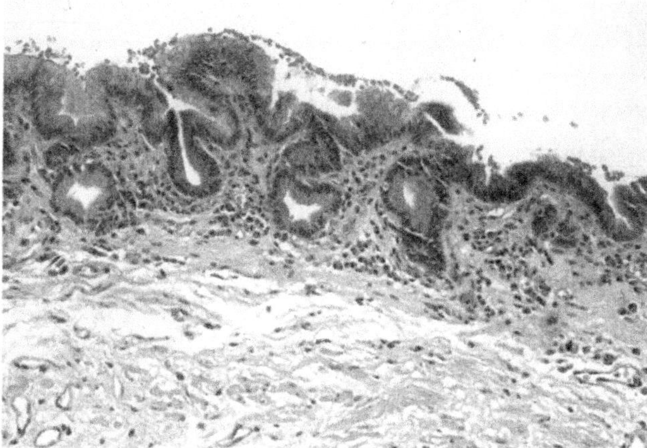

FIGURE 8.22 Cardiac mucosa consisting of only a surface layer and the foveolar pit. This area is lined by mucous cells. There is no glandular region deep to the foveolar pit.

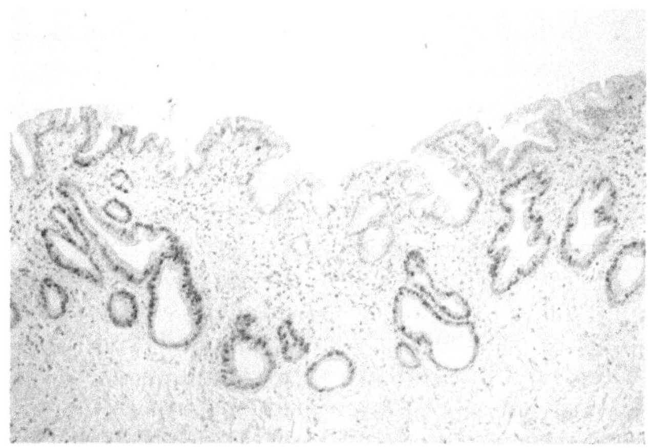

FIGURE 8.21 Ki67 in cardiac mucosa that has formed a foveolar pit. The Ki67-positive proliferative cells have moved to the base of the foveolar pit. The mucous cells at the surface are terminally differentiated, Ki67-negative cells.

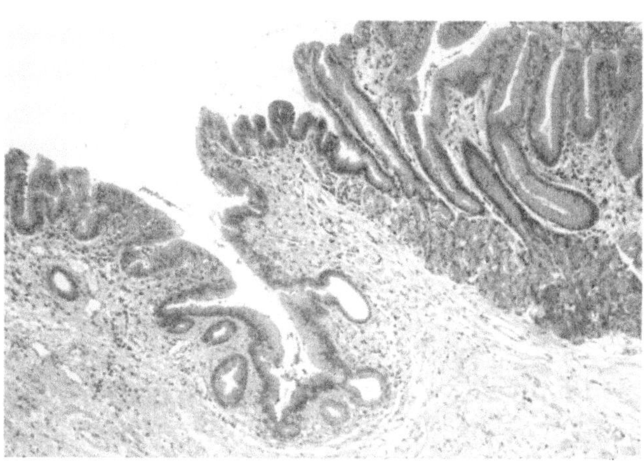

FIGURE 8.23 Cardiac mucosa, showing the development of glands under the foveolar region. The left side shows a purely foveolar epithelium with no glands; glands are present on the right side. (Cross reference: Color Plate 3.12)

beneath the foveolar region (Fig. 8.23). This is a fundamental change in which the dividing progenitor cells in the deep foveolar region direct the daughter cells to move downward to form glands rather than toward the surface to replace the surface cells. Glandular cells in cardiac mucosa remain as mucous cells, resembling the superficial glands that occur in fetal esophageal columnar epithelium. These glands consist of stable, nonmitotic, terminally differentiated secretory cells with a long life span; they are easily recognized as a Ki67-negative zone below the proliferative region in the deep foveolar pit (Fig. 8.24). This is the glandular type of cardiac mucosa.

In all of these types of cardiac mucosa, the progenitor cell pool is relatively accessible to the action of luminal molecules. The foveolar pit is open at the surface, and luminal molecules reach its depths relatively easily. This sets the stage for molecular interactions at the surface of the adult esophageal progenitor cell that has the potential to cause further genetic changes. These subsequent genetic switches alter the differentiation model of cardiac mucosa and may result in the development of specialized cells within cardiac mucosa. Many types of specialized cells appear in the metaplastic esophageal cardiac mucosa including parietal cells, goblet cells, Paneth cells, pancreatic

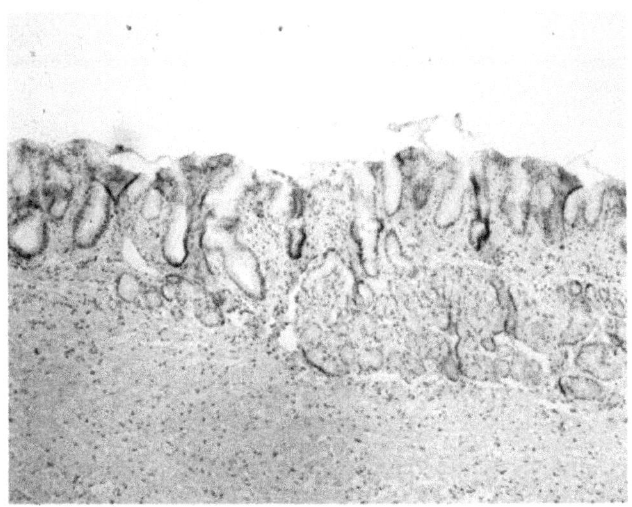

FIGURE 8.24 Cardiac mucosa, immunoperoxidase stain for Ki67. In the left, the mucosa consists only of a foveolar pit with the proliferative Ki67 cells at the base of the foveolar pit. In contrast, the center and right side show Ki67 negative glands deep to the foveolar region; the Ki67-positive proliferative cells are in the deep part of the foveolar pit.

cells, and neuroendocrine cells. This will be discussed in the next chapter.

Conclusion: Cardiac mucosa is damaged by components of gastroesophageal reflux; luminal molecules have ready access to the cardiac mucosal progenitor cell pool.

Clinical Effects of Cardiac Metaplasia in the Esophagus

Cardiac mucosa resembles other gastrointestinal columnar epithelia in that it has no intraepithelial nerve endings. Pain in cardiac mucosa is likely to be different from that in the highly sensitive squamous epithelium and resembles the pain associated with other columnar epithelia in the gastrointestinal tract (gastritis, ileitis, and colitis).

Columnar epithelium has been shown to produce a lower frequency of positive response to the Bernstein acid perfusion test, with a false-negative rate of 66% (34). There are also anecdotal reports of some patients with long-standing gastroesophageal reflux disease reporting spontaneous improvement in their pain, perhaps when they develop metaplastic columnar epithelium (35).

Fletcher et al. (36) reported a study designed to test the sensitivity of Barrett epithelium to acid and hypertonic solutions. This study consisted of (a) 7 healthy asymptomatic volunteers, (b) 14 patients with symptoms of gastroesophageal disease and no evidence of Barrett esophagus (8 had erosive esophagitis and 6 were endoscopically normal but had abnormal 24-hour pH tests), and (c) 14 patients with symptoms of reflux disease and endoscopic Barrett esophagus ranging in length from 5 to 15cm with intestinal metaplasia on biopsy. Of this group, 3 had proximal erosive esophagitis in the squamous epithelium.

Each subject underwent four provocative tests at weekly intervals via a catheter placed in the lower esophagus above the sphincter. Infusions were distilled water, hypertonic saline, hydrochloric acid at pH 1, and hypertonic saline acidified with hydrochloric acid pH 1. The patients recorded pain on a visual analog scale during the infusions.

Pain was produced only by the acid infusions in 4 of 7 healthy volunteers. In the non-Barrett GERD group, water infusion caused pain in 1 of 14 patients, the hypertonic saline caused mild pain in 3 of 14, and the two acid solutions caused moderate pain in all patients. In the Barrett esophagus group, water infusion produced mild pain in 2 of 14, the hypertonic saline caused pain in 10 of 14 (significantly more commonly than in the non-Barrett group), and the acid infusion caused pain in 13 of 14 patients but the intensity of the pain was significantly lower than the non-Barrett group. The symptoms were qualitatively different in the Barrett group, where patients experienced more frequent nausea in addition to the pain when compared to those in the non-Barrett GERD group.

This study is difficult to interpret because of the uncertainty relating to where the symptoms are actually originating. Patients with non-Barrett GERD can be assumed to have their symptoms relating to squamous epithelial nerve ending stimulation. However, patients with Barrett esophagus very likely have squamous damage (visible in three patients) and may have pain caused by stimulation of squamous epithelium. Thus, it is not possible to ascribe the pain to stimulation of the columnar epithelium. The decrease in pain sensitivity in the Barrett group compared to the non-Barrett GERD group strongly suggests that the columnar epithelia are less sensitive, but it is possible that the columnar epithelium is completely insensitive. The development of the new symptom of nausea in the Barrett group suggests that this is in some way related to stimulation of the columnar rather than squamous epithelium.

It should be noted that this study is defined by the presence of intestinal metaplasia in cardiac mucosa. There are no reports that study differences in pain sensitivity in non-Barrett columnar epithelia. There is no reason to believe that cardiac mucosal pain sensitivity is altered by the presence or absence of intestinal metaplasia, though this remains unproven.

Conclusions: Cardiac mucosa (and other metaplastic esophageal columnar epithelia) may be less pain sensitive than squamous epithelium.

SUMMARY

The following steps are involved in the first phase of reflux disease:

1. Reflux esophagitis, characterized by increased surface cell loss, compensatory hyperplasia with basal cell hyperplasia and papillary elongation, increased cell-cell separation, increased permeability of the epithelium to luminal molecules, cell damage with release of cytokines, and chemoattraction of neutrophils and eosinophils.
2. Columnar transformation of squamous epithelium, resulting in cardiac mucosa. This occurs as a result of a yet unknown genetic switch that causes the esophageal epithelial progenitor cell to change from squamous to columnar differentiation.

References

1. Chandrasoma PT, Der R, Ma Y, et al. Histology of the gastroesophageal junction: An autopsy study. Am J Surg Pathol 2000;24:402–409.
2. Marsman WA, van Sandyck JW, Tytgat GNJ, ten Kate FJW, van Lanschot JJB. The presence and mucin histochemistry of cardiac type mucosa at the esophagogastric junction. Am J Gastroenterol 2004;99:212–217.
3. Dresner SM, Griffin SM, Wayman J, Bennett MK, Hayes N, Raimes SA. Human model of duodenogastro-oesophageal reflux in the development of Barrett's metaplasia. Br J Surg 2003;90:1120–1128.
4. Csendes A, Maluenda F, Braghetto I, et al. Location of the lower esophageal sphincter and the squamocolumnar mucosal junction in 109 healthy controls and 778 patients with different degrees of endoscopic esophagitis. Gut 1993;34:21–27.
5. Glickman JN, Fox V, Antonioli DA, Wang HH, Odze RD. Morphology of the cardia and significance of carditis in pediatric patients. Am J Surg Pathol 2002;26:1032–1039.
6. Chandrasoma PT, Lokuhetty DM, DeMeester, TR, et al. Definition of histopathologic changes in gastroesophageal reflux disease. Am J Surg Pathol 2000;24:344–351.
7. Blot WJ, Devesa SS, Kneller RW, Fraumeni JF Jr. Rising incidence of adenocarcinoma of the esophagus and gastric cardia. JAMA 1991;265:1287–1289.
8. Karam SM. Lineage commitment and maturation of epithelial cells in the gut. Front Bio 1999;4:286–298.
9. Kiesslich R, Kanzler S, Vieth M, et al. Minimal change esophagitis: Prospective comparison of endoscopic and histological markers between patients with non-erosive reflux disease and normal controls using magnifying endoscopy. Dig Dis 2004;22:221–227.
10. Riddell RH. The biopsy diagnosis of gastroesophageal reflux disease, "carditis," and Barrett's esophagus, and sequelae of therapy. Am J Surg Pathol 1996;20(Suppl 1):S31–S51.
11. Takubo K, Honma N, Aryal G, Sawabe M, Arai T, Tanaka Y, Mafune K, Iwakiri K. Is there a set of histologic changes that are invariably reflux associated? Arch Pathol Lab Med 2005;129(2):159–163.
12. Lundell LR, Dent J, Bennett JR, et al. Endoscopic assessment of oesophagitis: Clinical and functional correlates and further validation of the Los Angeles classification. Gut 1999;45:172–180.
13. Allison PR, Johnstone AS. The oesophagus lined with gastric mucous membrane. Thorax 1953;8:87–101.
14. Tobey NA, Carson JL, Alkeik RA, Orlando RC. Dilated intercellular spaces: A morphological feature of acid reflux-damaged human esophageal epithelium. Gastroenterology 1996;111:1200–1205.
15. Caviglia R, Ribolsi M, Maggiano N, Gabbrielli AM, Emerenziani S, Guarino MP, Carotti S, Habib FI, Rabitti C, Cicala M. Dilated intercellular spaces of esophageal epithelium in nonerosive reflux disease patients with physiological esophageal acid exposure. Am J Gastroenterol 2005;100:543–548.
16. Villanacci V, Grigolato PG, Cestari R, Missale G, Cengia G, Klersy C, Rindi G. Dilated intercellular spaces as markers of reflux disease: Histology, semiquantitative score and morphometry upon light microscopy. Digestion 2001;64:1–8.
17. Tobey NA, Hosseini SS, Argore CM, Dobrucali AM, Awayda MS, Orlando RC. Dilated intercellular spaces and shunt permeability in non-erosive acid-damaged esophageal epithelium. Am J Gastroenterol 2004;99:13–22.
18. Veith M, Peitz U, Labenz J. What parameters are relevant for the histological diagnosis of gastroesophageal reflux disease without Barrett's mucosa? Dig Dis 2004;22:106–201.
19. Bowrey DJ, Williams GT, Clark GW. Histological changes in the oesophageal squamous mucosa: Correlation with ambulatory 24-hr pH monitoring. J Clin Pathol 2003;56:205–208.
20. Tummala V, Barwick KW, Sontag SJ, Vlahcevic RZ, McCallum RW. The significance of intraepithelial eosinophils in the histologic diagnosis of gastroesophageal reflux. Am J Clin Pathol 1987;87:43–48.
21. Isomoto H, Wang A, Mizuta Y, Akazawa Y, Ohba K, Omagari K, Miyazaki M, Murase K, Hayashi T, Inoue K, Murata I, Kohno S. Elevated levels of chemokines in esophageal mucosa of patients with reflux esophagitis. Am J Gastroenterol 2003;98:551–556.
22. Kanazawa Y, Isomoto H, Wen CY, et al. Impact of endoscopically minimal involvement on IL-8 mRNA expression in esophageal mucosa of patients with non-erosive reflux disease. World J Gastroenterol 2003;9:2801–2804.
23. Kuramochi H, Vallbohmer D, Uchida K, Schneider S, Hamoui N, Shimizu D, Chandrasoma PT, DeMeester TR, Danenberg KD, Danenberg PV, Peters JH. Quantitative, tissue-specific analysis of cyclooxygenase gene expression in the pathogenesis of Barrett's adenocarcinoma. J Gastrointest Surg 2004;8: 1007–1017.
24. DeMeester TR, Peters JH, Bremner CG, Chandrasoma P. Biology of gastroesophageal reflux disease: Pathophysiology relating to medical and surgical treatment. Annu Rev Med 1999;50: 469–506.
25. Oberg S, Peters JH, DeMeester TR, et al. Inflammation and specialized intestinal metaplasia of cardiac mucosa is a manifestation of gastroesophageal reflux disease. Ann Surg 1997; 226:522–532.
26. Hayward J. The lower end of the oesophagus. Thorax 1961; 16:36–41.
27. Barrett NR. The lower esophagus lined by columnar epithelium. Surgery 1957;41:881–894.
28. Paull A, Trier JS, Dalton MD, Camp RC, Loeb P, Goyal RK. The histologic spectrum of Barrett's esophagus. N Engl J Med 1976;295:476–480.

29. Chandrasoma PT, Der R, Ma Y, Peters J, DeMeester T. Histologic classification of patients based on mapping biopsies of the gastroesophageal junction. Am J Surg Pathol 2003;27:929–936.
30. Johns BAE. Developmental changes in the oesophageal epithelium in man. J Anat 1952;86:431–442.
31. Shields HM, Rosenberg SJ, Zwas FR, et al. Prospective evaluation of multilayered epithelium in Barrett's esophagus. Am J Gastroenterol 2001;96:3268–3273.
32. Der R, Tsao-Wei DD, DeMeester T, et al. Carditis: A manifestation of gastroesophageal reflux disease. Am J Surg Pathol 2001;25:245–252.
33. Olvera M, Wickramasinghe K, Brynes R, Bu X, Ma Y, Chandrasoma P. Ki67 expression in different epithelial types in columnar lined esophagus indicates varying levels of expanded and aberrant proliferative patterns. Histopathology 2005;47: 132–140.
34. Johnson DA, Winters C, Spurling TJ, et al. Esophageal acid sensitivity in Barrett's esophagus. J Clin Gastroenterol 1987; 9:23–32.
35. Ter RB, Castell DO. Gastroesophageal reflux disease in patients with columnar lined esophagus. Gastroenterol Clin N Amer 1997;26:549–563.
36. Fletcher J, Gillen D, Wirz A, McColl KEL. Barrett's esophagus evokes a quantitatively and qualitatively altered response to both acid and hypertonic solutions. Am J Gastroenterol 2003;98:1480–1483.

CHAPTER

9

The Pathology of Reflux Disease at a Cellular Level: Part 2—Evolution of Cardiac Mucosa to Oxyntocardiac Mucosa and Intestinal Metaplasia

Gastroesophageal reflux disease is extremely unusual as a human disease because it has three distinct phases that affect different types of epithelia. In the first phase, which we considered in the previous chapter, molecules in the refluxate cause damage to the squamous epithelium of the esophagus. Because acid is a major injurious agent to the squamous epithelium, all patients with exposure of the squamous epithelium to sufficient reflux to overcome the natural resistance of the epithelium will show changes. The most sensitive cellular change that results from reflux damage is columnar metaplasia, whereby squamous epithelium undergoes metaplasia to cardiac mucosa.

The presence of cardiac mucosa in the distal esophagus sets the stage for the second phase of the disease, which we describe in this chapter (Fig. 9.1). There are five things that may happen to cardiac mucosa: (a) it can reverse to squamous epithelium (unlikely without treatment); (b) it can remain as cardiac mucosa (very common); (c) it can develop parietal cells, becoming oxyntocardiac mucosa (very common); (d) it can develop goblet cells, becoming intestinal metaplasia, which defines Barrett esophagus (uncommon, but critical); (e) it can be damaged, producing changes of injury such as erosion, reactive hyperplasia of the epithelial cells, and inflammation. These changes are the result of interactions between molecules in the refluxate with cardiac mucosal cells. These molecules may cause direct cellular damage, usually to terminally differentiated surface cells, or they may cause genetic shifts by interacting with the deeper progenitor cells. The exact molecules that cause pathologic changes in cardiac mucosa are unknown (Table 9.1). There is very little research into this question because of the false dogma that dismisses cardiac mucosa as a normal epithelium.

The third phase of the disease begins with the occurrence of intestinal metaplasia in cardiac mucosa. This is believed to be the first point at which carcinogens act; only the intestinal type of epithelium is susceptible to the carcinogens in the refluxate. It is important to recognize that these carcinogens have always existed in the refluxate in a given patient, but they have no action unless and until intestinal metaplasia develops. Carcinogenesis in intestinal metaplasia will be the subject of a subsequent chapter.

HISTOLOGIC COMPOSITION OF COLUMNAR-LINED ESOPHAGUS

The Evolution and Interrelationship of Columnar Epithelial Types in the Esophagus

The three types of columnar epithelia that constitute columnar-lined esophagus do not each arise independently from squamous epithelium. Rather, they occur in an orderly sequence directed by changes in the genetic signals in the proliferating esophageal progenitor cell pool. When the signal changes from squamous to columnar, cardiac mucosa is produced,

TABLE 9.1 Phases of Gastroesophageal Reflux Disease

Phase	Epithelium	Molecule	Change
1	Squamous	H^+	Damage; erosion; ulcer, stricture
	Squamous	H^+	Increased permeability
	Squamous	?	Cardiac metaplasia
2	Cardiac	?	Inflammation—reflux carditis
	Cardiac	?	Erosion; increased proliferation
	Cardiac	H^+ (?)	Change to oxyntocardiac mucosa
	Cardiac	Bile salts (?)	Change to intestinal metaplasia
3	Intestinal	?	Inflammation; erosion
	Intestinal	?	Carcinogenesis

Note: These phases indicate the epithelium that is under attack, the likely causative molecule in the refluxate, and the cellular change that results from the event. Note that multiples of these epithelial types exist in a single patient and these different events may be occurring simultaneously in different areas of the esophagus.

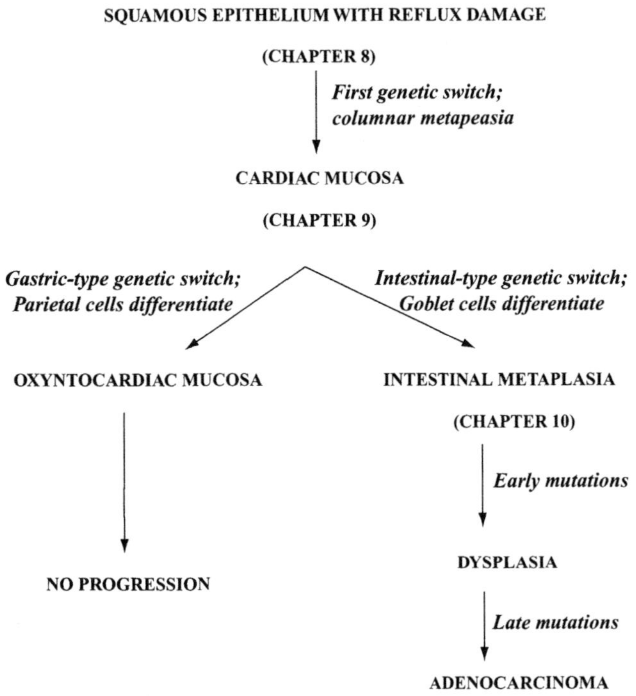

SQUAMOUS EPITHELIUM WITH REFLUX DAMAGE
(CHAPTER 8)
First genetic switch; columnar metapeasia
CARDIAC MUCOSA
(CHAPTER 9)
Gastric-type genetic switch; Parietal cells differentiate
Intestinal-type genetic switch; Goblet cells differentiate
OXYNTOCARDIAC MUCOSA
INTESTINAL METAPLASIA
(CHAPTER 10)
Early mutations
NO PROGRESSION
DYSPLASIA
Late mutations
ADENOCARCINOMA

FIGURE 9.1 The reflux to adenocarcinoma sequence. In this chapter, we consider phase 2 of the disease where cardiac mucosal changes occur, resulting in oxyntocardiac or intestinal metaplasia.

with subsequent evolution to oxyntocardiac and intestinal epithelial types when additional genetic signal changes occur. We studied the Ki67 staining patterns of columnar epithelial types in an attempt to understand the exact sequence of this multistep process.

This study (Chandrasoma, unpublished data) evaluated the Ki67 staining pattern in 26 patients who underwent esophagogastrectomy for either high-grade dysplasia or adenocarcinoma complicating Barrett esophagus. Ki67 (MIB-1) is an antigen that is expressed in cells that are in the cell cycle and is therefore a marker of proliferation. Nineteen patients had residual nondysplastic intestinal metaplasia, and all patients showed cardiac or oxyntocardiac mucosa in the tubular esophagus distal to the squamocolumnar junction. All cases had gastric oxyntic mucosa distal to the esophagus near the distal margin of resection of the esophagogastrectomy specimen.

In total, 129 areas of metaplastic columnar-lined esophagus were selected from sections taken from the tubular esophagus in the 26 esophagectomy specimens. These included oxyntocardiac mucosa (containing glands with a mixture of parietal and mucous cells deep to the isthmus; 43 areas), cardiac mucosa (containing only mucous cells without any parietal or goblet cells; 45 areas), and intestinal metaplastic epithelium with goblet cells (41 areas).

We examined 32 control areas of normal gastric oxyntic mucosa from the stomach to establish the normal components of glandular mucosa and the pattern of Ki67 expression. Gastric oxyntic mucosa consists of a surface layer of mucous cells, a foveolar pit composed of mucous cells, and a straight tubular gland composed of parietal and chief cells. Ki67 positive cells are restricted to the deep part of the foveolar pit and neck of the gland (Fig. 9.2). We classified columnar-lined esophagus based on this model of normal gastric mucosa into surface, foveolar, and glandular types of epithelia (Fig. 9.3).

When the columnar-lined epithelium consisted of a flat surface layer of mucous cells for a length of 2 mm, it was classified as a surface-only epithelium (surface type) (Figs. 9.4 and 9.5). This was a rare finding. In all other areas, the surface layer formed foveolar pits of varying length (foveolar type) (Figs. 9.6 and 9.7). When

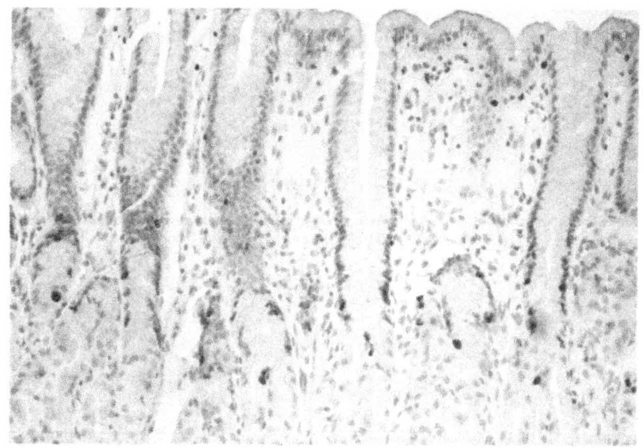

FIGURE 9.2 Immunoperoxidase stain for Ki67 of normal gastric oxyntic mucosa showing the proliferative cells in the deep foveolar region. The glands are below this level, and the acid secreted by the parietal cells drains into the foveolar pit, necessarily exposing the progenitor cells in this region continuously to acid.

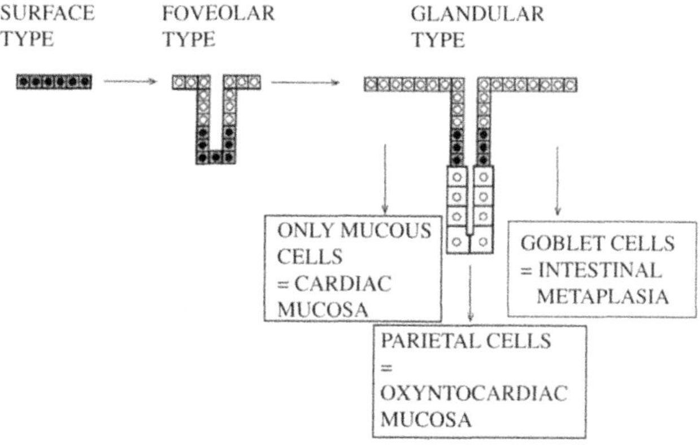

FIGURE 9.3 The evolution of cardiac mucosa, diagrammatic. Columnar transformation of squamous epithelium produces a single-layered epithelium composed of proliferative cells (dark cytoplasm and nuclei). This invaginates into a foveolar-type epithelium where the proliferative and stem cells are sequestered in the deep part of the foveolar pit. These surface and foveolar cells are only mucous cells. With the downward movement of cells from the proliferative region (arrows), glands are formed. If the glands remain mucous cell only, the epithelium remains within the definition of cardiac mucosa. The occurrence of parietal cells in the glands results in oxyntocardiac mucosa; the occurrence of goblet cells anywhere in the epithelium represents intestinal metaplasia.

the foveolar-type epithelium contained only mucous cells in cardiac mucosa or a combination of mucous and goblet cells in intestinal metaplastic mucosa, it was difficult to differentiate a true gland deep to the foveolar pit from the deepest area of a tortuous, tangentially sectioned foveolar pit. In cardiac mucosa and intestinal metaplastic mucosa, therefore, we used the Ki67 staining pattern to distinguish between tortuous and elongated foveolae from true glands

FIGURE 9.4 Early stage of columnar metaplasia. The columnar epithelium consists of a flat layer of mucous cells.

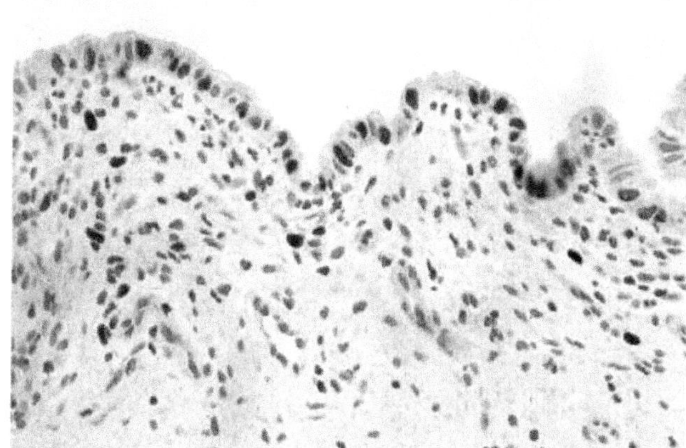

FIGURE 9.5 Immunoperoxidase stain for Ki67 in early metaplastic epithelium of surface-only type. A high percentage of cells show Ki67, indicating that the proliferative progenitor cells are on the surface.

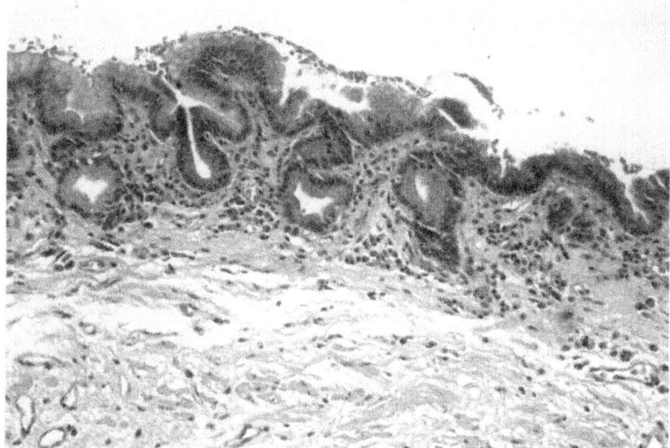

FIGURE 9.6 Cardiac mucosa of foveolar type. The surface epithelium has invaginated into short foveolar pits composed of mucous cells.

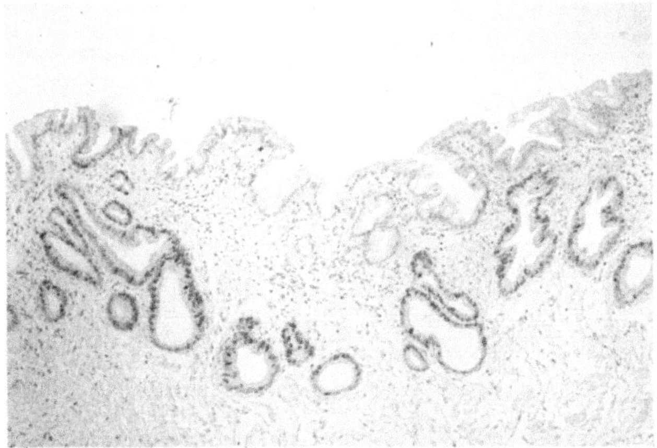

FIGURE 9.7 Immunoperoxidase stain for Ki67 in fovolar-type cardiac mucosa. The Ki67 positive proliferating cell pool has become sequested in the deep part of the foveolar pit. The surface epithelium is now Ki67 negative, indicating that they are terminally differentiated nonmitotic cells.

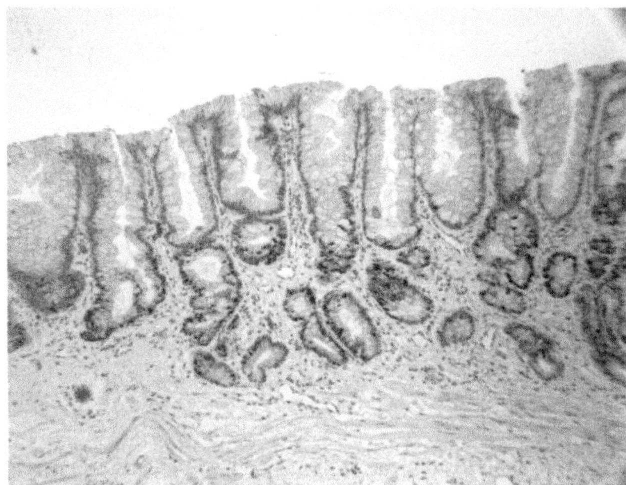

FIGURE 9.9 Immunoperoxidase stain for Ki67 showing Kl67 positive cells present to the deepest part of the foveolar pit. There are no Ki67 negative glands. This is therefore a foveolar-type epithelium.

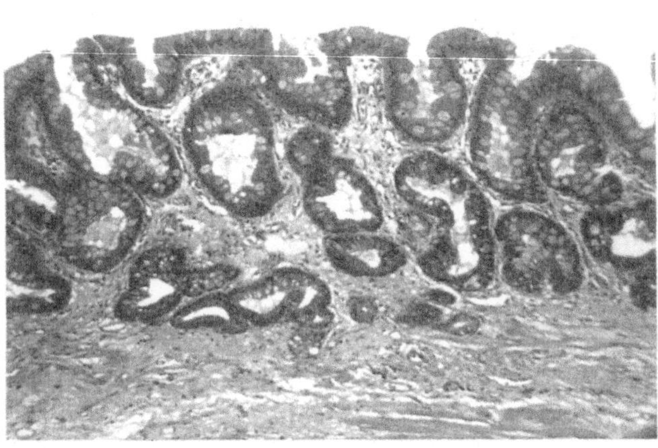

FIGURE 9.8 Better-developed columnar epithelium with goblet cells indicating intestinal metaplasia. The foveolar pit is long and complex. It is difficult to determine whether the deep "glands" are true glands or tortuous, tangentially sectioned foveolar pits.

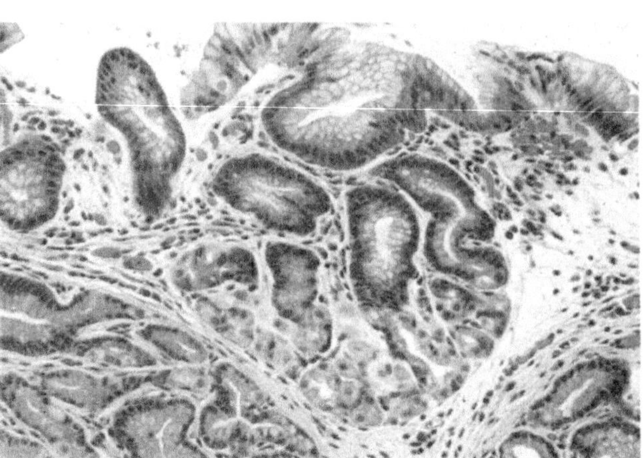

FIGURE 9.10 Oxyntocardiac mucosa, showing lobulated glands under the foveolar region composed of a mixture of parietal cells and mucous cells.

(Figs. 9.8 and 9.9). This decision was based on the observation that, in normal gastric oxyntic mucosa, the true glandular cells deep to the foveolar proliferative zone are differentiated, nonproliferative cells that do not express Ki67 (Fig. 9.2).

True glands could be easily recognized morphologically when they contained parietal or chief cells as was the case in oxyntocardiac mucosa (Fig. 9.10). In these instances, the parietal cell containing glandular region was confirmed as a region of Ki67 negativity below the proliferative deep foveolar zone (Fig. 9.11).

The 129 metaplastic columnar-lined epithelial areas in sections taken from the tubular esophagus could be divided into three patterns based on morphology and Ki67 staining (Table 9.2). Oxyntocardiac mucosa was always glandular because parietal cells do not normally occur in the surface layer and foveolar pit. Twenty-four of 129 areas (18.6%) were of the foveolar type, including 3 of 45 (6.7%) areas of cardiac mucosa, and 21 of 41 areas (51.2%) of nondysplastic intestinal metaplasia. The prevalence of a foveolar type was significantly greater in intestinal epithelium than cardiac mucosa (21/41 versus 3/45; $p < 0.001$ by the chi-square test).

In this study, Ki67 staining permitted us to follow the adult esophageal progenitor cell as it changed its differentiation from one epithelial type to another (Fig. 9.3). This permitted the development of a hypothetical model for the way in which these various epithelia evolved. The surface type, composed of Ki67 positive columnar epithelial cells in a single layer (Figs. 9.4 and 9.5), must be either the first epithelium that arises when squamous epithelium shifts to columnar metaplasia or an epithelium that regenerates over areas of erosion. In either event, the rarity of this epithelium (only 4 of 129 areas) suggests that this is a very transient epithelial type. This would be consistent with the general tendency for adult columnar epithelium to sequester the progenitor cells deeper in the mucosa. This epithelium appears to quickly invaginate into foveolar pits (Figs. 9.6 and 9.7), sequestering the Ki67 positive stem cells in the deep foveolar region away from the surface. The foveolar pits are at first rudimentary and then more elongated. Both surface and foveolar epithelial types are devoid of glands and therefore do not have parietal cells; they are therefore composed only of mucous cells and fall within the definition of cardiac mucosa or contain goblet cells and fall within the definition of intestinal metaplasia. We will provide evidence later in this chapter that intestinal metaplasia evolves from cardiac mucosa, which is the first epithelial type that arises from squamous epithelium.

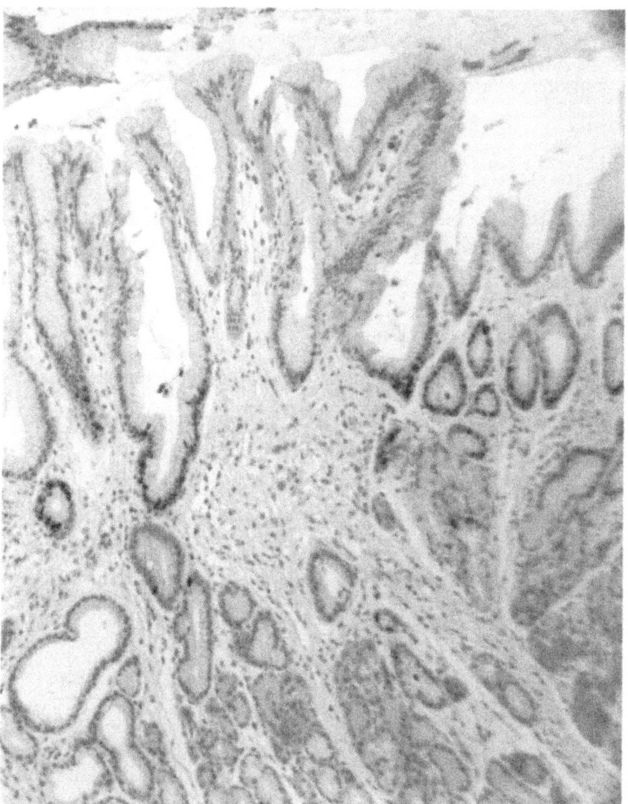

FIGURE 9.11 Glandular type columnar metaplastic epithelium. There are lobulated Ki67 negative glands under the Ki67 positive cells in the deep foveolar region. The glands contain a mixture of parietal and mucous cells (oxyntocardiac type).

The foveolar type of cardiac mucosa can remain as a purely foveolar epithelial type, or alternatively, and more commonly, glands develop under the foveolar region (Fig. 9.12). The development of glands represents a shift wherein the daughter cells of progenitor cell proliferation move downward to form glands in addition to moving up to the surface. The downward movement results in glandular cells, which develop secretory characteristics. These glandular cells are stable, terminally differentiated cells that generally have a long life span. They are nonproliferative and can be recognized as a Ki67 negative zone below the proliferative region of the epithelium in the deep foveolar pit (Figs. 9.3 and 9.11). Our data show that, in columnar metaplasia of the esophagus, the glands may remain as mucous cells (glandular type of cardiac mucosa) or develop a mixture of mucous and parietal cells and become oxyntocardiac epithelium (Fig. 9.12).

TABLE 9.2. Cardiac, Oxyntocardiac, and Intestinal Types of Columnar-Lined Esophagus

Type of epithelium[a]	Location	Surface type	Foveolar type	Glandular type	Total
Cardiac	Esophagus	4 (8.9%)	3 (6.7%)	38 (84.4%)	**45**
Oxyntocardiac	Esophagus	0	0	43	**43**
Nondysplastic IM (Barrett's)	Esophagus	0	21 (51.2%)	20 (48.8%)	**41**
Total	Esophagus	4 (3.1%)	24 (18.6%)	101 (78.3%)	**129**

[a]Classified on the basis of their composition into surface, foveolar, and glandular epithelial types.

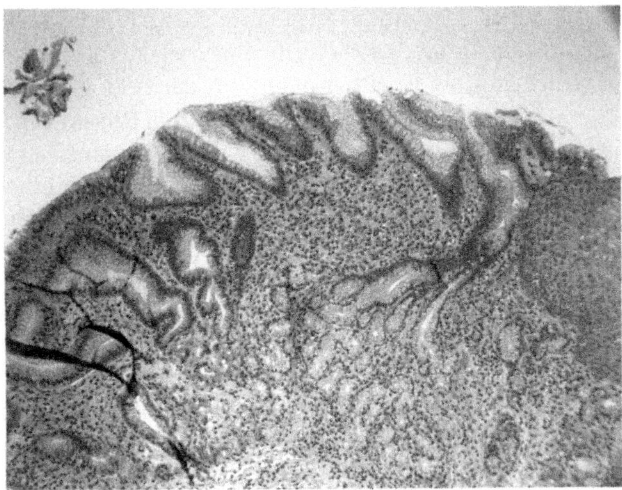

FIGURE 9.12 Columnar-lined esophagus of glandular type adjacent to squamous epithelium (*right*). The central part adjacent to the squamocolumnar junction contains glands with only mucous cells (1 to 2 foveolar complexes); this is cardiac mucosa. Parietal cell containing oxyntocardiac mucosa (*left*) is more distal. (Cross reference: Color Plate 3.8)

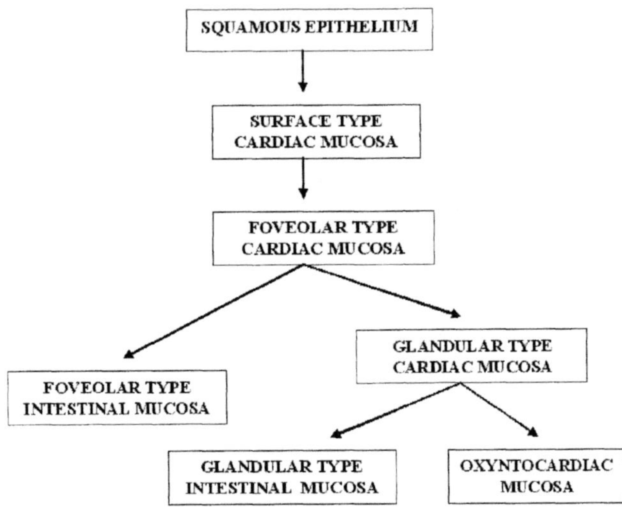

FIGURE 9.13 Schematic drawing of the evolution of the different types of columnar epithelia seen in the esophagus.

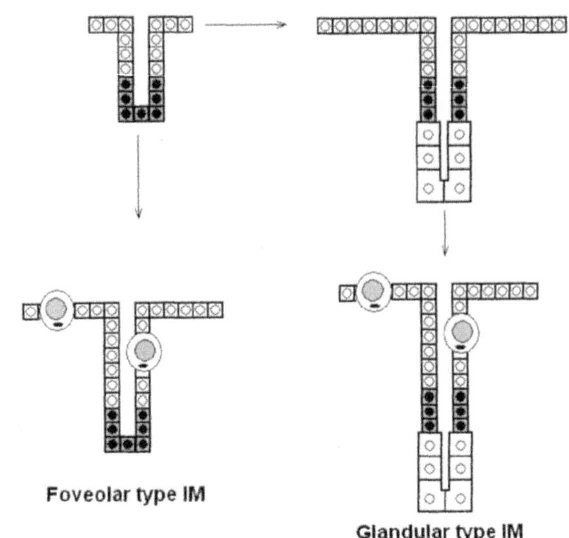

Figure 9.14 Evolution of cardiac mucosa into intestinal metaplasia. This occurs in both foveolar and glandular types of cardiac mucosa and is characterized by the occurrence of goblet cells (large round cells with gray vacuole), which usually populate the surface epithelium and foveolar region.

It therefore appears that oxyntocardiac mucosa must evolve from glandular type cardiac mucosa (Fig. 9.13).

Intestinal metaplasia represents the development of goblet cells in cardiac mucosa of either foveolar or glandular types (Fig. 9.14). Goblet cells are usually seen in the surface and foveolar region of the epithelium. The 41 areas of intestinal metaplasia studied showed nearly equal numbers of foveolar (21) and glandular (20) types of epithelia. This indicates that intestinal metaplasia develops either from foveolar or glandular types of cardiac mucosa. As described by Paull et al. (1) and confirmed in our previous studies (2), parietal cells and goblet cells do not coexist in the same foveolar-gland complex; this suggests that once parietal cells develop in a gland complex, they preclude the development of intestinal metaplasia.

In summary, the process of columnar metaplasia in the esophagus appears to be a complex but regularly sequenced process (Fig. 9.13). Cardiac mucosa develops from squamous epithelium, first appearing as a very transient surface-type epithelium. This quickly invaginates to form a foveolar pit. From this point, cardiac epithelium can develop glands, evolving into either glandular-type cardiac mucosa or oxyntocardiac mucosa, the latter characterized by the presence of parietal cells. Intestinal metaplasia occurs by the development of goblet cells in either the foveolar- or glandular-type of cardiac mucosa, but it never occurs in oxyntocardiac mucosa.

The Distribution of Columnar Epithelial Types in the Esophagus

If columnar-lined esophagus is defined as the segment of metaplastic esophageal columnar epithelium that is interposed between the stomach (which begins at the true gastroesophageal junction or

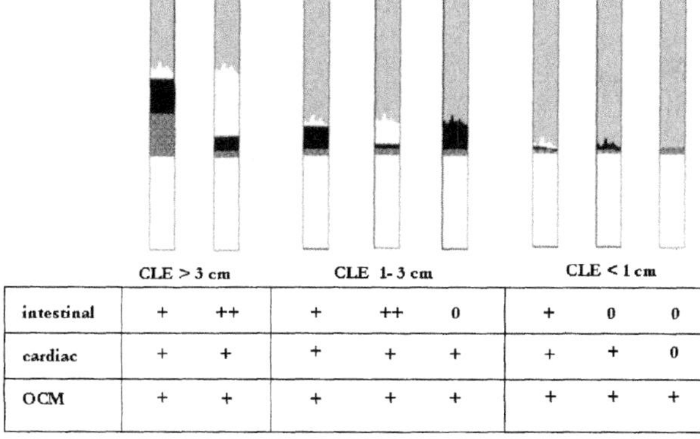

	CLE > 3 cm		CLE 1-3 cm			CLE < 1 cm		
intestinal	+	++	+	++	0	+	0	0
cardiac	+	+	+	+	+	+	+	0
OCM	+	+	+	+	+	+	+	+

FIGURE 9.15 Distribution of intestinal (white), cardiac (black), and oxyntocardiac (stippled) in columnar-lined esophagus. Long-segment (greater than 3 cm), short-segment (1 to 3 cm), and microscopic (less than 1 cm) columnar-lined esophagus can have very different amounts of these epithelial types in different patients. This can only be assessed by microscopic mapping. The true gastroesophageal junction is the line between oxyntocardiac mucosa and gastric oxyntic mucosa.

proximal limit of gastric mucosa) and the squamocolumnar junction, the following patterns can be recognized (Fig. 9.15):

1. Patients with a long (>3 cm) segment of columnar lined esophagus. The majority of such patients will have all three types of columnar epithelia that constitute columnar-lined esophagus (Fig. 9.15). Paull et al. (1) showed that the three epithelia are distributed in a constant manner with the intestinal metaplasia being nearest the squamocolumnar junction, cardiac mucosa being in the midregion, and oxyntocardiac mucosa nearest the gastroesophageal junction (Fig. 9.16). The amounts of these epithelial types present vary greatly in different individuals (Figs. 9.15 and 9.16).
2. Patients with short (1 to 3 cm) columnar-lined segments that are visible at endoscopy as short-segment disease. These patients have a lower prevalence of intestinal metaplasia. Those patients that have intestinal metaplasia have the same distribution of the three epithelia as seen in long-segment columnar-lined esophagus (Fig. 9.15). Patients in this group who do not have intestinal metaplasia will have cardiac mucosa proximally and oxyntocardiac mucosa distally.
3. Patients with columnar-lined esophagus of less than 1 cm (usually endoscopically invisible and detected only by microscopy) have a still lower prevalence of intestinal metaplasia. Those that have intestinal metaplasia will have the same distribution of the three epithelial types. Those that have no intestinal metaplasia may have a mixture of cardiac mucosa and oxyntocardiac mucosa or oxyntocardiac mucosa alone (Fig. 9.15). We will explore this epithelial distribution further.

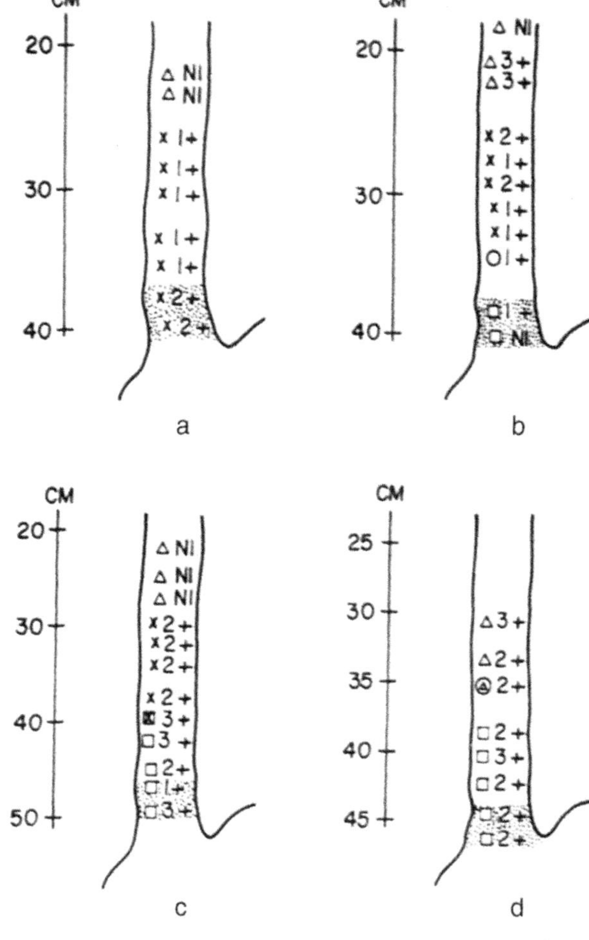

x SPECIALIZED COLUMNAR EPITHELIUM
□ GASTRIC FUNDAL EPITHELIUM
○ JUNCTIONAL EPITHELIUM
Δ SQUAMOUS EPITHELIUM
NI NO INFLAMMATION
1+ to 3+ DEGREE OF INFLAMMATION
CM CENTIMETERS FROM INCISORS
LOWER ESOPHAGEAL SPHINCTER

FIGURE 9.16 Paull et al.'s (1) Figure 4, showing the presence of oxyntocardiac (= fundic-type) epithelium in the distal part of the esophagus above the sphincter in patients with long-segment Barrett esophagus. Note that the distribution of the three epithelial types is very different in these four patients that have been illustrated.

Mechanism of Pathologic Changes in Reflux-Damaged Esophagus

In patients with reflux damage, pathologic changes are simultaneously occurring in squamous epithelium and all types of columnar metaplastic epithelia. It is likely that some components of the gastroesophageal refluxate are more involved in producing some kinds of changes than others. Acid, for example, is definitely involved in producing squamous epithelial injury. It is not known whether other molecules in the refluxate contribute to this. The actual molecules causing the genetic switches that result in cardiac metaplasia or injuring cardiac mucosa, or transforming cardiac mucosa into intestinal and oxyntocardiac mucosa, are unknown (Table 9.3).

It is naïve to believe that acid is responsible for every cellular event in every epithelial type within the reflux damaged esophagus, which is the basic assumption made when acid suppressive agents are used to treat reflux disease. What is of great concern is whether acid, which has a damaging effect on squamous epithelium, has a beneficial rather than deleterious effect on cardiac mucosa. If this is true, suppression of acid in patients with cardiac metaplasia may actually aggravate rather than prevent the disease progression through the second and third phases of this disease. While there is little direct evidence for a negative effect of acid on cardiac mucosa, this is largely an unknown and untested issue that is relevant because there is nothing that is known to explain the increase in reflux-induced adenocarcinoma that we have experienced since the 1970s.

The fact that the prevalence of intestinal metaplasia has increased greatly since the 1970s suggests that acid is not an important agent in the development of intestinal metaplasia in cardiac mucosa. Similarly, the increase in the incidence of adenocarcinoma since the 1970s suggests that acid does not play an important role in the development of malignant neoplasia in Barrett esophagus.

The question that needs urgent answer is whether acid suppression that is the mainstay of medical treatment of reflux disease contributes to the transformation of cardiac mucosa into intestinal metaplasia or promotes carcinogenesis in intestinal metaplastic epithelium. Whenever this question is raised in public forum, there is an aggressive defense of acid suppressive drugs by gastroenterologists that prevents further inquiry. However, this vehement defense is made without any evidence. There is evidence to suggest that acid suppression is associated with cancer. Lagergren et al. (3), in his epidemiologic study of adenocarcinoma of the esophagus and gastric cardia, reported:

> We compared the risk of esophageal adenocarcinoma among patients who used medication for symptoms of reflux at least five years before the interview with that among symptomatic persons who did not use such medication. *The odds ratio was 3.0 (95 percent confidence interval, 2.0 to 4.6) without adjustment for the severity of symptoms and 2.9 (95 percent confidence interval, 1.9 to 4.6) with this adjustment.* (pp. 827–829)

An odds ratio of 3.0 is a highly significant association; recently Merck voluntarily withdrew Vioxx from the market because it was shown to have an odds ratio of 1.9 in causing ischemic events. Unfortunately, Lageregren et al. (3) did not provide details in the article regarding the type of acid suppressive medications that were used.

Even more troubling is that gastroenterologists will dismiss the question that there is an association between use of acid suppressive medication and cancer when there is no evidence to prove that it does not. The fact is that we do not know whether acid suppression promotes cancer in this disease; there is some evidence to suggest that it does and no evidence that it does not. To support a $10 billion to 13 billion per year acid suppressive drug industry without the certainty of these data is certainly bad practice of medicine that goes against the fundamental edict, "Do no harm."

TABLE 9.3 Postulated Genetic Switches

Genetic switch	Caused by
Squamous to cardiac	Acid (?); other
Cardiac to oxyntocardiac	Acid (?); other
Cardiac to intestinal	Less acid pH (?); bile (?)

Note: These switches are the basis for metaplastic events that occur in the esophagus in reflux disease.

CARDIAC TO OXYNTOCARDIAC MUCOSA: THE BENIGN GENETIC SWITCH

In 1957, Barrett (4) recognized that the proximal region of the columnar-lined esophagus was lined by pure cardiac mucosa but that oxyntic (parietal) cells started appearing in the more distal region. Paull et al. (1), in their histologic classification of columnar-lined esophagus, referred to the mucosa in which there are glands containing a mixture of parietal and mucous cells as "gastric-fundic-type" mucosa (Fig. 9.10), distinguishing it by precise and reproducible histologic criteria from cardiac (= junctional) epithelium, which contains only mucous cells. Unfortunately, the succeeding medical literature shortened Paull et al.'s (1) term to "fundic" mucosa. Paull et al. (1) clearly showed that this epithelial type was seen in the esophagus above the lower esophageal sphincter and within the zone of the sphincter (Fig. 9.15). Because this has nothing to do with the gastric fundus, which is lined by pure gastric oxyntic mucosa, we prefer to use the term "oxyntocardiac mucosa" for this epithelium (5).

Oxyntocardiac mucosa is formed when cardiac mucosa develops a genetic signal that causes its progenitor cells to differentiate into parietal cells and move downward into glands in the deeper part of the mucosa below the proliferative progenitor cell region in the base of the foveolar pit (Fig. 9.10). Review of the embryology of the esophagus indicates that this event is aberrant for the esophagus. At no point during embryogenesis does the fetal stem cell in the esophagus produce parietal cells (6). Gastric mucosa, on the other hand, develops parietal cell containing glands very early in fetal life (7, 8). It therefore appears that oxyntocardiac transformation of cardiac mucosa represents the acquisition of an aberrant genetic signal that resembles the normal gastric genetic signal (Fig. 9.1).

Oxyntocardiac mucosa is morphologically different from gastric oxyntic mucosa; it shows chronic inflammation, has residual mucous cells in the glands, and the glands are frequently lobulated in contrast to the straight tubular glands of normal gastric mucosa (Fig. 9.10). The most important difference, however, is that it is located in the distal reflux-damaged epithelium above the true gastroesophageal junction, not in the stomach (Fig. 9.16). It contains gland ducts of the submucosal glands (Fig. 9.17) (9) and its submucosa contains esophageal glands (Fig. 9.18) (10, 11).

Studies indicate that the Sonic hedgehog gene directs epithelia in the gastrointestinal tract to differentiate into parietal cells (12). This gene is maximally expressed in normal adult gastric oxyntic mucosa. It is also expressed in heterotopic gastric mucosa in Meckel's diverticula and has been shown to be expressed in metaplastic columnar epithelia of the esophagus that contain parietal cells (i.e., oxyntocardiac mucosa). The genetic signal that is acquired by cardiac mucosa that results in oxyntocardiac mucosa appears to be the Sonic Hedgehog gene.

The proliferative characteristics of oxyntocardiac mucosa show that it is less proliferative than cardiac and intestinal epithelia, indicating that it is associated with a low damage environment (13). The "gastric type genetic signal" associated with the Sonic Hedgehog gene appears to be generated only in the low damage environment of the distal esophagus. Oxyntocardiac mucosa is to be regarded as the most resistant

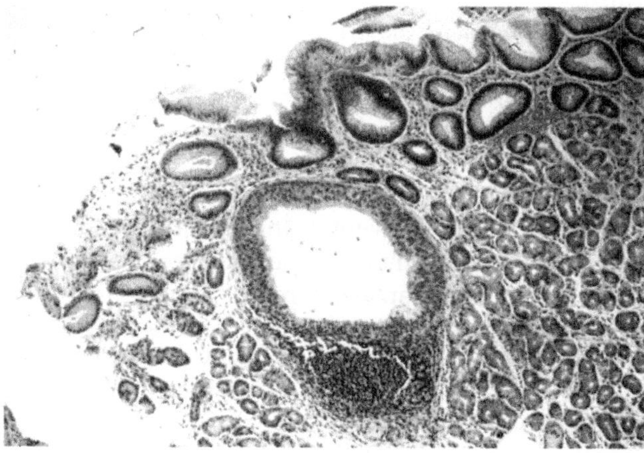

FIGURE 9.17 A submucosal gland duct traversing oxyntocardiac mucosa. This proves that oxyntocardiac mucosa is esophageal in this biopsy because submucosal glands and gland ducts are not normally found in gastric mucosa.

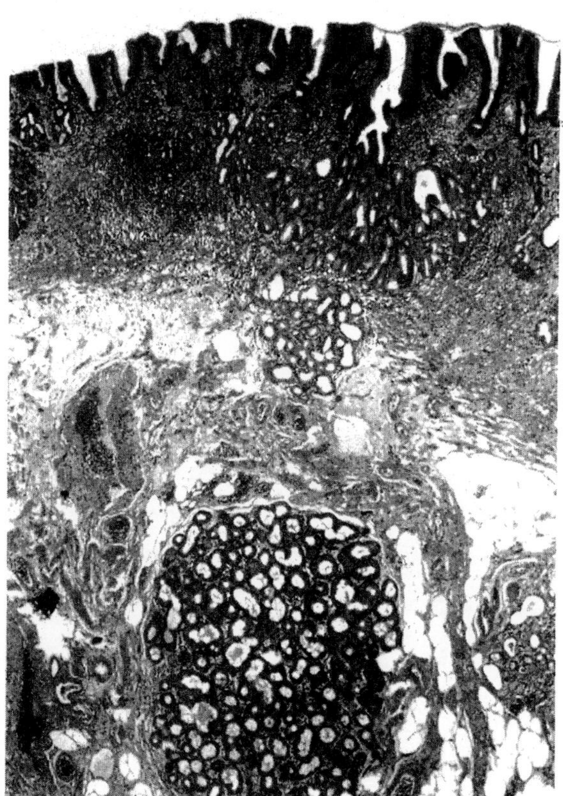

FIGURE 9.18 Section from esophagectomy specimen showing a lining of columnar epithelium of oxyntocardiac type and the presence of submucosal mucous glands. The latter identifies this area as esophagus because these glands are not found in the stomach.

esophageal epithelium to gastric refluxate; it is the least proliferative (Fig. 9.11), the least inflamed (Fig. 9.19), and the least likely to progress to intestinal metaplasia and adenocarcinoma. It is a benign epithelium. Conversion of other types of metaplastic columnar epithelia in the esophagus such as cardiac and intestinal epithelia to oxyntocardiac mucosa will decrease the risk of carcinoma. As such, it can be regarded as one mechanism of "histologic cure" of columnar metaplasia.

Oxyntocardiac mucosa is an important epithelial type in the columnar-lined esophagus. As first reported by Paull et al. (1) and confirmed by us (5), it is almost never associated with intestinal metaplasia in the same foveolar-gland complex. What this means is that when cardiac mucosa gets the gastric-type (?Sonic Hedgehog) genetic signal to produce parietal cells, it precludes the simultaneous development of the genetic signal (Cdx2) that causes intestinal metaplasia. The gastric-type genetic signal appears to block the occurrence of the intestinal-type genetic signal in the same foveolar-gland complex (Fig. 9.1).

Does Acid Activate the "Parietal Cell" Genetic Switch?

If the switch from cardiac to oxyntocardiac mucosa results in a metaplastic columnar epithelium in the esophagus that is not predisposed to intestinal metaplasia and adenocarcinoma, it is incredibly important to find out its mechanism. The answer to this question is not only unknown, it has never been thought of let alone investigated. As such, we can only hypothesize, using observation and logic, and leave the hypothesis to be subject to experimentation.

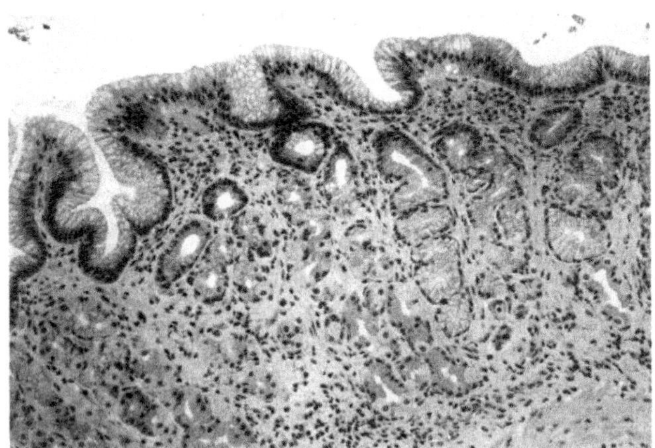

FIGURE 9.19 Oxyntocardiac mucosa with numerous parietal cells in the glands. Inflammation and foveolar hyperplasia are minimal. (Cross reference: Color Plate 3.14)

In the developing embryonic stomach, parietal cells develop in the earliest glands that form at the base of the foveolar pit at approximately the 17th week of gestation (8). This is clearly a programmed genetic change unassociated with any external stimulus. After the genetic signal has been established in gastric mucosa, it lasts throughout life in a manner that maintains the normal parietal cell mass in gastric oxyntic mucosa. Because parietal cells are long-lived stable cells, their turnover is slow. The genetic signal for the gastric epithelial progenitor cell to generate parietal cells is controlled at the required slow rate by some unknown mechanism.

The adult progenitor cells in gastric mucosa are found in the deepest region of the foveolar pit above the parietal cell containing glands (Fig. 9.2). The parietal cells secrete acid into the gland lumen, which drains into the base of the foveolar pit. The adult progenitor cells are continually exposed to this acid secretion. Let us hypothesize that this acid exposure of the adult progenitor cell in the gastric foveolar pit represents the continuing stimulus for the progenitor cell to differentiate into parietal cells in normal adult gastric mucosa.

The progenitor cells in metaplastic cardiac mucosa are located in the deep foveolar region (Fig. 9.7). Luminal molecules, including H^+ ions, have relatively easy access to these progenitor cells through the openings of the foveolar pit at the surface. When they are stimulated by acid in the refluxate, there is the possibility of a similar interaction between the acid and the progenitor cell to lead to the activation of the Sonic Hedgehog gene within cardiac mucosa to generate parietal cells in the glands (i.e., transform into oxyntocardiac mucosa). If this reasoning is correct, it would mean that the acid could also be the agent that causes cardiac mucosa to transform into oxyntocardiac mucosa. Because this transformation is benign and prevents the development of intestinal metaplasia and adenocarcinoma, acid may be protective in metaplastic columnar epithelium of the esophagus.

Does this possibly explain why adenocarcinoma rates are increasing? If this hypothesis is correct, acid suppression will prevent transformation of cardiac into oxyntocardiac mucosa and increase the probability of intestinal metaplasia. There is support for this in the finding of oxyntocardiac mucosa at a higher level in the esophagus in Paull et al.'s 1976 article (1) than is seen commonly today. In the four cases illustrated by Paull et al. (1), oxyntocardiac ("= gastric fundal") epithelium is shown in the distal esophagus above the sphincter zone in two patients (Fig. 9.16). In our study of 10 esophagectomy specimens, there was no oxyntocardiac mucosa in the tubular part of the esophagus in any patient, including those with long segments of columnar-lined esophagus (11); oxyntocardiac mucosa was restricted to the most distal part of the columnar-lined segment (Fig. 9.20).

If there has been a historic change in the amount and location of oxyntocardiac mucosa in the columnar-lined esophageal segment, there must be a reason. It seems logical to hypothesize that this change could result from a change in the pH environment in the esophagus. It is known that the pH in the esophagus in reflux patients progressively becomes more alkaline as the distance from the gastroesophageal junction increases. Normal patients without gastroesophageal reflux have a sharp pH gradient at the distal end of the lower esophageal sphincter; the neutral (approximately 7) pH of the esophagus is sharply separated from the high acidity (pH 1 to 3) of gastric contents by the sphincter. Reflux of acid gastric juice into the esophagus causes the length of this pH gradient to increase in a manner that is determined by the volume of reflux. With increasing volumes of reflux, the pH neutrality in the esophagus is pushed to increasingly proximal levels, and the esophagus is exposed to a range of pH varying between 1–3 at the gastroesophageal junction to neutral (pH 7) at the height of the refluxate column. The general increase in the use of acid suppressive medication (largely over the counter) is likely to have had an effect in producing an overall decrease in the acidity of gastric juice in symptomatic patients. The desired effect of acid suppressive drugs is the alkalinization of gastric juice from its norm of 1 to 3 to higher pH levels. For the same amount of reflux in a given

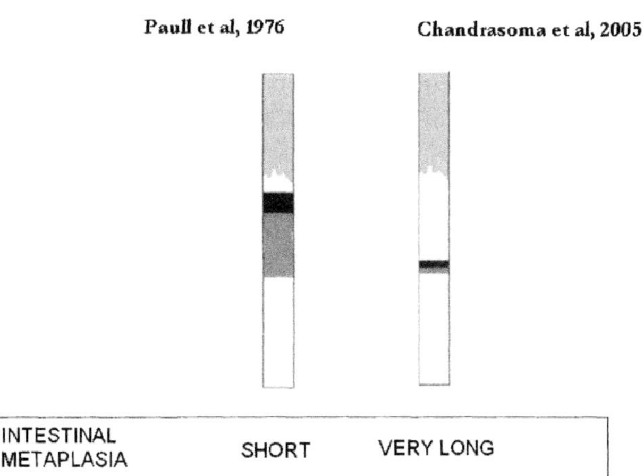

FIGURE 9.20 Historical differences in the distribution of the three epithelial types in columnar-lined esophagus. On the left is Paull et al.'s (1) data showing relatively small amounts of intestinal metaplasia (white) and larger amounts of oxyntocardiac mucosa (stippled) than Chandrasoma (11).

patient, the esophageal acidity would have decreased at any given level. If one assumes that acid is the stimulus that is responsible for converting cardiac to oxyntocardiac mucosa, this can explain how oxyntocardiac mucosa has decreased and is seen at a lower level in the columnar-lined segment in 2005 compared with 1976 (Fig. 9.20).

Like any other hypothesis, this needs to be thought of and tested. The fact that acid suppressive agents continue to be used freely—without evidence that this hypothesis and others not yet thought of that may indicate that acid has a negative effect in carcinogenesis are not correct—appears dangerous. The greatest dangers in science and medicine have always been the unknown and untested hypothesis. It is the reason why our ozone layer is in danger (no one thought of testing the effect of fluorocarbons to see whether they had an effect on the ozone layer); it was the reason why asbestos transformed from a wonder insulating agent to blight upon humanity (no one thought to test asbestos exposure for risk of mesothelioma). Is acid suppression the reason why reflux-induced adenocarcinoma is the most rapidly increasing cancer type in the Western world? Will omeprazole be the next Vioxx or thalidomide?

CARDIAC MUCOSA TO INTESTINAL METAPLASIA: THE SECOND GENETIC SWITCH

Metaplastic cardiac mucosa in columnar-lined esophagus is an active epithelium that is constantly reacting to the injurious effect of continuing gastroesophageal reflux. Its proliferative stem cells are located in the basal region of the foveolar pit where they multiply to renew the epithelium (Fig. 9.7). The rate of proliferation depends on the rate of cell loss resulting from reflux-induced damage. Study of proliferative rates using Ki67 staining in cardiac mucosa shows that there is a significantly higher proliferative rate in cardiac mucosa than both normal gastric oxyntic mucosa and oxyntocardiac mucosa (13).

Mechanism of Intestinal Metaplasia in Cardiac Mucosa

Luminal molecules have access to the progenitor cells in cardiac mucosa and can interact with them to cause further genetic changes. Of these, intestinal metaplasia is the most important because it seems necessary for carcinogenesis to progress. Intestinal metaplasia is characterized by the appearance of goblet cells in cardiac mucosa. A single goblet cell in cardiac mucosa satisfies the criterion for diagnosis of intestinal metaplasia. The density of goblet cells in cardiac mucosa is extremely variable (Figs. 9.21 and 9.22).

There is good evidence that the genetic switch that causes intestinal differentiation in cardiac mucosa is activation of the Cdx homeobox transcription gene system, which includes Cdx1 and Cdx2. These genes are suppressed in normal esophagus and stomach. However, they are expressed in the normal small and large intestine and are believed to be the genes that drive differentiation in these sites (14). Cdx1 and Cdx2 are expressed in most cases of intestinal metaplasia of the esophagus.

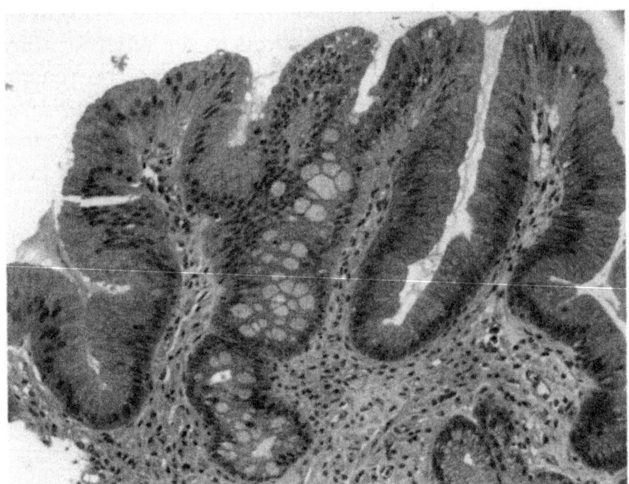

FIGURE 9.21 Intestinal metaplasia in cardiac mucosa, characterized by the occurrence of goblet cells in one of four foveolar complexes in cardiac mucosa. (Cross reference: Color Plate 4.1)

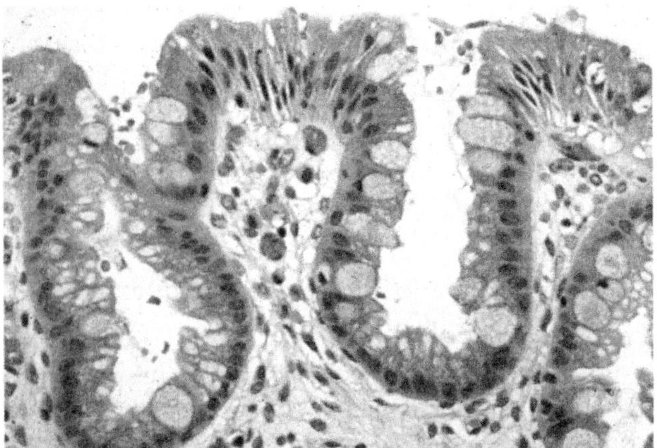

FIGURE 9.22 Intestinal metaplasia in cardiac mucosa with larger numbers of goblet cells; virtually every other cell is a goblet cell and all the foveolar complexes shown have intestinal metaplasia. (Cross reference: Color Plate 4.3)

In 1997, Silberg et al. (15) showed that Cdx1 is an intestine-specific transcription factor expressed early in intestinal development that may be involved in the regulation of proliferation and differentiation of intestinal epithelial cells. This study examined the pattern of Cdx1 expression in metaplastic and neoplastic tissue to provide insight into its possible role in abnormal differentiation. The study used human tissue samples from surgical specimens stained by immunohistochemistry using a polyclonal antibody against a peptide epitope of Cdx1.

Specific nuclear staining was found in epithelial cells of the small intestine and colon. Staining was maximal in the crypts of the small intestine and at the base of the colonic crypts and less at the tips of the villi and colonic surface. Normal stomach and esophagus, including both squamous epithelium and the submucosal glands, did not express Cdx1 protein. Areas of intestinal metaplasia in both stomach and esophagus showed intensely positive staining for Cdx1. All 10 cases of Barrett esophagus and 12 of 12 cases of gastric intestinal metaplasia showed nuclear staining for Cdx1 restricted to the areas of metaplasia. The authors concluded that Cdx1 may be important in the transition from normal gastric and esophageal epithelium to intestinal-type metaplasia.

Phillips et al. (16) used immunohistochemistry to detect Cdx2 in 134 esophageal specimens, including 62 with junctional-type (cardiac) epithelium, 13 of which had equivocal histologic features of Barrett's epithelium, 34 with Barrett's epithelium without dysplasia, and 38 with Barrett's epithelium and dysplasia or carcinoma (13 low-grade dysplasia, 19 high-grade dysplasias, and 6 adenocarcinomas). The 13 cases of junctional-type (cardiac) epithelium that was considered "equivocal for Barrett's epithelium" had no definitive goblet cells on the hematoxylin and eosin section, but had nongoblet columnar epithelial cells with cytoplasmic acid mucin in PAS-alcian blue-stained sections. Cdx2 was expressed in all cases of Barrett's epithelium, with diminution or focal loss of expression in some cases of high grade dysplasia and adenocarcinoma. Cdx2 was positive in 20/62 cases (30%) of junctional-type (cardiac) epithelium including 10 of 13 (77%) with equivocal histologic features of Barrett's epithelium.

These two studies suggest that Cdx1 and Cdx2 gene activation is involved in the transformation of cardiac mucosa to intestinal metaplasia. Expression of Cdx2 in a minority of specimens with cardiac mucosa raises the possibility of Cdx2 activation at an earlier point in the metaplastic process. The alternative explanation is that Cdx2 positive cardiac mucosa is in a lag phase between gene expression and phenotypic expression of intestinal metaplasia.

Vallbohmer et al. (17), from our group, evaluated the gene expression levels of Cdx2 in squamous, cardiac and Barrett's epithelium. Endoscopic biopsies from patients with gastroesophageal reflux disease were classified histologically into normal squamous (62 cases), cardiac (19 cases), oxyntocardiac (14 cases), and intestinal metaplasia (15 cases). Normal duodenum was used as a columnar control. After laser capture microdissection, Cdx2 mRNA expression levels were measured by quantitative real-time polymerase chain reaction.

Cdx2 gene expression levels were lowest in squamous epithelium (median 0.01; 25th to 75th percentile 0.01 to 0.05). There was a significant stepwise increase in Cdx2 expression from squamous to cardiac and oxyntocardiac to Barrett's epithelium. Cdx2 levels in cardiac mucosa (median 0.4; 25th to 75th percentile 0.3 to 0.71) were not significantly different from those in oxyntocardiac mucosa (median 0.76; 25th to 75th percentile 0.28 to 1.14), but they were significantly greater than in squamous epithelium. An approximately 10-fold higher Cdx2 expression was present in intestinal metaplasia when compared with cardiac mucosa. (median 6.72; 25th to 75th percentile 3.97 to 8.08). Control duodenal mucosa had the highest levels of Cdx2 expression, consistent with the fact that Cdx2 is normally expressed in duodenal mucosa. This study showed that Cdx2 gene expression is markedly upregulated at the point of conversion of cardiac mucosa to intestinal metaplasia. Expression of Cdx2 in cardiac mucosa at a lower level suggests that Cdx2 upregulation may occur at an earlier point than the actual appearance of goblet cells.

Conclusion: Cdx2 expression in the esophageal progenitor cell is an important genetic signal associated with the development of intestinal metaplasia in the esophagus.

Squamous Epithelium to Intestinal Metaplasia Is a Two-Step Process

There is little doubt that squamous epithelium does not progress directly to intestinal metaplasia without the intervening step of cardiac metaplasia (Fig. 9.1). In fact, when one looks at the different morphologic types of surface, foveolar, and glandular epithelia, the sequence of change from squamous to intestinal appears to be more diverse that even a simple multistep single pathway (Fig. 9.14). The pattern of Cdx2 expression in columnar epithelia of the esophagus also suggests that the metaplastic process is a multistep process: an intermediate columnar transformation to cardiac mucosa followed by intestinal metaplasia in the cardiac mucosa. There is evidence from clinical observations to support such a two-step metaplastic process.

Columnar-Lined Esophagus in Children Rarely Has Intestinal Metaplasia

Reflux-induced columnar-lined esophagus in children consists predominantly of cardiac mucosa without intestinal metaplasia. The frequent presence of small amounts of cardiac mucosa in children with reflux was well documented by Glickman et al. (18) who showed that small amounts of cardiac mucosa were present in 60 of 74 children (81%). In these children, they established that the probability of reflux symptoms was greater in children who had greater than 1 mm of cardiac mucosa compared with those with less than 1 mm, strongly suggesting that even very small amounts of cardiac mucosa were the result of reflux. Glickman et al. reported that the overall prevalence of goblet cells in these children was 15%; the amount of intestinal metaplasia was small, occupying less than 25% of the area in any biopsy specimen.

Hassall (19), in an excellent review from the British Columbia Children's Hospital in Vancouver, showed that the diagnosis of "Barrett esophagus" is often made in children without the presence of intestinal metaplasia. He reviewed 14 reports from the literature from 1984 to 1996, in which 119 children were reported to have Barrett esophagus. Only 43 of these had intestinal metaplasia.

Hassall reported his unpublished experience:

> Over a 12-year period (1985–1996), only 7 children were newly diagnosed as having Barrett esophagus (i.e., with specialized metaplasia), for a prevalence of 0.02% of all pediatric upper gastrointestinal endoscopies during that period. Their ages were 8–17 years (mean 14 years). In this unit, it is routine practice at endoscopy to document esophagogastric landmarks and take biopsy specimens from the Z line and tubular esophagus; use of this protocol makes it highly unlikely that Barrett esophagus would be missed. (p. 536)

Hassall also commented on his previously reported study from 1985 of 11 patients with "Barrett esophagus." Only 5 out of 11 of these patients had specialized mucosa with goblet cells. The other 6 patients had "biopsy specimens containing cardiac mucosa as proximal as 4 to 15 cm above the lower esophageal sphincter" (p. 536). Hassall suggested that it is likely that Barrett esophagus was missed because of sampling error of the biopsies. He considered less likely the possibility that a 10- or 15-cm columnar-lined esophagus in a child does not contain goblet cells. The youngest patient reported in the literature with intestinal metaplasia was 5 years old (20). Qualman et al. (20)

showed that pediatric patients usually had 25 or fewer goblet cells per square millimeter of Barrett mucosa compared with a mean of 57 cells per square millimeter of Barrett mucosa between the ages of 41 and 80 years.

In Hassall's review, most of the reported patients in the literature had long, obviously visible columnar-lined segments in the esophagus. The mean lengths in three reported studies where the length was documented were 4 cm (2), 7 cm (22), and 8.5 cm (23).

Children with columnar-lined esophagus almost always had significant comorbidities such as mental retardation, cerebral palsy, cystic fibrosis, repaired esophageal atresia, and chemotherapy for malignant disease (usually leukemia). Hassall estimated that 60% to 70% of children with Barrett esophagus have significant comorbid diseases. He commented that this situation is the opposite of that seen in adults with Barrett esophagus.

Hassall reaches the correct conclusion to explain the rarity of intestinal metaplasia in children even when long segments of columnar-lined esophagus are present: "Perhaps Barrett's esophagus evolves in some children by the development of goblet cell metaplasia in cardiac mucosa. In other words, perhaps cardiac mucosa in the tubular esophagus is a precursor of Barrett's esophagus in children" (p. 537).

It is clear from Hassall's review that long segments of columnar-lined esophagus are very uncommon in children and usually seen with other diseases. Even in children with very long segments of columnar-lined esophagus, intestinal metaplasia is uncommon, being seen in 2 of 13 cases in Dahms (21) where the columnar-lined segment was a mean of 4 cm, 5 of 11 cases in Hassall (22) with a mean length of 7 cm, and 2 of 2 cases in Hassall (23) with a mean length of 8.5 cm.

Glickman et al.'s (18) data show that when biopsies are done at the Z line in endoscopically normal children, microscopic (often <1 mm) segment of cardiac mucosa is common (60 of 74 patients). Glickman et al. reported a prevalence of intestinal metaplasia of 15% (11 of 74 patients) in this group, which is high. In Hassall's experience, the prevalence of intestinal metaplasia in children is 0.02% of all upper endoscopies with only 7 newly diagnosed cases of intestinal metaplasia during a 12-year period. The prevalence in these two studies seem very different; however, Glickman et al.'s 74 patients span a period of 9 years, and a total of 11 patients over this period is not greatly dissimilar to Hassall's 7 patients over 12 years. The small number of cases in Glickman's study suggests that, unlike Hassall, this group takes biopsies only in patients with visible endoscopic abnormality; this would create a bias toward finding intestinal metaplasia because intestinal metaplasia increases in frequency as the length of columnar-lined esophagus increases. The numbers strongly indicate that intestinal metaplasia is extremely uncommon in children despite the frequent presence of columnar-lined esophagus (cardiac and oxyntocardiac mucosa) in the distal esophagus.

The data in children strongly suggest that cardiac mucosa is the first type of columnar epithelium that occurs in patients with reflux disease. It likely precedes intestinal metaplasia by many years. Intestinal metaplasia becomes common only in adult patients, from which point its prevalence increases with age.

Changes in Postesophagectomy Patients

The second piece of evidence that intestinal metaplasia passes through the intermediate step of cardiac metaplasia is that after esophagogastrectomy, many patients develop cardiac mucosa in the esophagus above the anastomotic line. This progresses to intestinal metaplasia in some patients, often after many years. Cases are on record where this progresses to adenocarcinoma in patients whose initial esophagectomy was for benign disease.

The most conclusive article on this subject is that published by Dresner et al. (21), which is reviewed fully in Chapter 5 (Figure 9.23). In summary, Dresner et al. (24) performed serial endoscopic assessment, and systematic biopsy at the esophagogastric anastomosis was done in 40 patients who had esophagogastrectomy. Esophageal columnar epithelium developed at the anastomosis in 19 patients after the surgery. These revealed cardiac-type epithelium in 10 patients and intestinal metaplasia in 9; 7 patients followed serially

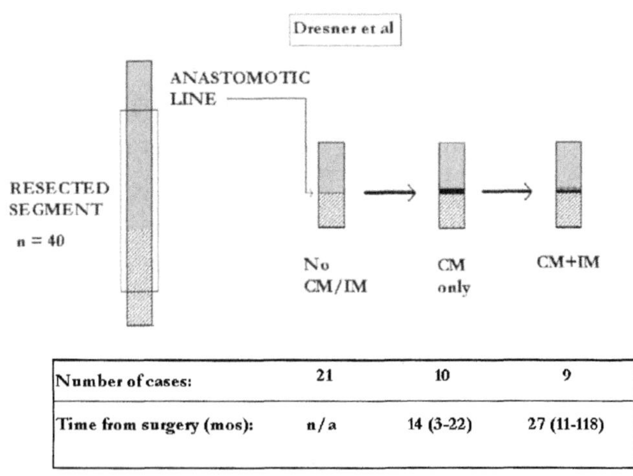

Number of cases:	21	10	9
Time from surgery (mos):	n/a	14 (3-22)	27 (11-118)

FIGURE 9.23 Summary of findings above the anastomotic line after esophagogastrostomy reported by Dresner et al. (24). The squamous epithelium undergoes sequential columnar metaplasia, first into cardiac and then into the intestinal type of epithelium.

showed progression from cardiac-type epithelium to intestinal metaplasia. The incidence of Barrett's metaplasia was similar irrespective of the histological subtype of the resected tumor. Patients with esophageal columnar epithelium had significantly higher acid ($p = 0.015$) and bilirubin ($p = 0.011$) reflux.

The authors concluded that severe duodenogastroesophageal reflux occurs following subtotal esophagectomy and provides an environment for the acquisition of Barrett's metaplasia via a sequence of cardiac epithelium and eventual intestinal metaplasia.

A similar study from our unit showed that the cardiac mucosa that developed in the esophagus after esophagectomy had genetic features that were similar to cardiac mucosa developing spontaneously in the esophagus (25). This was a retrospective study of 100 patients with esophageal carcinoma who had undergone esophagectomy with gastric tube reconstruction. The esophagectomy included the proximal stomach in all cases, ensuring removal of the gastroesophageal junctional region and all possible cardiac mucosa. The anastomosis was between squamous epithelium-lined esophagus proximally (the proximal margin of the esophagectomy contained squamous epithelium in all cases) and gastric oxyntic mucosa distally (the distal surgical margin of the esophagectomy specimen was at least 2 cm distal to the gastroesophageal junction and shown to contain gastric oxyntic mucosa in all cases).

Of these 100 patients, 20 had an endoscopic evaluation that included a biopsy of the remnant cervical esophagus taken from above the gastroesophageal anastomosis 9 months or more after the esophagectomy. The supra-anastomotic biopsies were examined for the presence of epithelial type; this is a study from the University of Southern California and the definitions used were those recommended for squamous, cardiac, oxyntocardiac, and intestinal.

Cardiac mucosa was present in the cervical esophagus above the anastomotic line in 10 out of 20 (50%) of these patients. The median interval between esophagectomy and postoperative biopsy showing cardiac mucosa was 36 months (range: 9 months to 42 years). Cardiac mucosa was found on the first endoscopy after esophagectomy in 9 patients; these were the only biopsies that were available for these patients. One patient had two postesophagectomy endoscopies with biopsy. At the first endoscopy, performed 15 months after esophagectomy, only squamous epithelium was present in the biopsy of the cervical esophagus. At the second endoscopy, performed 9 months later, cardiac mucosa was found in the cervical esophagus.

Four of the 10 patients with cardiac mucosa also had intestinal metaplasia with goblet cells in the supra-anastomotic biopsies. One of the 4 patients, a 57-year-old man, who had undergone esophagectomy at age 15 for an esophageal stricture secondary to ingesting a coin at 9 months of age, had Barrett's esophagus with dysplasia and an intramucosal adenocarcinoma in the remnant cervical esophagus 42 years after esophagectomy.

In a study from Sweden of 32 postesophagectomy patients who underwent 24-hour pH monitoring, bilirubin monitoring, and endoscopy with biopsy 3 to 10.4 years after surgery, 15 of 32 patients (46.9%) had developed metaplastic columnar mucosa in their cervical esophagus (26). The length of columnar metaplasia correlated with the degree of abnormality in the 24-hour pH study but not the bilirubin study, suggesting that acid was the main factor in causing columnar metaplasia. Intestinal metaplasia was found within the columnar-lined segment in three patients 8.5, 9.5, and 10.4 years after esophagectomy. All three patients with intestinal metaplasia had an abnormal 24-hour pH test and bilirubin test.

These three studies represent what is essentially an unwitting human experiment to evaluate the effects of reflux on squamous epithelium. The esophagogastrectomy first removes all possible cardiac mucosa at the junction and creates an anastomosis between esophageal squamous and gastric oxyntic mucosa. The surgery eradicates the normal sphincter mechanism and angle of His and sets the stage for postoperative reflux. This was shown to be present in all patients who developed columnar metaplasia above the anastomotic line in Oberg et al.'s (26) study by means of a 24-hour pH study with the pH electrode placed in the cervical esophagus 1 cm above the anastomotic line. Oberg et al. showed that there was a direct correlation between the length of the metaplastic segment and the percentage of time the cervical esophagus was exposed to acid.

Recognition of this two-step process provides a potential method of preventing reflux-induced adenocarcinoma of the esophagus. Cardiac mucosa is easily identified by biopsy in the first phase of the metaplastic sequence, which can last many years prior to the development of intestinal metaplasia. Transformation of cardiac mucosa to intestinal metaplasia must occur as a result of an interaction of a luminal molecule with the stem cells in cardiac mucosa to cause Cdx2 activation. If this molecule can be identified and inactivated, the Cdx2 activation and the resulting intestinal metaplasia in cardiac mucosa can be prevented. Because intestinal metaplasia is an essential precursor to reflux-induced carcinogenesis, prevention of Cdx2 activation will theoretically prevent the development of carcinoma.

Distribution of Intestinal Metaplasia in Columnar-Lined Esophagus

Intestinal metaplasia develops in cardiac mucosa in a highly consistent and nonrandom manner. This was shown by Paull et al. (1), who mapped the distribution of the three different types of epithelia beautifully and showed that the intestinal metaplasia was almost always seen in the most proximal segment of the columnar-lined esophagus (Fig. 9.16).

Literature Review

Paull A, Trier JS, Dalton MD, Camp RC, Loeb P, Goyal RK. The histologic spectrum of Barrett's esophagus. N Engl J Med 1976;295:476–480.

After classifying the three main histologic subtypes of columnar-lined esophagus (at the time synonymous with Barrett esophagus) with unerring accuracy, these authors went on to describe the distribution of these three epithelial types in their 11 patients. The authors used a highly controlled biopsy technique wherein biopsies were correlated with manometry; the lower esophageal sphincter was identified and biopsies taken from the lower esophageal sphincter zone and then at intervals of 1 to 3 cm from the body of the esophagus between the sphincter below and beyond the squamocolumnar junction above. These patients all had long-segment Barrett esophagus (short-segment disease was not recognized at the time). The total number of biopsies from the esophagus in the 11 patients was 122, a number that is not as high as present series but very likely an adequately representative sample.

In their results, the authors stated that all 11 patients had a segment of columnar-lined esophagus above the lower esophageal sphincter zone (it must be appreciated that the proximal limit of the manometric lower esophageal sphincter is normally 3 to 5 cm above the gastroesophageal junction; though the sphincter is shortened in patients with reflux disease, the presence of a columnar-lined esophagus above the sphincter zone indicates a total length that probably exceeds 3 cm). In 9 patients, the columnar-lined segment included an area of specialized columnar epithelium (= intestinal metaplasia; these 9 patients only would fit into the modern classification of Barrett esophagus). Of these 9 patients, the intestinal epithelium with goblet cells extended over a length of 4.0 to 14.5 cm in 7 patients but was present in only one biopsy in the other 2. In 5 of the 9 patients with intestinal metaplasia, either gastric-fundic-type (oxyntocardiac) or junctional-type (cardiac) was interposed between the intestinal epithelium and the lower esophageal sphincter zone. The interposed gastric-fundic-type (oxyntocardiac) epithelium extended over at least 2, 6, and 10 cm above the lower esophageal sphincter in 3 of these patients. In the 2 patients with junctional-type (cardiac) epithelium, but in whom no gastric-fundic-type (oxyntocardiac) epithelium could be demonstrated, the junctional (cardiac) epithelium extended at least 1 and 4 cm above the lower esophageal sphincter.

The 2 patients who did not have intestinal metaplasia had either a mixture of cardiac and oxyntocardiac mucosa or oxyntocardiac mucosa only. This showed "marked atrophic changes characterized by sparse and shortened glands containing a variable number of parietal cells" (p. 477).

In their discussion, the authors described the distribution of the epithelial types in the columnar-lined segment, diagrammatically illustrating detailed findings in 4 patients (Fig. 9.16). They stated: "When present, specialized columnar-type (intestinal) epithelium was always the most proximally located, and the gastric-fundic-type (oxyntocardiac) always the most distally located columnar epithelium. Junctional-type (cardiac) epithelium, when present, was interposed between the gastric-fundic-type (oxyntocardiac) and the specialized columnar (intestinal) type or squamous epithelium" (p. 479).

In a recent study (2) we confirmed Paull et al's basic distribution of epithelia, but showed that the amounts of oxyntocardiac and intestinal epithelia were very different (Figs. 9.16, 9.24). We will show later that this difference may be due to the use of more potent acid suppressive drugs.

Literature Review

Chandrasoma PT, Der R, Dalton P, Kobayashi G, Ma Y, Peters J, DeMeester T. Distribution and significance of epithelial types in columnar-lined esophagus. Am J Surg Pathol 2001;25:1188–1193.

In this study, we evaluated the distribution of the different epithelial types in 32 patients with a columnar-lined esophagus measuring a minimum of 2 cm (range: 2 to 16 cm; mean: 5.7 cm; median: 5 cm) who had intestinal metaplasia (IM) in at least one of their biopsies (i.e., patients satisfying the definition of long-segment Barrett esophagus). These patients had two to five biopsies per level at 1 to 2 cm measured intervals within the columnar-lined segment. The number of biopsies containing metaplastic columnar epithelium (oxyntocardiac and cardiac with and without intestinal metaplasia) in these 32 patients was 424 (range per patient: 5 to 37; mean: 13.3).

This permitted mapping of the different epithelial types within the columnar-lined esophagus. When intestinal metaplasia was present, the density of goblet cells in each biopsy was graded in the following manner: grade 0 = no goblet cells; grade 1 = less than one third of the glands contain goblet cells; grade 2 = one to two thirds of the glands contain goblet cells; grade 3 = more than two thirds of the glands contain goblet cells. An overall intestinal metaplasia grade was then assigned to each level based on the mean grade for all biopsies at that level. This permitted as accurate a mapping of the density of goblet cells within the columnar-lined esophagus as is possible.

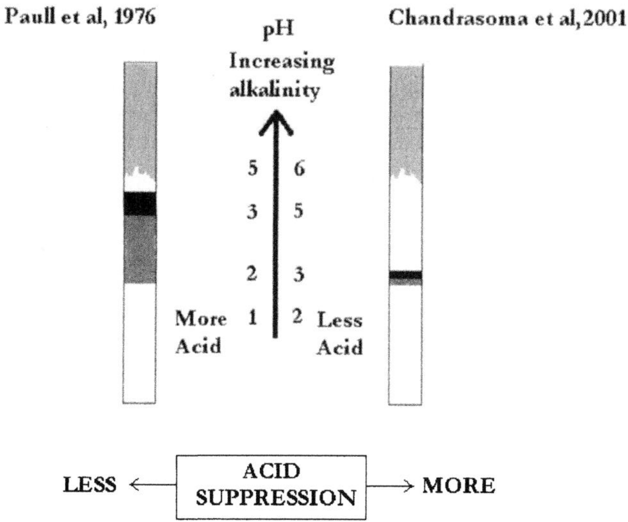

FIGURE 9.24 Possible explanation for historical increase in the amount of intestinal metaplasia and decrease in oxyntocardiac mucosa in columnar-lined esophagus. Increased use of acid suppressive drugs results in an increased alkalinity that favors intestinal metaplasia and inhibits development of oxyntocardiac mucosa.

The distribution of intestinal metaplasia in the 424 biopsies from the columnar-lined esophagus in these 32 patients was as follows:

	Number of patients with IM (%)	Number of biopsies with IM (%)	Number of biopsies without IM (%)
Total	32	311/424 (73%)	113/424 (27%)
Most proximal level	32 (100)	64/68 (94%)	4/68 (6%)
Most distal level	22 (69)	40/102 (39%)	62/102 (61%)

This shows that the intestinal metaplasia tended to favor the most proximal level of the columnar-lined segment. If biopsies were taken only at the most proximal level of a patient with long-segment Barrett esophagus (i.e., the columnar epithelium immediately distal to the squamocolumnar junction), intestinal metaplasia would have been detected in all patients and in 64 of 68 (94%) of the biopsies.

The mean density of goblet cells in the biopsies at the most proximal and most distal level, as determined by our grading system, also showed a significant difference:

	Grade 0	Grade 1	Grade 2	Grade 3
Most proximal level	0	5	6	21
Most distal level	10	8	13	1

The number of goblet cells also tended to be greatest in the most proximal level. Because the diagnosis of intestinal metaplasia is made when there is a single definitive goblet cell in cardiac mucosa, the likelihood of a false negative diagnosis is the least in biopsies of the most proximal level.

Cardiac mucosa was present at all levels in all patients. This finding differs from that shown in Paull et al.'s study (1). The reason for the difference is that we had many more biopsies at each level. In our general experience, it is not uncommon for cardiac and intestinal metaplastic epithelium to be present side by side in the proximal region of the columnar-lined esophagus.

The distribution of oxyntocardiac mucosa was the exact reverse of intestinal metaplasia; the number of parietal cells was maximal at the distal end and progressively decreased in the more proximal regions of the columnar-lined segment. This is identical to Paull et al.'s finding and the original descriptions of histology of the columnar-lined segment by Allison and Johnstone and Barrett.

This pattern of distribution suggests two things: (a) goblet cells first arise in cardiac mucosa in the most proximal region adjacent to the squamocolumnar junction, appearing to progressively extend downward to involve more cardiac mucosa in a contiguous manner; (b) the occurrence of intestinal metaplasia prevents further columnar metaplasia of squamous epithelium. If this were not true, one should expect alternation of cardiac and intestinal metaplasia within the columnar-lined segment, and this is rarely seen. If true, it would mean that while cardiac metaplasia of squamous epithelium increased progressively with cumulative reflux, the columnar metaplastic process would cease as soon as intestinal metaplasia occurred at the squamocolumnar junction. This would suggest that, once intestinal metaplasia (= Barrett esophagus) occurred, there would be no increase in the length of columnar-lined esophagus. In other words, long-segment Barrett esophagus requires a long-segment of cardiac metaplasia *before* intestinal metaplasia occurred within it. Short-segment Barrett esophagus would not progress to long-segment Barrett esophagus, and patients with microscopic segments of Barrett esophagus would never develop endoscopically visible Barrett esophagus. We will pursue the evidence for this and its significance later in this chapter.

Conclusion: In the columnar-lined segment, intestinal metaplasia occurs in the most proximal region adjacent to the squamocolumnar junction.

Columnar-Lined Esophagus Is Not Barrett Esophagus

One of the great and obvious errors made by many gastroenterologists is to equate columnar-lined esophagus with Barrett esophagus. The error is made by the following line of thinking: A segment of columnar-lined esophagus is seen at endoscopy; a biopsy is taken, which shows intestinal metaplasia; a diagnosis of Barrett esophagus is made; it is assumed that the entire columnar-lined segment consists of intestinal metaplasia. This is an obvious error that results from a failure to understand histology; it can never be made by anyone who systematically studies the microscopic pathology of biopsies that are taken from the columnar-lined esophagus.

The diagnosis of Barrett esophagus in a columnar-lined segment is made when there is one goblet cell in any biopsy from this segment. The number of goblet cells varies from one cell in one foveolar complex to being present in every foveolar complex (Figs. 9.25 and 9.26). The nonintestinal epithelia within the columnar-lined segment consist of cardiac, and oxyntocardiac mucosa. Because intestinal metaplastic, cardiac, and oxyntocardiac mucosa cannot be distinguished by endoscopy, the diagnosis of Barrett esophagus cannot be made by endoscopy. The entity seen at endoscopy is columnar-lined esophagus until a *reliable* method of defining intestinal metaplasia by endoscopy is developed; present chromo- and magnification endoscopy are far from reliable.

This error has another significant practical consequence. When a decision is made to ablate Barrett's epithelium, the objective must be to ablate epithelium at risk for malignancy. The only epithelium at risk for malignancy within columnar-lined esophagus is intestinal metaplasia. Failure to recognize the fact that the entire columnar-lined esophagus is not composed of intestinal metaplasia results in ablaters seeking to ablate the entire columnar-lined esophagus (27). Limiting the ablation to the actual area of intestinal metaplasia can be expected to shorten the time of the procedure and decrease the risk and complication rate of ablation while achieving its objective of removing the epithelium at risk. Ablating oxyntocardiac mucosa is not necessary. Ablaters can easily map the area of intestinal metaplasia by an appropriate biopsy protocol and limit their ablation to the area of intestinal metaplasia. This is possible because of the constant zonation of intestinal metaplasia, which begins at the squamocolumnar junction and extends distally *in a contiguous manner* for a variable distance (1, 2).

The second consequence of this error is a naive expectation that patients with a similar length of columnar-lined esophagus will have a similar risk of adenocarcinoma. This expectation is well stated by Spechler (28): "Although logic and indirect evidence suggest that the risk of cancer should vary with the extent of esophageal metaplasia, this contention has not been proved" (p. 836). The evidence at present is confusing at best with varying results in different studies. The present recommendation for equivalent surveillance is based on the belief that long- and short-segment Barrett esophagus have a similar risk of malignancy (29).

What is called long- and short-segment Barrett at the present time is not correct because it is defined by the endoscopic length of columnar-lined esophagus (Fig. 9.27). The great variation in the amount of intestinal metaplasia within the columnar-lined esophagus makes it critical to be precise in one's definition. If intestinal metaplasia is restricted to the proximal 0.5 mm of a 10 cm-long segment of columnar-lined esophagus, the correct designation for this is

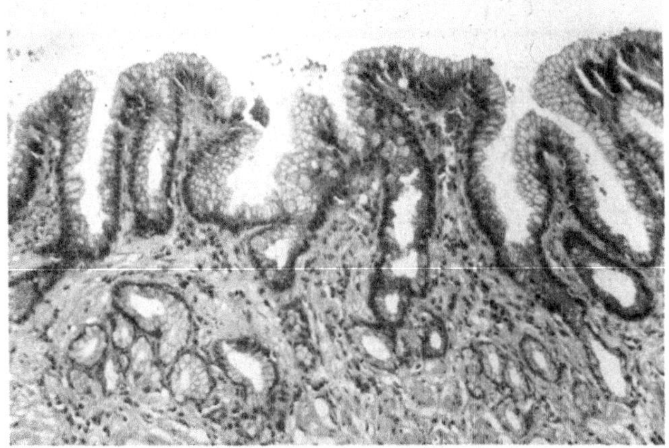

FIGURE 9.25 Early intestinal metaplasia in cardiac mucosa. Only one foveolar complex in the center shows goblet cells; the remainder of the epithelium is cardiac. This will have a low IM number.

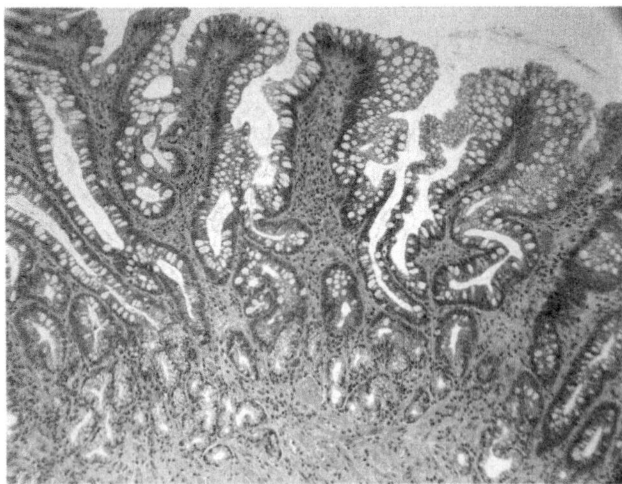

FIGURE 9.26 Extensive intestinal metaplasia in cardiac mucosa. Every foveolar complex shows goblet cells. This epithelium will have a higher IM number than the intestinal epithelium shown in Figure 9.25. (Cross reference: Color Plate 4.2)

"short-segment Barrett esophagus in long-segment columnar-lined esophagus" or better, "Barrett esophagus complicating severe gastroesophageal reflux disease" with a quantitation of the amount of intestinal metaplasia found. This patient will have less intestinal metaplasia than a patient with 1 cm of columnar-lined esophagus, which may not even be detectable at endoscopy, if that 1 cm-segment is predominantly intestinal metaplastic epithelium. If intestinal metaplasia is indeed the only epithelium with risk of cancer, the present endoscopy-based definitions of long- and short-segment Barrett esophagus are meaningless as a method of evaluating cancer risk. It can only be determined by an appropriate biopsy protocol that permits the mapping of the extent of intestinal epithelia within the columnar-lined segment. No studies in the literature have evaluated the risk of malignancy according to the amount of intestinal metaplasia within the columnar-lined segment.

Prevalence of Intestinal Metaplasia Is Proportional to the Length of Columnar-Lined Esophagus

There is an increasing prevalence of intestinal metaplasia with increasing length of columnar-lined esophagus (Fig. 9.28). When we mapped the different epithelial types in columnar-lined esophagus, 100% of patients with a columnar epithelium segment of greater than 5 cm had intestinal metaplasia, compared with 90% when the length was 3 or 4 cm, 70% when the length was 1 or 2 cm, and 15% with a length less than 1 cm (30). The direct correlation between intestinal metaplasia prevalence and length of columnar-lined esophagus is a well-documented and accepted fact with complete agreement in the literature.

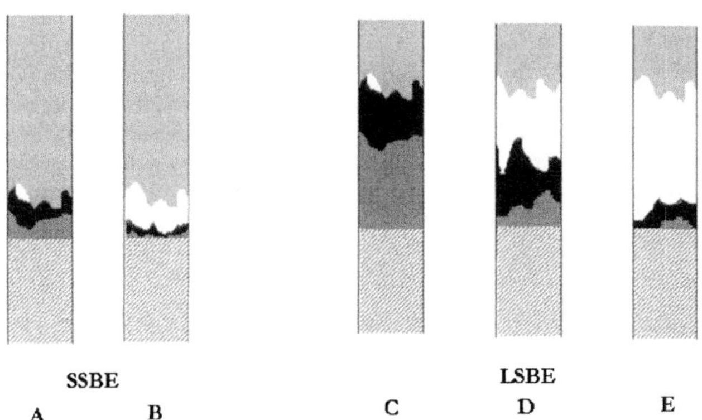

FIGURE 9.27 Patients with similar lengths of columnar-lined esophagus can have widely varying amounts of intestinal metaplasia (white). In this diagram, the patient (b) with short segment Barrett esophagus has more intestinal metaplasia the patient (c) with long-segment Barrett esophagus.

Literature Review

Chandrasoma PT, Der R, Ma Y, Peters J, DeMeester T. Histologic classification of patients based on mapping biopsies of the gastroesophageal junction. Am J Surg Pathol 2003;27:929–936.

This is a study of 959 patients in whom endoscopic biopsies were taken according to our standard protocol. The epithelial types at each level were classified by our standard definitions into squamous, cardiac, oxyntocardiac, intestinal, and gastric oxyntic. The length of columnar-lined esophagus was defined as the distance between the end of squamous epithelium and the proximal limit of gastric oxyntic mucosa; it was composed of oxyntocardiac mucosa and cardiac mucosa with and without intestinal metaplasia. The length of columnar-lined esophagus was determined as follows:

Group 1. Patients who did not have a columnar-lined epithelium in two measured biopsies. Because the minimum interval between biopsies was 1 cm, these patients included all patients without columnar-lined esophagus (i.e., only squamous and oxyntic mucosa) and patients who had less than 1 cm of columnar-lined esophagus.

Group 2. Patients who had columnar-lined esophagus in two measured biopsies that were 1 or 2 cm apart.

Group 3. Patients who had columnar-lined esophagus in two measured biopsies that were 3 or 4 cm apart.

Group 4. Patients who had columnar-lined esophagus in two measured biopsies that were 5 cm or greater apart. The following table shows the distribution of the five different epithelial types in the four groups. (*Note:* Patients who had intestinal metaplasia almost always had cardiac and oxyntocardiac mucosa as well; patients with cardiac mucosa without intestinal metaplasia almost always had oxyntocardiac mucosa as well.)

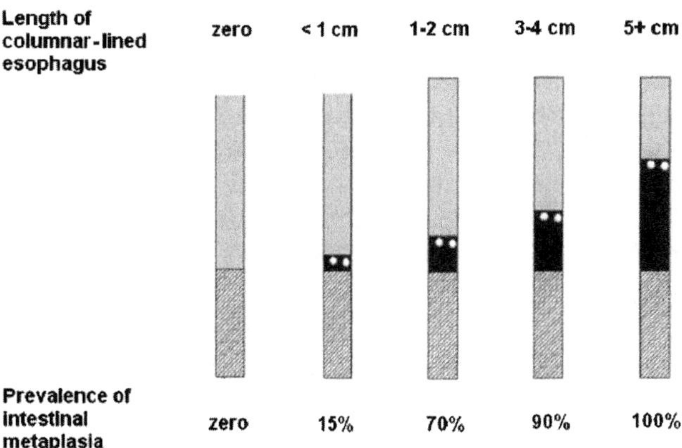

FIGURE 9.28 Diagram showing the prevalence of intestinal metaplasia in columnar-lined esophagus. As the length of columnar-lined esophagus increases, the prevalence of intestinal metaplasia increases. From Chandrasoma (24).

Group	Definition	Significance	Number (%)
1	**Abnormal columnar epithelium 0 to 0.9 cm**		**811**
1a	Only oxyntic and squamous epithelia	Normal	161 (19.9)
1b	+OCM only	Compensated reflux	158 (19.4)
1c	+CM without IM	Mild reflux disease	372 (45.9)
1d	+IM	Microscopic BE	120 (14.8)
2	**Abnormal columnar epithelium 1 to 2 cm**		**54**
2a	Only oxyntic and squamous epithelia	*	0 (0)
2b	+OCM only	*	1 (3.8)
2c	+CM without IM	Moderate reflux disease	15 (27.8)
2d	+IM	BE in moderate GERD	38 (70.4)
3	**Abnormal columnar epithelium 3 to 4 cm**		**38**
3a	Only oxyntic and squamous epithelia	*	0 (0)
3b	+OCM only	*	0 (0)
3c	+CM without IM	Severe reflux disease	4 (10.5)
3d	+IM	BE in severe GERD	34 (89.5)
4	**Abnormal columnar epithelium 5+ cm**		**56**
4a	Only oxyntic and squamous epithelia	*	0 (0)
4b	+OCM only	*	0 (0)
4c	+CM without IM	*	0 (0)
4d	+IM	BE in severe GERD	56 (100)

Abbreviations: OCM = oxyntocardiac mucosa; CM = cardiac mucosa; IM = intestinal metaplasia; BE = Barrett esophagus; GERD = gastroesophageal reflux disease.

*Too rare to attach a clinical significance.

This linear relationship between prevalence of intestinal metaplasia and length of columnar-lined esophagus suggests that intestinal metaplasia is not a random event. In some way, the likelihood of intestinal metaplasia appears to increase the further away the squamocolumnar junction moves from the true gastroesophageal junction (Fig. 9.28). This suggests that factors causing intestinal metaplasia are most active at points that are increasingly more proximal in the esophagus.

It is interesting to note that the distribution of intestinal metaplasia and oxyntocardiac mucosa within the columnar-lined segment are exactly opposite, which suggests that the factors that are involved in their etiology are probably related and opposite (Fig. 9.29). Earlier in this chapter, we suggested that oxyntocardiac mucosal transformation of cardiac mucosa may be caused by the high acid exposure of cardiac mucosa. It would then follow that the occurrence of intestinal metaplasia is favored by a lower level of acidity (higher pH).

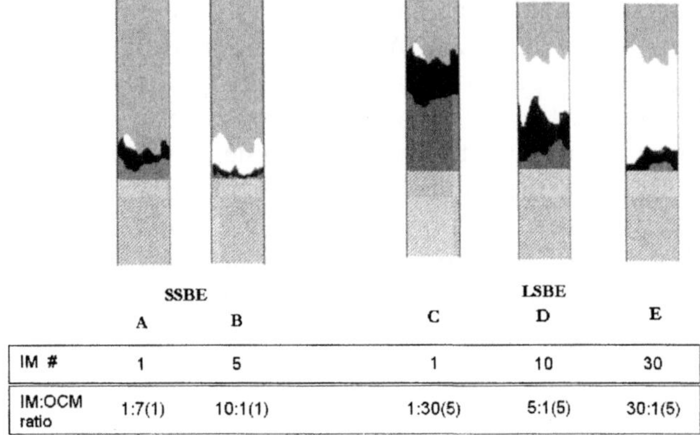

SSBE
LSBE
A
B
C
D
E
IM #
1
5
1
10
30
IM:OCM ratio
1:7(1)
10:1(1)
1:30(5)
5:1(5)
30:1(5)

FIGURE 9.29 Methods of quantitating the amount of intestinal metaplasia within a columnar-lined esophageal segment. The IM number refers to the amount of intestinal metaplasia found (number of biopsies with IM × number of foveolar complexes showing IM). The IM:OCM ratio divides the number of biopsies with IM by the number of biopsies with OCM and multiplies this by the length of columnar-lined esophagus.

Historical Changes in Distribution of Epithelial Types within Columnar-Lined Esophagus

Long-segment, columnar-lined esophagus was common from the time the entity was first described. Allison and Johnstone (31) reported that 21 of 125 patients had columnar-lined esophagus, extending to the level of the arch of the aorta. Hayward (32) encountered columnar-lined esophagus so frequently that he felt the need to define it as abnormal only when it exceeded 2 cm. Subsequent endoscopists increased the normal amount of columnar epithelium necessary for a diagnosis of columnar-lined esophagus to 3 cm. This is strong evidence to suggest that there has been no significant increase in the prevalence of columnar-lined esophagus.

It is more difficult to glean data from the literature regarding the changes in the distribution of epithelial types within columnar-lined esophagus since the mid-1960s. This is largely because of the following:

1. Cardiac and oxyntocardiac mucosa have been and still are regarded by most pathologists as "normal" and "gastric" even when they are found in the esophagus. As a result, cardiac and oxyntocardiac mucosa have been completely ignored in the literature.
2. The almost universal tendency in the literature to equate Barrett esophagus and columnar-lined esophagus has resulted in a paucity of accurate histologic mapping data in the vast majority of studies. This makes assessment of the amount intestinal metaplasia within columnar-lined esophagus impossible.
3. The numerous changes in the definition of Barrett esophagus make it impossible to glean the histology in many articles. Even today, there is no complete consensus that intestinal metaplasia is a sine qua non for the diagnosis of Barrett esophagus (33). Lagergren et al.'s influential article in 1999 (3) included patients with long segments of nonintestinalized cardiac mucosa described as "Barrett esophagus."

The histologic descriptions in Allison and Johnstone's 1953 article on columnar-lined esophagus (31) and Barrett's 1957 article (4) suggest that goblet cells were rarely seen even in patients who had columnar-lined esophagus to the arch of the aorta. Allison and Johnstone described one patient with goblet cells. The dominant epithelium in these early reports is cardiac mucosa with some mention of the presence of "a few parietal cells" in the distal part of the columnar-lined segment. In Paull et al.'s 1976 article (1) (Fig. 9.16), the maps of the three epithelial types in the esophagus show substantial amounts of oxyntocardiac mucosa fairly high in the body of the esophagus and relatively small amounts of intestinal metaplasia in many patients with very long segments of columnar-lined esophagus.

This reported distribution in the 1953–1976 period appears to be different from what we see at the present time. In our study of the distribution of histologic epithelial types in 32 patients with endoscopically visible columnar-lined esophagus (2), intestinal metaplasia was present in 64 of 68 biopsies from the most proximal level in all 32 patients. Intestinal metaplasia was present in 40 of 102 biopsies at the most distal level in 22 of 32 patients. In contrast, oxyntocardiac mucosa was absent at the proximal level in all 68 biopsies; it was present in the most distal level in 16 of 32 patients and in the retrograde (unmeasured) biopsy in another 3 patients. This is in striking contrast to Paull et al. (1), where only 1 of 11 patients had intestinal metaplasia extending into the lower esophageal sphincter zone. These data suggest that there is a much greater dominance of intestinal metaplasia over oxyntocardiac mucosa in the columnar-lined esophagus in today's patient compared with that shown in Paull et al.'s series (Figs. 9.16c and d, Fig. 9.24).

It is possible that one cause for the increasing incidence of adenocarcinoma is that there has been a shift in the direction of cardiac mucosal transformation from oxyntocardiac mucosa to intestinal metaplasia while the length of columnar-lined esophagus has remained constant. This will remain undetected by all tests other than careful mapping biopsies and correct histologic classification of the columnar-lined segment of esophagus.

Methods of Quantitating Intestinal Metaplasia in Columnar-Lined Esophagus

If it is true that intestinal metaplastic epithelium is the epithelium at risk for the development of cancer, it behooves us to define methods of accurately quantitating the amount of intestinal metaplasia (Fig. 9.29). Clearly, the presently used endoscopic classification into long- and short-segment Barrett esophagus is sufficiently flawed to be of little value in providing useful information.

If endoscopy cannot differentiate between the various epithelial types, quantitation demands a standardized biopsy protocol for columnar-lined esophagus (e.g., four quadrant biopsies with standardized biopsy forceps at 1- to 2-cm intervals extending from the squamocolumnar junction proximally to the true gastroesophageal junction defined as the proximal limit of gastric oxyntic mucosa distally).

The amount of intestinal metaplasia in these biopsies can be expressed in two ways.

Total Number of Foveolar Complexes Showing Intestinal Metaplasia (IM Number)

The most valuable information regarding cancer risk is the absolute number of foveolar complexes that show intestinal metaplasia. Theoretically, carcinogens act on the proliferative progenitor cells in these foveolar complexes that are marked phenotypically by the presence of intestinal metaplasia.

This number can be expressed by the formula: the number of biopsies that show intestinal metaplasia × the number of foveolar complexes in each biopsy that show intestinal metaplasia (Fig. 9.29). If one assumes that sampling of the entire columnar-lined esophagus is done in a uniformly random manner directed by protocol, this number will provide an accurate representation of the amount of intestinal metaplasia. This number will provide data that are comparable between different lengths of columnar-lined esophagus as long as the biopsy protocol used for sampling remains constant (Fig. 9.29). While it is unreasonable to actually count the involved foveolar complexes involved, the pathologist can make a quick estimate of this number based on the standard examination, possibly reaching an estimation in a 1 to 3 scale with reasonable accuracy and ease.

IM:OCM Ratio within a Segment of Columnar-Lined Esophagus

This will be another useful assessment of the risk of adenocarcinoma for columnar-lined esophagus because intestinal metaplasia is susceptible to carcinogenesis and oxyntocardiac mucosa (OCM) is not (Fig. 9.1). The most reproducible and accurate definition for the IM:OCM ratio will be the number of biopsies with IM divided by the number of biopsies with OCM (Fig. 9.29). Cardiac mucosa is not used in this definition because it is neither at risk for malignancy nor entirely benign because it can transform into intestinal metaplasia. The IM:OCM ratio has no meaning without a correction for the length of the columnar-lined segment (Fig. 9.29). The mathematical formula for this determination will be the number of biopsies showing IM divided by the number of biopsies showing OCM multiplied by the length of columnar-lined esophagus in cm.

It is likely that the IM:OCM ratio will provide the best indicator of risk in columnar-lined esophagus; the higher the IM:OCM ratio, the greater the risk for the same length of columnar-lined esophagus. This is particularly valuable when one assesses the treatment of columnar-lined esophagus because reversal of intestinal metaplasia as well as conversion of cardiac mucosa to oxyntocardiac mucosa potentially decreases cancer risk and both these will be reflected in the ratio.

It is unlikely, though, that any systematic biopsy protocol will ever be accepted as a standard protocol in any practice outside academic centers. The time commitment to taking 20 to 40 biopsies at upper endoscopy is so impractical that it is likely to limit any reasonable quantitation of intestinal metaplasia, particularly in long-segment columnar-lined esophagus.

Does Columnar-Lined Esophagus Increase in Length? Does Barrett Esophagus Increase in Length?

There is no doubt that columnar transformation of the esophageal squamous epithelium is a cumulative process. Children with reflux disease usually have very small lengths of cardiac mucosa (18) unless they have severe reflux associated with comorbid diseases, when long segments of cardiac mucosa develop (19). With increasing age and cumulative reflux, progressively longer segments of columnar-lined esophagus become more common (Fig. 9.30). In fact, the length of cardiac mucosa in the general population tends to increase with age, suggesting that the cardiac mucosal length in an individual is an expression of lifetime reflux damage of the squamous epithelium.

Literature Review

Cameron AJ, Lomboy CT. Barrett's esophagus: Age, prevalence, and extent of columnar epithelium. Gastroenterology 1992;103:1241–1245.

The title of this article (34) is misleading; the definition of "Barrett's esophagus" in this article is a greater than 3 cm

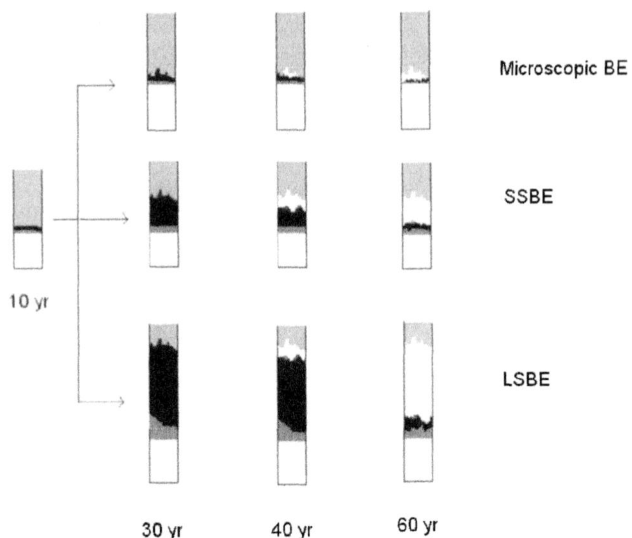

FIGURE 9.30 Evolution of columnar-lined esophagus with increasing age. In children (10 years—left), reflux is commonly manifested as short segments of nonintestinalized cardiac (black) and oxyntocardiac (stippled) mucosa. Depending on severity of reflux, this segment progressively elongates (shown to age 30 years in three patients) without intestinal metaplasia. When intestinal metaplasia occurs (age 40 years), elongation of the columnar segment stops. Further evolution is an increase in the amount of intestinal metaplasia within the columnar-lined segment.

length of columnar-lined esophagus, without any consideration of the presence of intestinal metaplasia. Over the 1976–1989 study period, 51,311 patients underwent upper gastrointestinal endoscopy. Of these, 377 patients had greater than or equal to 3-cm columnar epithelium in the esophagus and no carcinoma.

The prevalence of columnar-lined esophagus greater than 3 cm (the authors' Barrett esophagus) progressively increased with age to reach a plateau by the 7th decade. The data of prevalence of greater than 3 cm of columnar-lined esophagus according to age are shown in the table.

	0–9	10–19	20–29	30–39	40–49	50–59	60–69	70–79	80–89	>90
Number of pts undergoing endoscopy	176	679	2588	4671	6263	10,509	13,533	1004	2694	194
Number with Barrett esophagus	0	1	4	12	45	71	126	92	25	1
Percentage with Barrett esophagus	0	0.147	0.155	0.257	0.718	0.676	0.931	0.920	0.928	0.515

The increasing number of patients with columnar-lined esophagus greater than 3 cm with increasing age suggests that columnar metaplasia increases in length with age and reflects progressive and cumulative metaplasia of the squamous epithelium.

Unlike prevalence, the mean length of columnar epithelium did not increase with age in this population of long-segment columnar metaplasia. The mean length of columnar epithelium was similar in the different age groups, between 6 to 7 cm. No significant change in length was found in 21 patients followed up for a mean of 7.3 years (mean initial length, 8.29 +/− 0.85 cm; mean final length, 8.33 +/− 0.77 cm). The authors concluded that these data are consistent with a fairly rapid evolution of Barrett esophagus (i.e., greater than 3 cm columnar-lined esophagus) to its full length with little subsequent change.

The data are difficult to interpret because no information is given regarding the presence of intestinal metaplasia. It is tempting to assume that the 21 patients who were followed for 7.3 years had intestinal metaplasia because there was never any indication for follow-up of patients without intestinal metaplasia. If true, it would suggest that the lack of increase in length of the columnar segment occurred *after* intestinal metaplasia developed. However, such assumptions cannot be made.

There is controversy regarding the question of elongation of Barrett esophagus defined by the presence of intestinal metaplasia. Does short-segment Barrett esophagus evolve into long-segment Barrett esophagus? Few studies show any elongation of Barrett esophagus during surveillance. In fact, most studies that look at longitudinally sequential biopsies show either no change or decrease in the length of Barrett esophagus. A decrease is usually attributed to the effect of treatment (either acid suppression or antireflux surgery), while no change is taken to mean that the treatment has been ineffective in reducing the amount of Barrett esophagus. Few of these studies show any increase in the length of Barrett esophagus. The absence of increasing length of Barrett esophagus during surveillance strongly suggests that the length of Barrett esophagus remains constant.

The fact that the length of Barrett esophagus does not increase is very interesting because it means that patients progress from microscopic to short-segment to long-segment cardiac metaplasia *until* intestinal metaplasia occurs. When intestinal metaplasia occurs at the squamocolumnar junction, the further elongation of the columnar-lined esophagus stops (Fig. 9.30). Intestinal metaplasia appears to act as a barrier that prevents further columnar transformation of the squamous epithelium.

The constant zonation of epithelia with the intestinal metaplasia favoring the most proximal part of the columnar-lined segment supports this concept. The well-established zonation of epithelia can only be explained in two ways:

1. Cardiac transformation of squamous epithelium does not occur after intestinal metaplasia has developed. If cardiac transformation occurred after intestinal metaplasia has developed and there is a time interval before intestinal metaplasia develops in cardiac mucosa, there should be patients who have cardiac mucosa between the squamocolumnar junction and intestinal metaplasia. This is rarely seen.
2. The cardiac transformation of squamous epithelium that occurs after intestinal metaplasia has developed rapidly undergoes intestinal metaplasia with a very short lag phase.

The test for the correct answer between these two possibilities is whether the overall length of Barrett esophagus increases or not. If the length of Barrett esophagus remains constant in a given patient after it is detected, it means that intestinal metaplasia acts as a deterrent to further columnar transformation of squamous epithelium. If the length of Barrett esophagus progressively increases, the second explanation is correct. Though not to be regarded as proven, the present data suggest that Barrett esophagus does not elongate during surveillance after it is discovered. This suggests that the occurrence of intestinal metaplasia in cardiac mucosa acts as a brake for further columnar transformation of squamous epithelium by reflux (i.e., in some way, intestinal metaplasia protects the squamous epithelium against reflux-induced damage).

Although the evidence suggests that the length of columnar metaplasia does not increase after intestinal metaplasia has occurred, there is strong evidence that the *amount* of intestinal metaplasia progressively increases within the columnar-lined segment (Fig. 9.30). This is shown by studies in children (19), where even long segments of columnar-lined esophagus have either no or only very small amounts of goblet cells. In adults, long segments of columnar-lined esophagus have intestinal metaplasia involving a much larger area (1, 11, 30). The IM:OCM ratio and the IM number as defined earlier progressively increase with time. This increase is not constant; it is accentuated by any factor that causes the Cdx2-driven intestinal metaplasia switch in cardiac mucosa, and it is decreased by any factor that causes the Sonic Hedgehog gene–driven oxyntocardiac mucosa switch in cardiac mucosa.

Gender Differences in Barrett Esophagus

A strong gender difference exists in the patterns of Barrett esophagus. Long-segment Barrett esophagus is predominantly a male disease with a male-to-female ratio varying from 4–8:1 in different studies. The ratio decreases progressively as the length of Barrett esophagus decreases, reaching near equality between the sexes in microscopic or very short segment Barrett esophagus (presently falling under the term "intestinal metaplasia of the gastric cardia").

The gender difference present in varying lengths of Barrett esophagus is translated to the gender difference between adenocarcinoma of the lower esophagus, which is a dominantly male disease, compared with adenocarcinoma of the lower regions of the columnar-lined esophagus (presently falling within the definitions of "junctional adenocarcinoma" and "adenocarcinoma of the proximal stomach or gastric cardia"), where the male and female incidence is close to equal.

No explanations have ever been presented to explain this epidemiologically established gender difference. Only one study has described any gender difference in columnar-lined esophagus that may provide an explanation.

Literature Review

Der R, Tsao-Wei DD, DeMeester T, Peters J, Groshen S, Lord RVN, Chandrasoma P. Carditis: A manifestation of gastroesophageal reflux disease. Am J Surg Pathol 2001;25:245–252.

This is a study that we undertook to evaluate the relationship between the presence of inflammation in cardiac mucosa and gastroesophageal reflux disease as defined by an abnormal 24-hour pH study (35). During the statistical analysis of the data in this study by our highly skilled statisticians, an unexpected statistically significant difference between the sexes emerged. This is complex and requires a somewhat devious interpretation, which may be open to debate.

As reported in the study, a complex statistically significant relationship was found between the sex, severity of carditis, the presence of gastritis, and the percentage of time the pH < 4 in the 24-hour pH study (normal = 4.5%). The percentage of time the pH < 4 in the 24-hour study in these different subgroups is expressed in the table below:

	Chronic inflammation of cardiac mucosa			
	Mild (1+)		Severe (2+/3+)	
Gastritis	Male Mean (95% CI) (n)	Female Mean (95% CI) (n)	Male Mean (95% CI) (n)	Female Mean (95% CI) (n)
Yes	0	1.7 (0.3,9.6) (2)	5.5 (2.9,8.8) (16)	4.1 (0.3,12.2) (3)
No	5.2 (2.3,9.2) (11)	3.3 (1.2,6.5) (12)	9.1 (6.7,11.8) (37)	5.9 (3.6,8.7) (24)

These data suggest that overall, men with carditis had a statistically significant higher mean acid exposure than women. Acid exposure in the pH study is a marker of reflux; as such, these data indicate that at similar grades of carditis, men had more reflux. Another way of saying this is that women develop carditis and show evidence of damage (inflammation) at a lower level of reflux than males.

The reason for this gender difference is unknown. While it may be due to factors such as differences in diet and smoking, it is more likely to be related to a more fundamental difference such as hormones or pregnancy. Let us let our logical fantasy run amok here for a while. One difference between men and women as it may influence reflux disease is pregnancy. Let us assume that the intraabdominal changes associated with pregnancy caused an increased tendency to reflux. This would mean that women in their early reproductive life would be subjected to more reflux than men. Because evolution is geared primarily toward reproduction, it would seem logical that there would be a natural selection of women whose squamous epithelium was in some way more able to withstand the damage resulting from reflux. The conversion of squamous to cardiac mucosa not only produces a more resistant epithelium but also one that is less pain sensitive. The data in this study suggest that indeed women have a tendency to transform their squamous epithelium to cardiac mucosa with greater ease at a lower level of reflux than men.

If there was also a facilitation of the transformation of cardiac mucosa to intestinal metaplasia in women, and if the occurrence of intestinal metaplasia prevented further columnar metaplasia of squamous epithelium as we have suggested, then this gender difference in the squamous epithelium would explain why the predominant type of Barrett esophagus in women is the shortest segment Barrett esophagus.

We will be the first to agree that this is an unproven and somewhat convoluted explanation for the gender difference in Barrett esophagus. However, any attempt at explaining the difference is better than none; the preceding explanation is merely intended to begin the discussion of the subject.

Proliferative Patterns in Columnar-Lined Esophagus

We have indicated that the histologic appearance of the three types of columnar epithelia have different degrees of chronic inflammation and suggested that this relates to different levels of damage in these epithelia. In particular, the chronic inflammation in cardiac and intestinal types of epithelia is almost always significantly greater than that in oxyntocardiac mucosa.

Damage, though it is likely to be related to the degree of inflammation, is best assessed by the rate of cell loss in the epithelium secondary to exposure to the injury. The easiest method of estimating the rate of cell loss is to measure the normal compensatory reaction to this cell loss. In general, the higher the proliferative rate in an epithelium, the greater the rate of cell loss. Ki67 is an antigen that is expressed in the nuclei of cells that are in the mitotic cycle and gives a sensitive measure of the proliferative rate of the epithelium. Ki67 can be detected by immunoperoxidase technique using a specific antibody that has been well studied and is available commercially.

Literature Review

Olvera M, Wickramasinghe K, Brynes R, Bu X, Ma Y, Chandrasoma P. Ki67 expression in different epithelial types in columnar lined esophagus indicates varying levels of expanded and aberrant proliferative patterns. Histopathology 2005;47:132–140.

We undertook this study to characterize patterns of Ki67 expression in the different types of columnar metaplastic epithelia that occur in the esophagus as a result of gastroesophageal reflux disease.

Twenty six patients who underwent esophagogastrectomy for either high-grade dysplasia or adenocarcinoma complicating long-segment Barrett esophagus were selected for the study. Nineteen patients had residual nondysplastic intestinal metaplasia, and all patients showed cardiac or oxyntocardiac mucosa in the tubular esophagus distal to the squamocolumnar junction. All specimens had gastric oxyntic mucosa distal to the esophagus. Multiple sections were evaluated in each case, and a total of 241 representative noneroded areas that had specific epithelial types were selected for study. The selected areas were well oriented and had a sufficient length of a uniform epithelial type for evaluation. In squamous epithelium, a minimum of 2 mm was required, and in columnar lined mucosa, three adjacent foveolar-gland complexes composed of a uniform epithelial type were required for selection.

Forty areas of stratified squamous epithelium were studied. The columnar epithelium lined mucosae in sections taken from the tubular esophagus were defined as oxyntocardiac mucosa (43 areas), cardiac mucosa (45 areas), intestinal metaplastic epithelium with goblet cells (41 areas), intestinal metaplasia with low-grade dysplasia (9 areas), intestinal metaplasia with high-grade dysplasia (16 areas), and invasive adenocarcinoma (15 areas). Gastric mucosal samples were taken from the anatomic proximal stomach near the distal resection margin. These were all of the typical pure oxyntic type, containing straight tubular glands with parietal cells and chief cells but devoid of mucous cells deep to the isthmus of the pit-gland complex (32 areas). All areas of gastric oxyntic mucosa chosen were morphologically normal without significant inflammation, atrophy, or intestinal metaplasia; all were negative for *Helicobacter pylori.*

Serial sections were stained with hematoxylin and eosin and for Ki67 by an immunoperoxidase technique. Ki67 is an antigen that is expressed in cells that are in the cell cycle and is therefore a marker of proliferation. Staining was defined as strong and complete nuclear staining clearly recognizable at 10× magnification.

Metaplastic esophageal columnar epithelia had the basic structure of gastrointestinal tract epithelia, being composed of a surface epithelial layer, a foveolar pit region, and a region of glands deep to the foveolar region. Because selection excluded areas of erosion, all areas had a surface epithelial layer; 4 areas selected were composed only of a flat surface layer of cells. All other columnar epithelia had a foveolar region of varying length. Three areas had a rudimentary foveolar region that was too short to be divided in any meaningful manner. In all other epithelial areas, the foveolar pit region was divided into superficial, mid, and deep thirds and evaluated separately. True glands under the foveolar region were defined morphologically and by the fact that they were composed of Ki67 negative cells, similar to the Ki67 negative glands in normal gastric mucosa. The Ki67 positivity in each of these zones was scored as follows: 0 = no Ki67 positive cells; 1 = 1% to 10% Ki67 positive cells; 2 = 11% to 25% Ki67 positive cells; 3 = 26% to 50% Ki67 positive cells; 4 = >50% Ki67 positive cells.

The distribution of Ki67 positive cells in normal gastric oxyntic mucosa was used as a control to define normal proliferative activity in columnar-lined epithelia. In all 32 areas of gastric oxyntic mucosa studied, the Ki67 positive cells were restricted to the deep third of the foveolar pit without any Ki67 positive cells in the surface epithelial layer, the superficial and middle third of the foveolar pit, or the glands composed of parietal and chief cells deep to the foveolar pit.

We recognized two deviations from this normal pattern of Ki67 staining in the columnar epithelial types in the metaplastic columnar-lined esophagus (Fig. 9.31):

1. *Aberrant proliferation.* Aberrant proliferation is a measure of abnormal maturation. We will consider this in the discussion of dysplasia in Barrett esophagus.

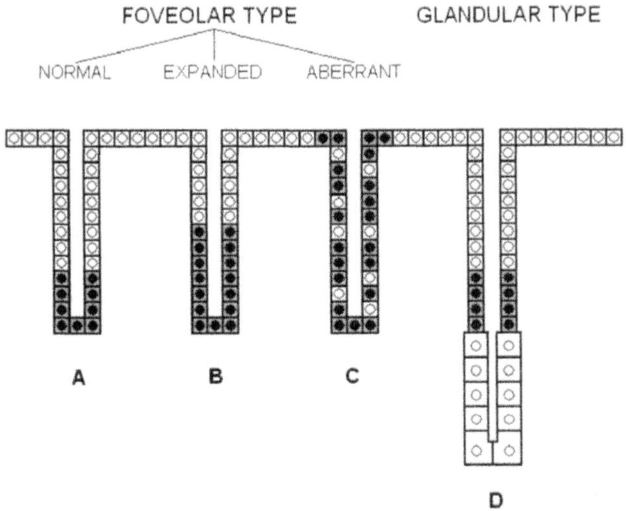

FIGURE 9.31 Patterns of Ki67 staining of columnar-lined esophagus. The normal pattern (a and d) shows limited positivity restricted to the deepest third of the foveolar pit in both foveolar- and glandular-type epithelia. Expansion of proliferation with normal maturation shows the presence of increased numbers of Ki67 positivite cells, restricted to the deep and middle thirds of the foveolar pit (b). Aberrant Ki67 staining shows Ki67 positivity extending to the superficial third of the pit and the surface (c). The expanded and aberrant patterns are shown only for the foveolar epithelium but equally apply to glandular epithelial types.

2. *Expanded proliferation.* This was defined by the presence of Ki67 positive cells in the middle third of the foveolar pit in addition to the deep third (Figs. 9.9 and 9.31). We quantitated this by taking the combined score for the mid and deep third of the foveolar pit and used a score greater than 5 to define expanded proliferation. Normal gastric oxyntic mucosa had a maximum mid + deep third score of 4 because no area showed staining in the middle third of the foveolar pit.

Oxyntocardiac mucosa in the esophagus showed a pattern of Ki67 staining that was similar to normal gastric mucosa. The mid foveolar third showed no Ki67 positive cells in 26 of 43 areas (60.5%); in the other 17 areas, Ki67 positivity expanded to the mid third only. The combined mid + deep third score was greater than 5 in 4 of the 43 areas studied (9.3%).

The 45 areas of cardiac mucosa studied consisted of 4 areas composed only of a flat surface layer of mucous cells and 3 areas where the foveolar pit was rudimentary and too short to permit evaluation of the criteria of expansion of normal proliferation. The other 38 areas of cardiac mucosa showed well developed Ki67 negative glands composed of mucous cells only without parietal cells below the proliferative foveolar region. A combined mid + deep third score of greater than 5 was seen in 11 of 38 areas (29%), significantly higher than both gastric oxyntic and oxyntocardiac mucosa ($p = 0.02$), indicating a higher frequency of expanded proliferation in cardiac mucosa.

Of the 41 areas of nondysplastic intestinal metaplasia studied, the combined mid + deep third score was greater than 5 in 26 areas (63.4%) (Fig. 9.6). This indicated a significantly higher frequency of expanded proliferation when compared with cardiac mucosa ($p = 0.02$).

The intensity of Ki67 staining, which recognizes the pool of cells in these epithelia that are actively proliferating, showed significant differences in these three epithelial types. The expanded pattern reflects increased proliferation as a compensatory response to damage. There was a significant stepwise increase in the frequency of expanded proliferation from normal gastric oxyntic mucosa to esophageal oxyntocardiac mucosa, esophageal cardiac mucosa, to nondysplastic intestinal metaplasia. Differences in proliferative activity between intestinalized and nonintestinalized columnar epithelium have been previously reported.

We used these differences in Ki67 expression to classify nondysplastic columnar epithelia in the esophagus into three damage levels: (a) low-damage oxyntocardiac mucosa, (b) intermediate-damage cardiac mucosa, and (c) high-damage intestinal epithelium.

These data suggest that parietal cell differentiation in cardiac mucosa is associated with a low-damage environment, whereas intestinal metaplasia occurs in a high-damage environment. It may also indicate that oxyntocardiac mucosa is more resistant to damage caused by reflux. In general, it appears that the more proximal segment of the columnar-lined esophagus has a tendency to express a greater level of damage as well as a greater tendency to undergo intestinal metaplasia. Because there is a strong relationship between the length of columnar-lined esophagus and the prevalence of intestinal metaplasia, it is highly likely that this intestinal metaplasia is a genetic switch occurring in a predictable manner, becoming more likely as the squamocolumnar junction moves increasingly cephalad. From a genetic standpoint, this means that factors causing the intestinal metaplasia switch (Cdx2 activation?) operate proximally in the esophagus much more than distally.

In contrast, the more distal segment of the columnar-lined esophagus expresses a lower level of damage and a greater tendency to activate the genetic switch (Sonic Hedgehog gene?) that induces parietal cell differentiation (i.e., become oxyntocardiac mucosa).

The Cause of the Cardiac Mucosa to Intestinal Metaplasia Switch

Recognizing these patterns of damage in different parts of the esophagus provides clues to etiology of these changes. Reflux is not a simple phenomenon. Rather, it is a highly complex phenomenon where multiple epithelial types are exposed to different molecules and patterns of reflux producing an infinite variety of interactions.

These interactions drive the cellular changes. It is crucial for us to understand these interactions and attempt to find their mechanisms because they will provide points of attack in the disease. Enhancement of the conversion of cardiac mucosa to oxyntocardiac mucosa, reversal of intestinal metaplasia, or inhibition of the development of intestinal metaplasia in cardiac mucosa will tend to decrease cancer risk. Unfortunately, these interactions have not been adequately studied because there has been no understanding of the processes involved. Even the fairly obvious concept that cardiac mucosa precedes intestinal metaplasia is relatively new and not completely accepted.

Data exist regarding characteristics of the type and composition of gastroesophageal reflux that may point to possible etiologic factors that promote the development of intestinal metaplasia in columnar-lined esophagus. The proliferative studies described earlier suggest that intestinal metaplasia occurs in a high-damage environment associated with high proliferative index.

Differences in pH along the Esophagus

Gastroesophageal reflux decreases in volume from the gastroesophageal junction to the proximal esophagus, at least in the upright position as a result of gravity. This is associated with a progressive decrease in the acidity (increasing pH) as one gets farther up the esophagus from the gastroesophageal junction.

Literature Review

Tharalson EF, Martinez SD, Garewal HS, Sampliner RE, Cui H, Pulliam G, Fass R. Relationship between rate of change in acid exposure along the esophagus and length of Barrett's epithelium. Am J Gastroenterol 2002;97:851–856.

This excellent report (36) from the University of Arizona in Tucson studied 17 patients with varying lengths of Barrett esophagus with 24-hour esophageal pH monitoring using a pH probe with four sensors located 5 cm apart. Barrett esophagus was defined as the presence of intestinal metaplasia on biopsy. Barrett's length was measured from the proximal margin of continuous Barrett's epithelium to the end of the tubular esophagus or the proximal margin of hiatal hernial folds. It should be noted how, even with highly experienced investigators such as this group, Barrett esophagus is diagnosed when any biopsy shows intestinal metaplasia and the entire length of endoscopic columnar-lined esophagus becomes "Barrett's length" despite absence of any statement regarding the presence of intestinal metaplasia in the entire extent. If histologic mapping is done as in Paull et al. (1), it will be found that a variable amount of the "Barrett's length" consists of nonintestinalized cardiac mucosa.

A reflux episode was defined as a decrease in pH < 4 and reflux time as the interval until the pH is >4. The data were reported for each of the four sensors as the percent time that the pH was <4 during the 24 hours of the test. The values for mean acid exposure expressed as mean percentage time pH < 4 (+ or – one standard deviation) at each sensor height (expressed as cm above the lower esophageal sphincter) are shown in the following table:

Sensor height (cm)	Total study time	Upright position	Supine position
1	**26.4**(+/−14.6)%	**24.2**(+/−11.4)%	**39.8**(+/−25.5)%
6	**23.2**(+/−15.8)%	**21.4**(+/−15.8)%	**25.8**(+/−20.8)%
11	**12.6**(+/−8.0)%	**10.2**(+/−8.0)%	**15.8**(+/−13.2)%
16	**6.7**(+/−6.9)%	**5.3**(+/−4.5)%	**8.4**(+/−12.0)%

The authors used a complicated mathematical calculation to show that there was a statistically significant relationship between the rate of change in acid exposure and the length of Barrett esophagus.

The data in the study are more interesting than the conclusion. They show that the acid exposure (and, potentially, exposure to all other molecules in the refluxate that accompanies the acid) progressively decreases from distal to proximal esophagus in patients with reflux. The severity of reflux determines the length of columnar-lined esophagus. The constant zonation of epithelial types within this columnar-lined segment has intestinal metaplasia occurring at the most proximal part of this columnar-lined segment. This indicates that factors that promote intestinal metaplasia are most active in the proximal region. Because all the molecular components are either equal to or lower in concentration in the proximal esophagus compared with distal, it becomes likely that the intestinal metaplasia results from a factor other than just the presence of a molecule. If a molecule in the refluxate was responsible for the development of intestinal metaplasia in cardiac mucosa, this would tend to be maximal in the distal esophagus, which is the region of highest concentration.

The use of multiple-level pH electrodes have shown that the pH tends to progressively increase more proximally in the esophagus in patients with columnar-lined esophagus. This suggests that the intestinalization of cardiac mucosa is a pH dependent phenomenon with higher pH promoting intestinal metaplasia. Similarly, it suggests that a lower pH stimulates parietal cell differentiation in cardiac mucosa because oxyntocardiac mucosa occurs in the most distal region of the columnar lined esophagus (Fig. 9.32). We again come to the rather frightening conclusion that increasing the pH of the gastric refluxate may promote intestinal metaplasia over oxyntocardiac transformation in cardiac mucosa.

Bile

Bile components commonly become admixed with gastric juice because of frequent duodenogastric reflux and may be found in the esophagus during gastroesophageal reflux. Components of bile, mainly bile acid and its metabolites, are believed to be involved in many points in the reflux-adenocarcinoma sequence including the generation of intestinal metaplasia. We will discuss bile more fully in the etiology of carcinogenesis in intestinal metaplasia where it is believed to have a maximal role.

Kazumori et al. (37) studied the effects of various bile acids on the expression of Cdx2 in cultured rat esophageal squamous epithelial cells. They found that out of 11 kinds of bile acids that were tested, only cholic acid and dehydrocholic acid dose-dependently increased Cdx2 promoter activity and Cdx2 protein production in cultured esophageal keratinocytes. Mutation analysis studies suggested that this bile acid-induced activation of the Cdx2 promoter acted via NF-kB binding sites on the cells. The rat esophageal keratinocytes that expressed Cdx2 produced intestinal type mucin, MUC2. The authors suggested that this may be one of the mechanisms of metaplasia in Barrett esophagus.

This is an animal study using rat keratinocytes in culture. If one recognizes that the interactions that are being tested are between bile acids and the esophageal progenitor cell, it is extremely relevant that the progenitor cell appears to have binding sites that bind specific bile acids and that this interaction causes Cdx2 activation and the production of MUC2, which is the typical mucin produced by intestinal metaplastic cells. This is an extremely artificial system, but it could explain how cardiac mucosa and intestinal metaplasia in human esophageal columnar epithelia express Cdx2. Human squamous epithelium does not express Cdx2; this experiment therefore suggests that columnar metaplasia occurs as soon as the bile acid–keratinocyte interaction occurs and Cdx2 is activated.

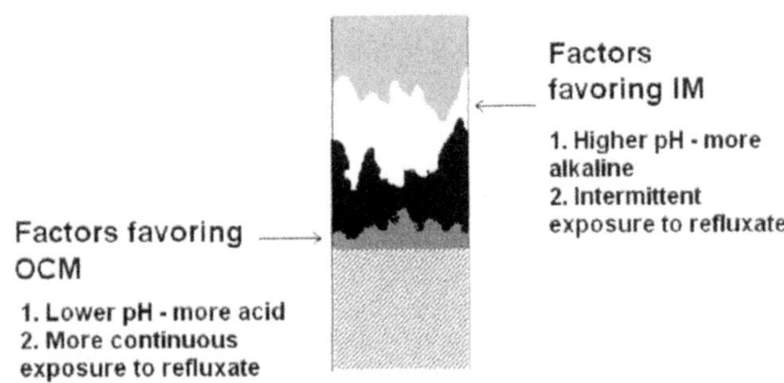

FIGURE 9.32 Factors involved in the differential evolution of cardiac mucosa into intestinal metaplasia in the more proximal region of the esophagus compared with evolution into oxyntocardiac mucosa in the more distal esophagus.

Patterns of Reflux: Intermittent (Pulse) versus Continuous

Another physical factor that maybe involved in the pathogenesis of intestinal metaplasia is the type of exposure of the epithelium to refluxate molecules. Fitzgerald et al. (38) used an *ex vivo* approach in which endoscopic biopsies were cultured in an organ culture system. Histologic examination of parallel samples confirmed that the biopsies showed intestinal metaplasia. Normal esophagus and duodenum were used as controls. The culture's tissues were exposed to acid (pH 3 to 5), either continuously or as a 1-hour acid pulse, and compared with cultures maintained at a pH of 7.4. Cell differentiation was assessed by villin expression, and cell proliferation was determined by tritiated thymidine incorporation and proliferative cell nuclear antigen expression.

Villin expression increased in Barrett tissue from a baseline of 25% of the samples to 50% and 83% of the samples after 6 and 24 hours of continuous acid exposure. Villin expression correlated with ultrastructural maturation of the brush border in the cells, confirming that this was a marker of cell differentiation. Villin expression did not increase with an acid pulse. This indicated that cell differentiation was promoted by continuous exposure to acid.

In contrast, continuous acid exposure blocked cell proliferation in Barrett mucosa compared with a 1-hour acid pulse, which promoted cell proliferation. This indicates that pulse exposure to acid promotes proliferation in Barrett epithelium.

This suggests that pulse type acid exposure is associated with a higher damage level to the cells of Barrett epithelium than continuous exposure. Pulse-type exposure is much more likely in the more proximal region of the esophagus. If this is a generic phenomenon for all columnar epithelia, it could explain the higher likelihood of the genetic change required for the development of intestinal metaplasia in cardiac mucosa. Olvera et al. (13) reported a stepwise significant increase in the proliferative activity (assessed by Ki67 expression) from oxyntocardiac to cardiac to intestinal metaplastic epithelia. The distribution of these epithelial types correlates exactly with the likelihood of pulse exposure in gastroesophageal reflux; the greatest likelihood of pulse exposure is most proximal where intestinal metaplasia occurs and the highest proliferation rates are seen; the least pulse exposure is in the most distal segment where oxyntocardiac mucosa occurs and the proliferation rates are lowest (Fig. 9.32).

Reversibility of Genetic Switches

We have suggested three genetic switches being responsible for the array of metaplasia that converts squamous epithelium to the different types of columnar epithelia (Table 9.4). The first causes the squamous epithelium to transform into the undifferentiated cardiac mucosa composed only of mucous cells. The evolution of the columnar mucosa goes in one of two directions from cardiac mucosa. The development of a "gastric-type genetic signal" associated with the Sonic Hedgehog gene occurs in the low-damage environment of the distal region and results in oxyntocardiac mucosa, which is characterized by the development of parietal cells. This is the benign pathway that is no longer susceptible to reflux-induced genetic transformations because it does not progress to intestinal metaplasia or adenocarcinoma. The second direction is the premalignant pathway. The development of intestinal type genetic signals, probably the homeobox Cdx genes, occurs in the high-damage environment of the proximal esophagus and results in the development of goblet cells, which defines intestinal (or Barrett-type) metaplasia. This is the only epithelium in columnar-lined esophagus that is at risk for developing adenocarcinoma.

Genetic switches result from expression and suppression of normal genetic pathways in the cell as a result of cell receptor interactions. They are reversible; removing or altering the receptor interactions can drive these differentiation pathways in different directions. If this is true, intestinal metaplasia can reverse back to cardiac mucosa if the Cdx gene activation is reversed; cardiac mucosa can revert to squamous epithelium if it loses the first columnar genetic switch or becomes oxyntocardiac mucosa if it can be made to acquire the "gastric-type genetic signal" associated with the Sonic Hedgehog gene. The ability to manipulate the columnar epithelia in the esophagus to move it away from intestinal metaplasia and toward the squamous and oxyntocardiac mucosa is equivalent to preventing adenocarcinoma. We believe that the most beneficial reversion is the generation of oxyntocardiac mucosa because it is a full thickness mucosal change. Squamous reepithelialization of the surface is frequently associated with the presence of residual glandular elements below the epithelium, and these have been known to progress to adenocarcinoma.

The best method of manipulating these genetic shifts in columnar-lined esophagus is to characterize the nature of the interactions that cause the genetic changes and neutralize them. Until these specific interactions are characterized, however, the only logical method of achieving reversal is to alter the damage environment or abolish reflux completely. Acid suppressive therapy is unlikely to do either; impedance studies have shown that the reflux persists, the refluxate becomes more alkaline, and the pulse effect does not change. The fact that the prevalence of intestinal metaplasia and adenocarcinoma has risen even as acid suppression has improved is evidence that acid is a relatively minor factor in these genetic

TABLE 9.4 Effect of acid suppressive drugs and anti-reflux surgery on the genetic switches that are responsible for epithelial metaplasias in the esophagus

Genetic switch	Caused by	Genetic factors	Effect of proton pump inhibitors	Effect of antireflux surgery
Squamous to cardiac	Acid (?)	Unknown	Prevention; reversal (?)	Prevention; reversal (?)
Cardiac to oxyntocardiac	Acid (?)	Sonic Hedgehog gene	Prevention	Promotion
Cardiac to intestinal	Alkalinity; bile (?)	Cdx-2	Promotion (?)	Prevention; reversal

switches. On the other hand, successful antireflux surgery abolishes all reflux into the esophagus by creating a new valve. We have observed the phenotypic expression of some of these reversals; intestinal metaplasia accompanying short-segment Barrett esophagus frequently reverses (39) and we have observed an increased amount of oxyntocardiac mucosa in post-fundoplication biopsies compared with preoperative biopsies. Both are highly beneficial changes that occur at a histologic level without any change in the endoscopic length of the columnar-lined esophagus.

The most promising point of attack in the attempt to prevent cancer is the stage of reflux disease before the development of intestinal metaplasia. The patients at risk in the future are those who have cardiac mucosa, which is detectable by screening and biopsy. There is a long lag phase before cardiac mucosa progresses to intestinal metaplasia in most patients. Recognition of the molecular component in the refluxate that drives Cdx2 activation and the "gastric-type genetic signal" can lead to the production of drugs that impact these molecules and drive differentiation of cardiac mucosa away from intestinal metaplasia and toward oxyntocardiac mucosa. This will theoretically prevent progression to adenocarcinoma. Of course, mass population screening is not cost-effective at the present time, but it may become so in the future if the incidence of adenocarcinoma continues its upward trend. For an individual patient undergoing endoscopy for any reason, the taking of biopsies at the squamocolumnar junction is a screening opportunity.

References

1. Paull A, Trier JS, Dalton MD, Camp RC, Loeb P, Goyal RK. The histologic spectrum of Barrett's esophagus. N Engl J Med 1976;295:476–480.
2. Chandrasoma PT, Der R, Dalton P, Kobayashi G, Ma Y, Peters J, DeMeester T. Distribution and significance of epithelial types in columnar lined esophagus. Am J Surg Pathol 2001;25: 1188–1193.
3. Lagergren J, Bergstrom R, Lindgren A, Nyren O. Symptomatic gastroesophageal reflux as a risk factor for esophageal adenocarcinoma. New Engl J Med 1999;340:825–831.
4. Barrett NR. The lower esophagus lined by columnar epithelium. Surgery 1957;41:881–894.
5. Chandrasoma PT, Lokuhetty DM, DeMeester, TR, et al. Definition of histopathologic changes in gastroesophageal reflux disease. Am J Surg Pathol 2000;24:344–351.
6. Johns BAE. Developmental changes in the oesophageal epithelium in man. J Anat 1952;86:431–442.
7. Salenius P. On the ontogenesis of the human gastric epithelial cells: A histologic and histochemical study. Acta Anat (Basel) 1962;50 (Suppl 46):S1–S70.
8. Menard D, Arsenault P. Cell proliferation in developing human stomach. Anat Embryol 1990;182:509–516.
9. Shi L, Der R, Ma Y, Peters J, DeMeester T, Chandrasoma P. Gland ducts and multilayered epithelium in mucosal biopsies from gastroesophageal-junction region are useful in characterizing esophageal location. Dis Esophagus 2005; (in press).
10. Sarbia M, Donner A, Gabbert HE. Histopathology of the gastroesophageal junction: A study on 36 operation specimens. Am J Surg Pathol 2002;26:1207–1212.
11. Chandrasoma P, Makarewicz K, Wickramasinghe K, et al. A proposal for a new validated histologic definition of the gastroesophageal junction. Hum Pathol 2006;37:40–47.
12. van den Brink GR, Hardwick JC, Nielsen C, Xu C, ten Kate FJ, Glickman J, van Deventer SJ, Roberts DJ, Peppelenbosch MP. Sonic hedgehog expression correlates with fundic gland differentiation in the adult gastrointestinal tract. Gut 2002; 51:628–633.
13. Olvera M, Wickramasinghe K, Brynes R, Bu X, Ma Y, Chandrasoma P. Ki67 expression in different epithelial types in columnar lined esophagus indicates varying levels of expanded and aberrant proliferative patterns. Histopathology 2005;47: 132–140.
14. Silberg DG, Swain GP, Suh ER, Traber PG. Cdx1 and Cdx2 during intestinal development. Gastroenterology 2000;119: 961–971.
15. Silberg DG, Furth EE, Taylor JK, Schuck T, Chiou T, Traber PG. CDX1 protein expression in normal, metaplastic, and neoplastic human alimentary tract epithelium. Gastroenterology 1997; 113:478–486.
16. Phillips RW, Frierson HF, Moskaluk CA. Cdx2 as a marker of epithelial differentiation in the esophagus. Am J Surg Pathol 2003;27:1442–1447.
17. Vallbohmer D, DeMeester SR, Peters JH, Oh D, Kuramochi H, Shimizu D, Hagen JA, Danenberg KD, Danenberg PV, DeMeester T, Chandrasoma PT. Cdx-2 expression in squamous and metaplastic columnar epithelia of the esophagus.
18. Glickman JN, Fox V, Antonioli DA, Wang HH, Odze RD. Morphology of the cardia and significance of carditis in pediatric patients. Am J Surg Pathol 2002;26:1032–1039.
19. Hassall E. Columnar lined esophagus in children. Gastroenterol Clin N Amer 1997;26:533–548.
20. Qualman SJ, Murray RD, McClung J, et al. Intestinal metaplasia is age related in Barrett's esophagus. Arch Pathol Lab Med 1990;114:1236–1240.
21. Dahms BB, Rothstein FC. Barrett's esophagus in children: a consequence of chronic gastroesophageal reflux. Gastroenterology 1984;86:318–323.
22. Hassall E, Weinstein WM, Ament ME. Barrett's esophagus in childhood. Gastroenterology 1985;89:1331–1337.
23. Hassall E. Barrett's esophagus: New definitions and approaches in children. J Pediatr Gastroenterol Nutr 1993;16:345–364.
24. Dresner SM, Griffin SM, Wayman J, Bennett MK, Hayes N, Raimes SA. Human model of duodenogastro-oesophageal reflux in the development of Barrett's metaplasia. Br J Surg 2003;90:1120–1128.
25. Lord RVN, Wickramasinghe K, Johansson JJ, DeMeester SR, Brabender J, DeMeester TR. Cardiac mucosa in the remnant esophagus after esophagectomy is an acquired epithelium with Barrett's like features. Surgery 2004;136:633–640.
26. Oberg S, Johnasson J, Wenner J, Walther B. Metaplastic columnar mucosa in the cervical esophagus after esophagectomy. Ann Surg 2002;235:338–345.
27. Sampliner RE. Ablative therapies for the columnar lined esophagus. Gastroenterol Clin N Amer 1997;26:685–694.
28. Spechler SJ. Barrett's esophagus. N Engl J Med 2002;346: 836–842.
29. Rudolph RE, Vaughan TL, Storer BE, et al. Effect of segment length on risk for neoplastic progression in patients with Barrett esophagus. Ann Intern Med 2000;132:612–620.

30. Chandrasoma PT, Der R, Ma Y, Peters J, DeMeester T. Histologic classification of patients based on mapping biopsies of the gastroesophageal junction. Am J Surg Pathol 2003;27:929–936.
31. Allison PR, Johnstone AS, Royce GB. Short esophagus with simple peptic ulceration. J Thorac Surg 1943;12:432–457.
32. Hayward J. The lower end of the oesophagus. Thorax 1961; 16:36–41.
33. Sharma P, McQuaid K, Dent J, Fennerty B, Sampliner R, Spechler S, Cameron A, Corley D, Falk G, Goldblum J, Hunter J, Jankowski J, Lundell L, Reid B, Shaheen N, Sonnenberg A, Wang K, Weinstein W. A critical review of the diagnosis and management of Barrett's esophagus: The AGA Chaicago Workshop. Gastroenterology 2004;127:310–330.
34. Cameron AJ, Lomboy CT. Barrett's esophagus: Age, prevalence, and extent of columnar epithelium. Gastroenterology 1992;103: 1241–1245.
35. Der R, Tsao-Wei DD, DeMeester T, et al. Carditis: A manifestation of gastroesophageal reflux disease. Am J Surg Pathol 2001;25:245–252.
36. Tharalson EF, Martinez SD, Garewal HS, Sampliner RE, Cui H, Pulliam G, Fass R. Relationship between rate of change in acid exposure along the esophagus and length of Barrett's epithelium. Am J Gastroenterol 2002;97:851–856.
37. Kazumori H, Ishihara S, Rumi MA, Kadowaki Y, Kinoshita Y. Bile acids directly augment caudal-related homeobox gene Cdx2 expression in esophageal keratinocytes in Barrett's epithelium. Abstract ID: S1674, DDW, 2005.
38. Fitzgerald RC, Omary MB, Triadafilopoulos G. Dynamic effects of acid on Barrett's esophagus. J Clin Invest 1996;98:2120–2128.
39. DeMeester SR, Campos GMR, DeMeester TR, Bremner CG, Hagen JA, Peters JH, Crookes PF. The impact of antireflux procedure on intestinal metaplasia of the cardia. Ann Surg 1998;228:547–556.

CHAPTER

10

Pathology of Reflux Disease at a Cellular Level: Part 3—Intestinal (Barrett) Metaplasia to Carcinoma

CARCINOGENESIS IN INTESTINAL METAPLASIA

The target cell for esophageal adenocarcinoma is the proliferative progenitor cell in intestinal metaplastic epithelium. Esophageal epithelia that do not have the genetic makeup that is expressed phenotypically as intestinal metaplasia (these include squamous epithelium, cardiac mucosa, and oxyntocardiac mucosa) are not susceptible to carcinogenesis (Fig. 10.1). It is almost as if the genetic switch that causes intestinal metaplasia opens a receptor on the surface that can now accept the carcinogen (Fig. 10.2). The carcinogens do not act on the goblet cells; these are usually terminally differentiated cells. They act on the proliferative cell pool that is present in the deep foveolar region of intestinal metaplastic epithelium (Fig. 10.3).

Carcinogenesis in intestinal metaplastic epithelium probably progresses through a series of irreversible genetic mutations in a manner comparable to the development of adenocarcinoma of the colon. The molecular pathway of carcinogenesis in Barrett esophagus is much less understood than that of colonic adenocarcinoma.

Irreversible carcinogenic mutations may or may not be expressed phenotypically; when expressed, the phenotypic change may follow the genetic mutation after a significant lag phase. In Barrett esophagus, the phenotypic expression of genetic changes of carcinogenesis is believed to progress through low-grade dysplasia, high-grade dysplasia, and invasive adenocarcinoma. There is evidence for such a progression from longitudinal studies in patients under long-term surveillance for Barrett esophagus (discussed later).

The exact molecular alterations associated with these changes are unknown as yet. Dysplasia has been shown to be associated with general genetic abnormalities such as aneuploidy, and there is some evidence for p53 mutations being involved in the process at least in some cases. A large number of other genetic mutations have been described in Barrett esophagus, dysplasia, and adenocarcinoma. Though these genetic changes have been shown to be present more frequently in these epithelia, neither a mechanism of carcinogenesis nor any consistent sequence of genetic change has been worked out. In the absence of reliable molecular markers that define the stage of carcinogenesis, reliance for clinical management of patients with Barrett esophagus must still be placed on the phenotypic expression of these changes (i.e., grade of dysplasia) (Fig. 10.4).

Assuming that carcinogenesis in gastroesophageal reflux disease is caused by luminal carcinogens, the agent or agents responsible for inducing carcinogenic transformations in the target cell must reside in the gastric juice. These agents are unknown at present and must gain access to the target cell by the occurrence of gastroesophageal reflux.

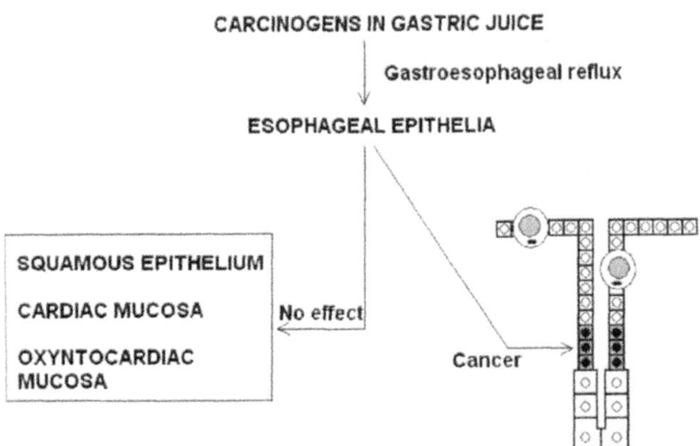

FIGURE 10.1 Carcinogenesis in the esophagus resulting from gastroesophageal reflux. The first requirement is the presence of carcinogenic molecules in the gastric contents. The second requirement is that gastric contents reach the esophageal epithelium via gasroesophageal reflux. The third requirement is that the esophageal epithelium has the appropriate target cell on which the carcinogens act. Evidence indicates that squamous epithelium, cardiac mucosa, and oxyntocardiac mucosa are not at risk. The carcinogens act only on the proliferating pool of cells in intestinal metaplasia.

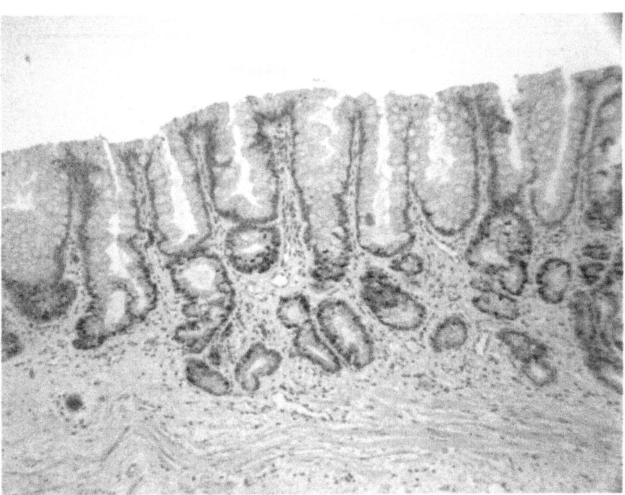

FIGURE 10.3 Immunoperoxidase stain for Ki67 in intestinal metaplastic epithelium. The target cell is the Ki67 proliferative cell pool in the deep foveolar region. The goblet cells, which are largely terminally differentiated nonproliferative cells toward the surface, simply serve as a marker for an epithelium that is receptive to the carcinogen.

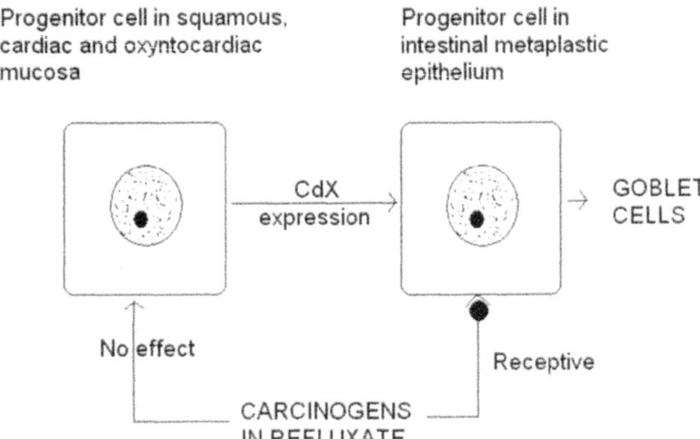

FIGURE 10.2 Schematic to show the effect of carcinogens in the gastric refluxate on the progenitor cell of the esophageal epithelium. When the progenitor cell occupies squamous, cardiac, and oxyntocardiac epithelia, the carcinogen has no effect. When this cell occupies intestinal metaplastic epithelium, it is receptive to the carcinogen. This figure shows the progenitor cell being altered by the CdX gene expression to induce goblet cell formation. When this happens, it appears that a receptor opens up on the progenitor cell to accept carcinogen interaction.

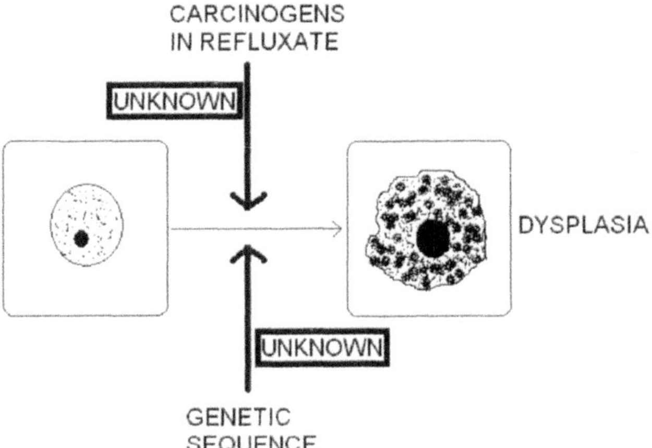

FIGURE 10.4 The process of neoplasia in the esophagus requires a carcinogen that likely produces a sequence of genetic mutations. Neither the carcinogens nor the genetic sequence are known. In the absence of this knowledge, we are left with detection of the phenotypic result of the mutations (i.e., dysplasia of increasing grade) to assess the progression of a patient through the carcinogenic sequence.

DOES ADENOCARCINOMA ARISE IN ESOPHAGUS WITHOUT INTESTINAL (BARRETT) METAPLASIA?

In the 1980s, Haggitt and Dean (1), Hamilton et al. (2), and McArdle et al. (3) provided evidence that esophageal adenocarcinoma developed mainly in the intestinal type of columnar-lined esophagus.

Literature Review

Hamilton SR, Smith RR, Cameron JL. Prevalence and characteristics of Barrett esophagus in patients with adenocarcinoma of the esophagus or esophagogastric junction. Hum Pathol 1988;19:942–948.

This is a histologic study from Johns Hopkins Hospital in Baltimore of 61 consecutive esophagogastrectomy specimens with adenocarcinoma of the esophagus and esophagogastric junction (including gastric cardia). At the time of the study, short-segment Barrett esophagus was not yet recognized and the definition of Barrett esophagus had still not changed to require intestinal metaplasia. The proximal limit of the rugal folds as being a criterion for the gastroesophageal junction had been recently reported but probably not in usage at the time this study was done.

> The gastroesophageal junction was identified on the basis of at least two of four possible gross and histopathologic landmarks: the most distal location of esophageal submucosal glands, the transition from the tubular configuration of the esophagus into the stomach, the most distal location of the squamous epithelium, and the change in orientation of the layers of muscularis propria in longitudinally oriented specimens. (2, p. 943)

In illustrations provided of two specimens, the gastroesophageal junction is marked well below the most distal location of the squamous epithelium and below the end of the tubular esophagus in the authors' Figure 1a (Fig. 10.5),

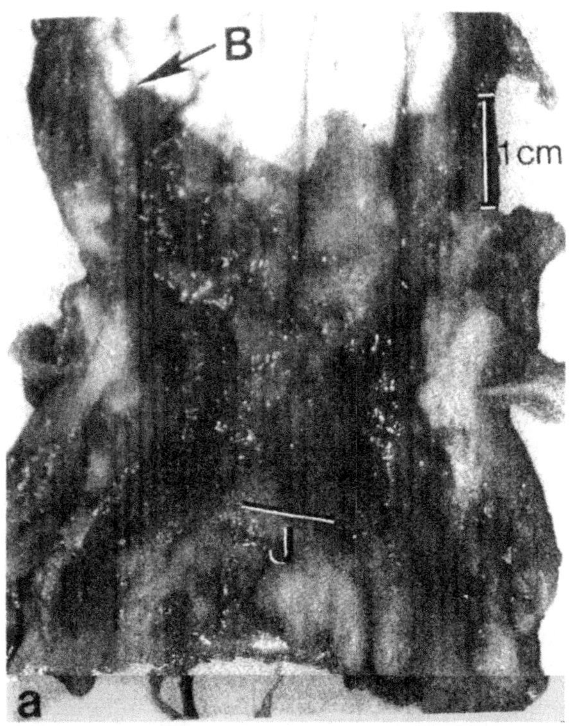

FIGURE 10.5 Figure 1a from Hamilton et al. shows a gross specimen of an adenocarcinoma arising in Barrett esophagus. The upper end of Barrett's mucosa (B) and the esophagogastric junction (J) are marked. Note that the esophagogastric junction is clearly distal to the point at which the tubular esophagus flares, and it appears to be close to the proximal limit of the rugal folds. Reproduced with permission from Hamilton SR, Smith RRL, Cameron JL. Prevalence and characteristics of Barrett's esophagus in patients with adenocarcinoma of the esophagus or esophago-gastric junction. Hum Pathol 1988;19:942–948.

making it probable that this location was determined by the presence of submucosal glands. In this specimen, it is quite clear that abnormal epithelium is present proximal to the rugal folds in the saccular part of the specimen distal to the end of the tubular esophagus. This is a clear example of end-stage dilated esophagus recognized as such by an astute pathologist using a more accurate definition of the gastroesophageal junction (the distal limit of the esophageal submucosal glands) rather than the incorrect definition of the end of the tubular esophagus.

Barrett esophagus was defined by the presence of columnar epithelium-lined mucosa in the esophagus. The mucosal types included (a) specialized Barrett's mucosa with incomplete intestinal metaplasia characterized by goblet cells that were interspersed with gastric-type columnar mucous cells (the authors are clearly describing intestinal metaplasia arising in cardiac mucosa), (b) dysplastic Barrett's mucosa, characterized by villiform mucosal architecture and dysplastic columnar epithelium with intestinal-type mucin, and (c) cardiac-type (junctional) Barrett's mucosa resembling gastric cardiac mucosa and lacking goblet cells. No cases with only fundic-type Barrett's mucosa were found. The length of Barrett esophagus was defined as the measured distance between the gastroesophageal junction and the most proximal Barrett's mucosa or ulcerated tumor. These descriptions are beautifully precise.

Barrett esophagus was found in 39 of 61 (64%) of the patients; 35 of these were identified on the basis of distinctive intestinal epithelium, 3 had dysplastic Barrett mucosa, and 1 had cardiac-type Barrett's mucosa. From their data, it appears that the authors will diagnose Barrett esophagus in the following circumstances: (a) when intestinal metaplasia is present, whatever the length of the segment, as shown by their Figure 2 (Fig. 10.6), where there are cases with 1 cm of Barrett's mucosa, and (b) in one patient with only cardiac mucosa, presumably because the length of columnar-lined esophagus exceeded 3 cm.

The authors provided a detailed figure showing the extent of Barrett's mucosa, tumor, and the epicenter of the tumor in these 39 patients (Fig. 10.6). The epicenter of the tumor is distal to the esophagogastric junction (i.e., in the stomach) in 4 patients (1 cm distal in 3 and 2 cm distal in 1 patient), at the EGJ in 7 patients, 1 cm above the EGJ in 9 patients, 2 cm above the EGJ in 9 patients, and greater than 2 cm above the EGJ in 17 patients to a maximum of 7 cm above the EGJ. The 4 patients with the tumor epicenter below the EGJ are depicted as tumor growing down into the stomach from Barrett mucosa in the esophagus. By present criteria, these would be defined as "adenocarcinoma of the gastric cardia."

From the preceding discussion it can be deduced that 22 of 61 patients (36%) did not have Barrett esophagus, which means that they had tumors in segments of columnar-lined esophagus that were less than 3 cm long and did not contain residual intestinal metaplasia. The details of these specimens are not provided except in the author's Table 2 where it is seen that the mean length from the epicenter of the tumor to the EGJ was 0 +/− 0.5 cm and that only 14 of 22 had

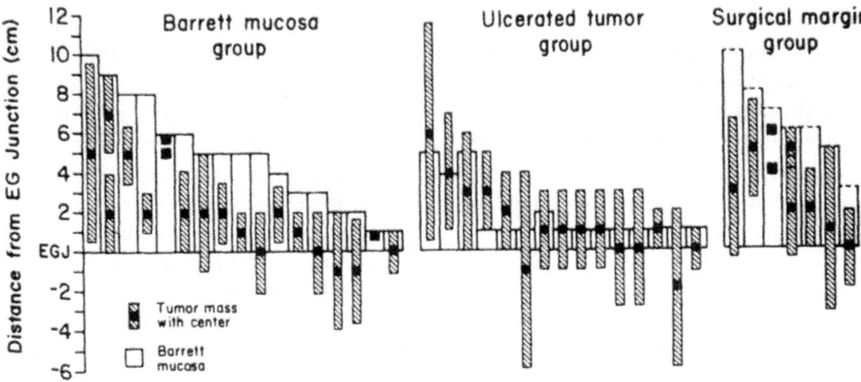

FIGURE 10.6 Figure 2 from Hamilton et al. shows extent of Barrett esophagus in each case (white rectangle) along with the center (black) and extent (striped) of adenocarcinoma. Four tumors were centered near the esophagogastric junction with four tumors having the epicenter below the junction. Reproduced with permission from Hamilton SR, Smith RRL, Camerson JL. Prevalence and characteristics of Barrett's esophagus in patients with adenocarcinoma of the esophagus or esophago-gastric junction. Hum Pathol 1988;19:942–948.

complete histologic examination. The latter fact leaves open the possibility that undetected residual intestinal metaplasia could have been present.

The patients with Barrett adenocarcinoma showed a striking predominance of white men (34 of 39; 87%), similar to patients having adenocarcinoma of the esophagus or esophagogastric junction without demonstrable Barrett esophagus (16 of 22; 73%), but in contrast to gastric adenocarcinoma cases examined at the hospital during this period (21 of 69; 30%). The authors conclude that most adenocarcinomas of the esophagus or esophagogastric junction are Barrett's carcinomas, rather than gastric cardiac cancers or other types of esophageal adenocarcinoma; most Barrett's adenocarcinomas occur in short segments of Barrett esophagus, which may be difficult to detect at endoscopy; and white men with Barrett esophagus may constitute a clinically identifiable at-risk group suitable for surveillance.

This research showed that *most* esophageal adenocarcinomas arose in the intestinal metaplastic type of Barrett esophagus. These authors are way ahead of their time because they included short-segment Barrett esophagus within the definition of Barrett esophagus and also recognized the occurrence of cancer in such patients. They have also suggested that adenocarcinomas of the proximal stomach are similar to adenocarcinomas of the esophagus and different from distal gastric adenocarcinomas. These pathologic data did not stimulate gastroenterologists to biopsy columnar-lined esophagus less than 3 cm long for 6 years after the publication of this article (4).

The fact that intestinal metaplasia is not always found in esophageal adenocarcinoma was also clearly reported in Lagergren et al. (5); Barrett esophagus was detected in 118 of 189 (62%) patients with esophageal adencoarcinoma. In this study from Sweden, the definitions had not changed much from Hamilton et al.; esophageal adenocarcinomas were defined as tumors having their epicenter greater than 2 cm above the gastroesophageal junction. The gastroesophageal junction was defined as the proximal limit of the rugal folds, but tumors with the epicenter less than 2 cm above the gastroesophageal junction (i.e., in the distal esophagus and therefore including most cases of short-segment Barrett esophagus) were included with adenocarcinomas of the gastric cardia. In this study also, Barrett esophagus included five patients who had nonintestinalized cardiac and oxyntocardiac mucosa greater than 3 cm proximal to the gastroesophageal junction. If we assume that all these patients had esophageal adenocarcinoma, the lowest possible incidence of intestinal metaplasia associated with esophageal adenocarcinoma in this study is 113 of 189 (60%). The prevalence of intestinal metaplasia in patients with adenocarcinoma of the gastric cardia (which includes the distal 2 cm of the esophagus) was not reported.

In a study of 74 esophagogastrectomy specimens at the University of Southern California in all cases of adenocarcinoma in the tubular esophagus and within 3 cm distal of the tubular esophagus, we found residual intestinal metaplasia in 48 (65%) (Chandrasoma, 2005, unpublished data). It seems fair to state that residual intestinal metaplasia is found in about two thirds of patients with reflux-induced adenocarcinoma.

Despite the fact that one third of patients who had esophageal adenocarcinoma were proven to not have intestinal metaplasia in relation to their tumor, it became generally accepted that *esophageal adenocarcinoma only arose in intestinal type columnar-lined esophagus.* There is no justification in the literature to reach this conclusion; like many things in medicine, this was probably the result of a consensus agreement among the experts in the field outside the formal literature.

Acceptance of the fact that all esophageal adenocarcinomas arose in intestinal metaplasia led to the following very significant changes:

1. The definition of Barrett esophagus shifted from the presence of a columnar-lined esophagus greater than 3 cm long in the tubular esophagus to requiring an additional criterion that there must be intestinal metaplasia in a biopsy taken from this columnar-lined esophagus (see the discussion of geographic differences presented later). The definition of Barrett esophagus shifted from being purely endoscopic to one that required a combination of endoscopic (>3 cm of columnar-lined esophagus) and histologic (presence of intestinal metaplasia) criteria. When short-segment Barrett esophagus was recognized, the requirement of intestinal metaplasia persisted.
2. Intestinal metaplasia became recognized as a premalignant epithelium. Patients with intestinal metaplasia were placed on long-term surveillance in the belief that any cancer arising in Barrett esophagus could be detected early and mortality could be prevented. Initially surveillance was restricted to patients with long-segment Barrett esophagus. When it was shown that short-segment Barrett esophagus also had an increased cancer risk, surveillance extended to this group of patients as well. Recommended surveillance protocols vary in the frequency of endoscopies from annually to every 5 years (6).
3. Patients with the nonintestinal types of columnar-lined esophagus (cardiac and oxyntocardiac epithelia) were removed from the equation completely. The presence of cardiac and oxyntocardiac epithelia in biopsies no longer meant anything. In the United States, these patients were quickly removed from the definition of Barrett esophagus, which required the presence of intestinal metaplasia. In Europe, as shown in Lagergren et al., this was not necessarily true; patients with more than 3 cm of columnar-lined esophagus without intestinal metaplasia still fell within the definition of Barrett esophagus. This geographic variation was not a huge discrepancy because the prevalence of intestinal metaplasia in columnar-lined esophagus greater than 3 cm was over 90%. In the United States, surveillance was not recommended for patients without intestinal metaplasia on a biopsy, whatever the length of columnar-lined esophagus.

At present, there is no effort to detect intestinal metaplasia at the junction even in patients with reflux who are endoscopically normal because of the recommendation that these patients should not be biopsied. However, if by chance a biopsy is taken from an endoscopically normal patient and it shows intestinal metaplasia, there is a tendency to place these patients on surveillance despite the absence of evidence that this is indicated. The fear of reflux-associated intestinal metaplasia as a precursor of adenocarcinoma is widespread.

Based on these above facts, it is not surprising that there is evidence of considerable confusion. In the American Gastroenterological Association Chicago meeting of experts (7), the statement (no. 1) that esophageal intestinal metaplasia documented by histology is a prerequisite criterion for the diagnosis Barrett esophagus did not meet with complete agreement. Only 5 of 18 experts accepted this statement completely, and 2 of 18 rejected it (were these the European experts?). The basis for rejection was that there was uncertainty about whether adenocarcinoma could arise in columnar-lined esophagus devoid of intestinal metaplasia—that is, in the cardiac and fundic (oxyntocardiac) types. This was felt to be an unproven fact, at least by the two experts who rejected it. These two experts are correct; this is an unproven fact. *While there is evidence that most adenocarcinomas arise in intestinal metaplastic Barrett esophagus, there is no proof in the literature that they never arise in nonintestinalized columnar-lined esophagus.*

In the absence of a surveillance protocol for patients with columnar-lined esophagus without intestinal metaplasia, longitudinal studies in this patient group are not available. If a patient has an endoscopic biopsy that shows columnar-lined esophagus without intestinal metaplasia and later presents with adenocarcinoma, there is no way to know whether there was an intermediate step of intestinal metaplasia unless the specimen contains residual intestinal metaplasia around the tumor. There is likely to be a significant number of patients without residual intestinal metaplasia, even if this was a precursor lesion; it is not uncommon for a cancer to completely destroy the precursor lesion (as colorectal carcinomas do to the adenomas they arise in).

A study by Jones et al. (8) on patients with endoscopic short-segment columnar-lined esophagus who do not have intestinal metaplasia on the index biopsy who go on to have a repeat endoscopy show that more than 20% of these patients are found to have intestinal metaplasia at the later endoscopy. This may represent one of two things: (a) sampling error at the first examination or (b) progression of the metaplastic sequence from cardiac to intestinal type of epithelium by the development of goblet cells. If the repeat endoscopy is immediate, sampling error is certain. With increasing intervals between the index and repeat endoscopy, the possibility of progression to intestinal metaplasia increases.

Longitudinal studies are available in patients who have had esophagogastrectomy. These studies are reviewed in Chapter 9 where we examine the evidence that reflux-induced adenocarcinoma is a multistep process that progresses from squamous epithelium through cardiac mucosa to intestinal metaplasia before carcinogenesis begins. Dresner et al. (9) showed that 19 of 40 patients developed columnar metaplasia in the esophagus above the anastomotic line (see Fig. 9.23). They found cardiac mucosa in 10 and intestinal metaplasia in 9 patients. Seven patients followed serially showed progression from cardiac-type epithelium to intestinal metaplasia. This is convincing evidence of a sequence of change that progresses from cardiac to intestinal metaplasia, but it does not prove that cancer does not arise in both types of epithelia.

It is not uncommon to see patients with adenocarcinomas occurring in dysplastic epithelium devoid of goblet cells and surrounded by oxyntocardiac (and possibly cardiac) epithelium. Is this evidence that adenocarcinoma can arise in columnar-lined esophagus without intestinal metaplasia? Two issues arise with this question:

1. Is it possible that the area of dysplasia from which the carcinoma arose actually represents intestinal metaplasia despite the absence of goblet cells? One of the features of dysplasia is a loss of cytoplasmic mucin and frequent loss of goblet cells. We do not hesitate to diagnose dysplastic intestinal metaplasia in epithelium devoid of goblet cells (Fig. 10.7). In Hamilton et al. (2) reviewed earlier, it appears that this is a tacit agreement among pathologists; the researchers defined dysplastic Barrett mucosa as "characterized by villiform mucosal architecture and dysplastic columnar epithelium with intestinal-type mucin" specifically omitting the requirement of goblet cells. This is, however, a proof by definition; we simply define all dysplasia in the esophagus as arising in intestinal metaplasia even if there are no goblet cells. We will be the first to admit that proof by definition is the worst kind of proof.
2. Is it possible that adenocarcinoma and dysplasia destroy the intestinal metaplasia from which they arise? Testing this question requires the use of indirect data. We (10) studied 74 patients who underwent esophagogastrectomy for adenocarcinoma of the distal esophagus (38), gastroesophageal junction (25), and gastric cardia (11). The tumors were classified as to location by the relationship of the epicenter of the tumor to a line drawn across the end of the tubular esophagus. The tumor was measured for maximum mucosal diameter and depth of gross invasion. The mucosal types present immediately surrounding the tumor on all sides were examined.

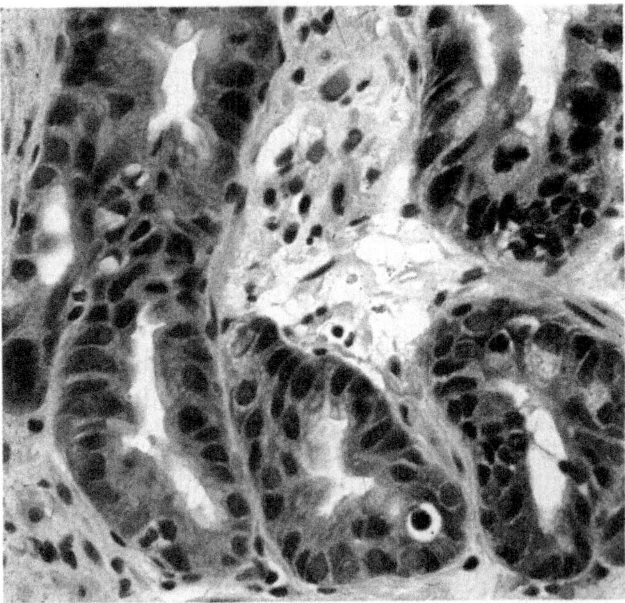

FIGURE 10.7 Dysplasia in Barrett esophagus showing depletion of mucin and an increased nuclear:cytoplasmic ratio typical of dysplasia and resulting in the absence of goblet cells. The presence of goblet cells is not necessary to define intestinal metaplasia when the epithelium is dysplastic.

Residual intestinal metaplasia, characterized by the presence of goblet cells, was found in 48 of the 74 patients (65%). Intestinal metaplasia was seen in 33 of 38 patients with distal esophageal tumors (87%), 10 of 25 patients with junctional tumors (40%), and 5 of the 11 patients with gastric cardiac tumors (45%). Cardiac and oxyntocardiac types of mucosa were present around the tumor in all cases. Cardiac and oxyntocardiac epithelial types were the only epithelial types found in the 26 patients who did not have residual intestinal metaplasia. Gastric oxyntic mucosa, when present, was found only at the distal edge of the tumors, representing extension of the tumor into the part of the stomach lined by oxyntic mucosa.

The prevalence of residual intestinal metaplasia was greatest in the tumors that did not infiltrate through the muscle wall of the esophagus. All 14 patients with intramucosal adenocarcinoma (100%), 7 of 8 patients with submucosal tumors (88%), and 3 of 4 patients with intramural tumors (75%) (i.e., a total of

24 of 26 or 92% of patients with tumors restricted to the wall of the esophagus) had residual intestinal metaplasia, compared with 24 of 48 patients who had transmural tumors (50%) ($p < 0.01$). In another study of 78 patients who had esophagectomy for intramucosal adenocarcinoma, we found that all these patients had residual intestinal metaplasia (Oh et al., unpublished data).

The likelihood of finding residual intestinal metaplasia decreased with increasing tumor size. All eight tumors with a size of 1 cm or less had intestinal metaplasia; the locations of these tumors were distal esophageal (five cases) and junctional (three cases). There were no tumors 1 cm or less that were classified as gastric cardiac carcinomas. A total of 31 of 41 tumors (76%) with a size of 4 cm or less had intestinal metaplasia. This contrasted with 17 of 33 patients (52%) with tumors greater than 4 cm in size with residual intestinal metaplasia. The smallest lesion that did not have residual intestinal metaplasia was 1.4 cm in greatest diameter.

The data in this study show that esophageal adenocarcinoma is surrounded by metaplastic columnar epithelia of the esophagus in all patients, essentially proving the concept that these tumors arise in columnar-lined esophagus.

It also shows that residual intestinal metaplasia was not found in the area immediately surrounding the tumor in 26 of 74 patients (35%). In these patients, the tumor was surrounded by cardiac and oxyntocardiac epithelia without goblet cells. The thoroughness of the histologic study makes it unlikely that the absence of intestinal metaplasia can be explained by sampling error.

There are two possible explanations for the absence of residual intestinal metaplasia around the tumor. The first possibility is that the adenocarcinoma in the 26 patients who did not have intestinal metaplasia arose in cardiac and oxyntocardiac mucosa without goblet cells. The second possibility is that the adenocarcinoma arose in intestinal metaplasia but destroyed the precursor epithelium by its growth. This is similar to the well-accepted explanation as to why residual benign adenomas are frequently not present at the edge of colorectal carcinomas.

The correct test to determine which of these possibilities is correct is as follows: If adenocarcinomas developed in columnar epithelia without intestinal metaplasia with any significant frequency, there would be no difference in the prevalence of residual intestinal metaplasia in small and large tumors. If, on the other hand, adenocarcinomas developed in intestinal metaplasia and destroyed the precursor lesion as it grew, the prevalence of residual intestinal metaplasia would bear a strong relationship with tumor size and stage. This is dictated by the fact that destruction of the precursor lesion becomes more common as the size of the tumor increases.

The data in the study show a clear and significant relationship between the prevalence of intestinal metaplasia and the size and stage of the tumor. Tumors that were less than 1 cm in size and intramucosal tumors invariably showed residual intestinal metaplasia. The probability of intestinal metaplasia being absent increased with increasing tumor size and stage and was common only in tumors greater than 4 cm in size that showed transmural invasion.

This study provides strong support for the concept that intestinal metaplasia is a necessary precursor lesion for adenocarcinoma of the esophagus. Carcinogenic factors do not act in columnar-lined esophagus until intestinal metaplasia develops (Figs. 10.1 and 10.2). The restriction of the definition of Barrett esophagus to patients who have intestinal metaplasia is correct if the intent is to use Barrett esophagus as a marker of a cancer risk. The practice of not recommending surveillance for patients with columnar-lined esophagus that does not have histologic intestinal metaplasia is correct.

Although cardiac mucosa does not progress to adenocarcinoma, it is the epithelium within columnar-lined esophagus in which intestinal metaplasia arises, frequently after a lag phase of many years. As such, cardiac mucosa is the precursor of intestinal metaplasia and cannot be completely ignored as being irrelevant. A repeat endoscopy to rule out the possibility of sampling error is one need (8). The other need is to ensure, maybe after an interval of several years, that this patient has not progressed to develop intestinal metaplasia.

Conclusion: Intestinal metaplasia is a necessary precursor lesion for the occurrence of adenocarcinoma.

NATURAL HISTORY OF DYSPLASIA: SURVEILLANCE FOR BARRETT ESOPHAGUS

Does Nondysplastic Barrett Esophagus Progress through Dysplasia to Adenocarcinoma?

The statement that patients progress from nondysplastic Barrett esophagus to dysplastic Barrett esophagus (first low grade and then high grade) to invasive adenocarcinoma is true (Fig. 10.8). There is strong

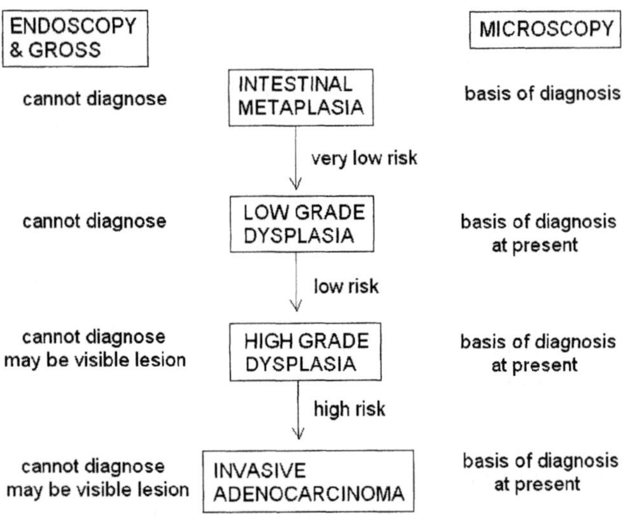

FIGURE 10.8 The sequence of low-grade dysplasia, to high-grade dysplasia, to invasive adenocarcinoma that likely exists in Barrett esophagus. The diagnosis of the level of progression is based on microscopic examination of biopsies, often of random biopsies, and, in the higher areas of the spectrum, of lesions visible at endoscopy. The risk of progression increases with increasing degrees of dysplasia.

evidence for the existence of such a sequence in patients who have had serial longitudinal biopsies while on long-term surveillance for Barrett esophagus. However, this sequential progression is rare in patients with Barrett esophagus; only very few patients with non-dysplastic Barrett esophagus in the index biopsy progress to dysplasia and adenocarcinoma. The risk of progression increases when low-grade dysplasia is found in the index biopsy and increases even further when the index biopsy shows high-grade dysplasia. However, not even all patients with high-grade dysplasia progress to adenocarcinoma, at least within the periods of follow-up that have been reported.

Literature Review

Weston AP, Badr AS, Hassanein RS. Prospective multivariate analysis of clinical, endoscopic, and histological factors predictive of the development of Barrett's multifocal high-grade dysplasia and adenocarcinoma. Am J Gastroenterol 1999;94:3413–3419.

This is a well-designed study of 108 patients with Barrett esophagus followed from the index biopsy for a period ranging from 12 to 101 months (a total of 361.8 patient-years). The baseline biopsy showed no dysplasia in 80 patients, low-grade/indefinite dysplasia in 20 patients, and unifocal high-grade dysplasia in 8 patients. Unifocal high-grade dysplasia was diagnosed when a single biopsy specimen harbored high grade dysplasia; multifocal high-grade dysplasia requires more than one biopsy from different Barrett segments to contain high-grade dysplasia. Exclusion criteria included the presence of adenocarcinoma, multifocal high-grade dysplasia, or dysplasia-associated lesion or mass (DALM) at or within 6 months of initial diagnosis. The finding of multifocal high-grade dysplasia, adenocarcinoma, or DALM was an indication for surgery or ablation.

Overall, 5 patients developed multifocal high-grade dysplasia and 5 developed adenocarcinoma.

Three of the 80 patients who had no dysplasia at the index biopsy progressed to multifocal high-grade dysplasia (1 patient) or adenocarcinoma (2 patients). Two of these patients had low-grade dysplasia before high-grade dysplasia and adenocarcinoma were found, showing full progression within the surveillance interval.

Four of the 20 patients with an initial diagnosis of low-grade dysplasia progressed to unifocal high-grade dysplasia; 1 each of these progressed to multifocal high-grade dysplasia and adenocarcinoma, with the other 2 unifocal high-grade dysplasia patients reversing to no dysplasia.

Five of the 8 patients with a diagnosis of unifocal high-grade dysplasia at the initial diagnosis subsequently developed multifocal high-grade dysplasia (3 patients) or adenocarcinoma (2 patients) during follow-up 14 to 84 months after the index biopsy.

Weston et al.'s study (11) clearly shows the malignant potential of Barrett esophagus. One point of difference from colonic adenocarcinoma in the phenotypic expression of mutations in Barrett esophagus is that an adenoma phase where there is a localized polypoid lesion that is visible endoscopically is uncommon. Most dysplasia in Barrett esophagus occurs in a flat mucosa that cannot be differentiated endoscopically from the adjacent nondysplastic columnar epithelium (Fig. 10.8). The phenotypic expression of Barrett's dysplasia is similar to that of dysplasia in chronic ulcerative colitis. Weston et al. have extrapolated to patients with Barrett esophagus the entity of dysplasia-associated lesion/mass (DALM) that has been shown to be a high-risk lesion in chronic ulcerative colitis. In chronic ulcerative colitis, DALM is a significant risk indicator even in the presence of low-grade dysplasia. There is no evidence in the esophagus that a visible lesion with low-grade dysplasia has a significant association with cancer. The presence of DALM in Barrett esophagus should not be an indication for esophagectomy unless there is high-grade dysplasia.

Conclusion: There is no doubt that patients show progression from nondysplastic, low-grade dysplastic, and high-grade dysplastic Barrett esophagus to adenocarcinoma.

Do All Adenocarcinomas Pass through a Dysplastic Phase?

There is no proof that the sequence of low-grade to high-grade dysplasia precedes all adenocarcinomas arising in Barrett esophagus. Over 90% of patients with adenocarcinoma do not have a prior history of Barrett esophagus. These patients have invasive adenocarcinoma at the index examination and in a significant number of these patients, there is no evidence of either dysplastic or nondysplastic Barrett esophagus around the tumor. In our study of 74 esophagectomy specimens, there was no residual intestinal metaplasia in 26 (35%).

What remains uncertain is whether some patients can skip parts of this orderly sequence, progressing directly from nondysplastic Barrett esophagus to adenocarcinoma without dysplasia or passing directly from low-grade dysplasia to adenocarcinoma without high-grade dysplasia. This is an impossible question to answer in a setting where the vast majority of patients with adenocarcinoma have never had a prior endoscopy or biopsy. Present belief, without any definite evidence, is that increasing grades of dysplasia precede invasive carcinoma in most, if not all, cases.

There is a strong possibility that this belief will change. In patients with carcinoma of the uterine cervix, the initial belief of an orderly sequence has been replaced by one that recognizes that invasive cancer may arise rapidly in some patients, passing from low-grade dysplasia in one examination to invasive carcinoma in the next. This is not proof in such cases that there was no high-grade dysplasia before invasive carcinoma occurred; it is a practical statement that the progression can be so quick in some patients that the surveillance interval may be too long. If this is true, it becomes impossible to be complacent about low-grade dysplasia; in the uterine cervix, recognition of this fact has resulted in the more aggressive eradication of low-grade dysplasia. The uterine cervix is not the esophagus. Eradication can be achieved by a cone biopsy of the cervix; eradication of disease is much more complicated at the present time in the esophagus. However, the refinement of endoscopic ablative techniques may raise the possibility of more aggressive eradication of low-grade Barrett dysplasia in the future.

Conclusion: There is no proof that all Barrett adenocarcinomas follow a sequential progression from nondysplastic to low-grade dysplastic to high-grade dysplastic Barrett esophagus to adenocarcinoma. This is likely, though the rate of change is highly variable.

Is There a Pattern of Distribution of Dysplasia and Adenocarcinoma in Barrett Esophagus?

The present belief is that dysplasia and adenocarcinoma develop randomly in a segment of columnar-lined esophagus. Based on this belief, the recommendation for biopsy protocol during surveillance endoscopy for Barrett esophagus is to (a) take four-quadrant random biopsies at 1- to 2-cm intervals using a jumbo forceps; (b) biopsy any visible lesion if any is present. This standard biopsy protocol is additionally based on the present belief that dysplasia in Barrett esophagus cannot be distinguished from nondysplastic intestinal metaplasia by endoscopy, even with newer chromoendoscopic and magnification techniques. If the ability to localize dysplasia at endoscopy improves, this protocol can change. In esophageal squamous carcinoma, Lugol's iodine has been shown to be useful in directing biopsies to areas of high risk for early neoplastic change in esophageal squamous epithelium.

A study by McArdle et al. (3) examined the entire mucosal surfaces of seven esophagectomy specimens resected for high-grade dysplasia (3 of 7) or early invasive intramucosal carcinoma (4 of 7). The specimens were examined by longitudinal sections extending from squamous epithelium to gastric mucosa. In all, 358 slides from the 7 esophagectomies were examined and reconstructed to provide a mapping of the entire region. Barrett esophagus was defined by the presence of goblet cells and dysplasia graded according to standard criteria.

The average length of Barrett's mucosa measured from the tissue sections (therefore, defined by intestinal metaplasia) ranged from 1.9 to 7.3 cm. The study showed the presence of an equal likelihood of high-grade dysplasia or invasive carcinoma occurring throughout the length of Barrett's epithelium with a slightly higher incidence within the central portions (Figs. 10.9 and 10.10). Foci of carcinoma appeared within fields of Barrett's epithelium and adjacent to areas of dysplasia, supporting a dysplasia-carcinoma sequence. The amount of dysplastic epithelium was related to the surface area of Barrett's epithelium, but there was no association between the extent of dysplasia and the likelihood of finding carcinoma.

The authors concluded that this study supports the current standard of practice for clinical surveillance of patients with Barrett's esophagus by uniformly distributed endoscopic biopsy of the complete length. In addition, the presence of any degree of dysplasia may be an indication for close clinical follow-up.

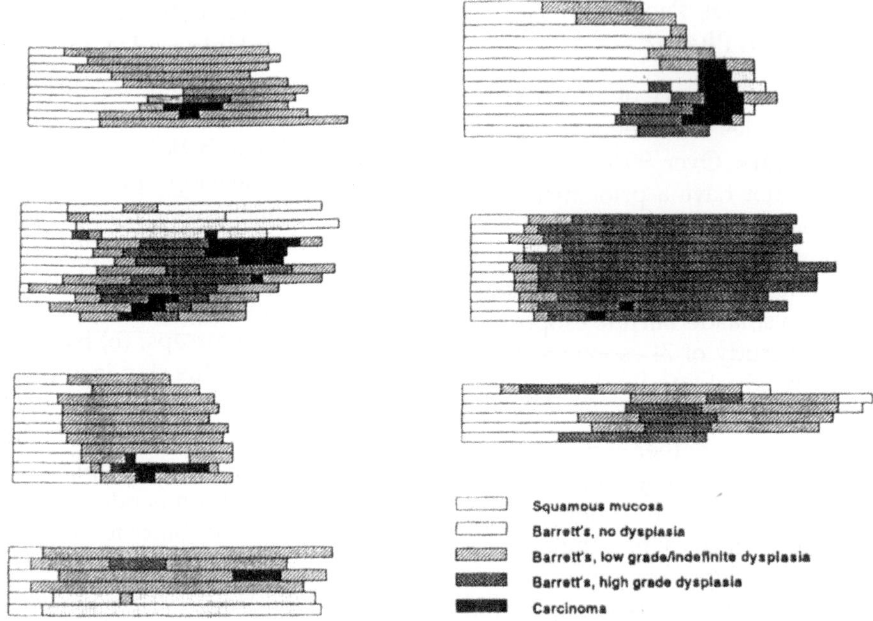

FIGURE 10.9 Complete histologic mapping of esophagectomy specimens in seven patients with high-grade dysplasia or early adenocarcinoma complicating Barrett esophagus. The low- and high-grade dysplasia and adenocarcinoma are seen in a random distribution. Reproduced with permission from McArdle JE, Lewin KJ, Randall G, et al. Distribution of dysplasias and early invasive carcinoma in Barrett's esophagus. Hum Pathol 1992;23:479–482.

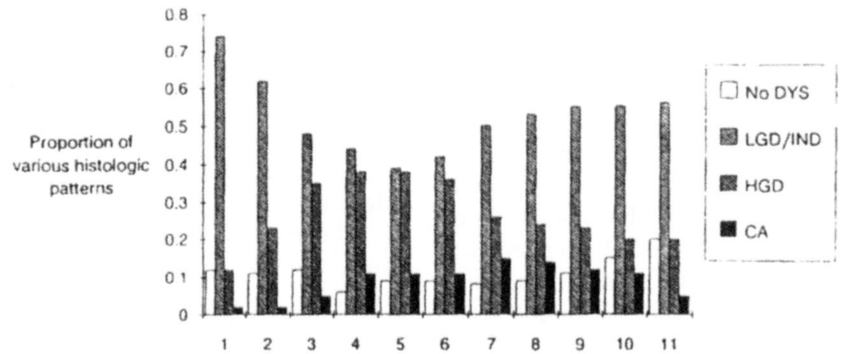

FIGURE 10.10 Same information in Figure 10.9 expressed in graphic form, with each number in the horizontal axis representing 10% of the length of Barrett esophagus and the vertical axis showing the percentages of nondysplastic, low- and high-grade dysplastic, and cancerous epithelium found at each level. Reproduced with permission from McArdle JE, Lewin KJ, Randall G, et al. Distribution of dysplasias and early invasive carcinoma in Barrett's esophagus. Hum Pathol 1992;23:479–482.

In the detailed study by Hamilton et al. (2) of 61 esophagectomy specimens, the cancers were predominantly centered at or within 2 cm of the gastroesophageal junction (Fig. 10.6).

Our experience with regard to the distribution of dysplasia and adenocarcinoma in the esophagus is similar. In a study of 78 patients with intramucosal adenocarcinoma (12), a greater than 3-cm segment of columnar-lined esophagus with intestinal metaplasia was present in 57 patients, a less than 3-cm visible segment of columnar-lined segment was present in 8 patients, and no visible columnar-lined segment with intestinal metaplasia immediately distal to the endoscopically normal squamocolumnar junction was present in 13 patients. The adenocarcinomas were found distributed throughout the area of the columnar-lined esophagus. This meant that all 13 patients without a visible columnar-lined esophagus had tumors at or distal to the normal endoscopic squamocolumnar junction (presently classified as "adenocarcinoma of the cardia" occurring in "cardiac intestinal metaplasia"). Those with less than 3 cm of columnar-lined esophagus had tumors in the distal esophagus within 3 cm of the gastroesophageal junction (i.e., proximal limit of rugal folds), at the junction, or within 2 cm distal to the junction. Those with greater than 3 cm of columnar-lined esophagus had tumors throughout the long segment of columnar-lined esophagus with a somewhat greater tendency for tumors to be in the distal esophagus rather than the proximal regions of the columnar-lined segment (Fig. 10.11).

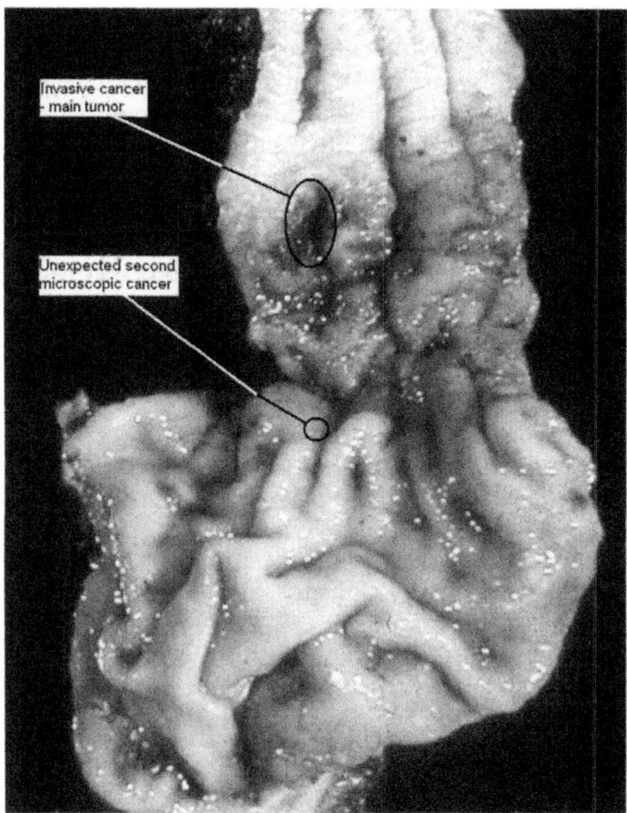

FIGURE 10.11 Esophegectomy specimen in a patient with adenocarcinoma arising in Barrett esophagus. The main cancer is seen as an ulcerated mass in the distal esophagus in the more proximal part of the columnar-lined segment. A second early (intramucosal) adenocarcinoma was present 1 cm distal to the endoscopic gastroesophageal junction. This was an unexpected microscopic finding. (Cross reference: Color Plate 3.5)

Care is necessary when interpreting data from studies that use resected specimens, such as our study above. It is possible that the distribution will be skewed if there is a tendency for cancers in some locations to present at an earlier stage and be more amenable to resection. Surveillance which is presently limited to Barrett esophagus in visible segments of columnar lined esophagus will tend to select patients with longer segments, resulting in a bias of early resectable cases in this group.

Conclusion: Dysplasia and adenocarcinoma arise randomly within the columnar-lined segment in a distribution that favors the distal esophagus and junctional region.

How Many Patients with Barrett Esophagus without High-Grade Dysplasia Progress to Adenocarcinoma?

The more serious uncertainty relates to the number of patients within each group who progress on to adenocarcinoma. How many patients with nondysplastic intestinal metaplasia, low-grade dysplasia, and high-grade dysplasia will ultimately develop adenocarcinoma? Information regarding this question is available, but there is no consensus.

Another serious gap in our knowledge is the time frame for the development of adenocarcinoma. What is the rate of progression from low-grade dysplasia to high-grade dysplasia to adenocarcinoma? It is likely that the time frame is variable, but it is necessary to know the minimum time it takes to progress from low to high grade and high grade to adenocarcinoma. Lack of certainty regarding these questions makes patient management difficult.

Table 10.1 summarizes the results of several long-term surveillance studies on patients with Barrett esophagus, depending on whether their index biopsy was nondysplastic, had low-grade dysplasia, or had high-grade dysplasia. These studies have led to the present consensus that patients who are detected as having Barrett esophagus without dysplasia or with dysplasia classified as indefinite or low grade are at very low risk for developing adenocarcinoma even after many years of surveillance. In these studies, only 15 of 929 patients (1.6%) followed for a significant period (the mean follow-up in these studies was greater than 3 years) developed adenocarcinoma. The presently accepted estimate for cancer risk in a patient with Barrett esophagus is 0.5% per year.

Reid et al. reported that 5 of 129 patients with nondysplastic Barrett esophagus, 1 of 79 patients who were indefinite for dysplasia, and 3 of 43 patients with low-grade dysplasia in their index endoscopy devel-

TABLE 10.1 The Incidence of Adenocarcinoma in Patients with Nondysplastic, Low-Grade Dysplastic, and High-Grade Dysplastic Barrett Esophagus

Study	Number of patients	Follow-up (mean yrs)	No dysplasia	LGD	HGD
Miros (13)	124	3.6	0/111	1/10 (10%)	2/3 (67%)
Hameetman (14)	50	5.2	2/43 (5%)	3/6 (50%)	0/1
O'Connor (15)	136	4.2	1/125 (1%)	1/10 (10%)	0/1
Weston (11)	108	3.3	2/80 (2.5%)	1/20 (5%)[a]	2/8 (25%)
Reid (16)	327	5	6/208 (3%)[a]	3/43 (7%)	45/76 (59%)
Montgomery (17)	125	—	4/66 (6%)[a]	4/26 (15%)	20/33 (61%)
Schnell (18)	1057	7.3	0/230	10/748 (1.3%)	12/75 (16%)

[a]Includes "indefinite for dysplasia."

oped adenocarcinoma on surveillance. The 5-year cumulative cancer incidence for this group without high-grade dysplasia was 4% (95% confidence interval = 1.6, 9). Reid et al. reported the use of flow cytometric data as an independent risk indicator in these patients. This research will be reviewed fully later but, in summary, these workers reported that patients with Barrett esophagus without high-grade dysplasia in their index biopsy had no risk of cancer if flow cytometry showed neither aneuploidy or increased 4N fractions. This compared with a 5-year cumulative cancer incidence of 28% for this group if aneuploidy or increased 4N fraction was present. The authors suggested that flow cytometry be used routinely in patients with Barrett esophagus as a method of determining surveillance protocols. This view has not been generally accepted and flow cytometry is not widely used, probably because of the need for complex specimen handling and expense.

One problem relating to elucidating the natural history of low-grade dysplasia and making decisions based on the presence of low-grade dysplasia is the lack of reproducibility of this diagnosis. We will discuss this factor in Chapter 14.

Conclusion: The risk of adenocarcinoma occurring within 5 years is low in nondysplastic and low-grade dysplastic Barrett esophagus and high in high-grade dysplastic Barrett esophagus.

Is Long-Term Surveillance for Barrett Esophagus Justified?

At present, patients with both long- and short-segment Barrett esophagus are placed on lifelong endoscopic surveillance with random biopsies at intervals ranging from 1 to 3 years. The rarity of adenocarcinoma on follow-up of patients with nondysplastic Barrett esophagus has led to questions being raised about the cost-effectiveness of the long-term surveillance of patients with Barrett esophagus. Cost-effectiveness must take into account not only the detection of cancer but also whether the cancers detected during surveillance were cured by treatment. In the preceding studies (Table 10.1), there is no information about how many of these patients were cured of their cancer.

In a long-term follow-up study of surveillance for a mean of 4.4 years in 143 patients with nondysplastic Barrett esophagus, MacDonald et al. (19) reported that surveillance detected only one patient with asymptomatic cancer, and this patient died of the ensuing surgery. This study suggested that surveillance was a futile and costly exercise in patients with nondysplastic Barrett esophagus, although the criticism can be made that the sample size and length of follow-up was too small for definite conclusions.

When one adds into this equation the fact that only 5% of adenocarcinomas are diagnosed in patients with a prior diagnosis of Barrett esophagus (20), it is quite clear that the practice of long-term surveillance of patients with nondysplastic Barrett esophagus will not impact the incidence rates of reflux-induced adenocarcinoma. Most patients who are diagnosed with adenocarcinoma of the esophagus are found outside the setting of surveillance.

Spechler (21), in a review of the present evidence relating to the cost-effectiveness of endoscopic surveillance in the *New England Journal of Medicine,* stated:

> A report from the United Kingdom estimated that the cost of detecting one case of esophageal cancer with the use of endoscopic surveillance was approximately $23,000 among male patients with Barrett's esophagus and $65,000 among female patients [22]. A U.S. study estimated that the cost was approximately $38,000, which was lower than the cost of surveillance mammography ($55,000 for each breast cancer detected; [23]). Both studies, however, used a considerably higher incidence of cancer to calculate surveillance costs than the current estimate of 0.5 per cent per year.

> A statistical-probability model has also been used to evaluate surveillance strategies in a simulated cohort of 10,000 middle aged patients with Barrett's esophagus [6]. Assuming an annual incidence rate of esophageal cancer of 0.4 percent, this analysis suggested that endoscopic surveillance every five years was the preferred strategy, costing $98,000 per quality-adjusted year of life gained. Using a different decision model, another group estimated the incremental cost effectiveness of endoscopic surveillance performed every other year at approximately $17,000 per year of life saved [24]. . . . It is important to appreciate the limitations of these computer models, which incorporate multiple layers of soft data and questionable assumptions. (p. 838)

Spechler's analysis indicates that there is a considerable degree of uncertainty and a lack of conviction about the presently recommended surveillance protocols for patients with nondysplastic Barrett esophagus. This lack of conviction was obvious among the panel of experts at the Chicago workshop on Barrett esophagus (7).

There is evidence that the small number of patients who develop adenocarcinoma while on surveillance have earlier stage cancers at the time of detection and have a better overall likelihood of survival than patients who present with symptoms of cancer for the first time (25). There is therefore the conflict between a public health recommendation supporting surveillance that is based on cost-effectiveness and a recommendation for surveillance in the individual patient. Because there are data suggesting that the individual patient may benefit from surveillance if he or she develops cancer, most gastroenterologists will place patients with Barrett esophagus on a long-term surveillance protocol. The basis for this decision is that the cost and discomfort of surveillance for all those who do not develop cancer is justified by the decreased mortality in the patients who develop cancer. If nothing else, in the absence of a strong and generally accepted recommendation to deviate from the recommended surveillance protocol, practicing gastroenterologists will incur significant liability risk if they do not place their patients with Barrett esophagus under some recommended surveillance protocol.

In a practice setting where the presence of nondysplastic Barrett esophagus is an indication for surveillance, the only practical question that arises when low-grade dysplasia or "indefinite for dysplasia" is present in a biopsy is whether there is any evidence that suggests the need to change the surveillance interval. There are few data to suggest that this action is necessary, but it is not uncommon for gastroenterologists to decrease the surveillance interval in patients with low-grade dysplasia.

There is no universally accepted standard surveillance protocol in Barrett esophagus. In practice, the actual surveillance interval is a combination of physician commitment to surveillance and patient compliance. In the University of Washington, Seattle, Barrett Esophagus Project, the actual median surveillance interval in 309 patients in one study during the 1983–1988 period was 25 months for patients with a baseline diagnosis of nondysplastic Barrett esophagus, 18 months for patients with a baseline diagnosis of indefinite for dysplasia or low-grade dysplasia, and 5 months for patients with a baseline diagnosis of high-grade dysplasia (26).

Conclusion: Surveillance is probably beneficial in detecting early cancer in the individual patient, but its cost-effectiveness is questionable and it is not likely to impact the incidence of esophageal adenocarcinoma.

The Natural History and Management of Patients with High-Grade Dysplasia

The natural history of high-grade dysplasia is controversial. At one extreme, high-grade dysplasia is taken as an indication for an immediate surgical or ablative procedure to remove at least the area of high-grade dysplasia. At the present time, the most common method of removing high-grade dysplasia is some kind of limited esophagectomy (transhiatal and vagal sparing esophagectomies are becoming favored). This has the disadvantage of significant morbidity and low but not insignificant mortality. It can be only justified if the risk of incident cancer in patients with a diagnosis of high-grade dysplasia is significant or if the likelihood of progression to cancer in the near future is high.

At the other extreme, a diagnosis of high-grade dysplasia is an indication for intense surveillance. Intense surveillance includes frequent endoscopy with extensive biopsy sampling. The Seattle protocol calls for jumbo four-quadrant biopsies at 1-cm intervals in these patients. In patients with a long segment of columnar-lined esophagus, this is an extremely tedious and time-consuming procedure requiring the taking of as many as 60 biopsies in extreme cases. It is important to recognize that following a patient with high-grade dysplasia without a commitment to this intensive biopsy protocol is a dangerous and unproven practice that can have significant liability risk.

There are two separate issues to be resolved in the patient who is found to have high-grade dysplasia in a biopsy. The first is the possibility that there is an invasive cancer that has been missed by the biopsy sampling. The second is the question of the imminence and amount of risk of invasive carcinoma developing in the patient. Another question that must be answered in relation to the second issue is whether patients who develop invasive cancer during surveillance can be cured or whether they are at significant risk of dying from the cancer.

Literature Review

Peters JH, Clark GWB, Ireland AP, Chandrasoma P, Smyrk TC, DeMeester TR. Outcome of adenocarcinoma arising in Barrett's esophagus in endoscopically surveyed and nonsurveyed patients. J Thorac Cardiovasc Surg 1994;108:813–821.

This is a study of 17 patients who were referred from endoscopic surveillance programs for management of high-grade dysplasia or adenocarcinoma developing in Barrett esophagus compared with 35 patients who had a newly recognized Barrett's adenocarcinoma and who had not been in a surveillance program. The referral diagnosis in the surveyed group was adenocarcinoma in 6 and high-grade dysplasia in 11. After repeat endoscopy with aggressive biopsy, 2 additional patients with adenocarcinoma were identified. Of the 9 patients who underwent esophagectomy for high-grade dysplasia, 5 had invasive adenocarcinoma in the esophagectomy specimen, which had been missed before the operation, despite the fact that the median number of biopsy specimens obtained per 2 cm of Barrett's mucosa was 7.8 (range: 1.5 to 15.0). Overall, 13 patients in the surveyed group had adenocarcinoma, 12 staged early and one staged intermediate by the WNM classification. Surveyed patients were operated on at an earlier stage than the nonsurveyed patients (10 early, 14 intermediate, and 11 late-stage tumors; $p < 0.01$). Despite the presence of adenocarcinoma in 13 of the 17 surveyed patients, their survival was significantly better than that of the nonsurveyed group ($p < 0.05$).

This study concluded that (a) patients referred from surveillance programs for Barrett's esophagus have a better outcome and earlier stage tumors than nonsurveyed patients and (b) because multiple biopsy endoscopy does not exclude the presence of adenocarcinoma, continued surveillance of high-grade dysplasia is dangerous and potentially destructive to surveillance efforts. The results of this study suggest that the risk of concurrent invasive cancer is sufficiently high in patients with high-grade dysplasia that resection was indicated. The indication for esophagectomy was not related to the natural history of high-grade dysplasia; this factor could not be assessed in this group because esophagectomy was performed.

The rationale for a nonsurgical approach to the diagnosis of high-grade dysplasia comes from studies that show that a significant number of patients with a diagnosis of high-grade dysplasia do not progress to adenocarcinoma.

1. In Weston et al. (11), the presence of unifocal high-grade dysplasia (high-grade dysplasia in one area) was taken as an indication for surveillance, whereas multifocal high-grade dysplasia or any dysplasia associated with a lesion/mass was an indication for surgery. This permitted evaluation of the natural history of unifocal high-grade dysplasia. Five of eight patients (63%) with unifocal dysplasia progressed to develop multifocal high-grade dysplasia or adenocarcinoma within 14 to 84 months.
2. In Reid et al. (16), the 5-year cumulative cancer incidence among 76 patients with baseline high-grade dysplasia was 59%. The relative risk of cancer for a patient with baseline high-grade dysplasia compared to those without high-grade dysplasia in their index biopsy was 28 (95% confidence interval 13; 63).
3. In a study of 100 patients with a diagnosis of high-grade dysplasia by Buttar et al. (27), 32% of patients developed esophageal adenocarcinoma during an 8-year follow-up period.
4. Schnell et al. (18) from Dr. Sontag's group reviewed a cohort of 1099 patients with a diagnosis of Barrett esophagus during a 20-year period; 36,251 esophageal mucosal specimens obtained from these patients were reviewed. Seventy-nine of 1099 patients (7.2%) initially had high-grade dysplasia (34 prevalent) or subsequently developed high-grade dysplasia (45 incident) without evidence of cancer. Patients with a diagnosis of high-grade dysplasia were placed on an intensive follow-up protocol. Dr. Sontag's intensive searching protocol consisted of endoscopy every 3 months with extensive biopsy in line with the Seattle protocol. Under this protocol, 4 of 79 patients developed cancer during the first year. Of the 75 patients with high-grade dysplasia who remained without detectable cancer after the first year of intensive searching, 12 (16%) developed cancer during a mean 7.3-year surveillance period: 11 of the 12 who were compliant were considered cured with surgical or ablation therapy. Of the remaining 63 patients with high-grade dysplasia, cancer did not develop during the surveillance period.

These authors concluded from these data that high-grade dysplasia without cancer in Barrett esophagus follows a relatively benign course in the majority of patients. In the patients who eventually progress to cancer during regular surveillance, surgical resection is curative. They concluded that surveillance endoscopy with biopsy is a valid and safe follow-up strategy for Barrett patients who have high-grade dysplasia without cancer.

The rationale of surveillance of patients with high-grade dysplasia includes the fact that a nonsurgical approach is feasible because the prognosis of patients who develop adenocarcinoma during surveillance is excellent. In Weston et al. (11), three of the five patients who developed "frank adenocarcinoma" had esophagectomy; pathologic staging showed these to be T1N0M0 in one and TisN0M0 in two (this is difficult to understand because Tis refers to a noninvasive adenocarcinoma, not "frank adenocarcinoma"). One of the other patients with cancer (clinically staged as T1N0M0) had photodynamic therapy to ablate the disease. The final patient refused follow-up until he developed metastatic disease and died. The five patients with multifocal high-grade dysplasia in Weston et al.'s study all underwent ablative therapy. In Schnell et al., 11 of 12 patients who were compliant underwent surgery or ablation therapy and were cured.

Care must be used in interpreting these studies. The authors tend to minimize the issue of patients with high-grade dysplasia who have invasive carcinoma in some unsampled area of the esophagus. In the Peters et al. study reviewed earlier, five of the nine patients who underwent esophagectomy for high-grade dysplasia had invasive adenocarcinoma in the esophagectomy specimen These had been missed by endoscopic biopsy, despite the fact that the median number of biopsy specimens obtained per 2 cm of Barrett's mucosa was 7.8 (range 1.5 to 15.0). The studies that follow patients with high-grade dysplasia find a method to exclude patients who have adenocarcinoma missed during the index endoscopy. Thus, in Schnell et al., 4 of 79 patients developed cancer in the first 6 months and were not included in the 16% incidence of cancer in the series. In Weston et al., all patients (number not stated) who developed multifocal high-grade dysplasia or cancer within six months of the index biopsy were excluded without any information given about these patients. These patients must be included if the study is to be evaluated correctly. The decision to follow these patients after a diagnosis of high-grade dysplasia precluded immediate surgery. If these patients with undetected adenocarcinoma had an advancement in their cancer stage during the period of intense study that was aimed at detecting the concurrent cancer (6 months in Weston et al. and 12 months in Schnell et al.), the decision to follow patients with high-grade dysplasia is a serious mistake. Not providing these data by excluding these patients does not permit evaluation of this crucial question. Before these authors get to the point of surveillance (6 months in Weston et al. and 12 months in Schnell et al.), sufficient damage may have already been done to patients inappropriately excluded from the study to make the question of follow-up for high-grade dysplasia moot.

Esophageal adenocarcinoma is a serious disease with an overall mortality of around 80% at the present time. This extremely virulent cancer progresses rapidly to death. If the four patients who were detected with cancer within the first year of surveillance in Schnell et al.'s study died or had a diminished likelihood of survival as a result of the delay, the authors' conclusions are questionable.

The other point of caution in these studies is the diagnosis of high-grade dysplasia. In Weston et al., the diagnosis of high-grade dysplasia was subject to "confirmation as high grade dysplasia by independent review" (p. 3415), whatever this means. There was no review in Schnell et al. In the discussion on diagnosing dysplasia in Chapter 14, we will consider the importance of variation in the pathologic diagnosis of high-grade dysplasia in terms of reproducibility and accuracy. There is a concern about whether the diagnosis of high-grade dysplasia is accurate and specific in the Schnell et al.'s study (22). It was not routinely subject to review by a second pathologist. This is not a problem if the pathologist had a proven record of accuracy in diagnosing high-grade dysplasia, but this is not documented by any validation studies. There is objective cause for questioning the pathologic diagnosis: the number of patients developing high-grade dysplasia among 1099 patients with Barrett esophagus is 79 (7.2%). This is a higher percentage than is encountered in most other studies in a primary care setting. In Table 10.1, the incidence of high-grade dysplasia in the cited studies was as follows: Miros et al. (13): 1 of 124 (0.8%); Hameetman et al. (14): 1 of 50 (2%); O'Connor et al. (15): 1 of 136 (0.7%); and Weston et al. (11): 8 of 108 (7.4%). In a tertiary care setting, which tends to drain complicated patients, the prevalence of patients with high-grade dysplasia tends to be much higher, as shown by Reid et al. (16): 76 of 327 (23.2%); and Montgomery et al. (17): 33 of 125 (26.4%).

There is a natural tendency for a pathologist working with a gastroenterologist who places patients with high-grade dysplasia on a stringent surveillance protocol to have a lower specificity of diagnosis of high-grade dysplasia than a pathologist working with a surgeon (such as Dr. Chandrasoma) for whom high-grade dysplasia is an indication for esophagectomy. The latter tends to demand stricter criteria and greater specificity than the former. We will show in Chapter 14 that the diagnosis of high-grade dysplasia is liable to marked variation between expert pathologists even in a study setting that is away from a patient care environment where only histologic criteria for high-grade dysplasia are being evaluated.

The rationale for the surveillance for high-grade dysplasia is that it prevents a surgical procedure with significant mortality and morbidity in those patients who do not progress to adenocarcinoma (or some other end point that precipitates surgery in a given institution such as Weston et al.'s multifocal high-grade dysplasia and dysplasia-associated lesion/mass). The critical question is one of numbers. It would seem that the risk is less worthwhile in Weston et al.'s study where 5 of 8 patients with unifocal high-grade dysplasia and all patients (unknown number) with multifocal high-grade dysplasia present at or within 6 months of the index endoscopy were candidates for surgery; the surveillance protocol prevented surgery in only 3 patients with high-grade dysplasia. In Reid et al. (16), 59% of patients with high-grade dysplasia developed cancer within 5 years; it is not known whether these patients required a more radical esophagectomy at the time of diagnosis of cancer as compared with the time of the original diagnosis of high-grade dysplasia. The esophagus was saved in 41% of the patients at the end of the follow-up period. In contrast, Schnell et al. (18) reported that 63 of 79 patients with high-grade dysplasia were without invasive cancer after a mean 7.3 years of follow-up. This is a much higher rate of prevention of esophagectomy and a surveillance protocol for high-grade dysplasia that potentially damages 16 of 79 patients by possibly precipitating a more radical surgery.

Surgeons are trained to think differently than gastroenterologists. While gastroenterologists opt for preserving the esophagi in those who will not get cancer even if it involves some risk to those that will develop cancer, surgeons tend to opt for removing all esophagi to optimize the survival in those who develop cancer by treating them at the earliest possible time. The latter is a well-recognized principle of oncologic surgery; patients have prostatectomy for very early prostate cancer despite evidence that many of these early cancers will not kill all patients within their natural lifetime; patients have excision of very early noninvasive breast cancers without "waiting for invasion to develop."

The crucial question here is the morbidity associated with the surgical procedure. Esophagectomy has the reputation of being a procedure with significant mortality and high morbidity. Surgeons will counter this argument by suggesting that the surgery required for high-grade dysplasia is not the standard esophagectomy. Transhiatal, vagal sparing esophagectomy, which can even be done via laparoscopy in some centers, has minimal risk of operative mortality and much lower morbidity than standard esophagectomy. As such, waiting for invasion to develop during surveillance may result in a conversion from a low-morbid transhiatal vagal sparing esophagectomy to a more radical esophagectomy with the associated increased risk and morbidity.

Conclusion: High-grade dysplasia is a high-risk lesion. The safest treatment is removal of the high-grade dysplasia, but this potentially hurts patients who are not destined to develop adenocarcinoma in the near future.

Can Other Factors in High-Grade Dysplasia Predict the Presence of Concurrence of Cancer or a Higher Risk within the Group?

The safest method for managing high-grade dysplasia is immediate esophagectomy. This is justified by many studies that show the presence of unexpected invasive cancer in the resected esophagus in at least a few of these patients. Any expectant protocol will place these patients at increased risk of death and extended esophagectomy with significantly higher morbidity. However, it is important to refine this recommendation in an attempt to stratify risk accurately among patients with high-grade dysplasia. The goal is to define characteristics that indicate a high risk and then subject these patients to esophagectomy while placing patients at lower risk for cancer on an appropriate surveillance protocol.

It appears that some authorities already use these variables to make this decision; in Weston et al.'s study, the presence of multifocal high-grade dysplasia and a dysplasia-associated lesion/mass are indications for abandoning surveillance and proceeding to surgery. This is not the case in Schnell et al. or Reid et al., where the end point for surveillance requires the detection of invasive cancer.

The following study (28) in our unit sheds some light on the two criteria of multifocal versus unifocal high-grade dysplasia and high-grade dysplasia with and without a visible lesion. The aim of this study was to determine the endoscopic, histologic, and demographic features associated with the presence of concurrent undetected invasive cancer in 31 patients who underwent esophagectomy for a diagnosis of high-grade dysplasia.

The presence of a visible lesion associated with the high-grade dysplasia was noted; these included nodules, polyps, plaques, and erosions. High-grade dysplasia was categorized as unilevel if the dysplasia was limited to a single level of biopsy and as multilevel if more than one level was involved (this is identical to Weston et al.'s definition of unifocal and multifocal). Patients were divided into two groups according to the presence or absence of cancer in the resected specimens, and these variables were compared.

The prevalence of coexisting invasive cancer in these patients was 45% (14 of 31). Of these 31 patients, 9 had a visible lesion. Seven of the 9 (78%) with a visible lesion compared to 7 of the 22 (32%) patients without a visible lesion had cancer in the resected specimens ($p = 0.019$). Of the 22 patients without a visible lesion, 10 had multilevel high-grade dysplasia and 12 had unilevel involvement. Six of 10 patients with multilevel high-grade dysplasia compared to 1 of 12 patients with unilevel high-grade dysplasia had cancer in the resected esophagus ($p = 0.009$).

We concluded that in these 31 patients with high-grade dysplasia, either a visible lesion on endoscopy or high-grade dysplasia in multiple biopsy levels was present in 19 patients (61%). Concurrent invasive cancer was present in 13 of these 19 patients (68%). Of the 12 patients who had no visible lesion and unilevel dysplasia, only 1 patient had invasive cancer in the resected specimen.

A single pathologist, Dr. Chandrasoma, interpreted the biopsies in all these patients. Biopsies in which a diagnosis of intramucosal adenocarcinoma was even suggested are not included in this group of patients. Dr. Chandrasoma tends to be aggressive in suggesting the presence of intramucosal adenocarcinoma in patients with high-grade dysplasia because of his bias that these patients are best treated by surgery. This group therefore represents patients who had normal gland contours, an absence of significant crowding and gland architecture distortion, and absent dilatation of glands with cytologic criteria of high-grade dysplasia.

This study suggests that Weston et al.'s protocol in which surveillance is abandoned immediately on discovery of a visible lesion or the presence of high-grade dysplasia at multiple levels is an excellent one. The 68% prevalence of invasive cancer in these patients makes it dangerous to place these patients on a surveillance protocol. The low risk of invasive cancer in the group with no visible lesion and the presence of unilevel high-grade dysplasia suggest that surveillance may be reasonable.

If one accepts that surveillance is a reasonable option for patients with high-grade dysplasia without a visible lesion and the presence of only unilevel dysplasia, it is still important to search for factors that may predict a high likelihood of cancer developing during surveillance in this group.

Literature Review

Reid BJ, Levine DS, Longton G, Blount P, Rabinovich PS. Predictors of progression to cancer in Barrett's esophagus: Baseline histology and flow cytometry identify low- and high-risk patient subsets. Am J Gastroenterol 2000;95:1669–1676.

In this study, the authors routinely subjected their biopsy specimens to flow cytometry in addition to histology, dividing each biopsy into two halves for this concurrent assessment. Of 75 patients with baseline high-grade dysplasia on the biopsy, 49 (65%) had baseline aneuploidy, increased 4N, or both. Patients who had aneuploidy or increased 4N in baseline (prevalent) and incident high-grade dysplasia had 3-year cancer incidences of 61% (95% confidence interval = 44, 78) and 58% (95% C.I. = 16, 99), respectively, compared to 42% (95% C.I. = 23, 68) and 7.7% (95% C.I. = 1.1, 43) for baseline and incident high-grade dysplasia without cytometric abnormalities, respectively. The authors concluded that these biomarkers assessed by flow cytometry were not statistically significant as predictors of cancer in either prevalent or incident cases of high-grade dysplasia.

It therefore seems that flow cytometric criteria are not useful in identifying a subgroup of patients with high-grade dysplasia who have a higher likelihood of developing cancer during surveillance. The only available strategy in these patients is regular endoscopy with extensive biopsy sampling.

Based on the available evidence, a strategy can be developed for the management of high-grade dysplasia (Fig. 10.12) that is aimed at minimizing mortality in patients who go on to develop cancer and maximizing the likelihood of retaining the esophagus in patients who do not progress to cancer. It must be recognized at the outset that these objectives cannot be met completely with any management strategy. The bias of any strategy should be to preserve life and prevent cancer; advanced esophageal adenocarcinoma is a terrible disease that must be prevented at any reasonable cost. It is also recognized that this is an evolving strategy. As alternatives to esophagectomy such as ablation and endoscopic mucosal resection are developed, they may replace esophagectomy as the appropriate treatment in this overall strategy.

Conclusion: There is near consensus that multifocal high-grade dysplasia and high-grade dysplasia with a visible lesion should have esophagectomy. Unifocal high-grade dysplasia can be treated with intense surveillance protocols or esophagectomy.

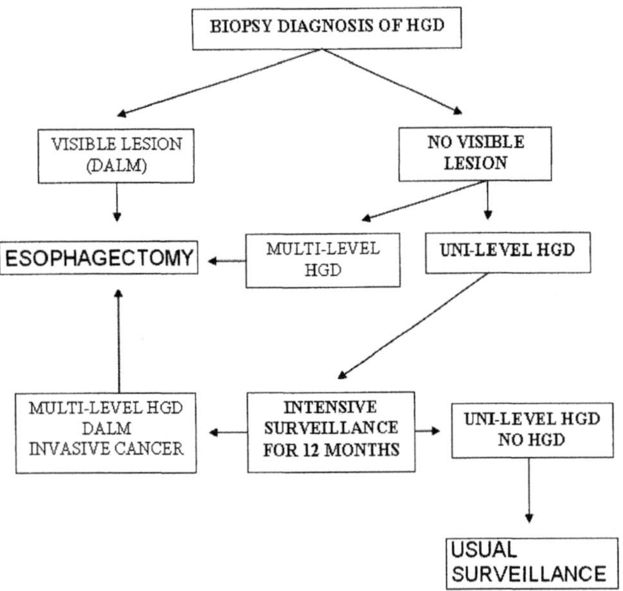

FIGURE 10.12 Algorithm for management of patients with high-grade dysplasia if it is decided that surgery is not the treatment of choice after a diagnosis. The risk of prevalent invasive cancer is so high in patients with multilevel high-grade dysplasia or high-grade dysplasia associated with a visible lesion that some intervention (surgery or ablation if the patient is not a suitable candidate for surgery) is essential.

DOES DYSPLASIA REVERSE?

Because the changes of low-grade dysplasia are mimicked closely by nondysplastic processes, there is likely to be a significant false positive diagnosis rate even with the most experienced pathologists. This is not pathologist error; it is an inherent error based on using dysplasia to define a genetic change. The two are not exactly concordant; it is possible to have genetic changes without cytologic abnormality, and many cytologic abnormalities may not be associated with the critical genetic changes that are the precursor of adenocarcinoma. This why it is anticipated that molecular tests will supplant dysplasia as the criterion that defines the level of progression in the cancer sequence.

In this setting, the concept of "reversal" of dysplasia after treatment must be treated with skepticism. True dysplasia is not likely to reverse if the genetic abnormality it is associated with is irreversible. As stated in Reid et al.'s (16) original criteria: "The diagnosis of low grade dysplasia and high-grade dysplasia is based on . . . criteria that suggest neoplastic transformation of the columnar epithelium" (p. 1670). Neoplastic transformation is, by the fundamental definition of neoplasia, an irreversible process. Reversal of dysplasia most likely means that a false positive diagnosis of dysplasia has disappeared. This is generally true only of low-grade dysplasia. With increasing severity of dysplasia, and particularly when the criteria for high-grade dysplasia are present, the number of conditions mimicking dysplasia decreases and false positive diagnoses become increasingly less frequent. As such, high-grade dysplasia is unlikely to reverse with any treatment other than ablation or surgical removal.

Conclusion: Theoretically, dysplasia does not reverse. Practically, reversal probably indicates a false positive microscopic diagnosis of true dysplasia.

THEORETICAL CONSIDERATIONS RELATING TO ADENOCARCINOMA DISTRIBUTION IN THE COLUMNAR-LINED ESOPHAGUS

The random occurrence of dysplasia within a segment of columnar-lined esophagus is logical if one considers carcinogenesis as an interaction between a target cell within the columnar-lined esophagus and a carcinogen in the refluxate. The target cell in intestinal metaplasia tends to be most prevalent in the most proximal part of the columnar-lined segment, but the carcinogen is most likely to be present closest to the gastroesophageal junction. The variation in the maximum target cell concentration (in the proximal region) and the maximum point of carcinogenic activity (in the distal region) would make the point of occurrence of dysplasia and carcinoma totally random and unpredictable.

Dysplasia and Carcinoma in Long-Segment Barrett Esophagus

A patient with long-segment Barrett esophagus is part of a diverse population based on (a) the length of columnar-lined esophagus, which can vary from 3 to 25 cm, and (b) the histologic composition of the columnar epithelium. The number of target cells at risk in the columnar-lined esophagus will depend on the number of foveolar-gland complexes that show intestinal metaplasia (Fig. 10.13). We have suggested that this be quantitated by standardized biopsy sampling (the IM Number or IM:OCM ratio; see Figure 9.29). In Chapter 9, we showed that intestinal metaplasia tends to begin in the most proximal region of the columnar-lined esophagus in patients who develop Barrett esophagus (29, 30). Once it has appeared at the squamocolumnar junction, the amount of intestinal metaplasia increases by contiguous downward extension

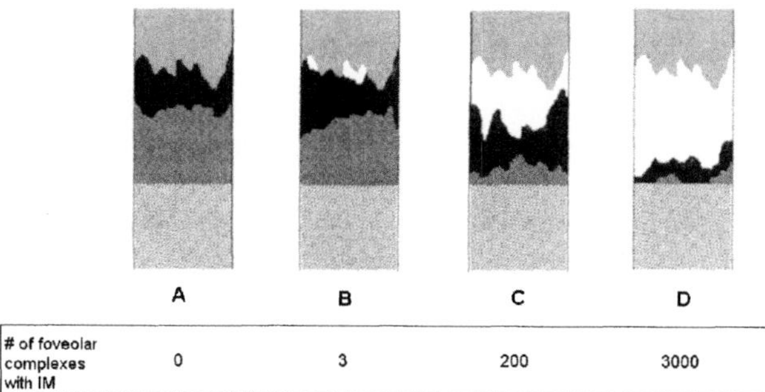

FIGURE 10.13 Variation in the amount of intestinal metaplasia in four patients with identical lengths of columnar-lined esophagus. The patients in parts (b), (c), and (d) are classified as Barrett esophagus of the identical length (long segment if >3 cm, short segment if <3 cm). The amount of intestinal metaplasia (solid white areas) progressively increases from part (a) to part (d). The number of foveolar complexes showing intestinal metaplasia (the "IM number") is shown as a possible method of risk assessment.

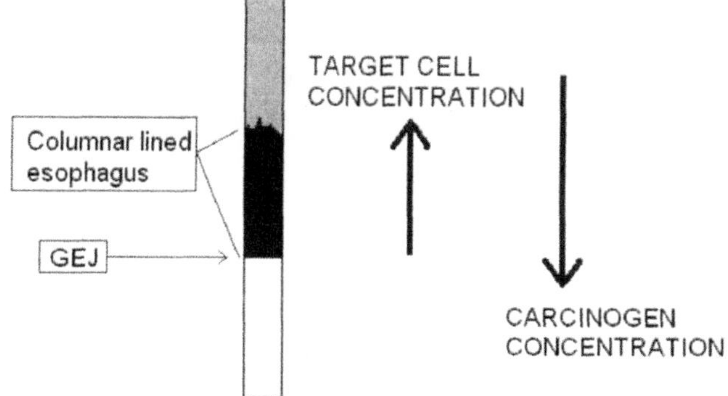

FIGURE 10.14 The tendency to produce intestinal metaplasia (the target cell for carcinogenesis) in a columnar-lined segment of esophagus is highest at the most proximal region. The carcinogenic environment is maximal in the most distal esophagus if it is assumed that the carcinogen is a molecule present in gastric contents.

from the squamocolumnar junction with little tendency to skip areas.

If one assumes that the carcinogen that is responsible for dysplasia and adenocarcinoma in intestinal metaplasia is a luminally acting carcinogen that is present in the gastric refluxate, it is likely that the concentration of the carcinogen is maximal in the most distal esophagus and becomes progressively lower as one proceeds higher in the esophagus (Fig. 10.14). Carcinogenic efficacy will cease at some point in the distal esophagus based on the amount of reflux in the patient, the concentration of the carcinogen in the gastric contents, and the potency of the carcinogen. It is possible for patients with intestinal metaplasia to have varying risk of cancer depending on how far down their intestinal metaplasia extends and how far up the carcinogenic effect extends in the esophagus (Fig. 10.15).

The distribution of dysplasia and adenocarcinoma in Barrett esophagus tends to favor the central and more distal part of a columnar-lined segment (2, 3; see Figs. 10.6, 10.9, and 10.10). In Hamilton et al.'s study of 61 esophagectomy specimens (2), the mean length of the tumor epicenter from the gastroesophageal junction was 2 cm; more than half of the patients with segments of Barrett esophagus 3 cm or greater (Fig. 10.6) had the epicenter of the cancer within 2 cm of the gastroesophageal junction. The fact that the most proximal region is relatively spared can only be explained by the fact that carcinogenic activity rarely reaches that high in the esophagus. The upper regions of long segments of columnar-lined esophagus, where intestinal metaplasia is most prevalent, may actually be a largely risk-free area for cancer because the carcinogenic activity simply does not reach that region in most people.

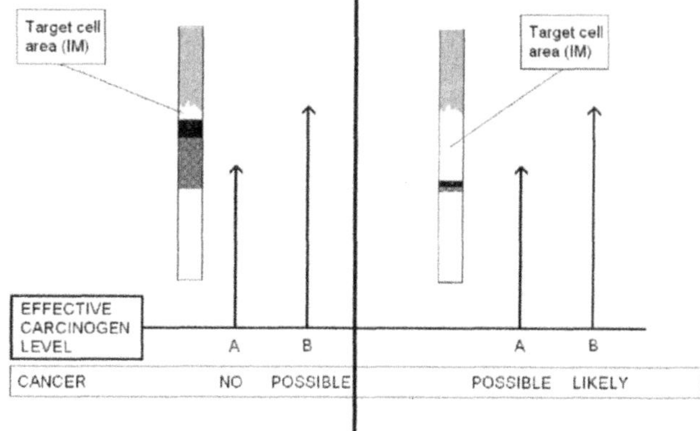

FIGURE 10.15 Two patients with identical lengths of columnar-lined esophagus but with different amounts of intestinal metaplasia (solid white). Two levels of effective carcinogen activity in the refluxate (a and b) are shown by vertical arrows. With level A effectiveness, the carcinogen never reaches the target cell (intestinal metaplasia) in the patient on the left; cancer does not occur in this patient. In the patient on the right, both levels of carcinogenic activity will access target cells with consequent cancer risk.

It is interesting to follow the life of a patient who develops reflux-induced columnar-lined esophagus (Fig. 10.16). At birth, with normal development, the entire esophagus is lined by squamous epithelium. With cumulative reflux, cardiac metaplasia occurs and progressively increases in length. The development of intestinal metaplasia in this columnar-lined esophagus is dependent on the length of the columnar-lined esophagus and the factors that favor intestinal metaplasia in the patient. After intestinal metaplasia develops, there is no further elongation of the columnar-lined segment. However, the intestinal metaplasia extends progressively downward to involve increasing amounts

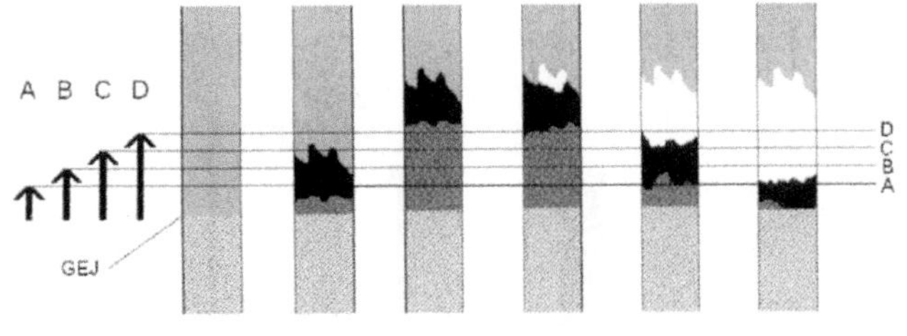

AGE (yrs)	0	20	30	40	50	60
CANCER RISK						
A	NO	NO	NO	NO	NO	NO
B	NO	NO	NO	NO	NO	YES
C	NO	NO	NO	NO	NO	YES
D	NO	NO	NO	NO	YES	YES

FIGURE 10.16 The lifetime change in a patient who develops long-segment Barrett esophagus. At birth, there is no columnar-lined esophagus. With reflux beginning early in life, the columnar metaplastic segment (black = cardiac mucosa; stippled = oxyntocardiac mucosa) progressively increases (to age 30 in this patient). Intestinal metaplasia (white) and a diagnosis of Barrett esophagus occurs at age 40. The amount of intestinal metaplasia progressively increases with age. Four levels of effective carcinogenic activity are shown by the vertical arrows (A,B,C,D). The possibility of cancer exists only when effective carcinogenic activity occurs in areas of intestinal metaplasia.

Paull et al, 1976
Chandrasoma et al, 2005
Target cell area (IM)
Target cell area (IM)
EFFECTIVE CARCINOGEN LEVEL

CANCER	NO	POSSIBLE
ACID SUPPRESSIVE DRUGS	INEFFECTIVE	VERY EFFECTIVE
pH	MORE ACID	LESS ACID

FIGURE 10.17 Diagram showing how changes in the distribution and amount of intestinal metaphasia resulting from acid suppression has altered the probability of cancer. For the same level of carcinogenic activity, the present acid-suppressed patient (*right*) is at risk whereas the historic patient without acid suppression (*left*) is not.

of the columnar-lined esophagus. The risk of cancer in this patient depends on the carcinogenic milieu of the esophagus, which progressively decreases as one proceeds up the esophagus. In Figure 10.16, four different carcinogenic levels (a, b, c, and d) are shown in four different patients. Dysplasia and cancer will occur in these patients whenever an area of intestinal metaplasia meets an adequately carcinogenic milieu (Fig. 10.16). This will explain the increasing risk of adenocarcinoma in Barrett esophagus with age as well as the typical distribution of adenocarcinoma in the more distal parts of the esophagus. It should be recognized that the population varies infinitely in how much columnar epithelium is generated by reflux, how much intestinal metaplasia occurs, and the level of carcinogenicity that exists in the gastric refluxate.

We have presented evidence in Chapter 9 (see Fig. 9.20) that the prevalence of intestinal metaplasia has changed significantly in modern times compared to the 1950s when Allison and Johnstone (31) first reported the entity and the 1970s (30). Today's patient with a long segment of columnar-lined esophagus tends to have intestinal metaplasia extending much lower in the esophagus than the typical patient in Paull et al. (3, 10). We have suggested that this change is likely due to the fact that the efficacy and wide usage of acid suppressive drugs has created a less acid environment in the esophagus that has favored the conversion of cardiac mucosa to intestinal metaplasia. For a given level of carcinogenic activity, today's patient is much more likely to be at risk than the patient in historic times (Fig. 10.17). This simple shift in the amount of intestinal metaplasia since the 1960s has the capability of explaining the increasing incidence of adenocarcinoma of the esophagus.

Dysplasia and Carcinoma in Short-Segment Barrett Esophagus

Adenocarcinomas arising in Barrett esophageal lengths of less than 3 cm must occur in the distal esophagus within 1 to 2 cm of the gastroesophageal junction. Many of these tumors have their epicenter at the junction and are classified as junctional adenocarcinomas. Some have their epicenter distal to the gastroesophageal junction and are called "adenocarcinomas of the gastric cardia" (5).

There is controversy as to whether the risk of adenocarcinoma in short-segment Barrett esophagus is similar or less than that for long-segment Barrett esophagus. We will consider this controversy in detail later. However, the fact that the distal part of the esophagus is favored as the site of cancer in long-segment Barrett esophagus suggests that this is the high-risk cancer environment. It should not be a surprise, then, to find that the cancer risk in short-segment Barrett esophagus is similar to that in long-segment Barrett esophagus (Fig. 10.18). In fact, when one com-

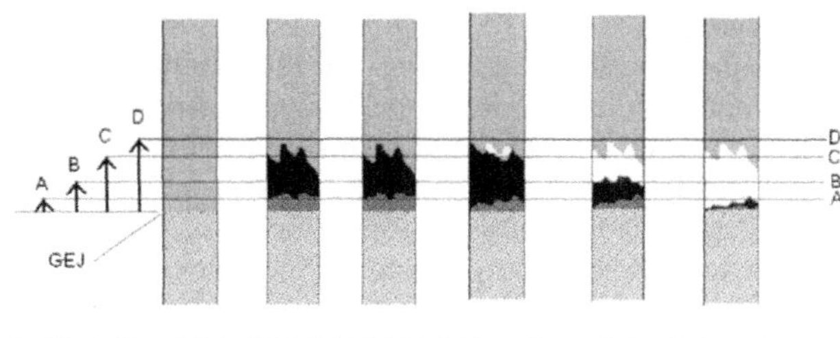

AGE (yrs)	0	20	30	40	50	60
CANCER RISK						
A	NO	NO	NO	NO	NO	YES
B	NO	NO	NO	NO	YES	YES
C	NO	NO	NO	YES	YES	YES
D	NO	NO	NO	YES	YES	YES

FIGURE 10.18 Lifetime change in a patient who develops short-segment Barrett esophagus. The figure is identical to Figure 10.16 except for the length of Barrett esophagus and the potential cancer risk with different levels of carcinogenic activity.

pares the time line diagram of patients with long- and short-segment Barrett esophagus, there is a suggestion that the number of patients at risk may be greater because of the higher concentration of intestinal metaplasia within the high carcinogen milieu of the more distal esophagus (Fig. 10.18).

Dysplasia and Carcinoma in Microscopic Barrett Esophagus

Microscopic Barrett esophagus is a condition that is presently ignored. By defining Barrett esophagus as the occurrence of intestinal metaplasia in a visible segment of columnar-lined esophagus, gastroenterologists fail to identify this group. This group has no endoscopic abnormality (i.e., no visible columnar-lined esophagus) until they develop a visible lesion caused by high-grade dysplasia or adenocarcinoma. This lesion will be distal to the endoscopic gastroesophageal junction, and these patients will be classified by present definition as "adenocarcinoma of the gastric cardia."

We suggested in Chapter 7 and will provide strong evidence in Chapter 11 that this region immediately distal to the endoscopically defined gastroesophageal junction actually represents reflux-damaged esophagus in patients with reflux disease (10). Lagergren et al. (5) showed that "adenocarcinoma of the gastric cardia" has an association with symptomatic reflux (discussed later). The increase in the incidence of adenocarcinoma of the gastric cardia has paralleled that of adenocarcinoma of the distal esophagus. The predominant viewpoint is that adenocarcinoma of the gastric cardia is part of the spectrum of reflux-induced Barrett esophagus.

The overall number of adenocarcinomas arising in the "gastric cardia" is similar to those arising in the distal esophagus. This should not be surprising, despite the fact that the maximum amount of target epithelium in microscopic Barrett esophagus is much less than in long- and short-segment Barrett esophagus. Microscopic Barrett esophagus (presently called "intestinal metaplasia of the cardia") is much more prevalent in the population than long- and short-segment Barrett esophagus (32). Also, the extent of intestinal metaplasia in microscopic Barrett esophagus is exposed to the highest concentration of carcinogenic molecules because of its distal location.

Conclusion: The distribution of adenocarcinoma that favors the distal esophagus appears largely dictated by a higher carcinogenic milieu in this region. This means that the shortest segments of Barrett esophagus have significant risk.

ASSESSMENT OF CANCER RISK IN BARRETT ESOPHAGUS

Accurate assessment of cancer risk in patients with Barrett esophagus must await the discovery of the components in the refluxate that are carcinogenic and the genetic mutational sequence involved in carcinogenesis. Until these factors are elucidated, different parameters must be evaluated to stratify risk of adenocarcinoma in Barrett esophagus.

Demographic Risk Factors

Adenocarcinoma of the esophagus is predominantly a disease of Caucasian males over the age of 40 who are in the higher socioeconomic strata of the population. The disease is common in Western Europe, North America, Australia, and New Zealand. These demographic groups are also those most likely to have gastroesophageal reflux disease of all types including Barrett esophagus.

Symptomatic Reflux

In a classic epidemiologic study from the Karolinska Institute in Sweden that was reported in the *New England Journal of Medicine* in 1999, Lagergren et al. (5) demonstrated that symptomatic gastroesophageal reflux was a risk factor for adenocarcinoma of the esophagus and gastric cardia.

Literature Review

Lagergren J, Bergstrom R, Lindgren A, Nyren O. Symptomatic gastroesophageal reflux as a risk factor for esophageal adenocarcinoma. N Engl J Med 1999;340:825–831.

In this nationwide, population-based, case control study in Sweden, the authors interviewed 189 patients with esophageal adenocarcinoma and 252 patients with adenocarcinoma of the cardia for information on the subjects' history of gastroesophageal reflux. Controls included 820 subjects from the general population and 167 patients with esophageal squamous carcinoma. The odds ratios were calculated by logistic regression.

Symptomatic gastroesophageal reflux was defined as the presence of heartburn, regurgitation, or both occurring at a frequency of at least once per week. To avoid reverse causality (i.e., to avoid collecting data on reflux caused by adenocarcinoma), the authors disregarded symptoms that had occurred less than 5 years before the interview. To further evaluate the effect of severity of symptoms, the symptoms were graded with scores for heartburn only (1 point), regurgitation only (1 point), heartburn and regurgitation combined (1.5 points), nightly symptoms (yes = 2 and no = 0), frequency of symptoms (scale of 0 = once a week, 1 = 2 to 6 times a week, 2 = 7 to 15 times a week, and 3 = more than 15 times a week). The location of the tumor and histologic type were verified by endoscopic data and histologic examination.

Among patients with symptomatic reflux, as compared with persons without symptoms, the odds ratios were 7.7 (95% C.I., 5.3 to 11.4) for esophageal adenocarcinomas. Persons with severe reflux (4.5 points or higher) had a risk of esophageal adenocarcinoma that was 20 times higher than asymptomatic persons. Among persons with reflux duration more than 20 years and severe symptoms, the adjusted odds ratio was 43.5 (95% C.I., 18.3 to 103.5) as compared to asymptomatic persons.

Symptomatic reflux was also associated with a smaller increased risk of adenocarcinoma of the cardia; the odds ratio was 2.0 (95% C.I., 1.4 to 2.9). The risk increased with the increasing severity of symptoms and the duration of symptoms but not to the same extent as esophageal adenocarcinoma. Persons who reported both severe reflux symptoms and long duration of symptoms had an odds ratio of 4.4 (95% C.I., 1.7 to 11.0) of developing adenocarcinoma of the cardia.

The authors justifiably concluded that there was a strong and probably causal relation between gastroesophageal reflux and esophageal adenocarcinoma and a weak association with adenocarcinoma of the cardia. These conclusions have wide acceptance.

Acid Suppressive Medication Use

Lagergren et al.'s methods do not include any questions directed at the use of acid suppressive medications by their patient or control group. It is therefore surprising to find the following statement in the body of the results section:

> We compared the risk of esophageal adenocarcinoma among persons who used medication for symptoms of reflux at least five years before the interview with that among symptomatic persons who did not use such medication. The odds ratio was 3.0 (95 percent confidence interval, 2.0 to 4.6) without adjustment for the severity of symptoms and 2.9 (95 percent confidence interval, 1.9 to 4.6) with this adjustment. (pp. 827–829)

These data suggest that patients who suppress acid and alter the pH of their gastric refluxate have a higher risk of developing adenocarcinoma (Table 10.2). This can be explained by two events:

1. The decreased acid milieu of patients using acid suppression favors the conversion of cardiac mucosa to intestinal metaplasia. We have suggested that there is strong evidence for this conclusion (see Chapter 9).

TABLE 10.2 Differences between Patients with Reflux with and without Acid Suppression

	Without acid suppression	With acid suppression
Gastric pH	1–3	3–6 (usually)
24-hour pH test	Abnormal	Normal
Symptoms of reflux	Present	Controlled
Erosive esophagitis	Present	Healed
Impedance test	Reflux present (acid)	Reflux present (less acid)
Intestinal metaplasia in CLE	Inhibited	Promoted
OCM in CLE	Promoted	Inhibited
Bile acids in stomach	Precipitated; inactive	Soluble; un-ionized; carcinogenic
Esophageal adenocarcinoma	Low risk	Increased risk

2. The change in the pH resulting from acid suppression potentiates the carcinogenic milieu of the gastric refluxate. We will consider this possibility later in the chapter.

Not all the data in the literature provide evidence for an increased incidence of adenocarcinoma with acid suppressive drug use. Farrow et al. (33) reported data from a large population-based case-control study. Patients newly diagnosed with esophageal adenocarcinoma (n = 293), esophageal squamous cell carcinoma (n = 221), gastric cardia adenocarcinoma (n = 261), or noncardia gastric adenocarcinoma (n = 368) in three areas with population-based tumor registries were the study groups. Controls (n = 695) were chosen by random digit dialing and from Health Care Financing Administration rosters. Data were collected using an in-person structured interview. The authors reported that the risk of esophageal adenocarcinoma increased with frequency of reflux symptoms; the odds ratio in those reporting daily symptoms was 5.5 (95% C.I. 3.2 to 9.3). Ever having used H2 blockers was unassociated with esophageal adenocarcinoma risk (odds ratio 0.9, 95% C.I. 0.5 to 1.5). The odds ratio was 1.3 (95% C.I. 0.6 to 2.8) in long-term (4 or more years) users but increased to 2.1 (95% C.I. 0.8 to 5.6) when use in the 5 years prior to the interview was disregarded. Risk was also modestly increased among users of antacids. Neither reflux symptoms nor the use of H2 blockers or antacids was associated with risk of the other three tumor types. The authors concluded that individuals with long-standing reflux disease are at increased risk of esophageal adenocarcinoma, whether or not the symptoms are treated with H2 blockers or antacids.

Farrow et al.'s study (33), despite finding no statistically significant association between acid suppression and esophageal adenocarcinoma, seems to suggest a trend that there may be such a relationship if the statistical power of the study population is increased. An odds ratio of 2.1 with a 95% confidence interval of 0.8 to 5.6 is certainly not proof that acid suppression is not related to adenocarcinoma. Also, the study tested only H2 receptor antagonists, not proton pump inhibitors.

While it is true that there is no absolute proof that acid suppressive drugs increase the risk of esophageal adenocarcinoma, there is certainly no proof that they do not have such an effect. This is troubling when one recognizes that the mainstay of all medical treatments of symptomatic reflux disease at the present time is acid suppression, that these drugs are available over the counter without prescription, and that these drugs are aggressively marketed to the population for the empiric treatment of heartburn.

Obesity

Several epidemiological studies have identified obesity, measured by a high body mass index (BMI), as an independent risk factor for adenocarcinomas of the esophagus and gastric cardia (34–41). Two of the positive studies indicate that BMI in the more distant past is a stronger predictor of risk than recent BMI (5, 34).

Bu et al. (42) in our unit studied the relationship of body mass index and the histologic types of epithelia found in columnar-lined esophagus in a case-control study. A total of 174 Barrett esophagus patients, 333 patients who had cardiac mucosa without intestinal metaplasia, and 274 controls (patients whose biopsies showed neither intestinal metaplasia nor cardiac mucosa) were included in this study. Multivariate logistic regression methods were used to estimate odds ratios for Barrett esophagus or cardiac mucosal metaplasia associated with body mass index. Linear regression analysis was employed to examine the relationship between the length of columnar-lined esophagus and body mass index.

A strong dose-response relationship between body mass index and the presence of Barrett esophagus was observed ($p = 0.0004$) (Table 10.2). The multivariate-adjusted odds ratio for Barrett esophagus comparing obese (BMI $\geq 30\,kg/m^2$) to lean individuals (BMI $< 22\,kg/m^2$) was 3.3 (95% C.I. = 1.6–6.7) (Table 10.3).

A substantial, although more modest, dose-dependent relationship was observed for the relationship between body mass index and risk of cardiac mucosal metaplasia ($p = 0.03$) (Table 10.3). The multivariate-adjusted odds ratio for obese (BMI $\geq 30\,kg/m^2$) versus lean individuals (BMI $< 22\,kg/m^2$) was 1.8 (95% C.I. = 1.04 to 3.1) (Table 10.3).

Linear regression analysis showed that the length of columnar-lined esophagus (adjusted for height) increased with increasing body mass index ($p = 0.04$) in the 103 cases with measured columnar-lined esophagus (86 Barrett esophagus cases and 17 cases of cardiac mucosa without Barrett esophagus). A positive relationship was also found between unadjusted length of columnar-lined esophagus and weight ($p = 0.03$).

The stepwise increased association between body mass index and cardiac metaplasia and Barrett esophagus shown in this study suggests that obesity acts at an early stage of the reflux → adenocarcinoma sequence. It suggests that obesity promotes reflux and causes squamous transformation into cardiac mucosa of increasing length. Other factors that lead to intestinal metaplasia, dysplasia, and adenocarcinoma would then act independently, affecting later stages in the sequence.

TABLE 10.3 **Body Mass Index and Risk for Barrett Esophagus and Cardiac Mucosal Metaplasia**

Variables	Cases	Controls	Age- and sex-adjusted odds ratio and 95% C.I.	Trend p-value
Barrett esophagus				
BMI (kg/m^2)[a]				
Quartile I (<22–low)	18	61	1 (reference)	0.0004
Quartile II (22–24.9)	37	69	1.2 (0.6–2.5)	
Quartile III (25–29.9)	71	100	1.6 (0.9–3.1)	
Quartile IV (>30–high)	48	44	3.3 (1.6–6.7)	
Cardiac mucosal metaplasia				
BMI (kg/m^2)[a]				
Quartile I (<22–low)	56	61	1 (reference)	0.03
Quartile II (22–24.9)	77	69	1.2 (0.7–2.0)	
Quartile III (25–29.9)	127	100	1.3 (0.8–2.1)	
Quartile IV (>30–high)	73	44	1.8 (1.04–3.1)	

[a]Quartile cut points per World Health Organization criteria.

Hiatal Hernia

Weston et al. (11), in their study of 108 patients with Barrett esophagus, showed that the incidence of multifocal high-grade dysplasia or invasive adenocarcinoma was associated with the presence of a hiatal hernia, the length of Barrett esophagus and the presence of dysplasia. Hiatal hernia is a well-recognized factor in all elements of gastroesophageal reflux disease including adenocarcinoma.

Intestinal Metaplasia of the Esophagus (Barrett Esophagus)

Once a patient develops Barrett esophagus, ethnic, age, and sex differences become less important, being dwarfed by the importance of Barrett esophagus as a risk factor. In patients with Barrett esophagus, therefore, the risk indicators are mainly determined by factors related to the intestinal metaplasia that is the precursor lesion for adenocarcinoma.

An individual who has intestinal metaplasia has a low overall risk of cancer; the best estimate of cancer incidence in Barrett esophagus is 0.5% per year (22). In most surveillance studies, the incidence of cancer in Barrett esophagus is very low when expressed as numbers per patient-years of follow-up. In Weston et al.'s study (11) of 108 patients with Barrett esophagus without multifocal high-grade dysplasia, the incidence of cancer was 1 per 71.9 patient-years. If the 41 patients in this study with short-segment Barrett esophagus were excluded, the cancer incidence increases to 1 per 49.4 patient-years. If patients with unifocal high-grade dysplasia were removed along with patients with short-segment Barrett esophagus, the incidence becomes 1 per 76.5 patient-years.

However, when looked at from the opposite perspective, every patient who develops adenocarcinoma of the esophagus comes from the pool of patients who have Barrett esophagus. There is no other pathway for adenocarcinoma of the esophagus.

This combination of facts makes this a difficult disease to treat from a practical standpoint. While it is impossible to ignore an individual patient who has intestinal metaplasia because of the proven cancer risk, it stretches resources to keep all patients with intestinal metaplasia under surveillance, and the cost-effectiveness of such surveillance protocols becomes questionable.

The carcinogenic mechanism in reflux-induced adenocarcinoma is unique. Let us postulate that every patient has a unique carcinogenic potential for reflux-induced adenocarcinoma of the esophagus resident in his or her gastric contents (Fig. 10.19). The carcinogenic potential of the refluxate is irrelevant in patients who have no intestinal metaplasia, irrespective of the severity of reflux-induced damage. Patients who have a high carcinogeniuc refluxate tend to progress rapidly through increasing dysplasia to cancer as soon as intestinal metaplasia develops. On the other hand, patients with a low carcinogenic milieu tend to have stable reflux-induced columnar metaplasia and intestinal metaplasia without progressing to dysplasia for long periods and with a very low incidence of cancer. Clinical experience supports such a hypothesis. At one extreme are patients with a long history of symptom-

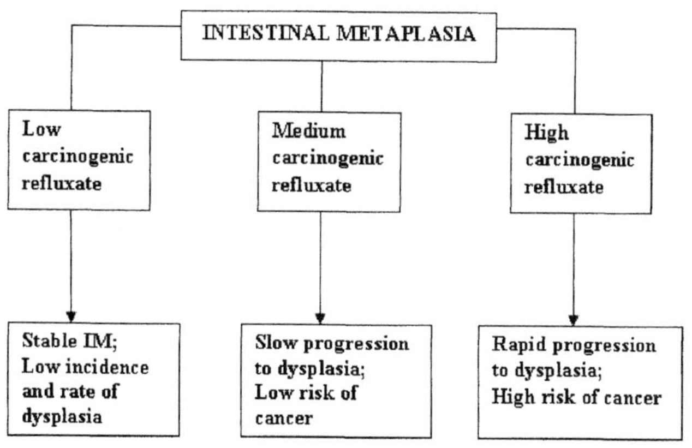

FIGURE 10.19 The probability of changes associated with severe reflux in patients who have different levels of carcinogenic activity in their refluxate. Rate of progression to dysplasia and cancer are depicted as depending on the level of carcinogenicity with intestinal metaplasia depicted as an independent event.

atic reflux disease and long-segment Barrett esophagus who remain stable without progressing to dysplasia and cancer, suggesting that this may be a population in whom reflux damage is severe but the carcinogenicity of the gastric refluxate is low. At the other extreme, many patients who present with adenocarcinoma have never had symptoms of reflux and have very short segments of Barrett esophagus. Because both symptoms and length of columnar metaplasia are directly related to the severity of reflux, these patients likely have relatively low reflux damage but a high carcinogenicity in their refluxate. If the major factor for the development of adenocarcinoma in reflux disease is the carcinogenic potential of the refluxate, it is not surprising that present methods of surveillance have little impact on the incidence of esophageal adenocarcinoma.

Long- and Short-Segment Barrett Esophagus

The division of Barrett esophagus into long- and short-segment diseases was the product of historical error. Until the early 1990s, the designation of Barrett esophagus was limited by the use of an incorrect definition to patients with columnar-lined esophagus greater than 3 cm with intestinal metaplasia on biopsy. When intestinal metaplasia was found to be present in the distal 3 cm of the esophagus, this was called "short-segment Barrett esophagus." The intent was to distinguish it from the classical disease because its significance in terms of its risk of malignancy was not established until later. The present consensus is that the division of Barrett esophagus into short- and long-segment diseases is artificial and that they represent a continuum without any justification for separation by length (7).

There is some conflict in the literature regarding the risk of malignancy in short-segment Barrett esophagus. In Weston et al.'s study, 10 of 108 patients developed multifocal high-grade dysplasia or adenocarcinoma. This broke down to zero of 51 patients with a Barrett length of less than 2 cm, 1 of 28 patients with a length of 2 to 6 cm, and 9 of 39 patients with a length greater than 6 cm. Though the numbers are small, they suggest that there is a relationship between the length of Barrett esophagus and the risk of developing cancer.

In 1993, Menke-Pluymers et al. (43), in a retrospective case control study that compared 96 patients with benign Barrett esophagus and 62 patients with adenocarcinoma, referred to the Rotterdam Esophageal Tumor Study Group between 1978 and 1985 and reported that the length of columnar-lined esophagus correlated with cancer risk. They found that a doubling of the length of columnar-lined esophagus increased risk 1.7 times. In 2002, Avidan et al. (44) also reported that the risk of both high-grade dysplasia and adenocarcinoma increased with the increasing length of Barrett esophagus.

These studies are contradicted by the data in the largest and best-controlled study by Rudolph et al. (26). This was a prospective cohort study from the University of Washington, Seattle, which consisted of a total of 309 patients with Barrett esophagus. Barrett segment length was defined as "the distance between the esophagogastric junction and the squamocolumnar junction. The esophagogastric junction was defined as the endoscopic lower esophageal sphincter or, if this was not apparent, the location at which the tubular esophagus joined the proximal margin of the gastric folds" (26, p. 613). The patients were monitored for progression to adenocarcinoma by repeated endoscopy with biopsy for an average of 3.8 years. Cox proportional hazards analysis was used to calculate adjusted relative risks and 95% confidence intervals.

In the full cohort of 309 patients, the incidence of cancer was somewhat greater among those with longer segments. When all baseline histologic diagnoses were included, the incidence of cancer was 2.5 per 100 person-years for a Barrett length of less than 3 cm, 2.8 for a length of 3 to 6 cm, 3.2 for a length of 7 to 10 cm, and 7.0 for a length greater than 10 cm. After adjustment for histologic diagnosis at study entry, the authors reported that segment length was not related to risk for cancer in the full cohort ($p > 0.2$ for trend). When patients with high-grade dysplasia at baseline were excluded, however, a nonsignificant trend was observed; based on a linear model, a 5-cm difference in segment length was associated with a 1.7-fold (95% C.I., 0.8-fold to 3.8-fold) increase in cancer risk. Important was also their finding that 7 of 83 patients with a Barrett length of less than 3 cm developed cancer.

Based on their data, the authors concluded that the risk for esophageal adenocarcinoma in patients with short-segment Barrett esophagus was not substantially lower than that in patients with longer segments. Although the results suggest a small increase in risk for neoplastic progression with increasing segment length, this did not reach statistical significance. They concluded that "until further data are available, the frequency of endoscopic surveillance should be selected without regard to segment length" (26, p. 612).

This study provided evidence that the development of adenocarcinoma is either only slightly related or not related to the length of Barrett esophagus. More important, it showed a significant number of cancers arising in patients with short-segment Barrett esophagus. Based on this study, the consensus viewpoint at present is that there should be an identical surveillance protocol for all patients with Barrett esophagus, irrespective of length.

Amount of Intestinal Metaplasia within Columnar-Lined Esophagus

The expectation that there should be an increase in the incidence of cancer with increasing lengths of Barrett esophagus is somewhat misplaced. It is caused by the common misconception that in patients with Barrett esophagus, the entire columnar-lined esophagus is lined by intestinal metaplasia. Barrett esophagus is diagnosed by the presence of *any* intestinal metaplasia within it; after it is so diagnosed, its length is defined by the length of columnar-lined esophagus as determined by endoscopic measurement. Histologic examination shows that there is often a considerable range in the amount of intestinal metaplasia and the density of goblet cells within areas of intestinal metaplasia in different patients with an equal length of columnar-lined esophagus. If the only epithelium at risk for cancer in columnar-lined esophagus is the intestinal type, this variation makes it likely that there will be an absence of a strong correlation between cancer incidence and the length of Barrett esophagus (Fig. 10.13). Two segments of columnar-lined esophagus of equal length will be expected to have greatly different risks if they had predominant intestinal metaplasia or oxyntocardiac mucosa (Fig. 10.13).

We suggested in Chapter 9 that it is logical that the risk and incidence of cancer in Barrett esophagus will be best predicted by some estimate of the amount of intestinal metaplasia (see Fig. 9.29). There are no studies in the literature that assess cancer risk and incidence as a measure of the amount of intestinal metaplasia. This is a serious omission resulting from a failure to understand the cellular nature of columnar-lined esophagus.

Dysplasia

The recognition of morphologic changes of dysplasia is presently the most accurate method of stratifying patients with intestinal metaplasia with regard to their cancer risk. Dysplasia is believed to progress from low grade through high grade before invasive adenocarcinoma develops. The rate of progression of dysplasia and the frequency with which different grades of dysplasia progress to adenocarcinoma probably vary significantly in different people. However, as a general rule, higher grades of dysplasia are associated with a greater likelihood and imminence of adenocarcinoma.

Because dysplasia is not endoscopically visible and is a random event within the intestinal metaplastic segment, the only way to detect it is by a systematic biopsy protocol with adequate sampling of the entire region. This usually requires four-quadrant biopsies at 1 to 2 cm intervals within the involved segment. Though this is a relatively small sample, evidence suggests that it is effective in detecting dysplasia. Any protocol less than this has the potential to miss dysplasia by sampling error.

The diagnostic criteria for dysplasia will be considered in Chapter 14. These criteria are more difficult to apply with reproducibility in the lower grades but become more certain with the higher grades. However, it must be recognized that the morphologic assessment of dysplasia, though the best present indicator of cancer risk in Barrett esophagus, is far from perfect. Molecular determinants of cancer risk are urgently needed to improve patient care in this situation.

CLINICALLY USEFUL MOLECULAR TESTS

The final answer to the question of predicting cancer risk in Barrett esophagus awaits elucidation of specific genetic changes in the process of carcinogenesis. This is still in a relatively primitive stage of understanding, and we are probably left with dysplasia as the best indicator of cancer risk for many years to come.

As a molecular definition of carcinogenesis becomes better understood, it is widely expected that the detection of specific genes or patterns of multiple gene expression as seen with gene array studies will supplant histologic grading of dysplasia as the preferred method of predicting the risk of malignancy in patients with Barrett esophagus. This is, however, by no means certain. Even when the carcinogenic pathway is elucidated in the future, it is not certain that genetic testing will replace dysplasia as a diagnostic criteron. In colorectal carcinoma, for example, the genetic changes associated with carcinogenesis have been well known for many years, but biopsy assessment of adenomas and dysplasia in ulcerative colitis still remains the mainstay of clinical care. Genetic abnormalities are evaluated only for determining familial risk in patients with familial types of colon cancer such as familial polyposis coli and hereditary nonpolyposis colorectal carcinoma. One of the reasons for this is the expense associated with molecular testing compared with histologic examination of a biopsy. The increasing use of molecular markers is expected to drive down costs associated with such tests, but this must be matched by proof that molecular criteria for diagnosis has higher specificity and sensitivity than grading of dysplasia in a biopsy.

No universally accepted molecular markers are mandated for clinical diagnostic use in Barrett esophagus. Molecular studies that assess genetic markers remain research modalities. These include specific markers as well as combinations of genes that are detected by testing panels of several genes by gene array technology.

Patients with Barrett esophagus are managed entirely by histologic grading of dysplasia, which represents today's gold standard. When the diagnosis of high-grade dysplasia is made on morphologic examination, these molecular markers are not necessary because the morphologic diagnosis is not altered by any finding (i.e., negative flow cytometry, negative p53, or absent aberrant Ki67 staining do not negate the diagnosis of high-grade dysplasia). When there is no evidence of any dysplasia, these molecular markers are not tested; the frequency of positivity is very low, its significance is unknown, and the expense is enormous. *The only situations where these markers may be clinically useful at the present time are (a) in patients with high-grade dysplasia, where their presence may indicate that it is dangerous to maintain patients on surveillance without surgery, and (b) in patients with low-grade dysplasia.* Although the expression of any of these markers does not alter the diagnosis of low-grade dysplasia, the presence of aneuploidy, an increased 4N fraction, strong and specific p53 expression, and aberrant Ki67 patterns may provide independent indicators of possible danger and heighten surveillance of this subgroup. None of these indicators is presently used, but there is some suggestion in the literature that they may be useful (Table 10.4).

Aneuploidy and Increased (Greater Than 6%) 4N Fraction

The detection of aneuploidy and increased 4N fraction greater than 6% has been reported to be of value in predicting cancer in a study by Reid et al. from the University of Washington, Seattle.

Literature Review

Reid BJ, Levine DS, Longton G, Blount P, Rabinovich PS. Predictors of progression to cancer in Barrett's esophagus: Baseline histology and flow cytometry identify low- and high-risk patient subsets. Am J Gastroenterol 2000;95:1669–1676.

This study evaluated the significance of two flow cytometric abnormalities—aneuploidy and increased (>6%) 4N—in 327 patients with Barrett esophagus followed prospectively for a mean of 3.9 years (median 2.4 years) repre-

TABLE 10.4 Molecular Tests That May Supplement the Grade of Dysplasia in Assessing Risk of Cancer in Barrett Esophagus

Test	Method	Indication	Comment
Aneuploidy	Flow cytometry of biopsy	Low-grade dysplasia	Positive test = increased risk
Increased (>6%) 4N fraction	Flow cytometry of biopsy	Low-grade dysplasia	Positive test = increased risk
p53	Immunohistochemistry	Low-grade dysplasia	Strong positive = increased risk
p53	Immunohistochemistry	High-grade dysplasia	Strong positive = surgery over surveillance
Aberrant Ki67 expression	Immunohistochemistry	Low-grade dysplasia	Positive = increased risk

senting 1200 patient-years of follow-up. One biopsy from each level of columnar-lined esophagus was divided for histology and flow cytometry.

In 322 patients who had baseline flow cytometry, 41 cancers were diagnosed, including 34 within 5 years of the baseline endoscopy. Of the 322 patients, 241 (75%) had neither baseline aneuploidy nor increased 4N fractions. Twelve of 241 patients in this group progressed to cancer; 48 of 322 (15%) had baseline increased 4N fraction; and 18 of 48 progressed to cancer. The relative risk of cancer for patients with increased 4N fraction compared to those without was 7.5 (95% C.I. = 4.0, 14) ($p \leq 0.001$); 53 of 322 (16%) patients had aneuploidy in the baseline biopsy; 17 of 53 developed cancer. The relative risk of cancer in patients with baseline aneuploidy compared to those without was 5.0 (95% C.I. = 2.7, 9.4) ($p \leq 0.001$).

When the findings on histology were combined with flow cytometric abnormalities, researchers found out the following: (a) Of 120 patients who were negative for dysplasia, 110 (92%) had neither aneuploidy nor increased 4N fraction. Only two progressed to cancer, compared with 1 of 10 patients with aneuploidy or increased 4N ($p = 0.05$). (b) Of 83 patients whose biopsies were indefinite for dysplasia, 72 (87%) had neither aneuploidy or increased 4N fraction. Only 1 of 72 of these patients progressed to cancer compared with 1 of 11 with aneuploidy or increased 4N fraction ($p = 0.26$). (c) Of 44 patients with low-grade dysplasia on biopsy, 33 (75%) had neither aneuploidy nor increased 4N fraction. None of these 33 patients progressed to cancer, compared with 4 of 11 who had aneuploidy or increased 4N fraction ($p \leq 0.001$).

These data show that the flow cytometrically defined abnormalities of aneuploidy and increased 4N fraction are valuable in identifying a subset of patients with low-grade dysplasia who are at increased risk for cancer. The data are not so convincing for patients who are negative for dysplasia or indefinite for dysplasia. In a patient with low-grade dysplasia, the absence of both aneuploidy and increased 4N fraction strongly suggests that cancer is much less likely to develop.

Flow cytometry requires special handling of specimens at the time of biopsy and considerable added expense. This precludes its use in every case in standard practice outside the academic environment. However, the results suggest that this test is justified in patients with low-grade dysplasia. The protocol can be adjusted such that patients with a diagnosis of low-grade dysplasia can have special arrangements made for appropriate specimen handling to permit flow cytometry at the time of their next surveillance endoscopy. The presence of aneuploidy or increased 4N fraction can precipitate heightened surveillance in this high-risk group, whereas the absence of both these parameters can reduce intensity of surveillance (Table 10.4; Fig. 10.20).

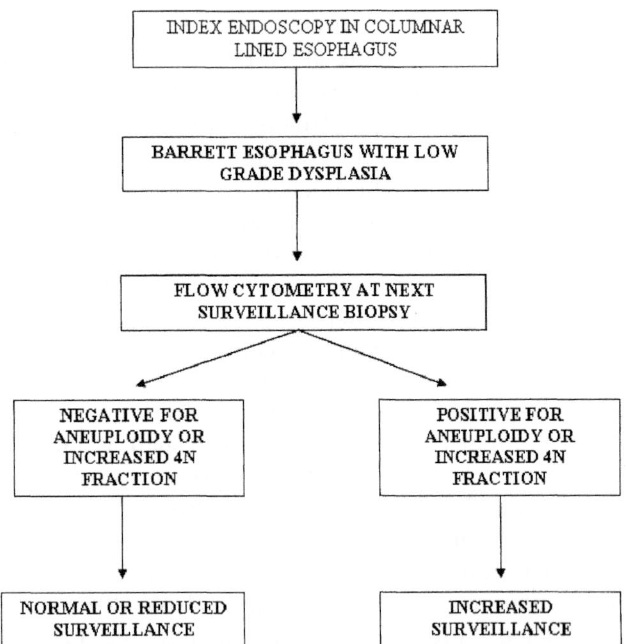

FIGURE 10.20 Algorithm for management of patients with low-grade dysplasia using flow cytometric data to direct differences in intensity of surveillance.

Expression of p53 by Immunohistochemistry

The p53 gene is a ubiquitous tumor suppressor that is involved in cell cycle regulation in cells. Mutations of p53 have been shown to be involved in many types of malignant neoplasm. The p53 gene normally responds to DNA damage occurring in a cell during replications. DNA damage activates a mechanism that upregulates p53, which acts via p53 regulated factors such as p21, bcl-2, and bax-1 to cause an arrest of the cell cycle in the G1 phase and either produce apoptosis or DNA repair. The action of p53 in the normal cell is therefore to maintain normal genetic stability in the face of DNA damage. Mutation of the p53 gene results in a nonfunctional p53, which fails to arrest mitosis, fails to repair DNA, and fails to produce apoptotic death of the cell, setting the stage for continued proliferation of the genetically damaged cell and potentially causing uncontrolled neoplastic proliferation.

Many types of mutations in the p53 gene complex exist. Some of these can be detected by simple immunohistochemical staining for p53 in routine biopsy material. Strong expression of p53 in a cell is evidence of the presence of some type of mutated p53 within the cell. Wild-type p53 normally upregulated in cells does not produce strong positivity with p53 staining by immunohistochemistry. While genetic testing provides a more accurate assessment of the exact nature of the p53 abnormality, immunoperoxidase staining for p53 is a simple and relatively cheap method that has good correlation with p53 mutation.

Kim et al. (45) reported that p53 expression was present in over 80% of high-grade dysplasia and adenocarcinoma compared with highly significantly lower expression levels in nondysplastic Barrett esophagus and low-grade dysplasia. Similar findings have been reported in many other studies. In a review by Ireland et al. (46), in which researchers collected data from multiple studies, the p53 positivity rate in the different patient groups were as follows: (a) Barrett esophagus without dysplasia: 15 of 283 (5.3%); (b) low-grade dysplasia: 31 of 156 (19.9%); (c) high-grade dysplasia: 61 of 93 (65.6%); and (d) invasive adenocarcinoma: 143 of 221 (64.7%).

The value of p53 as an independent indicator of higher risk in patients with low-grade dysplasia was evaluated by Weston et al. (47), who followed 48 Barrett esophagus patients with low-grade dysplasia for a mean of 41.2+/−22.5 months. Five of 48 patients progressed to multifocal high-grade dysplasia or cancer; 12 of 48 had persistent low-grade dysplasia, and 31 of 48 regressed to no dysplasia. Immunoperoxidase staining for p53 in the index biopsy was positive and localized to areas of low-grade dysplasia in 4 of 31 patients who regressed, 3 of 12 who persisted, and 3 of 5 who progressed to developing multifocal high-grade dysplasia or cancer. The authors concluded that p53 expression in low-grade dysplasia is a significant risk factor for progression to higher grades of dysplasia and cancer and can be used to identify a population of low-grade dysplasia patients with a higher risk that required increased surveillance or intervention (47).

Our yet unpublished experience is similar. The p53 gene is frequently expressed in high-grade dysplasia and adenocarcinoma in a strong and highly specific manner with only the affected cells staining. In cases where we performed the p53 stain, we reported the presence of p53 positivity and suggested that it may represent an independent risk indicator in addition to the reported high-grade dysplasia (Table 10.2). This conclusion is open to interpretation but may precipitate surgery over intense surveillance in a borderline patient with unifocal high-grade dysplasia. We have rarely encountered strong positive staining in low-grade dysplasia but would agree that if it is found, it is a cause for concern that the patient requires heightened surveillance. We consider the data as inconclusive but justify the performance of immunoperoxidase staining for p53 in patients in whom dysplasia is encountered.

Ki67 Patterns and Level of Expression

Proliferative patterns in esophageal columnar epithelia provide useful information. This is easily assessed by immunoperoxidase staining of routine biopsy material using Ki67, an antibody that detects a nuclear antigen that is expressed in cells that are in the cell cycle (i.e., in all phases of the cell cycle—G1, S, G2—other than the resting G0 phase). In a study of patterns of Ki67 expression in metaplastic epithelia of the esophagus, we showed that two patterns were recognizable (48). The first was an expanded normal pattern where the number of proliferative cells was increased but their distribution, with restriction to the deep and midfoveolar region, was normal (Figs. 10.3 and 10.21). A pattern of expanded proliferation was seen in all nine cases of low-grade dysplasia and 12 of 16 areas of high-grade dysplasia. Expanded proliferation was therefore present in 21 of 25 (84%) areas of dysplastic epithelium, which was a significantly higher frequency than nondysplastic columnar epithelia of all types.

Increased proliferation also occurs in non-neoplastic reactive epithelia as a result of compensation for cell damage. The overlap between the increased proliferation that occurs in reactive change and dysplasia is such that it cannot be used from diagnosis in an individual case. Also, as seen in this study, there is no significant quantitative difference between proliferative activity in low- and high-grade dysplasia.

Two modifications using proliferative studies may prove useful in discriminating reactive from dysplas-

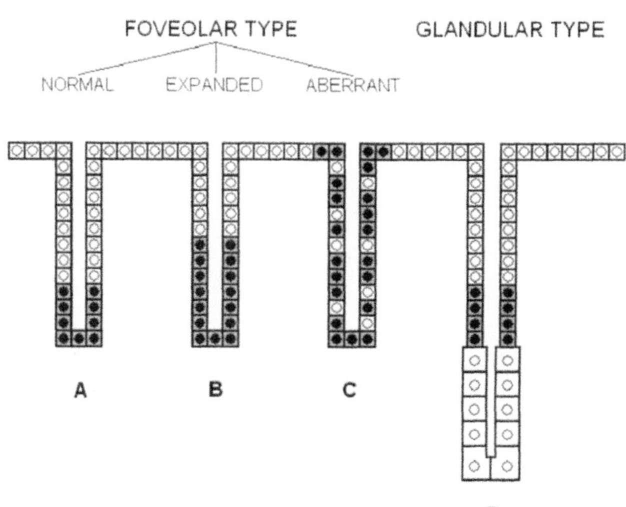

FIGURE 10.21 Diagram of Ki67 staining patterns in columnar epithelia, showing a normal pattern (a and d), an expanded normal pattern (b), and an aberrant pattern (c). The expanded normal and aberrant patterns are only shown for the foveolar epithelium, but they are the same for glandular epithelium.

tic and neoplastic Barrett esophagus. First, as reported by Whittles et al. (49), assessment of the proliferation to apoptosis ratio shows significant differences between nonintestinalized columnar-lined esophagus, nondysplastic Barrett esophagus, dysplastic Barrett esophagus, and adenocarcinoma. This suggests that increased proliferation occurs as compensation to increased apoptosis in normal epithelia, producing a balanced ratio. In contrast, with increasing dysplasia the proliferation is in excess of apoptosis, suggesting that it is a primary, genetics-driven occurrence rather than a compensatory phenomenon.

The other method is the use of the distribution of the proliferative cells as shown by their Ki67 positivity within the epithelium (48). We defined an aberrant pattern of Ki67 expression in columnar-lined esophagus, hypothesizing that the mutation-driven proliferation of dysplastic and neoplastic cells is more likely to show abnormal maturation than normal (Fig. 10.22). Aberrant proliferation was defined by the presence of Ki67-positive cells in the surface epithelial layer or the superficial third of the foveolar pit (Fig. 10.20). None of the 43 areas of esophageal oxyntocardiac mucosa, 10 of 45 areas of cardiac mucosa, 7 of 41 (17.1%) areas of nondysplastic intestinal metaplasia, and 3 of 9 (33.3%) areas of intestinal metaplasia with low-grade dysplasia showed an aberrant Ki67 proliferation pattern. In contrast, the 16 areas with high-grade dysplasia showed an aberrant pattern with surface and superficial foveolar positivity in 14 of 16 (87.5%) areas, indicating a frequency of aberrant proliferation that was significantly higher than low-grade dysplasia and all nondysplastic columnar epithelia ($p \leq 0.001$). The pattern in invasive carcinoma was always aberrant and quantitatively more aberrant than high-grade dysplasia (Fig. 10.23).

We concluded from these data that while Ki67 staining is of no practical value in differentiating low-grade dysplasia from reactive changes, there is a strong suggestion that it may be useful in differentiating low-grade from high-grade dysplasia. The difference in the frequency of aberrant proliferation as defined in this study between low-grade and high-grade dysplasia was extremely significant.

The fact that aberrant proliferation is much more commonly seen in high-grade dysplasia as compared with low-grade dysplasia does not mean that this pattern is of diagnostic value in an individual case where the morphologic features are borderline between low- and high-grade dysplasia. We now perform Ki67 stain routinely in cases of Barrett dysplasia and are testing various combinations of proliferative activity and aberrant proliferation in an attempt to devise criteria that will help in the practical differentiation of low- from high-grade dysplasia. These studies are in progress and have not yet shown definitive findings; however, the results show considerable differences between low-grade and high-grade dysplasia and make us optimistic that Ki67 staining patterns may provide valuable criteria to discriminate low-grade from high-grade dysplasia in the future.

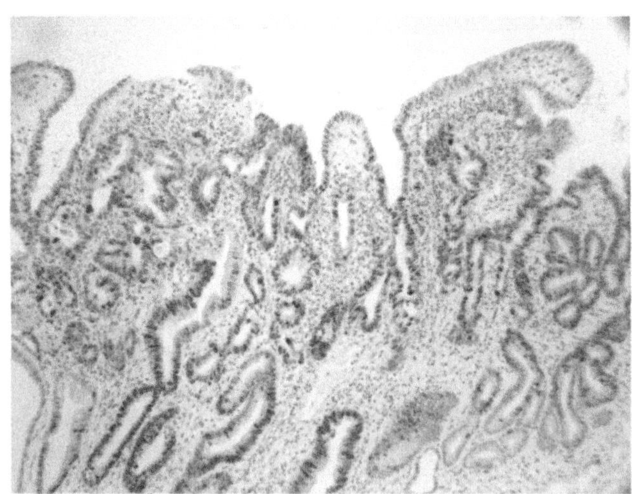

FIGURE 10.22 Immunoperoxidase stain for Ki67 in high-grade dysplastic intestinal metaplasia showing aberrant proliferation pattern with Ki67 expression in the superficial foveolar region and surface epithelium.

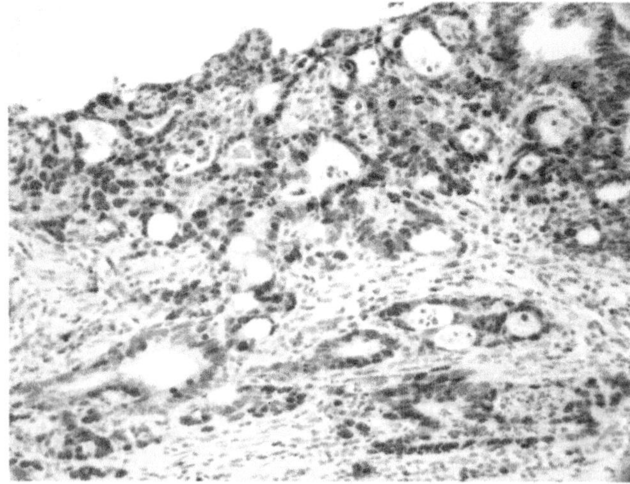

FIGURE 10.23 Immunoperoxidase stain for Ki67 in invasive adenocarcinoma showing marked expansion and aberration of staining pattern. The majority of the cells at all levels show Ki67 positivity.

CAUSE OF CARCINOGENESIS IN BARRETT ESOPHAGUS

The carcinogens involved in Barrett esophagus are uncertain. From a theoretical standpoint, it seems certain that Barrett adenocarcinoma is the result of a luminally acting carcinogen. There is almost nothing to suggest that a blood-borne carcinogen is involved. If the carcinogen is luminally acting, the likelihood is that it will be present in gastric contents. The only other possible sources are saliva and mucous secretion of esophageal glands. The carcinogenic agent is highly likely to be an endogenous molecule that is secreted into the intestinal lumen and is found in the stomach. Nothing in the literature suggests that any exogenous compound in food and drink causes Barrett adenocarcinoma.

In summary, while the carcinogens are unknown, it is presently believed and highly likely to be luminally acting endogenous molecules that are secreted into the foregut and find their way into the stomach. When gastroesophageal reflux occurs, these molecules act on intestinal metaplastic epithelium in the esophagus to cause neoplastic genetic transformation. The carcinogen does not act on squamous, cardiac, and oxyntocardiac epithelia in the esophagus.

Salivary Nitrogenous Compounds

The salivary secretion contains significant amounts of nitrate, derived from exogenous dietary nitrite absorbed from food and endogenous nitrate. In the mouth, the normal oral bacterial flora rapidly convert 10% to 90% of this secreted salivary nitrate to nitrite. Under fasting conditions, significant nitrite concentrations can be found in the saliva and esophagus (50). When nitrite passing down the esophagus encounters gastric acid, it is converted to nitrous oxide and potent nitrosating compounds that are able to react with organic nitrogenous compounds in food to form the potentially carcinogenic N-nitroso compounds. These compounds are proven experimental carcinogens that have been used to induce gastric and esophageal cancers in animals.

McColl (50) postulated that reflux-induced adenocarcinoma is caused by generation of these carcinogenic compounds when saliva meets gastric acid. Normally, without significant reflux, the site of this saliva-gastric acid interaction is the gastric cardia. In patients with reflux, the location of this interaction is in the distal esophagus and causes cancer therein.

McColl (50) suggested that the increase in the incidence of adenocarcinoma of the distal esophagus and gastric cardia is related to the increased dietary content of nitrates resulting from the increased use of nitrogenous fertilizer. For this hypothesis to be true, carcinogenesis must be an extremely rapid and highly localized phenomenon. The carcinogen must act within seconds of being produced and be deactivated immediately before it can diffuse to immediately adjacent areas. Otherwise it is impossible to explain how this process remains localized to the distal esophagus and gastric cardia without involving the rest of the stomach, where the cancer incidence has declined since the 1970s. This is a highly improbable model of a molecule with carcinogenic activity.

Acid

There is almost no evidence, clinical or experimental, that suggests that H^+ ions in gastric juice are capable of causing genetic mutations. The only pathologic change that has been shown to be caused by acid has related to cellular injury, which in turn results in increased proliferative activity. Increasing the proliferative activity of an epithelium has a secondary effect in facilitating mutations; because mutations occur during cell division, anything that increases proliferative activity will increase the chance of random mutations.

It is highly unlikely that acid is capable of causing genetic changes. H^+ ions are the smallest and simplest particles in nature and it is highly unlikely that cellular receptors have developed to recognize H^+. In general, cell surface receptors have complex structures that are complementary to large molecules with complex tertiary structures that lend specificity to the molecule-receptor interaction.

Despite the lack of evidence that acid is directly involved in carcinogenesis and the theoretical objections to its action as a molecule capable of interacting with cell surface receptors to cause genetic mutations, most people regard acid as the main factor in the etiology of reflux-induced adenocarcinoma of the esophagus. The emphasis of treating reflux disease is acid suppression. There is a naive expectation that removing acid from the equation will prevent all the complications of reflux disease including Barrett esophagus and adenocarcinoma. Much research is being done to determine whether acid suppression causes regression of Barrett esophagus.

At the same time that this unrelenting focus on acid suppression has led to the development of increasingly effective acid suppressive drugs from simple antacids to H2 receptor antagonists to proton pump inhibitors, the incidence of adenocarcinoma of the esophagus has increased sixfold in the United States from 1975 to 2000 (51), and evidence exists that the risk of developing esophageal adenocarcinoma is significantly increased in symptomatic patients who take these drugs compared with symptomatic people who do not (39).

In Chapter 9, we provided evidence that acid suppression, by producing a more alkaline environment in the esophagus, may promote the development of intestinal metaplasia in cardiac mucosa and inhibit the conversion of cardiac mucosa to the benign oxyntocardiac mucosa (see Chapters 8 and 9).

Acid in reflux does not cause cancer; acid suppression promotes the development of intestinal metaplasia in cardiac mucosa and promotes the development of adenocarcinoma in intestinal metaplasia.

Bile

Bile frequently enters the stomach in many people by duodenogastric reflux. Bile components, notably the bile acids cholic, taurocholic, and glycocholic acid and their deoxygenated metabolites, have been considered to be important in causing severe esophageal injury including erosive esophagitis, strictures, and Barrett esophagus (52).

In addition, bile salt metabolites have been shown to have capability in entering epithelial cells and binding with cellular receptors in esophageal progenitor cells to activate many metabolic pathways in experimental models using cell lines. Jaiswal et al. (53) showed that the conjugated bile salt glycochenodeoxycholic acid was highly effective in activating the P13 kinase/Akt signaling pathway in a Barrett adenocarcinoma cell line. This pathway is known to promote cell proliferation and inhibit apoptosis. Jaiswal et al. (54) studied the effect of bile salts on normal esophageal squamous cells and non-neoplastic Barrett's cells. Normal esophageal squamous cells exposed to 5 minutes of bile salts did not increase cell numbers. In Barrett's epithelial cells, bile salt exposure increased cell number by 31%, increased phosphorylated p38 and ERK levels by twofold to threefold, increased BrdU incorporation by 30%, and decreased UV-induced apoptosis by 15% to 20%. The authors concluded that bile salt exposure induces a non-neoplastic Barrett's cell line to proliferate by activation of both the ERK and p38 MAPK pathways and suggested that this could be a potential mechanism whereby bile reflux may facilitate the neoplastic progression of Barrett esophagus.

The effect of bile salts in activating Barrett's epithelial cells by changes in genetic pathways has also been shown in experimental animals. Stamp (55) showed that rats who were gastrectomized and jejunostomized to allow bile acids to reflux into the esophagus developed many carcinomas in 50 weeks, while other modifications that kept bile out of the esophagus did not produce any cancers. The authors concluded that bile acids refluxing into the esophagus of humans should also cause cancers, especially in Westernized societies with their high-fat diets, which provide an abundant supply of bile. They showed that bile acids can enter the model OE33 cells and activate the oncogene c-myc at pH 4, the gene complex NF-kappaB at pH 6.5, and induce cellular proliferation. In their discussion, the authors suggested that acid suppression therapy used to treat patients with Barrett esophagus will solubilize free bile acids and some of the glycine conjugates, allowing them to enter the epithelial cells. These intracellular bile acids have the ability to produce cellular changes that induce carcinogenesis. Therefore, the authors concluded, techniques to keep bile acids out of the stomach, or prevent them from reacting, must be developed, but until then, acid suppression therapy should be restricted, not promoted.

Penetration of bile salt metabolites into epithelial cells of the esophagus is a critical factor in carcinogesis as this appears to be a prerequisite to bile acids activating a variety of metabolic pathways that lead to increased proliferation of the cells. Study of bile salt metabolism shows that there is a critical pH range, between 3 and 6, in which bile acids exist in their soluble, un-ionized form, can penetrate cell membranes, and accumulate within mucosal cells (Fig. 10.24). At a lower pH, bile acids are precipitated, and at a higher pH, bile acids exist in their noninjurious ionized form (56).

These experimental data on bile acid metabolism and cellular activity provide a frightening scenario. The entry of bile into the stomach by duodenogastric reflux is a very common if not universal phenomenon. In a patient with normal gastric acidity in the 1 to 3 pH range, bile acids precipitate into harmless insoluble molecules (Fig. 10.25). At a pH above 6, the bile acids remain in an ionized form, which precludes their entry into cells. Between a pH of 3 and 6, the bile acids and their metabolites are converted to un-ionized soluble molecules that can penetrate cell membranes and enter epithelial cells.

Acid suppression, particularly when it is incomplete as in the vast majority of patients who take these medications, both over the counter and prescribed, has the effect of increasing gastric pH into the critical 3 to 6 pH range where it promotes the generation of carcinogenic molecules from the endogenous bile acids when these are present in the stomach. Even in the highly controlled setting with high-dosage proton pump inhibitors, it is rarely possible to achieve the total acid suppression that keeps the gastric pH consistently above 6. It is therefore

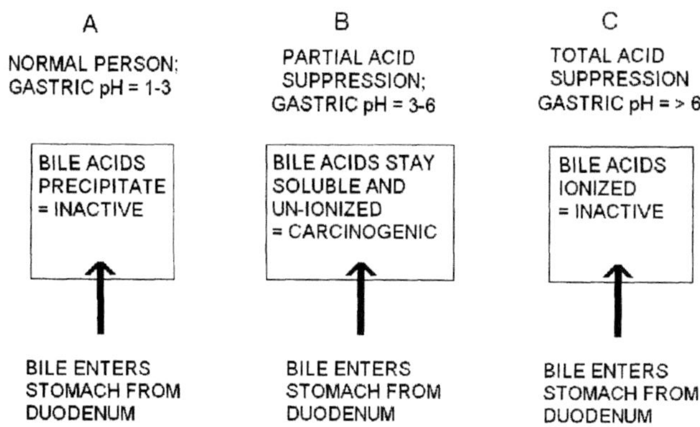

FIGURE 10.24 Effect of gastric pH on bile acids that have entered the stomach as a result of duodenogastric reflux. In a normal person with a gastric pH in the 1 to 3 range (a), the bile acids are inactivated by precipitation. In the partially acid suppressed state commonly present in patients on acid suppressive drugs (b), the gastric pH is in the 3 to 6 range and bile acid metabolites are in the soluble, un-ionized state where they have been shown to enter Barrett epithelial cells and induce carcinogenesis. In patients who are totally acid suppressed (c), bile acids are in the soluble ionized state where they are not active as carcinogens.

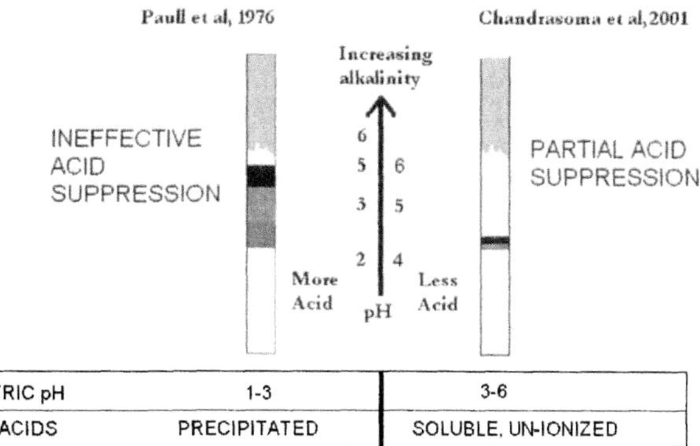

GASTRIC pH	1-3	3-6
BILE ACIDS	PRECIPITATED	SOLUBLE, UN-IONIZED
CARCINOGENICITY	NO	YES
CANCER RISK	LOW	INCREASED

FIGURE 10.25 The effect of partial acid suppression (*right*) where the gastric pH is in the 2 to 6 range compared with ineffective (i.e., gastric pH in the 1 to 3 range) acid suppression (*left*). The increased pH of the refluxate induces increased intestinal metaplasia in the acid-suppressed patient. The conversion of bile acids to carcinogens increases the likelihood of conversion of the intestinal metaplasia to cancer. Note that the two patients are the typical patients in Paull et al. (31) and Chandrasoma et al. (10), suggesting that these differences can explain the increased incidence of adenocarcinoma that has occurred during this period.

highly likely that the majority of patients who take acid suppressive therapy have an increased amount of carcinogenic bile acid derivatives in their stomach. These bile acid derivatives are not carcinogenic to gastric epithelial cells, esophageal squamous cells, and metaplastic esophageal columnar epithelial cells in cardiac and oxyntocardiac mucosae. However, these molecules enter the cells of intestinal epithelium in the esophagus, causing transformations that likely induce carcinogenesis.

Esophageal impedance studies show that acid suppression does not significantly reduce either the number of reflux episodes or their volume. It only increases the pH sufficiently to prevent acid-induced symptoms in the esophagus. As it causes symptomatic improvement, we have suggested in Chapters 8 and 9 that it increases the conversion of metaplastic cardiac mucosa to intestinal metaplasia and inhibits the conversion of cardiac mucosa to oxyntocardiac mucosa (Fig. 9.24). This dramatically increases the population of target cells as intestinal metaplasia increases both in prevalence and quantity within the columnar-lined esophageal segment. The increased alkalinity of the gastroesophageal refluxate that has been rendered carcinogenic by acid suppression has simultaneously created a fertile environment in the esophagus for its action.

In the patient who has intestinal metaplasia, continued reflux allows the entry of these soluble bile acid metabolites and appears to induce alterations in genetic and metabolic pathways, which lead to proliferation of these cells and inhibit apoptosis. Though the exact pathway causing carcinogenesis is unknown, it is likely that this is the mechanism of carcinogenesis in Barrett esophagus.

The loop whereby acid suppression increases the likelihood of cancer in a patient with Barrett esophagus is now completed. The increasing effectiveness of acid suppression would also explain the increasing incidence of reflux-induced adenocarcinoma since the mid-1970s (Fig. 10.25).

References

1. Haggitt RC, Dean PJ. Adenocarcinoma in Barrett's epithelium. *In:* Spechler SJ, Goyal RK (eds), Barrett's Esophagus: Pathophysiology, Diagnosis and Management. Elsevier Science, New York, 1985, pp. 153–166.
2. Hamilton SR, Smith RRL, Camerson JL. Prevalence and characteristics of Barrett's esophagus in patients with adenocarcinoma of the esophagus or esophago-gastric junction. Hum Pathol 1988;19:942–948.
3. McArdle JE, Lewin KJ, Randall G, et al. Distribution of dysplasias and early invasive carcinoma in Barrett's esophagus. Hum Pathol 1992;23:479–482.
4. Spechler SJ, Zeroogian JM, Antonioli DA, Wang HH, Goyal RK. Prevalence of metaplasia at the gastroesophageal junction. Lancet 1994;344:1533–1536.
5. Lagergren J, Bergstrom R, Lindgren A, Nyren O. Symptomatic gastroesophageal reflux as a risk factor for esophageal adenocarcinoma. N Engl J Med 1999;340:825–831.
6. Provenzale D, Schmiit C, Wong JB. Barrett's esophagus: A new look at surveillance based on emerging estimates of cancer risk. Am J Gastroenterol 1999;94:2043–2053.

7. Sharma P, McQuaid K, Dent J, Fennerty B, Sampliner R, Spechler S, Cameron A, Corley D, Falk G, Goldblum J, Hunter J, Jankowski J, Lundell L, Reid B, Shaheen N, Sonnenberg A, Wang K, Weinstein W. A critical review of the diagnosis and management of Barrett's esophagus: The AGA Chicago Workshop. Gastroenterology 2004;127:310–330.
8. Jones TF, Sharma P, Daaboul B, Cherian R, Mayo M, Topalovski W, Weston AP. Yield of intestinal metaplasia in patients with suspected short segment Barrett's esophagus (SSBE) on repeat endoscopy. Dig Dis Sci 2002;47:2108–2111.
9. Dresner SM, Griffin SM, Wayman J, Bennett MK, Hayes N, Raimes SA. Human model of duodenogastro-oesophageal reflux in the development of Barrett's metaplasia. Br J Surg 2003;90:1120–1128.
10. Chandrasoma P, Wickramasinghe K, Ma Y, DeMeester T. Is intestinal metaplasia a necessary precursor lesion for adenocarcinomas of the distal esophagus, gastroesophageal junction and gastric cardia? Dis Esophagus (in press).
11. Weston AP, Badr AS, Hassanein RS. Prospective multivariate analysis of clinical, endoscopic, and histological factors predictive of the development of Barrett's multifocal high grade dysplasia or adenocarcinoma. Am J Gastroenterol 1999; 94:3413–3419.
12. Oh D. The clinial biology and surgical therapy of intramucosal adenocarcinoma of the esophagus. (Submitted for publication.)
13. Miros M, Kerlin P, Walker N. Only patients with dysplasia progress to adenocarcinoma in Barrett's esophagus. Gut 1991; 32:1441–1446.
14. Hameetman W, Tytgat GNJ, Houthoff HJ, et al. Barrett's esophagus: Development of dysplasia and adenocarcinoma. Gastroenterology 1989;96:1248–1256.
15. O'Connor JB, Falk GW, Richter JE. The incidence of adenocarcinoma and dysplasia in Barrett's esophagus. Am J Gastroenterol 1999;94:2037–2042.
16. Reid BJ, Levine DS, Longton G, et al. Predictors of progression to cancer in Barrett's esophagus: Baseline histology and flow cytometry identify low- and high-risk patient subset. Am J Gastroenterol 2000;95:1669–1676.
17. Montgomery E, Goldblum JR, Greenson JK, et al. Dysplasia as a predictive marker for invasive carcinoma in Barrett's esophagus: A follow-up study based on 138 cases fro a diagnostic variability study. Hum Pathol 2001;32:379–388.
18. Schnell TG, Sontag SJ, Chejfec G, et al. Long-term nonsurgical management of Barrett's esophagus with high grade dysplasia. Gastroenterology 2001;120:1607–1619.
19. MacDonald CE, Wicks AC, Playford RJ. Final results from 10 year cohort of patients undergoing surveillance for Barrett's oesophagus: Observational study. Br Med J 2000;321: 1252–1255.
20. Dulai GS, Guha S, Kahn KL, Gornbein J, Weinstein WM. Preoperative prevalence of Barrett's esophagus in esophageal adenocarcinoma: A systematic review. Gastroenterology. 2002;122:26–33.
21. Spechler SJ. Barrett's esophagus. N Engl J Med 2002;346:836–842.
22. Wright TA, Gray MR, Morris AI, et al. Cost effectiveness of detecting Barrett's cancer. Gut 1996;39:574–579.
23. Streitz JM Jr, Ellis FH Jr, Tilden RL, Erickson RV. Endoscopic surveillance of Barrett's esophagus: A cost-effectiveness comparison with mammographic surveillance for breast cancer. Am J Gastroenterol 1998;93:911–915.
24. Sonnenberg A, Soni A, Sampliner RE. Medical decision analysis of endoscopic surveillance of Barrett's esophagus to prevent oesophageal adenocarcinoma. Aliment Pharmocol Ther 2002;16:41–50.
25. Peters JH, Clark GWB, Ireland AP, Chandrasoma P, Smyrk TC, DeMeester TR. Outcome of adenocarcinoma arising in Barrett's esophagus in endoscopically surveyed and nonsurveyed patients. J Thorac Cardiovasc Surg 1994;108:813–821.
26. Rudolph RE, Vaughan TL, Storer BE, et al. Effect of segment length on risk for neoplastic progression in patients with Barrett esophagus. Ann Intern Med 2000;132:612–620.
27. Buttar NS, Wang KK, Sebo TJ, et al. Extent of high grade dysplasia in Barrett's esophagus correlates with risk of adenocarcinoma. Gastroenterology 2001;120:1607–1619.
28. Tharavej C, Bremner CG, Chandrasoma P, DeMeester TR. Predictive factors of coexisting cancer in Barrett's high grade dysplasia. (Submitted for publication.)
29. Chandrasoma PT, Der R, Dalton P, Kobayashi G, Ma Y, Peters J, DeMeester T. Distribution and significance of epithelial types in columnar lined esophagus. Am J Surg Pathol 2001;25: 1188–1193.
30. Paull A, Trier JS, Dalton MD, Camp RC, Loeb P, Goyal RK. The histologic spectrum of Barrett's esophagus. N Engl J Med 1976;295:476–480.
31. Allison PR, Johnstone AS. The oesophagus lined with gastric mucous membrane. Thorax 1953;8:87–101.
32. Rex DK, Cummings OW, Shaw M, Cumings MD, Wong RKH, Vasudeva RS, Dunne D, Rahmani EY, Helper DJ. Screening for Barrett's esophagus in colonoscopy patients with and without heartburn. Gastroenterology 2003;125:1670–1677.
33. Farrow DC, Vaughan TL, Sweeney C, Gammon MD, Chow WH, Risch HA, Stanford JL, Hansten PD, Mayne ST, Schoenberg JB, Rotterdam H, Ahsan H, West AB, Dubrow R, Fraumeni JF Jr, Blot WJ. Gastroesophageal reflux disease, use of H2 receptor antagonists, and risk of esophageal and gastric cancer. Cancer Causes Control 2000;11:231–238.
34. Wu AH, Wan P, Bernstein L. A multiethnic population-based study of smoking, alcohol and body size and risk of adenocarcinoma of the stomach and esophagus. Cancer Causes Control 2001;12:721–732.
35. Brown LM, Swanson CA, Gridley G, Swanson GM, Schoenberg JB, Greenberg RS, Silerman DT, Pottern LM, Hayes RB, Schwartz AG. Adenocarcinoma of the esophagus: Role of obesity and diet. J Natl Cancer Inst 1995;87:104–109.
36. Vaughan TL, Davis S, Kristal A, Thomas DB. Obesity, alcohol, and tobacco as risk factors for cancers of the esophagus and gastric cardia: Adenocarcinoma versus squamous cell carcinoma. Cancer Epidemiol Biomarkers Prev 1995;4:85–92.
37. Ji BT, Chow WH, Yang G, McLaughlin JK, Gao RN, Zheng W, Shu XO, Jin F, Fraumeni JF Jr, Gao YT. Body mass index and the risk of cancers of the gastric cardia and distal stomach in Shanghai, China. Cancer Epidemiol Biomarkers Prev 1997;6:481–485.
38. Chow WH, Blot WJ, Vaughan TL, Risch HA, Gammon MD, Stanford JL, Dubrow R, Schoenberg JB, Mayne ST, Farrow DC, Ahsan H, West AB, Rotterdam H, Niwa S, Fraumeni JF Jr. Body mass index and risk of adenocarcinomas of the esophagus and gastric cardia. J Natl Cancer Inst 1998;90:150–155.
39. Lagergren J, Bergstrom R, Nyren O. Association between body mass index and adenocarcinoma of the esophagus and gastric cardia. Ann Intern Med 1999;130:883–890.
40. Cheng KK, Sharp L, McKinney PA, Chilvers CED, Cook-Mozaffari P, Ahmed A, Day NE. A case-control study of esophageal adenocarcinoma in women: A preventable disease. Br J Cancer 2000;83:127–132.
41. Lagergren J. Increased incidence of adenocarcinoma of the esophagus and cardia. Reflux and obesity are strong and inde-

pendent risk factors according to the SECC study. Lakartidningen 2000;97:1950–1953.

42. Bu X, Ma Y, Der R, DeMeester T, Bernstein L, Chandrasoma PT. Body mass index is associated with Barrett esophagus and cardiac mucosal metaplasia. Dig Dis Sci 2005; in press.
43. Menke-Pluymers MB, Hop WC, Dees J, van Blankenstein M, Tilanus HW. Risk factors for the development of an adenocarcinoma in columnar-lined (Barrett) esophagus: The Rotterdam Esophageal Tumor Study Group. Cancer. 1993;72:1155–1158.
44. Avidan B, Sonnenberg A, Schnell TG, Chejfec G, Metz A, Sontag SJ. Hiatal hernia size, Barrett's length, and severity of acid reflux are all risk factors for esophageal adenocarcinoma. Am J Gastroenterol. 2002;97:1930–1936.
45. Kim R, Clarke MR, Melhem MF, Young MA, Vanbibber MM, Safatle-Ribeiro AV, Ribeiro U Jr, Reynolds JC. Expression of p53, PCNA, and C-erbB-2 in Barrett's metaplasia and adenocarcinoma. Dig Dis Sci 1997;42:2453–2462.
46. Ireland AP, Clark GW, DeMeester TR. Barrett's esophagus. The significance of p53 in clinical practice. Ann Surg 1997;225: 17–30.
47. Weston AP, Banerjee SK, Sharma P, Tran TM, Richards R, Cherian R. p53 protein overexpression in low grade dysplasia (LGD) in Barrett's esophagus: Immunohistochemical marker predictive of progression. Am J Gastroenterol 2001;96: 1321–1323.
48. Olvera M, Wickramasinghe K, Brynes R, Bu X, Ma Y, Chandrasoma P. Ki67 expression in different epithelial types in columnar lined esophagus indicates varying levels of expanded and aberrant proliferative patterns. Histopathology 2005;47:132–140.
49. Whittles CE, Biddlestone LR, Burton A, Barr H, Jankowski JA, Warner PJ, Shepherd NA. Apoptotic and proliferative activity in the neoplastic progression of Barrett's oesophagus: A comparative study. J Pathol 1999;187:535–540.
50. McColl KEL. When saliva meets acid: Chemical warfare at the oesophagogastric junction. Gut 2005;54:1–3.
51. Pohl H, Welch HG. The role of overdiagnosis and reclassification in the marked increase of esophageal adenocarcinoma. J Natl Cancer Inst 2005;19:142–146.
52. Nehra D, Howell P, Williams CP, Pye JK, Beynon J. Toxic bile acids in gastro-oesophageal reflux disease: Influence of gastric aciditiy. Gut 1999;44:598–602.
53. Jaiswal K, Tello V, Lopez-Guzman C, Nwariaku F, Anthony T, Sarosi GA. Bile salt exposure causes phosphtidyl-inositol-3-kinase-mediated proliferation in a Barrett's adenocarcinoma cell line. Surgery 2004;136:160–168.
54. Jaiswal K, Lopez-Guzman C, Souza RF, Spechler SJ, Sarosi GA Jr. Bile salt exposure increases proliferation through p38 and ERK-MAPK pathways in non-neoplastic Barrett's cell line. Am J Physiol Gastrointest Liver Physiol, 2006;290:G335–342.
55. Stamp DH. Bile acids aided by acid suppression therapy may be associated with the development of esophageal cancers in Westernized societies. Med Hypothesis 2002;59:398–405.
56. Kauer WK, Stein HJ. Bile reflux in the constellation of gastroesophageal reflux disease. Thorac Surg Clin 2005;15:335–340.

CHAPTER

11

Pathology of Reflux Disease at an Anatomic Level

So far, we have considered the cellular changes of reflux disease without giving any consideration to changes in the anatomy of the distal esophagus. To do this, we have used a diagrammatic device that is essentially a vertical section passing across the distal esophagus into the stomach (Fig. 11.1). This diagrammatic device is reproduced by a vertical histologic section taken from the distal esophagus into the proximal stomach at autopsy or in an esophagectomy specimen. At endoscopy, it is reproduced by taking a biopsy sample (ideally, a four-quadrant biopsy) at measured horizontal levels passing from the distal esophagus to the proximal stomach.

When this is done, it is clear that the entire normal esophagus is lined by squamous epithelium and the entire stomach is lined by oxyntic mucosa (Fig. 11.1). We have defined reflux disease by the presence of an interposed metaplastic columnar-lined esophagus composed of cardiac mucosa with or without intestinal metaplasia and oxyntocardiac mucosa (Fig. 11.2). In Chapter 6 we showed that these are always abnormal epithelia resulting from reflux-induced metaplasia of the esophageal squamous epithelium. As such, these mucosal types are always esophageal.

Obvious problems arise when the information in this diagrammatic vertical section is applied to real life. There is no problem whatsoever in the normal person, but normal people are rarely encountered at endoscopy. They are encountered at autopsy. Because autopsy studies of this region were not reported until our study of 2000 (1), the concept of normalcy had developed incorrectly by the examination of patients with pathologic abnormality. When normal autopsies were studied, it was clear that normal people had no columnar-lined esophagus. In these normal patients, the entire squamous-lined esophagus was tubular, traversed the diaphragm, and had a 2- to 3-cm abdominal length before it entered the stomach. The left side of this point of entry formed an acute angle of His with the stomach because the gastric fundus moved sharply up and to the left to come to lie under the left diaphragm (Fig. 11.3). The entire stomach had a saccular shape and was lined by oxyntic mucosa, which begins at the squamocolumnar junction in normal people who have no columnar-lined esophagus (Fig. 11.4). In normal people, therefore, the gastroesophageal junction is defined by the squamocolumnar junction, the end of the tubular esophagus, and the proximal limit of the rugal folds. However, of these three criteria that define the gastroesophageal junction, only the proximal limit of gastric oxyntic mucosa remains constant.

Problems arise in the abnormal patient with reflux damage to the lower esophagus. Cardiac epithelium, not infrequently complicated by intestinal metaplasia, and oxyntocardiac epithelium are often found distal to the end of the tubular esophagus. Because this is a common definition of the end of the esophagus (i.e., the gastroesophageal junction), this epithelium was placed in the proximal stomach (i.e., distal to the gastroesophageal junction). When the gastroesophageal junction became defined as the proximal limit of the rugal folds, the same thing happened. In some patients, cardiac mucosa with and without intestinal metaplasia and oxyntocardiac mucosa were found distal to the proximal limit of the rugal folds, (i.e., in the proximal stomach) (Fig. 11.5).

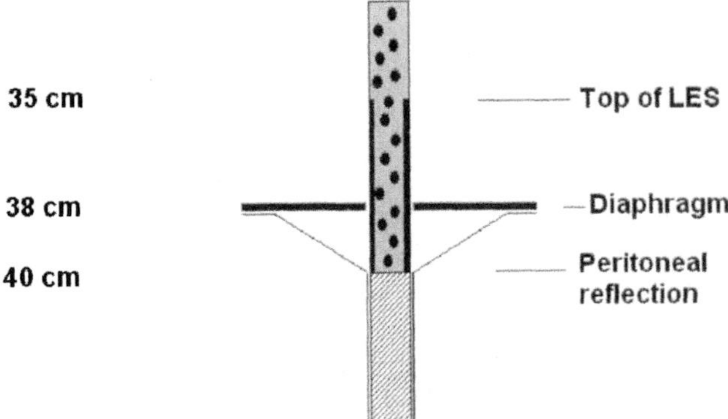

FIGURE 11.1 Diagrammatic vertical section to show normal esophagus and stomach. The esophagus passes through the diaphragm and enters the stomach at the gastroesophageal junction, represented by the peritoneal reflection. The mucosal junction is where the squamous epithelium of the esophagus meets gastric oxyntic mucosa. The wall of the entire abdominal esophagus (approximately 2 to 3cm) and the distal 2 to 3cm of the thoracic esophagus constitutes the lower esophageal sphincter (black wall). The esophagus contains submucosal glands (black dots). The distances on the left are from the incisor teeth; they are the average distances, which show considerable individual variation and therefore are unreliable to define landmarks.

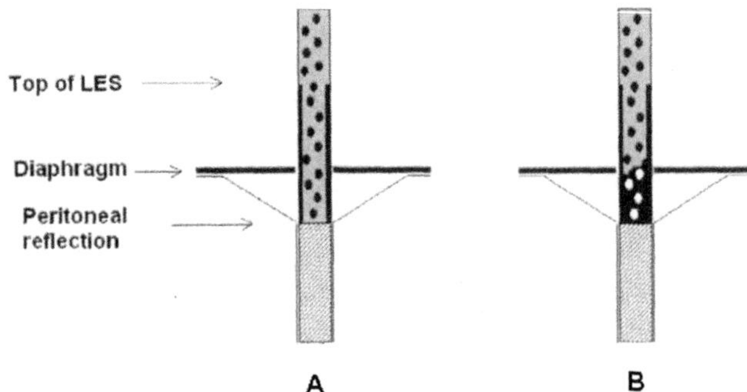

FIGURE 11.2 Abnormal reflux-damaged esophagus as shown in this vertical section (*right*). This shows a columnar-lined esophagus (black) interposed between the gastric oxyntic mucosa (striped; which remains constant) and the squamocolumnar junction (which has moved cephalad to a degree that is directly proportional to the severity of reflux). The peritoneal reflection remains constant in that it is at the gastroesophageal junction. It may, however, change in position anatomically if there is esophageal shortening (not shown).

An obvious discrepancy now exists (Fig. 11.5). By our diagrammatic vertical section, cardiac mucosa with and without intestinal metaplasia and oxyntocardiac mucosa were always a metaplastic columnar esophageal epithelium induced by reflux; by the endoscopic definitions presently in use, these epithelia were located in the proximal stomach. Either the literature presented in the previous chapters, which stated that cardiac mucosa with and without intestinal metaplasia and oxyntocardiac mucosa represented columnar-

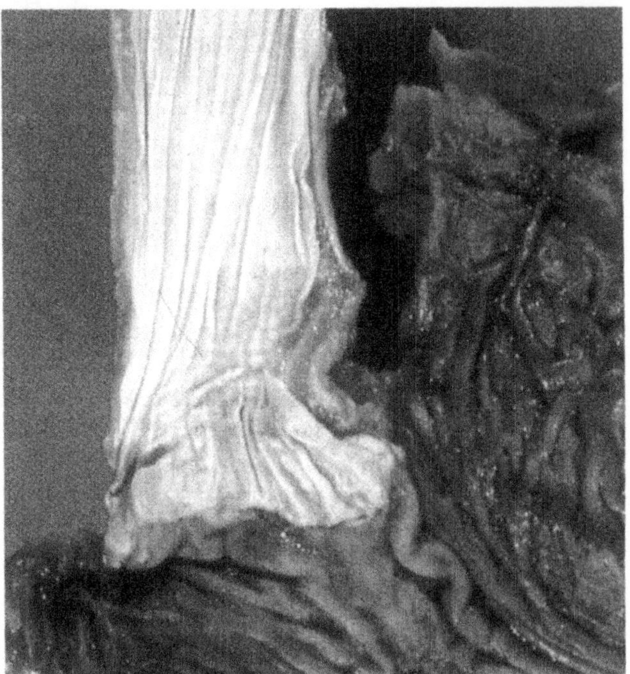

FIGURE 11.3 Normal anatomy of the distal esophagus and proximal stomach showing the entire tubular esophagus lined by squamous epithelium with a horizontal Z line. This differs from the diagrammatic representation in that the tubular esophagus flares out into the saccular stomach at the gastroesophageal junction and the left side (greater curvature) of the stomach makes the acute angle of His at the junction with the wall passing upward and to the left to form the fundus of the stomach. (Cross reference: Color Plate 1.1)

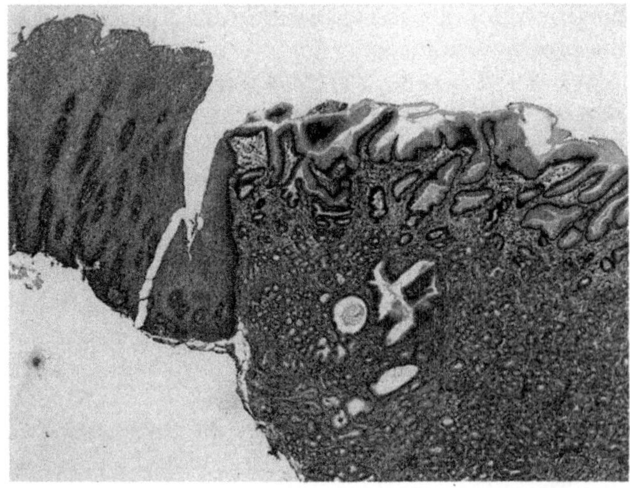

FIGURE 11.4 Normal histology of the gastroesophageal junction. This shows squamous epithelium of the esophagus (*left*) transitioning into gastric oxyntic mucosa with parietal cell containing glands (*right*). There is no cardiac or oxyntocardiac mucosa (i.e., no metaplastic columnar lined-esophagus).

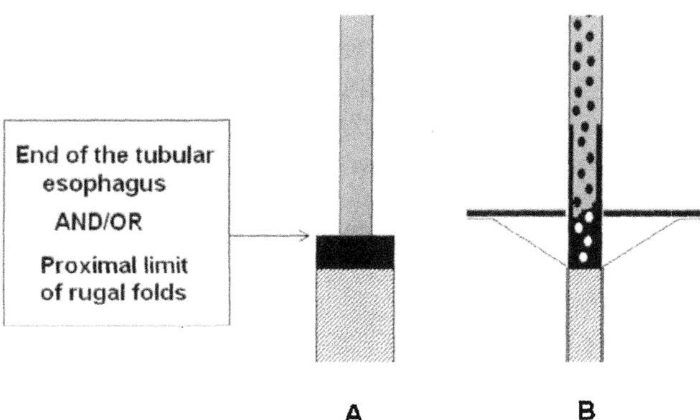

FIGURE 11.5 The discrepancy that now exists. (a) Cardiac mucosa distal to the end of the tubular esophagus and proximal limit of rugal folds. By the criteria used at present, there is no abnormality. The cardiac mucosa found distal to the endoscopic gastroesophageal junction (end of tubular esophagus or proximal limit of rugal folds) is considered to be normal gastric cardiac mucosa. (b) In the diagrammatic vertical section, cardiac mucosa is always esophageal and always abnormal.

lined esophagus, was wrong, or the present definitions of the gastroesophageal junction were wrong.

What Is the Presently Held View Regarding the Anatomy and Histology of the Gastroesophageal Junction?

Of note is what has transpired since the mid-1990s regarding the concept of normalcy in this region. As recently as 1997, the medical world believed that there was at least 2 to 3 cm of cardiac mucosa normally straddling the gastroesophageal junction. This belief has changed dramatically, drifting slowly toward the truth that we believe, which is that there is no normal cardiac mucosa between the squamous epithelium of the esophagus and gastric oxyntic mucosa. One attempt at describing what mainstream pathologists accept as the anatomy of this region was a review article from Harvard published in 2005 (2). This is now the contrary opinion to our hypothesis and belief.

Literature Review

Odze RD. Unraveling the mystery of the gastroesophageal junction: A pathologist's perspective. Am J Gastroenterol 2005;100:1853–1867.

This review presents what is now the contrary viewpoint to our concepts, as presented in this book that (a) there is normally no gastric cardia is composed of cardiac mucosa and (b) cardiac mucosae with and without intestinal metaplasia and oxyntocardiac mucosa are metaplastic esophageal epithelia resulting from reflux-induced damage to the esophageal squamous epithelium.

Dr. Odze (2) defined the "true" gastric cardia as "the area of mucosa located distal to the anatomic gastroesophageal junction (defined as the proximal limit of the gastric folds) and proximal to the portion of the stomach (corpus) that is composed entirely of oxyntic glands" (p. 1854). In the abstract, the true gastric cardia is stated to be "an extremely short segment (<0.4 mm) of mucosa that is typically composed of pure mucous glands, or mixed mucous/oxyntic glands that are histologically indistinguishable from metaplastic mucinous columnar epithelium of the distal esophagus" (p. 1853). The less than 0.4 mm is probably a typographic error; in the body of the paper, Odze appears to be taking the figure of Kilgore et al.'s autopsy study (3) of 0.1 to 0.4 cm as the sole basis for his definition of the extent of the "true" gastric cardia. Odze failed to recognize the flaws in Kilgore et al. (see Chapter 6). The reason why he favors these data over our autopsy data, which show cardiac mucosa to be absent in 56% of persons (1), is that "this [our] study . . . utilized a highly referral based population with GERD and lacked a symptom-free control group" (p. 1856). This is blatantly false; ours was an autopsy study where only three patients had any mention of symptoms of reflux during life and is the strongest normal control population in the literature.

Odze's review is a series of contradications consequent on the inability to recognize the obvious fact that there is no "true gastric cardia" and it is ridiculous to define a "normal" organ by a minute extent of less than 0.4 cm. These contradictions are as follows:

1. Odze's "true" gastric cardia is not lined by cardiac mucosa. While accepting Kilgore et al. (2000) in ascribing a length of less than 0.4 cm to the true gastric cardia, Odze rejected Kilgore et al.'s finding that this area is lined by cardiac mucosa. Using data from their own unit (Glickman et al. [4]), they defined the mucosa that lines the "true" gastric cardia as being either pure cardiac mucosa (81% in Glickman et al., which is a clinical pediatric population with a high prevalence of reflux) or oxyntocardiac mucosa (19% in Glickman et al.).
2. Odze contradicted his claim regarding the universal existence of "true" gastric cardia by stating: "The type and length of epithelium in the true gastric cardia may also vary in different portions of the cardia within individual patients. A paradoxical observation is that a small proportion of patients have mucosa composed of pure oxyntic glands only, without mucous glands, in the true gastric cardia" (p. 1854). This is not possible; Odze defined the distal limit of his "true" gastric cardia as the proximal limit of pure oxyntic mucosa. These patients, therefore, have no "true" gastric cardia by his definition. Odze covered this absence of normalcy by using the amazing artificial definitional device of *not ascribing a lower size limit to his "true" gastric cardia*. Less than <0.4 cm can include zero!
3. He defined the upper size limit of "true" gastric cardia as 0.4 cm. This is not correct except in Kilgore et al.'s normal autopsy population. It is not an error when standard his-

tology textbooks (5) state that 2 to 3 cm of cardiac mucosa may exist in the "proximal stomach." Jain et al. (6) showed by extensive biopsy that cardiac mucosa was present in decreasing prevalence 1 and 2 cm distal to the endoscopic gastroesophageal junction. In Sarbia et al. (7), a study of elderly patients who had esophagectomy for esophageal tumors, the maximum length of pure cardiac mucosa distal to the gastroesophageal junction was 1 to 15 mm (median 5 mm) and the maximum combined length of cardiac and oxyntocardiac mucosa was 5 to 28 mm (median 11 mm). We (Chandrasoma et al., [8], also in a study of elderly patients who had esophagectomy for neoplasms) found a maximum cardiac and oxyntocardiac mucosa distal to the proximal limit of rugal folds varying between 3.1 and 20.5 mm. Odze's definition of "true" gastric cardia as 0.4 cm has the possible effect or creating a new disease when cardiac mucosa expands *toward the antrum* to an extent greater than 0.4 cm. In many patients with this potential "new" entity of "expansion of the true gastric cardia," the oxyntic mucosa is normal (9).

4. Odze used the proximal limit of the rugal folds as the sole criterion that defines the proximal limit of the "true" gastric cardia. This would suggest that Odze believes in the accuracy of assessing this criterion endoscopically with the precision required to define an extent less than 0.4 cm. Nothing is further from the truth. He stated: "At the time of endoscopy, it is difficult, if not impossible, to detect a minor proximal displacement of the Z-line" (p. 1854). If this is true, and if zero to 0.4 cm comes within the definition of his "minor proximal displacement," Odze's definition of the proximal limit of "true" gastric cardia is *based on a criterion that is within his own recognized margin of error.*
5. In his discussion of the differential diagnosis of intestinal metaplasia of this region, Odze stated: "As it is difficult, if not impossible, to know the precise anatomic location (i.e., esophagus or stomach) of a particular biopsy obtained from the gastroesophageal junction at the time of endoscopy . . . clinicians have relied on pathologists to help establish the correct location and etiology of the inflammatory condition in mucosal biopsies from the gastroesophageal junctional region" (p. 1862). This is blatantly false. The present practice among pathologists, including Odze, is to ascribe the anatomic location to a biopsy by its relationship to the endoscopic gastroesophageal junction; if the biopsy is distal to this, it is gastric cardia by Odze's own definition. Odze cannot use histology to refute this endoscopically defined cardiac location because, by his own admission, "the histologic characteristics of the short segment of mucosa located above the anatomic gastroesophageal junction in these individuals (with a proximally displaced squamocolumnar junction) are similar to that of the true gastric cardia" (p. 1854). Odze had, therefore, only the clinician's definition of the gastroesophageal junction to enable him to differentiate the esophagus from the true gastric cardia.

It is interesting that Odze used our definitions in two important places (Fig. 11.6): (a) For the first time, a pathologist other than those in our group has used the histologically defined proximal limit of gastric oxyntic mucosa in a definition. Odze uses this point to limit his "true" gastric cardia distally. In our definition, the proximal limit of gastric oxyntic mucosa is the end of the esophagus. The distance separating our two opinions is zero to 0.4 cm and heading inexorably to the truth of zero. (b) Odze was stating our position when he noted that the location of an endoscopic biopsy must be determined by histology rather than anatomy and endoscopy. He cannot do that without agreeing with our position that cardiac mucosa and oxyntocardiac mucosa are *always metaplastic esophageal epithelia (9). It is only when a pathologist holds that belief that he or she can tell the endoscopist that a biopsy distal to the endoscopic gastroesophageal junction that contains cardiac mucosa is esophageal and not gastric. Otherwise, one is stuck with the 0.4 cm of "true" gastric cardia.*

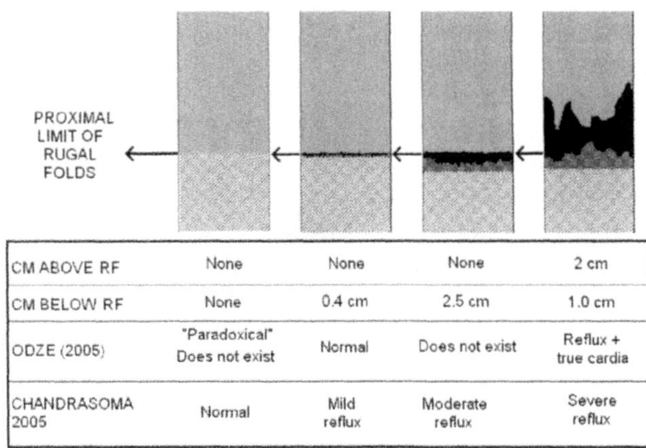

CM ABOVE RF	None	None	None	2 cm
CM BELOW RF	None	0.4 cm	2.5 cm	1.0 cm
ODZE (2005)	"Paradoxical" Does not exist	Normal	Does not exist	Reflux + true cardia
CHANDRASOMA 2005	Normal	Mild reflux	Moderate reflux	Severe reflux

FIGURE 11.6 Comparison between latest present belief (2) and our belief (8, 9). The arrows point to the endoscopic proximal limit of rugal folds (RF). We believe there is normally no cardiac (black) and oxyntocardiac (stippled) mucosa, and, when these are found, they always represent reflux-induced columnar-lined esophagus. Increasing amounts of cardiac and oxyntocardiac mucosa separating the squamous epithelium from gastric oxyntic mucosa (right three esophagi) indicate increasing amounts of reflux-induced damage. Odze believes that the complete absence of cardiac and oxyntocardiac mucosa is "paradoxical" or does not exist (*left*); believes that less than 0.4 cm of cardiac and oxyntocardiac mucosa below the proximal limit of rugal folds is normal; ignores patients with greater than 0.4 cm of cardiac mucosa below the proximal limit of rugal folds; and when columnar-lined mucosa exists above the proximal limit of rugal folds, regards the 0.4 cm of the cardiac mucosa below the rugal folds to be normal gastric cardia.

What Is the Evidence That the End of the Tubular Esophagus or the Proximal Limit of the Rugal Folds Defines the True Gastroesophageal Junction?

The Association of Directors of Anatomic and Surgical Pathology (10) recommended that the end of the tubular esophagus be used as the criterion for the gastroesophageal junction in dissections of esophagectomy specimens. The proximal limit of the gastric rugal folds was described in 1987 by McClave (11) as a useful endoscopic landmark. It has since become the universal endoscopic definition of the gastroesophageal junction.

There is almost no evidence to validate these endoscopic definitions of the gastroesophageal junction. In the American Gastroenterological Association Chicago Workshop attended by 18 experts on Barrett esophagus (12), statement 7 stated: "The proximal margin of the gastric folds is a reliable endoscopic marker for the gastroesophageal junction" (p. 315). Review of the nature of the evidence in the literature that supported this statement received a grade of IV (i.e., the evidence available consisted of "opinions of respected authorities based on clinical experience, descriptive studies, or reports of expert committees" [p. 315]). This indicates that the basis of the proximal limit of the rugal folds being a definitional marker for the true gastroesophageal junction is opinion, not fact. Despite the lack of hard evidence, the subgroup support for this statement was a grade C ("there is poor evidence to support the statement, but recommendations may be made on other grounds").

In the discussion under the statement, repeated verbatim:

> All workshop members were in agreement that a reliable endoscopic determination of the gastroesophageal junction (GEJ) is needed to make a diagnosis of Barrett esophagus and to record the length of columnar epithelium in the esophagus. A widely used endoscopic marker of the GEJ is the proximal margin of the rugal folds. Other markers of the GEJ include the level at which the tubular esophagus meets the wider, sac-like stomach and the pinch at the end of the tubular esophagus. Given its widespread use by experts in the field, the group uniformly favored using the proximal margin of the gastric folds to identify the GEJ but recognized that there are scant data that validate it. No studies have compared the upper margin of the gastric folds with other markers, such as the level at which the tubular esophagus meets the wider, sac-like stomach. However, the group agreed that this still represented the best landmark recognized at endoscopy. (p. 315)

In the individual voting of their level of support for the statement, 14 of the experts accepted it with some reservation and 4 accepted it with major reservation. None of the experts accepted this statement completely and none rejected it.

This is not a strong vote of confidence for the reliability of the endoscopic definition of the gastroesophageal junction (the proximal limit of the rugal folds). In fact, it suggests that this definition has been accepted by a consensus of experts without any validation or proof. It is interesting that these experts have only the end of the tubular esophagus to use for validating this definition. The end of the tubular esophagus has also not been validated as the true gastroesophageal junction. The important question that should have been asked is whether it is possible to reliably identify the gastroesophageal junction by any available endoscopic criterion. Attempting to do what is impossible is certain to cause error. To recognize that the endoscope is not capable of identifying the gastroesophageal junction accurately appears to be anathema to this group.

This is not a minor problem. If there is an error in identifying the true gastroesophageal junction by endoscopy, every other definition of diseases in this region becomes suspect (Fig. 11.6). The very concept of normal endoscopic appearance becomes suspect; patients presently regarded as normal endoscopically may actually be quite abnormal and at risk for Barrett esophagus and adenocarcinoma.

IDENTIFICATION AND VALIDATION OF THE TRUE GASTROESOPHAGEAL JUNCTION

In the preceding chapters, we suggested that the correct definition of the true gastroesophageal junction is the proximal limit of gastric oxyntic mucosa. In the normal person, this coincides with the squamocolumnar junction at the point at which the tubular esophagus ends. In patients with reflux, a metaplastic columnar-lined esophageal segment is interposed proximal to the proximal limit of gastric oxyntic mucosa as the squamocolumnar junction moves cephalad.

What is the anatomic configuration of this segment of columnar-lined esophagus? Present belief is that the metaplastic columnar esophageal epithelium is located in the tubular esophagus, which is one endoscopic definition of the gastroesophageal junction. Present belief is also that it is proximal to the proximal margin of the rugal folds, which is the other endoscopic definition of the gastroesophageal junction. The latter definition assumes that all epithelia in mucosa with rugal folds is gastric, irrespective of its histologic type. By this definition, if cardiac mucosa with and without intestinal metaplasia or oxyntocardiac mucosa is found in rugated mucosa, it is gastric (2). The literature ignores the situation when the proximal limit of the rugal folds does not coincide with the end of the tubular esophagus; by default, it would seem that the proximal limit of the rugal folds takes precedence in defining the gastroesophageal junction (Fig. 11.7).

To settle this question, we must go back to Allison and Johnstone's article published in 1953 (13) in which they proved to Barrett that what Barrett was calling

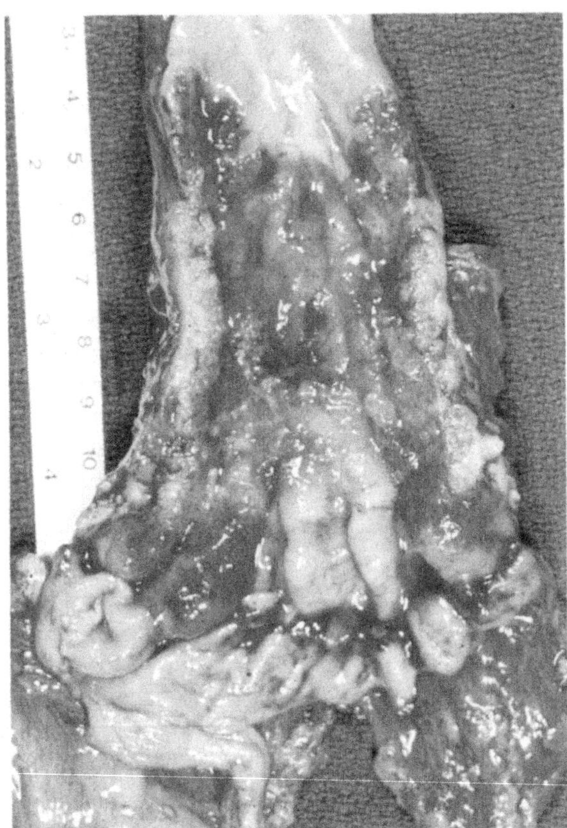

FIGURE 11.7 Discrepancy between point of flaring of the tubular esophagus (seen to be gradual and not very clear) and the proximal limit of rugal folds (not very well demarcated). (Cross reference: Color Plate 3.2)

tubular stomach was actually a columnar-lined esophagus. Dr. D. H. Collins, Allison's pathologist, describes his masterful assessment of the esophagectomy specimen from case 1: "the oesophagus was separated with a knife from the stomach along the line of the peritoneal reflection. A vertical slice was then made through the centre of the reconstituted specimen, i.e., up the posterior wall, and three vertically contiguous blocks were prepared" (13, p. 94). Dr. Collins was exactly reproducing our diagrammatic vertical section across the distal esophagus and proximal stomach. He used the line of the peritoneal reflection, which is a validated definition of the gastroesophageal junction, to separate the esophagus from the stomach.

Allison and Johnstone then examined the part above the peritoneal reflection (i.e., the esophagus) and below it (i.e., the stomach) and defined the following criteria as being indicative of the esophagus rather than stomach: "that it has no peritoneal covering, that the musculature is that of the normal oesophagus, that there may be islands of squamous epithelium within it, that there are no oxyntic cells in the mucosa, and that in addition to gastric glands there are present typical oesophageal mucous glands" (p. 87). In 1953, at the first moment columnar esophagus was described, it was known to Allison, Johnstone, and their pathologist, that "the stomach below the anatomical junction with the oesophagus is lined by gastric mucosa of fundal type.... Cardiac glands and cardiac gastric mucosa do not appear until 0.6 cm up the anatomical oesophagus" (13, p. 94). What we are trying to prove in this book was proved by Allison and Johnstone at the very inception of the recognition of the columnar-lined esophagus!

In Chapter 4, we assessed these criteria in detail and concluded that the most reliable marker of the esophagus is the presence of submucosal glands. According to standard histology texts, submucosal glands are a feature of the esophagus and are not found in the stomach (14). Embryologic studies show that submucosal glands develop in the esophagus after the surface epithelium has changed from fetal columnar epithelium to stratified squamous epithelium. According to Johns (15), "the deep glands do not appear until the epithelium has become stratified squamous in character, and they develop as outgrowths from this epithelium" (p. 438). Goetsch et al. (16) reported that submucosal glands were invariably found in the esophagus and that their numbers varied from 60 to 741 in the adult esophagus. These submucosal glands remain when the esophagus undergoes columnar metaplasia and their ducts can be seen in mucosal biopsies in columnar-lined esophagus (Fig. 11.2).

Literature Review

Chandrasoma P, Makarewicz K, Ma Y, DeMeester T. A proposal for a new validated histologic definition of the gastroesophageal junction. Human Pathol 2006;37:40–47.

In this study of the gastroesophageal junctional region in esophagogastrectomy specimens, we correlated the distal limit of esophageal submucosal glands with the surface epithelial type, the point of flaring of the tubular esophagus, and the proximal limit of the rugal folds in an attempt to establish the true definition of the gastroesophageal junction.

Methods

We selected 10 esophagogastrectomy specimens based on the following criteria: (a) the absence of significant distortion of the gastroesophageal junctional region by tumor; (b) the presence of a well-defined point of flaring of the tubular esophagus that coincided with the proximal limit of the rugal folds (Fig. 11.8); and (c) the presence of grossly and

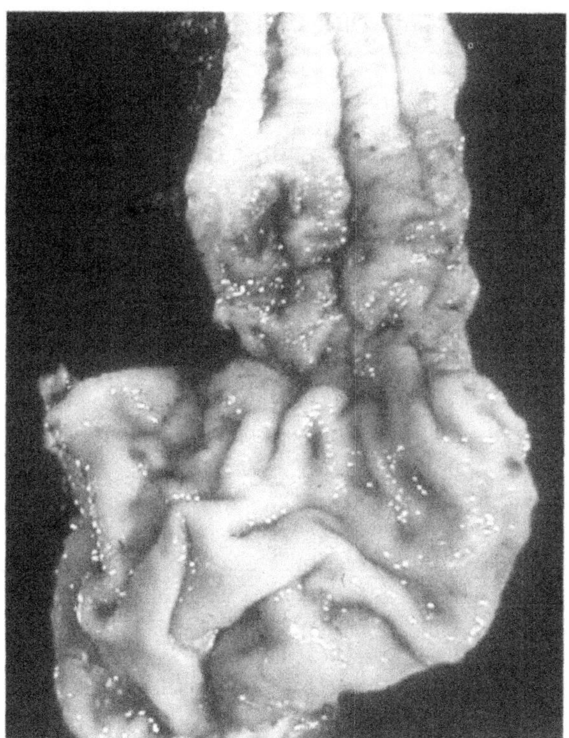

FIGURE 11.8 Esophagogastrectomy specimen of a patient with adenocarcinoma arising in long-segment Barrett esophagus. The tubular esophagus flares at a distinct point. The rugal folds reach the end of the tubular esophagus. The distal esophagus is lined by columnar epithelium containing an ulcerated tumor. The squamocolumnar junction is seen as an irregular line several centimeters up the tubular esophagus. (Cross reference: Color Plate 3.5)

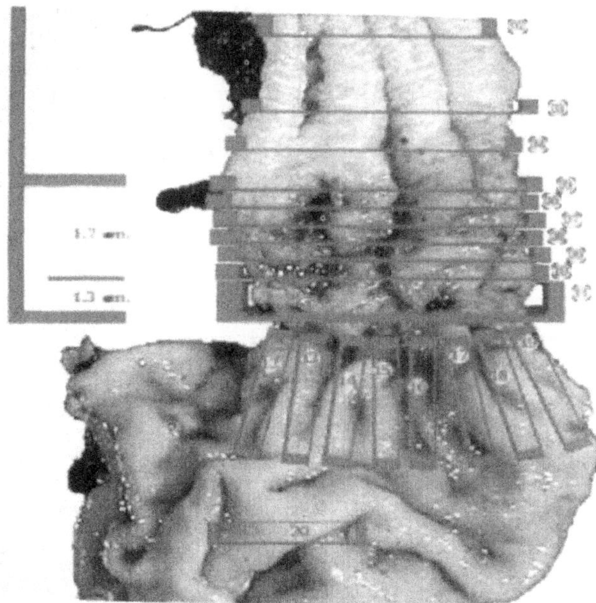

FIGURE 11.9 Specimen in Figure 11.8 separated by a line drawn across the end of the tubular esophagus, which is also the proximal limit of the rugal folds. The method of sectioning of the specimen is shown. The horizontal sections are taken from the columnar-lined esophagus at 5-mm intervals. The area distal to the presently accepted gastroesophageal junction is completely cut in using vertical sections passing from the junction distally.

microscopically normal gastric oxyntic mucosa at the distal resection margin of the specimen. These included 8 patients with Barrett esophagus (1 patient with high-grade dysplasia and 7 patients with adenocarcinoma) and 2 patients with squamous carcinoma of the esophagus.

After gross examination, each specimen was separated by a horizontal line made at the end of the tubular esophagus where it flared out into the gastric pouch (Fig. 11.9). Our selection criteria were such that this line was also the proximal limit of the rugal folds. When the lower esophagus had a columnar lining, the columnar-lined tubular esophagus was completely cut in using horizontal sections at 0.5 cm intervals (Fig. 11.9). Random sections were also taken from the more proximal squamous-lined tubular esophagus. The proximal part of the pouch distal to the tubular esophagus was cut in completely using vertical sections passing from the end of the tubular esophagus to the maximum distance permitted by the limitation of the size of standard cassettes for histologic processing (approximately 2.5 cm) (Fig. 11.9). Random sections were taken from the more distal gastric pouch nearer the distal surgical margin.

The full thickness sections were examined for the lining epithelial types and the presence of submucosal mucous glands. The length of the different epithelial types in the pouch was measured with an ocular micrometer and the epithelial types mapped for each case (Table 11.1). The presence and location of submucosal mucous glands in the tubular esophagus and the distal pouch were noted. The location of the glands in the pouch was measured from the end of the tubular esophagus.

Results

Submucosal glands were present in the tubular esophagus in all 10 cases. The number of glands varied considerably between patients. When the tubular esophagus was lined by glandular epithelium, the number of glands present was similar, in general, to that seen in the squamous-lined esophagus.

Submucosal glands were present in the vertical sections taken through the proximal pouch in 8 patients. The two patients in whom glands were not found were (a) 1 patient (case 9) who had only one submucosal gland in the entire esophagogastrectomy specimen and (b) the patient with squamous carcinoma who had a mean cumulative length of cardiac + oxyntocardiac + intestinal epithelium of 0.31 cm distal to the tubular esophagus. The number of glands in the pouch varied between 0 and 66 in the different patients, the variation being a function of the number of glands present in the esophagus in that patient and the mean cumulative length of cardiac + oxyntocardiac + intestinal epithelium in the pouch (Table 11.1).

TABLE 11.1 Clinical and Pathologic Characteristics in the 10 Patients

	1	2	3	4	5	6	7	8	9	10
Sex/age	M/62	F/56	F/55	M/58	M/47	M/56	M/65	M/58	F/60	M/61
Tumor type	Adeno	SCC	Adeno	Adeno	SCC	Adeno	HGD	Adeno	Adeno	Adeno
Mean CM + OCM + IM length in tubular esophagus (cm)	3.0	0	5.5	4.5	0	<0.5	2.0	<0.5	<0.5	5.0
Mean CM + OCM + IM length in pouch (cm)	1.03	0.31	1.05	1.13	0.43	1.10	1.68	1.60	1.39	2.05
Submucosal glands under SM in tube (total number)	115	59	6	9	4	15	17	17	0	104
Submucosal glands under CM, OCM, IM in tube	114	N/A	53	38	N/A	2	52	3	1	124
Submucosal glands under CM and IM in pouch	18	0	13	4	6	1	25	12	0	40
Submucosal glands under OCM in pouch	7	0	6	0	1	0	0	1	0	26
Submucosal glands under Oxyntic mucosa in pouch	0	0	0	0	0	0	0	0	0	0

Adeno = Adenocarcinoma; SCC = squamous cell carcinoma; HGD = high-grade dysplasia; GEJ = gastroesophageal junction; SM = squamous mucosa; CM = cardiac mucosa; OCM = oxyntocardiac mucosa; IM = intestinal metaplasia; N/A = not applicable.

All patients had cardiac and oxyntocardiac mucosa between the distal end of the tubular esophagus and the gastric oxyntic mucosa in their vertical pouch sections. In the patients with squamous carcinoma, the cumulative cardiac + oxyntocardiac + intestinal epithelium length in the pouch was 0.31 cm and 0.43 cm. The 8 patients with Barrett esophagus had cumulative cardiac + oxyntocardiac + intestinal epithelium lengths ranging from 1.03 to 2.05 cm in the pouch distal to the tubular esophagus.

Intestinal metaplasia was present in the vertical pouch sections at least focally in all 8 patients with Barrett esophagus and in 1 of the 2 patients with squamous carcinoma.

The epithelial types overlying submucosal glands in the pouch sections included cardiac mucosa (8 cases), oxyntocardiac mucosa (3 cases), and intestinal metaplasia (5 cases) (Figs. 11.10, 11.11, and 11.12). The density of submucosal glands in the pouch was similar to that in the tubular esophagus in areas of the pouch that was lined by cardiac mucosa and intestinal metaplasia, and it was slightly less under oxyntocardiac mucosa. No submucosal glands or gland ducts were found in any part of the pouch that was lined by oxyntic mucosa, either in the vertical sections of the proximal pouch or in any of the numerous random sections of the distal pouch. There was a very close (<0.5 cm) correlation between the most distal submucosal gland and the distal limit of cardiac + oxyntocardiac + intestinal epithelium present in the vertical pouch sections (Fig. 11.13).

This study shows that submucosal glands are present distal to both the end of the tubular esophagus and the proximal limit of the rugal folds in 8 of 10 patients. Of the 2 patients who did not have submucosal glands, 1 patient had a very small number of submucosal glands in the entire specimen and the other was a patient with squamous carci-

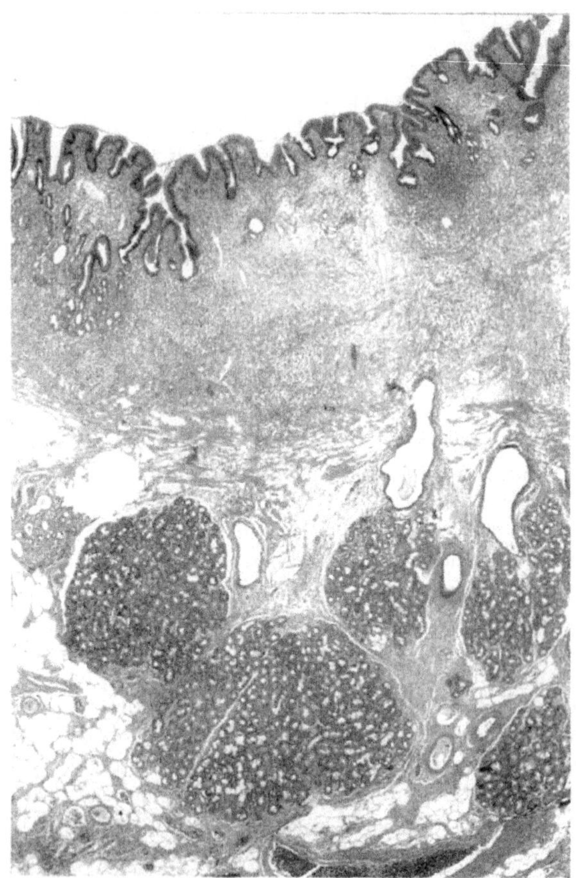

FIGURE 11.10 Section from the saccular region showing the cardiac and oxyntocardiac mucosal lining of the surface and a lobulated submucosal gland in the submucosa.

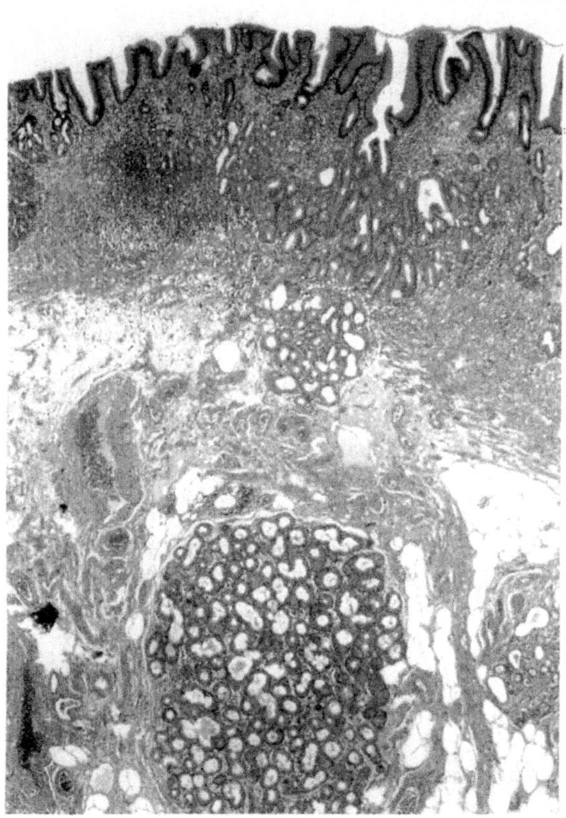

FIGURE 11.11 Higher power of section from the saccular region showing oxyntocardiac mucosal lining of the surface and a lobulated submucosal gland in the submucosa.

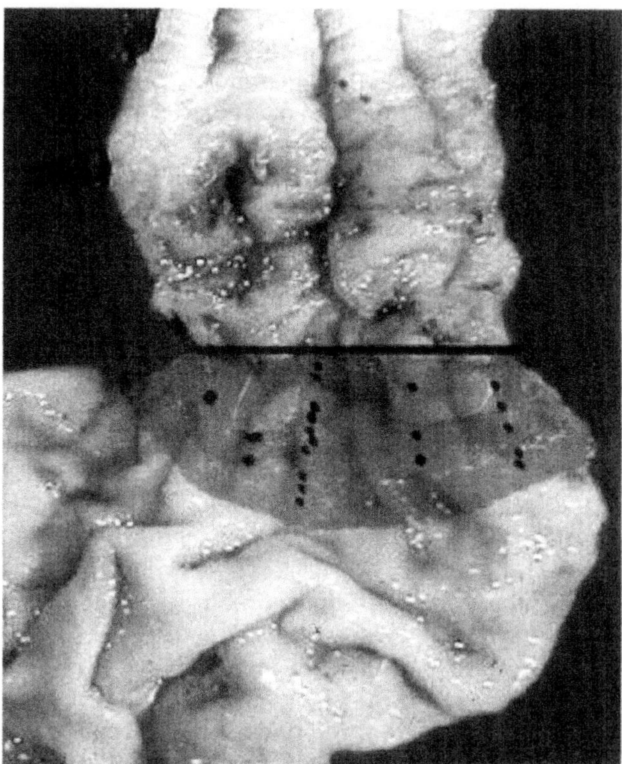

FIGURE 11.12 Specimen in Figure 11.8 with histology of the vertical sections across the proximal part of the saccular region drawn in, showing cardiac mucosa with and without intestinal metaplasia and oxyntocardiac mucosa extending distal to the presently accepted gastroesophageal junction for 2.05 cm (shaded area). The location of the submucosal glands is shown as black dots. (Cross reference: Color Plate 3.6)

noma who had only 0.31 cm of cardiac and oxyntocardiac mucosa. If it is true that submucosal glands are a marker for esophageal location, these data indicate that esophageal tissue is present distal to the presently accepted endoscopic definitions of the gastroesophageal junction.

The amount of esophageal tissue distal to the end of the tubular esophagus can be assessed by two criteria:

1. *The distance to which cardiac mucosa with and without intestinal metaplasia and oxyntocardiac mucosa extended.* These are, according to our criteria, metaplastic esophageal epithelia that end at the proximal limit of gastric oxyntic mucosa, which is our definition of the true gastroesophageal junction. By this definition, the amount of esophageal tissue distal to the end of the tubular esophagus was 0.31 to 2.05 cm in these patients.
2. *The distance to which submucous glands were present distal to the end of the tubular esophagus.* This distance coincided with the length of metaplastic epithelia within 0.5 cm in all cases. It was slightly less in all cases; never did submucosal glands extend to the area covered by gastric oxyntic mucosa, confirming that submucosal glands are not found in the stomach. There was evidence to suggest that the most distal 0.5 cm of the esophagus often contained fewer glands than the remainder; this

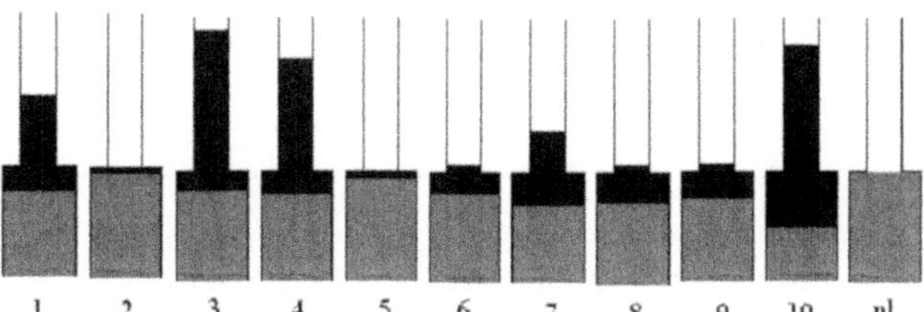

FIGURE 11.13 Summary of findings in the ten esophagogastrectomy specimens compared to our concept of normal (nl). The columnar metaplastic epithelium is solid black between the squamous epithelium (white) and gastric oxyntic mucosa (gray). All specimens show columnar metaplastic epithelium distal to the end of the tubular esophagus (and proximal limit of rugal folds) ranging in length from 0.31 to 2.05 cm. Eight out of 10 specimens also contain columnar metaplastic epithelium in the distal tubular esophagus. The submucosal glands are shown as white dots; their extent is seen to correspond almost exactly to the distribution of the columnar metaplastic epithelium. Submucosal glands were never seen overlying oxyntic mucosa. Submucosal glands under squamous epithelium are not shown in this diagram.

was suggested by the fact that the number of glands was similar under squamous, cardiac, and intestinal epithelia but lower under oxyntocardiac mucosa. Oxyntocardiac mucosa is usually the most distal type of epithelium in the columnar-lined esophagus.

The use of the presence of metaplastic columnar epithelia is more accurate than using the submucosal glands, which are randomly distributed in varying numbers in different people. While their presence proves an esophageal location, their absence does not indicate that a given point is not esophageal. As such, the use of submucosal glands to define the end of the esophagus would be expected to place the true gastroesophageal junction more proximally than its true position, as was the finding in this study. Submucosal glands are valuable in validating the gastroesophageal junction but are not sufficiently constant to be used to define it.

This study validates the definition of the proximal limit of gastric oxyntic mucosa as the true gastroesophageal junction. It also proves that the present endoscopic definitions of the gastroesophageal junction (proximal limit of the rugal folds and end of the tubular esophagus) place the gastroesophageal junction more proximally than its true location. In these 10 patients, the magnitude of the error varied between 0.31 and 2.05 cm. Stated in a different way, in these 10 patients, using the endoscopic definitions of the gastroesophageal would have resulted in 0.31 to 2.05 cm of the distal esophagus being erroneously designated as the proximal stomach. Sarbia et al. (7) found this segment to be as long as 28 mm; we suggest that the figure of 3 cm be used as the maximum possible extent of this segment of cardiac mucosa that can be found distal to the end of the tubular esophagus (Fig. 11.14).

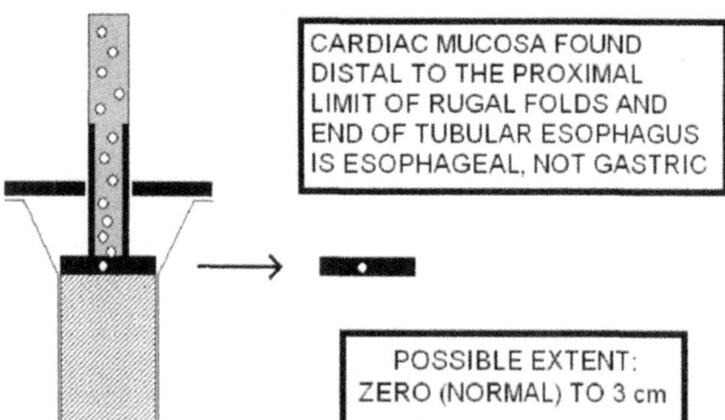

CARDIAC MUCOSA FOUND DISTAL TO THE PROXIMAL LIMIT OF RUGAL FOLDS AND END OF TUBULAR ESOPHAGUS IS ESOPHAGEAL, NOT GASTRIC
POSSIBLE EXTENT: ZERO (NORMAL) TO 3 cm

FIGURE 11.14 A patient with approximately 1 cm of cardiac mucosa distal to the tubular esophagus. The presence of submucosal glands under this epithelium shows that this is esophageal, not gastric as presently believed. The maximum extent of abnormal, dilated esophagus found distal to its tubular segment is a maximum of 3 cm.

THE ANATOMIC CHANGES ASSOCIATED WITH SLIDING HIATAL HERNIA

We have recognized the obvious alteration of the anatomy of the lower esophagus and proximal stomach that is associated with a sliding hiatal hernia since the earliest times. They are well described in Allison's 1948 report (17) (Fig. 11.15).

The changes that characterize a sliding hiatal hernia are as follows (Fig. 11.16):

1. The esophagus has become shortened and ends in the thorax. The stomach, with a sac of peritoneum, is drawn up into the mediastinum as a hernia.
2. There are invariable changes in the lower esophageal sphincter. In all cases, and by definition, the sphincter ends in the thorax. There is no abdominal component of the lower esophageal sphincter. In most patients, there is a reduced resting

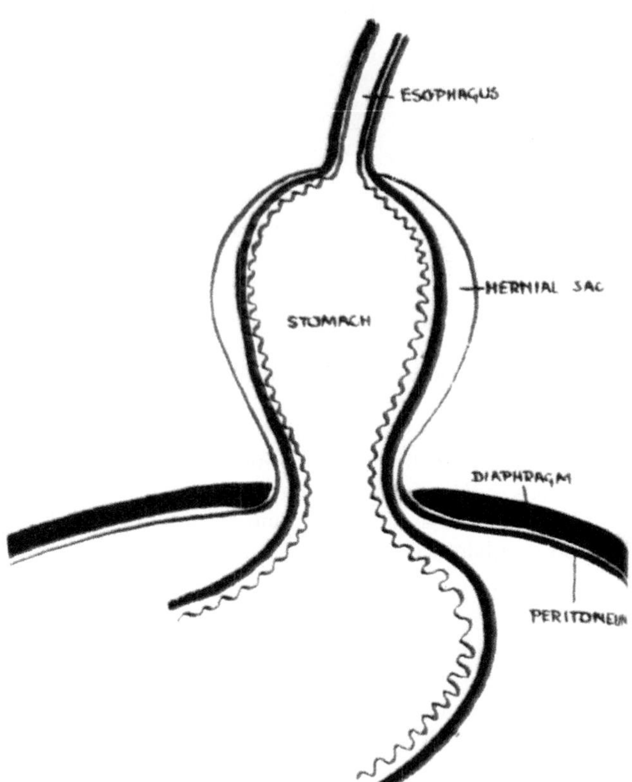

FIGURE 11.15 Sliding hiatal hernia, as illustrated by Allison in (17). Note that the distended stomach is entirely covered by a peritoneal sac that reaches the end of the tubular esophagus. Allison showed squamous epithelium (straight line) lining the entire tubular esophagus and gastric mucosa (serrated line) lining the entire saccular hernia. At this time in history, columnar-lined esophagus had not yet been described. This situation is very idealized and quite uncommon. Most patients with sliding hiatal hernia have severe malfunction of the lower esophageal sphincter and reflux-induced columnar-lined esophagus.

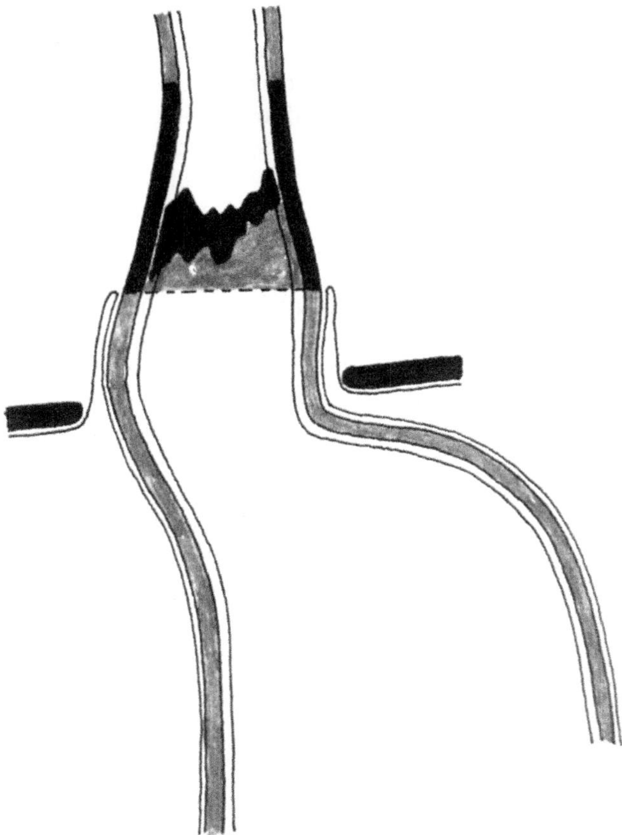

FIGURE 11.16 More typical situation in hiatal hernia. The anatomy is represented in a manner that is identical to Allison (see Fig. 11.15), but a segment of cardiac and oxyntocardiac mucosa is shown above the proximal limit of oxyntic mucosa (broken line) within the hiatal hernia. The upper part of the "hiatal hernia" consists of dilated esophagus. This can only be defined by histology.

sphincter pressure as well as considerable sphincter shortening.

3. There is complete obliteration of the normal acute angle of His at the gastroesophageal junction. Obliteration of the angle of His results in a straightening of the left border of the esophagus and stomach to resemble the right side. As described by Allison in 1948 (17): "In the patient with short oesophagus the picture is quite different (to the normal endoscopic appearance), for in the absence of stenosis the instrument is passed down from oesophagus to stomach without impediment or deflection. The viscus is lined by oesophageal mucosa at one level and by gastric at the next; the level is higher than would be anticipated in the normal" (p. 36). The esophagus and stomach have changed into an appearance that resembles a bottle of chardonnay with a gradually tapering continuity between the esophagus and the stomach. Once this anatomic change has occurred, shortening of the esophagus can pull the stomach up into the thorax as a sliding hiatal hernia. As long as an acute angle of His is present, the stomach cannot slide into the thorax.

When a sliding hiatal hernia is present, endoscopic assessment becomes more difficult (18). The end of the tubular esophagus and the point where it dilates has become indistinct, and the end of the tubular esophagus cannot be used as a reliable marker for the gastroesophageal junction. In this situation, the only marker at endoscopy for the gastroesophageal junction is the proximal limit of the rugal folds. As indicated previously this is not an accurate marker. We have observed cardiac mucosa and oxyntocardiac mucosa in biopsies taken distal to the proximal limit of the rugal folds, suggesting that some part of the dilated organ distal to the proximal limit of the rugal folds is actually dilated esophagus that is mistaken as the stomach by the use of an incorrect definition of the gastroesophageal junction. As in normal patients, the true gastroesophageal junction in patients with sliding hiatal hernia can be defined by the proximal limit of gastric oxyntic mucosa (Fig. 11.16).

The anatomic changes associated with a sliding hiatal hernia represent the end result of the most severe abnormality that is associated with reflux. It is quite amazing that we recognize normal anatomy and the altered anatomy of a sliding hiatal hernia without any consideration of the intermediate steps that must exist between these two extremes.

STAGES BETWEEN NORMAL AND SLIDING HIATAL HERNIA: THE REFLUX-DAMAGED DISTAL ESOPHAGUS—DEFINING THE END-STAGE ESOPHAGUS

Gastroesophageal reflux first affects the most distal part of the esophagus that is closest to the stomach. An incredible aspect of this disease is that the medical community, from the very beginning, has ignored this part of the esophagus. Short segments of columnar-lined esophagus were called normal for four decades. At present, we do not recognize Barrett esophagus until we can see it endoscopically in the *tubular esophagus.* If, as we have shown, the true gastroesophageal junction is significantly (up to 2.05 cm in the 10 patients in our study [8]) distal to the end of the tubular esophagus, we are calling up to 2.05 cm of reflux-damaged esophagus "normal" and completely ignoring it. We are slow to learn from the mistakes of the past.

The changes in the anatomy of the distal esophagus resulting from reflux damage have been greatly underestimated. There is a naive tendency to believe that landmarks in the normal person are retained in the patient whose distal esophagus has been damaged by reflux. The end of the tubular esophagus is believed to mark the gastroesophageal junction even in patients with severe reflux damage to the esophagus. In actuality, reflux damage to the distal esophagus produces devastating alteration to the anatomy, and we have not appreciated this fully.

Normal Anatomy and Physiology (also see Chapter 4)

Consider the anatomy of the distal 2 to 3cm of the esophagus; this is the abdominal segment of the esophagus that is normally completely part of the lower esophageal sphincter. This is normally tubular in shape and lined by squamous epithelium. It enters the abdomen through the diaphragmatic hiatus. The true gastroesophageal junction in the normal person is at the end of the tubular esophagus where the esophagus flares into the saccular stomach, which is entirely lined by gastric oxyntic mucosa (Fig. 11.17).

From a functional and anatomic standpoint, the distal 10cm of the esophagus must be recognized as being composed of three different segments (Fig. 11.18):

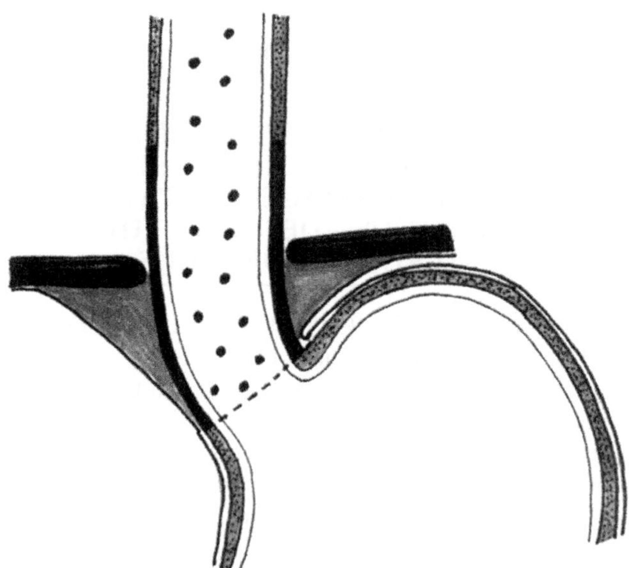

FIGURE 11.17 Normal anatomy showing the lower esophageal sphincter (black wall) in the distal thoracic and entire abdominal esophagus. The dotted line is the normal gastroesophageal junction defined by the end of esophageal squamous epithelium (white above line), the proximal limit of gastric oxyntic mucosa (white below line), the peritoneal reflection, and the end of the tubular esophagus. Submucosal glands extend to the end of the esophagus (black circles).

1. *The body of the esophagus in the thorax above the lower esophageal sphincter.* This has no resting muscle tone and its intraluminal pressure is intrathoracic (–5mmHg) with variation during respiration typical of an intrathoracic organ (i.e., more negative during inspiration).
2. *The intrathoracic part of the lower esophageal sphincter.* This has a resting pressure of 15 to 20mmHg dependent on the muscle tone of the lower esophageal sphincter. The variation with respiration is identical to the body of the esophagus. The intraluminal pressure in the intrathoracic part of the sphincter will be the same as the body of the esophagus (–5mmHg) if the sphincter muscle tone became lost for any reason.
3. *The intraabdominal part of the lower esophageal sphincter.* This has the resting pressure of 15 to 20mmHg dependent on the muscle tone of the lower esophageal sphincter. The variation with respiration is the reverse of that seen in the body of the esophagus and intrathoracic part of the sphincter. The variation of pressure is that of an abdominal organ, becoming more positive with inspiration. It is identical to the pattern seen in the stomach. It should be recognized that the intraluminal pressure in the intraabdominal part of the sphincter will be the same as the stomach (+5mmHg) if the sphincter muscle tone became lost for any reason. At the distal limit of the lower esophageal sphincter, the stomach is entered. This has no resting tone, but

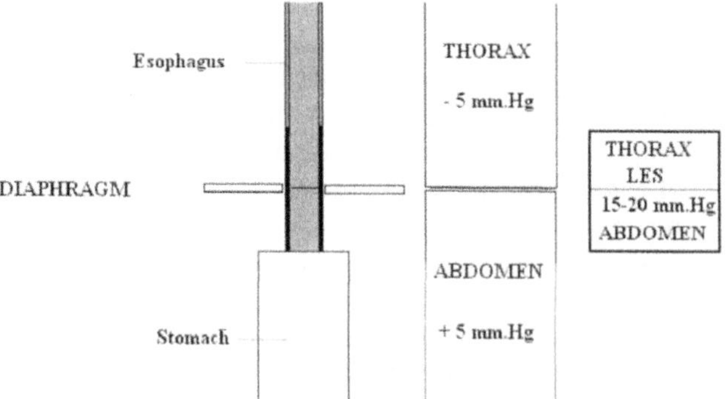

FIGURE 11.18 Functional anatomy showing the high pressure zone of the lower esophageal sphincter (15 to 20mm Hg) in the distal thoracic esophagus and entire abdominal esophagus. The intraluminal pressure in the normal esophagus proximal to the diaphragm (respiratory inversion point) is thoracic (–5mm Hg), whereas that in the abdominal esophagus is abdominal (+5mm Hg). The dilatory effect of this positive intraluminal pressure is normally inhibited by the high tonic resting pressure of the sphincter.

the intraluminal gastric pressure is +5 mm Hg typical of an intraabdominal organ.

The intraabdominal part of the esophagus is entirely composed of the intraabdominal part of the lower esophageal sphincter. In a normal person, this is about 2 to 3 cm long and ends at the normal gastroesophageal junction. The peritoneal reflection of the undersurface of the diaphragm passes down to the gastroesophageal junction enclosing a cylindrical mass of connective tissue that surrounds the intraabdominal esophagus (Fig. 11.17).

On the left side, the end of the esophagus makes an acute angle with the fundus of the stomach, which passes upward and to the left to come into relation with the undersurface of the left diaphragm (Fig. 11.17). This acute angle of His is established early in embryologic development (see Chapter 3) as a result of differential growth rates between the greater and lesser curvatures of the fetal stomach (19).

The intraabdominal part of the esophagus and the angle of His are believed crucial in the prevention of gastroesophageal reflux when there are sudden increases in intraabdominal pressure during straining, exercise, coughing, and so on. Any increase in intraabdominal pressure tends to cause the angle of His to increase the valvelike action of the normal gastroesophageal junction, permitting the esophagus to effectively resist the tendency to reflux caused by increased intraabdominal pressure. The greater the pressure increase in the abdomen, the greater the accentuation of this mechanism (20). The strain associated with increased intraabdominal pressure increase is not an important mechanism of abnormal reflux in patients with gastroesophageal reflux disease (21).

The Mechanism of Gastroesophageal Reflux

A large amount of data exists regarding the mechanisms of reflux and the changes in the lower esophageal sphincter that are associated with the occurrence of reflux. The sphincter normally relaxes to permit the passage of food during swallowing. Because this is associated with the onward passage of food through the sphincter, and because the sphincter returns to its normal high pressure immediately following the passage of the bolus of food, this relaxation does not cause reflux.

Strain-Induced Gastroesophageal Reflux

A potential mechanism for reflux is an increase in the pressure within the abdominal cavity that occurs during straining and coughing. The abdominal segment of the sphincter is crucial in preventing reflux during times of increased abdominal pressure. With a normal abdominal sphincter segment and a normal acute angle of His, any increase in abdominal pressure would tend to accentuate the sphincter action, thereby resisting gastroesophageal reflux (20).

In patients with loss of the abdominal sphincter component and loss of the normal angle of His, such as in a patient with a hiatal hernia, any increase in abdominal pressure will tend to cause reflux (22). Strain-induced gastroesophageal reflux is a minor factor that operates only in patients with severe reflux disease, usually those with a hiatal hernia.

Transient Lower Esophageal Sphincter Relaxations (tLESRs); Belching

A mechanism of sphincter relaxation exists that is designed for the physiologic venting of excessive swallowed air (i.e., belching). Belching occurs in people with a normal sphincter by a mechanism that is crucial to understand; these are called transient lower esophageal sphincter relaxations (tLESRs). Normal belching is preceded by overdistension of the stomach by excessive swallowed air resulting in an increased intragastric pressure. As the stomach dilates and intragastric pressure increases, physical forces develop that pull down the distal esophagus, causing it to be "taken up" into the distended gastric fundus, shortening the abdominal segment of the sphincter and decreasing the acute angle of His. This process continues until a critical shortening occurs; when the abdominal component of the sphincter is somewhere between 1 and 2 cm, sudden sphincter relaxation occurs and the swallowed air is vented as a belch. The venting of air decreases intragastric pressure, reduces gastric distension, and restores the anatomy of the sphincter to its normal state of effectiveness as a barrier to reflux. Belching represents a situation where a sufficiently large amount of air accumulates in the stomach to overcome the normal sphincter.

It has been hypothesized that tLESRs that are caused by gastric distension are triggered by some tension receptor in the proximal stomach (23). No anatomic basis for such receptors has been identified. The ability to explain the occurrence of tLESRs by a purely physical mechanism whereby sphincter relaxation follows shortening of the abdominal component of the sphincter obviates the need for hypothesizing the presence of receptors. The mechanism for tLESRs is a transient shortening of the abdominal segment of the lower esophageal sphincter.

Abnormal tLESR Frequency Is the Main Mechanism for Abnormal Gastroesophageal Reflux

The present belief is that tLESRs precipitated by gastric distension represent the dominant mechanism of abnormal gastroesophageal reflux associated with reflux disease (23, 24). This belief suggests that an increase in the frequency of tLESRs beyond the normal physiologic level associated with belching causes gastroesophageal reflux of sufficient severity to produce reflux disease. The frequency of tLESRs has also been shown to be greatly increased in patients with hiatal hernia.

The increased frequency of tLESRs in patients with reflux disease and hiatal hernia is easy to explain. Both these groups of patients have evidence of permanent sphincter damage, often with permanent shortening of the abdominal component of the lower esophageal sphincter. In a patient with a permanently shortened sphincter, the degree of gastric distension needed to reach the length at which transient sphincter relaxation is precipitated is decreased proportionately to the degree of shortening. This means that, with a constant dietary intake, the tLESR frequency is inversely related to the length of the abdominal component of the lower esophageal sphincter; the shorter the sphincter, the greater the likelihood and frequency of tLESRs.

Ismail et al. (25) showed that there was a close relationship between the amount of gastric distension necessary to overcome the sphincter and the anatomy of the gastroesophageal junctional region. The intragastric pressure at which the sphincter opened in response to distension by air during endoscopy was higher in patients with an intact angle of His when compared to patients with a hiatal hernia who had lost their acute angle of His.

This mechanism was shown to be accurate in a brilliant study from the University of Chicago.

Literature Review

Kahrilas PJ, Shi G, Manka M, Joehl RJ. Increased frequency of transient lower esophageal sphincter relaxation induced by gastric distension in reflux patients with hiatal hernia. Gastroenterology 2000;118:688–695.

In this study, intraluminal pressures from within the esophagus, esophagogastric junction (EGJ), and stomach, along with distal esophageal pH, were measured during gastric air distension. Intraluminal pressures were measured with a 17-lumen catheter, positioned to permit pressure recordings at 1-cm intervals in the distal esophagus above and including the lower esophageal sphincter (LES) and the proximal stomach. Esophageal pH was recorded by a catheter positioned 5 cm above the proximal margin of the LES. Air was infused by a catheter exiting 4 cm proximal to the most distal pressure recording site at 15 mL/min.

Three subject groups were evaluated: Eight healthy asymptomatic volunteers constituted the first group. There were fifteen patients with symptomatic reflux disease (heartburn equal to or greater than three times per week controlled by acid suppression) without Barrett's epithelium. The subjects were endoscoped and a clip placed at the level of the normal squamocolumnar junction (SCJ). These patients were divided into those without hiatal hernia (group 2) and with a hiatal hernia (group 3) by the relationship of the clip at the SCJ to the diaphragmatic hiatus. A hiatal hernia was defined as present when the clip at the SCJ was at least 1 cm proximal to the hiatus. The 8 healthy volunteers and 7 patients with symptomatic reflux disease had no hiatal hernia; 8 of 15 patients with symptomatic reflux had hiatal hernia.

A reflux event was defined as either an abrupt decrease in esophageal pH to less than 4 for at least 5 seconds or, if the esophageal pH was already less than 4, a further abrupt decrease of at least 1 pH unit for at least 5 seconds. The manometric tracing was analyzed to characterize the basal EGJ pressure, the length of the EGJ high pressure zone, and the pressure activity associated with each reflux event. tLESRs were defined using the criteria of Holloway et al.: (a) absence of swallowing for 4 s before to 2 s after the onset of LES relaxation, (b) relaxation rate of > or =1 mmHg/s, (c) time from onset to complete relaxation of < or =10 s, and (d) nadir pressure of < or =2 mmHg. Each reflux event was then classified as tLESR-related, swallow-related, strain-related, or free.

The following results were reported:

1. Intragastric air infusion significantly increased the number of reflux episodes and the esophageal acid exposure time in all subject groups. During air infusion, the 15 patients with symptomatic reflux had significantly more reflux events and acid exposure time than the 8 healthy controls (Fig. 11.19, 11.20, and 11.21).
2. The predominant mechanism of reflux events was tLESRs in the normal controls and nonhernia patients in both the baseline period (75% and 83%) and during the air infusion (70% and 83%). Among the hernia patients, swallowing-induced reflux accounted for 89% of the reflux events during baseline recording with tLESRs responsible for only 7%. During air infusion, tLESRs became the dominant mechanism, accounting for 52% of reflux events (Fig. 11.19).
3. During baseline recording, all three subject groups had a similar number of tLESRs. With intragastric air infusion, normal controls and nonhernia GERD patients had an increase in the tLESR frequency of 4.0 and 4.5 per hour, respectively ($p < 0.05$), and hernia patients had a median increase in tLESR frequency of 9.5 per hour ($p < 0.001$). The hernia patients had significantly more

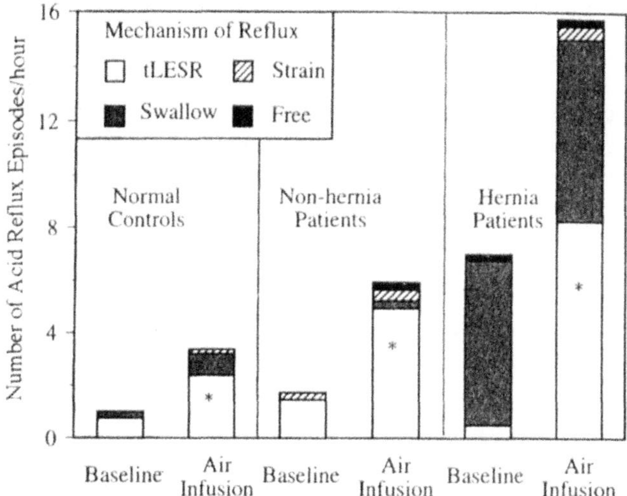

FIGURE 11.19 The number of acid reflux events related to different mechanisms during baseline recording and during a period of intragastric air infusion among normal controls, nonhernia GERD patients, and hernia patients. Air infusion resulted in a significant increase in reflux events only by the tLESR mechanism. Reproduced with permission from Kahrilas PJ, Shi G, Manka M, Joehl RJ. Increased frequency of transient lower esophageal sphincter relaxation induced by gastric distension in reflux patients with hiatal hernia. Gastroenterology 2000;118:688–695.

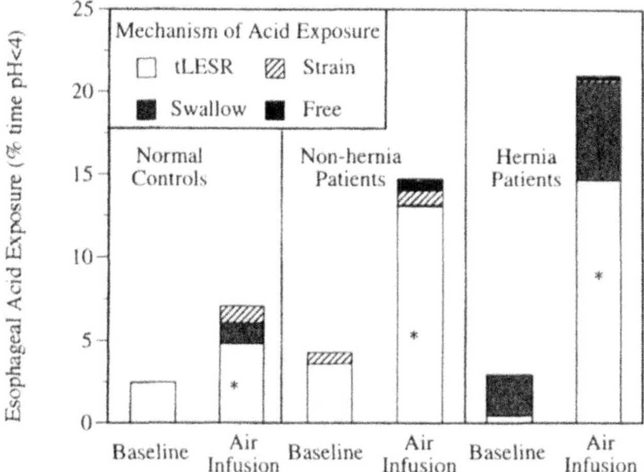

FIGURE 11.20 Esophageal acid exposure time associated with different mechanisms of reflux during baseline recording and during a period of intragastric air infusion among subject groups. Air infusion significantly increased acid exposure time resulting from increased tLESRs in all groups. Reproduced with permission from Kahrilas PJ, Shi G, Manka M, Joehl RJ. Increased frequency of transient lower esophageal sphincter relaxation induced by gastric distension in reflux patients with hiatal hernia. Gastroenterology 2000;118:688–695.

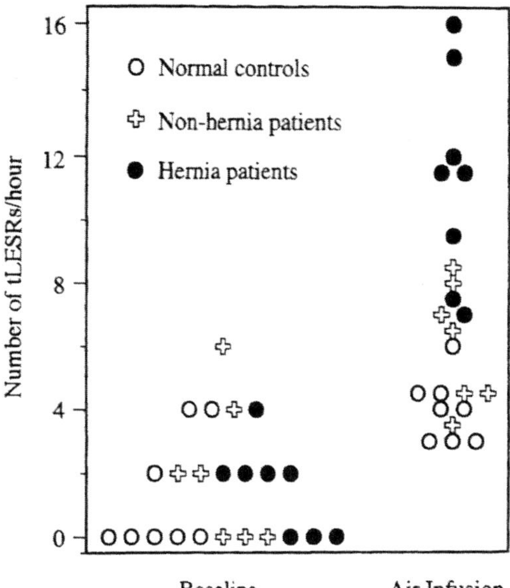

FIGURE 11.21 Individual data on the frequency of tLESRs in normal controls, nonhernia GERD patients, and hernia patients during baseline recording and during a period with intragastric air infusion. Air infusion increased tLESR frequency in all groups. Both patient groups had a significantly higher tLESR frequency than the control group during air infusion, and the hernia patients had a significantly higher frequency than nonhernia patients. Reproduced with permission from Kahrilas PJ, Shi G, Manka M, Joehl RJ. Increased frequency of transient lower esophageal sphincter relaxation induced by gastric distension in reflux patients with hiatal hernia. Gastroenterology 2000;118:688–695.

tLESRs than normal controls ($p < 0.001$) and nonhernia GERD patients ($p < 0.01$). The authors concluded: "The data suggest that these subject groups represent a continuum in the diminution of the threshold for the elicitation of tLESRs in response to gastric distension" (21, p. 688) (Fig. 11.21).

4. The study permitted measurement of the length of the high-pressure zone (i.e., LES) by the number of adjacent side holes in the manometric catheter concurrently residing in the high-pressure zone. During the baseline recording, the high-pressure zone (LES) was significantly longer in the normal controls, showing a stepwise decrease in length in nonhernia GERD patients and hernia GERD patients (Fig. 11.22). Intragastric distension associated with air infusion resulted in a gradual shortening of the high-pressure zone in all three groups; this shortening became significant 20 to 30 minutes after the beginning of distension (Fig. 11.22).
5. The shortening of the high-pressure zone preceded the tLESR, suggesting that "the distal EGJ was opened as an early event in triggering a tLESR" (21, p. 694).
6. During intragastric air infusion, there was a gradually increased intragastric pressure which was then restored to baseline after a tLESR. The threshold for increased intragastric pressure for triggering tLESRs was similar

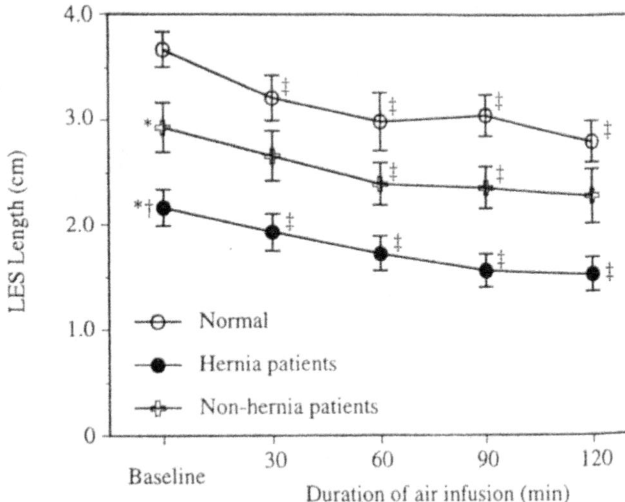

FIGURE 11.22 Esophagogastric junction (= lower esophageal sphincter) length during recordings averaged longer than 30-minute periods. This shows that sphincter length at baseline is longest in normal patients and less in nonhernia GERD and still less in hernia patients. All three groups show a similar decrease in the lower esophageal sphincter length during air infusion. Reproduced with permission from Kahrilas PJ, Shi G, Manka M, Joehl RJ. Increased frequency of transient lower esophageal sphincter relaxation induced by gastric distension in reflux patients with hiatal hernia. Gastroenterology 2000;118:688–695.

among the three subject groups, suggesting that it was the perturbed anatomy of the EGJ (LES) associated with hiatal hernia and reflux disease that was of critical importance in precipitating tLESRs.

It is interesting that in their discussion, Kahrilas et al. made the following statement: "It is possible that with hiatal hernia the distal area of the EGJ (LES) is already partly distended" (21, p. 694). This is what I have described as end-stage dilated esophagus as an early anatomic change caused by reflux disease and represents a merging of the physiologic data of Kahrilas et al. and our histologic data, leading to the same conclusion.

The Earliest Cellular Changes Caused by Reflux

The sequence of changes that occurs with a meal is likely to be similar to that induced by air infusion in Kahrilas et al.'s experiment. With most meals, the rise in intragastric pressure is not great because of the distensibility of the stomach. With an excessively heavy meal, however, gastric distension reaches its limit and then the pressure rises. The increased intragastric pressure causes the sphincter to shorten, resulting in sudden relaxation (tLESR), causing postprandial reflux. The likelihood of tLESRs is increased in patients with a shortened baseline length of the lower esophageal sphincter.

The earliest pathologic change in reflux disease occurs when the most distal part of the esophagus becomes exposed to gastric luminal contents. The human anatomy with a lower esophageal sphincter that has a resting pressure well in excess of intragastric pressure and the valvular effect of the angle of His is designed to prevent this occurrence.

For the first time is biological history, humans have developed a situation in which excess food has become easily available in the more developed parts of the world. Since the mid-1950s, overeating and obesity have rapidly become the most common human problem in the United States. The areas of the world where this ready availability of food is maximal—Western Europe, North America, and Australia—closely match the geographical distribution of highest prevalence of Barrett esophagus and adenocarcinoma of the esophagus. Reflux is primarily a disease of affluent Caucasians.

Overeating is a common occurrence in affluent countries. One of the most incredible things Dr. Chandrasoma noticed when he immigrated to the United States was the massive size of food portions in restaurants compared with those in Asia. Though unproved and conjectural, it is likely that overeating is the event that precipitates the reflux changes. Once it is set in motion, reflux becomes a self-perpetuating vicious cycle phenomenon.

Obesity has been shown to be strongly associated with Barrett adenocarcinoma of the esophagus (26, 27; see Chapter 10). Bu et al. (28) from our unit (see Chapter 10) showed that the effect of obesity is seen in the earliest stages of the reflux to adenocarcinoma sequence. Obesity was significantly associated with the presence of cardiac mucosa at the junction compared to controls who had no cardiac mucosa; the association with obesity increased when intestinal metaplasia was found. Because severity of reflux is associated with intestinal metaplasia, these data suggest that the severity of reflux is related to the amount of obesity. It is suggested that the mechanism whereby obesity is associated with reflux is secondary rather than primary (i.e., obesity does not cause reflux). The cause of obesity, which is the ingestion of frequent large meals, is the likely mechanism whereby these patients have increased reflux.

Gastric distension beyond its usual capacity causes an increase in the intragastric pressure and causes deformation of its points of junction with the duodenum at the pylorus and the esophagus at the gastroesophageal junction. Gastric contents include food and swallowed air; in general, the air accumulates as a bubble in the proximal stomach. There is a normal mechanism of belching where the lower esophageal sphincter relaxes to permit venting of swallowed air into the esophagus to relieve gastric distension. Belching is thought to occur as a result of shortening of the intraabdominal part of the lower esophageal sphincter. This results when overdistension of the stomach deforms the distal part of the gastroesophageal junction, pushing open the sphincter to a point where the shortening is sufficient to cause a sudden relaxation of the sphincter (21). The air is vented, relieving the gastric overdistension and reversing the deformation of the sphincter.

During the deformation of the gastroesophageal junction that accompanies gastric distension, the squamous epithelium lining the lower part of the sphincter must become temporarily exposed to gastric luminal contents (Fig. 11.23). When an air bubble exists, there is no problem because the exposed squamous epithelium does not come into contact with liquid gastric contents. However, if gastric overdistension occurs when there is no air bubble due to overeating after

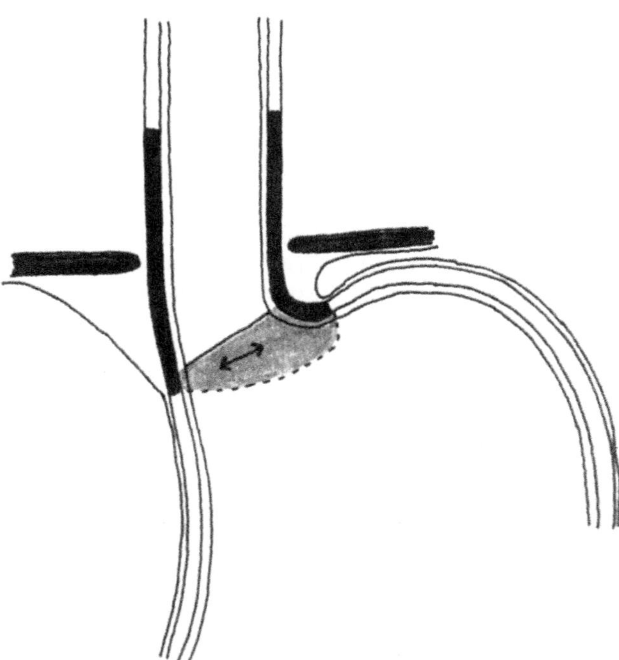

FIGURE 11.23 The change in anatomy resulting from gastric distension. The dilatory pressure at the gastroesophageal junction causes the most distal part of the sphincter to "open" and "be taken up" into the stomach. The functional sphincter length decreases and the gastroesophageal junction (dotted line) moves into the stomach. Part of the esophageal squamous epithelium (solid gray area) in this "taken up" distal esophagus becomes exposed to gastric contents and susceptible to acid-induced damage. This is a temporary phenomenon; when gastric distension decreases, the anatomy reverts to normal. Any squamous epithelial damage that has occurred by the temporary exposure can heal.

belching has vented the air, the exposed squamous epithelium is exposed to gastric luminal contents. This may be especially damaging because gastric pH studies show that there tends to be a pocket of highly acidic material in the proximal stomach in the postprandial phase (29).

The stage is now set for the first reflux damage. The interaction between gastric acid contents and the squamous epithelium during the period of temporary gastric overdistension can damage the squamous epithelium. No change is likely when these gastric overdistension episodes are mild and separated in time sufficiently to allow healing of any damage that has occurred. In this situation, the resistance of the squamous epithelium and the ability to repair mild damage is sufficient to prevent disease (Fig. 11.23).

With severe and frequently repeated episodes of gastric overdistension, squamous epithelial damage occurs, leading to the changes described in Chapter 8, characterized by dilated intercellular spaces, increased permeability of the epithelium, and entry of luminal molecules (acid and all others) into the epithelium. Stimulation of nerve endings by luminal molecules (acid and others) causes heartburn. Interactions of luminal molecules with the proliferating stem cell pool in the basal region of the squamous epithelium cause change in the genetic differentiating signal and lead to columnar metaplasia (Fig. 11.24).

Columnar (cardiac) metaplasia of the esophagus alters the neural complex of the epithelium, interfering with pain generation (though cardiac mucosa may be less sensitive to pain, patients can continue to have pain due to damage in adjacent squamous epithelium) and possibly some reflex neural mechanism that is responsible for maintenance of sphincter tone. By whatever mechanism they result from, early reflux changes in the distal esophagus are associated with sphincter shortening beginning in the most distal part of the sphincter. In Oberg et al.'s study (30), significant shortening of the abdominal part of the lower esophageal sphincter was present in endoscopically normal patients who had cardiac or oxyntocardiac mucosa distal to the squamous epithelium. In Kahrilas et al. (20), baseline sphincter length was shortened progressively in patients with reflux who did not have a hiatal hernia and hiatal hernia patients (Fig. 11.22).

Dilatation of the Sphincter-less Abdominal Esophagus

The loss of the high resting tone of the distal few millimeters of the esophagus has the theoretical potential to result in dilatation of this part of the esophagus. The intraluminal pressure of the abdomi-

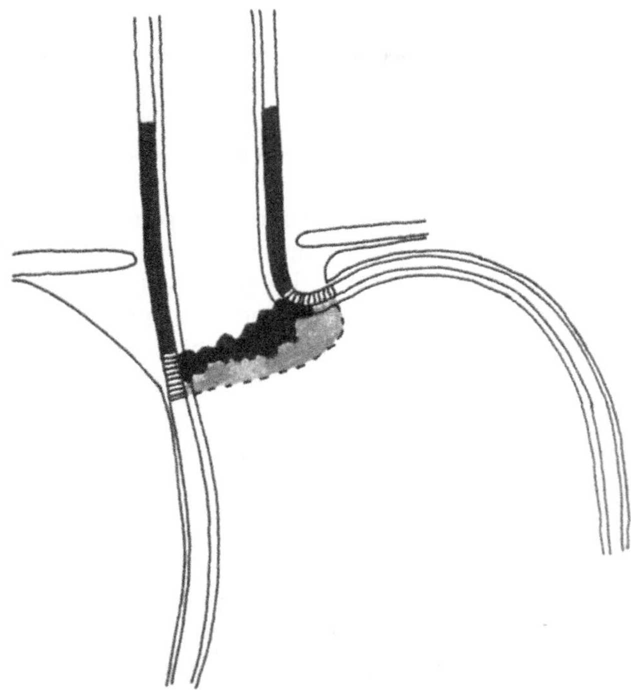

FIGURE 11.24 With repeated and frequent exposures of the distal esophageal squamous epithelium, damage does not heal sufficiently and gastric molecules gain access to the basal region where it induces columnar metaplasia. Cardiac (black) and oxyntocardiac (gray) metaplasia results and is associated with a permanent loss of sphincter tone in this region (striped wall). This part of the esophagus now remains "taken up" and permanently dilated. The abdominal esophagus is shortened and ends closer to the diaphragm, and the angle of His has become less acute. The peritoneal reflection and submucosal glands remain unaltered. The cardiac mucosa evolves within this region; in this diagram there is a mixture of cardiac mucosa and oxyntocardiac mucosa. This diagram actually shows a fairly severe degree of damage with an approximately 50% (1 to 1.5cm) reduction in the length of the abdominal sphincter. With less severe damage, this change is millimetric and extremely difficult to detect by endoscopy or gross examination or even by manometry. The only marker for this is the presence of columnar metaplastic esophageal epithelium. This patient is now at risk for developing intestinal (Barrett's) metaplasia in the cardiac mucosa.

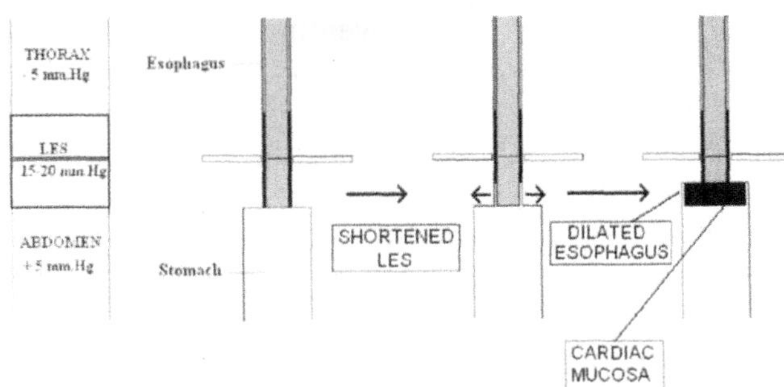

FIGURE 11.25 Mechanism of dilatation of the reflux-damaged distal esophagus. The loss of sphincter tone in the distal abdominal esophagus permits the normally masked positive intraluminal pressure to cause progressive dilatation whenever the intragastric pressure increases. This ultimately results in a permanently dilated esophagus lined by metaplastic columnar epithelium (black). The esophageal wall in the dilated segment contains the destroyed lower esophageal sphinter and submucosal glands.

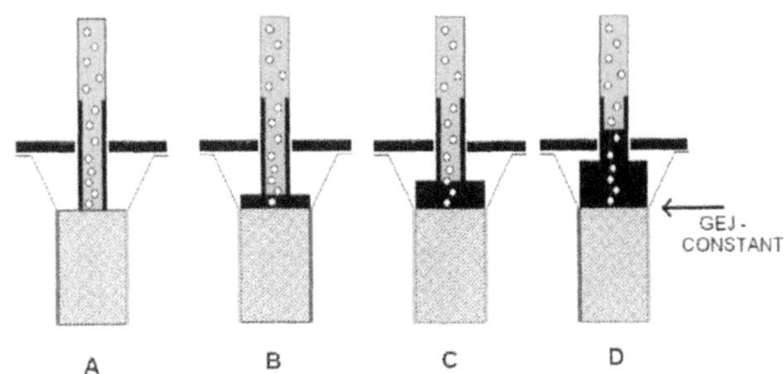

FIGURE 11.26 The stages of change occurring with increasing reflux-induced damage. (a) Normal. (b) A 5-mm shortening of the lower esophageal sphincter with consequent dilatation and columnar metaplasia of the distal esophagus. (c) A 1- to 1.5-cm shortening of the lower esophageal sphincter with dilatation and columnar metaplasia. (d) Severe reflux damage associated with complete loss of the abdominal part of the sphincter showing columnar metaplasia of the distal tubular esophagus associated with free reflux into the esophagus. In all cases, columnar metaplastic esophageal epithelium is shown as a solid black area between squamous epithelium (gray) and gastric oxyntic mucosa (striped gray). Note that the original gastroesophageal junction (arrow) remains unaltered if defined by reliable criteria such as the proximal limit of oxyntic mucosa, the peritoneal reflection, and the distal limit of submucosal glands (circles, white, and black). The rugal folds are not shown; they are irrelevant and unreliable.

nal esophagus is positive (+5mmHg) and is now not counteracted by the resting tone of the sphincter. With every meal and every rise in intragastric pressure, particularly with overdistension and deformation of this region, the sphincter-less part of the distal esophagus undergoes dilatation. With repeated postprandial dilatations, permanent dilatation may occur. This dilatation will result in a proximal migration of the end of the tubular esophagus equivalent to the length of sphincter that has been lost. The most distal part of the esophagus, which has undergone columnar metaplasia, now assumes the contour of the stomach and is located distal to the end of the tubular esophagus (Fig. 11.25).

This change has a maximal effect on the left side of the gastroesophageal junction. As more and more of the abdominal sphincter becomes lost, the tubular esophagus will end closer and closer to the diaphragmatic hiatus (Fig. 11.26). The junction of the tubular esophagus and the saccular organ will no longer be the gastroesophageal junction; it is the junction

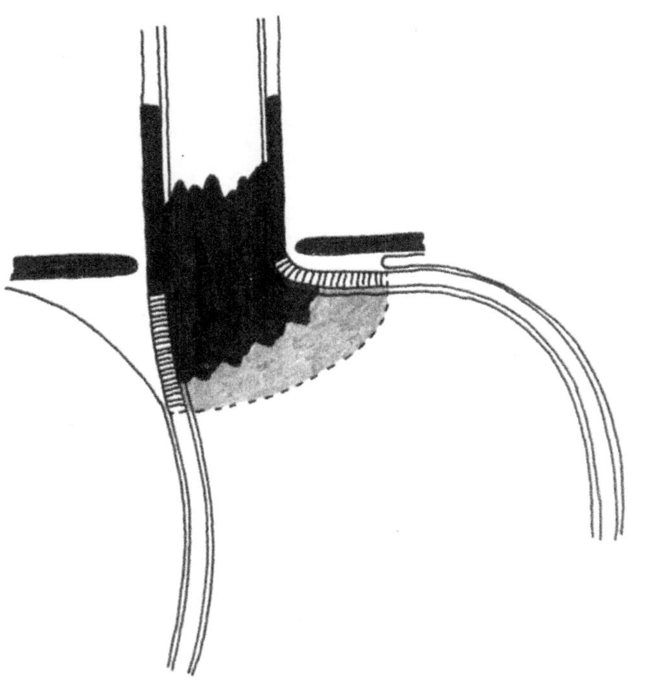

FIGURE 11.27 Severe reflux disease with near complete loss of the abdominal part of the sphincter. The esophagus curves to the left at an almost right angle immediately distal to the diaphragm without any acute angle of His. The peritoneal reflection and submucosal glands remain unaltered. The lower part of the esophagus is permanently dilated, and its epithelium is permanently exposed to gastric contents and has undergone columnar metaplasia (shown as a mixture of cardiac mucosa (black) and oxyntocardiac mucosa (gray)). The sphincter is severely incompetent, resulting in severe gastroesophageal reflux. In this case, the reflux has caused columnar metaplasia of the distal tubular esophagus in the thorax. Even this severely damaged esophagus is frequently called "not abnormal" by present criteria. The columnar epithelium in the tubular esophagus will be visible at endoscopy, but if biopsies do not show intestinal metaplasia (absent in this patient), no diagnosis is attached to the patient. Horribly advanced reflux disease is being completely missed.

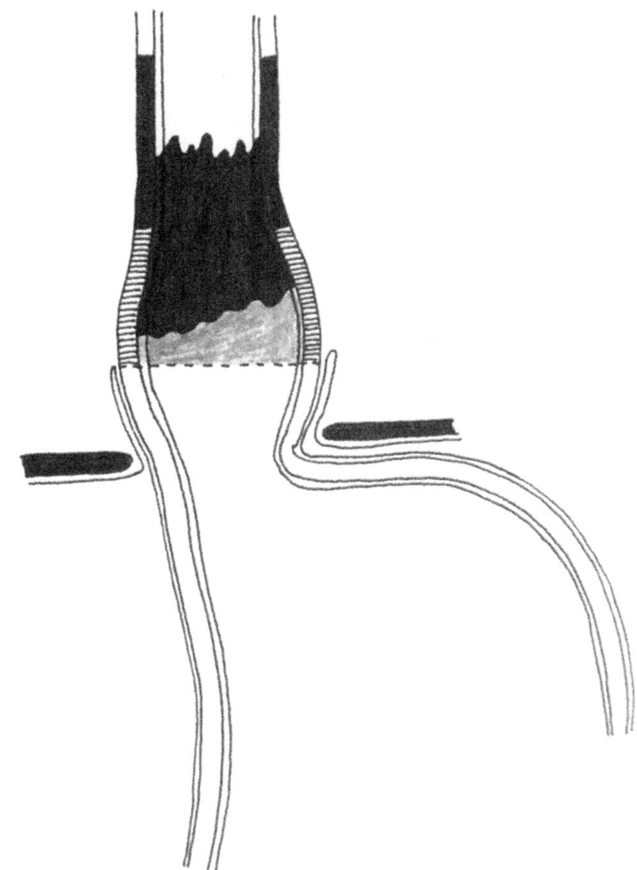

FIGURE 11.28 A sliding hiatal hernia. This is identical in all respects to Figure 11.27 except for the fact that the esophagus has shortened, pulling the stomach into the thorax as a sliding hiatal hernia. This carries the peritoneal reflection with it. The gastroesophageal junction is still marked by the peritoneal reflection, the proximal limit of oxyntic mucosa, and the submucosal glands. This drawing is also essentially identical to Figure 11.16 except for the fact that the lost part of the abdominal part of the sphincter is shown as a striped area.

between the esophagus that has the abdominal sphincter intact and the esophageal segment that has become dilated because it has lost the sphincter. The dilated distal esophagus becomes continuous with the curvature of the stomach. The angle of His will become progressively less acute. When the entire abdominal sphincter has been lost, the left side of the esophagus passes to the left at a right angle from the end of the tubular esophagus to become continuous with the stomach (Fig. 11.27). At this stage, there is total sphincter failure, severe gastroesophageal reflux, and damage to the body of the esophagus. If this damage is associated with shortening, the obliteration of the acute angle of His has set the stage for a sliding hiatal hernia to occur (Fig. 11.28).

Between the normal stage and the complete loss of the abdominal sphincter, there is a millimetric progression of dilatation of the distal reflux-damaged esophagus. In the vast majority of patients in the population, the cumulative lifetime reflux damage of the distal esophagus is less than 1 cm (Fig. 11.25). This correlates with the autopsy data, where the vast majority of persons will have less than 5 mm of metaplastic esophageal columnar epithelia (cardiac with and without intestinal metaplasia and oxyntocardiac epithelia) between the squamous epithelium and gastric oxyntic mucosa (9). A lower esophageal sphincter that has been shortened by less than 0.5 cm is still functionally competent except in the postprandial phase when gastric distension causes further sphincter shortening. This majority of the population is susceptible to increased postprandial gastroesophageal reflux.

The recognition of this anatomic abnormality is difficult. At surgery, the dilated esophagus is contained within the subdiaphragmatic connective tissue above the peritoneal reflection. It is difficult to appreciate a subtle decrease in the length of the abdominal part of the esophagus. Surgeons will theoretically be most likely to appreciate this as a significant difference in the anatomy when they operate on patients with reflux disease (i.e., when they perform a fundoplication or esophagectomy for adenocarcinoma) and patients with achalasia (i.e., when they perform a myotomy). The abdominal part of the esophagus will be much more easy to access in patients with achalasia because the entire abdominal length of the sphincter is retained (unless the patient has had numerous prior dilatations and has severe iatrogenically induced sphincter damage, which is a common situation at the time patients with achalasia undergo surgery).

At endoscopy, mild degrees of this abnormality will not be detectable. The tubular esophagus will tend to flare out into the stomach almost as usual except that it will be slightly closer to the diaphragmatic hiatus; it is unlikely that this can be reliably differentiated from the normal situation until a significant hiatal hernia develops. With the present definitions used at endoscopy, the dilated reflux-damaged few millimeters of the distal esophagus will almost certainly be misinterpreted as stomach, particularly if the metaplastic esophageal segment develops rugal folds, as we have shown it can (9). The only situation that is possible to recognize at endoscopy is if the columnar-lined esophagus remains as a flat, abnormal appearing mucosa between the proximal limit of the rugal folds and the end of the tubular esophagus. This situation is never reported in the endoscopic literature. Flat columnar epithelium is always described as being in the tubular esophagus (Fig. 11.29). This may be because endoscopy is performed on a constantly moving organ and creates its own artifact, making it difficult to judge the exact location of millimetric abnormalities.

Characteristics of the Dilated Reflux-Damaged End-Stage Esophagus

The dilated part of distal esophagus becomes functionally like the proximal stomach, dilating with meals and collapsing as the stomach empties. It shows functional characteristics of a reservoir. The columnar mucosa stretches with distension after meals and collapses into folds when the stomach is collapsed. This segment can therefore develop mucosal rugal folds similar to those in the stomach. Rugal folds do not characterize any particular histologic type of epithelium; it characterizes a distensible epithelial-lined organ. The only reason why rugal folds will not be seen is if there is atrophy or inflammation, just like in gastric mucosa. We have called this process end-stage reflux-damaged distal esophagus; an alternate term that we have suggested is "gastricization."

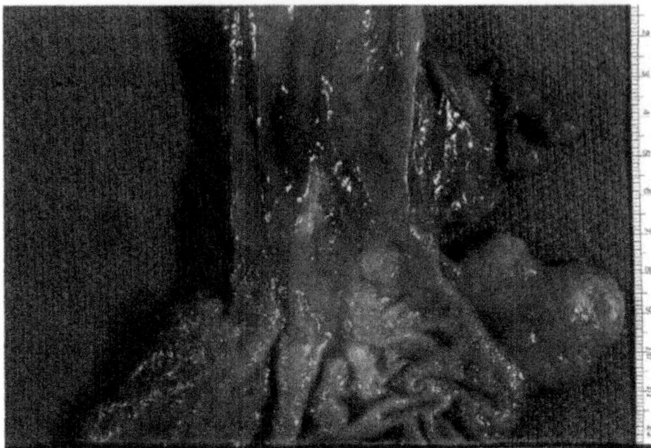

FIGURE 11.29 Esophagectomy specimen showing a short segment of flat mucosa between the serrated squamocolumnar junction and the gastroesophageal junction. The dilated end-stage esophagus cannot be recognized; it is of a variable length under rugated mucosa and can be defined only by histology by the presence of metaplastic columnar esophageal epithelium. The true gastroesophageal junction is the proximal limit of gastric oxyntic mucosa and can be determined only by histology. (Cross reference: Color Plate 3.1)

However much of it functions like the stomach, this dilated end-stage esophagus *is still the esophagus.* It can be recognized as such because it is lined by metaplastic columnar epithelia of the esophagus (cardiac mucosa, with and without intestinal metaplasia, and oxyntocardiac mucosa), not gastric oxyntic mucosa. It also contains submucosal glands and gland ducts that open into the epithelial surface, a feature that characterizes it as esophagus and not stomach (1, 7, 8).

The extent to which the distal esophagus can undergo end-stage dilatation is limited by the length of the abdominal part of the sphincter (2 to 3 cm). The thoracic part of the sphincter has a resting intraluminal pressure that is similar to thorax (i.e., −5 mm Hg). The pressure decreases further during inspiration. There is no mechanism of increasing intraluminal pressure in the thoracic part of the esophagus; this happens only when there is esophageal obstruction such as in achalasia when the esophageal luminal pressure proximal to the obstruction increases and causes massive dilatation of the esophagus. Loss of the high resting tone in the intrathoracic part of the sphincter, which occurs in severe reflux-induced damage, does not therefore cause significant dilatation of this part of the esophagus.

It is interesting to review the findings in Jain et al.'s study (6) that was published in abstract form only.

This is an endoscopic study with extensive biopsy sampling in 31 patients. "At endoscopy, both the squamocolumnar junction (the Z line) and the anatomic esophagogastric junction (EGJ) were identified, and biopsy specimens were taken from the columnar epithelium at each of four quadrants in the following locations: (a) the Z line, (b) the EGJ, (c) the gastric cardia 1 cm below the EGJ, and (d) the gastric cardia 2 cm below the EGJ." The endoscopic findings in this population are as follows: "The Z-line and EGJ were located at the same level in 25 patients, whereas in 6 the Z-line was located >1 cm proximal to the EGJ." The frequency of finding the different epithelial types in at least one of the four biopsy specimens obtained at the various locations is tabulated as follows:

Biopsy site	Number of patients	Cardiac epith	Fundic epith	Intestinal met
EGJ	31	11 (35%)	21 (68%)	10 (32%)
1 cm below EGJ	29	4 (14%)	23 (79%)	7 (24%)
2 cm below EGJ	30	1 (3%)	30 (100%)	2 (7%)

This study, in a population that is largely endoscopically normal, shows that the prevalence of cardiac mucosa decreased from 11 of 31 (35%) at the gastroesophageal junction, to 4 of 29 (14%) 1 cm distal to the gastroesophageal junction, to 1 to 30 (3%) 2 cm distal to the gastroesophageal junction. The interpretation of these data was that cardiac mucosa was infrequent but present at the gastroesophageal junction and in the proximal 2 cm of the stomach. It is difficult to call this "normal" because the prevalence is so low. According to the interpretation and definitions that we have suggested, 4 out of 29 (14%) of these patients had 1 cm dilated end-stage esophagus mistaken for proximal stomach and 1 out of 30 (3%) had 2 cm of dilated end-stage esophagus (Fig. 11.26).

The data relating to intestinal metaplasia are more difficult to interpret because it is not known whether the intestinal metaplasia was esophageal (i.e., arising in cardiac mucosa) or gastric (i.e., arising in oxyntic mucosa). The fact that the prevalence of intestinal metaplasia decreased from 10 of 31 (32%) at the gastroesophageal junction to 7 to 29 (24%) 1 cm distal and 2 of 7 (14%) 2 cm distal to the gastroesophageal junction makes it likely that most of these were esophageal; the prevalence would be expected to remain the same throughout the gastric body in chronic atrophic gastritis with intestinal metaplasia. It is likely that most of these patients would have been classified as having "intestinal metaplasia of the gastric cardia." These patients likely had intestinal metaplasia in the dilated end-stage esophagus. It should be noted that, in this study, the distance of the dilated end-stage esophagus distal to the end of the tubular esophagus is within the theoretical limit of 2 to 3 cm that we have predicted. It is extremely unusual to find cardiac mucosa greater than 3 cm distal to the tubular esophagus.

In rare patients, squamous epithelium covers part of the mucosal surface of the end-stage dilated esophagus. Fass and Sampliner (31) reported the presence of extensions of squamous epithelium distal to the proximal limit of the rugal folds in 16 of 547 (3%) of patients. In 12 patients, this consisted of a solitary tongue of squamous epithelium; 4 patients had two tongues and two of these had in addition two squamous islands within the cardia. The extension of squamous epithelium distal to the proximal limit of the rugal folds measured 0.5 cm in length in 4 patients, 1 cm in 3 patients, 1.5 cm in 3 patients, 2 cm in 3 patients, and 3 cm in 3 patients.

These patients were all males, 14 Caucasian, with a mean age of 61.2 years. Most patients had evidence of gastroesophageal reflux disease; heartburn was present in 12 patients, 8 reported acid regurgitation, 10 had complications including strictures and Barrett esophagus, and 13 patients were on acid suppressive medications. Hiatal hernia was present in all patients and the squamous epithelial extension was always limited to the hernia sac.

The authors' interpretation of this phenomenon is interesting. Because the squamous epithelium is distal to the endoscopic gastroesophageal junction, it is designated as being in proximal stomach. This, and the strong association with gastroesophageal reflux disease, leads to the conclusion: "acid reflux may be one of the culprits responsible for the development of squamous epithelium in the proximal stomach" (p. 31). This is incomprehensible; how can reflux of gastric acid into the esophagus cause the development of squamous epithelium in the proximal stomach? In their abstract, they conclude: "This mucosal abnormality may represent an esophageal mucosal response to proximal gastric injury" (p. 27). Their purported esophageal mucosal injury is in the proximal stomach and there is no evidence of a proximal gastric injury in their patients!

One has only to recognize that Fass and Sampliner are accurately describing partial columnar metaplasia in the end-stage dilated esophagus caused by reflux disease. They provide one microphotograph of the squamocolumnar junction which shows cardiac mucosa distal to the squamous epithelium. By our definitions, the squamous epithelium is proximal to the gastric oxyntic mucosa and therefore esophageal. It is also worthwhile nothing that the maximal extent of squamous epithelium found is 3 cm, which is our theporetical limit for the end-stage dilated esophagus.

Diagnostic Errors Resulting from the Dilated End-Stage Esophagus

A patient with end-stage dilatation of the reflux damaged esophagus is a source of significant error when present definitions of the gastroesophageal junction are applied. It is distal to both presently accepted definitions of the gastroesophageal junction. It is distal to the end of the tubular esophagus, which is the recommended definition of the Association of Directors of Anatomic and Surgical Pathology for gross dissection of esophagogastrectomy specimens. Pathologists will, according to this definition, erroneously designate the dilated end-stage esophagus as the proximal stomach or gastric cardia. Gastroenterologists who follow the American Gastroenterology Association practice guidelines (32) will designate the dilated end-stage esophagus as proximal stomach (gastric cardia).

The error results in several extremely important practical misconceptions:

1. The present definition of the normal endoscopic appearance will fail to recognize the dilated end-stage esophagus. Because the dilated end-stage esophagus will be classified as "normal stomach," it will not be biopsied; any Barrett's epithelium present in it will be missed and if this has a risk of malignancy, patients will develop adenocarcinoma of this segment without any possibility that a premalignant lesion will be detected. The error has prevented evaluation of any premalignant risk of intestinal metaplasia occurring in the dilated end-stage esophagus because it is ignored and not biopsied.
2. Intestinal metaplasia of the dilated end-stage distal esophagus will be incorrectly classified as "intestinal metaplasia of the gastric cardia." Because the length of dilated end-stage esophagus is variable from 0 to 3 cm depending on the severity of reflux damage, there is no way to know whether a biopsy at any point distal to the proximal limit of the rugal folds represents end-stage dilated esophagus or true stomach. It is therefore not surprising that studies using biopsies from the anatomically defined gastric cardia will find that intestinal metaplasia of the gastric cardia are etiologically related to both gastroesophageal reflux disease and *Helicobacter pylori* gastritis. When this biopsy is taken "just distal to the end of the tubular esophagus" (33), it will have less gastric admixture than if it is taken "within 5 mm from the squamocolumnar junction" (34). It is not surprising that the prevalence of *H. pylori* infection associated with what the authors call intestinal metaplasia of the cardia was 38% in Rex et al. (33) and 88% in Goldblum et al. (33).
3. Adenocarcinoma of the dilated end-stage esophagus will be misclassified as "adenocarcinoma of the gastric cardia" or "adenocarcinoma of the proximal stomach." We now have an explanation as to why adenocarcinoma of the gastric cardia has epidemiologic similarity to adenocarcinoma of the distal esophagus and why they have been shown to have an association with symptomatic gastroesophageal reflux disease. They are the same entity. They have differences based on the fact that adenocarcinoma of the dilated end-stage distal esophagus ("gastric cardia") occurs in very short segments of intestinalized columnar-lined esophagus. Short segments of columnar-lined esophagus result from mild reflux-induced damage (and therefore the patient may frequently be asymptomatic, be as likely female than male, have no endoscopic evidence of reflux disease, and have a normal or slightly abnormal 24-hour pH test).

Recognizing the Dilated End-Stage Esophagus and the True Gastroesophageal Junction

Gross examination and endoscopy are incapable of identifying the dilated end-stage segment of the distal esophagus. The only methods of defining its true extent are the following:

1. *To use the fact that it is lined by metaplastic esophageal columnar epithelia (cardiac mucosa with and without intestinal metaplasia and oxyntocardiac mucosa).* This requires that it be accepted that these mucosal types are never normal and never gastric. We are close to this at present; most authorities cling to a very small, <0.4 cm, zone of native "true" gastric cardia, lined by either cardiac or oxyntocardiac mucosa as a normal gastric epithelium (2). This prevents the criterion of the presence of cardiac and oxyntocardiac mucosa from being applied to define reflux disease.
2. *To use the presence of the esophageal submucosal gland complex.* In resection specimens, the glands themselves can be identified; in mucosal biopsies, the presence of gland ducts is useful. However, glands are sporadically and variably present in esophagi, and their absence is not meaningful. In one study

we showed that approximately 15% of biopsies containing metaplastic esophageal columnar epithelia contain gland ducts (35). This frequency is the measure of the average number of submucosal glands in esophagus divided by the volume of sampling by biopsy.

The identification of the true gastroesophageal junction is also beyond the technical capability of either endoscopy or gross examination because gastric oxyntic mucosa cannot be distinguished from metaplastic esophageal columnar epithelia. They both may or may not have rugal folds; rugal folds do not signify epithelial type, they signify a distensible organ; the absence of rugal folds does not signify epithelial type, it signifies the presence of inflammation and atophy in any columnar epithelial type. Histologic mapping to delineate the proximal limit of gastric oxyntic mucosa is necessary to define the true gastroesophageal junction.

RELATIONSHIP OF THE DILATED END-STAGE ESOPHAGUS TO COLUMNAR-LINED ESOPHAGUS

Patients with dilated end-stage esophagus fall into two categories.

Patients with Only Dilated End-Stage Esophagus

These are patients with the earliest evidence of reflux damage involving the most distal 1 to 10 mm of the esophagus (Figs. 11.24, 11.25, 11.26b, and 11.26c). They have reflux damage of the squamous epithelium leading to columnar metaplasia, loss of length of the abdominal sphincter, and dilatation of the sphincterless abdominal esophagus (end-stage esophagus). The shortening of the sphincter is not adequate to compromise its function; sufficient sphincter function remains to prevent free reflux into the body of the esophagus. These patients will therefore have no pathologic changes above the end-stage dilated segment of esophagus. We suggest the term "intrasphincteric reflux disease" for this entity (see Chapter 12).

In our study of esophagectomy patients reviewed here (8), there were two patients with squamous carcinoma. These patients had no columnar-lined esophagus and had a dilated end-stage esophagus measuring 0.31 and 0.43 cm distal to the end of the tubular esophagus.

In a similar study of esophagectomy specimens, Sarbia et al. (11) looked for submucosal glands in 36 patients with squamous carcinoma where, in all the cases, the entire tubular esophagus was lined by squamous epithelium and transitioned into "the brown-red gastric mucosa." (The authors really meant brown-red columnar epithelium.) They sectioned the entire junction by longitudinal sections and measured the minimal and maximal lengths of cardiac mucosa and oxyntocardiac mucosa found distal to the squamous epithelium. They reported that cardiac mucosa was present along the entire circumference of the squamocolumnar junction in 20 patients, in part of the circumference in 15 patients, and was completely absent in one patient. In the one patient without cardiac mucosa, oxyntocardiac mucosa was present distal to the squamous epithelium. The minimum length of cardiac mucosa ranged between 0 and 3 mm (median 1 mm); the maximum length ranged between 1 and 15 mm (median 5 mm). The combined length of cardiac plus oxyntocardiac mucosa was a minimum of 1 to 12 mm (median 4 mm) and the maximum length was between 5 and 28 mm (median 11 mm).

Sarbia et al. then evaluated the relationship of these epithelia to submucosal glands which define the esophagus. In eight cases, cardiac mucosa or oxyntocardiac mucosa was situated over submucosal mucous glands. In a ninth case, CM/OCM was situated over squamous epithelium-lined ducts. Thus, Sarbia et al. proved that in 25% of their patients, the cardiac and oxyntocardiac mucosae were esophageal in location by virtue of the fact that they were associated with the esophageal submucosal gland-duct complexes. Because these glands are distributed sporadically in the esophagus, the absence of submucosal glands does not prove that the location is not esophageal, particularly when the length of tissue is in the 5 to 10 mm range.

Sarbia et al.'s study show that all 36 patients had a dilated end-stage esophagus distal to the end of the tubular esophagus if it is believed that cardiac and oxyntocardiac mucosa are esophageal epithelia. The median combined length of cardiac + oxyntocardiac mucosa was 4 to 11 mm. In 25% of these cases, the presence of submucosal glands under these epithelia proved that they were located in the esophagus.

The vast majority of the general population belongs in this category of mild reflux disease where the only manifestation of reflux is end-stage damage of the most distal few millimeters of the esophagus (Fig. 11.24). Its frequency is a manifestation of the universal occurrence of reflux in humans, akin to carbon pigment in the lungs of urban dwellers and atherosclerosis of the abdominal aorta in developed countries. This is shown by the fact that in most autopsy studies, the majority of the patients have an overall length of columnar metaplastic epithelia of less than 5 mm when measured from the squamocolumnar junction. Unfortunately, unlike anthracosis and mild atherosclerosis of the abdominal aorta, this minimal reflux disease can progress through intestinal metaplasia to adenocarcinoma.

This most common manifestation of gastroesophageal reflux is completely undetected by present endoscopic and histologic criteria. It can only be recognized when the medical community agrees that cardiac mucosa is the defining criterion of reflux disease rather than a normal finding and biopsies are taken distal to what is perceived incorrectly to be the endoscopic gastroesophageal junction. At the present time these patients will not undergo biopsy if the American Gastroenterology Association practice guideline recommendations are followed because they are endoscopically normal (32). If they do undergo biopsy and cardiac or oxyntocardiac mucosa is found distal to the endoscopic gastroesophageal junction, they will be deemed to have "normal" gastric cardiac mucosa (2). If intestinal metaplasia is found, they will be deemed to have "intestinal metaplasia of the gastric cardia." In a subsequent chapter, we will review an article by Rex et al. (33) that shows the relatively high frequency of this condition if routine biopsies are taken.

Patients with Dilated End-Stage Esophagus Distal to a Visible Columnar-Lined Esophagus

As the length of abdominal sphincter becomes damaged further and its shortening increases beyond 10 to 15 mm, the length of effective abdominal sphincter that remains can no longer sustain normal sphincter function. These patients have decreased baseline sphincter length and have a high frequency of transient lower esophageal sphincter relaxation and increased acid exposure of the lower esophagus whenever there is gastric distension (21). These patients will develop reflux changes in the body of the esophagus. The 24-hour pH test will become abnormal, and impedance will show increased free liquid reflux into the esophagus detected by devices placed above the proximal margin of the lower esophageal sphincter. Columnar metaplasia occurs in the tubular esophagus to a length that correlates with the severity of reflux. Endoscopy becomes abnormal when there is either erosion of the squamous epithelium or a visible segment of columnar-lined tubular esophagus (Fig. 11.29).

Most if not all patients with a visible columnar-lined esophagus will have dilated end-stage esophagus distal to it (Figs. 11.22, 11.26d, and 11.29). In our study of esophagectomy specimens reviewed earlier (8), the eight patients with columnar metaplasia in the tubular esophagus had a dilated end-stage esophageal segment measuring 1.03 to 2.05 cm distal to the end of the tubular esophagus (Fig. 11.13).

If the present definitions of the endoscopic junction are used, this dilated end-stage esophagus will be misinterpreted as proximal stomach because it is distal to the end of the tubular esophagus and distal to the proximal limit of the rugal folds. This will result in an underestimation of the length of columnar-lined epithelium by 1.03 to 2.05 cm in these eight patients. It is apparent that the error increases as the length of columnar-lined esophagus increases, because the greater loss of the abdominal sphincter (which will determine the length of dilated end-stage esophagus) will be associated with a greater likelihood of reflux disease.

An important piece of evidence supports the fact that the visible columnar-lined esophagus and the dilated end-stage esophagus represent a single pathologic unit caused by reflux. The zonation of the three types of epithelia first described by Paull et al. (36) and confirmed by us (37) show that the three types of epithelia tend to be distributed in a consistent manner with the intestinal metaplasia being proximal, cardiac mucosa being central, and oxyntocardiac mucosa being distal. This zonation pattern is maintained throughout the columnar-lined segment that includes the columnar epithelia of the dilated end-stage esophagus (Fig. 11.30). When both the visible columnar lined

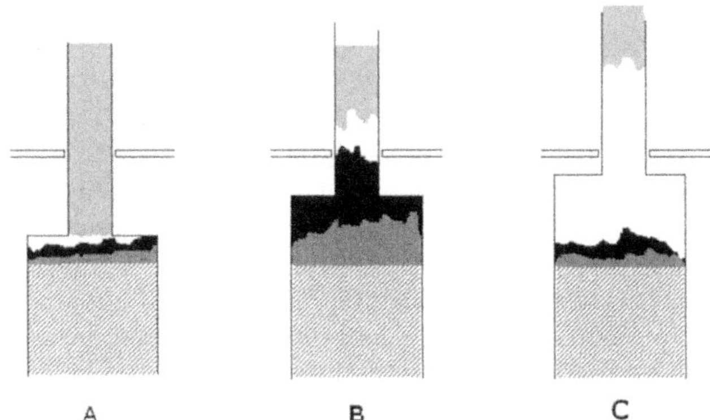

FIGURE 11.30 Three patients with columnar-lined esophagus, showing the constant zonation of the three epithelial types: intestinal metaplasia (solid white) is proximal, cardiac mucosa (black) is in the midregion, and oxyntocardiac mucosa (stippled) is distal. (a) The entire segment of columnar-lined esophagus is in the dilated end-stage esophagus. (b and c) There is a visible columnar-lined segment in the tubular esophagus with the dilated end-stage esophagus composed of nonintestinalized and intestinalized epithelium, respectively. Note that as the columnar-lined segment in the tubular esophagus lengthens, the dilated end-stage esophagus becomes larger, both indicative of the severity of reflux disease.

esophagus and columnar lining of the end-stage esophagus are regarded as one entity, it is found that intestinal metaplasia is seen adjacent to the squamocolumnar junction, progressively involving an increasing amount of the columnar-lined segment and frequently extending into the dilated end-stage esophagus. In many of these patients, oxyntocardiac mucosa will be adjacent to gastric oxyntic mucosa and limited to the end-stage dilated segment of the esophagus.

THE DISCREPANCY HAS NOW DISAPPEARED

The recognition that the present definitions of the gastroesophageal junction at endoscopy as the proximal limit of the rugal folds and at gross examination as the end of the tubular esophagus are incorrect is the key to solving the discrepancy that was described at the beginning of this chapter and in Figure 11.5. In a patient with columnar-lined distal esophagus that dilates to take the contour of the stomach or develops rugal folds, use of these definitions will place the gastroesophageal junction proximal to the true gastroesophageal junction (Fig. 11.26).

When reliable criteria for the true gastroesophageal junction are used, the problem resolves completely (Fig. 11.31). The true gastroesophageal junction is defined by one of three definitive criteria:

1. *The histologically defined proximal limit of gastric oxyntic mucosa.* The only source of inaccuracy with this definition is the difficulty differentiating gastric oxyntic mucosa from esophageal oxyntocardiac mucosa when the latter has numerous parietal cells. This is a relatively minor problem, but it may create an error of 1 to 2 millimeters. This error is not clinically relevant because esophageal oxyntocardiac mucosa is not at risk for malignancy. It will, however, cause a slight underestimation of the length of columnar-lined esophagus.
2. *The distal limit of esophageal submucosal glands.* This is not exactly the true gastroesophageal junction because these glands are distributed sporadically. It is an important landmark because it clearly establishes a point that must represent esophagus because these glands are not present in the stomach. The true gastroesophagel junction must be *at or distal* to this point. In our study, reviewed earlier, this point was always less than 0.5 cm proximal to the histologically defined proximal limit of gastric oxyntic mucosa.
3. *The peritoneal reflection.* This remains a constant marker but its exact correlation with a point in the mucosa is difficult to establish because of technical reasons (see Chapter 4).

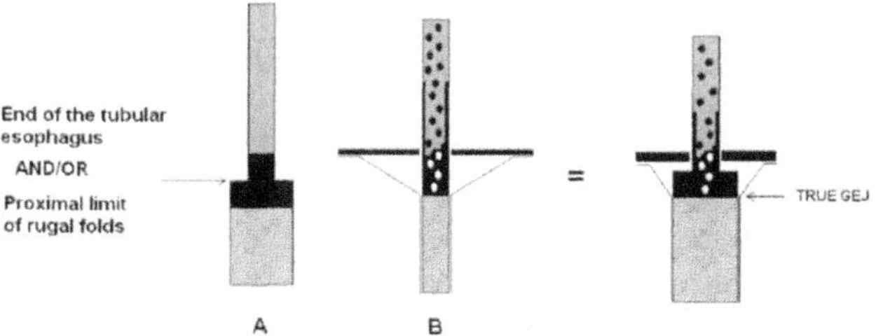

FIGURE 11.31 Removal of the discrepancy presented in Figure 11.5. The problem resolves when the true gastroesophageal junction is recognized by the proximal limit of gastric oxyntic mucosa, the distal limit of submucosal glands, or the peritoneal reflection. In reflux disease, this has significant separation from the presently recognized gastroesophageal junction (end of tubular esophagus and proximal limit of rugal folds). This separation is the discrepancy; it is also the area of esophagus that is misclassified as "gastric cardia" and ignored by present clinical practice.

Of these three constant markers of the true gastroesophageal junction, the only one that is available for clinical use is the proximal limit of gastric oxyntic mucosa, which can be established by an appropriate system of endoscopic biopsy. It is only when this is used to define the true gastroesophageal junction and the present definitions are discarded that we will be finally able to define the true boundaries of the esophagus and stomach (Fig. 11.31). The use of the present definitions of the endoscopic junction results in the misclassification and ignoring of the distal reflux-damaged esophagus.

ANATOMY AND HISTOLOGY OF A SLIDING HIATAL HERNIA

The presence of the acute angle of His between the end of the abdominal esophagus and the fundus of the stomach depends largely on the length of the abdominal esophagus (Fig. 11.17). As long as this angle is maintained, it is extremely unlikely for the esophagus to slide into the thorax as a sliding hiatal hernia. If the distal esophagus dilates as a result of reflux, the length of the tubular part of the abdominal esophagus becomes shorter and the angle of His less acute.

When the sphincter action of the entire 2 to 3 cm of the abdominal esophagus is lost and the dilated end-stage esophagus reaches its maximum, the acute angle of His no longer exists (Fig. 11.27). The tubular esophagus ends at the diaphragmatic hiatus, and the left side passes at an almost right angle to become continuous with the dilated end stage esophagus, which flows into the fundus of the stomach. This anatomic configuration sets the stage for the occurrence of a hiatal hernia. Extreme loss of lower esophageal sphincter length and severe reflux are often associated with sliding hiatal hernia because severe reflux causes fibrosis of the esophageal wall with longitudinal contraction causing esophageal shortening (Fig. 11.28). The intermediate anatomic abnormality between the normal state (with 2 to 3 cm of abdominal esophagus with a normal sphincter and an acute angle of His) and a sliding hiatal hernia (characterized by absence of an abdominal sphincter and an obliterated acute angle of His) is the dilated end-stage esophagus. The only difference between a dilated end-stage esophagus with and without a sliding hiatal hernia is the presence of esophageal shortening in the former; the organs remain in their normal position if there is no hiatal hernia.

When a sliding hiatal hernia occurs, it is difficult to identify the exact location of the true gastroesophageal junction (Fig. 11.32). Radiologic assessment of hiatal hernia, which depends on recognizing the point of dilation of the tubular esophagus, often overestimates the size of the hernia because it tends to include the dilated end-stage distal esophagus as part of the hernia. Endoscopic assessment is also inaccurate for the same reason; the end-stage dilated esophagus has rugal folds and is mistaken as stomach. It has been our experience that biopsies taken within the hiatal hernia show it to be a mixture of end-stage esophagus (lined by metaplastic esophageal columnar epithelia in its more proximal part) and stomach (lined by gastric oxyntic mucosa in the more distal part) (Fig. 11.28). The only way to correctly assess these structures is by mapping this area histologically to locate the proximal limit of gastric oxyntic mucosa.

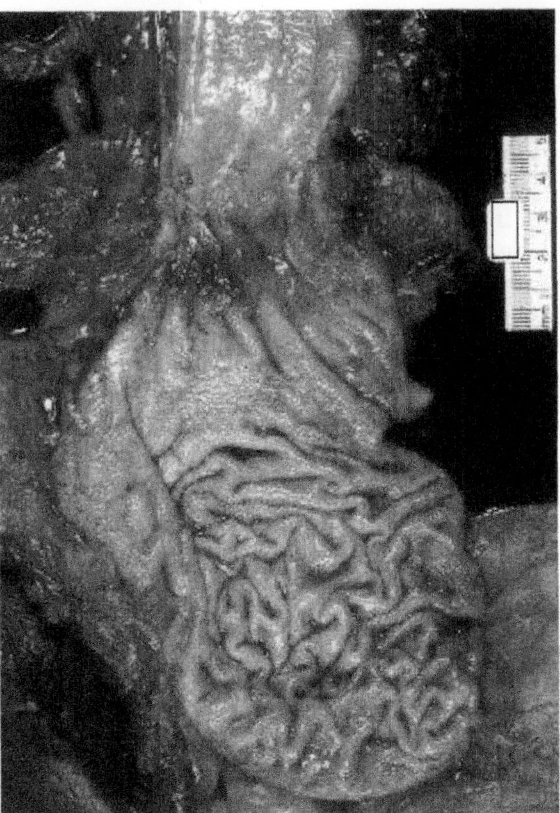

FIGURE 11.32 Esophagectomy specimen showing an ulcerated adenocarcinoma at the squamocolumnar junction. A large dilated segment is present distal to the squamocolumnar junction that is lined by flat epithelium before the rugated epithelium is reached more distally. The differentiation between dilated end-stage esophagus and hiatal hernia can only be made by the location of the peritoneal reflection, the distal limit of submucosal glands, or the proximal limit of gastric oxyntic mucosa. (Cross reference: Color Plate 5.2)

THE GREAT HISTORICAL MISUNDERSTANDING CONTINUES TO THE PRESENT

In Chapter 2, we considered how serious misunderstanding has caused confusion in the diagnosis and management of reflux disease. Though there is slow progress toward the truth, two huge mistakes persist at the present time.

Error 1: Underestimation of the Magnitude of the Problem by Designating Reflux-Damaged Esophagus as Part of the Stomach

There has always been a tendency to recognize only the most severe disease caused by reflux and to ignore the less severe forms of the disease because of the use of an incorrect definition of the gastroesophageal junction and false definitions of normalcy. This has resulted in a continuous underestimation of the magnitude of the problem.

Before 1953, long segments of columnar-lined esophagus were denied. By declaring that the esophagus ended at the squamocolumnar junction, Barrett made columnar-lined esophagus a gastric disease (38). When it was first described, Allison and Johnstone (13) suggested that it was not a disease at all; rather it was a congenital anomaly in which the lower esophagus was lined by "gastric mucous membrane." When it was proved that columnar-lined esophagus was acquired as a result of gastroesophageal reflux, Hayward, in 1961 (39), rushed to limit the disease to those with very long segments (2 to 3 cm) of columnar lining in the distal tubular esophagus. This was achieved by the simple expediency of declaring without proof that 2 (or 3) cm of the distal esophagus was normally lined by columnar epithelium. Until the 1990s, all patients who developed Barrett's adenocarcinoma in shorter segments than 3 cm paid the penalty for this mistake.

In the 1990s, short-segment Barrett esophagus was recognized (18). However, the medical community devised artificial methods to limit the magnitude of the problem. By declaring without any proof that the proximal limit of the gastric rugal folds was the true gastroesophageal junction, we found a method of ignoring Barrett esophagus in the most distal esophagus. The device they used here was similar to that used by Barrett: simply call the columnar-lined distal esophagus gastric by using a false and unproved definition of the gastroesophageal junction (the proximal limit of the rugal folds is as wrong and unproven a definition of the gastroesophageal junction as Barrett's squamocolumnar junction).

The degree to which the problem is underestimated today is potentially massive. In our series of mapping biopsies in a population of 959 patients biased toward having reflux disease, 650 (68%) patients had a columnar lined esophagus measuring less than 10 mm (40). In our autopsy study, 44% of patients had cardiac mucosa and 100% had oxyntocardiac mucosa, almost always less than 10 mm (1). These patients are likely to be regarded as "endoscopically normal," and in a normal practice setting where American Gastroenterology Association guidelines were followed, they would not have undergone biopsy. In 959 patients in our study (40), 120 (13%) had intestinal metaplasia, 372 (39%) had cardiac mucosa, and 158 (17%) had oxyntocardiac mucosa measuring less than 10 mm. This information would not be available without biopsy. The number of patients in that study with intestinal metaplasia in the dilated end-stage esophagus (120) is only slightly less than the total number of patients (128) who had intestinal metaplasia in a visible (1 cm or greater) length of columnar-lined esophagus (40).

If intestinal metaplasia in end-stage esophagus is precancerous with even a small risk of malignancy, failing to biopsy the endoscopically normal population is a continuation of the mistake of underestimating this disease. Patients who develop cancer from intestinal metaplasia in this end-stage esophagus pay the penalty for this error (Fig. 11.33). The incidence of "adenocarcinoma of the proximal stomach" is similar to or greater than adenocarcinoma of the esophagus. These cases are presently misclassified as gastric rather than esophageal cancers because of a faulty gross definition of the gastroesophageal junction recommended to pathologists. This results in the underreporting of reflux-induced esophageal adenocarcinoma and a failure to appreciate the true magnitude of the problem.

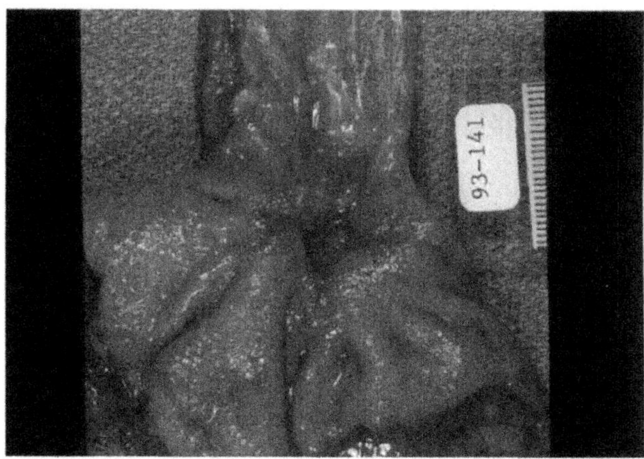

FIGURE 11.33 Adenocarcinoma of the area distal to the end of the tubular esophagus. This will be almost universally classified as a tumor of the proximal stomach. However, it was shown to contain metaplastic esophageal columnar epithelium both at its epicenter and distal margin, characterizing this as an adenocarcinoma of the distal dilated end-stage esophagus. (Cross reference: Color Plate 5.3)

Error 2: Metaplastic Columnar Epithelium of the Esophagus Is Designated as Normal or as Gastric

It is amazing that we are at the same crossroads that the medical community faced in 1950. At that time, Barrett had proclaimed (without proof) that the gastroesophageal junction ended at the squamocolumnar junction. Diseases occurring distal to the squamocolumnar junction were classified as gastric diseases. In 1953, Allison and Johnstone (13) proved that this was incorrect; they showed Barrett and the medical community the undoubted proof that there was a columnar-lined segment of esophagus between the squamocolumnar junction and the stomach. This removed the incorrect definition of the gastroesophageal junction.

Since 1953, there has been no validated gastroesophageal junction. We have used two definitions of the gastroesophageal junction at gross examination and endoscopy. These are the end of the tubular esophagus and the proximal limit of the rugal folds. These have been declared to be the gastroesophageal junction by edict in a manner that is not different than Barrett's edict that the gastroesophageal junction was the squamocolumnar junction. By this edict, everything distal to the junction is gastric. Just as Allison and Johnstone (13) proved to the medical community that Barrett's definition of the gastroesophageal junction was wrong, we have shown that esophageal submucosal glands are present distal to the gastroesophageal junction as defined by the end of the tubular esophagus and the proximal limit of the rugal folds. Just as Barrett's tubular stomach was proved to be a reflux-damaged columnar-lined esophagus in 1953, we have proved that what is called "gastric cardia" today is actually dilated end-stage, reflux-damaged, columnar-lined esophagus. The similarity of the error, the result of the error, and the method used in proving the existence of the error is absolutely the same today as it was in 1953. The only difference is that Allison and Johnstone corrected the error in the proximal part of the columnar-lined esophagus; we are correcting the error in the distal part and for the first time developing a method of recognizing the entire columnar-lined esophagus, which ends at the histologically defined proximal limit of gastric oxyntic mucosa. This definition will not change in the future; we have reached the final end point defined by the truth.

References

1. Chandrasoma PT, Der R, Ma Y, et al. Histology of the gastroesophageal junction: An autopsy study. Am J Surg Pathol 2000;24:402–409.
2. Odze RD. Unraveling the mystery of the gastroesophageal junction: A pathologist's perspective. Am J Gastroenterol 2005;100:1853–1867.
3. Kilgore SP, Ormsby AH, Gramlich TL, et al. The gastric cardia: Fact or fiction? Am J Gastroenterol 2000;95:921–924.
4. Glickman JN, Fox V, Antonioli DA, Wang HH, Odze RD. Morphology of the cardia and significance of carditis in pediatric patients. Am J Surg Pathol 2002;26:1032–1039.
5. Owen DA. Stomach. *In:* Sternberg SS (ed), Histology for Pathologists, second edition. Lippincott-Raven Publishers, Philadelphia, 1997. pp 481–493.
6. Jain R. Aquino D, Harford WV, Lee E, Spechler SJ. Cardiac epithelium is found infrequently in the gastric cardia. Gastroenterology 1998;114:A160 (Abstract).
7. Sarbia M, Donner A, Gabbert HE. Histopathology of the gastroesophageal junction: A study on 36 operation specimens. Am J Surg Pathol 2002;26:1207–1212.
8. Chandrasoma P, Makarewicz K, Ma Y, DeMeester T. A proposal for a new validated histologic definition of the gastroesophageal junction. Human Pathol 2006;37:40–47.
9. Chandrasoma P. Controversies of the cardiac mucosa and Barrett's esophagus. Histopathol 2005;46:361–373.
10. Association of Directors of Anatomic and Surgical Pathology. Recommendations for reporting of resected esophageal adenocarcinomas. Am J Surg Pathology 2000;31:1188–1190.
11. McClave SA, Boyce HW Jr, Gottfried MR. Early diagnosis of columnar lined esophagus: A new endoscopic diagnostic criterion. Gastrointest Endosc 1987;33:413–416.
12. Sharma P, McQuaid K, Dent J, Fennerty B, Sampliner R, Spechler S, Cameron A, Corley D, Falk G, Goldblum J, Hunter J, Jankowski J, Lundell L, Reid B, Shaheen N, Sonnenberg A, Wang K, Weinstein W. A critical review of the diagnosis and management of Barrett's esophagus: The AGA Chicago Workshop. Gastroenterology 2004;127:310–330.
13. Allison PR, Johnstone AS. The oesophagus lined with gastric mucous membrane. Thorax 1953;8:87–101.
14. De Nardi FG, Riddell RH. Esophagus. *In:* Sternberg SS (ed), Histology for Pathologists. Raven Press, New York, 1992, pp 524–526.
15. Johns BAE. Developmental changes in the oesophageal epithelium in man. J Anat 1952;86:431–442.
16. Goetsch E. The structure of the mammalian oesophagus. Amer J Anat 1910;10:1–40.
17. Allison PR. Peptic ulcer of the oesophagus. Thorax 1948;3: 20–42.
18. Spechler SJ, Zeroogian JM, Antonioli DA, Wang HH, Goyal RK. Prevalence of metaplasia at the gastroesophageal junction. Lancet 1994;344:1533–1536.
19. Liebermann-Meffert D, Duranceau A, Stein HJ. Anatomy and embryology, Chapter 1. *In:* Orringer MB, Heitmiller R (eds), The Esophagus, Vol 1, Zudeima GD, Yeo CJ. Shackelford's Surgery of the Alimentary Tract, 5th ed. Saunders, Philadelphia, 2002. pp 3–39.
20. DeMeester TR, Peters JH, Bremner CG, Chandrasoma P. Biology of gastroesophageal reflux disease: Pathophysiology relating to medical and surgical treatment. Annu Rev Med 1999;50: 469–506.
21. Kahrilas PJ, Shi G, Manka M, Joehl RJ. Increased frequency of transient lower esophageal sphincter relaxation induced by gastric distension in reflux patients with hiatal hernia. Gastroenterology 2000;118:688–695.
22. Sloan S, Rademaker AW, Kahrilas PJ. Determinants of gastroesophageal junction incompetence: Hiatal hernia, lower esophageal sphincter, or both? Ann Intern Med 1992;117: 977–982.
23. Mittal RK, Holloway RH, Penagini R, Blackshaw LA, Dent J. Transient lower esophageal sphincter relaxation. Gastroenterology 1995;109(2):601–610.
24. Dodds WJ, Dent J, Hogan WJ, Helm JF, Hauser R, Patel GK, Egide MS. Mechanisms of gastroesophageal reflux in patients with reflux esophagitis. N Engl J Med 1982;307:1547–1552.
25. Ismail T, Bancewicz J, Barlow J. Yield pressure, anatomy of the cardia and gastroesophageal reflux. Br J Surg 1995;82:943–947.
26. Wu AH, Wan P, Bernstein L. A multiethnic population-based study of smoking, alcohol and body size and risk of adenocarcinoma of the stomach and esophagus. Cancer Causes Control 2001;12:721–732.
27. Lagergren J, Bergstrom R, Nyren O. Association between body mass index and adenocarcinoma of the esophagus and gastric cardia. Ann Intern Med 1999;130:883–890.
28. Bu X, Ma Y, Der R, DeMeester T, Bernstein L, Chandrasoma PT. Body mass index is associated with Barrett esophagus and cardiac mucosal metaplasia. Dig Dis Sci 2005; in press.
29. Katzka DA, Gideon RM, Castell DO. Normal patterns of acid exposure at the gastric cardia: A functional midpoint between the esophagus and stomach. Am J Gastroenterol 1998;93: 1236–1242.
30. Oberg S, Peters JH, DeMeester TR, et al. Inflammation and specialized intestinal metaplasia of cardiac mucosa is a manifestation of gastroesophageal reflux disease. Ann Surg 1997;226:522–532.
31. Fass R, Sampliner RE. Extension of squamous epithelium into the proximal stomach: A newly recognized mucosal abnormality. Endoscopy 2000;32:27–32.
32. Sampliner RE. Practice guidelines on the diagnosis, surveillance, and therapy of Barrett's esophagus. Am J Gastroenterol 1998;93:1028–1031.
33. Rex DK, Cummings OW, Shaw M, Cumings MD, Wong RKH, Vasudeva RS, Dunne D, Rahmani EY, Helper DJ. Screening for

Barrett's esophagus in colonoscopy patients with and without heartburn. Gastroenterology 2003;125:1670–1677.

34. Goldblum JR, Vicari JJ, Falk GW, et al. Inflammation and intestinal metaplasia of the gastric cardia: The role of gastroesophageal reflux and *H. pylori* infection. Gasteroenterology 1998;114: 633–639.
35. Shi L, Der R, Ma Y, Peters J, DeMeester T, Chandrasoma P. Gland ducts and multilayered epithelium in mucosal biopsies from gastroesophageal-junction region are useful in characterizing esophageal location. Dis Esophagus 2005;18:87–92.
36. Paull A, Trier JS, Dalton MD, Camp RC, Loeb P, Goyal RK. The histologic spectrum of Barrett's esophagus. N Engl J Med 1976;295:476–480.
37. Chandrasoma PT, Der R, Dalton P, Kobayashi G, Ma, Y, Peters JH, DeMeester TR. Distribution and significance of epithelial types in columnar lined esophagus. Am J Surg Pathol 2001;25:1188–1193.
38. Barrett NR. Chronic peptic ulcer of the oesophagus and "oesophagitis." Br J Surg 1950;38:175–182.
39. Hayward J. The lower end of the oesophagus. Thorax 1961; 16:36–41.
40. Chandrasoma PT, Der R, Ma Y, Peters J, DeMeester T. Histologic classification of patients based on mapping biopsies of the gastroesophageal junction. Am J Surg Pathol 2003;27:929–936.

CHAPTER

12

Reflux Disease Limited to the Dilated End-Stage Esophagus: The Pathologic Basis of NERD

GASTROESOPHAGEAL REFLUX VERSUS REFLUX DISEASE

A misconception about the cellular changes of gastroesophageal reflux disease is that it always occurs as a result of a failure of the normal lower esophageal sphincter mechanism. As Allison (1), in the first accurate description of reflux esophagitis in 1948, stated: "A failure of this (sphincter) mechanism will allow acid to reach the esophagus, and in time this leads inevitably to inflammation and ulceration" (p. 39).

In Allison's time, the sphincter mechanism was not clearly understood; it was thought to be a valvelike effect at the gastroesophageal junction. Within this framework of knowledge, Allison was correct. If the sphincter was a valve limited to the gastroesophageal junction, any pathology above the valve (junction) caused by reflux would be considered abnormal. However, it is now known that the lower esophageal sphincter has a normal length of 3 to 5 cm around the distal thoracic and entire abdominal part of the esophagus (2). The lower esophageal sphincter is part of the esophagus; it is normally tubular and lined by squamous epithelium.

Complete failure of sphincter function requires severe damage. Zaninotto et al. (3) demonstrated that a permanently defective sphincter is characterized by one (or any combination) of the following three abnormalities: an average resting pressure of less than 6 mm Hg, an average overall length of 2 cm or less, and an average length of the abdominal component of the sphincter of 1 cm or less. The normal length of the abdominal component of the sphincter is 2 to 3 cm. This means that over 1 cm of the abdominal part of the sphincter can be destroyed with retention of relatively normal sphincter function. Free reflux into the body of the esophagus therefore occurs only at a relatively advanced stage of sphincter destruction; an abnormal 24-hour pH test and an abnormal impedance test are relatively late manifestations of gastroesophageal reflux disease. The first pathologic changes of reflux must occur in the most distal few millimeters of the esophagus, within its abdominal intrasphincteric part. These patients can therefore manifest symptoms and cellular changes of reflux disease in the squamous epithelium lining the most distal esophagus without having free reflux or evidence of sphincter failure (Fig. 12.1). The primary cause of the changes that we ascribe to reflux disease is exposure of the esophageal squamous epithelium to gastric luminal contents.

This is a very important but unrecognized distinction between the pathophysiology of reflux and the pathophysiology of cellular changes associated with reflux disease. Changes equivalent to reflux disease can occur in the most distal esophagus without any sphincter dysfunction or abnormal reflux if the squamous epithelium becomes exposed to gastric contents. We will show later in this chapter that this change, which we will call "reflux disease limited to the dilated end-stage esophagus," is the earliest and most common manifestation of reflux disease. It is unrecognized at this time except when it is associated with clinical symptoms of reflux. A patient with symptomatic reflux who has unrecognized endoscopic abnormalities will be defined as having "nonerosive reflux disease" (NERD).

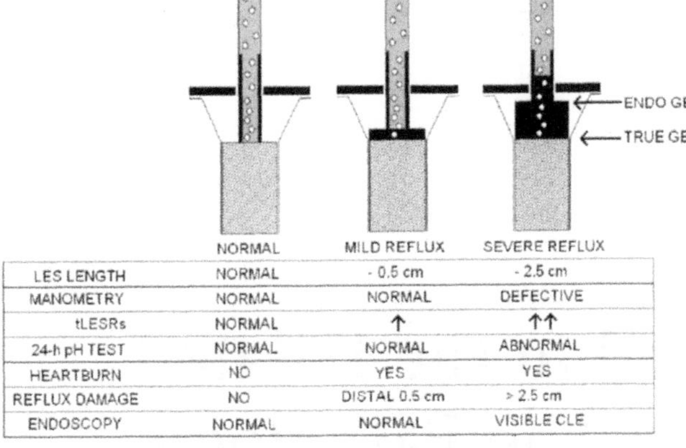

	NORMAL	MILD REFLUX	SEVERE REFLUX
LES LENGTH	NORMAL	- 0.5 cm	- 2.5 cm
MANOMETRY	NORMAL	NORMAL	DEFECTIVE
tLESRs	NORMAL	↑	↑↑
24-h pH TEST	NORMAL	NORMAL	ABNORMAL
HEARTBURN	NO	YES	YES
REFLUX DAMAGE	NO	DISTAL 0.5 cm	> 2.5 cm
ENDOSCOPY	NORMAL	NORMAL	VISIBLE CLE

FIGURE 12.1 The stages of reflux disease. On the left is the normal person without any reflux-induced changes. In the middle is mild reflux change with no evidence of abnormal free reflux into the esophageal body, a sphincter that is 0.5 cm shorter than normal and 0.5 cm of metaplastic columnar epithelium limited to the area of sphincter loss. On the right is a patient with severe reflux with greater sphincter shortening and free reflux into the body with metaplastic columnar epithelium that has extended into the esophagus above the area of sphincter damage.

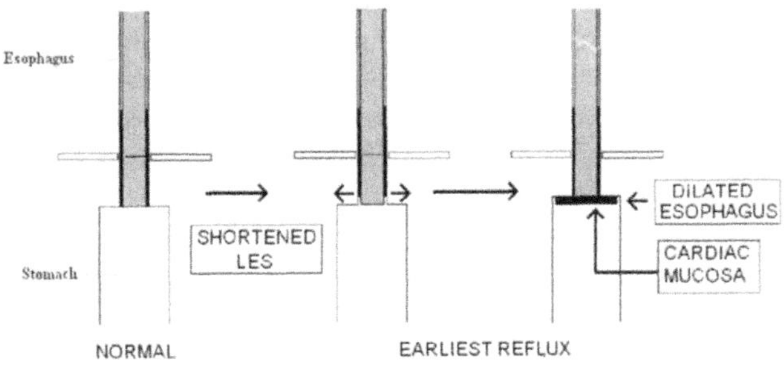

FIGURE 12.2 Pathogenesis of earliest reflux changes. The sphincter shortening has resulted in dilatation of this part of the esophagus. The squamous lining, which is now exposed to gastric contents, has undergone cardiac metaplasia limited to this dilated segment.

In the previous chapter, we considered the pathophysiology of gastroesophageal reflux, which gave us insight into the mechanisms involved in the earliest changes of this disease (Fig. 12.2). This resulted in the characterization of two entities associated with reflux disease:

1. *Reflux disease associated with sphincter failure and free reflux into the body of the esophagus.* This is the only type of reflux disease that we recognize using the present criteria of definition and practice guidelines of the American Gastroenterology Association (4). In these patients, the extent of change in the esophagus is greater than the length of sphincter loss.
2. *Reflux disease limited to the dilated end-stage esophagus.* This is presently unrecognized but represents the earliest and most common type of reflux disease; it can only be diagnosed by histologic definitions that we have recommended. In these patients, the extent of change is the same as the length of sphincter loss (Figs. 12.1 and 12.2). Reflux disease limited to the dilated end-stage esophagus always precedes reflux disease associated with sphincter failure and free reflux, but it frequently exists alone.

REFLUX DISEASE LIMITED TO THE DILATED END-STAGE ESOPHAGUS

Reading the literature leads one to believe that the absolutely earliest mechanism producing reflux is the occurrence of tLESRs. This is true. However, when one evaluates the mechanism that produces tLESRs, it is seen that the lower esophageal sphincter shortens temporarily whenever the stomach overdistends (5). The overdistension causes an increase in the intragastric pressure, which in turn causes the distal esophagus to be "taken up" into the fundus during this process. It is only when this process has resulted in a sufficient degree of shortening of the sphincter that tLESRs are precipitated and a reflux event occurs. In Kahrilas et al.'s study (5), significant sphincter shortening was seen 20 to 30 minutes after the beginning of gastric distension and preceded the occurrence of the tLESR.

The Pathogenesis of Reflux Disease Limited to the Dilated End-Stage Esophagus

What happens in that period where the gastric distension and intragastric pressure increase has produced a temporary shortening of the lower esophageal sphincter that is not of sufficient severity to cause sphincter failure and a tLESR? It is important to understand what "shortening of the sphincter" means. Where does the shortened part of the sphincter go?

The mechanism is best illustrated by taking a standard round balloon where the long neck is the sphincter and the collapsed part is the empty stomach (Fig. 12.3a). As the balloon is blown up, the collapsed part fills to capacity; there is little resistance to this phase of filling because there is no significant pressure increase (Fig. 12.3b). This is equivalent to a normal meal and illustrates the reservoir function of the stomach. When air is blown in excess of its capacity, the balloon distends and pressure increases; with increasing distension, the neck shortens by a mecha-

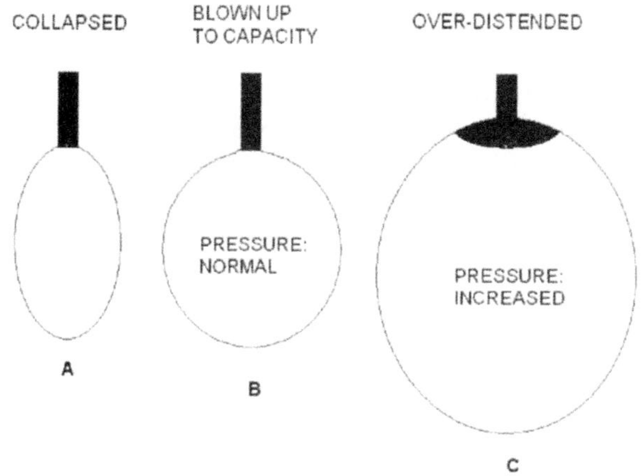

FIGURE 12.3 (a) Experiment using a standard round balloon where the long neck is the sphincter and the collapsed part is the empty stomach. (b) As the balloon is blown up, the collapsed part fills to capacity; there is very little resistance to this phase of filling and there is no significant pressure increase. (c) When air is blown in excess of its capacity, the balloon distends and pressure increases; with increasing distension, the neck shortens by a mechanism in which the most distal part of the neck is "taken up" into the contour of the stomach to an extent that is directly related to the amount of distension.

nism in which the most distal part of the neck is "taken up" into the contour of the stomach to an extent that is directly related to the amount of distension (Fig. 12.3c). Anyone who has blown up a balloon will remember how the neck of the balloon shortens as the balloon distends. The neck has not disappeared; it is now part of the contour of the distended part of the balloon.

There is another situation in the human body where a muscular organ is taken up; during the first stage of labor, as uterine contractions increase the pressure in the uterine body, the cervix is progressively taken up into the body, causing shortening and finally effacement of the cervix that is associated with dilatation.

Let us assume that the gastric distension with a reasonably heavy meal has caused a taking up of the distal 2 mm of the lower esophageal sphincter. Shortening or taking up of the sphincter means that the most distal 2 mm of the esophagus has taken the contour of the fundus of the stomach (Fig. 12.4a). Because the entire esophagus is lined by squamous epithelium, this must mean that *the most distal 2 mm of*

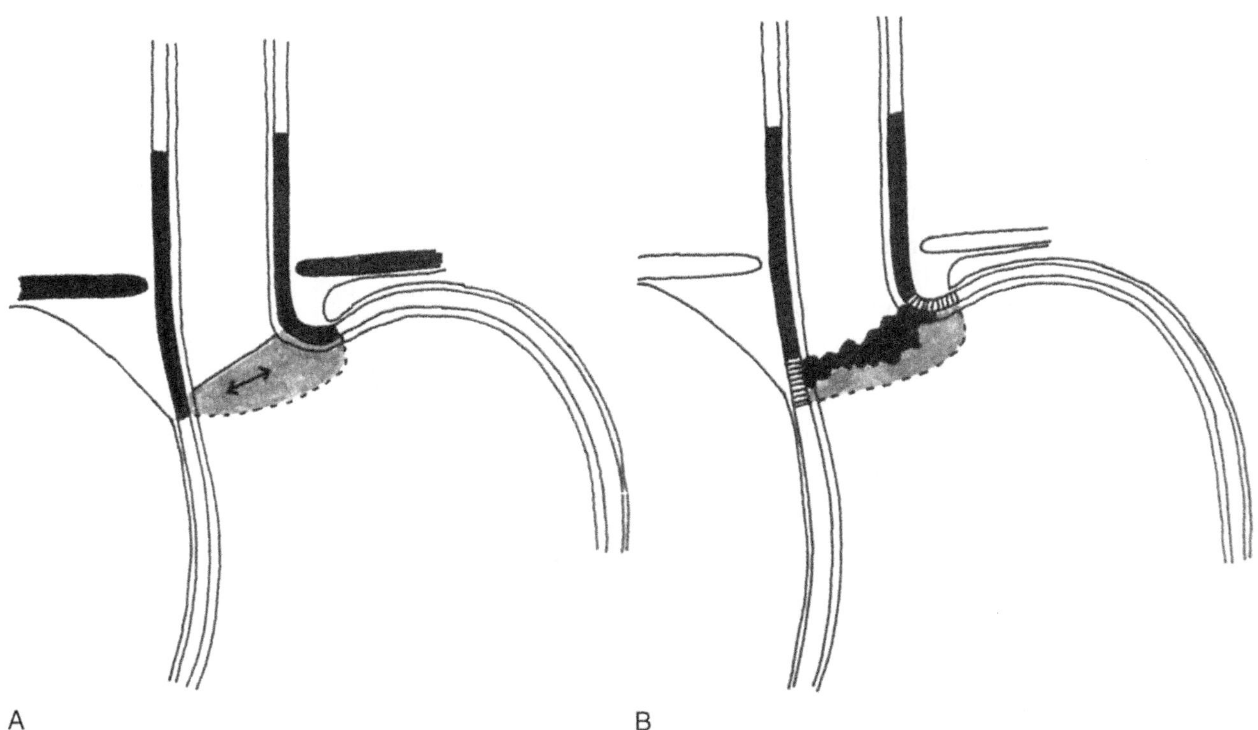

FIGURE 12.4 (a) The effect of overdistension of the stomach after a heavy meal. The increased gastric pressure has caused the distal few millimeters of the lower esophageal sphincter to "be taken up," assuming the contour of the stomach. The squamous epithelium lining this distal few millimeters of the esophagus (solid gray area above the true gastroesophageal junction–dotted line) is now exposed to gastric contents and can undergo damage. (b) Earliest reflux disease. The damage has resulted in the permanent sphincter loss (striped wall), dilatation of the area of sphincter damage, and replacement of squamous epithelium by cardiac (black) and oxyntocardiac (gray) mucosa. *Note:* The drawing shows a relatively long segment of exposed squamous epithelium for ease of visualization. In practice, millimetric areas are exposed.

the esophageal squamous epithelium is now exposed to gastric luminal contents.

This exposure continues throughout the period of gastric overdistension. If a tLESR does not occur, the squamous epithelium remains exposed to gastric luminal contents until the distension is relieved by gastric emptying. Any cause of delayed gastric emptying will tend to increase the exposure. In Kahrilas et al.'s study (5) with air infusion, which is an artificially created gastric overdistension, the shortening was of sufficient severity to produce a tLESR only after 20 to 30 minutes of air infusion. In real life, the shortening of the sphincter in a normal person after a heavy meal is likely to be smaller (in the 1- to 2-mm range), not associated tLESRs, and last a longer time. Significant exposure of the squamous epithelium to gastric luminal contents is therefore likely to be a common occurrence in people who ingest heavy meals.

Exposure of esophageal squamous epithelium to gastric contents causes pathologic changes of reflux disease. The squamous epithelium is damaged every time the stomach distends sufficiently to shorten the sphincter; when the stomach empties, the shortening reverses and the damage may heal. However, when gastric distension is frequent as a result of constant overeating, the squamous epithelial damage can become permanent (Fig. 12.4b). The cellular changes described in Chapter 8 become established. There is separation of squamous cells (dilated intercellular spaces), increased permeability of the epithelium, penetration of the epithelium by molecules in the gastric juice, stimulation of nerve endings (causing postprandial heartburn), alteration of the neural mechanisms (causing permanent sphincter shortening), and finally columnar metaplasia to produce cardiac mucosa. The permanent 2-mm loss of sphincter length results in dilation of this 2 mm of esophagus (end-stage esophagus), which permanently takes up the contour of the stomach. Reflux disease has occurred in the distal 2 mm of the esophagus without abnormal reflux (Fig. 12.4b). The most distal 2 mm of the esophagus has ceased to function as the tube that transmits food to the stomach; it has taken the shape of the stomach and become part of the reservoir.

From this point, pathologic progression of reflux disease becomes a vicious cycle. The person with a 2-mm dilated end-stage esophagus has a 2-mm permanent shortening of the lower esophageal sphincter and a decrease in the acuity of the angle of His (Fig. 12.4b). This person is at increased risk for changes associated with gastric distension, further temporary (followed by permanent) shortening of the lower esophageal sphincter, increasing the frequency of tLESRs and abnormal reflux. This is the patient group with reflux disease (symptomatic) without hiatal hernia in Kahrilas et al.'s study (5).

When permanent shortening has caused the abdominal segment of the lower esophageal sphincter to reach the critical 1-cm length commonly seen in patients with an abnormal 24-hour pH study, free reflux into the body of the esophagus by the tLESR mechanism occurs much more freely and is not restricted to the postprandial period. It is only at this stage that reflux into the body of the esophagus severe enough to cause an abnormal 24-hour pH study and an abnormal impedance study.

Marked shortening of the abdominal part of the sphincter is commonly associated with severe reflux (Fig. 12.5) and esophageal shortening, which cause sliding hiatal hernia (Fig. 12.6). In Kahrilas et al.'s study (5), patients with hiatal hernia were extremely susceptible to tLESRs induced by gastric distension.

Reflux Disease Limited to Dilated End-Stage Esophagus Is the Pathological Basis of Nonerosive Reflux Disease

It is incredible that, until recently, no attempt has been made to characterize this earliest manifestation

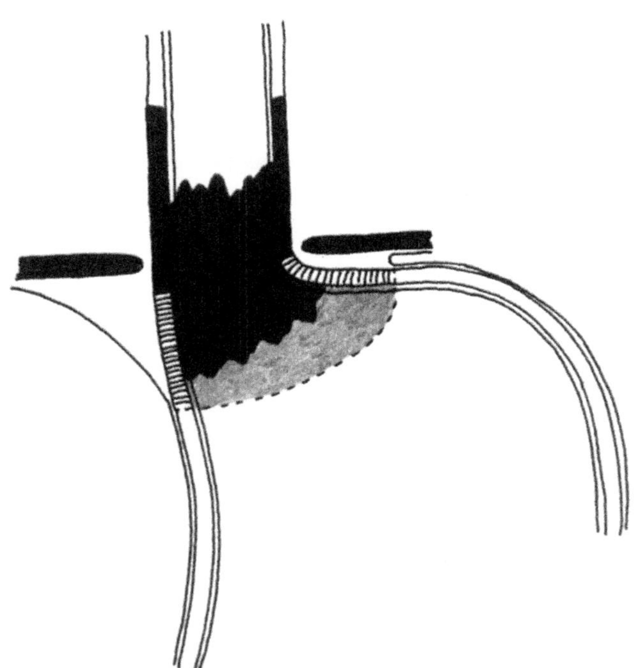

FIGURE 12.5 Severe reflux disease showing a segment of columnar-lined esophagus (black = cardiac and gray = oxyntocardiac mucosa) involving both the dilated end-stage esophagus and the tubular esophagus. The true gastroesophageal junction (proximal limit of oxyntic mucosa) is shown as a broken line at the distal limit of metaplastic columnar esophageal epithelium.

of reflux disease. Throughout history, the medical community has looked at reflux in a most illogical manner. It is like dipping a capillary tube (the esophagus) into a beaker of acid (gastric contents); the contents rise to a variable extent depending on the physical properties but will always involve the most distal part first (Fig. 12.7). Instead of first looking for changes at the most distal esophagus, we have looked for changes at much higher levels, ignoring the point of maximum attack. It makes no sense to place a pH electrode or an impedance probe 5 cm and 3 cm, respectively, above the proximal limit of the lower esophageal sphincter (which is 8 to 10 cm proximal to the end of the esophagus, assuming a lower esophageal sphincter length of 5 cm), or defining Barrett esophagus as a change 3 cm above the gastroesophageal junction. It is almost as if the medical community is frightened by the enormity of the problem of gastroesophageal reflux disease that it ignores its most common pathologic changes.

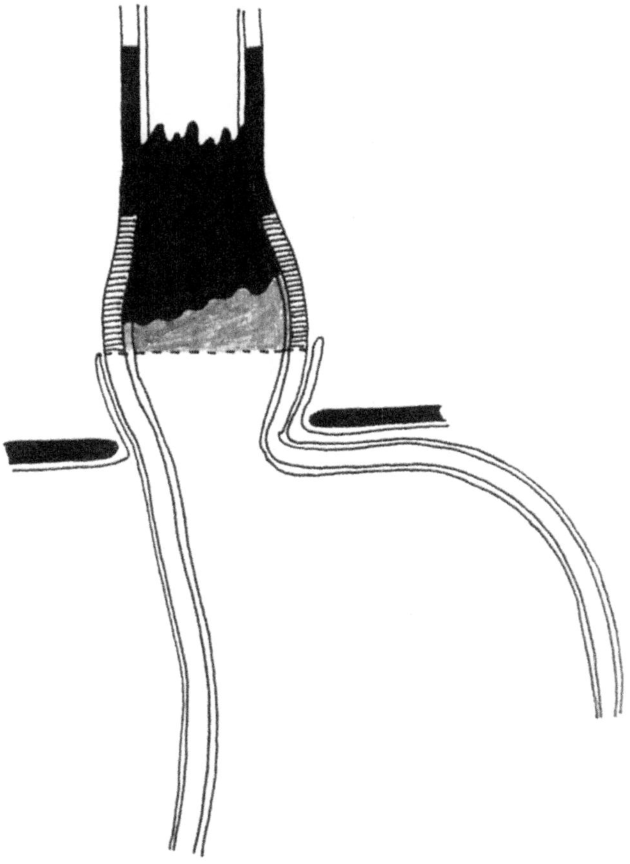

The largest population of patients with reflux disease falls into the category of patients who have symptoms of reflux but do not have any endoscopic abnormality. These patients are designated as having NERD. Even with restrictive definitions requiring very specific symptoms at high frequency (in Kahrilas et al. [5], symptomatic reflux is defined as the presence of heartburn more than three times per week; in Lagergren et al. [6], it was heartburn or regurgitation more than one time per week), this includes a large number of people. If one includes patients who experience heartburn without restricting the symptom by the requirement of a given frequency, around 40% of the adult American population suffers from heartburn. This is the population that is the target of direct marketing advertising for over-the-counter acid suppressive drugs that are ubiquitous on television in the United States (Fig. 12.8). Viewers are led to believe that they will be cured of their "acid reflux disease" when they take these drugs. These people have reason to believe this is true because these agents are extremely effective at controlling heartburn. The suppression of acid increases gastric pH and decreases the stimulation of nerve endings in the squamous epithelium that is responsible for heartburn. While reflux continues, the person on acid suppression is relieved of symptoms.

The patient whose heartburn persists despite over-the-counter acid suppressive drugs seeks medical care, presenting to an internist or gastroenterologist. Failure of simple treatment is likely the result of inadequate

FIGURE 12.6 The loss of the angle of His resulting from dilatation of the abdominal segment of the esophagus sets the stage for a sliding hiatal hernia. These patients have severe functional loss of sphincter action with severe reflux into the body of the esophagus. If the damage resulting from this damage causes shortening of the esophagus, there is little resistance to the stomach being pulled up into the thorax.

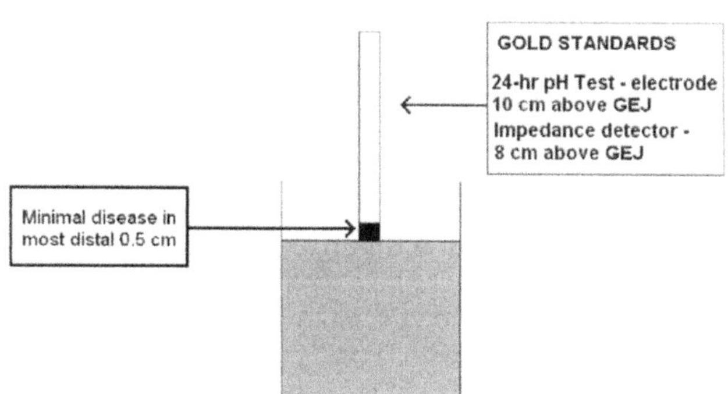

FIGURE 12.7 Esophagus shown as a capillary tube dipping into a beaker of acid (representing the stomach). As acid is drawn up, the first changes must occur in the most distal region. We have constantly looked at a much higher point in the esophagus to define reflux disease, ignoring the point of maximal involvement. The present gold standards for diagnosis of reflux (24-hour pH testing and impedance) place the recording devices 10 and 8 cm above the true gastroesophageal junction.

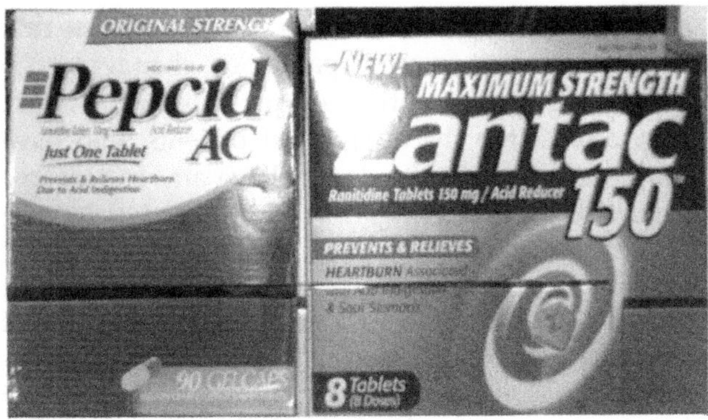

FIGURE 12.8 Over-the-counter acid suppressive agents proclaiming their efficacy in preventing and relieving heartburn.

acid suppression or the fact that the damaged squamous epithelium is sensitive to nonacid agents that continue to reflux into the esophagus.

These patients face one of two outcomes:

1. They are treated empirically by "prescription strength" acid suppressive drugs. Many of these patients respond to higher doses of proton pump inhibitors (40 mg or 80 mg of omeprazole or newer agents maybe required), often used in combination with H2 receptor antagonists. Response to these regimens is around 95%. Response requiring higher dosages and longer duration is probably related to healing of the damaged squamous epithelium. The dilatation of the intercellular spaces reverses, and the nerve endings become protected against nonacid reflux as a result of healing.
2. They are subject to "the gold standard" of diagnostic testing to document that their symptoms are caused by reflux, which is a 24-hour pH test, impedance test, or both. This test is based on the assumption that reflux must be associated with loss of sphincter function and looks for evidence of reflux at a point 5 cm above the upper limit of the lower esophageal sphincter (Fig. 12.7). This is approximately 10 cm above the site of first damage in the most distal millimeter of the sphincter, assuming sphincter length to be 5 cm. Newer tests with impedance catheters placed 3 cm above the upper limit of the lower esophageal sphincter similarly look for free reflux into the body of the esophagus 8 cm above the end of the esophagus. *Patients who have a normal 24-hour pH test or a normal impedance study are designated as "not having reflux disease."* The wise physician will recognize that these "gold standard" tests have a significant false negative rate and will treat these patients for reflux symptoms despite the negative test. This is a strong argument to not do these diagnostic tests in the first place.

The majority of patients who are placed on long-term acid suppressive drug therapy never have an upper endoscopy. In a study of pharmacy billing data for two insurers within a large eastern Massachusetts provider network, 4684 of 168,727 patients (2%) were prescribed chronic (>90 day) acid suppressive drugs; 47% were taking H2 receptor antagonists and 57% were taking proton pump inhibitors (4% were taking both). Diagnostic testing was uncommon in these patients, with only 19% of the patients on acid suppressive drugs having undergone esophagogastroduodenoscopy within the prior two years (7).

The majority of patients who have an upper endoscopy will have no detectable endoscopic abnormality. Only 30% of patients with reflux symptoms will have erosive esophagitis, and approximately 10% will have a visible columnar-lined esophagus (8). The patients who have no endoscopic abnormality fall within the definition of NERD. It is not that these patients have no manifestation of reflux disease; it is that the manifestations of reflux disease are missed by present definitions and practice guidelines. Physicians are defining away and ignoring the earliest pathologic changes of reflux disease. Just like they ignored short-segment Barrett esophagus for more than three decades by defining that columnar-lined esophagus less than 3 cm was normally present in the distal esophagus, they are now ignoring the changes in the dilated end-stage esophagus by defining these as normal (Fig. 12.1).

The importance of this error is that the occurrence of adenocarcinoma does not depend on the amount of reflux damage (Fig. 12.9). It probably depends largely on the characteristics of the refluxate that promote intestinal metaplasia and carcinogenesis. Thus, patients with reflux disease limited to the dilated end-stage esophagus can develop intestinal metaplasia if the refluxate factors promote the genetic switch that causes goblet cell differentiation in cardiac mucosa. If the characteristics in the refluxate promote carcinogenesis in intestinal metaplasia, dysplasia and adenocarcinoma can develop in this short segment of abnormal metaplastic epithelium.

Reflux Disease Limited to the Dilated End-Stage Esophagus: An Explanation for Many Contradictions

There are many serious contradictions in what is presently believed about the anatomy, endoscopy, and

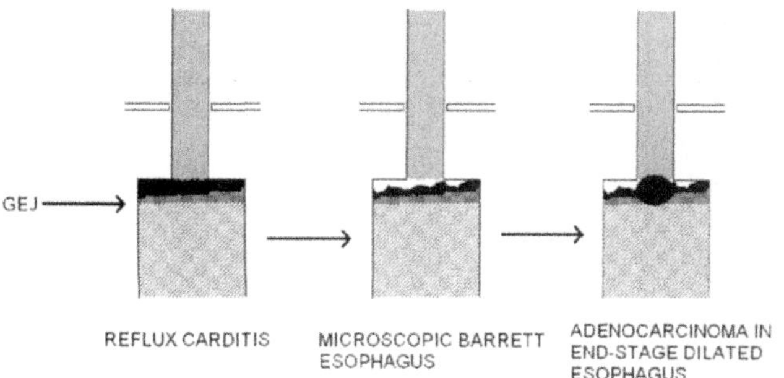

FIGURE 12.9 The spectrum of reflux disease limited to the end-stage esophagus. The cardiac metaplasia (black) leads to intestinal metaplasia (white), which is at risk for developing adenocarcinoma (black circle). This disease will be misclassified as "normal gastric cardia," "intestinal metaplasia of the gastric cardia," and "adenocarcinoma of the gastric cardia" by present definitions. History repeats itself: a reflux-induced esophageal disease is once again mistaken as "normal" and "gastric."

histology of the gastroesophageal junction and the pathologic changes of reflux disease. These should raise at least the following questions:

1. *Patients with classical symptoms of reflux commonly have no endoscopic abnormality. This is called NERD.*

   Question: Is it really true that patients with symptomatic reflux exist without any endoscopic abnormality, or are we missing some pathology in these patients?

   Answer: Most patients with nonerosive reflux disease represent the population with intrasphincteric reflux disease limited to a dilated end-stage esophagus. This is not diagnosed because endoscopically normal patients are not biopsied; even if they are biopsied, the criterion that defines this entity (cardiac mucosa) is incorrectly called "normal gastric mucosa." At endoscopy, the wrong place (the squamous epithelium) is being studied for evidence of disease by sophisticated modalities like chromoendoscopy and magnification endoscopy. The pathology is distal to the squamocolumnar junction; it is completely ignored and defined away as normal.
2. *Cardiac mucosa is frequently found distal to the endoscopic and gross gastroesophageal junction and is called "normal gastric mucosa." However, when the data for the significance of the presence of cardiac mucosa in this location are critically evaluated, even 1 mm of cardiac mucosa correlates with the presence of reflux disease* (9).

   Question: Does this mean that the proximal stomach is injured in gastroesophageal reflux disease, or are we misinterpreting something?

   Answer: We are mistaking the dilated end-stage esophagus for the proximal stomach because we use an incorrect definition of the gastroesophageal junction.
3. *When isolated intestinal metaplasia is found distal to the endoscopic gastroesophageal junction with the remainder of the stomach being absolutely normal, a diagnosis of "intestinal metalaplasia of the gastric cardia" is made. Studies suggest that this is sometimes related to gastroesophageal reflux disease.*

   Question: Does this mean that there is a disease of the proximal stomach caused by gastroesophageal reflux that is limited to the most proximal few millimeters of the stomach, or are we misinterpreting something?

   Answer: We are mistaking the dilated end-stage esophagus for the proximal stomach because we are using an incorrect definition of the gastroesophageal junction.
4. *Epidemiologists find a relationship between adenocarcinoma of the gastric cardia and symptomatic gastroesophageal reflux disease.*

   Question: Does this mean that reflux disease causes gastric adenocarcinoma, or are we misinterpreting something?

   Answer: We are mistaking adenocarcinoma arising in the dilated end-stage esophagus for adenocarcinoma of the proximal stomach. It is no real surprise that the epidemiology of "adenocarcinoma of the gastric cardia" parallels adenocarcinoma of the esophagus and is completely different than adenocarcinoma of the stomach.

Definition and Diagnosis of Reflux Disease Limited to Dilated End-Stage Esophagus

Reflux disease limited to the dilated end-stage esophagus is defined by the presence of metaplastic columnar esophageal epithelium (cardiac mucosa with and without intestinal metaplasia, and oxyntocardiac mucosa) between the squamocolumnar junction at the end of the tubular esophagus and the proximal limit of gastric oxyntic mucosa (Fig. 12.10) (i.e., the true gastroesophageal junction). The metaplastic columnar epithelium may be under rugated mucosa or may be so small in extent that it cannot be detected by endoscopy or gross examination. The patients do not have reflux disease by all traditional tests (no erosive esophagitis, no visible columnar-lined esophagus, the 24-hour pH test and impedance test is within normal limits). The patients may have symptoms (symptomatic reflux disease) or be asymptomatic.

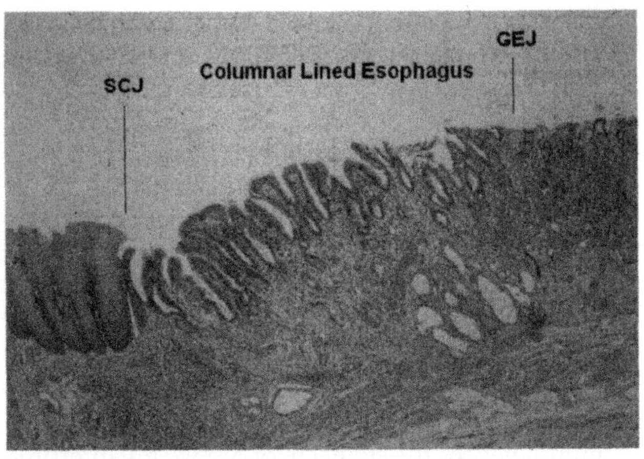

FIGURE 12.10 Early reflux-induced change. The squamocolumnar junction (left) has moved proximally as a result of cardiac metaplasia (center). The area of cardiac mucosa between the proximal limit of gastric oxyntic mucosa (right; gastroesophageal junction) and the distal limit of squamous epithelium is the earliest manifestation of reflux disease. The cardiac mucosa here is long (about 5 mm) and shows foveolar hyperplasia and chronic inflammation.

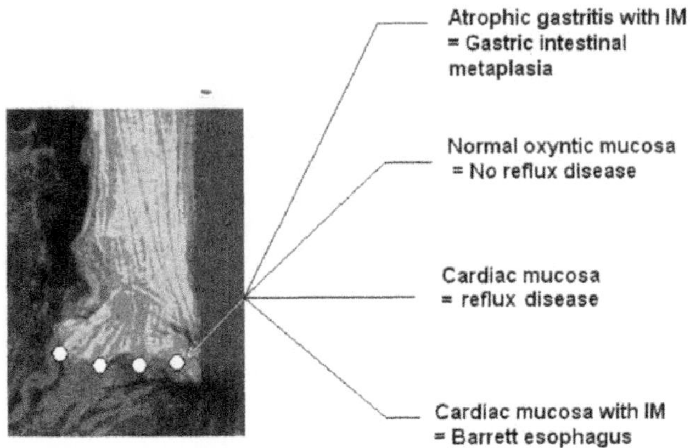

FIGURE 12.11 Classification of patients who have a normal endoscopic/gross appearance is based on a biopsy immediately distal to the squamocolumnar junction (ideally one that straddles the squamocolumnar junction). The variation in the type of columnar epithelium defines normal (oxyntic mucosa), atrophic gastritis (intestinal metaplasia in oxyntic mucosa), reflux disease (cardiac mucosa), or Barrett esophagus (intestinal metaplasia in cardiac mucosa). The patient with oxyntocardiac mucosa adjacent to squamous epithelium is not shown in this figure. (Cross reference: Color Plate 1.1)

Most of these patients never have symptoms of reflux disease during life; they can be identified at autopsy as that group of patients who have a very small amount of cardiac and oxyntocardiac mucosa between the squamous epithelium and gastric oxyntic mucosa (10). Unfortunately, some of these asymptomatic patients go on to develop Barrett esophagus and reflux-induced adenocarcinoma and can be detected only by screening or when symptomatic cancer arises.

This diagnosis is precluded by presently accepted definition. When the incorrect endoscopic definition of the gastroesophageal junction (end of the tubular esophagus or proximal limit of rugal folds) is used, reflux disease limited to dilated end-stage esophagus will be misinterpreted as being "normal stomach" or gastric pathology. Reflux disease limited to dilated end-stage esophagus will be misinterpreted as a normal histologic finding by all pathologists who believe that cardiac mucosa is a normal gastric epithelium. In fact, present practice guidelines (4) guarantee that this entity will not be found unless the endoscopist rejects the recommendation to not biopsy endoscopically normal patients.

These patients will have no visible endoscopic abnormality (Fig. 12.11). The dilated end-stage esophagus will be indistinguishable from normal gastric mucosa by endoscopy and gross examination. It will be distal to the end of the tubular esophagus and can occur in mucosa that shows rugal folds. The only method of diagnosis is by taking biopsies distal to the normal-appearing squamocolumnar junction (Fig. 12.11). The existence and stratification of the disease is based on the type of glandular epithelium found distal to the normal-appearing squamocolumnar junction (Table 12.1).

TABLE 12.1 Clinical Features of Reflux Disease Limited to End-Stage Esophagus

Symptoms	Atypical, episodic, classical, or absent
Endoscopy	Normal; tubular esophagus squamous lined, SCJ horizontal, rugal folds not separated from Z line
24-hour pH test	Mildly abnormal or within normal limits, slightly increased tLESRs, increased post-prandial reflux
Impedance test	Mildly abnormal or within normal limits, slightly increased tLESRs, increased postprandial reflux
Response to acid suppressive agents	Usually excellent symptom relief
Manometry	Usually within normal limits, may have detectable mild shortening
Histology (biopsy, resection, autopsy)	Cardiac mucosa between squamocolumnar junction and gastric oxyntic mucosa
Complications	Intestinal metaplasia (Barrett esophagus), dysplasia, adenocarcinoma

Epithelium found distal to squamocolumnar junction	Interpretation
Gastric oxyntic mucosa only (Fig. 12.12)	No evidence of gastroesophageal reflux
Oxyntocardiac mucosa (Fig. 12.13)	Compensated reflux
Cardiac mucosa (Figs. 12.14 and 12.15)	Intrasphincteric reflux disease limited to dilated end-stage esophagus
Intestinal metaplasia in cardiac mucosa (Fig. 12.16)	Barrett esophagus in dilated end-stage esophagus

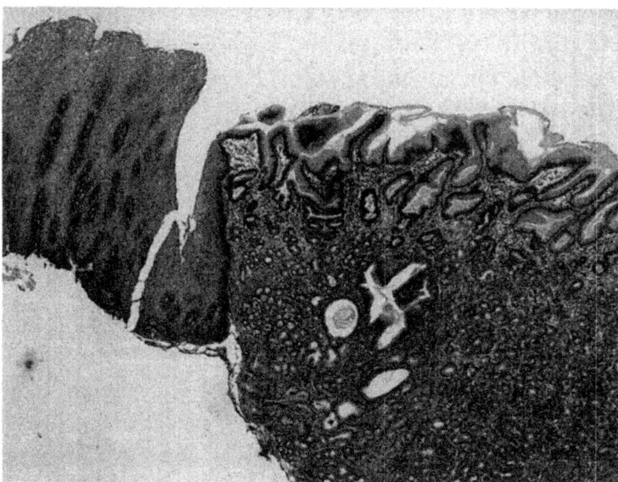

FIGURE 12.12 Biopsy straddling the squamocolumnar junction showing squamous epithelium directly transitioning into gastric oxyntic mucosa. There is no metaplastic esophageal columnar epithelium.

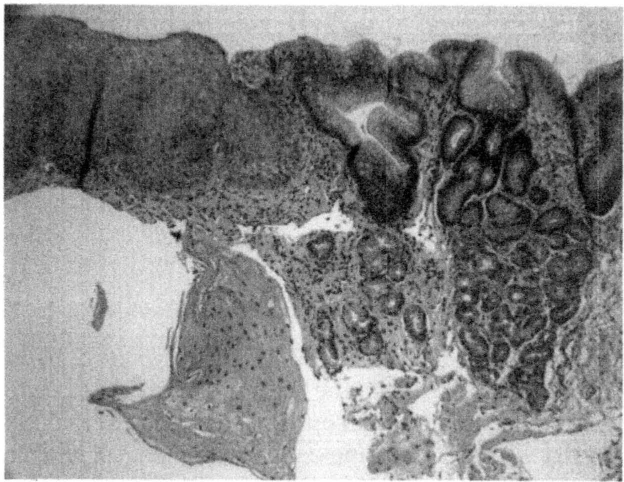

FIGURE 12.13 Biopsy straddling the squamocolumnar junction showing squamous epithelium directly transitioning into oxyntocardiac mucosa. Two foveolar-gland complexes are shown. The one adjacent to the squamous epithelium shows a few parietal cells; the one on the right shows numerous parietal cells. There is no cardiac mucosa.

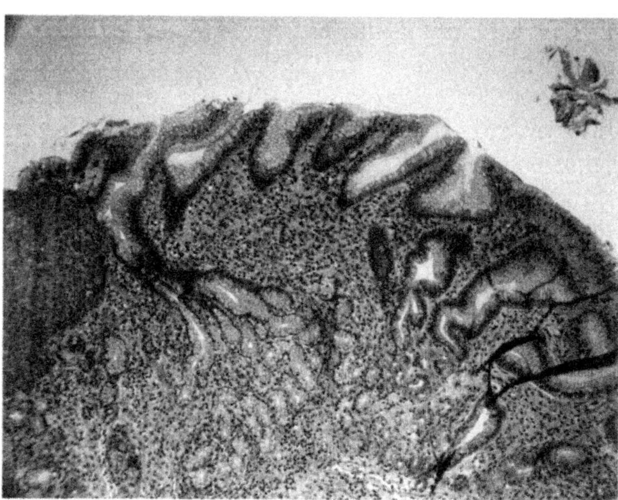

FIGURE 12.14 Biopsy straddling the squamocolumnar junction showing squamous epithelium directly transitioning into cardiac mucosa. The single foveolar-gland complex adjacent to the squamous epithelium consists only of mucous cells. The gland immediately next to this contains parietal cells (oxyntocardiac mucosa). This is cardiac mucosa whose extent is a single foveolar-gland complex. (Cross reference: Color Plate 3.8)

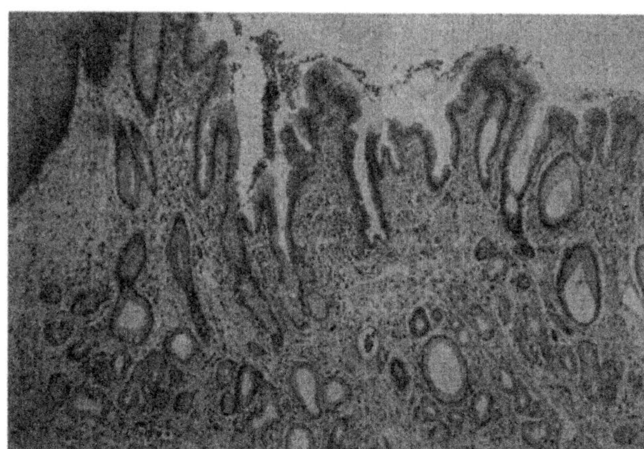

FIGURE 12.15 Higher power of Figure 12.10 showing a longer segment of cardiac mucosa distal to squamous epithelium. The cardiac mucosa shows reactive hyperplasia of the foveolar region, distorted glands, and chronic inflammation.

The epithelia found in a biopsy taken from immediately distal to the squamocolumnar junction (ideally, the biopsy should straddle the junction and contain the squamous-glandular transition to avoid sampling error) shows the same epithelial types that occur in columnar-lined esophagus if the patient has reflux disease in a dilated end-stage esophagus. If the patient does not have reflux disease, the glandular epithelium will be gastric oxyntic mucosa (Fig. 12.12). The

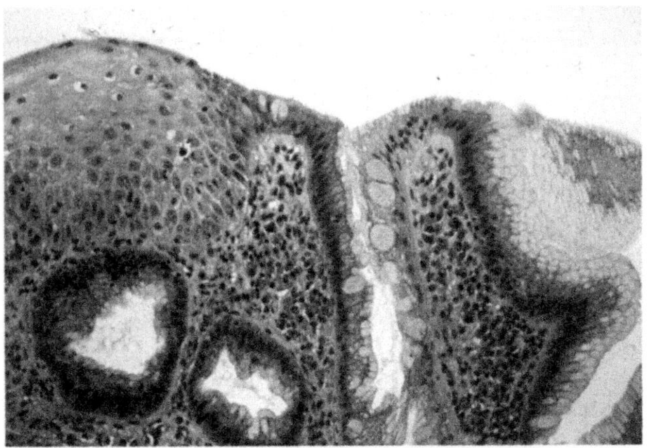

FIGURE 12.16 Biopsy straddling the squamocolumnar junction showing squamous epithelium directly transitioning into a very short segment of intestinal metaplasia, consisting of goblet cells limited to one foveolar complex followed by cardiac mucosa. This is the shortest segment of Barrett esophagus that is possible.

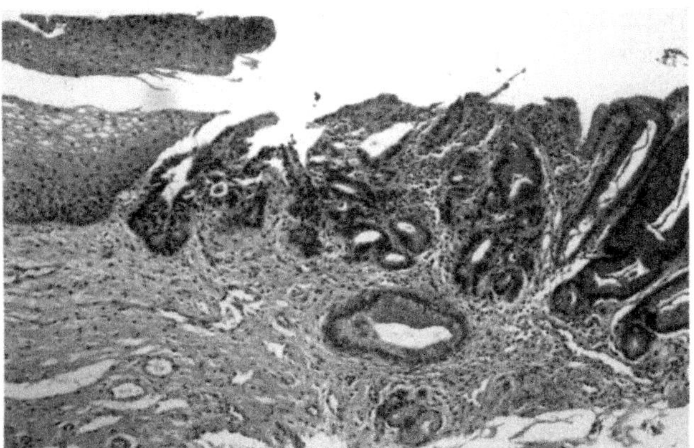

FIGURE 12.17 High-grade dysplasia occurring in reflux disease limited to end-stage esophagus showing a limited dysplastic area between squamous epithelium and nondysplastic Barrett epithelium. This patient underwent endoscopic mucosal resection for high-grade dysplasia "of the gastroesophageal junction." (Cross reference: Color Plate 5.4)

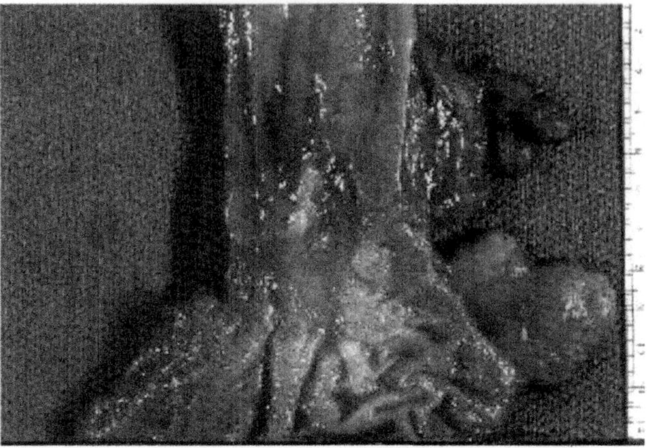

FIGURE 12.18 Esophagectomy specimen showing a short segment of columnar-lined esophagus between the irregular, proximally displaced squamocolumnar junction and the proximal limit of the rugal folds. Gross examination does not permit accurate identification of the proximal limit of gastric oxyntic mucosa, which may be at or up to 2cm distal to the proximal limit of the rugal folds.

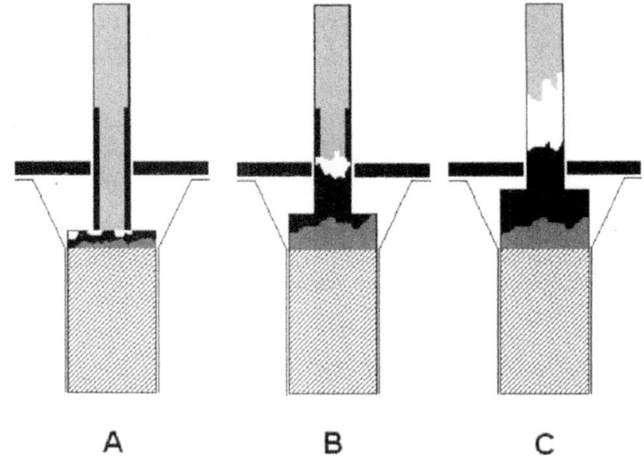

FIGURE 12.19 Different lengths of columnar-lined esophagus with the typical zonation. Intestinal metaplasia (white) is proximal, cardiac mucosa (black) is intermediate, and oxyntocardiac mucosa (stippled) is distal. (a) This zonation is maintained in reflux disease of all lengths, including disease limited to the dilated end-stage esophagus. (b and c) Short- and long-segment Barrett esophagus.

presence of intestinal metaplasia in cardiac mucosa in this biopsy is Barrett esophagus (Fig. 12.16) even though it does not fit the present definition of Barrett esophagus, which requires intestinal metaplasia in an *endoscopically visible* columnar-lined esophagus. We should not make the historically repeated mistake of using an incorrect definition of this disease to ignore it when it is present. Patients will still go on to develop complications from this disease even as we claim it does not exist (Fig. 12.17).

It should be noted that when more severe damage to the sphincter occurs, the dilated end-stage esophagus will increase in length but will become associated with evidence of free reflux into the body of the esophagus (Figs. 12.18, 12.19, and 12.20; see also Figs. 12.1 and 12.5). In these patients, the disease is not restricted to the dilated end-stage esophagus; evidence of reflux disease such as erosive esophagitis and a visible columnar-lined esophagus will be present in the distal tubular esophagus. It is only in this relatively advanced stage of the disease that an endoscopic diagnosis of reflux disease is possible with present criteria.

No proof is necessary for the viewpoint that a microscopic phase of reflux disease must exist before

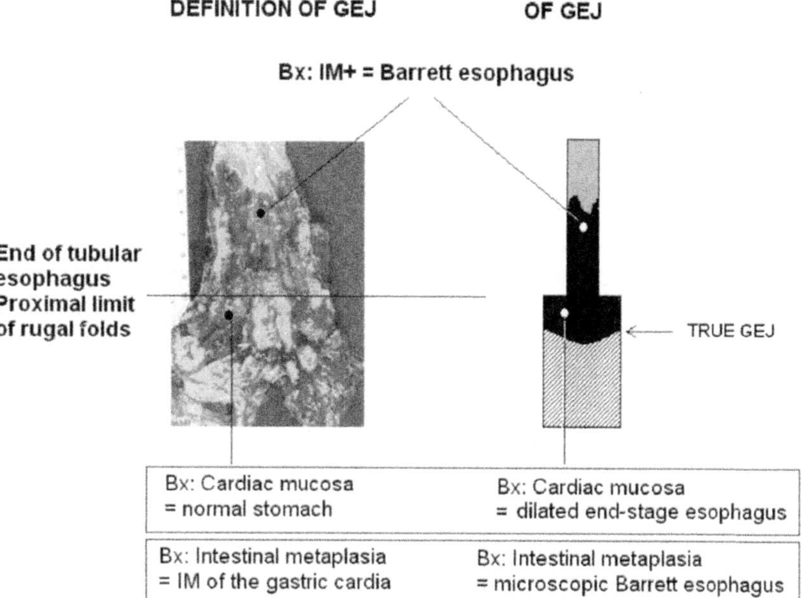

FIGURE 12.20 Esophagectomy specimen with a line drawn across the end of the tubular esophagus. This line is chosen because there appear to be rugal folds extending to this line in the center. There is abnormal, flat mucosa above and below this line representing a long segment of columnar-lined esophagus. If a biopsy taken from the tubular esophagus shows intestinal metaplasia, a diagnosis of Barrett esophagus is made. The interpretation of the presence of cardiac mucosa and intestinal metaplasia distal to the line of the gastroesophageal junction will depend on whether you believe the presently accepted definitions (= normal gastric cardia and intestinal metaplasia of the gastric cardia) or the new suggested definition (= the distal region of columnar metaplasia with and without intestinal metaplasia involving dilated end-stage esophagus). (Cross reference: Color Plate 3.2)

the visible changes used to diagnose reflux disease. This statement is proved by the fact of physics that the highest level of optical resolution of the naked eye and endoscope is less than the optical resolution of the microscope. One cannot progress to a visible change in the esophagus without passing through a microscopic change. Depending on the visible change to define reflux disease and Barrett esophagus *must mean that one misses the microscopic change.* The microscopic phase even has a name applied to it: nonerosive reflux disease. All I am doing is defining the cellular and anatomic nature of this microscopic phase as reflux disease limited to the dilated end-stage esophagus. Nonerosive reflux disease equates to normal endoscopic and gross features, but does not equate to normal histology. The answer to finding the pathology is not looking at the squamous epithelium for phantom changes; it is to biopsy the 0.5 cm area distal to the "endoscopically normal" squamocolumnar junction.

Patients with morphologic evidence of reflux disease in the body of the esophagus (erosive esophagitis and visible columnar-lined esophagus) always have an endoscopically undetected dilated end-stage esophagus distal to the visible columnar-lined esophagus. The only effect is endoscopy will underestimate the length of esophagus that appears to be involved because the dilated end-stage esophagus is regarded as stomach. If adenocarcinoma arises in this segment, it will not be detected during surveillance because this is not regarded as part of the abnormal columnar-lined segment from which biopsies are taken.

Distribution of Epithelia in Reflux Disease Limited to the Dilated End-Stage Esophagus

In 1976, Paull et al. (11) showed that the three epithelial types in columnar-lined esophagus had a constant zonation with the intestinal (specialized) type being most proximal, the cardiac (junctional) type in the middle, and the oxyntocardiac (fundic) type most distal. One of the most powerful arguments for the existence of reflux disease limited to a dilated end-stage esophagus is that this zonation is only applicable when the entire segment of columnar epithelium from the squamocolumnar junction to the proximal limit of gastric oxyntic mucosa is taken.

This zonation is maintained with all lengths of columnar-lined esophagus (Fig. 12.19). In patients with long segments of Barrett esophagus, it is not uncommon for a large part of the tubular esophagus to be lined by intestinal or intestinal + cardiac epithelia with the oxyntocardiac +/− cardiac mucosa being limited to the area distal to the end of the tubular esophagus and often in mucosa containing rugal folds (Fig. 12.20). In patients with short-segment Barrett esophagus, intestinal metaplasia is frequently restricted to the proximal region in the tubular esophagus, cardiac mucosa frequently crosses the end of the tubular esophagus, and oxyntocardiac mucosa is often restricted to the part of the esophagus distal to the end of its tubular part. In patients with reflux disease limited to the dilated end-stage esophagus, all three epithelial types are distal to the end of the tubular esophagus and in rugated mucosa, but the zonation is maintained. In these patients the intestinal, cardiac, and oxyntocardiac epithelia are frequently present over a very short segment. We have seen cases where all three epithelia are present in a single biopsy piece (Fig. 12.21).

Ignoring the dilated end-stage esophagus distal to the visible columnar-lined esophagus is a problem when intestinal metaplasia is present in that segment. Such intestinal metaplasia is missed at surveillance by any biopsy protocol that uses the proximal limit of the rugal folds and can progress to dysplasia and adenocarcinoma without being detected even by rigorous surveillance (Fig. 12.22).

Clinical Features of Reflux Disease Limited to Dilated End-Stage Esophagus (Table 12.1)

The patient with reflux disease limited to a dilated end-stage esophagus may or may not have symptoms. The majority do not fit definitions used for "classical reflux disease" symptoms (heartburn or regurgitation at least once per week is the most common definition used); a significant number will have episodic, ill-defined, and mainly postprandial symptoms; some will have classical symptoms; many will be asymptomatic. Asymptomatic patients will be detectable at autopsy by the presence of very short segments of oxyntocardiac mucosa and cardiac mucosa with and without intestinal metaplasia (10).

The presence of heartburn indicates squamous epithelial damage sufficient to cause an increased permeability that permits molecules to enter the epithelium and stimulate nerve endings therein. Cessation of pain after such damage can occur without any treatment by two mechanisms: (a) if the damage to the squamous epithelium is temporary and heals with a return of the normal lack of permeability of the epithelium or (b) if columnar metaplasia occurs in the area of squamous epithelial damage. If, as is likely, columnar epithelium

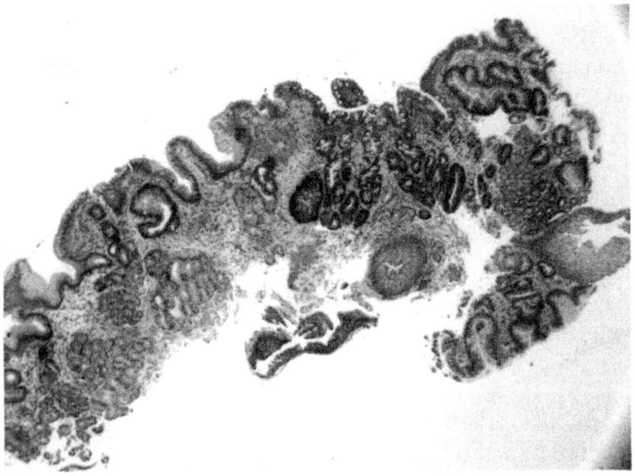

FIGURE 12.21 Single biopsy from a patient with intrasphincteric reflux disease with Barrett esophagus showing intestinal, cardiac, and oxyntocardiac mucosal type. Note the presence of a squamous-lined duct in the mucosa indicating the esophageal location of this biopsy. (Cross reference: Color Plate 3.4)

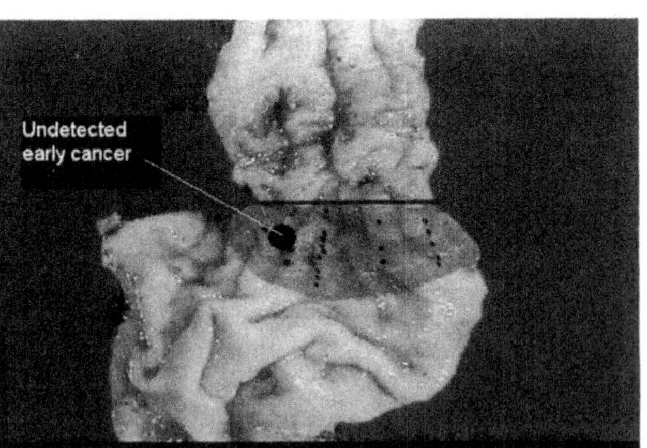

FIGURE 12.22 Esophagectomy specimen in patient with adenocarcinoma in the distal esophagus arising in long-segment Barrett esophagus. Histologic examination of the region distal to the end of the tubular esophagus and the proximal limit of the rugal folds showed an early adenocarcinoma arising in intestinal (Barrett-type) metaplasia (black circle) that was unsuspected preoperatively. This was in an area showing esophageal submucosal glands, identifying the cancer as esophageal in location. (Cross reference: Color Plate 3.6)

is less pain sensitive when exposed to gastric contents, the typical heartburn will cease and the patient is "cured" of symptoms at the expense of columnar metaplasia. Heartburn will recur only when the squamous epithelium above the metaplastic segment undergoes damage. Such episodes may be temporally separated by significant periods.

These patients are extremely common in the population. Many patients probably have ill-defined, intermittent, mild, and often ignored symptoms that come under the nonspecific terms of "dyspepsia" or "acid indigestion." They do not seek medical care and represent the population targeted for over-the-counter acid suppressive medications in ubiquitous television advertisements. When they seek care, they are frequently treated by primary care physicians with prescribed acid suppression to control their symptoms without having any endoscopy and diagnostic testing. These are the patients with early reflux that fall through the cracks and present for the first time with reflux-induced adenocarcinoma as their first serious manifestation of reflux disease. Because their symptoms do not fall into definitions of "classical symptomatic reflux disease," they are designated as "asymptomatic" or "control" groups in most studies.

Endoscopically, these patients are normal by present definitions. The tubular esophagus is completely lined by squamous epithelium, and the rugal folds extend to this point (Fig. 12.11). When these patients have classical symptoms, they are designated as having NERD.

The 24-hour pH study may be either normal or abnormal, depending on whether or not the increase in tLESRs resulting from sphincter shortening has resulted in the test reaching the threshold of abnormality for the test (usually defined as a pH of <4 for 4.5% of the 24-hour period). In a study of 105 endoscopically normal patients with cardiac mucosa who had a 24-hour pH test at our institution, Der et al. (12) reported that 64 (61%) of 105 patients had an abnormal test defined as a pH of less than 4 for less than 4.5% of the 24-hour period; 41 of 105 patients (39%) had a pH study that was within the normal range. The 39% of patients with cardiac mucosa and a normal 24-hour pH test can be classified as having reflux disease *only* by the criterion of the histologic presence of cardiac mucosa (reflux carditis). These data indicate that the presence of cardiac mucosa is a more sensitive indicator of reflux disease than a 24-hour pH test.

Manometric abnormalities, namely a shortening of the abdominal segment of the lower esophageal sphincter, are present in all patients with reflux disease limited to the dilated end-stage esophagus. This does not mean that the abnormality can be detected. The sphincter length has a normal variation in different people, resulting in a range of normalcy (e.g., the abdominal sphincter length is normally 2 to 3 cm). If there is a 1-mm shortening, the sphincter length is likely to remain within normal limits. Manometry does not have the sensitivity to detect the minimal shortening associated with reflux disease limited to a very short segment of dilated end-stage esophagus. This was shown in a study by Oberg et al. (10) from our institution, where Dr. Chandrasoma was the pathologist who examined all the slides. The study looked at 334 patients who had a normal endoscopic appearance. Of these, 246 (74%) had either cardiac or oxyntocardiac mucosa in their biopsies; the other 88 patients (26%) had only squamous and gastric oxyntic mucosa. The patients with cardiac and oxyntocardiac mucosa had a greater likelihood of having lower esophageal sphincter abnormalities compared with the patients who did not have these mucosal types. The most common sphincter abnormality was a shortening of the abdominal length of the sphincter. In this study, many patients with cardiac mucosa did not have a detectable sphincter abnormality. These are the patients in whom the test was not sufficiently sensitive to detect the minimal sphincter shortening.

If patients with reflux disease limited to the dilated end-stage esophagus have a shortening of the abdominal segment of the sphincter, they should be more susceptible than normal people to reflux occurring in the postprandial state when the stomach is distended. Because their abdominal sphincter segment is permanently shortened, the degree of gastric overdistension and increase in intragastric pressure necessary to cause transient lower esophageal sphincter relaxation and reflux is less. These patients should therefore be expected to have a higher frequency of transient sphincter relaxations and postprandial reflux disease.

Kahrilas et al.'s study (5) showed this was true. When the stomach was distended artificially by air infusion, the frequency of tLESRs increased 4.0, 4.5, and 9.5 per hour in normal controls, nonhernia GERD patients, and GERD patients with hiatal hernia. These data suggested that there was a continuum from normal to nonhernia GERD to GERD with hernia in the diminution of the threshold for the precipitation of tLESRs in response to gastric distension. Kahrilas et al.'s GERD patient group was composed predominantly of patients who had erosive esophagitis; only 4 of 15 did not have erosive esophagitis and would have fallen into the definition of nonerosive reflux disease.

Kahrilas et al.'s study (5) also demonstrated that the number of acid reflux events and the time of acid exposure increased with gastric distension. These data are summarized in the following table from their article:

	Number of acid reflux events		Percentage of time with pH < 4	
	Baseline	Air infusion	Baseline	Air infusion
Normal controls	0 (0–2)	4 (1.8–4.3)[a]	0 (0–5.6)	5.4 (3.1–10.4)[a]
Nonhernia GERD	0 (0–3)	6 (2.4–10.6)[a,b]	0 (0–4.9)	14.6 (5.6–16.6)[a,b]
GERD with hernia	1 (0–7)	12.8 (6.5–26.3)[a,b,c]	1.8 (0–5.8)	22.7 (14–24.8)[a,b,c]

Note: Data are presented as median (interquartile range).
[a] $P < 0.05$ air infusion vs baseline.
[b] $P < 0.05$ versus normal controls.
[c] $P < 0.07$ versus nonhernia GERD patients.

Our group (Tharavej et al. unpublished data) evaluated the relationship between histologic subtypes at the squamocolumnar junction in patients with a normal 24-hour pH study with the likelihood of postprandial reflux. This study analyzed 177 consecutive patients who had a 24-hour pH study that had a normal composite score. This group was selected from 662 patients who underwent a 24-hour pH test during the period that spanned from January 2001 to August 2004. The histologic type of columnar epithelium adjacent to the squamous epithelium was classified by our standard histologic criteria. The total percentage time pH of less than 4, percentage time pH less than 4 two hours after a standardized meal, and the prevalence of increased pre-/postprandial esophageal acid exposure ratio were calculated in patients according to the histologic epithelial type that was found. Of the 177 patients, 24 had intestinal metaplasia, 107 had cardiac mucosa, 32 had oxyntocardiac mucosa, and 14 had only oxyntic mucosa. The pH study parameters tested (expressed as median value) are given for each histologic group in the following table:

	IM (24)	CM (107)	OCM (32)	OM (14)	p value
Total percentage of time pH < 4	1.5	1.6	1.1	0.9	ns
Percentage pH < 4 two hours after a standardized meal	4.7[a]	4.2[b]	1.8[c]	1.1[d]	p = 0.01
Prevalence of increased pre-/postprandial esophageal acid exposure ratio	7/24[a]	24/107[b]	2/32[c]	0/4[d]	p = 0.02

Statistically, [a] versus [b] = not significant; [a] versus [c], [a] versus [d], [b] versus [c], and [b] versus [d] were all significant with p = 0.05.

This study shows a significant increase in reflux during the postprandial period in patients with reflux disease limited to the dilated end-stage esophagus (defined by the presence of cardiac mucosa with and without intestinal metaplasia and a normal 24-hour pH test).

Prevalence of Reflux Disease Limited to Dilated End-Stage Esophagus

Because it is defined by the presence of a histologic criterion (the presence of cardiac mucosa with and without intestinal metaplasia) distal to the squamocolumnar junction in patients who are endoscopically normal, there are a considerable amount of data in the literature regarding the prevalence of reflux disease in dilated end-stage esophagus. These data have to be extracted from the studies, which generally report these data as the prevalence of these histologic types in "normal gastric cardia." Any report that accurately documents histologic types will provide this information; unfortunately, accurate histologic documentation is not common even in the best studies.

In our autopsy study of 18 normal patients, we reported that cardiac mucosa was absent in 10 of 18 of the patients (56%), a 44% prevalence of reflux disease limited to dilated end-stage esophagus (10). The maximum cardiac mucosal length was less than 3 mm in all patients, indicating that in this "normal" autopsy population, the amount of reflux disease was very small.

As the likelihood of reflux disease increases, studies will report increasing prevalence or length of reflux disease in dilated end-stage esophagus. In Jain et al.'s endoscopic study (14), in which 25 of 31 patients had normal endoscopic features, the frequency of finding the different epithelial types in at least one of the four biopsy specimens obtained at the various locations is tabulated as follows:

Biopsy site	Number of patients	Cardiac epith	Fundic epith	Intestinal met
EGJ	31	11 (35%)	21 (68%)	10 (32%)
1 cm below EGJ	29	4 (14%)	23 (79%)	7 (24%)
2 cm below EGJ	30	1 (3%)	30 (100%)	2 (7%)

In this clinical population, the prevalence of reflux disease in dilated end-stage esophagus (defined by cardiac mucosa distal to the endoscopic EGJ) is 35%; 14% and 3% of patients have a length of reflux disease involving the dilated end-stage esophagus measuring 1 cm and 2 cm (see Fig. 11.26).

Marsman et al. (15), in a study of patients undergoing clinically indicated endoscopy who had a normal endoscopic gastroesophageal junction, reported a prevalence of 62% of patients with cardiac mucosa (= reflux disease in dilated end-stage esophagus) distal to the endoscopic esophagogastric junction. This study required the presence of the actual squamocolumnar junction and was therefore very precise. The study does not permit evaluation of the length of the segment of reflux disease in dilated end-stage esophagus.

Glickman et al. (9), in an endoscopic study of 74 clinical pediatric patients, a significant number of whom had clinical reflux, reported that 19% did not have cardiac mucosa; 81% had reflux disease (i.e., cardiac mucosa) distal to the endoscopic gastroesophageal junction, which often measured less than 1 mm. This reflects the fact that children, with a short cumulative exposure to reflux, are much more likely to have shorter segments of damaged esophagus.

Sarbia et al. (16), in a study on 36 older patients undergoing esophagectomy for squamous carcinoma, showed that 20 patients had a circumferential segment of reflux disease (cardiac mucosa) in end-stage esophagus, 15 patients had partial circumferential disease, and 1 patient did not have any. This higher prevalence with age reflects the increased cumulative exposure of the esophagus to reflux during life. Sarbia et al. also showed that this cardiac mucosa-lined segment distal to the end of the tubular esophagus contained esophageal glands in the submucosa, confirming end-stage esophagus rather than stomach.

Evolution of Columnar-Lined Esophagus in Dilated End-Stage Esophagus

The metaplastic columnar epithelium in the dilated end-stage esophagus is damaged by gastric contents.

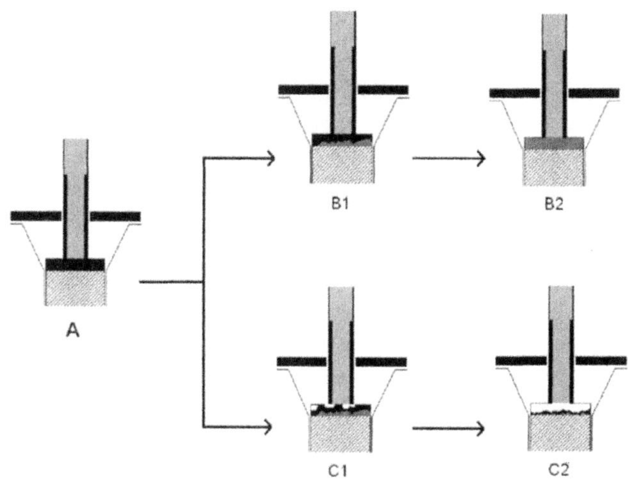

FIGURE 12.23 Evolution of epithelial types in reflux disease limited to the dilated end-stage esophagus. The same patterns of epithelia are seen as in columnar-lined esophagus involving the tubular esophagus. (a) Cardiac mucosa differentiates into oxyntocardiac mucosa (stippled) with increasing numbers of parietal cell–containing glands (b1 and b2) or into intestinal metaplasia (white) with increasing numbers of goblet cells (c1 and c2).

The potential exposure of this epithelium to gastric contents is almost continuous because it has lost the protection of the lower esophageal sphincter, which has been permanently destroyed. It is different than in the tubular esophagus, which is at least partially protected by the damaged sphincter, and exposure to refluxate is more likely to be intermittent.

The columnar epithelia in the end-stage esophagus develop by the same metaplastic sequence as in the tubular esophagus (Fig. 12.23). Cardiac mucosa arises from squamous epithelium and then evolves into either the benign oxyntocardiac mucosa or intestinal metaplasia. Because the tendency to intestinal metaplasia is related to the length of columnar epithelium, these patients have a much lesser tendency toward intestinal metaplasia. This means that the percentage of patients who have intestinal metaplasia in reflux disease limited to the end-stage esophagus is expected to be much lower than the percentage of patients with intestinal metaplasia in longer segments of visible columnar-lined esophagus. However, because reflux disease in dilated end-stage esophagus is much more common than visible columnar-lined esophagus, it is likely that the total number of patients who have intestinal metaplasia in reflux disease limited to end-stage esophagus is far greater than the number of patients presently diagnosed as having Barrett esophagus.

In our study of 959 patients with mapping biopsies (these included endoscopically normal patients because of our standard protocol), 811 patients (85%) had a columnar metaplastic epithelial length less than 1 cm (17). Most of these patients were classified as being endoscopically normal and had their metaplastic epithelia in dilated end-stage esophagus. Of these 811 patients, 120 (14.8%) had intestinal metaplasia. This compares with a 128 out of 148 (87%) prevalence of intestinal metaplasia in patients with a columnar-lined esophagus of greater than 1 cm.

This population is not a normal population; it is one from a foregut surgery unit with a strong bias toward having referrals for severe reflux uncontrolled by medical treatment and patients with Barrett esophagus referred for consideration of antireflux surgery. Despite this, the total number of patients (120 versus 128) in the two groups with markedly different percentages for prevalence of intestinal metaplasia (87% in patients with greater than 1 cm of columnar-lined esophagus versus 14.8% in patients with a less than 1 cm of columnar-lined esophagus) was nearly equal. In a more normal population, it is likely that the prevalence of intestinal metaplasia in the dilated end-stage esophagus will be much higher than in the tubular esophagus when expressed as a total number.

A similar argument can be extended to the incidence of adenocarcinoma arising in reflux disease limited to end-stage esophagus. The risk of cancer in this segment is determined by two opposing factors: (a) the amount of epithelium at risk (intestinal metaplasia) is lower than in the longer segments of Barrett esophagus in the tubular esophagus; and (b) the at-risk epithelium is immediately accessible to any carcinogen because of its location, much more so than intestinal metaplasia higher in the tubular esophagus. The interplay of these two factors results in the observed overall number of cancers that occur, which is similar for adenocarcinoma of the "gastric cardia" (= adenocarcinoma in the dilated end-stage esophagus) as it is for adenocarcinomas of the distal tubular esophagus.

A NEW LOOK AT INTESTINAL METAPLASIA OF THE GASTRIC CARDIA

Intestinal metaplasia of the gastric cardia (CIM) is defined as the finding of intestinal metaplasia in a biopsy taken at or distal to the endoscopic gastroesophageal junction. We made a case in Chapter 4 to not use the word "cardia" because it permits the fence-sitter to have a large place to sit on. If intestinal metaplasia is found distal to the gastroesophageal junction, it is gastric intestinal metaplasia. It is difficult to explain how a gastric disease can be localized to the most proximal 5 mm of the stomach when the remainder of the stomach is absolutely normal. It is difficult to explain how gastric intestinal metaplasia can be etiologically related to gastroesophageal reflux disease. The use of the term "cardia" permits the fence-sitter to hide behind an ambiguous term. There are only two organs in this region: the esophagus and the stomach; there is no third organ called the cardia. Intestinal metaplasia is either esophageal (= Barrett esophagus) or gastric.

"Intestinal metaplasia of the cardia" is usually diagnosed only in patients who are endoscopically normal (i.e., who do not have Barrett esophagus by the present definition). In studies that simultaneously evaluate the distal stomach, intestinal metaplasia of the cardia may be isolated to the "cardia" or be associated with diffuse gastritis.

Do Barrett Esophagus and Intestinal Metaplasia of the Gastric Cardia Coexist?

By definition, intestinal metaplasia of the gastric cardia is the occurrence of intestinal metaplasia in a biopsy taken distal to a normal appearing gastroesophageal junction. In the literature, intestinal metaplasia of the gastric cardia is generally not recognized in patients who have a visible segment of Barrett esophagus.

The distal limit of a visible columnar-lined esophagus is the endoscopically defined gastroesophageal junction. The method of biopsy recommended in these patients is to sample the endoscopically visible columnar-lined segment at 1- to 2-cm intervals and stop at the endoscopic gastroesophageal junction. This biopsy protocol ensures that any coexisting "intestinal metaplasia of the gastric cardia" is not found. In practice, because of definitional criteria and standard biopsy technique, *the diagnosis of intestinal metaplasia of the gastric cardia is never made in patients with Barrett esophagus.* Why? Does it mean that the two diseases never coexist? Surely, that is impossible, considering the high frequency of intestinal metaplasia of the gastric cardia in the population. In Rex et al. (18), 122 of 940 (or 12.9%) patients had intestinal metaplasia of the gastric cardia. In 7 of these 122 patients, coexistent Barrett esophagus was present in the tubular esophagus.

If a biopsy is routinely taken distal to the endoscopic gastroesophageal junction in a patient with Barrett esophagus, as was done in the screening study by Rex et al. (18), this would be a biopsy of the "gastric cardia" by present definition. If intestinal metaplasia is found in this biopsy, it must be classified as "intestinal metaplasia of the gastric cardia" *even though it is continuous with and histologically identical to the intestinal metaplasia in the columnar-lined esophagus.* One reaches the somewhat improbable conclusion in these patients that a histologically identical and continuous lesion represents two pathologic entities divided by the presently defined gastroesophageal junction: Barrett esophagus in that part above the junction and intestinal metaplasia of the gastric cardia below the line (Fig. 12.20).

In Figure 12.22, the intestinal metaplasia of a long segment of Barrett esophagus in the tubular part of the esophagus extends into the area distal to the end of the tubular esophagus and the proximal limit of "gastric" rugal folds. What does this patient have? We believe that this patient has one disease characterized by a long segment of columnar-lined esophagus between the squamocolumnar junction and gastric oxyntic mucosa (the true gastroesophageal junction). According to this interpretation, all intestinal metaplasia is in the esophagus and constitutes Barrett esophagus. This can only be true if the present definition of the gastroesophageal junction as the end of the tubular esophagus or the proximal limit of the rugal folds is incorrect. If one applies the present definition of the gastroesophageal junction, this patient must have two diseases (Fig. 12.20): (a) Barrett esophagus, which terminates at the end of the tubular esophagus or the proximal limit of the rugal folds, and (b) intestinal metaplasia of the cardia. When this is limited to the proximal few millimeters of the stomach with the remainder of the stomach being normal, as is commonly the case, the interpretation approaches the ridiculous because there is no histologic difference between the intestinal metaplasia in the proximal 2 mm of the stomach and the long segment of Barrett esophagus in the tubular esophagus. In practice, this situation is avoided by not taking biopsies from the mucosa distal to the endoscopic gastroesophageal junction. This is dangerous because adenocarcinoma may occur in this ignored segment (Fig. 12.22).

The only rational explanation for this situation is that the present definition of the gastroesophageal junction is incorrect and the correct definition of the true gastroesophageal junction is the proximal limit of gastric oxyntic mucosa defined by histology. When the true gastroesophageal junction is recognized, it becomes obvious that there is one long segment of columnar-lined esophagus involving the tubular esophagus and the dilated end-stage esophagus (Figs. 12.20 and 12.22). There are not two diseases; there is one long segment of columnar-lined esophagus interposed between the squamocolumnar junction and gastric oxyntic mucosa (the true gastroesophageal junction). The only reason for the incorrect belief that there are two diseases is the fact that an incorrect line is drawn to define the gastroesophageal junction (Figs. 12.20 and 12.22).

Is There Never Flat Mucosa between the End of the Tubular Esophagus and the Proximal Limit of the Rugal Folds?

The other problem about the diagnosis of intestinal metaplasia of the gastric cardia is that in the best articles on the subject, the definitions are too good to be true. They are presented with a total clarity that does not exist in reality. Anything that is too perfect is suspect of being contrived. The definition of normal endoscopy has the entire tubular esophagus lined by squamous epithelium with the gastric rugal folds lining the entire proximal stomach and extending to the horizontal squamocolumnar junction. Any tongues of flat columnar epithelium extending into the tubular esophageal squamous epithelium are recognized as abnormal (if intestinal metaplasia is present and the tongues are less than 3 cm, this is noncircumferential short-segment Barrett esophagus).

According to this scheme of definition, there is no situation in which the rugal folds fail to completely reach the squamocolumnar junction at the end of the tubular esophagus in the normal patient. In the patient with columnar-lined esophagus, the literature makes it appear that the rugal folds always extend to the horizontal line marking the end of the tubular esophagus. This is not correct. The proximal limit of the rugal folds is not infrequently concave with the rugal folds reaching the end of the tubular esophagus in some parts of the circumference but failing to do so in others (Fig. 12.24). What does one do in a situation where the proximal limit of the rugal folds does not extend all the way to the end of the tubular esophagus? This does not fit any criterion of definition. *The only option is to ignore that this situation exists.*

The rational explanation for the finding of flat and rugated mucosa is that the presence or absence of rugal folds has nothing to do with the epithelial type. The presence of rugal folds indicates a distensible organ. The stomach is normally lined by rugated mucosa. However, the reflux-damaged end-stage esophagus also distends with the stomach and can acquire rugal folds. The absence of rugal folds in columnar epithelium is the result of inflammation and atrophy. Rugal folds are commonly absent in metaplastic columnar esophageal epithelium because it is commonly inflamed and atrophic. The absence of rugal folds in gastric mucosa is a recognized endoscopic feature of chronic atrophic gastritis. The epithelial

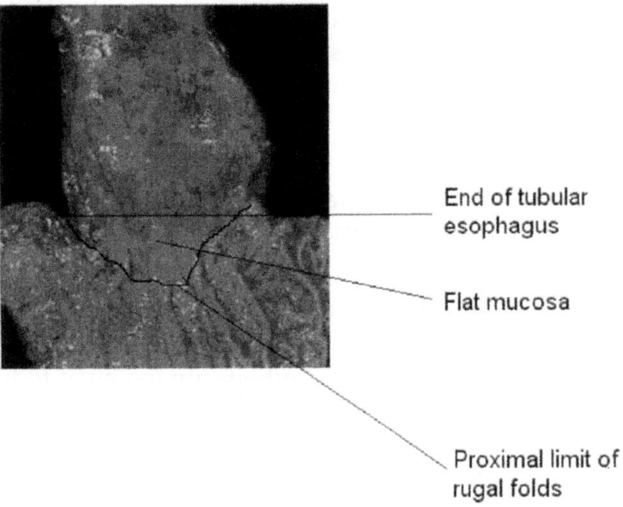

FIGURE 12.24 Esophagectomy specimen (see Fig. 4.14 for full size of this picture) in a patient with long-segment Barrett esophagus with the flat epithelium extending into the area distal to the end of the tubular esophagus. The proximal limit of the rugal folds is a concave line. The flat area between the end of the tubular esophagus and the proximal limit of the rugal folds contained cardiac and oxyntocardiac mucosa, representing columnar epithelium in dilated end-stage esophagus. (Cross reference: Color Plate 3.3)

type cannot be defined endoscopically because rugal folds are not specific for epithelial type; rugal folds do not indicate that the epithelium is gastric or esophageal, they simply mean that the epithelium is not sufficiently inflamed and atrophic to lose the folds.

In practice, anything short of a definitely visible columnar-lined segment in the tubular esophagus will fall under the designation of "normal." This creates problems in studies that try to define the gastroesophageal junction with millimetric precision. We will demonstrate this tendency by reviewing three articles on intestinal metaplasia of the gastric cardia that we have selected as being the best on this subject:

Literature Review

Rex DK, Cummings OW, Shaw M, Cumings MD, Wong RKH, Vasudeva RS, Dunne D, Rahmani EY, Helper DJ. Screening for Barrett's esophagus in colonoscopy patients with and without heartburn. Gastroenterology 2003;125:1670–1677.

This is a study of 961 patients screened by upper endoscopy to evaluate the prevalence of intestinal metaplasia. At endoscopy, the following landmarks were identified: (a) the squamocolumnar junction (SCJ); (b) the esophagogastric junction (EGJ) defined as the proximal end of the gastric folds where the tubular esophagus meets the stomach; and (c) the diaphragmatic hiatus. Long-segment Barrett esophagus (LSBE) was defined as columnar epithelium with specialized intestinal metaplasia (SIM) of greater than or equal to 3 cm proximal to the EGJ, whereas short-segment Barrett esophagus (SSBE) was defined as columnar epithelium with SIM of less than 3 cm above the EGJ.

Four-quadrant biopsy samples were taken from the tubular esophagus at least every 2 cm in the case of circumferential segments and at least one sample from each tongue in the case of tongues. *Two samples were taken from the gastric cardia (defined as the proximal edge of the gastric folds, just distal to the end of the tubular esophagus).*

The total number of patients with Barrett esophagus was 65 of the 961 (6.8%). Twelve of 961 (1.2%) had long-segment Barrett's esophagus; 53 of 961 (5.5%) had short-segment Barrett esophagus. Of the 940 patients with evaluable cardia tissue, intestinal metaplasia (IM-cardia) was identified in 122 (12.9%). There were 7 patients in whom intestinal metaplasia was identified in both tubular esophagus (Barrett esophagus) and in the gastric cardia. Because these 7 patients were included in the total of 122 patients diagnosed as having intestinal metaplasia of the cardia, these authors seem to be saying that Barrett esophagus can coexist with intestinal metaplasia of the cardia. But this is contrary to the standard definition of intestinal metaplasia of the cardia that requires a normal endoscopic appearance.

Of the 122 patients with intestinal metaplasia of the cardia, 109 had testing for *Helicobacter pylori*; of these, 41 (38%) were positive and 68 (62%) were negative. Unfortunately, the distal gastric biopsies were tested only by Pyloritek; no histologic features are given. It is amazing that in such a well-designed study, the researchers did not feel the need to provide histologic data in distal gastric biopsies, which would add so much to the value of the study.

Another defect in the study is that no histologic data of epithelial type (i.e., cardiac versus oxyntic epithelium) is provided except the presence or absence of intestinal metaplasia. If the authors had provided data on the epithelial type in which intestinal metaplasia was found, and data related to the histology in distal gastric biopsies, it would have been simple to divide these 122 patients into three groups:

1. Patients with only reflux disease with Barrett esophagus. The intestinal metaplasia would have been in cardiac epithelium *found in biopsies from either the tubular esophagus or the "gastric cardia," H. pylori* would be absent, and gastric mucosa distal to the "cardia" would have been normal.
2. Patients with only *H. pylori*-induced atrophic pangastritis involving the proximal stomach. These patients would have had the intestinal metaplasia in atrophic gastric oxyntic mucosa, and there would have been no cardiac mucosa.
3. Patients with both. In these patients, the intestinal metaplasia may have been in cardiac mucosa (Barrett esophagus) or atrophic oxyntic mucosa; the patients would have been *H. pylori* positive and had distal gastritis.

In the absence of this information it is an interesting exercise to attempt to glean the etiology of the intestinal metaplasia in these 122 patients. The 68 patients who are *H. pylori* negative are likely to be in group 1 and their IM-cardia likely to be Barrett esophagus in reflux disease involving the dilated end-stage esophagus; this is almost certain in the 23 of 109 patients who had erosive esophagitis. The 41 of 109 patients who were *H. pylori* positive belong to either group 2 or group 3, determined by whether the intestinal metaplasia is in cardiac mucosa (group 3) or atrophic oxyntic mucosa (group 2).

Literature Review

Goldblum JR, Vicari JJ, Falk GW, Rice TW, Peek RM, Easley K, Richter JE. Inflammation and intestinal metaplasia of the gastric cardia: The role of gastroesophageal reflux and *H. pylori* infection. Gastroenterology 1998;114:633–639.

In this study from the Cleveland Clinic, patients with evidence of columnar-lined esophagus were excluded. Gastric cardia biopsy specimens were obtained with the endoscope in the retroflexed position within 5 mm from a normal-appearing squamocolumnar junction. The gastric specimens, including the one from the cardia, were evaluated for inflammation using the Sydney system, the degree of intestinal metaplasia, the degree of glandular atrophy, and the degree of *H. pylori* colonization. Unfortunately the epithelial type (i.e., cardiac versus oxyntic) found in the biopsy of the cardia is not recorded.

Intestinal metaplasia was identified in 8 of 85 patients (9%) in the gastric cardiac biopsy. This consisted of 6 of 27 (22%) in the control group without GERD symptoms and 2 of 58 (3%) in the GERD group (difference highly significant; p = 0.01). The detailed analysis of the 8 patients with intestinal metaplasia is given in the following table:

Age/sex	GERD symptoms	Carditis	HP+ in cardia	HP serology	Antral/ fundic IM
42/M	No	Yes	Yes	+	Yes
73/M	No	Yes	No	+	Yes
50/M	No	Yes	Yes	+	Yes
74/M	No	Yes	Yes	NA	Yes
70/M	No	Yes	Yes	+	No
86/F	No	Yes	Yes	NA	Yes
25/M	Yes	Yes	No	–	No
84/M	Yes	Yes	Yes	+	Yes

The 6 control group patients (no GERD symptoms) all had *H. pylori* infection, and 5 of 6 had multifocal chronic atrophic pangastritis. One of the 2 patients in the GERD group had *H. pylori* infection and multifocal chronic atrophic pangastritis. The other had no evidence of *H. pylori* infection and no evidence of multifocal chronic atrophic gastritis.

This study appears to be dominated by patients with *H. pylori* pangastritis. Of the 8 patients with intestinal metaplasia of the cardia, only 1 patient had no evidence of *H. pylori* infection and is likely to have had Barrett esophagus as a manifestation of reflux disease in the end-stage esophagus; 6 of the other 7 patients (5 controls and 1 with symptomatic GERD) had *H. pylori*-associated multifocal chronic atrophic pangastritis. Whether they had associated reflux disease cannot be determined without knowledge of the epithelial type in the biopsy from the "gastric cardia." One of the control patients is interesting; this patient had *H. pylori* gastritis, but the intestinal metaplasia was restricted to the cardia; in our experience, patients who have this combination of findings typically have Barrett esophagus coexisting with nonatrophic *H. pylori* gastritis.

Literature Review

Hirota WK, Loughney TM, Lazas DJ, Maydonovitch CL, Rholl V, Wong RKH. Specialized intestinal metaplasia, dysplasia and cancer of the esophagus and esophagogastric junction: Prevalence and clinical data. Gastroenterology 1999;116:277–285.

In this study from Walter Reed Army Medical Center, 889 patients were enrolled from January 1995 to September 1996. The following esophageal landmarks were documented: diaphragmatic hiatus, most proximal tip of the gastric fold (EGJ; Note: this wording has a subtle difference to the usual proximal limit of the rugal folds; is there a difference between tip and limit?) and esophageal SCJ. The diagnosis of EGJ-SIM (specialized intestinal metaplasia of the esophagogastric junction) was defined by the presence of a normal-appearing SCJ, devoid of any tongues of pink columnar-lined epithelium above the endoscopically defined EGJ but associated with SIM on antegrade biopsy specimens distal to the SCJ. This is the definition of a very careful person. Note that we are told that the SCJ was normal (i.e., horizontal), without tongues of columnar epithelium extending upward. *We are not given information on the relationship of the SCJ to the proximal limit of the rugal folds.* Do the authors want us to assume that this was coincident with the SCJ or are they saying that it was not necessarily coincident with the SCJ and they really did not know what to do with that data? Were they seeing flat columnar epithelium between the normal SCJ and the proximal tip of the gastric folds as we have suggested exists in some esophagectomy specimens? The authors then stated that the SIM that defined EGJ-SIM was found on antegrade biopsy specimens *distal to the SCJ*. Why do the authors not say that the biopsies were taken distal to their defined EGJ which is the proximal tip of the rugal folds?

Also interesting is the new term—"EGJ-SIM"—which the authors introduced instead of the more usual "intestinal metaplasia of the gastric cardia." It is even more ambiguous than "intestinal metaplasia of the gastric cardia." The EGJ is not a structure; it is an imaginary line drawn between the esophagus and the stomach. An imaginary horizontal line has no length; it cannot therefore have a disease.

Of the 889 patients, 151 had specialized intestinal metaplasia; 40 had long-segment Barrett esophagus, 64 had short-segment Barrett esophagus, and 47 had EGJ-SIM. The term "specialized intestinal metaplasia" is generally applied to the specific histology of Barrett esophagus; it really describes intestinal metaplasia occurring in cardiac mucosa rather than the gastric intestinal metaplasia of atrophic gastritis. The fact that the location of these biopsies was defined as "immediately distal to the SCJ" rather than by their relation to the proximal limit of the rugal folds suggests that they would have been likely to contain metaplastic columnar epithelium if such epithelium existed.

Selected characteristic of these patients are given in the following table:

Feature	Reference	EGJ-SIM	SSBE	LSBE
Number of patients[a]	738 (83%)	47 (5.3%)	64 (7.2%)	40 (4.5%)
Sex; M : F	394/344	25/22	45/19	35/5
White race	485 (66%)	31 (66%)	55 (86%)	40 (100%)
Heartburn	62%	59%	83%	63%
H. pylori at EGJ	64 (8.8%)	10 (21.3%)	3 (4.7%)	1 (2.5%)

[a]Numbers should not be used to indicate prevalence of the subgroups because this number included patients undergoing surveillance endoscopy for Barrett esophagus. The corrected prevalence of EGJ-SIM was 47/833 (5.6%), SSBE 6.0%, and LSBE 1.6%.

The most interesting aspects of these three studies are their similarities and differences. They all examine intestinal metaplasia occurring immediately distal to the normal endoscopically defined gastroesophageal junction (we hope). The relevant comparisons are shown in the table below.

The comparison shows that the featured characteristics are quite divergent in these three studies, all from highly reputable gastroenterology departments and all considered to be of sufficient excellence to be published in *Gastroenterology*. (a) *The prevalence difference* is strange; the Hirota et al. study was in patients presenting with symptoms that represented a clinical indication for upper endoscopy, whereas the Rex et al. study was a screening population who did not have an indication for upper endoscopy. Why does the screening population have a higher prevalence of intestinal metaplasia of the gastric cardia (12.9% versus 5.6%)? (b) *The gender difference* ranges from 53% males in Hirota et al. to 88% male in Goldblum et al. (19). (c) The difference in esophagitis between the studies is not easy to compare, but the data show a 25% GERD rate in Goldblum et al., lower than the 59% rate in Hirota et al. The percentage of patients with esophagitis is lower in Rex et al. than in Hirota et al.; this is expected because Rex et al. is a screening population and Hirota et al. is a clinical population. (d) The prevalence of *H. pylori* infection was very high in Goldblum et al., low in Hirota et al. (21%), and intermediate (38%) in Rex et al. The prevalence of *H. pylori* infection in a given population varies considerably with the socioeconomic status of the population and cannot be compared easily.

There should not be such a divergence in the data relating to such important characteristics in these three studies. Is it possible that the differences are the result of the slight differences in methodology? The data discrepancies are perfectly explained if the location of the biopsy was most distal in Goldblum et al.

Feature	Goldblum et al. (19)	Hirota et al. (8)	Rex et al. (18)
Definition	IM in retroflexed biopsy within 5 mm from the normal appearing SCJ	IM in an antegrade biopsy distal to the normal appearing SCJ	IM in a biopsy at the proximal edge of gastric folds just distal to the end of tubular esophagus
Population	Normal endoscopy; no CLE	All patients having upper endoscopy	Screened population; no indication for upper endoscopy
Prevalence of IM-cardia/EGJ	8/85 (9%)	47/833 (5.6%)	122/940 (12.9%)
Sex: M : F (Male %)	7/1 (88%)	25/22 (53%)	76/46 (68%)
White race	8/8 (100%)	31/47 (66%)	96/112 (86%)
Heartburn in patients with IM-cardia/EGJ	2/8 (25%)	20/34 (59%)	Not given
Esophagitis		7/34 (21%)	23/112 (15%)
H. pylori infection	7/8 (88%)	10 (21.3%)	41/109 (38%)

and most proximal in Hirota et al. with Rex et al. being intermediate, albeit by a few millimeters. A few millimeters in this area makes a world of a difference; the difference between whether the biopsy is from the esophagus or stomach. If this is true, intestinal metaplasia would be much more likely to be esophageal in Hirota et al. (= Barrett esophagus, more associated with heartburn, and less associated with *H. pylori* infection) and gastric in Goldblum et al. (and therefore atrophic pangastritis, less associated with heartburn, and more associated with *H. pylori* infection), and mixed in Rex et al. All of the data fall into place with incredible precision. What is really happening here is that these three authors are using the incorrect, variable (but all acceptable) definitions of the gastroesophageal junction. They are accessing a variable point distal to the incorrect gastroesophageal junction. The likelihood of whether the point of biopsy is esophageal or gastric depends entirely on the distance of the biopsy distal to the endoscopic EGJ. If one looks back at the definitions, it is quite likely that this is true; Hirota et al.'s antegrade biopsies, which are immediately distal to the SCJ are likely to be more proximal than Rex et al.'s biopsies which are from the proximal edge of the gastric folds, and these are likely to be more proximal than Goldblum et al.'s retroflexed biopsies taken within 5 mm from the SCJ. These three studies all seem to study the same thing. They are, however, different because they used different methodology and definitions; definitions are at the whim of the author. Slight differences in methodology can reflect millimetric differences that dictate whether there is a sampling bias toward esophagus or stomach (Table 12.2).

TABLE 12.2 Differences in the Location of Biopsies Used to Define "Intestinal Metaplasia of the Gastric Cardia" or IM of the EG Junction

Reference	Location of target biopsies
Rex et al. (18)	Gastric cardia (defined as the proximal edge of the gastric folds, just distal to the end of the tubular esophagus)
Goldblum et al. (19)	Retroflexed biopsy of the gastric cardia within 5 mm from a normal-appearing squamocolumnar junction
Hirota et al. (8)	Antegrade biopsies distal to the normal appearing SCJ
Hackelsberger et al. (20)	Immediately distal to the SCJ located within 2 cm proximal to the EGJ (proximal limit of the rugal folds)
Voutilainen et al. (21)	Within 2 cm proximal to the proximal limit of the rugal folds

Two studies from Europe suggest that the definitions used may be quite different from the definitions used in the above preceding American studies (Table 12.2). Hackelsberger et al. (20) divided his patients by endoscopic features into the following groups: group I: the squamocolumnar junction described as unremarkable in location (= within 2 cm above the proximal limit of the rugal folds which is their definition of the gastroesophageal junction) and appearance (relatively straight); group II: the squamocolumnar junction described as suggestive or typical of Barrett esophagus. The biopsy specimens in patients from group I were taken from the columnar mucosa immediately below the SCJ. The data in this study have nothing to do with intestinal metaplasia of the esophagogastric junction or gastric cardia because, contrary to the title of the article, the biopsy location is potentially up to 2 cm proximal to the defined gastroesophageal junction.

In a study from Finland (21), a similar definition of the gastroesophageal junction is used: "The normal squamocolumnar mucosal junction or Z-line is normally located within 2 cm of the proximal edge of the gastric folds.... Esophagogastric junctional biopsies were obtained within 2 cm of the proximal edge of the gastric folds." (p. 914).

It appears that the European studies are looking at biopsies up to 2 cm above the endoscopic gastroesophageal junction, whereas the American studies are looking at biopsies at or up to 5 mm distal to the endoscopic gastroesophageal junction. A great deal of care is therefore needed when interpreting the data in these studies.

None of these studies provide data that can be interpreted in the light of the new data that we have presented regarding the anatomic changes associated with reflux disease. If patients commonly have a segment of end-stage esophagus as an early manifestation of reflux disease, and the length of this segment increases with the severity of reflux and age, it is impossible to assign an esophageal or gastric location to a biopsy taken distal to the endoscopically defined gastroesophageal junction. In patients without reflux disease, this biopsy will contain gastric mucosa; in patients with reflux disease in the dilated end-stage esophagus, it will contain metaplastic esophageal columnar epithelium. Without defining the biopsies by histology into the different epithelial types, there is no meaning to the data. Most of these studies show that what is being examined is etiologically related to both gastroesophageal reflux and *H. pylori* gastritis. The reason for this is that some biopsies will have esophagus and others will have stomach; it does not mean that "intestinal metaplasia of the cardia" is caused by these two conditions.

Interpretation of Intestinal Metaplasia Distal to the Endoscopic Gastroesophageal Junction

Articles that report on intestinal metaplasia of the cardia must be carefully evaluated for their methods. Ideally, these articles should not be permitted to enter the medical literature unless there is clear information on the types of mucosa that are involved using precise definition, information regarding the histology of the tubular esophagus proximal to the "cardia," and information regarding the histology of the stomach distal to the "cardia." Without such information, the interpretation of data in articles on the subject is meaningless and represents only an interesting guessing game. If one reviews the literature, not a single report on intestinal metaplasia of the gastric cardia satisfies these criteria of an evaluable paper.

The diagnosis of intestinal metaplasia in a biopsy taken from a columnar-lined segment proximal to the endoscopic gastroesophageal junction is not controversial—it represents Barrett esophagus. Thus, the finding in Voutilainen et al. (21) that intestinal metaplasia of the esophagogastric junction (defined as within 2 cm proximal to the EGJ) is associated with reflux disease and not *H. pylori* infection is easy to understand; this is really a study of short-segment Barrett esophagus, and no one doubts that short-segment Barrett esophagus is caused by reflux disease.

The interpretation of intestinal metaplasia in a biopsy taken distal to the endoscopic gastroesophageal junction is dependent on the exact location of the biopsy and the type of epithelium present in the biopsy. "Intestinal metaplasia of the cardia" as presently defined represents either Barrett esophagus occurring in reflux disease of the dilated end-stage esophagus (= intestinal metaplasia occurring in esophageal metaplastic cardiac epithelium) or *H. pylori* (or autoimmune)–induced chronic pangastritis with multifocal atrophy involving the most proximal stomach (= intestinal metaplasia occurring in atrophic gastric oxyntic mucosa) (Fig. 12.2). The distinction between the two cannot be made if any nonhistologic definition of the "cardia" is used, because reflux disease in the dilated end-stage esophagus is invisible to the endoscope. This means that a biopsy taken from a point (usually 5 mm) distal to the endoscopically defined gastroesophageal junction may be esophagus in a patient with reflux disease limited to end-stage esophagus (Fig. 12.25a) or gastric mucosa in a patient with no reflux disease (i.e., no columnar-lined esophagus; Fig. 12.25b). When intestinal metaplasia is found in columnar metaplastic esophageal (cardiac) mucosa in patient A, the correct diagnosis is Barrett esophagus. When intestinal metaplasia is found in patient B, the correct diagnosis is gastric intestinal metaplasia (chronic atrophic gastritis). The distinction can be made only by histologic criteria that permit distinction of Barrett type intestinal metaplasia occurring in esophageal cardiac mucosa from gastric intestinal metaplasia occurring in gastric oxyntic mucosa (see Chapter 5).

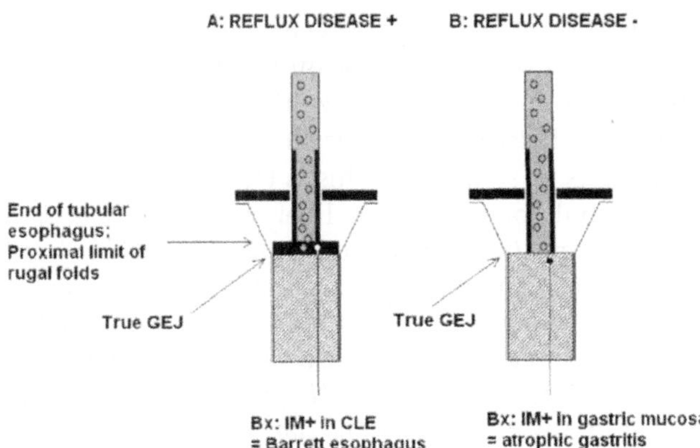

FIGURE 12.25 Interpretation of a biopsy taken from a point distal to the presently defined endoscopic gastroesophageal junction (end of tubular esophagus or proximal limit of rugal folds). Whether this biopsy is from esophagus or stomach depends entirely on whether or not the patient has reflux disease. In the patient with reflux disease that is severe enough that the dilated end-stage esophagus reaches the point biopsied, the biopsy will be esophageal (*left*); otherwise the biopsy is gastric (*right*). The presence of intestinal metaplasia will be Barrett esophagus if it occurs in metaplastic esophageal epithelium (cardiac mucosa = black; *left*) and atrophic gastritis if it occurs in gastric oxyntic mucosa (striped; *right*).

Three separate diagnostic possibilities exist when intestinal metaplasia is found in a biopsy from this region in a patient who is endoscopically normal (i.e., without a visible columnar-lined segment in the tubular esophagus) (Table 12.3)

Is There an Increased Risk of Adenocarcinoma in Barrett Esophagus Limited to Dilated End-Stage Esophagus?

There is little doubt that Barrett esophagus restricted to the dilated end-stage esophagus is the source of adenocarcinoma of this region. No other disease entity can be responsible. The epidemiologically demonstrated increase in the incidence of "adenocarcinoma of the gastric cardia" that has paralleled the increase in incidence of esophageal adenocarcinoma (discussed later) makes this certain. The incidence change is com-

TABLE 12.3 Differential Diagnosis of Intestinal Metaplasia in a Biopsy Taken Distal to the Endoscopic Gastroesophageal Junction in an Endoscopically Normal Patient

1. Barrett esophagus in reflux disease limited to the dilated end-stage esophagus without gastric pathology:	
Pathologic findings:	Reflux changes in squamous epithelium +/−; intestinal metaplasia in cardiac mucosa, limited to the area within 3 cm distal to the endoscopic GEJ; normal gastric oxyntic and pyloric mucosa
2. Atrophic gastritis without reflux disease of the esophagus	
Pathologic findings:	Normal squamous epithelium; no cardiac mucosa; intestinal metaplasia in atrophic gastric oxyntic mucosa; diffuse involvement of the stomach by atrophic gastritis
3. Simultaneous Barrett esophagus and atrophic gastritis	
Pathologic findings:	Reflux changes in squamous epithelium +/−; intestinal metaplasia of cardiac mucosa, limited to the area within 3 cm distal to the GEJ; atrophic gastritis with intestinal metaplasia involving gastric oxyntic and pyloric mucosa diffusely

pletely the opposite of gastric cancer incidence trends. It can be stated with reasonable certainty that the approximately 8000 patients per year who develop adenocarcinoma that is presently classified as "gastric cardiac" have Barrett esophagus limited to dilated end-stage esophagus as the basis of their cancer (discussed later).

The only controversy is in determining the magnitude of the risk in the individual patient. From a theoretical standpoint, if reflux-induced cancer results from the interaction of a target cell (the stem cell in intestinal metaplasia) and a carcinogenic factor in the refluxate, the risk of cancer in Barrett esophagus is expected to be proportional to the number of target cells and the carcinogenicity of the refluxate. There is a certainty that patients with intestinal metaplasia restricted to reflux disease limited to dilated end-stage esophagus have a small number of susceptible target cells. However, the target cells are in the area of the damaged esophagus that is at the highest concentration of the luminally acting carcinogen in the gastric contents. There are therefore two opposing factors that will determine the actual cancer risk. The risk is unknown. To determine the risk would require the following:

1. *The recognition of criteria to correctly diagnose patients with Barrett esophagus in reflux disease limited to the dilated end-stage esophagus.* Present criteria do not permit this. As shown in the three articles reviewed, patients presently diagnosed as having "intestinal metaplasia of the cardia" fall into two etiologic groups: one with Barrett esophagus involving esophageal (cardiac) mucosa and the other with gastric intestinal metaplasia occurring in gastric (atrophic oxyntic) mucosa. The exact admixture of these groups varies with the population studies and the methodology, making it difficult to evaluate the correct cancer risk. For example, in any study of intestinal metaplasia designed with methods adopted in Goldblum et al., the high likelihood of a predominance of atrophic gastritis caused by *H. pylori* infection would tend to dilute the risk of Barrett adenocarcinoma. In contrast, a study of intestinal metaplasia of the cardiac designed with Hirota et al.'s criteria, which had 59% of patients with heartburn (probable Barrett esophagus) and only 21% with *H. pylori*, positivity would tend to identify a more accurate cancer risk. However, it is only when we adopt the histologic basis for differentiating Barrett esophagus from atrophic gastritis will accurate data emerge regarding the true incidence of adenocarcinoma in Barrett esophagus arising in reflux disease limited to the the dilated end-stage esophagus.
2. *A change in the present practice guidelines.* Without routinely taking biopsies to identify the population who has Barrett esophagus in reflux disease limited to dilated end-stage esophagus, these patients will remain an ignored population and data will never accumulate sufficiently to determine the magnitude of the risk. This is particularly true because of the likelihood that the risk is small; very large populations need to be studied with long follow-up before researchers have the adequate statistical power to recognize a small risk.

Although the predicted risk of adenocarcinoma in Barrett esophagus is small, the number of patients with this earliest manifestation of Barrett esophagus is likely to be very high in the general population. This situation is similar to that in the colon. Recognizing the lesions at high risk for colorectal carcinoma—such as familial polyposis coli, villous adenoma, and chronic ulcerative colitis—is very effective in preventing cancer in these individual patients. However, these interventions do not affect the overall incidence of colorectal cancer because most cancers arise in low-risk but extremely prevalent lesions (adenomas). It is only with a screening protocol that detects and removes adenomas that the incidence of colorectal cancer will be affected.

In Barrett esophagus, long-segment disease is at high risk but rare. Interventions designed to prevent cancer in these patients will not impact the overall incidence of reflux-induced adenocarcinoma. The low-risk lesion responsible for most reflux-induced adenocarcinoma is likely to be the extremely common Barrett esophagus involving the dilated end-stage esophagus.

A decision to institute a screening protocol to detect Barrett esophagus is a public health issue that has to be based on cost-effectiveness. Before such a decision is made, it is critical to accurately assess the magnitude of the problem by appropriate study. The recommendations for biopsy at the present time ensure that this problem will be ignored because patients will not be biopsied. In the setting of this total ignorance, no logical recommendations can be made.

At present, the first recognized manifestation of Barrett esophagus restricted to dilated end-stage esophagus is the appearance of ulcers or nodular mass lesions in the "proximal stomach" or "gastric cardia" (Figs. 12.26 and 12.27). These tumors are misclassified as gastric adenocarcinomas rather than esophageal adenocarcinomas. This results in an underestimation of reflux-induced esophageal adenocarcinoma in epidemiologic studies, decreasing the numbers of patients who die every year from this disease. Present data indicate that 8000 patients die every year from esophageal adenocarcinoma. It is estimated that an even larger number die from "proximal gastric adenocarcinoma." If these are correctly classified as esophageal adenocarcinomas, the incidence of reflux-induced adenocarcinoma is likely to more than double. Because analysis of the cost-effectiveness of screening is largely dependent on incidence data, it is quite possible that the problem of Barrett adenocarcinoma is already serious enough to consider mass screening.

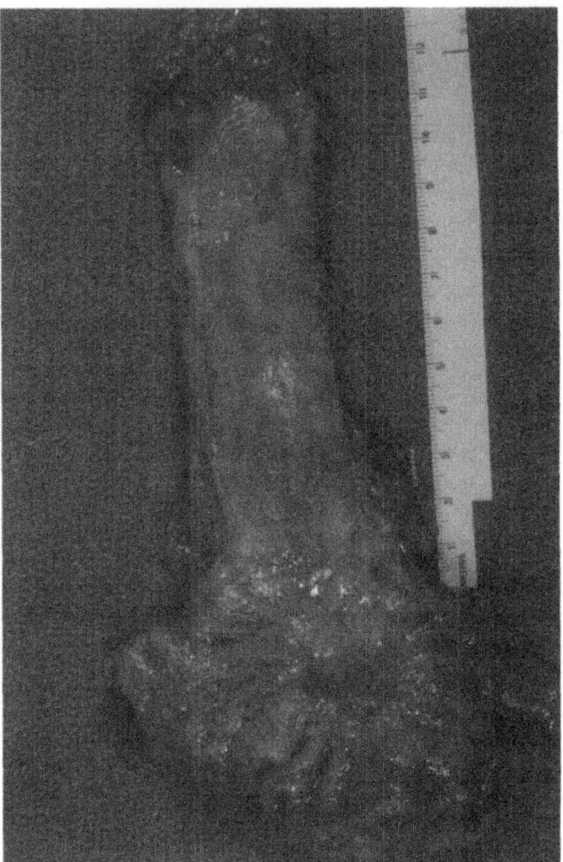

FIGURE 12.26 Esophagectomy specimen showing an ulcerated carcinoma immediately distal to the squamous-lined tubular esophagus. The rugal folds here are the result of contraction related to the tumor. This is an adenocarcinoma arising in reflux disease limited to dilated end-stage esophagus arising in metaplastic columnar esophageal epithelium.

A NEW LOOK AT ADENOCARCINOMA OF THE GASTRIC CARDIA

Adenocarcinoma of the gastric cardia is an adenocarcinoma that arises in the region immediately distal to the defined gastroesophageal junction. When the end of the tubular esophagus or the proximal limit of the rugal folds is used to define the gastroesophageal junction, adenocarcinoma of the gastric cardia refers to tumors arising within 3 cm of this point. In locating the point of origin of the tumor, it is assumed that the tumor has a uniform growth in all directions; the relationship of the epicenter of the tumor is used to define its point of origin. If this is within 3 cm distal to the

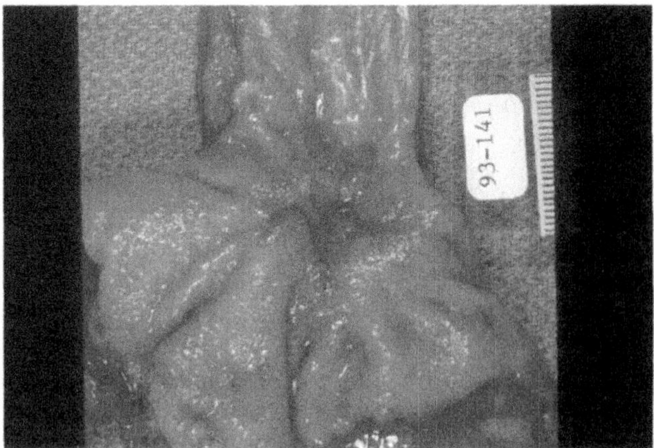

FIGURE 12.27 Esophagectomy specimen showing an ulcerated adenocarcinoma arising in reflux disease limited to dilated end-stage esophagus. The tubular esophagus is lined entirely by squamous epithelium; the rugal folds have been distorted by the tumor. (Cross reference: Color Plate 5.3)

gastroesophageal junction, the tumor is defined as an adenocarcinoma of the gastric cardia. If the epicenter is greater than 3 cm from the gastroesophageal junction, it falls under the designation of an usual gastric adenocarcinoma.

The epidemiology of adenocarcinoma of the gastric cardia closely resembles that of adenocarcinoma of the esophagus and is completely opposite to carcinoma of the stomach more distal than the 3 cm designated as "gastric cardia." Adenocarcinoma of the gastric cardia has paralleled the rise in incidence of esophageal adenocarcinoma since the mid-1970s. In contrast, noncardiac adenocarcinoma of the stomach and squamous carcinoma of the esophagus have declined in incidence during this period (see Chapter 1). There is a tendency to interpret this evidence as suggesting that adenocarcinoma of the esophagus and adenocarcinoma of the gastric cardia represent one epidemiologic entity. The evidence for this is strong.

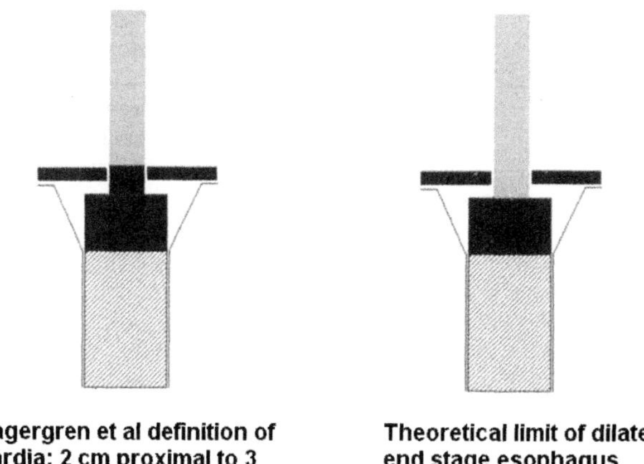

Lagergren et al definition of cardia: 2 cm proximal to 3 cm distal to GEJ

Theoretical limit of dilated end stage esophagus = 3 cm

FIGURE 12.28 The definition of the region in which a tumor is classified as "adenocarcinoma of the gastric cardia" in Lagergren et al. (6). This includes the region 2 cm proximal to and 3 cm distal to the gastroesophageal junction (left). It should be noted that the region 3 cm distal to the presently defined gastroesophageal junction is the theoretical limit of dilated end-stage esophagus distal to the end of the tubular esophagus (right).

Adenocarcinoma of the Gastric Cardia Is Associated with Symptomatic Reflux

One of the most remarkable of present beliefs is that adenocarcinoma of the gastric cardia has an epidemiologically proven association with symptomatic gastroesophageal reflux disease. This was shown in the excellent study from Sweden by Lagergren et al. (6). In this important research that established the strong epidemiologic relationship between symptomatic gastroesophageal reflux disease and esophageal adenocarcinoma, evidence is also presented for a definite but lesser association with adenocarcinoma of the cardia.

The authors' definition of an adenocarcinoma of the gastric cardia is interesting: "For a case to be classified as a cancer of the gastric cardia, the tumor had to have its center within 2 cm proximal, or 3 cm distal, to the gastroesophageal junction" (p. 826) (Fig. 12.28). Tumors arising within 2 cm proximal to the defined gastroesophageal junction are esophageal adenocarcinoma arising in short segment Barrett esophagus. The inclusion of these cases in the "gastric cardiac adenocarcinoma" group would tend to admix esophageal characteristics to an unknown extent. This likely admixture is evident in the authors' statement regarding the classification of esophageal squamous carcinomas: "Squamous cell carcinomas were classified as esophageal even if the location was the gastric cardia" (p. 826)

"Among patients with recurrent symptoms of reflux, as compared with persons without such symptoms, the odds ratios ratios were 7.7 (95% confidential interval, 5.3 to 11.4) for esophageal adenocarcinoma and 2.0 (95% C.I. 1.4 to 2.9) for adenocarcinoma of the cardia" (p. 825). It is unknown what effect the possible admixture of adenocarcinoma arising in the distal esophagus resulting from the unorthodox definition has on the odds ratio for adenocarcinoma of the cardia; it will tend to decrease the risk.

The authors classified the severity of reflux symptoms by using the following grading system that assigned a "reflux score" to each patient (heartburn only—1 point; regurgitation only—1 point; heartburn and regurgitation combined—1.5 points; nightly symptoms—yes = 2, no = 0; frequency of symptoms—once a week = 0, 2 to 6 times a week = 1, 7 to 15 times a week = 2, more than 15 times a week = 3).

The results showed that 685 of 820 controls (84%), 76/189 (40%) patients with esophageal adenocarcinoma, 187/262 (71%) of patients with adenocarcinoma of the gastric cardia, and 142/167 (85%) patients with esophageal squamous carcinoma were asymptomatic. The odds ratios of symptomatic patients based on the reflux symptom score and duration of reflux symptoms in patients with esophageal adenocarcinoma and adenocarcinoma of the gastric cardia are shown in Table 12.4.

The evidence for a relationship between symptomatic reflux (frequency, severity, and duration) and esophageal adenocarcinoma is very convincing. The relationship between symptomatic reflux and adenocarcinoma of the cardia is less convincing, particularly with the admixture of distal esophageal adenocarcinoma in this group.

TABLE 12.4 Reflux Symptom Score and Duration of Reflux Symptoms Related to Odds Ratios for Adenocarcinoma of the Esophagus and Gastric Cardia

	Esophageal adenocarcinoma (total number: 189)		Adenocarcinoma of gastric cardia (total number: 262)	
Symptom severity (score)	Number (%)	Odds ratio (95% C.I.)	Number (%)	Odds ratio (95% C.I.)
No symptoms	76 (40)	1.0	187 (71)	1.0
Score 1–2 pts	10 (5)	1.4 (0.7–3.0)	30 (11)	1.7 (1.0–2.9)
Score 2.5–4 pts	39 (21)	8.1 (4.7–16.1)	27 (10)	1.8 (1.0–3.2)
Score 4.5–6.5 pts	64 (34)	20.0 (11.6–34.6)	18 (7)	2.8 (1.6–5.0)
Symptom duration				
No symptoms	76 (40)	1.0	187 (71)	1.0
<12 years	31 (16)	7.5 (4.2–13.5)	19 (7)	1.6 (0.9–2.9)
12–20 years	42 (22)	5.2 (3.1–8.6)	34 (13)	1.8 (1.1–2.9)
>20 years	40 (21)	16.4 (8.3–28.4)	22 (8)	3.3 (1.8–6.3)

When these factors were added, the authors concluded: "Among patients with long-standing and severe symptoms of reflux, the odds ratios were 43.5 (95% C.I., 18.3 to 103.5) for esophageal adenocarcinoma and 4.4 (95% C.I., 1.7 to 11.0) for adenocarcinoma of the cardia. The risk of esophageal squamous carcinoma was not associated with symptomatic reflux (odds ratio 1.2; 95% C.I., 0.7 to 1.9)" (p. 825).

The data in this study prove that adenocarcinomas with an epicenter that is more than 2 cm above the defined gastroesophageal junction have a strong likelihood of having symptomatic reflux in a cumulative sense (i.e., severity × duration). Patients with adenocarcinomas arising within 2 cm proximal to and 3 cm distal to the defined gastroesophageal junction have a lesser likelihood of reflux.

In Lagergren et al.'s study, what is called esophageal adenocarcinoma has an epicenter that is more than 2 cm above the definition of the gastroesophageal junction. These will be patients whose adenocarcinoma developed in a segment of columnar-lined esophagus at least 2 cm long (very close to long-segment Barrett esophagus).

It is unfortunate that this study combines adenocarcinoma of the cardia (i.e., tumors arising within 3 cm distal to the defined gastroesophageal junction) with adenocarcinoma in the distal 2 cm of the esophagus. It would have been interesting to see how adenocarcinomas of the distal 2 cm of the esophagus (i.e., adenocarcinoma arising in short-segment Barrett esophagus) compared with adenocarcinoma within 3 cm distal to the gastroesophageal junction.

Is Adenocarcinoma of the Gastric Cardia Equal to Adenocarcinoma of the Dilated End-Stage Esophagus?

Lagergren et al.'s study (6) suggests that as the distance of the origin of the tumor from the gastroesophageal junction becomes more proximal, patients have an increasing severity of reflux. If the definition of the gastroesophageal junction is true, this increasing association should begin at the gastroesophageal junction and increase progressively at all points proximal to it. If, in fact, as this article suggests, the association of adenocarcinoma of this region with reflux disease begins 3 cm distal to the defined gastroesophageal junction, it strongly suggests that the definition of the gastroesophageal junction that is used is 3 cm proximal to the true gastroesophageal junction. This is the theoretical limit that we have set for the length of the dilated end-stage esophagus (Fig. 12.25). This would strongly suggest that what is being called "adenocarcinoma of the gastric cardia" in this article consists of two entities: (a) adenocarcinoma arising in short-segment Barrett esophagus (those tumors where the center is located within 2 cm proximal to the defined gastroesophageal junction) and (b) adenocarcinoma arising in dilated end-stage esophagus (those tumors where the center is located with 3 cm distal to the defined gastroesophageal junction). The true gastroesophageal junction is the point at which the association of tumors with an increasing severity of reflux begins; by the definition applied in this tumor, this is 3 cm distal to the defined gastrointestinal junction.

We have suggested that the true gastroesophageal junction is distal to the presently defined junction (either the end of the tubular esophagus or the proximal limit of gastric rugal folds) by the length of the dilated end-stage esophagus. This can be recognized histologically because it is lined by metaplastic esophageal columnar epithelia (cardiac mucosa with and without intestinal metaplasia and oxyntocardiac mucosa). The theoretical limit of this segment is the length of the abdominal part of the lower esophageal sphincter, which is approximately 2.5 to 3 cm.

In the following study, we looked at the relationship of the epicenter of tumors classified by present criteria (the end of the tubular esophagus) into adenocarcinoma of the distal esophagus, gastroesophageal junction, and gastric cardia to the histologic epithelial type.

Literature Review

Chandrasoma P, Wickramasinghe K, Ma Y, DeMeester T. Is the esophagus the origin of most adenocarcinomas of the distal esophagus, "gastroesophageal junction" and "gastric cardia"? Unpublished data.

This study consists of 74 patients who underwent esophagogastrectomy for adenocarcinoma of the distal esophagus, gastroesophageal junction, and gastric cardia (proximal stomach), so classified by the relationship of the epicenter of the tumor to the grossly defined gastroesophageal junction. The gastroesophageal junction was defined as a line drawn between the end of the tubular esophagus and the saccular stomach as recommended by the Association of Directors of Anatomic and Surgical Pathology. The specimen was sectioned extensively in a manner that permitted evaluation of the mucosal types present immediately proximal to the tumor, immediately distal to the tumor, and at the lateral edge of the epicenter of the tumor. The last determination was not possible in 10 cases where the tumor involved the full circumference of the esophagus.

The epithelial types were classified into squamous, cardiac, oxyntocardiac, intestinal, and oxyntic. The type of epithelium at the proximal, distal, and central lateral edge of each tumor was recorded.

Results

1. When the end of the tubular esophagus was used as the defining criterion for the gastroesophageal junction, 38 of these tumors (51.4%) were classified as distal esophageal, 25 (33.8%) as junctional, and 11 (14.8%) as gastric cardiac.
2. Of the 38 distal esophageal cancers, 30 were men and 8 (21%) were women. Four of the 25 patients with junctional tumors (16%) and 4 of the 11 patients with gastric cardiac tumors (36%) were women. The mean age was 64.1 years (median: 66 years; range: 31 to 86 years). The tumor size ranged from grossly invisible intramucosal carcinomas to large tumors exceeding 15 cm in greatest dimension. Histologically, they were all pure adenocarcinomas. The depth of invasion of the tumor was intramucosal in 14 cases (18.9%), submucosal in 8 cases (10.8%), intramural in 4 cases (5.4%), and transmural in 48 cases (64.9%).
3. The epithelial type at the proximal edge of the tumor was determined by examining the section immediately above the upper edge of the tumor (Table 12.1). No cases had oxyntocardiac or oxyntic mucosa at the proximal edge of the tumor. In the 38 distal esophageal tumors, squamous epithelium was present in 16 (42%), intestinal metaplasia was present in 20 (53%), and cardiac mucosa was present in 2 (5%) cases. In the 25 junctional tumors, squamous epithelium was found in 20 cases (80%), 3 of the cases (12%) had intestinal, and 2 cases (8%) had cardiac mucosa. In the 11 gastric cardiac tumors, 7 cases (64%) had squamous epithelium and 2 cases (18%) each had intestinal and cardiac mucosa.
4. The epithelial type at the lateral edge of the epicenter of the tumor was determined by examination of the section across the center of the tumor in the 64 cases where the tumor was not circumferential. None of the tumors had oxyntic mucosa at the epicenter (Table 12.1). Of the 34 noncircumferential distal esophageal tumors, 2 (6%) had squamous epithelium, 25 (73%) had intestinal metaplasia, 5 (15%) had cardiac mucosa, and 2 (6%) had oxyntocardiac mucosa at the epicenter. Of the 21 noncircumferential junctional tumors, 3 (14%) had squamous epithelium, 7 (33%) had intestinal metaplasia, 9 (43%) had cardiac mucosa, and 2 (10%) had oxyntocardiac mucosa at the epicenter. Of the 9 noncircumferential gastric cardiac tumors, 2 (22%) had intestinal, and 7 (78%) had cardiac mucosa at the epicenter of the tumor.
5. Ten patients had circumferential tumors, which made it impossible to determine the epithelium at the epicenter. By presently accepted criteria, 4 of these were classified as distal esophageal, 4 as junctional, and 2 as proximal gastric. Examination of the epithelium at the distal limit of these tumors showed metaplastic glandular epithelium in 7 patients (cardiac mucosa in 3 cases and oxyntocardiac mucosa in 4 cases). Three patients had oxyntic mucosa at the distal tumor edge. These three tumors were all large (4 cm, 6.5 cm and 11 cm), transmural, had positive lymph nodes, had lymphovascular invasion, undermining of proximal squamous epithelium and distal oxyntic mucosa, and had satellite nodules in the mucosa away from the main tumor.
6. The epithelial type at the distal edge of the tumor was determined by examining the vertical section taken across the distal edge of the tumor. In the 38 distal esophageal tumors, the epithelium at the distal edge was intestinal in 12 cases (32%), cardiac in 13 cases (34%), oxyntocardiac in 7 cases (18%), and oxyntic in 6 cases (16%). In the 25 junctional tumors, the epithelium at the distal edge was intestinal in 2 cases (8%), cardiac in 12

cases (48%), oxyntocardiac in 5 cases (20%), and oxyntic in 6 cases (24%). In the 11 proximal gastric tumors, 2 cases (18%) each had intestinal and cardiac mucosa, 3 cases (27%) had oxyntocardiac mucosa, and 4 cases (36%) had oxyntic mucosa.

7. The 16 cases with oxyntic mucosa at the distal edge of the tumor were characterized by being large tumors (the sizes ranged from 2 to 11 cm; mean size: 6 cm; median size: 5.5 cm), that were deeply invasive (all 16 tumors were transmural) with evidence of lymph node involvement (all 16 tumors showed lymph node involvement). These tumors also showed frequent undermining of the proximal squamous epithelium (15 cases), undermining of distal oxyntic mucosa (10 cases), lymphovascular involvement (14 cases), and the presence of satellite tumor nodules in squamous or oxyntic mucosa (12 cases) (Table 12.5).

This study provides strong evidence that the present classification of tumors in this region into esophageal and gastric cardiac (with or without a junctional category) is incorrect and based on the incorrect definition of the gastroesophageal junction on gross examination. When the dilated end-stage esophagus is recognized, most, if not all, of these tumors become esophageal. In the present study, 58 of 74 cases had metaplastic esophageal columnar epithelia at *the distal limit* of the tumor, making them entirely within the esophagus. The 16 patients whose tumors had oxyntic mucosa at the distal edge all had advanced tumors, making it likely that these tumors had grown into the proximal stomach by extension. In 13 of the 16 patients who had non-circumferential tumors, the epithelium at the epicenter of the tumor was metaplastic esophageal columnar epithelium; by presently accepted criteria, these tumors would also be classified as esophageal. Only in the three patients who had circumferential tumors with oxyntic mucosa at the distal edge of the tumor is there any doubt about whether the tumor was esophageal or gastric; the fact that these were all large, advanced tumors favors an esophageal origin to these as well.

A New Classification Method for Adenocarcinoma of This Region

We are now in a position to define adenocarcinomas of this region into three subtypes within the overall designation of esophageal adenocarcinoma (Fig. 12.29):

1. *Adenocarcinoma arising in long-segment Barrett esophagus.* These will be defined by the presence of a visible columnar-lined segment in the tubular esophagus that is greater than 2 cm (or 3 cm, depending on which definition of long-segment Barrett esophagus is used). The cancer can have its epicenter anywhere in this segment based on its random occurrence within the area of intestinal metaplasia.

   The only change from present terminology is that cancers occurring in the dilated end-stage

TABLE 12.5 Epithelial Types Found at Proximal Edge, Distal Edge, and Lateral Edge at the Epicenter of Tumors Classified by Presently Accepted Criteria

Epithelial type	Distal esophageal tumors (n = 38)	Gastroesophageal junctional tumors (n = 25)	Proximal gastric tumors (n = 11)
At proximal edge of tumor			
Squamous	16	20	7
Intestinal	20	3	2
Cardiac	2	2	2
Oxyntocardiac	0	0	0
Oxyntic	0	0	0
At the lateral edge of epicenter of tumor			
Squamous	2	3	0
Intestinal	25	7	2
Cardiac	5	9	7
Oxyntocardiac	2	2	0
Oxyntic	0	0	0
Circumferential tumor	4	4	2
At the distal edge of tumor			
Squamous	0	0	0
Intestinal	12	2	2
Cardiac	13	12	2
Oxyntocardiac	7	5	3
Oxyntic	6	6	4

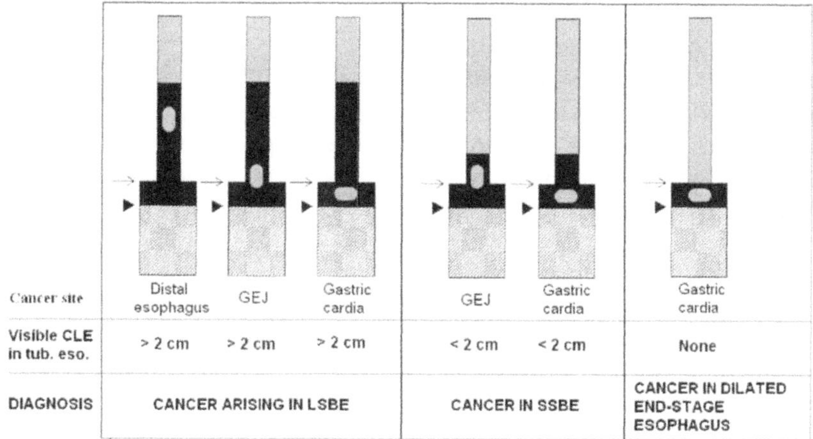

FIGURE 12.29 Classification of adenocarcinoma of the esophagus. The epicenter of all these tumors has columnar metaplastic epithelium (solid black). They arise in long-segment Barrett esophagus (*left*; defined by the presence of greater than 2 cm of columnar-lined tubular esophagus), short-segment Barrett esophagus (*center*; less than 2 cm of columnar lined tubular esophagus), or reflux disease limited to dilated end-stage esophagus (*right*; no visible columnar-lined tubular esophagus). The location of the tumor may be anywhere within the columnar-lined segment.

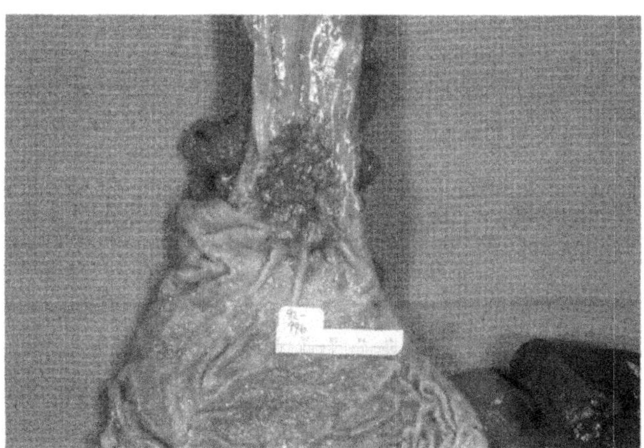

FIGURE 12.30 Adenocarcinoma arising in the distal tubular esophagus. Note the relatively flat mucosa lining the dilated end-stage esophagus between the end of the tubular esophagus and the proximal limit of rugal folds. (Cross reference: Color Plate 5.1)

esophagus in the distal region of the long segment of Barrett esophagus will cease to be classified as adenocarcinoma of the gastric cardia. They will be classified as esophageal adenoarcinoma arising in long-segment Barrett esophagus based on the criterion that their epicenter is in histologically defined columnar-lined esophagus and therefore above the true gastroesophageal junction.

This group of tumors will have demographic and epidemiologic features of long-segment Barrett esophagus. There will be a strong male predominance, and the patients will have a high likelihood of severe symptomatic reflux over a long duration and have a low probability of being "asymptomatic" (i.e., heartburn less than once per week; in Lagergren's study, 40% of this group were "asymptomatic").

2. *Adenocarcinoma arising in short-segment Barrett esophagus.* These will be defined by the presence of a columnar-lined esophagus that is visible and involves less than 2 (or 3) cm of the distal tubular esophagus. The tumor may have its epicenter in the distal 2 (or 3) cm of the tubular esophagus or in the dilated end-stage esophagus below the visible segment of columnar-lined esophagus. Because intestinal metaplasia favors the more proximal segment of columnar-lined segment, tumors in this group will tend to be in the area just above or straddling the end of the tubular esophagus (Fig. 12.30).

   This group will have the epidemiologic and demographic features of short-segment Barrett esophagus. They will have a lower male predominance, have a lower likelihood of severe symptomatic reflux, and be asymptomatic more frequently.
3. *Adenocarcinoma arising in dilated end-stage esophagus.* These patients will have no visible columnar-lined esophagus; the tumor's epicenter will be distal to the end of the tubular esophagus and the tumor will be called a "tumor of the gastric cardia" by present criteria (Figs. 12.26 and 12.27). Such tumors are defined by the presence of columnar metaplastic esophageal epithelium at the epicenter. They are differentiated from gastric adenocarcinoma, which arise in and have gastric oxyntic mucosa at the

epicenter of the tumor. These tumors can grow downward across the true gastroesophageal junction (when the distal margin will be gastric oxyntic mucosa) or grow upward into the tubular esophagus. The epicenter of the tumor can be accurately determined preoperatively by appropriately taken biopsies and by appropriate sectioning of the resected specimen. There is a slight possibility of error in using the epicenter of the tumor, which assumes that the tumor grows equally in all directions. It is more certain to use the epithelium at the distal edge of the tumor; if this is metaplastic esophageal epithelia, the diagnosis is certain. However, this method will exclude esophageal tumors that grow into the stomach.

This group will have the epidemiologic and demographic features of reflux disease limited to the dilated end-stage esophagus. There will be no gender predominance; these patients have a much lower likelihood of severe symptomatic reflux and will frequently be asymptomatic. In Lagergren et al.'s study (6), 71% of patients with "adenocarcinoma of the gastric cardia" (which by their definition was combined adenocarcinoma in short-segment Barrett esophagus and adenocarcinoma in dilated end-stage esophagus) were asymptomatic.

Adenocarcinoma may arise in the dilated end-stage esophagus in patients with long- and short-segment Barrett esophagus in addition to those with reflux disease limited to the dilated end-stage esophagus. These patients will show columnar metaplastic epithelium above the tumor, while those with cancer arising in reflux disease limited to the dilated end-stage esophagus will have the entire tubular esophagus lined by squamous epithelium (Figs. 12.26 and 12.27).

References

1. Allison PR, Johnstone AS. The oesophagus lined with gastric mucous membrane. Thorax 1953;8:87–101.
2. DeMeester TR, Peters JH, Bremner CG, Chandrasoma P. Biology of gastroesophageal reflux disease: Pathophysiology relating to medical and surgical treatment. Annu Rev Med 1999;50: 469–506.
3. Zaninotto G, DeMeester TR, Schwizer W, et al. The lower esophageal sphincter in health and disease. Am J Surg 1988;155:104–111.
4. Sampliner RE. Practice guidelines on the diagnosis, surveillance, and therapy of Barrett's esophagus. Am J Gastroenterol 1998;93:1028–1031.
5. Kahrilas PJ, Shi G, Manka M, Joehl RJ. Increased frequency of transient lower esophageal sphincter relaxation induced by gastric distension in reflux patients with hiatal hernia. Gastroenterology 2000;118:688–695.
6. Lagergren J, Bergstrom R, Lindgren A, Nyren O. Symptomatic gastroesophageal reflux as a risk factor for esophageal adenocarcinoma. N Engl J Med 1999;340:825–831.
7. Jacobson BC, Ferris TG, Shea TL, Mahlis EM, Lee TH, Wang TC. Who is using chronic acid suppression therapy and why? Am J Gastroenterol 2003;98:51–58.
8. Hirota WK, Loughney TM, Lazas DJ, Maydonovitch CL, Rholl V, Wong RKH. Specialized intestinal metaplasia, dysplasia, and cancer of the esophagus and esophagogastric junction: Prevalence and clinical data. Gastroenterology 1999; 116:277–285.
9. Glickman JN, Fox V, Antonioli DA, Wang HH, Odze RD. Morphology of the cardia and significance of carditis in pediatric patients. Am J Surg Pathol 2002;26:1032–1039.
10. Chandrasoma PT, Der R, Ma Y, et al. Histology of the gastroesophageal junction: An autopsy study. Am J Surg Pathol 2000;24:402–409.
11. Paull A, Trier JS, Dalton MD, Camp RC, Loeb P, Goyal RK. The histologic spectrum of Barrett's esophagus. N Engl J Med 1976;295:476–480.
12. Der R, Tsao-Wei DD, DeMeester T, et al. Carditis: A manifestation of gastroesophageal reflux disease. Am J Surg Pathol 2001;25:245–252.
13. Oberg S, Peters JH, DeMeester TR, et al. Inflammation and specialized intestinal metaplasia of cardiac mucosa is a manifestation of gastroesophageal reflux disease. Ann Surg 1997;226: 522–532.
14. Jain R. Aquino D, Harford WV, Lee E, Spechler SJ. Cardiac epithelium is found infrequently in the gastric cardia. Gastroenterology 1998;114:A160 (Abstract).
15. Marsman WA, van Sandyck JW, Tytgat GNJ, ten Kate FJW, van Lanschot JJB. The presence and mucin histochemistry of cardiac type mucosa at the esophagogastric junction. Am J Gastroenterol 2004;99:212–217.
16. Sarbia M, Donner A, Gabbert HE. Histopathology of the gastroesophageal junction. A study on 36 operation specimens. Am J Surg Pathol 2002;26:1207–1212.
17. Chandrasoma PT, Der R, Ma Y, Peters J, DeMeester T. Histologic classification of patients based on mapping biopsies of the gastroesophageal junction. Am J Surg Pathol 2003;27:929–936.
18. Rex DK, Cummings OW, Shaw M, Cumings MD, Wong RKH, Vasudeva RS, Dunne D, Rahmani EY, Helper DJ. Screening for Barrett's esophagus in colonoscopy patients with and without heartburn. Gastroenterology 2003;125:1670–1677.
19. Goldblum JR, Vicari JJ, Falk GW, et al. Inflammation and intestinal metaplasia of the gastric cardia: The role of gastroesophageal reflux and *H. pylori* infection. Gasteroenterology 1998;114: 633–639.
20. Hackelsberger A, Gunther T, Schultze V, Manes G, Dominguez-Munoz JE, Roessner A, Malfertheiner P. Intestinal metaplasia at the gastro-oesophageal junction: Helicobacter pylori gastritis or gastro-oesophageal reflux disease? Gut 1998;43:17–21.
21. Voutilainen M, Farkkila M, Juhola M, et al. Specialized columnar epithelium of the esophagogastric junction: Prevalence and associations. Am J Gastroenterol 1999;94:913–918.

CHAPTER

13

Definition of Gastroesophageal Reflux Disease and Barrett Esophagus

In Chapters 8 and 9, we detailed the cellular events that occur in the preneoplastic phase of gastroesophageal reflux disease. These are summarized in Table 13.1: Any one of these cellular changes or even a combination of changes can be potentially used to define reflux disease. Before deciding on any specific definitive criteria, it is useful to determine the characteristics of an ideal definition for reflux disease. The first characteristic must be that it is a cellular abnormality. All human diseases are ideally defined by a cellular abnormality. While a 24-hour pH study or an impedance test may be the gold standard for the *diagnosis of gastroesophageal reflux*, the possibility that there are individual variations in epithelial susceptibility or resistance to reflux makes it necessary that the *defining criterion for reflux disease must be a cellular change*. An abnormal 24-hour pH test cannot define reflux disease just as coronary atherosclerosis cannot define ischemic heart disease and smoking cannot define lung cancer. An etiologic factor, however specific and powerful, can only define a risk for the disease, not the disease itself.

The ideal cellular abnormality used to define reflux disease will have a highly specific causal relationship with reflux and be extremely sensitive. This means that if a patient has the defining criterion, the diagnosis of reflux is definite (no other disease is possible), and if the diagnostic criterion is absent, the patient has no reflux disease.

Table 13.1 shows that only one squamous epithelial change (classical symptoms caused by the stimulation of intraepithelial nerve endings by acid) reaches a high level of specificity, albeit with only intermediate sensitivity. In contrast, columnar metaplasia of all types is 100% specific for reflux. Despite the fact that this knowledge has existed since the 1950s, columnar metaplasia has never been used as a diagnostic criterion for reflux disease. The reason for this is that it has never been considered to be 100% specific because of the false dogma that cardiac mucosa is a normal epithelium in this region. Once this dogma is removed, columnar metaplasia approaches the ideal definition of reflux disease with 100% specificity and very high sensitivity (1).

PRESENTLY USED CRITERIA FOR DEFINING GASTROESOPHAGEAL REFLUX DISEASE

Amazingly, there is no standardized, universally accepted definition of gastroesophageal reflux disease. Without a standardized definition, researchers can devise any criterion to define reflux disease. The fact that the definition of reflux disease varies in different studies makes it impossible to assess the literature in any meaningful and scientific manner.

An attempt to define gastroesophageal reflux disease at the Genval Workshop in 1999 (2) led to the acceptance of the following definition: *The term "gastro-oesophageal reflux disease" (GORD, reflux disease) should be used to include all individuals who are exposed to the risk of physical complications from gastro-oesophageal reflux, or who experience clinically significant impairment of health related well being (quality of life) due to reflux related symptoms, after adequate reassurance of the benign nature of their symptoms* (p. 51). This is a perfect definition, and we will show that it is achievable but not by presently accepted criteria.

TABLE 13.1 Cellular Changes That Occur as a Result of Gastroesophageal Reflux

Epithelium involved	Change	Sensitivity	Specificity
Squamous	Increased Ki67 expression	Very high	Low
Squamous	Basal cell hyperplasia	Intermediate	Low
Squamous	Papillary elongation	Intermediate	Low
Squamous	Abnormal vascular pattern in mucosa	High	Low
Squamous	Intraepithelial neutrophils	Intermediate	Low
Squamous	Intraepithelial eosinophils	Intermediate	Intermediate
Squamous	Dilated intercellular spaces	High	Low
Squamous	Increased permeability	High	Low
Squamous	Chemokines	High	Unknown
Squamous	Esophageal pain	Intermediate	High
Squamous	Erosion	Low	Intermediate
Squamous	Ulceration	Low	Intermediate
Squamous	Strictures	Low	Intermediate
Squamous	Genetic changes	Intermediate	Unknown
Squamous	Columnar metaplasia	Near 100%	100%
Cardiac mucosa	Intestinal (Barrett-type) metaplasia	Low	100%
Intestinal epithelium	Adenocarcinoma	Very low	100%

There is no presently accepted definitive criterion or set of criteria that will identify "all individuals who are exposed to the risk of physical complications of reflux." Presently accepted definitions work if reflux-induced adenocarcinoma is excluded as a physical complication of reflux, but then the disease really ceases to be a major problem. The control of pain, erosions and ulcers, strictures, and the improvement of quality of life are easily achieved with either good acid suppression or antireflux surgery. It is the adenocarcinoma that is out of control and kills people with reflux. Reflux-induced adenocarcinoma is possible in any person irrespective of whether or not he or she has any presently used criterion to define reflux disease; it can occur in asymptomatic people, people without erosive esophagitis, and people without Barrett esophagus as presently defined. We have no basis for reassuring anyone that they are not at risk for adenocarcinoma. Therefore, the Genval Workshop definition includes everyone in the population.

If we substitute the word "adenocarcinoma" for "physical complications from gastro-oesophageal reflux" the statement beautifully defines what is needed by a definition of reflux disease: (a) The presence of the criterion needs to include all individuals (with or without symptoms) who are at risk for adenocarcinoma; and (b) its absence needs to be able to reassure people (with or without symptoms) that they have no risk of developing adenocarcinoma. We will show in this chapter that presently used definitive criteria fail in both these regards. We will also show that the use of "reflux carditis" as the defining criterion of reflux disease achieves these objectives perfectly.

Classical Symptoms

Classical symptoms are a commonly used defining criterion. If heartburn and regurgitation are used at a frequency more than occasional, it is a very specific criterion of reflux disease (Table 13.1). However, there is no standard symptom combination or symptom frequency to define reflux disease. For example, Spechler et al. (3) from Harvard used heartburn equal to or greater than 1 day per week to define symptomatic GERD. However, Kahrilas et al. (4) at the University of Chicago defined symptomatic reflux as heartburn equal to or greater than three times per week controlled by acid suppression, whereas at the Cleveland Clinic, Goldblum et al. (5) defined "classic GERD" by the presence of heartburn and/or acid regurgitation at least twice per week for at least 6 months. Even studies from the same institution at different times use different definitions. Thus, in their 2002 report from Harvard, Glickman et al. (6) used "symptoms related to GERD (heartburn, regurgitation, or dysphagia)" to define their GERD population without any reference to frequency of symptoms. What this means is that the GERD patients at Harvard may be controls in the University of Chicago and the Cleveland Clinic or even at Harvard a few years earlier. *The definition of reflux disease based on classical symptoms is not reproducible because there is no standard.*

It is well known that many patients presenting with serious complications of reflux disease such as adenocarcinoma have never had classical symptoms; in Lagergren et al. (7), 40% of patients with esophageal adenocarcinoma did not have reflux symptoms. One cannot define reflux disease by symptoms and at the same time recognize "asymptomatic reflux disease." *Classical symptoms have an unacceptably low sensitivity to define reflux disease.*

Endoscopic Criteria: Erosive Esophagitis

Erosive esophagitis is even less sensitive for reflux disease than classical symptoms (Genval Workshop, 1999). Only about 50% of symptomatic patients have erosive esophagitis (Fig. 13.1); the others are said to have "nonerosive reflux disease." Erosions are less specific as a single criterion than symptoms because they may result from infections, pills, and corrosives, suggesting that the use of erosive esophagitis is worse than classical symptoms and therefore useless as a criterion to define reflux.

Erosive esophagitis is used to grade the severity of reflux disease. Of the several classifications that exist, the most commonly used for grading erosive esophagitis is the Los Angeles classification (Table 13.2).

The grades in the Los Angeles classification have been shown to have an excellent correlation with esophageal acid exposure as assessed by the 24-hour pH test (8). The correlation with the 24-hour pH test is best with the higher Los Angeles grades; with grades A and B, the pH study may be normal in nearly half the patients. This can be due to a lack of specificity of grades A and B erosive esophagitis or a lack of sensitivity of the 24-hour pH test in the diagnosis of reflux. The presence of grade C and D erosive esophagitis has a high specificity for the diagnosis of reflux disease.

Recently, the use of magnifying endoscopes shows the presence of abnormal vascular pattern in squamous epithelium that indicates abnormal reflux (9). These correlate with the increased vascularity seen histologically in the elongated papillary ridges. This is likely to be more sensitive but less specific than erosive esophagitis; hypervascularity is a nonspecific pathologic response to many types of injuries.

Though the evidence is sometimes confusing, it is likely true that the detection of erosive esophagitis is a highly predictive indicator of the presence of reflux damage of the squamous epithelium in the presence of classical reflux symptoms and when other causes of erosions in the esophagus are excluded. In the immunocompetent patient, nonreflux causes of erosions are rare and include corrosive ingestion and pills that have fairly specific clinical features that are distinct from reflux disease. In the immunocompromised patient, a variety of infections and nonspecific ulceration of the esophagus by HIV infection complicate the value of erosions as a diagnostic test for reflux, probably to the extent of making it practically useless.

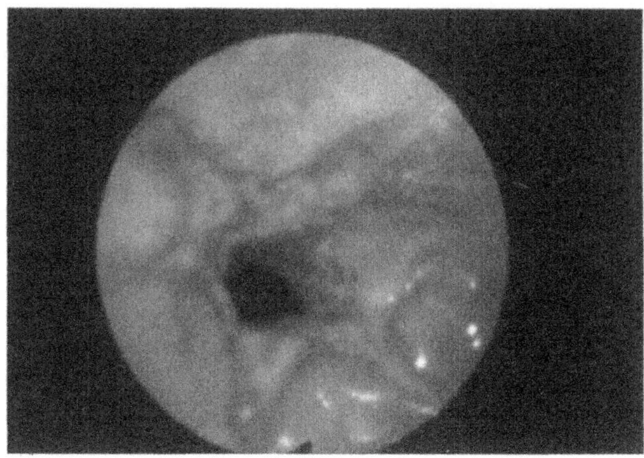

FIGURE 13.1 Erosive esophagitis at endoscopy. This has a very low sensitivity as a diagnostic criterion for reflux disease. (Cross reference: Color Plate 2.1)

TABLE 13.2 Los Angeles Classification for Grading Erosive Reflux Esophagitis

Grade A: One or more mucosal breaks no longer than 5 mm, none of which extends between the tops of the mucosal folds
Grade B: One or more mucosal breaks more than 5 mm long, none of which extends between the tops of two mucosal folds
Grade C: Mucosal breaks that extend between the tops of two or more mucosal folds but involve less than 75% of the esophageal circumference
Grade D: Mucosal breaks that involve 75% or greater of the esophageal circumference

Conclusion: Erosive esophagitis is a criterion of low sensitivity in the diagnosis of reflux. Its specificity is high in immunocompetent patients when combined with the presence of classical reflux symptoms and low in immunocompromised patients.

The Bernstein Test

The Bernstein test consists of exposing the distal esophagus to acid instilled at the time of endoscopy. A positive test is indicated by the precipitation of heartburn (ideally, similar to the patient's presenting symptom). The patient in whom the acid produces no symptom). The patient in whom the acid produces no symptomatic effect is a negative test. The Bernstein test has a high specificity for the diagnosis of reflux; its sensitivity is not known but likely to be low. The Bernstein test is not commonly used at this time.

The basis for the Bernstein test is that it detects invisible increased permeability of the squamous epithelium to acid. In patients with dilated intercellular spaces resulting from reflux, the instilled acid penetrates the epithelium to stimulate nerve endings therein, essentially reproducing the mechanism of symptom generation in reflux.

Therapeutic Response of Symptoms to Acid Suppression

The presence of highly effective acid suppressive drugs such as proton pump inhibitors makes a therapeutic test for reflux feasible. In this test, the patient with reflux is placed on the maximum dose of proton pump agents for a defined period. If the patient's symptoms reverse with this regimen, it is highly likely that the patient's symptoms are caused by abnormal acid exposure of the esophagus. The test is highly specific if the placebo effect of the drug can be excluded.

Unfortunately, a negative test does not exclude abnormal reflux. Approximately 5% of patients with reflux will not respond to the maximal proton pump inhibitor regimen despite the fact that they have reflux disease (i.e., the sensitivity of the test is 95% with a false negative rate of 5%). Failure to control symptoms with proton pump inhibitors is most often related to the fact that nonacid reflux continues in these patients and nonacid molecules in the refluxate continue to precipitate pain in the damaged squamous epithelium (nonacid reflux).

Despite its lack of total specificity, the acid suppression therapeutic test is a valuable test for demonstrating abnormal reflux. It is simple and cheap and requires no special equipment or discomfort. As such, it is commonly used by gastroenterologists as a practical method of confirming the diagnosis of reflux disease. Obviously, the test cannot be used to check for asymptomatic reflux disease.

Histologic Criteria in the Squamous Epithelium

Changes in the squamous epithelium associated with reflux include erosion, ulceration, basal cell hyperplasia, papillary elongation, intraepithelial eosinophils, and neutrophils (Fig. 13.2) are all neither very

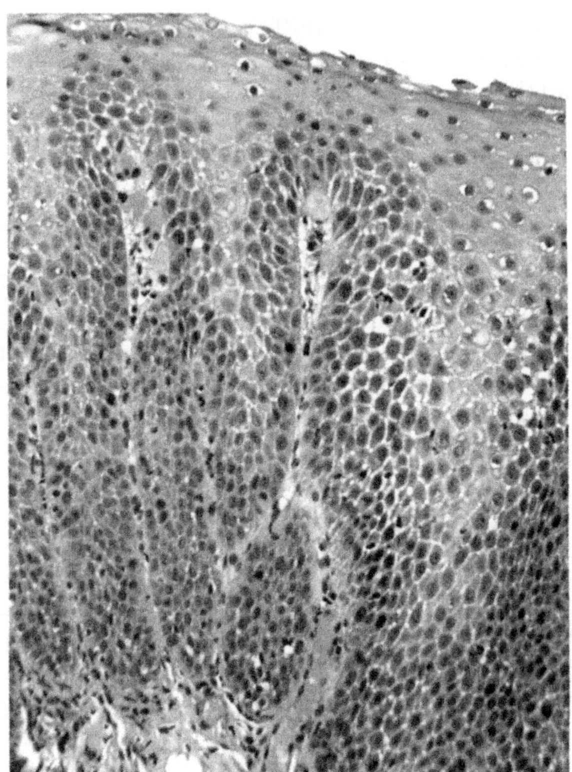

FIGURE 13.2 Histologic criteria for a diagnosis of reflux esophagitis in a biopsy from squamous-lined distal esophagus, showing basal cell hyperplasia, dilated intercellular spaces, and rare eosinophils in the epithelium. This has a low sensitivity as a diagnostic criterion for reflux disease. (Cross reference: Color Plate 2.2)

specific nor very sensitive and are essentially worthless for the diagnosis of reflux disease.

Of these criteria, the presence of intraepithelial eosinophils is probably the most sensitive marker for reflux disease. The following table, attributed to Rodger Haggitt, is present in Riddell's review article (10) and shows the correlation between reflux disease as measured by papillary length and the number of neutrophils and eosinophils:

		Number of neutrophils per hpf				Number of eosinophils per hpf			
Papillary length	**Number of biopsies**	**0**	**1–5**	**6–10**	**>10**	**0**	**1–5**	**6–10**	**>10**
>0.19 mm	60	44	10	3	3	25	20	7	8
<0.19 mm	31	30	1	0	0	23	8	0	0

The study suggests that if reflux disease is accurately determined by the criterion of a papillary length greater than 0.19 mm, the presence of eosinophils is more specific (35 of 60 true positives and only 8 of 31 false positive) and sensitive (25 of 60 false negative and 23 of 31 true negative) than neutrophils for the diagnosis of reflux. The problem is that the criterion used to define reflux disease (papillary length greater than 0.19 mm) is not a standard or proven criterion for the diagnosis of reflux disease. Tummala et al. (11) reported that eosinophils are present in 4 of 12 "normal controls." This article is often quoted as evidence that the finding of eosinophils in small numbers is not a reliable criterion for the diagnosis of reflux, despite the fact that there is no method of defining a "normal control" population.

The correct conclusion is that of all the criteria that occur in the squamous epithelium, intraepithelial eosinophils are the most specific, assuming that the rare entity of eosinophilic esophagitis has been excluded. There is no evidence that eosinophils occur "normally" in adult esophageal squamous epithelium because the "normal controls" in these studies cannot be guaranteed not to have reflux. As such, we use the presence of any intraepithelial eosinophils as the most reliable criterion for the histologic diagnosis of reflux disease. Even when used like this, its sensitivity is so low as to make it an essentially worthless criterion for the diagnosis of reflux disease. The specificity of eosinophils and neutrophils is also low; patients with allergic eosinophilic esophagitis have large numbers of intraepithelial eosinophils in the absence of reflux, and many infections are associated with neutrophils.

At the present time, the earliest recognized cellular change that indicates abnormal reflux is the presence of dilated intercellular spaces (Fig. 13.3). This results in increased permeability of the epithelium permitting refluxate molecules to permeate into the epithelium and stimulate pain-sensitive nerve endings in the epithelium. The resulting heartburn is often present in patients whose dilated intercellular spaces are detectable only at electron microscopy. As such, in many patients, heartburn is the earliest clinically apparent change associated with abnormal reflux, but is not accompanied by any visible change in squamous epithelium in routine specimens.

No single or combination of findings in squamous epithelium can be used for the diagnosis of a patient with reflux disease (12).

Conclusion: Histologic criteria of reflux esophagitis in the squamous epithelium are of such low sensitivity and specificity as to be worthless in defining reflux disease.

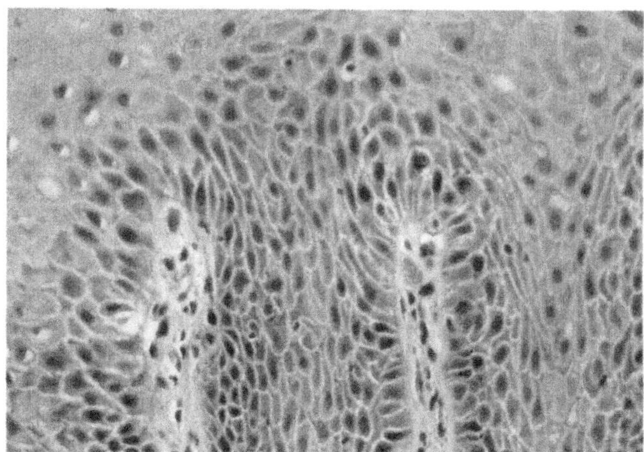

FIGURE 13.3 Basal region of the squamous epithelium in reflux-induced damage showing separation of epithelial cells ("dilatation of intercellular spaces"). (Cross reference: Color Plate 2.3)

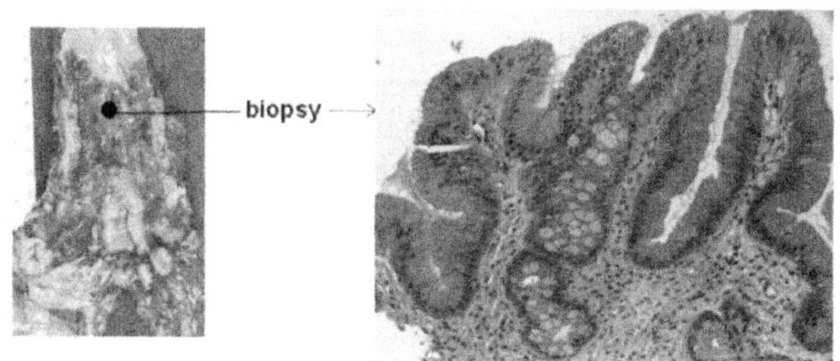

FIGURE 13.4 Barrett esophagus is presently defined as the presence of intestinal metaplasia in a biopsy taken from an abnormal columnar-lined esophagus visualized at endoscopy. (Cross references: Color Plates 3.2 and 4.1)

Barrett Esophagus

Table 13.1 suggests that columnar metaplasia of the squamous epithelium is a 100% specific and highly sensitive change produced by gastroesophageal reflux. At present, the only type of columnar metaplasia that is recognized as definitive for gastroesophageal reflux is Barrett esophagus, which is defined as the presence of intestinal metaplasia in a biopsy taken from an endoscopically visible segment of columnar-lined esophagus (Fig. 13.4). The present definition of Barrett esophagus is flawed. By the edict of definition, we are ignoring Barrett esophagus from the time of its onset to the time it becomes endoscopically visible (see Chapter 12).

The presence of Barrett esophagus is 100% specific for reflux disease. It is not produced by any other disease. However, it has an extremely low sensitivity for the diagnosis of reflux disease. Hirota et al. (13) showed that the prevalence of Barrett esophagus in patients with symptomatic chronic gastroesophageal reflux disease was only 3% to 5% (long segment) and 10% to 15% (short segment) for an overall prevalence of at most 20%. Rex et al. (14), in a screening study, showed an even lower prevalence of Barrett esophagus among symptomatic patients; in this study, the total prevalence of Barrett esophagus and long-segment Barrett esophagus was 8.3% and 2.6%, respectively. The prevalence of Barrett esophagus in asymptomatic patients was 5.6% and 0.36% for Barrett esophagus and long-segment Barrett esophagus. If the presence of Barrett esophagus is used as a criterion for reflux, the 80% to 90% of patients with symptomatic reflux disease will be classified as negative for reflux and 5% of asymptomatic patients will be classified as having reflux disease. *The use of Barrett esophagus is too insensitive to be of value as a criterion for the diagnosis of reflux disease.*

Barrett esophagus has value as a marker for adenocarcinoma risk. The reflux-adenocarcinoma sequence passes through Barrett esophagus (see Chapter 10). It is generally believed that only the intestinal type of columnar metaplasia in the esophagus is susceptible to carcinogenesis (15). If true, as we believe it is, Barrett esophagus has a 100% specificity as a marker for cancer risk. The sensitivity of Barrett esophagus as a cancer risk marker is also likely to be 100% as long as microscopic Barrett esophagus in dilated end-stage esophagus is included in the definition; the absence of Barrett esophagus indicates absence of risk. Unfortunately, only 5% of patients who develop esophageal adenocarcinoma have a prior diagnosis of Barrett esophagus (16). This makes Barrett esophagus an ineffective practical risk indicator. However, it is potentially valuable because it can be detected by an appropriate screening protocol.

Summary statement regarding present criteria for defining reflux disease: All presently used criteria to define reflux disease—symptoms, erosive esophagitis, histologic changes in squamous epithelium, and Barrett esophagus—are useless either because of low sensitivity or low specificity. A better criterion is desperately needed.

DEFINITIONS BASED ON QUANTITATING GASTROESOPHAGEAL REFLUX

The present gold standards for the diagnosis of reflux are considered to be the 24-hour pH test and newer multichannel impedance studies. These represent a group of tests that accurately quantitate gastroesophageal reflux. The 24-hour pH test uses acid as the marker to quantitate reflux; the Bilitec test uses bilirubin as the marker. Impedance evaluates the volume and type of reflux.

These tests are highly accurate in quantitating reflux. However, they are evaluating the etiology of reflux disease, not its cellular effect. No disease is accurately assessed by the etiologic agent. This is because there is significant individual variation between different people in the response to a given quantity of reflux. Defining reflux disease by the amount of reflux is like defining lung cancer by the number of cigarettes smoked.

The Ambulatory 24-Hour pH Test

If a pH electrode is placed in the distal esophagus, almost all patients will show reflux of gastric juice into the esophagus, detected by intermittent acid spikes in the electrode (17). For most people in whom reflux is infrequent and of low volume and in whom the esophageal mechanisms that clear the refluxed gastric juice are functioning normally, this is asymptomatic. Abnormal reflux and reflux disease are therefore not definable by the simple presence or absence of gastroesophageal reflux. Reflux is not abnormal, however, when there is no cellular damage or disease (detectable or not). In these patients, the reflux is so small that it is resisted by the squamous epithelium.

The ambulatory 24-hour pH test is customarily performed by placing a pH-sensitive electrode under manometric guidance at a point in the esophagus that is 5 cm above the proximal limit of the lower esophageal sphincter. The electrode provides a continuous tracing of lower esophageal pH during a 24-hour period. Ambulatory 24-hour pH testing is generally regarded as the gold standard for the objective diagnosis of abnormal reflux.

An acid spike in the pH electrode must be regarded as the most easily measured component of the gastric refluxate. It is one of a large number of molecules that may be part of gastric contents; these normally include swallowed saliva, food, and other secretions of the esophagus (mucus) and stomach (mucus, pepsin, rennin). Duodenogastric reflux is also a relatively common occurrence in humans. The entry of alkaline duodenal contents neutralizes gastric acid to varying degrees. Many patients take over-the-counter antacids that may also neutralize gastric acid, albeit relatively ineffectively unless well controlled. For these reasons, acid is not the most constant element in gastric refluxate, and its measurement in the esophagus may not be the most accurate marker of reflux. This is certainly true in patients who are on effective acid-suppressive medication; these patients have a gastric pH in the 3 to 7 range and reflux (marked as a $pH < 4$) may not be detected by a pH-sensitive electrode placed in the lower esophagus. Patients who are on proton pump inhibitors need to be taken off acid suppression for 2 weeks for the 24-hour pH study to provide an accurate measurement of reflux.

Studies with standard 24-hour pH testing have documented that symptoms associated with reflux occur when the pH is less than 4 for a period exceeding 4.5% of the 24-hour period (Jamieson et al.). It should be recognized that 4.5% of a 24-hour period is 64.8 minutes. This means that esophageal epithelium exposed to a pH of less than 4 for less than 1 hour per day is usually not associated with symptoms.

The tracing obtained in the 24-hour testing provides many other parameters. These can be quantitated to provide more detailed assessment of acid exposure. The most commonly used of these combined parameters is the DeMeester score, which uses a combination of six components of the pH tracing (total time pH < 4, total % supine time pH < 4, total % upright time pH < 4, number of reflux episodes, number of reflux episodes lasting more than 5 minutes, longest episode). The DeMeester score tends to be used more frequently by surgeons, and the percentage time pH less than 4 (which is the most important and best studied component) is used by most gastroenterologists. Both provide a result that is generally equivalent in terms of diagnostic accuracy for quantitating reflux.

The tracing can be compared with patient symptoms to correlate spikes of reflux with the patient's symptoms as recorded during the day. This provides information as to whether the acid reflux episode as recorded in the tracing is the cause of the patient's symptoms.

More recently, implantation of a pH-sensitive capsule in the esophageal mucosa at the same location has permitted pH monitoring without a tube by detecting continuous wireless transmission from the implanted capsule. The capsule records pH for the duration of the battery life or until it is sloughed off the mucosa, whichever is shorter. Normally, 24 to 72 hours of pH recording is obtained by these implanted capsules. The advantages of the Bravo capsule are that it records data over a longer time and it does not require a nasal tube to be kept for 24 hours. The latter permits the patient to lead a more normal existence for the test period than with the nasal tube. One interesting result of the Bravo capsule test is that significant differences in the reflux patterns have been found in the first 2 days of the test. In approximately 20% of patients, the test is normal on one day and abnormal on the next. There is a suggestion that the use of a 48-hour period of assessment may improve sensitivity of detecting abnormal reflux by 15%.

Standard pH monitoring may not be the optimum method of assessing reflux. The positioning of the catheter at a level 5 cm above the proximal limit of the lower esophageal sphincter may detect only the most severe cases of reflux. This level will be exposed to acid when the sphincter effectiveness has been lost so completely as to permit free reflux into the body of the esophagus. There is a suspicion that lesser degrees of reflux may cause damage to the distal esophagus without the standard 24-hour pH study being abnormal. Studies using pH electrode placements lower in the esophagus and within the high-pressure sphincter zone have shown results that are significantly different from those shown with the standard pH electrode placement.

Conclusion: The 24-hour ambulatory pH test is considered the gold standard for defining abnormal reflux. However, it is by no means perfect and does not define a cellular abnormality of reflux disease.

Interpretation of an Abnormal 24-Hour pH Test

It is a common mistake to conclude that the correlation of any lesion (symptoms, erosive esophagitis, Barrett esophagus) with an abnormal 24-hour pH study is evidence that that lesion is etiologically related to acid. Acid is simply a marker of reflux; correlation with 24-hour acid exposure is a statement of correlation with any molecule that may be present in the refluxate that accompanies the acid. It is correct to say that the lesion involved correlates with the amount of reflux, but not with the amount of acid. This is an important distinction because acid suppression will only cause the reversal of a lesion that is caused by acid, not one caused by a molecule that accompanies the acid in the refluxate.

Let us examine the excellent article from the highly regarded Arizona gastroenterology group to demonstrate the subtlety of this distinction (18). In their introduction, the authors made the following statements: "(a) . . . acid reflux has been implicated as the major contributing factor in the development of Barrett esophagus. Patients with Barrett's mucosa are known to have a greater degree of esophageal acid exposure than patients with erosive esophagitis, those with non-erosive reflux disease, and normal control subjects" (p. 851). *(This is true; acid reflux = reflux measured by a 24-hour pH study. Any component in the refluxate including acid can be the cause of Barrett esophagus.)* "(b) Once intestinal metaplasia develops to its greatest extent, reduction in acid exposure with proton pump inhibitors does not lead to a significant regression of Barrett's mucosa" (p. 851). *(This is true. There are two possible reasons for this. First, it may be that the change is irreversible. Second, the change is caused not by acid but by a component in the refluxate that accompanies the acid.)* The authors concluded from these two statements that "it seems that acid exposure may play a greater role in determining the appearance of Barrett esophagus than in maintaining it" (p. 851). This is not a valid conclusion; the fact that patients with Barrett's mucosa more frequently have an abnormal 24-hour pH test does not mean that acid exposure determines the appearance of Barrett esophagus. The pH test is simply a *measure of reflux*. The correct conclusion is that the appearance of Barrett's mucosa is determined by the amount of reflux; any component in the refluxate including acid and any other molecule in the refluxate can produce the correlation.

In our article on carditis (19), we showed that the severity of chronic inflammation in carditis correlates with acid exposure levels and make a similar erroneous conclusion that carditis is caused by acid exposure. Correlation with an abnormal 24-hour pH test does not mean that inflammation in cardiac mucosa is caused by acid. In fact, the likelihood is that it is not caused by acid because carditis is often seen at its most severe level in patients who are on acid suppressive agents. We have found that cardiac mucosal inflammation is decreased after antireflux surgery. The fact that inflammation in cardiac mucosa is unabated with acid suppression but decreases after reflux has been completely stopped by an antireflux surgical procedure is strong evidence that a component of the refluxate that is not acid is the injurious agent. Similarly, because reversal of Barrett epithelium has been shown to occur after antireflux surgery, it is likely that Barrett esophagus is actually caused by an injurious molecule other than acid.

Let us take a similar study that uses the presence of erosive esophagitis instead of Barrett esophagus. The severity of erosive esophagitis correlates with an abnormal 24-hour pH study. However, unlike Barrett esophagus and carditis, effective acid suppression almost always causes a significant reversal of erosive esophagitis. This means that acid itself and not some other component of the refluxate is probably the cause of erosive esophagitis.

Conclusion: An abnormal 24-hour pH test is highly specific for abnormal gastroesophageal reflux. Correlation of any lesion with an abnormal 24-hour pH does not mean it is caused by acid; it can be caused by any molecule in the gastric refluxate.

Bilitec Probes

Bilitec probes, which detect bilirubin, are only useful to detect the presence of molecules derived from bile in the esophagus. A positive Bilitec study indicates potential exposure of the esophagus to proximal duodenal, pancreatic, and biliary secretions in addition to gastric contents (20). All these molecules enter the duodenum with bilirubin at or before the ampulla of Vater. Bilirubin itself is not harmful to the esophagus; the accompanying molecules are suspected of being important to pathogenesis of the cellular injury.

Gastroesophageal reflux in a patient can be divided into pure gastroesophageal reflux (Bilitec negative) or duodenogastroesophageal reflux (Bilitec positive). As with the use of acid to detect gastric juice refluxing into the esophagus, the use of bilirubin simply detects the presence of duodenal juice in the refluxate and provides evidence for coexisting duodenogastric reflux across the pylorus. If it is shown that Bilitec positivity correlates with any lesion in the esophagus, it does not prove that bilirubin is the cause of that lesion; any molecule in duodenal juice that accompanies bilirubin can be the injurious agent.

Conclusion: An abnormal Bilitec test is highly specific for duodenogastroesophageal reflux. Correlation of any lesion with an abnormal Bilitec test does not mean it is caused by bilirubin; it can be caused by any molecule in the duodenogastric refluxate.

Multichannel Impedance Studies

Impedance studies are becoming recognized as a valuable method of assessing reflux. Unlike pH electrodes, impedance technology detects the occurrence of reflux episodes, the height to which reflux extends, and the effectiveness of clearing of the refluxed gastric juice from the esophagus. It can differentiate between reflux of air in belching and reflux of fluid. It cannot discriminate between the various chemical constituents in the refluxate, and does not quantitate the volume of reflux.

Impedance technology is particularly important in assessing reflux in patients who are on acid suppressive treatment (21). Even when their symptoms are controlled, patients continue to reflux gastric contents into the esophagus. In patients who have adequate acid suppression, such reflux episodes will not be detected by the 24-hour pH test because the pH of the refluxate is not acid. Impedance studies show that some patients who continue to have symptoms after effective acid suppression develop symptoms as a result of nonacid reflux episodes detected by impedance studies.

Conclusion: Impedance technology provides accurate information regarding the occurrence of reflux episodes irrespec-

tive of its chemical composition. It does not define reflux disease by any cellular abnormality.

Alkaline Reflux

The gastroesophageal refluxate contains many chemicals other than acid. These include endogenous secretory products of the stomach such as pepsin, duodenal contents such as the secretory products of the intestine, bile, and pancreas that frequently enter the stomach via duodeno-gastric reflux, products of bacteria that may colonize the stomach, and exogenous chemicals ingested as food. Because many of these alkaline substances are now considered important in the pathogenesis of reflux-induced damage, the pure determination of acid exposure is recognized as being insufficient.

Alkaline substances in the stomach neutralize the acid, bringing its pH toward normal and sometimes tilting to the alkaline range. When the pH electrode records a pH greater than 7, alkaline reflux can be diagnosed. However, when the pH is 7 or less, there is the possibility of alkaline reflux that is difficult to prove. The use of Bilitec probes, which detect chromogenic molecules in bile such as bilirubin, show that duodenal contents frequently enter the esophagus, producing complex changes in the pH of the refluxate that are difficult to assess by the tracing obtained with a pH electrode. If these alkaline molecules cause damage to the esophagus, their presence may not be adequately assessed by the 24-hour pH study. For the sake of argument, let us assume that there is a strongly injurious alkaline molecule in the gastric refluxate and this has neutralized the pH to within the normal range (5 to 7). Reflux will not be detected by the pH electrode, which is programmed to consider any pH less than 4 to be non-acid. The term "weak acid reflux" is being increasingly used for patients who have reflux episodes with the refluxate in the 4–7 pH range that falls neither into acid or alkaline reflux. In patients who are on acid suppressive drug therapy, weak acid reflux and damage to the esophagus by the injurious alkaline molecules continue despite a normal pH study. This can be best detected by a combined and simultaneous multichannel impedance and pH study.

The combination of continuing damage to the esophagus by nonacid components of the refluxate is particularly likely to be true in patients who have atrophic gastritis and patients who are on acid suppressive medications. In these patients, the pH study may not show acid reflux, while other injurious molecules in gastric juice continue to reflux and cause damage in the esophagus. If symptoms in reflux disease are primarily related to acid, these patients may be considered "cured" by antireflux treatment as their esophagus progresses toward adenocarcinoma.

Impedance studies have shown continuing reflux at significant volumes in patients who are on acid suppressive drugs indicating that the esophageal epithelium continues to be subject to other molecules in the gastric refluxate after acid secretion has been suppressed (21).

Conclusion: Alkaline components of reflux may produce ongoing esophageal injury in patients who are "cured" by acid suppression if cure is assessed as relief of symptoms and/or normalization of a 24-hour pH study.

PROPOSED NEW CRITERION FOR DEFINING GASTROESOPHAGEAL REFLUX DISEASE

Since the mid-1970s, the most common threat to life for patients with gastroesophageal reflux disease has become Barrett esophagus and adenocarcinoma. It therefore seems appropriate to make adenocarcinoma the end point in the diagnosis of reflux disease and seek a criterion in the reflux to adenocarcinoma sequence to define the disease. At present, the only accepted criterion is Barrett esophagus with intestinal metaplasia. We have shown that while it is an excellent marker of increased risk of adenocarcinoma, Barrett esophagus cannot be used to define reflux disease because of its low sensitivity.

In the previous two chapters, we described the cellular events of gastroesophageal reflux disease (Fig. 13.5, Table 13.1).

The transformation of normal squamous epithelium to Barrett esophagus occurs through a series of changes, often taking many years. The early steps in the series of changes that involve the squamous epithelium are relatively nonspecific, representing injury changes that are caused by many esophageal diseases (12). For example, dilated intercellular spaces and intraepithelial eosinophils are present in allergic (eosinophilic) esophagitis; chemokine release, increased permeability, and squamous epithelial genetic changes such as Cox2 expression are untested for specificity but are unlikely to become definitive criteria because they require complex testing outside of simple histologic examination of biopsies.

Table 13.1 shows that cardiac metaplasia, recognized by the presence of cardiac mucosa in a biopsy, is 100% specific for reflux disease. We have detailed the evidence in the literature that supports this viewpoint in Chapter 6 (Table 6.2). Cardiac mucosa results from a

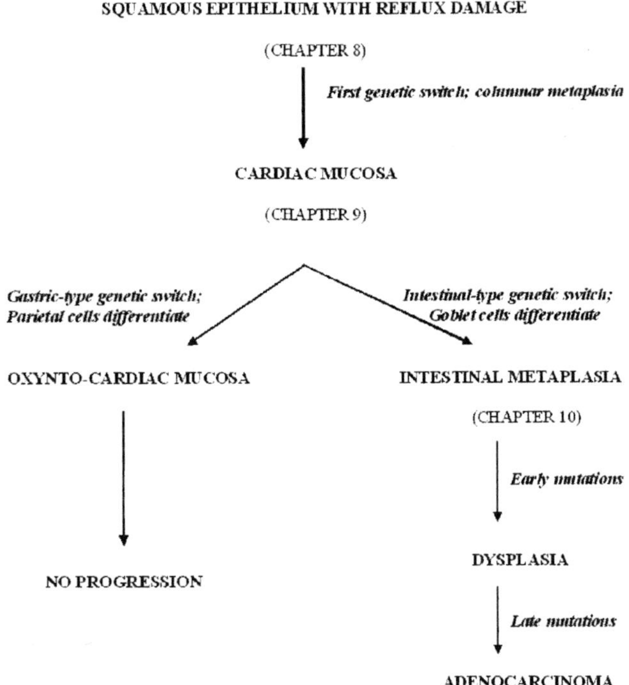

FIGURE 13.5 The reflux-adenocarcinoma sequence. The occurrence of columnar metaplasia of squamous epithelium requires a complex series of events that are specifically produced by the interaction of esophageal epithelia with molecules in the gastroesophageal refluxate.

genetic switch resulting from an interaction between molecules in the refluxate with cell receptors in adult stem cells in the basal region of the squamous epithelium (Fig. 13.5). The exact nature of this interaction is unknown. However, the resulting conversion of squamous epithelium to cardiac mucosa is absolutely specific for reflux disease. Cardiac metaplasia of squamous epithelium has not been described with any other esophageal disease.

The problem with using cardiac mucosa as a diagnostic criterion for gastroesophageal reflux disease in the past has been that cardiac mucosa has been regarded as a normal epithelium in the distal esophagus, proximal stomach, or both. Most textbooks of histology describe a normal extent of cardiac mucosa around 2 to 3 cm (22, 23). It is interesting that there is no primary reference in the literature for this claim. Initially localized to both the distal esophagus and proximal stomach (24), cardiac mucosa came to be regarded as a normal epithelium in the proximal stomach when the endoscopic definition of the normal condition demanded that the entire tubular esophagus be lined by squamous epithelium (25).

In the 1990s, we first suggested that cardiac mucosa was absent in normal people and that its presence at the squamocolumnar junction was a marker for reflux disease (26). Since that time, the amount of cardiac mucosa considered "normal" by the medical community has shrunk dramatically. The present belief is that a very small, probably around 1 mm segment of cardiac mucosa is normally present (6). In a review on the subject from Harvard, Odze (27) suggested that "the true gastric cardia" should be defined as a normal structure that extended between the proximal limit of the gastric rugal folds to the beginning of gastric oxyntic mucosa, and that its extent was less than 0.4 cm. According to Odze, the "true gastric cardia" so defined was composed of either cardiac mucosa or oxyntocardiac mucosa and could rarely be absent. The opinion of the medical community is inexorably tending toward the truth that cardiac mucosa is normally absent; the contrary opinion has shrunk to less than 0.4 cm with zero being considered possible.

As long as cardiac mucosa is regarded as a normal structure, it is not possible to use it as a criterion to define the pathologic entity of reflux disease. In Chapter 6, we reviewed the data in the literature that clearly show that cardiac mucosa does not normally exist in either the esophagus or in the stomach. The reader is referred back to this chapter, but the summary statement of this chapter is shown here as Table 13.3.

If cardiac mucosa is not a normal structure, its presence is 100% specific and highly sensitive for reflux disease, and it is the source of all metaplastic columnar epithelia in the esophagus, it is the perfect definition of gastroesophageal reflux. We have strongly recommended that it be so used (1).

While all metaplastic columnar epithelia are 100% specific for reflux disease, only two, cardiac mucosa and intestinal metaplasia, are in the reflux-adenocarcinoma sequence (Fig. 13.5). Oxyntocardiac mucosa is a benign epithelium that does not progress to intestinal metaplasia and adenocarcinoma (1). As such, the definition of reflux disease can be made more meaningful if the defining criterion is restricted to cardiac mucosa, with and without intestinal metaplasia. The presence of oxyntocardiac mucosa, though caused by and diagnostic of reflux disease, is irrelevant in terms of the adenocarcinoma that we are using as the end point of our definition. We have suggested the term "compensated reflux" for oxyntocardiac mucosa (1).

There are two other practical advantages for excluding oxyntocardiac mucosa from the definition of reflux disease:

1. Oxyntocardiac mucosa is almost universally present if the entire circumference of the gastroesophageal junction is examined (28), reflecting the almost universal presence of gastroesophageal reflux in

TABLE 13.3 Summary Statement Developed in Chapter 6 from Review of Evidence Relating to Cardiac Mucosa

1. Cardiac mucosa is not universally present at the squamocolumnar junction in fully developed humans (over 1 year old).
2. When present, cardiac mucosal length is less than 5 mm in the majority of patients, and often less than 1 mm.
3. The prevalence and length of cardiac mucosa tend to increase with age.
4. The prevalence of cardiac mucosa is lowest in normal autopsy populations (Chandrasoma et al.—44%) and biopsies obtained from a population of patients with a bias toward having a normal gastroesophageal junction at endoscopy (Jain et al.—35%).
5. The prevalence of cardiac mucosa is higher in populations where there are patients with clinical evidence of reflux. When endoscopy is normal, cardiac mucosa is present in 60% to 62% (Marsman et al., Chandrasoma et al.). When endoscopically visible columnar epithelium is present, the prevalence of cardiac mucosa approaches 100%.
6. The presence of cardiac mucosa at the junction, even when endoscopy is normal and the length of cardiac mucosa is in the 1-mm range, has a strong association with gastroesophageal reflux.
7. The length of cardiac mucosa has a direct correlation with the severity of reflux.
8. When evaluated by objective criteria for defining the gastroesophageal junction such as the peritoneal reflection, the angle of His, or the association with esophageal submucosal glands, cardiac mucosa is esophageal in location, not gastric.
9. In patients who have had esophagogastrectomy, cardiac mucosa develops by metaplasia of squamous epithelium above the anastomotic line. This columnar metaplasia correlates with the occurrence of duodenogastroesophageal reflux.

If one adds all the above into a conclusive statement, cardiac mucosa is an abnormal epithelium that occurs in the esophagus due to reflux-induced columnar metaplasia of the stratified squamous epithelium.

humans (28). As such, it will be too sensitive as a definitive criterion to have practical meaning. In contrast, cardiac mucosa is present in 50% or less of the general population. This is based on our autopsy study, where 44% of patients with complete sampling of the junction had cardiac mucosa (28), and Jain et al.'s study (29), which showed cardiac mucosal presence at the esophagogastric junction in 35% of their patients in endoscopic biopsies.

2. Oxyntocardiac mucosa can be difficult to differentiate from normal gastric oxyntic mucosa when it contains large numbers of parietal cells. As such, diagnostic error is possible. This is not the case with cardiac mucosa, where the absence of a single parietal cell makes it a highly reproducible and accurate diagnosis.

We therefore propose the definitions of gastroesophageal reflux disease that are displayed in Table 13.4.

REFLUX CARDITIS: A NEW ENTITY THAT DEFINES GASTROESOPHAGEAL REFLUX DISEASE WITH PERFECTION

When cardiac mucosa is found, it is always histologically abnormal. This was noted by the earliest researchers. Allison and Johnstone (30) described it as follows: "the rather villous type of cardiac mucosa, its lack of depth, and a diffuse fibrosis of the submucosa . . . suggest healing of previous shallow ulcerations" (p. 94).

Riddell, in a review article in 1996, wrote:

> Because of the increasing use of biopsies immediately on the squamous side of the Z-line . . . some specimens inevitably contain tissue from the cardiac mucosa immediately distal to the Z-line. Some of these have contained an excess of chronic and sometimes acute inflammatory changes,

TABLE 13.4 Definitional Value of Different Types of Metaplastic Columnar Epithelia in Gastroesophageal Reflux Disease

Definition	Sensitivity	Specificity	Risk of cancer
Cardiac mucosa (reflux carditis)[a]	High	100%	Very low[b]
Intestinal metaplasia (Barrett esophagus)	Low	100%	Low
Oxyntocardiac mucosa	100% (too sensitive)	100%	None

[a] The terms "cardiac mucosa" and "reflux carditis" are synonymous because cardiac mucosa is always inflamed (10, 19).

[b] Cardiac mucosa is not transformed to cancer directly by carcinogens. Its presence indicates that the patient is at risk of developing intestinal metaplasia and progressing in the reflux-adenocarcinoma sequence.

> which were not only unaccompanied by inflammatory changes elsewhere in the stomach, as might be expected with *Helicobacter pylori* gastritis, but inflammation appeared to be completely limited to the gastric cardia, and therefore was appropriately termed "gastric carditis." Preliminary studies suggest that this may be a more sensitive marker for GERD than inflammatory changes in the squamous mucosa, as judged by correlation with 24-h pH studies [Riddell cited the abstract from our unit by Clark et al. in 1994, which was the first suggestion that carditis was a marker for reflux disease]. This therefore indicates that . . . an additional biopsy be deliberately taken on the cardiac side of the Z-line. Moreover, in some patients with GERD this may be the only abnormal biopsy specimen (personal observation). (p. 532)

Riddell's observations are exactly the same as ours. The only difference is that he assigned a gastric location to the cardiac mucosa that is found immediately distal to the Z line. In effect, he used the squamocolumnar junction to define the gastroesophageal junction despite the fact that this was disproved in 1953. What Riddell termed *"gastric* carditis" is actually very unlikely to be gastric if it is an abnormality associated with gastroesophageal reflux disease. Again, to state the obvious: reflux causes abnormalities in the esophagus, not in the stomach. Riddell was describing carditis associated with reflux disease.

In a study by Der et al. of 141 consecutive patients with cardiac mucosa (reviewed later), there was not a single area of cardiac mucosa devoid of chronic inflammation. This was true even when the immediately adjacent oxyntic mucosa was completely normal without any inflammation. Cardiac mucosa without chronic inflammatory cells in the lamina propria does not exist in patients who have not had antireflux surgery. This has been proven true in all the many thousands of biopsies we have examined over the years.

There are two possible explanations for the invariable presence of chronic inflammatory cells in the lamina propria of cardiac mucosa: (a) lymphocytes, plasma cells, and eosinophils are normally present in cardiac mucosa, in a manner similar to duodenal and colonic lamina propria, or (b) it represents chronic inflammation and therefore carditis. Evidence for chronic inflammation rather than normal lamina propria lymphocytes is seen in Oberg et al.'s 1997 study (from our institution) (31), where the presence of any cardiac and oxyntocardiac mucosa, without consideration of inflammation, was shown to be associated with gastroesophageal reflux disease. In the study by Der et al. (19) (from our institution), reviewed in detail later, the degree of chronic inflammation in cardiac mucosa was shown to correlate with the severity of gastroesophageal reflux as assessed by a 24-hour pH test.

The presence of cardiac mucosa is therefore equivalent to carditis because cardiac mucosa is always inflamed. There is no "histologically normal" cardiac mucosa. When defined by histologic criteria, the presence of cardiac mucosa (= carditis) is associated with gastroesophageal reflux; the term "reflux carditis" is appropriate. Reflux carditis results from columnar metaplasia of the squamous epithelium of the esophagus followed by continued damage of the metaplastic columnar epithelium by the refluxate.

It is interesting to compare the degree of inflammation in cardiac mucosa with oxyntocardiac and oxyntic mucosa, which are found very close together, frequently at the same level in a biopsy sample from the region of the true gastroesophageal junction. In patients who have no gastritis, the inflammation in cardiac mucosa is always significantly greater than that in gastric oxyntic mucosa. In fact, the difference is frequently dramatic in its suddenness: gastric mucosa devoid of inflammation immediately adjacent to severely inflamed cardiac mucosa in biopsies taken at the same level. In general, inflammation in oxyntocardiac mucosa is mild, less than coexisting cardiac mucosa and more than oxyntic mucosa (unless gastritis is also present). The amount of inflammation in oxyntocardiac mucosa seems to correlate with the number of parietal cells: the greater the number of parietal cells, the lesser the inflammation.

Definition of Reflux Carditis

At the University of Southern California, we have used the diagnostic term "reflux carditis" since the 1990s. When defined in the manner we define it, it is a highly sensitive and specific histologic marker for gastroesophageal reflux (1). The presence of reflux carditis indicates that gastroesophageal reflux has been of sufficient severity during the life of the patient to have caused the squamous epithelial injury to result in columnar metaplasia.

Reflux carditis is defined as the finding of cardiac mucosa between the squamocolumnar junction and gastric oxyntic mucosa (Fig. 13.6). It is not necessary to say "inflamed cardiac mucosa" because cardiac mucosa is always inflamed. These biopsies may come to the pathologist labeled as "esophagus," "gastroesophageal junction," "gastric cardia," "proximal stomach," or "1 to 3 cm distal to the gastroesophageal junction." The endoscopic landmarks are superceded by the histologic finding of cardiac mucosa. In Chapter 11, we showed how anatomic changes in the distal esophagus that is damaged by reflux result in misinterpretation of the exact location of tissue samples at endoscopy and gross examination.

Our preferred term for this entity is not "reflux carditis." Because cardiac mucosa is always esophageal, we believe that the use of the term "carditis" results in confusion with gastric pathology as long as the term "gastric cardia" exists to describe the proximal stomach. We would prefer the term "reflux esophagitis, cardiac mucosal type" emphasizing that this is esophageal pathology. However, initial attempts to use this term resulted in such virulent antagonism that it led us to compromise back to "reflux carditis." It is our hope that when the dust clears and the concepts presented here are accepted, then the entity that we now describe as "reflux carditis" will be recognized simply as a criterion of reflux disease and be called "reflux esophagitis, cardiac mucosal type" distinct from reflux esophagitis involving the squamous epithelium.

Histologic Features of Reflux Carditis

The histologic features of reflux carditis are those of cardiac mucosa: it is an epithelium in the distal esophagus (including dilated end-stage esophagus) consisting of mucous cells without parietal or goblet cells (Fig. 13.6). The mucous cells may line a flat surface, a thin epithelium with rudimentary foveolar pits, or an epithelium that contains glands (see Chapter 9). The foveolar region commonly shows reactive features such as elongation, serration, and hyperdistension of the cells with mucin ("pseudogoblet cells") (Figs. 13.7 to 13.10). The glands are lobulated and often vary greatly in size and form (Fig. 13.10). There is variable, often severe chronic inflammation in the lamina propria with lymphocytes, plasma cells, and eosinophils (Figs. 13.11 and 13.12). Neutrophils are uncommon and indicate secondary pathology such as erosion (Figs. 13.13 and 13.14) or secondary infection in patients who have a coexisting *H. pylori* gastritis (Fig. 13.15). The lamina propria often shows fibrosis and the muscularis mucosae is hyperplastic, extending into the lamina propria as vertical smooth muscle fibers (Fig. 13.16). The severity of the chronic inflammation and foveolar hyperplasia correlates with the severity of reflux. Erosion of the surface with granulation tissue may be seen (Fig. 13.14).

It is incredible that pathologists over the years have seen these obviously pathologic changes in cardiac mucosa and either completely ignored them or regarded them as being normal. The term "carditis"

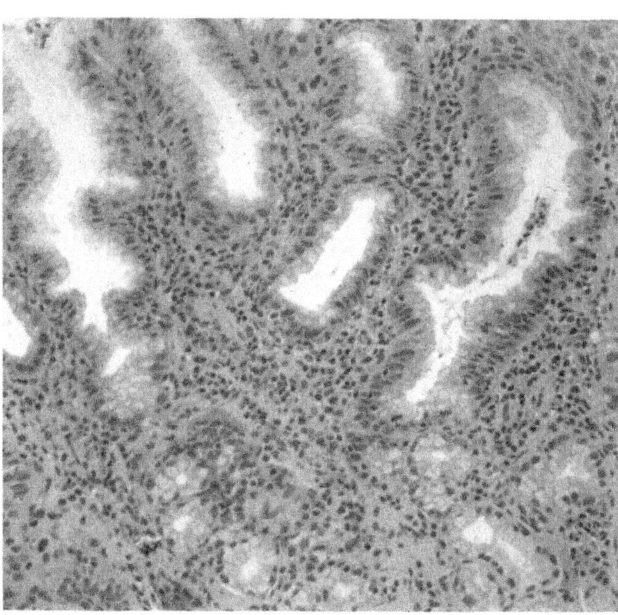

FIGURE 13.7 Cardiac mucosa with markedly elongated and serrated foveolar region with marked chronic inflammation. (Cross reference: Color Plate 3.11)

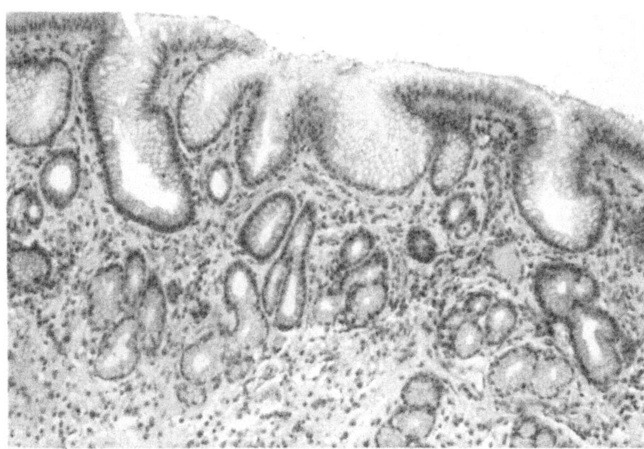

FIGURE 13.6 Cardiac mucosa showing a glandular epithelium consisting of mucous cells only lining the surface, foveolar region, and irregular glands. There is mild chronic inflammation in the lamina propria. No parietal or goblet cells are present.

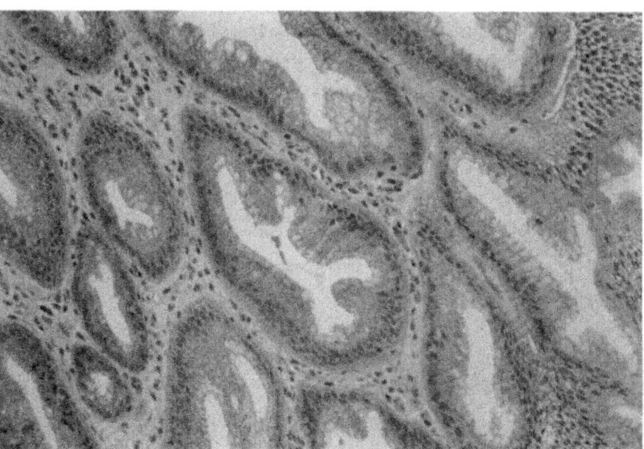

FIGURE 13.8 Tangentially sectioned foveolar region of cardiac mucosa showing luminal serration resembling the changes seen in a hyperplastic polyp.

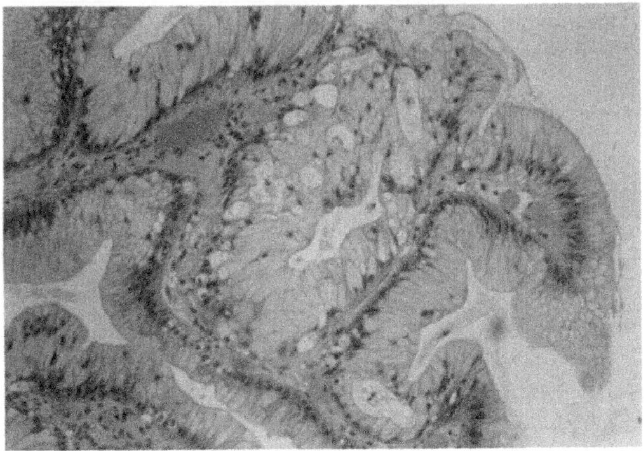

FIGURE 13.9 Mucin distension of cytoplasm of cardiac mucosal cells ("pseudogoblet cells").

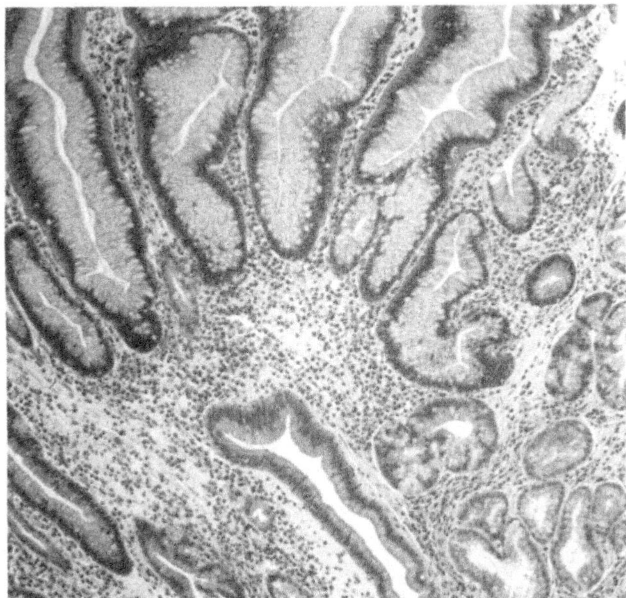

FIGURE 13.11 Cardiac mucosa showing severe chronic inflammation and foveolar hyperplasia.

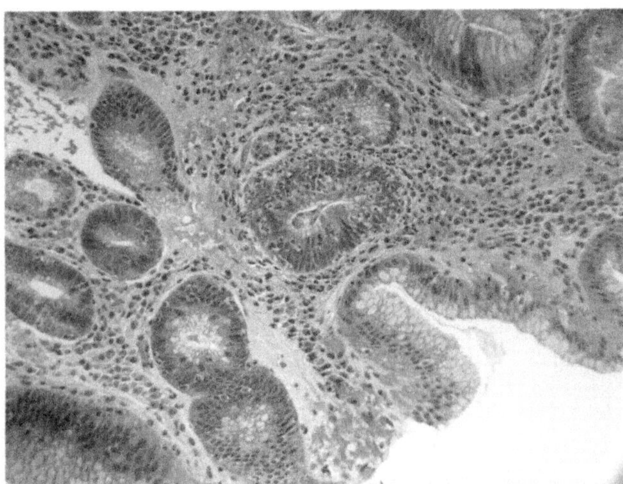

FIGURE 13.13 Cardiac mucosa showing focal neutrophil infiltration of lamina propria and a foveolar-gland complex. This was associated with *H. pylori* pangastritis.

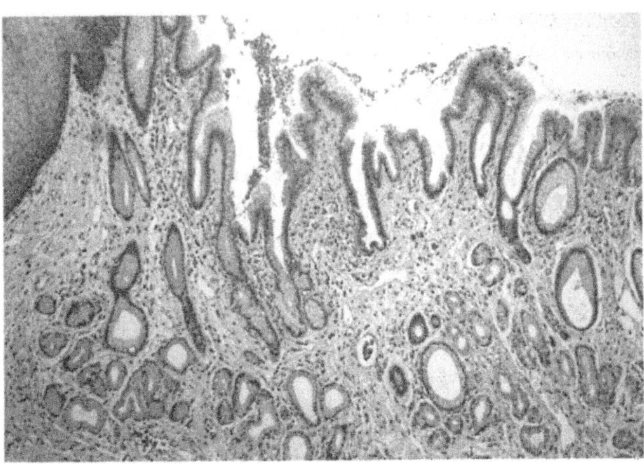

FIGURE 13.10 Cardiac mucosa showing irregularly distributed mucous glands with a suggestion of lobulation under the hyperplastic foveolar region. There is marked chronic inflammation, fibrosis, and smooth muscle in the lamina propria. (Cross reference: Color Plate 3.10)

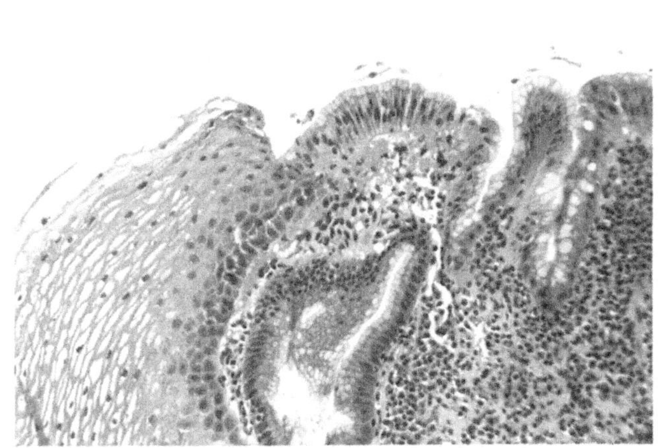

FIGURE 13.12 Squamocolumnar junction showing cardiac mucosa with severe chronic inflammation consisting of lymphocytes, plasma cells, and eosinophils. (Cross reference: Color Plate 3.7)

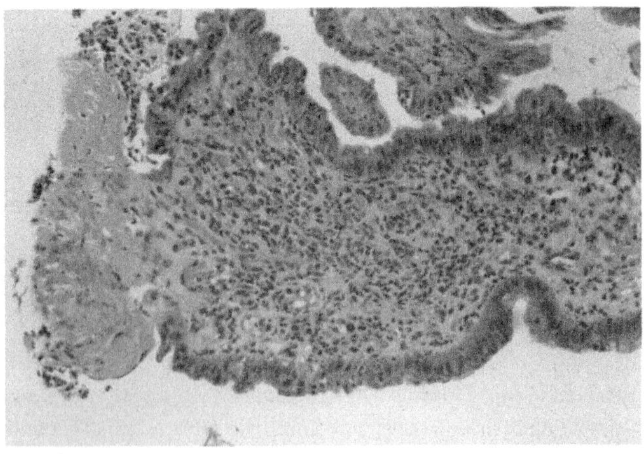

FIGURE 13.14 Cardiac mucosa showing focal surface erosion.

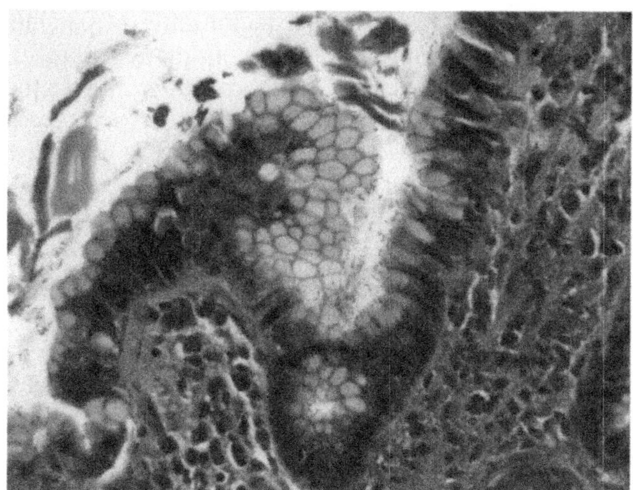

FIGURE 13.15 Giemsa stain showing numerous *H. pylori* in cardiac mucosa. The presence of *H. pylori* in cardiac mucosa is a secondary phenomenon rather than an etiologic relationship. *H. pylori* infects cardiac mucosa only in those people who have cardiac mucosa as a result of reflux disease; it does not cause cardiac mucosa to develop.

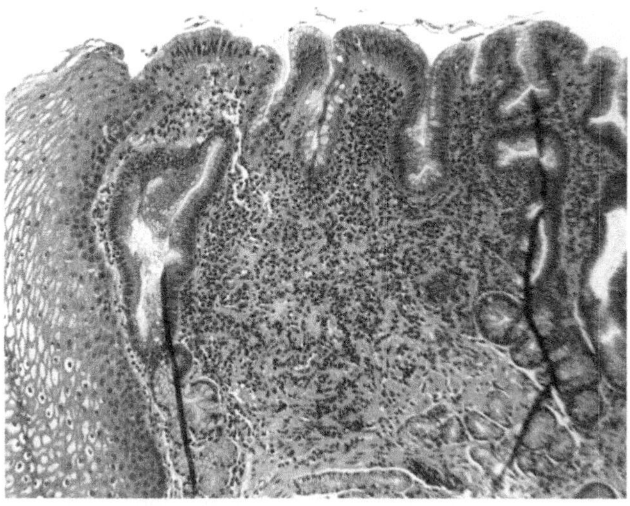

FIGURE 13.16 Cardiac mucosa showing marked chronic inflammation and hyperplasia and irregularity of the muscularis mucosae.

was not encountered in the literature before we suggested that it may be a marker for gastroesophageal reflux disease (26). It still does not commonly appear in pathology reports. We have used this term as a diagnostic criterion of reflux disease at the University of Southern California since the 1990s based on evidence that we have that the presence of this histologic feature has a high correlation with an abnormal 24-hour pH test (31).

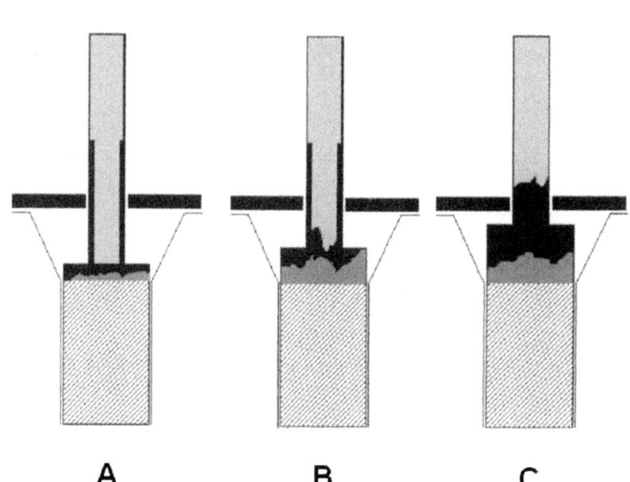

FIGURE 13.17 Appearance of columnar-lined esophagus without intestinal metaplasia with progressive elongation of the segment (black = cardiac mucosa; stippled = oxyntocardiac mucosa). (a) Reflux disease limited to end-stage dilated esophagus. (b) Short-segment (<2 cm) columnar-lined esophagus visible as columnar metaplasia in the distal tubular esophagus. (c) Long-segment (>2 cm) columnar-lined esophagus.

Gross Pathologic and Endoscopic Features of Reflux Carditis

Grossly and endoscopically, reflux carditis may or may not be visible. When it is invisible, it is present in rugated epithelium distal to the squamocolumnar junction in dilated end-stage esophagus (Fig. 13.17) (see Chapter 12). When visible, it is seen as a flat epithelium between the proximal limit of the rugal folds and the squamocolumnar junction. It may form tongues extending into the squamous epithelium or be circumferential in the tubular esophagus (Fig. 13.17). It may be interposed between the Z line at end of the tubular esophagus and the proximal limit of the rugal folds, occupying the most proximal part of the saccular region distal to the tubular esophagus, which is the dilated end-stage esophagus. In rare cases, the reactive cardiac mucosa forms small inflamed polypoid lesions. Reflux carditis is the histologic finding associated with the majority of non-neoplastic polyps found in the distal esophagus.

Visible reflux carditis is equivalent to endoscopic columnar-lined esophagus. Many gastroenterologists make the mistake of equating columnar-lined esophagus with Barrett esophagus at endoscopy. Endoscopy has no capability to differentiate the three epithelia—intestinal, cardiac, and oxyntocardiac—that constitute the histologic spectrum of columnar-lined esophagus.

It is only when biopsies show intestinal metaplasia in any part of the columnar-lined esophagus that the diagnosis of Barrett esophagus can be made. In patients with intestinal metaplasia (and therefore Barrett esophagus), the entire columnar-lined segment of the esophagus often does not have intestinal metaplasia (32, 33). In these cases, the nonintestinalized columnar-lined esophagus consists of reflux carditis and oxyntocardiac mucosa (34). In patients without intestinal metaplasia, the columnar-lined esophagus consists only of reflux carditis and oxyntocardiac mucosa. The prevalence of intestinal metaplasia increases as the length of columnar-lined esophagus increases, approaching 90% when the length is 3 cm and nearly 100% when the length is 5 cm (34).

To make a diagnosis of reflux carditis, biopsies must be taken. When there is any deviation from normal endoscopy such as a visible flat columnar epithelium in the tubular esophagus, tongues of columnar epithelium in the distal squamous epithelium, an irregular Z line, or flat epithelium distal to the squamocolumnar junction before rugal folds appear, the probability of finding reflux carditis in a biopsy is high. These represent the subtle and at present unrecognized endoscopic features of reflux carditis.

Reflux carditis can also be invisible to the endoscope. This occurs distal to the squamocolumnar junction in a patient who is endoscopically normal. The area involved by reflux carditis may be so small that its presence between the proximal limit of the rugal folds and squamocolumnar junction is undetected because of the limited resolution capacity of the endoscope. Depending on the observer, it is likely that 5 to 10 mm of flat columnar epithelium between the squamocolumnar junction and the proximal limit of the rugal folds may be interpreted as normal. It is possible that the concept of endoscopic normalcy will change when reflux carditis becomes recognized as a criterion of diagnosis of reflux disease. At present, endoscopists concentrate on the squamous epithelium when they look for evidence of minimal reflux; when they recognize that the changes are distal to the squamous epithelium, new and more sensitive endoscopic criteria for the diagnosis of reflux may emerge. Reflux carditis may also be present under epithelium that shows rugal folds representing the technical error in the endoscopic definition of the gastroesophageal junction as the proximal limit of the rugal folds. This error can be over 2 cm (see Chapter 11).

Clinical Features of Reflux Carditis

Patients with reflux carditis may or may not have symptoms of reflux. At the present time, gastroenterologists take biopsies of this region only in patients with a clinical suspicion of reflux. In many instances, therefore, the diagnosis of reflux carditis confirms the clinical suspicion of reflux. *We have never understood why so many academic clinicians resist having a sensitive histologic criterion that confirms their clinical suspicion. All the gastroenterologists we report biopsies for are extremely happy when we confirm their clinical suspicion of reflux with the diagnosis of reflux carditis. It is infinitely better than "no significant pathologic lesion."*

A significant number of patients who are asymptomatic will have cardiac mucosa in their biopsies. In our autopsy study, it was 44% (28). It is clear that many patients with significant reflux disease are asymptomatic or have mild and atypical symptoms that do not fall under the label of "classical symptomatic reflux." This is a group of patients who are very important because the first time they present with symptoms of any kind are when they develop serious complications of reflux disease. Today, this is commonly advanced-stage adenocarcinoma of the esophagus when the patient develops dysphagia. Also being recognized in this group are extraesophageal manifestations such as laryngitis, asthma, chronic cough, and idiopathic pulmonary fibrosis (35).

Reflux carditis is valuable in the diagnosis of atypically symptomatic reflux disease because of its high sensitivity as a criterion for reflux disease. If an atypically symptomatic patient does not have reflux carditis in an appropriate biopsy, it is highly unlikely that the atypical symptoms are caused by reflux. However, the presence of reflux carditis does not prove whether the atypical symptoms are caused by or simply coincident with reflux disease.

The presence of reflux carditis is not associated with any known risk of carcinoma; however, it is the earliest specific step in the reflux-adenocarcinoma sequence (Fig. 13.5). Patients with cardiac mucosa have an epithelium that is at risk for developing intestinal metaplasia and progressing in the reflux-adenocarcinoma sequence. The greater value of this diagnostic criterion is that the absence of both cardiac mucosa and intestinal metaplasia can be used to reassure the patient that there is little danger of adenocarcinoma. This is a significant population (approximately 50% of people). If a 40-year-old patient has no reflux carditis in an adequately sampled junction at endoscopy, it is highly unlikely that cardiac mucosa will develop and progress through intestinal metaplasia to cancer during his or her lifetime. Cameron (36) showed that the final length of columnar-lined esophagus is established relatively quickly and changes little after the fourth decade.

The objectives for the proposed definition of gastroesophageal reflux disease in the Genval Workshop report (2) have been met by the criterion of reflux carditis. The presence of reflux carditis in a biopsy *includes all people (symptomatic or not) who are at risk for developing adenocarcinoma.* The absence of reflux carditis in an appropriate biopsy set (appropriate = no sampling error) *can be used to reassure people (symptomatic or not) that they are not at risk for developing adenocarcinoma.*

Reflux Carditis and the 24-Hour pH Test

Most patients with reflux carditis have an abnormal 24-hour pH study. In Oberg et al. (31), the presence of cardiac or oxyntocardiac mucosa in patients who were endoscopically normal was predictive of the greater likelihood of an abnormal 24-hour pH test compared with patients who did not have these mucosal types.

However, the presence of reflux carditis is a more sensitive criterion for reflux disease than an abnormal 24-hour pH study, because the level of acid exposure in the 24-hour pH test that was used to define abnormal reflux was calibrated to the presence of symptoms (37). Therefore, it should not be surprising that patients with asymptomatic reflux disease can have reflux carditis but have a normal 24-hour pH test.

In the study from our unit by Der et al. (19) of 141 patients with reflux carditis on biopsy, a 24-hour pH study was available in 105 patients. Of these, 64 (61%) had an abnormal test defined as a pH of less than 4 for greater than 4.5% of the 24-hour period, 41 of 105 (39%) had a pH study that was within the normal range. Among these patients, the lowest percentage time for the pH less than 4 was 2.1%. This means that the 31 patients who had normal 24-hour pH studies had an acid exposure of between 30 and 60 minutes per day in the esophagus. While this may or may not cause symptoms, it should not be surprising that it causes cellular damage to the squamous epithelium leading to cardiac metaplasia.

It should be also noted that the pH electrode is located in the esophageal body 5 cm above the upper limit of the lower esophageal sphincter. This is approximately 8 to 10 cm above the gastroesophageal junction, given a sphincter length of 3 to 5 cm. It should not be a surprise that there can be reflux-induced changes in the most distal few millimeters of the esophagus when the pH study is normal. We have suggested that the most distal few millimeters of the esophageal squamous can be damaged in the absence of free reflux when it becomes exposed to gastric luminal contents during periods of gastric overdistension and in patients who have short segments of dilated end-stage esophagus resulting from sphincter damage. These are the bases for patients who have reflux carditis with a normal 24-hour pH test. The new concept of "reflux disease limited to a dilated end-stage esophagus" that we developed in Chapter 12 is defined by the presence of reflux carditis and a negative 24-hour pH test.

Significance of Reflux Carditis

Reflux carditis is a 100% specific histologic marker for gastroesophageal reflux. No other etiologic agent is known to produce this change. Reflux carditis is infinitely more specific than present squamous epithelial criteria for the diagnosis of gastroesophageal reflux. Reflux carditis is much more sensitive than Barrett esophagus for the diagnosis of gastroesophageal reflux. It is less sensitive than the presence of oxyntocardiac mucosa.

The sensitivity of reflux carditis for the diagnosis of reflux disease is less than 100%. This is due to several factors:

1. *True false negative, type 1 (patients with reflux disease who do not have any metaplastic columnar epithelia).* True false negatives may occur in the very early stage of reflux disease where reflux has caused changes in squamous epithelium without causing columnar metaplasia. This is rare and is most likely restricted to young children at the onset of their lifelong reflux disease; most adult patients with evidence of esophagitis in the squamous epithelium will have reflux carditis. Many patients with reflux carditis have normal residual squamous epithelia. The sensitivity of carditis for the diagnosis of reflux disease is far greater than the sensitivity of squamous epithelial changes recognized at present as criteria for reflux disease.
2. *True false negative, type 2 (patients with reflux disease who have only oxyntocardiac mucosa as a columnar metaplastic type).* This occurs when all the cardiac mucosa resulting from columnar metaplasia converts to oxyntocardiac mucosa. This is a common occurrence in the asymptomatic population. In our autopsy study (28), small amounts of oxyntocardiac mucosa were present in some part of the circumference of the squamocolumnar junction in the 56% of patients who did not have cardiac mucosa. In Marsman et al. (38), oxyntocardiac mucosa was found distal to the squamocolumnar junction in the 38% of patients who did not have cardiac mucosa. Patients who have only oxyntocardiac mucosa generally have less than 1 cm of columnar-lined esophagus (34) and are likely to be endoscopically normal. When the columnar epithelial length reaches 1 cm, there is almost always some cardiac mucosa in the columnar-lined esophagus (34).
3. *Sampling error (where inadequate sampling fails to biopsy cardiac mucosa that is present in the patient).* In evalu-

ating sensitivity, it is important to recognize that there is a significant danger of sampling error by endoscopic biopsy (these are false negatives resulting from inadequate sampling). The absence of cardiac mucosa in a biopsy does not necessarily mean its absence in the patient; the greater the sampling, the greater the level of confidence that absence of cardiac mucosa is a true negative result. Biopsies should always try to straddle the squamocolumnar junction. If the actual junction is present in a single tissue sample and the squamous epithelium transitions to oxyntocardiac or oxyntic mucosa, it is certain that cardiac mucosa is absent, at least in the biopsied part of the circumference.

Differential Diagnosis of Reflux Carditis

There is no confusion or difficulty with the diagnosis of reflux carditis. The diagnosis is accurate, precise, and reproducible. It is the presence of a columnar epithelium in the esophagus, including the proximal 3 cm of the "endoscopic proximal stomach" that is composed only of mucous cells; there are no parietal cells or goblet cells. Paneth cells, neuroendocrine cells, and pancreatic (serous) cells are permissible. Problems may arise when only part of the mucosa is available in a biopsy or the biopsy is tangentially sectioned (see Chapter 4).

Carditis resembles inflamed antral mucosa, but this should not be a problem as long as biopsy locations are labeled accurately. There is little chance that an endoscopist will mistake antral location for dilated end-stage esophagus.

To make a diagnosis of reflux carditis, one must accept the concept that there is no "normal" cardiac mucosa in the proximal stomach. Until that happens, the diagnosis of reflux carditis cannot be made because the histologic features of reflux carditis that occur as columnar metaplasia in what everyone recognizes as the esophagus (i.e., above the endoscopic gastroesophageal junction) are identical to those seen in "normal" cardiac mucosa (27). The pathologist who believes that a "normal" gastric cardia lined by "normal" cardiac mucosa exists is entirely dependent on the endoscopist to tell him or her that the biopsy is from the esophagus and not from the stomach.

The only condition that may theoretically cause confusion with reflux carditis is atrophic gastritis that complicates *H. pylori* gastritis and autoimmune gastritis. Atrophic gastritis can involve the entire stomach in either a diffuse or multifocal distribution and can therefore involve the most proximal part of the stomach. When oxyntic mucosa undergoes atrophy, parietal cells decrease and can completely disappear; the mucosa then consists of a foveolar region composed of mucous cells only (pseudopyloric metaplasia). Atrophic gastritis can therefore theoretically result in an inflamed "mucous cell only" epithelium. This epithelium is seen throughout the stomach in these patients; just because the most proximal part of it involves the region immediately distal to the gastroesophageal junction does not mean that it is any different than the similar mucosa in the rest of the stomach. This is an uncommon occurrence. In practice, atrophic gastritis sufficient to result in complete absence of parietal cells commonly has intestinal metaplasia (Fig. 13.18). The more difficult and significant distinction is between cardiac mucosa with intestinal metaplasia (Barrett esophagus) and atrophic gastritis, which we will consider in the next chapter under the diagnosis of Barrett esophagus. The presence of a mucous-cell-only epithelium without intestinal metaplasia in this region is essentially diagnostic of reflux carditis.

The Present State of the Art

The state of the art with regard to this area is pathetic. Despite all the data relating to the fact that cardiac mucosa is frequently absent and that its presence correlates with reflux disease, most gastroenterologists do not comprehend cardiac mucosa. To them there are three entities: squamous esophagus, which is damaged by reflux, normal gastric mucosa including cardiac mucosa, and Barrett esophagus (intestinal metaplasia). Review of the best research shows this

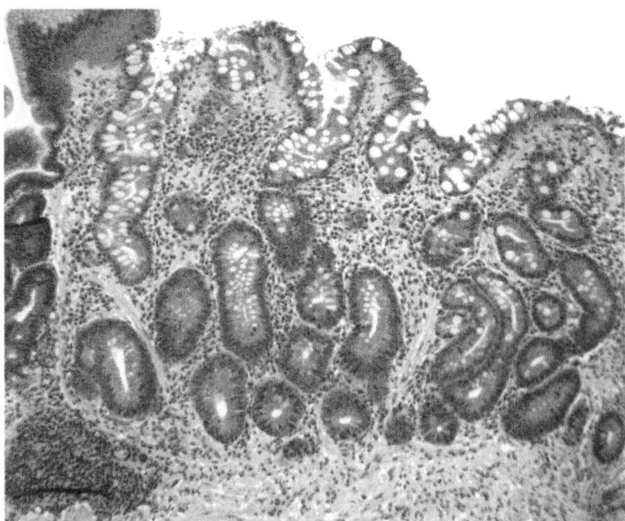

FIGURE 13.18 Atrophic gastritis showing flat mucosal surface with extensive intestinal metaplasia. The foveolar region shows no serration or hyperplastic features, and the muscularis mucosa is a flat layer at the base of the mucosa.

repeatedly; there is never any mention of histology except the recognition of intestinal metaplasia. Rex et al. (14) presented an excellent screening study to evaluate the prevalence of Barrett esophagus. In this study, 164 of 961 patients had columnar mucosa 5 to 30 mm in length in the esophagus; 53 of 164 of these (32%) had intestinal metaplasia. This was in the form of tongues in 48 and circumferential short-segment Barrett esophagus in 5 patients. The researchers reported that 111 of 164 (68%) of patients with "endoscopic short-segment columnar-lined esophagus" did not have intestinal metaplasia. These 111 patients were not considered any further in the study; their histology, clinical features, or demographics were not mentioned. They were ignored or considered normal. If intestinal metaplasia is not present, the thinking seems to be, no histologic finding has any significance. It is so irrelevant that it is not even worth recording in studies.

The ignored 111 patients (68%) with tongues of columnar mucosa 5 to 30 mm long that are devoid of intestinal metaplasia will all have metaplastic columnar epithelium in the form of cardiac and oxyntocardiac mucosa. This ignored histology is the best diagnostic criterion of reflux disease. The only reason why it is ignored is because gastroenterologists equate cardiac mucosa to normal gastric mucosa *even when it is found in what they have defined endoscopically as being above the gastroesophageal junction and even when it is the only change seen in a patient who has symptoms of reflux disease!*

In their discussion, despite the fact that the authors did not report the clinical findings in these 111 patients without intestinal metaplasia, the authors stated: "The absence of association of heartburn with Barrett esophagus overall was *the result of a prevalence of heartburn among those with SSBE [short-segment Barrett esophagus] in "tongues" of columnar mucosa that was no different than those without BE*" (14, p. 1674). The authors had evidence that tongues of columnar mucosa have an equal significance in terms of reflux symptoms irrespective of whether or not intestinal metaplasia is present. This supports our viewpoint that the severity of reflux correlates best with the length of columnar metaplasia of squamous epithelium and not the occurrence of intestinal metaplasia.

It is only the lack of understanding of histology that prevents these gastroenterologists from recognizing the obvious: In reflux disease, the squamous epithelium has undergone cardiac metaplasia, producing first tongues and then circumferential columnar epithelium in the distal esophagus. The amount of this metaplastic change, expressed as length, is related to the severity of cumulative reflux damage. The primary metaplastic epithelium arising from squamous epithelium is cardiac mucosa; short-segment columnar-lined esophagus with cardiac mucosa always precedes Barrett esophagus, which is the occurrence of intestinal metaplasia in this reflux-damaged metaplastic cardiac epithelium.

The present state of the art completely ignores metaplastic cardiac epithelium as a criterion for reflux disease even though it has been always accepted as a manifestation of reflux disease. We are staring at the best diagnostic criterion of reflux disease in every biopsy we take and calling it "normal cardiac mucosa." When this disease is unraveled, this will rank as one of the least intelligent achievements of physicians in the past century.

Review of Literature on Carditis

The literature relating to carditis is controversial, mainly because the topic is changing rapidly and many articles use information that is outdated by the time they reach publication. Only since about 2005 has it become generally recognized that cardiac mucosa is either absent (we presented evidence in Chapter 6 that this is the correct interpretation) or present only in very small amounts (around 1 to 4 mm; the present belief held by most authorities (27). The nature of this small amount of cardiac mucosa is also controversial; we believe that cardiac mucosa is an esophageal columnar epithelium resulting from reflux-induced metaplasia of squamous epithelium. We have presented evidence that the alternate opinion that cardiac mucosa is a normal or native gastric epithelium is no longer tenable.

Based on either of these viewpoints, it is no longer reasonable to take a biopsy 0.5 cm distal to the endoscopic gastroesophageal junction junction (the most common distance used to define the cardia) and regard cardiac mucosa in that biopsy as normal. When recognizable cardiac mucosa is present in the 3 to 4 mm of mucosal biopsy, it is obvious that the amount of cardiac mucosa very likely exceeds what is now accepted as normal (1 to 4 mm) in the majority if not all cases. In fact, Marsman et al. (38) showed that if the biopsy does not contain the squamocolumnar junction, it is very likely to miss any cardiac mucosa that may be present.

The study of carditis is a confusing problem that requires precise methodology to elucidate its cause, for the following reasons:

1. Cardiac mucosa is an abnormal esophageal columnar epithelium resulting from reflux-induced columnar metaplasia of the squamous epithelium.

2. Reflux-induced damage of the cardiac mucosa includes reactive changes (villiform surface, foveolar hyperplasia, smooth muscle proliferation) and *chronic* inflammation. Reflux does not cause acute inflammation in cardiac mucosa (19). The use of the Sydney classification to grade carditis is inappropriate in this setting. The Sydney classification was designed for gastritis where *H. pylori* causes both acute and chronic inflammation.
3. Most patients with cardiac mucosa have isolated inflammation of cardiac mucosa. Gastric oxyntic mucosa is normal, even when found immediately adjacent to inflamed cardiac mucosa. "Carditis" is therefore commonly an isolated pathologic abnormality unassociated with gastritis.
4. In patients without cardiac mucosa (and therefore without evidence of reflux disease), *H. pylori* gastritis frequently results in a pangastritis; the gastric oxyntic mucosa immediately distal to squamous epithelium undergoes inflammation. This is gastritis; it is not carditis. However, if carditis is defined by the presence of inflammation in a biopsy from endoscopically defined "cardia," these cases of pangastritis will fall within that definition.
5. Patients with reflux carditis who develop *H. pylori* gastritis frequently show extension of the infection from the stomach into the cardiac mucosa (Fig. 13.15). The presence of *H. pylori* in cardiac mucosa does not mean that there is an etiologic relationship; it is a secondary infection. When cardiac mucosa is secondarily infected with *H. pylori*, the inflammation is aggravated; acute inflammation appears and the severity of the chronic inflammation increases (Fig. 13.13). Use of the Sydney classification in carditis will always show a positive correlation between *H. pylori* and grade of inflammation because reflux does not cause acute inflammation.
6. Reflux disease, and therefore carditis, and *H. pylori* infection of the stomach are common diseases. Therefore, they can coexist in the same patient.

In evaluating articles relating to carditis, the important questions to be answered are the following:

1. What is the definition of carditis? If the definition is not inflammation of histologically defined cardiac mucosa, the data tend to be confusing and of little value (Table 13.5).
2. Does the study separately evaluate chronic and acute inflammation? If the study uses the Sydney classification where the presence of acute inflammation defines maximum severity, the data will produce confusing results if the grade of inflammation is correlated with carditis.
3. What is the absolute (not percentage) prevalence of *H. pylori* infection and associated carditis? If *H. pylori* has a significant correlation with carditis, the prevalence of *H. pylori* must be much higher than in patients without carditis. If the prevalence of *H. pylori* is similar in patients with and without carditis, it is highly unlikely that *H. pylori* causes carditis.

Of these questions, the one of overwhelming importance is the definition of carditis. We will first consider studies where the authors define carditis by histologic criteria as "inflammation of cardiac mucosa"; these studies produce data that can be evaluated even if the interpretations and conclusions are incorrect. The second group consists of studies where the authors

TABLE 13.5 Definition of "Carditis" in Selected Articles (all reviewed in detail in the text)

Reference	Definition of carditis	Anatomic location of biopsy
Spechler et al. (3)	Inflammation in gastric cardiac-type mucosa (histologic)	Not recorded
Der et al. (19)	Inflammation in cardiac mucosa (histologic)	Not recorded
Glickman et al. (6)	Inflammation in mucosa histologically next to squamous epithelium	Squamocolumnar junction; no visible columnar-lined esophagus at endoscopy
Goldblum et al. (5)	Inflammation in a biopsy within 0.5 cm distal to SCJ	Endoscopy normal; SCJ not separated from proximal limit of rugal folds
Wieczorek et al. (44)	Inflammation in biopsy immediately distal to squamous epithelium	No endoscopic evidence of columnar-lined esophagus
Lembo et al. (45)	Inflammation in columnar epithelium adjacent to Z line scored >2 by complex system	Z line in patients whose Z line was contiguous with the lower esophageal sphincter
Polkowski et al. (46)	Inflammation restricted to cardiac mucosa in proximal 2 cm of stomach with rest of stomach being normal	Esophagectomy specimens; no columnar-lined esophagus

define carditis anatomically as "the presence of inflammation in a biopsy taken from the anatomically defined cardia." These studies either completely ignore histology or use histology as a secondary criterion. The data in these studies are generally uninterpretable. It is a general rule that the first group will show a relationship between carditis and reflux and the second group will show a relationship between carditis and *H. pylori* or both *H. pylori* and reflux.

Carditis Defined by Histologic Criteria

We would like to consider one abstract by Spechler et al. from 1997 (39) to illustrate the importance of precise methodology. This abstract is chosen because we have a great deal of confidence in the data produced by this group and there is a subtle but very important discrepancy between the data and the authors' interpretation.

Literature Review

Spechler SJ, Wang HH, Chen YY, Zeroogian JM, Antonioli DA, Goyal RK. GERD vs. *H. pylori* infections as potential causes of inflammation in the gastric cardia. Gastroenterology 1997;112:A297.

The aim of this study was to evaluate the potential contribution of gastroesophageal reflux disease and *H. pylori* infection in carditis (defined as "gastric cardiac-type mucosa"; this study was from the Harvard laboratory supervised by Dr. Goyal which is one laboratory where one can be sure that this was indeed cardiac mucosa). All adult patients having elective endoscopy over a 6-month period were invited to participate, irrespective of the indication for endoscopy. Symptoms of GERD (defined as heartburn ⩾1 day/wk), endoscopic esophagitis (⩾grade 1), histologic esophagitis (no definition given) and histologic evidence of *H. pylori* infection were correlated with the grade of carditis on a scale of 0 to 3 using the updated Sydney system. Of the 156 patients in the study, 116 (74%) had cardiac-type mucosa. The correlation between grade of carditis and other features evaluated were as follows:

Grade of carditis	0 (n = 9)	1 (n = 58)	2 (n = 35)	3 (n = 14)
Heartburn	43%	47%	38%	23%
Endoscopic esophagitis	38%	38%	38%	8%
Histologic esophagitis	50%	42%	51%	43%
H. pylori infection	0%	9% (n = 5)	31% (n = 11)	57% (n = 8)

The only statistically significant correlation was that found between grade of carditis and the prevalence of *H. pylori* infection.

The authors reached the following conclusion: "Carditis was associated significantly with *H. pylori* infection, but not with symptoms or signs of GERD. These findings do not support an important role for GERD in the pathogenesis of inflammation in the gastric cardia" (p. A297).

Let us ask the three important questions outlined earlier. The definition of carditis is accurate; it is defined by histologic criteria by a highly reliable group of workers. The most telling datum in this study is the number of patients who had evidence of *H. pylori* infection—a total of 24 of the 116 patients (21%). This is within the range of *H. pylori* infection in many general hospital populations and suggests that the *H. pylori* prevalence in this group of patients with carditis is not greatly dissimilar to that in the general population under study. It is incredible that these highly experienced researchers suggest a causal relationship between carditis and *H. pylori* when the prevalence of *H. pylori* is 21%. It is unfortunate that they do not provide data on the 40 patients who did not have cardiac mucosa in the biopsies. Did these patients have a similar or different *H. pylori* prevalence? This is a crucial test.

In contrast, there is a substantial reason to believe that most if not all patients with carditis had reflux disease. The incidence of histologic esophagitis in the squamous epithelium is in the 40% to 50% range, which is the usual range in populations with proven reflux disease. It is to be noted that Jain et al. (S.J. Spechler was the senior author; this was after he had moved to Texas from Harvard) reported a year later that only 35% of a more normal population had cardiac mucosa in biopsies of the gastroesophageal junction. The 74% of patients with cardiac mucosa in this study is much higher than the 35% in their subsequent study; this is most rationally explained by the fact that this group of patients has a higher incidence of reflux disease. Another crucial datum is missing: What was the prevalence of reflux in the 40 patients who did not have cardiac mucosa in their biopsies?

The conclusion that the authors reach ("Carditis was associated significantly with *H. pylori* infection, but not with symptoms or signs of GERD (p. A297)") is not supported by their data or methodology. The correct conclusion is "The severity of inflammation in cardiac mucosa, when present, is associated significantly with *H. pylori*, but not with symptoms and signs of GERD." This association is based on the error that the authors use the Sydney classification to grade carditis; acute inflammation defines the highest grade in this classification. The highest grade of carditis with the Sydney classification will always correlate with *H. pylori* because reflux alone is not associated with acute inflammation. If the authors correlate the grade of chronic inflammation, they will find a relationship with reflux in those patients who are *H. pylori* negative. If they look at the grade of total inflammation or presence of acute inflammation, they will find a strong correlation with *H. pylori* infection.

Unfortunately, this abstract never reached publication as an article. There is much potentially valuable information in this study. Our study of carditis illustrates a different way of looking at the same problem:

Literature Review

Der R, Tsao-Wei DD, DeMeester TR, Peters JH, Groshen S, Lord RVN, Chandrasoma P. Carditis: A manifestation of gastroesophageal reflux disease. Am J Surg Pathol 2001;25:245–252.

This is a study of 141 patients who had cardiac mucosa in a biopsy sample from the gastroesophageal junctional region. Carditis was defined as the presence of inflammation in histologically defined cardiac mucosa irrespective of any relationship to the endoscopically defined gastroesophageal junction. Carditis was classified as acute (defined by the presence of neutrophils) and chronic (defined by the presence of plasma cells, lymphocytes, and eosinophils in the lamina propria). Chronic inflammation was graded as follows: none; mild = chronic inflammatory cells not significantly expanding interglandular spaces; and severe = chronic inflammatory cells focally expanding interglandular spaces.

Many of these patients also had a distal gastric biopsy, which sampled the antrum and body. The distal gastric biopsy was evaluated for gastritis and *H. pylori* infection with Giemsa stain. Using statistical analysis, we searched for associations between acute carditis (present or absent), chronic carditis (absent, mild, or severe), gastritis in the distal gastric biopsy (present or absent), *H. pylori* infection (positive or negative), and severity of reflux as defined by the percentage time the pH is less than 4 in a 24-hour pH test (classified as less than or equal to 4.5 and >4.5).

All 141 patients with cardiac mucosa had chronic inflammation in the lamina propria, graded as mild in 34 and severe in 107 patients. There was no instance where the cardiac mucosa had no inflammation; this was true even when an immediately adjacent piece of gastric oxyntic mucosa was absolutely normal. This indicated that carditis was localized to the area immediately distal to the squamocolumnar junction in patients who did not have gastritis. Acute carditis, characterized by the presence of neutrophils in the lamina propria, was present in 26 of 141 patients. Thirty patients had distal gastritis; in 111 patients the distal gastric biopsy showed no significant inflammation. Twenty patients were positive for *H. pylori.*

A 24-hour pH test was done in 105 patients; in these, 64 had an abnormal test defined as a pH of less than 4 for greater than 4.5% of the 24-hour period. In the other 41, the 24-hour pH test was normal. Fifteen of 25 patients (60%) with mild carditis had a normal 24-hour pH test compared with a significantly lower 26 of 80 patients (32%) with severe carditis. Patients who had severe chronic inflammation in cardiac mucosa had a significantly higher acid exposure by the 24-hour pH test as compared with those with mild chronic inflammation. There was no correlation between reflux severity and acute inflammation in cardiac mucosa. There was a strong association between the presence of acute inflammation in cardiac mucosa and gastritis and *H. pylori* infection ($p < 0.001$). Patients with gastritis had a weakly significant association with chronic inflammation in cardiac mucosa ($p = 0.042$). This was most likely a secondary association resulting from the fact that all patients with acute inflammation in cardiac mucosa had severe chronic inflammation.

We interpret these data to indicate that chronic inflammation in cardiac mucosa is strongly associated with reflux. Acute inflammation is not common in cardiac mucosa unless there is a concomitant distal gastritis caused by *H. pylori* infection.

The study indicates that the type of inflammation associated with reflux in cardiac mucosa is chronic inflammation with lymphocytes, plasma cells, and eosinophils. Neutrophils are not associated with reflux carditis unless there is a secondary event in the epithelium such as erosion or infection of the cardiac mucosa by *H. pylori* when there is an associated gastritis.

It is important in the study of one disease to not use grading systems developed for another disease. This is a common mistake that pathologists make when they evaluate carditis; most studies use the Sydney grading system. The Sydney grading system was developed for gastritis and it works well for *H. pylori* gastritis because both the active and chronic components of the inflammation are caused by *H. pylori.* Thus, the most severe grade that has neutrophils correlates well with activity of the *H. pylori* infection. Confusion results when this system is used for grading carditis. The most severe reflux injury of cardiac mucosa is a severe chronic inflammation devoid of acute inflammation. Acute inflammation results from erosion or a superimposed *H. pylori* infection. This is well shown in Der et al.'s study where there was no difference in acid exposure by the 24-hour pH study in patients with and without acute inflammation in cardiac mucosa. There was a highly significant correlation between the presence of acute inflammation in cardiac mucosa and *H. pylori* infection. When the Sydney system is used to grade carditis, the most severe grades will be seen in patients who have *H. pylori* gastritis because the highest grade in the system is defined by the presence of neutrophils. This gives the artificial impression to the unwary, that *H. pylori* causes inflammation of cardiac mucosa. The most severe reflux carditis is defined by the amount of chronic inflammation; acute inflammation in cardiac mucosa is irrelevant in the assessment of reflux severity. If we had used the Sydney system grading, our results would have provided a correlation between grade of carditis and *H. pylori* infection that would have been identical to that reached by Spechler et al. The prevalence of *H. pylori* in our study population was 20 of 141 (14%), which is approximately the same as our *H. pylori* prevalence in the entire population seen in the Foregut Surgery Department, which consists largely of affluent Caucasian patients.

Probably the best study of the significance of cardiac mucosa at the gastroesophageal junction is the study of pediatric patients from the Harvard group.

Literature Review

Glickman JN, Fox V, Antonioli DA, Wang HH, Odze RD. Morphology of the cardia and significance of carditis in pediatric patients. Am J Surg Pathol 2002;26:1032–1039.

This study correlated the histologic findings of 74 patients at the squamocolumnar junction with the clinical and pathological features. The patients were less than 19 years old, had no endoscopically visible columnar-lined esophagus, and had the squamocolumnar junction in a biopsy specimen. The patients included a spectrum of clinical abnormalities; 22 (30%) patients had "symptoms related to GERD (heartburn, regurgitation, or dysphagia)"; 28 (38%) had active esophagitis in the squamous epithelium; 9 of 69 patients (13%) who had biopsies of the distal stomach that showed gastritis; 5 (7%) patients had *H. pylori* infection.

The authors reported:

> Patients who had pure mucous glands located both within and > 1 mm from the squamocolumnar junction showed . . . an increased prevalence of active esophagitis (55% vs 21%, p = 0.04) compared with patients who had a combination of pure mucous glands within 1 mm of the squamocolumnar junction and either mixed mucous/oxyntic glands or pure oxyntic glands > 1 mm from the squamocolumnar junction. (6, p. 1035)

This is a convoluted way of saying that the presence of greater than 1 mm of cardiac mucosa had a significantly greater association with reflux than those with less than 1 mm of cardiac mucosa. Patients with greater than 1 mm of cardiac mucosa also had more severe inflammation than those with less than 1 mm of cardiac mucosa, indicating that inflammation increased with increasing lengths of cardiac mucosa.

The authors correlated the type of inflammatory cells in cardia mucosa with active esophagitis, distal gastritis, and *H. pylori* infection. They reported: "a significantly greater proportion of patients who had eosinophils in cardia mucosa had active esophagitis in comparison with those without eosinophils (54% vs 18%, p = 0.002). Furthermore, patients who had >10 lymphocytes per high power field in the cardia had a higher prevalence of chronic gastritis and *H. pylori* infection" (6, p. 1035).

This is an excellent study in terms of the data produced. The authors showed that eosinophils are the best criteria in cardiac mucosa in terms of its correlation with reflux. This is superior to Der et al.'s study, where we showed that the correlation with reflux was present when all chronic inflammatory cells were considered together.

The data are powerful. Of the 74 patients with carditis, only 13% had distal gastritis; in the other 87% the carditis was strictly isolated to a few millimeters distal to the squamocolumnar junction. Only 5 patients (7%) had evidence of *H. pylori* infection. Unfortunately, the absolute belief the authors had that the mucosa distal to the squamous epithelium is gastric made them reach the conclusion that the cardiac mucosa is gastric even though they proved beyond doubt that carditis is associated with gastroesophageal reflux. The only relationship that the authors found between carditis and *H. pylori* infection and distal gastritis was that the number of lymphocytes is greater in cardiac mucosa when there is coexisting *H. pylori* gastritis. This was the result of a double pathology in the cardiac mucosa where the superimposed *H. pylori* infection extending into the cardiac mucosa aggravated the inflammation caused by the reflux. The same association was found in Der et al. and Spechler et al. Glickman et al. correctly evaluated each element of inflammation separately without using the Sydney classification leading to correct conclusions.

Carditis Defined by Anatomic Criteria

In contrast to carditis defined by histologic criteria that have a strong and unique association with reflux, carditis defined by anatomic criteria as "inflammation occurring in a biopsy distal to the endoscopically defined gastroesophageal junction" will find an association with both gastroesophageal reflux and *Helicobacter pylori* infection. This confusion results from the use of the proximal limit of the rugal folds to define the gastroesophageal junction at endoscopy. In Chapter 10 we showed that the proximal limit of the rugal folds can be more than 2 cm proximal to the true anatomic gastroesophageal junction in patients with severe reflux disease.

If it is true that the proximal limit of the rugal folds is variably proximal to the true gastroesophageal junction, a biopsy from the anatomic cardia distal to the endoscopically defined gastroesophageal junction may sample cardiac mucosa in patients with reflux disease and gastric oxyntic mucosa in patients without reflux in a totally unpredictable manner. Inflammation in these biopsies may therefore result from both true carditis (caused by reflux) and false-positive carditis (really gastritis, which is caused by *H. pylori* or autoimmunity). Three important studies—Genta et al. (40), Hackelsberger et al. (41), and Goldblum et al. (42)—found a relationship between carditis and *H. pylori* infection. Of these, Goldblum et al. is reviewed here.

Literature Review

Goldblum JR, Vicari JJ, Falk GW, Rice TW, Peek RM, Easley K, Richter JE. Inflammation and intestinal metaplasia of the gastric cardia: The role of gastroesophageal reflux and *H. pylori* infection. Gastroenterology 1998;114:633–639.

"The aim of this study was to evaluate the relationship between cardia inflammation, *H. pylori* infection, and cardia intestinal metaplasia in patients with and without gastroesophageal reflux disease" (5, p. 633). The study population, consisting of patients from the Cleveland Clinic Foundation and the Nashville Department of Veterans Affairs Medical Center, were selected by the presence of (a) "classic GERD" defined by the presence of heartburn and/or acid regurgitation at least twice per week for at least 6 months, (b) absence of evidence of columnar-appearing mucosa in the tubular esophagus (Barrett's esophagus) or carcinoma.

In their *Methods* section, the authors described their biopsy technique. "Gastric cardia biopsy specimens were obtained with the endoscope in a retroflexed position within 5 mm from a normal appearing squamocolumnar junction" (5, p. 634). The fact that a point within 5 mm from the squamocolumnar junction was called "gastric cardia" shows that their endoscopic definition of the gastroesophageal junction was the squamocolumnar junction. The reader is left to assume that the proximal limit of the rugal folds coincided exactly with the squamocolumnar junction by the authors' statement that there was no columnar-appearing mucosa in the tubular esophagus. Biopsies were also taken from the distal esophagus, gastric antrum, and fundus in addition to the cardia. *H. pylori* infection was diagnosed when either histology or serology was positive.

In their *histologic analysis*, the authors indicated that all gastric specimens (including the gastric cardia biopsy) were evaluated for mononuclear inflammation and neutrophils (activity) by Dixon's modification of the Sydney system of classification of gastritis, for intestinal metaplasia, degree of gland atrophy, and degree of *H. pylori* colonization. They defined "carditis" as the presence of inflammation in the gastric cardia biopsy. There is no attempt whatsoever to classify the epithelial types seen in the biopsy of the "gastric cardia" by histologic criteria, despite the fact that the lead author of this article is a highly regarded gastrointestinal pathologist. The only thing that this article does is study the presence, severity, and type of inflammation, the presence of intestinal metaplasia, and the presence of *H. pylori* infection in the columnar epithelium at a point within 5 mm of the squamocolumnar junction.

Of 85 patients in the study, 58 had GERD and 27 were controls (patients undergoing endoscopy for reasons other than classic GERD symptoms, surveillance for Barrett's, peptic ulcer disease, or dyspepsia). The prevalence of *H. pylori* was similar in the GERD group (24 of 58, or 41%) and the control group (13 of 27, or 48%). In the vast majority of the patients with *H. pylori* infection of the cardia in both the GERD and the control group, the organism was also present in the distal gastric biopsies, indicating pangastritis.

The prevalence of carditis did not differ in the two groups. In the control group, 11 of 27 patients (41%) had carditis; all of these patients had *H. pylori* infection. In the GERD group, 23 of 58 patients (40%) had carditis; of these, 22 had evidence of *H. pylori* infection. This means that 16 of 27 patients in the control group and 35 of 58 patients in the GERD group had no inflammation in the cardiac biopsy. Histologic esophagitis was found in 19 of 58 patients in the GERD group (33%) and 2 of 27 in the control group (7%).

In their discussion, the authors concluded: (a) "We found the prevalence of carditis in patients with GERD (40%) to be similar to that of controls.... Thus carditis does not seem to be a marker of GERD in our population" (5, p. 636). (b) "On the other hand, all 11 patients with carditis in the control group and 22 of 23 patients with carditis in the GERD group had evidence of *H. pylori* infection... suggesting a strong link between *H. pylori* infection and carditis" (p. 637). (c) "We found evidence of a pangastritis involving the antrum and fundus, as well as the cardia in most of our *H. pylori* infected patients" (p. 637).

If carditis is defined by the criteria used in the Cleveland Clinic article as inflammation of the mucosa within 5 mm of the squamocolumnar junction, we agree completely that the inflammation will result largely from *H. pylori* infection. By their own autopsy data, the mean length of cardiac mucosa is 1.8 mm (43); it is therefore probable that their biopsy 5 mm from the squamocolumnar junction did not sample cardiac mucosa. It should not be a surprise that proximal gastritis (their "carditis") is caused by *H. pylori* infection. It is important to note that no patient with *H. pylori* infection had inflammation limited to the proximal stomach.

The data in this study are meaningless: (a) the control group may have been normal or may have been composed of patients with asymptomatic reflux disease; (b) patients with *H. pylori* gastritis may or may not have had concurrent reflux disease; (c) patients with reflux disease may or may not have had concurrent *H. pylori* infection. The data are uninterpretable because the histologic mucosal type that was present in the biopsies is not given. If the definition of carditis had been limited to inflammation in cardiac mucosa, as in the studies by Spechler et al. (3), Der et al. (19), and Glickman et al. (6), the data would have provided valuable information. This is very easy to do because the histologic definitions are precise. Why were these data not provided? This is the effect of the false dogma that cardiac mucosa is a normal epithelium. Those conducting this study had such a strong belief that cardiac mucosa is a normal gastric epithelium that they did not even report the epithelial types. If that is true, there is no point in the study. Goldblum et al. (5) were looking for changes of reflux disease in the stomach, not the esophagus. They might as well have examined a biopsy 5 cm distal to the gastroesophageal junction. It is highly likely that a biopsy in that location would have shown *H. pylori* gastritis in 11 of 27 control patients and 22 of the 58 GERD patients. There is only one GERD patient who had carditis without *H. pylori* infection that likely had reflux carditis (i.e., inflammation in histologic cardiac mucosa. Stating the obvious is important: gastroesophageal reflux causes esophageal and not gastric pathology. *The only value of a study of biopsies 0.5 cm distal to the endoscopic gastroesophageal junction is to determine how frequently this location contains esophageal tissue (cardiac mucosa) that proves the presence of reflux-induced dilated end-stage esophagus.* The alternate conclusion is that gastroesophageal reflux causes proximal gastritis, which is improbable. When histologically defined inflammation of cardiac mucosa is studied, as in the study by Glickman et al. (6), the association with *H. pylori* disappears and a strong association with gastroesophageal reflux emerges.

The research by Goldblum et al. (5) was designed to study how frequently *H. pylori* gastritis extends into the proximal stomach. This study, as well as the research by Genta et al. and Hackelsberger et al., shows that the vast majority of patients with *H. pylori* gastritis in the distal stomach have a pangastritis with evidence of involvement of the proximal stomach. *H. pylori* infection is almost never limited to the proximal stomach.

We also examine another study from the Harvard group (44).

Literature Review

Wieczorek TJ, Wang HH, Antonioli DA, Glickman JN, Odze RD. Pathologic features of reflux and *Helicobacter pylori*–associated carditis: A comparative study. Am J Surg Pathol 2003;27:960–968.

The authors introduced the topic by defining the gastric cardia:

> The gastric cardia is the most proximal anatomic region of the stomach. It is located immediately distal to the gastroesophageal junction in patients without Barrett's esophagus. Recent studies have shown that the cardia is much shorter in length than originally thought, consisting of, at most 3–5 mm of columnar mucosa containing pure mucous, *or a mixture of mucous and oxyntic,* glands. The cardia often contains chronic inflammation, in which case the term "chronic carditis" is used. The etiology of carditis is most often the result of either gastroesophageal reflux disease or *Helicobacter pylori* infection. (p. 960)

The authors defined carditis as inflammation distal to the squamocolumnar junction as long as there is no visible Barrett esophagus endoscopically. The authors went on to state: "Distinction between GERD-associated and *H. pylori* associated carditis by clinical and endoscopic methods is difficult. . . . Therefore the aim of this study was to evaluate and compare the histologic features of the gastric cardia in patients with reflux or *H. pylori* as the most likely cause of their carditis" (p. 960). Like Goldblum et al., the authors were testing a hypothesis that makes no sense; gastroesophageal reflux should not be a cause of pathology distal to the gastroesophageal junction.

The study group consisted of 85 adult patients between 1993 and 2001 who were (a) identified by the presence of the term "cardia" or "carditis" in the biopsy report and (b) had biopsy of both the squamous epithelium of the distal esophagus as well as distal stomach (corpus or antrum). Patients with endoscopic evidence of Barrett esophagus were excluded. A remarkable aspect of this research is the small number of patients studied—only 85 patients over the 8-year study period. This suggests an unacceptable case selectivity.

> The cardia was defined as the area of mucosa located immediately distal to (and within 3–5 mm of the anatomic gastroesophageal junction. The anatomic gastroesophageal junction was defined as the most proximal limit of the gastric folds. Because patients with Barrett esophagus were excluded, the gastroesophageal junction corresponded to the location of the squamocolumnar junction.

It is surprising how much credibility is given to the ability of the proximal limit of the rugal folds to define the gastroesophageal junction within a precision of 3 to 5 mm.

Patients were subdivided into three groups:

1. Thirty patients with probable reflux-associated carditis were defined by the presence of clinical symptoms of GERD, histologic evidence of active esophagitis in the squamous epithelium, *and* a biopsy of the distal stomach that showed absence of inflammation and *H. pylori* infection. This is incredible. All their patients with GERD had active esophagitis in the squamous epithelium and none of these patients had coexisting reflux disease and *H. pylori* gastritis.
2. Twenty-five patients with carditis most likely secondary to *H. pylori* infection, defined by the absence of symptoms of GERD, the finding of a normal squamous epithelium, and the presence of *H. pylori* in any biopsy.
3. Thirty controls who did not have symptoms of GERD, showed normal squamous epithelium and was negative for *H. pylori.*

Among the features that were evaluated by histology were the type and quantity of glandular epithelium in the cardia classified as mucous, mixed mucous/oxyntic, oxyntic, pancreatic acinar and the quantity of inflammation with individual cell types graded separately.

In their table, the authors' data indicate that pure mucous epithelium (= cardiac mucosa) was found in 44 patients and mixed mucous/oxyntic mucosa (= oxyntocardiac mucosa) was found in 41 patients.

The data reported in the descriptive part of the results are extremely difficult to interpret analytically; their Table 1 is the only source of intelligible information. There was no statistical significance in the prevalence of pure cardiac mucosa in the three groups (63% in the reflux group, 48% in the *H. pylori* group, and 43% in the control group). The only features that reach statistical significance between the *H. pylori* and reflux groups were male predominance in the reflux group, multilayered epithelium in the reflux group, and greater overall inflammation, neutrophils, and plasma cells in the *H. pylori* group. Unfortunately, the authors even managed to lose the significant finding in their previous study that the number of eosinophils in carditis correlated significantly with reflux (6).

This study has little meaning. The objective is to recognize histologic differences in three groups that are defined by criteria that are meaningless. The following criteria were used to classify patients into the three groups:

1. *The presence of clinical symptoms of GERD.* Because reflux disease can be present without classical symptoms of GERD, using these symptoms to differentiate between a reflux and nonreflux group is scientifically not valid.
2. *The presence of histologic evidence of reflux in the squamous epithelium.* This is meaningless. The sensitivity of histologic esophagitis in reflux disease is around 50%. The absence of reflux changes in squamous epithelium therefore does not mean there is no reflux; it is scientifically invalid to separate reflux and nonreflux controls by using the criterion of reflux esophagitis in squamous epithelium.
2. *The presence or absence of* H. pylori *infection is without merit.* The presence of *H. pylori* gastritis in no way means that the patient does not also suffer from reflux. When the objective is defined in terms that are meaningless, it should not be a surprise that the data produced by the study are essentially worthless.

Literature Review

Lembo T, Ippoliti AF, Ramers C, Weinstein WM. Inflammation of the gastroesophageal junction (carditis) in patients with symptomatic gastro-esophageal reflux disease: A prospective study. Gut 1999;45:484–488.

This is a study (46) of 30 patients being evaluated for possible gastroesophageal reflux symptoms (heartburn, regurgitation, epigastric pain, or extraesophageal manifestations). Only patients who had endoscopy, esophageal motility, and ambulatory pH monitoring were included. Patients were included only if they had a normal appearing esophagus and if the Z line appeared to be contiguous with the lower esophageal sphincter. Biopsies were taken from the Z line, sampling squamous and columnar epithelium. In addition to the Z line samples, specimens were also taken 1 to 2 cm below the Z line and from the more distal stomach. The biopsies were assessed histologically for the type of epithelium (squamous, squamous and columnar, or columnar alone), epithelial abnormalities in surface and pit epithelium, inflammation, and presence of goblet cells.

It should be noted that the definitions are not histologic. "Cardia" mucosa is any columnar mucosa that is found distal to the Z line, not a histologically defined epithelium containing only mucous cells. "Cardia mucosa" is permitted to contain parietal cells. All but one patient had at least one biopsy that contained the actual squamocolumnar junction. Also interesting is that these authors used the distal limit of the lower esophageal sphincter to define the true gastroesophageal junction.

"Carditis" was defined by a method that is unique for this article and the authors. It differs from any other definition of carditis in the literature. "A carditis score was calculated as follows: the sum of the neutrophil score (0–3) and the monocyte score (0–3), plus the greater of the surface or pit score for epithelial abnormalities. A carditis score greater than 2 was considered positive" (p. 485). We look upon this definition with amazement. The only thing it signifies is that anyone in the world can define carditis any way they want. It is unlikely that this definition is reproducible by anyone other than the authors.

The authors reported that 96% of patients had a positive carditis score; this is not surprising in a group of patients being evaluated for clinical reflux. The carditis scores were significantly higher in columnar epithelium immediately adjacent to the squamous epithelium than in columnar epithelium that was 1 to 2 cm distal to the Z line. This indicates that many patients had isolated inflammation in the region immediately distal to the Z line that decreased sharply within 1 to 2 cm. There is no mention of the epithelial type that was involved; all columnar epithelium fell under the broad classification of "cardia mucosa."

H. pylori gastritis was present in 4 of 30 (13%) patients. The mean carditis score was similar at the Z line in patients who were positive and negative for *H. pylori*. These 4 patients had significantly higher mean carditis scores in the biopsy 1 to 2 cm distal to the Z line than the *H. pylori* negative patients. These patients had *H. pylori* pangastritis.

The 24-hour pH test was normal in 14 of 30 patients and abnormal in 16 of 30; the carditis scores at the Z line were similar in the two groups. In a group of patients being evaluated for clinical symptoms of reflux, the finding of such a low rate of abnormality in the 24-hour pH test is surprising.

The authors concluded that mucosal injury in the columnar epithelium is highly localized to the region adjacent to the squamocolumnar junction in patients with gastroesophageal reflux disease. To us, this is clearly reflux disease limited to end-stage esophagus. The authors were very willing to ascribe this reflux-induced abnormality to a change in the proximal gastric mucosa. In their discussion, the authors proposed a unique mechanism for this carditis: "wear and tear," a consequence of "gastric contents lapping at the shores of the oesophagus in health" (p. 487). They ascribed this "wear and tear" phenomenon to explain regenerative changes in the squamous epithelium in volunteers and state: the same case can be made for carditis, perhaps even more so. We are glad we have never seen this explanation repeated in any other literature.

Though there are no usable standard histologic definitions in this study, the data suggest that in a population with a high bias toward reflux disease, there is a small area of columnar epithelium that shows inflammation that rapidly decreases over a 1 to 2 cm distance. This is the "isolated carditis" group and has a very strong association with reflux. In this study, only 13% of the patients had evidence of *H. pylori* gastritis, which is a normal prevalence rate for an affluent Caucasian population in Los Angeles. (Across town, the prevalence of *H. pylori* in our similarly affluent

University Hospital population is 14%; across the street, at the indigent-care Los Angeles County General Hospital, the *H. pylori* prevalence is close to 35%, similar to nearly 40% in the Cleveland Clinic population reported by Goldblum et al. [5]). Despite the lack of histologic definitions, this study characterizes reflux-induced pathology in the columnar epithelium distal to the Z line. Surely, this is esophageal, not gastric. To state the obvious again: gastroesophageal reflux causes disease in the esophagus, not in the stomach.

Literature Review

Polkowski W, van Lanschot JJB, ten Kate FJW, Rolf TM, Polak M, Tytgat GNJ, Obertop H, Offerhaus GJA. Intestinal and pancreatic metaplasia at the esophagogastric junction in patients without Barrett's esophagus. Am J Gastroenterol 2000;95:617–625.

This is a study (47) of the gastroesophageal junctional region of 76 esophagogastrectomy specimens in patients for conditions other than Barrett esophagus. The authors provided the following definitions: "The gastroesophageal junction was defined as the site of transition of the tubular esophagus into the stomach, characterized by a sharply demarcated transition line between squamous and columnar epithelium" (p. 618). This is stated to be a "normally located squamocolumnar junction" without evidence of Barrett esophagus. There is no mention of the relationship of the rugal folds to this line. Longitudinal sections passing from the squamous epithelium and including "the proximal 2 cm of the gastric mucosa" were examined. The material was archival and the number of sections of the esophagogastric junction available was 1 to 6 (mean 1.4) per patient.

An interesting part of the study is in the data the authors provided on the inflammation found in the mucosa immediately distal to the squamocolumnar junction in these 76 patients. They reported that "inflammatory changes in the gastric mucosa was present in 49 patients (64%) out of the total study group." This means that 27 patients did not have any inflammatory changes. It is unfortunate that the histologic type of these 27 patients is not reported. If it is true that cardiac mucosa is always inflamed, it is highly likely that these patients did not have histologic cardiac mucosa distal to the squamous epithelium. However, this must remain conjectural in the absence of these data.

The authors did not classify the mucosae by histologic criteria. They, however, employed criteria that permit partial definition of histologic type:

> When inflammatory changes were present only in the cardia (defined as the proximal 2 cm of gastric mucosa, bearing cardiac type glands) it was scored as "carditis." When inflammation was seen in the cardia and in the distal gastric resection margin, the inflammatory changes in the cardia were interpreted as an extension of a more generalized gastritis/pangastritis. (p. 619)

Reporting on the findings of the 49 patients who had inflammation in the mucosa distal to the squamocolumnar junction, the articles stated, "in 27 of these patients (35%) generalized pangastritis was found, whereas carditis was seen in 22 patients (29%)" (p. 620).

The definition of an isolated "carditis" without distal gastritis, in our experience, has an absolute correlation with histologically defined carditis. In the 141 patients reported by Der et al. (19), 111 had an isolated carditis that would fall into the definition of carditis used in this study. These authors also indicated this by stating that their patients with carditis had mucosa bearing cardiac-type glands. They reported: "*H. pylori* was observed in eight cases with a generalized pangastritis, whereas none of the patients with carditis had *H. pylori* ($p = 0.006$)" (p. 620). In their conclusions, the authors wrote: "Carditis may be a distinct inflammatory condition of the gastric mucosa that is not related to *H. pylori* infection."

This study provides excellent data. It suffers only in that the authors had implicit faith in their ability to define the gastroesophageal junction in these specimens. As a result, everything distal to the squamocolumnar junction is "gastric mucosa." The authors did not consider the possibility that their gastroesophageal junction was incorrectly defined. It is also interesting in this regard that the proximal limit of the rugal folds is never mentioned in this article; the authors have essentially reverted to the squamocolumnar junction and the end of the tubular esophagus to define the gastroesophageal junction.

Based on their results, 27 patients had no inflammation in the mucosa distal to the squamocolumnar junction. These most likely represent patients who had no histologic cardiac mucosa. In the 22 patients falling under the authors' definition of carditis, inflammation was restricted to the 2 cm distal to the squamocolumnar junction; this occurred in histologically defined cardiac-type mucosa. None of these patients had evidence of *H. pylori* infection, leading the authors to correctly conclude that carditis was not related to *H. pylori* infection. These patients would fall into what we classify as reflux carditis occurring in dilated end-stage esophagus irrespective of the presence of absence of any other evidence of reflux. The 27 patients who fell within the definition of pangastritis in this study will, based on our experience, fall into two categories:

1. Patients with pangastritis without histologic cardiac mucosa at the junction. This group will have no gastroesophageal reflux and will have etiologic associations similar to gastritis with a predominance of *H. pylori* infection.
2. Patients with pangastritis and histologic cardiac mucosa (our definition of carditis) at the junction. This group will consist of patients with both *H. pylori* gastritis and reflux carditis.

Unfortunately, the study provides no data regarding the histologic epithelial types distal to the squamous epithelium in the patients with pangastritis. This precludes accurate analysis of how many of the 27 patients classified as pangastritis had pure pangastritis (inflammation in oxyntic mucosa) or a mixture of pangastritis and reflux carditis (inflammation in cardiac mucosa devoid of parietal cells and showing reactive changes).

Summary Statement on Definition of Reflux Disease

1. Columnar metaplasia of the esophagus, defined as the presence of cardiac mucosa with or without intestinal metaplasia and oxyntocardiac mucosa, is a unique histologic feature that indicates reflux-induced cellular damage with 100% sensitivity and 100% specificity.
2. The use of reflux carditis is the best histologic criterion to define reflux.
3. Reflux carditis should be used to define gastroesophageal reflux disease; it is accurate, reproducible, and cheap because it requires no special technology other than standard biopsy.
4. Future studies on reflux disease must demand a histologic definition of epithelia.

A NEW DEFINITION OF BARRETT ESOPHAGUS

The definition of Barrett esophagus has evolved through history (Table 13.6). When it was originally described by Allison and Johnstone (30) and confirmed by Barrett (47), it was regarded as a congenital anomaly in which the distal esophagus had a columnar lining. No definition was required in terms of length because it was not a disease. When, however, it became recognized that columnar-lined esophagus was caused by gastroesophageal reflux, it became necessary to define the disease. In 1961, Hayward (24) limited the diagnosis of Barrett esophagus to the presence of columnar-lined esophagus in the distal esophagus exceeding 2 cm in length. This led to the tacit belief that less than 2 cm of columnar lining was normal in the distal esophagus.

TABLE 13.6 Evolution of the Definition of Columnar-Lined (Barrett) Esophagus

Year	Definition/change
1953	Columnar-lined segment between squamocolumnar junction and stomach (hiatal hernia) (believed to be a congenital anomaly)
1957	Barrett coins the name "columnar-lined esophagus" (= Barrett esophagus)
1961	Barrett esophagus = endoscopic presence of any columnar epithelium above 2 (3) cm of the distal tubular esophagus
1976	Histologic classification into three types; junctional (cardiac), fundic type (oxyntocardiac), and specialized (intestinal); no change in definition
1980s	Definition of BE required intestinal metaplasia in a biopsy in addition to the 2/3 cm rule; other types ceased to the BE
1990s	BE = intestinal metaplasia in any visible segment of columnar epithelium in any part of the tubular esophagus; divided into LSBE and SSBE
2000s	BE = intestinal metaplasia in any visible segment of columnar epithelium in any part of the tubular esophagus; one entity, not divided into LSBE and SSBE
2006	BE = intestinal metaplasia of the esophagus (= intestinal metaplasia of cardiac mucosa)

Until the mid-1980s, Barrett esophagus was an endoscopic diagnosis based entirely on the criterion of a columnar-lined esophagus being present in the distal esophagus greater than 2 (or 3, by some authorities) cm above the gastroesophageal junction (at that time, the end of the tubular esophagus). Hayward provided no evidence to back his definition. Based on Paull et al.'s histologic study, the endoscopically defined Barrett esophagus was classified into three histologic subtypes: junctional (= cardiac), fundic (= oxyntocardiac), and specialized (= intestinal). The criterion for diagnosis, however, remained endoscopic. At the time, the presence of Barrett esophagus indicated that the patient had reflux disease. The incidence of adenocarcinoma was very low.

When the incidence of adenocarcinoma of the esophagus began increasing in the 1980s, it was recognized that it was the specialized (= intestinal) type of columnar-lined esophagus that led to cancer (15). This resulted in the diagnosis of Barrett esophagus shifting to require the criterion of histologically defined intestinal metaplasia in a biopsy taken from a columnar-lined segment greater than 2 (or 3) cm above the end of the tubular esophagus.

The device of an artificial definition that 2 (or 3) cm of columnar-lined esophagus was normal resulted in the recommendation that these patients need not have biopsy. This resulted in a delay of nearly three decades before Spechler et al.'s classical study (39) where they biopsied the epithelium immediately distal to the squamocolumnar junction in patients with "normal" amounts of columnar-lined esophagus.

Literature Review

Spechler SJ, Zeroogian JM, Antonioli DA, Wang HH, Goyal RK. Prevalence of intestinal metaplasia at the gastro-oesophageal junction. Lancet 1994;344:1533–1536.

In the introduction, the authors summarized the state of the art as it existed in 1994:

> Columnar epithelium in the oesophagus has a characteristic red colour and velvet-like texture that contrasts sharply with the pale, glossy appearance of adjacent squamous epithelium. Although endoscopic examination can usually distinguish columnar epithelium from squamous epithelium in the oesophagus, the several types of columnar epithelia cannot be differentiated on endoscopic appearance alone. The distinction between specialized columnar epithelium (SCE) and gastric columnar epithelium can be made only by histological examination of biopsy specimens. Gastric columnar epithelium may normally line a short segment of the distal oesophagus. Therefore, endoscopists usually diagnose Barrett's oesophagus only when they see columnar epithelium extending well above the gastro-oesophageal junction . . . short segments of SCE in the distal oesophagus are not recognized as abnormal. . . . Endoscopists usually do not obtain biopsy specimens from columnar epithelium that involves only a short segment of the distal oesophagus, and therefore the characteristics of patients who have short segments of SCE at the gastro-oesophageal junction have not been well defined. (p. 1533)

The patients included all adults scheduled for elective endoscopy. For patients with no apparent hiatus hernia, the gastroesophageal junction was defined as the point at which the tubular distal esophagus flared to become a sacklike structure (the stomach). For patients with a hiatal hernia, the junction was defined as the proximal margin of the gastric folds in the hernia pouch. The extent of columnar epithelium in the distal esophagus was the distance between the junction and the most proximal extension of columnar epithelium.

Two biopsy specimens were obtained from the columnar epithelium at the squamocolumnar junction to seek SCE and one biopsy specimen 2 cm above the squamocolumnar junction to seek esophagitis.

There is confusion regarding what the authors considered to be the gastroesophageal junction. After defining the gastroesophageal junction precisely by endoscopic criteria, the authors used a biopsy from the squamocolumnar junction to fulfill their stated aim of evaluating the prevalence and characteristics of intestinal metaplasia at the gastroesophageal junction. The biopsies were not from the gastroesophageal junction; they were from the most proximal point of the columnar lined esophagus.

The patients were divided into two groups:

1. Those with endoscopically apparent Barrett esophagus (n = 16). These patients had columnar epithelium extending greater than 3 cm above the gastroesophageal junction. All had intestinal metaplasia on biopsy.
2. Those without endoscopically apparent Barrett esophagus. These were all the patients who had less than 3 cm of columnar-lined esophagus. This group was divided into patients who had intestinal metaplasia (SCE) (n = 26) and those without intestinal metaplasia (n = 114).

The 26 patients with intestinal metaplasia are described in detail:

> In 9 (35%) the epithelial squamocolumnar junction and the gastro-oesophageal junction were at the same level so that the squamocolumnar junction appeared as a relatively straight line at the very end of the oesophagus. In 17 patients (65%), columnar epithelium extended up to 3 cm into the distal oesophagus; in these cases the squamocolumnar junction appeared as a wavy or zigzag line, occasionally with eccentric tongues of columnar epithelium projecting as far as 3 cm up the oesophagus. (p. 1534)

There is no description of the appearance in the 114 patients who did not have intestinal metaplasia in the biopsy. This most likely means that the authors regarded these patients as having "normal gastric mucosa" in their distal esophagus.

The authors compared the patients with endoscopic Barrett esophagus and without endoscopic Barrett esophagus, the latter divided into those with and without SCE. The findings are summarized in the following table:

	Endoscopic Barrett's	**No endoscopic Barrett's**	
Feature	**SCE on Bx (n = 16)**	**SCE on Bx (n = 26)**	**No SCE on Bx (n = 114)**
M:F (ratio)	10/6 (1.7:1)	17/9 (1.9:1)	49/65 (0.8:1)
White race	16 (100%)	26 (100%)	99 (87%)
Heartburn	11 (69%)	14 (54%)	62 (54%)
Any GORD symptom	12 (86%)	15 (58%)	72 (63%)
Esophagitis (endoscopic/histologic)	8 (50%)	19 (73%)	50 (44%)

It is interesting how the patients without endoscopic Barrett esophagus who did not have intestinal metaplasia are completely ignored. If only the authors had classified the epithelial types found in these patients, the authors may have come up with valuable data. The fact that many patients in this group had evidence of reflux (63% had symptoms of reflux, and 44% had esophagitis in the squamous epithelium) should have led them to seek features that would have permitted separation of those with evidence of reflux and without. This would have led them to recognize the importance of cardiac mucosa as a criterion of reflux. Unfortunately, the opportunity was missed.

The main focus of this study was that intestinal metaplasia was found at the squamocolumnar junction in 26 of 142 patients (18%) without endoscopically apparent Barrett esophagus. The researchers concluded: "Adults frequently have unrecognized segments of SCE at the gastro-oesophageal junction; this may underlie the rising frequency of cancer of the gastro-oesophageal junction in the USA and Europe" (p. 1533).

No one has ever suggested that intestinal metaplasia anywhere in this region is "normal"; as such, this was recognized as an abnormality.

The findings in Spechler et al.'s study (39) resulted in several changes: (a) the recognition of short-segment

Barrett esophagus; (b) a change in the definition of "normal," wherein any columnar lining in the tubular esophagus was considered sufficiently abnormal at endoscopy to require biopsy; (c) a change in the definition of Barrett esophagus to include patients with short-segment Barrett esophagus; (d) the recommendation for a routine biopsy of patients with short segments of columnar lining in the distal esophagus visualized endoscopically; and (e) an accumulation of data to prove that short-segment Barrett esophagus was a premalignant lesion. Because patients with short-segment Barrett esophagus *were not routinely biopsied before 1994, the accumulation of data relating to this entity was delayed for three decades.*

The present definition of Barrett requires the presence of intestinal metaplasia in a biopsy taken from any endoscopically visible columnar lining in the tubular esophagus above the endoscopic gastroesophageal junction (= end of the tubular esophagus or, when this is not clearly defined, the proximal limit of the rugal folds) (25). This makes two assumptions, both false:

1. *That the endoscope is capable of detecting all metaplastic columnar lining that may be present in the esophagus. It is self-evident that this assumption is incorrect.* There is clearly a stage between the onset of columnar metaplasia of the esophagus (which begins with one cell) and the point at which it becomes visible at endoscopy. This is determined by the physics of resolution of the endoscope. The present definition ensures that those patients with subendoscopic or microscopic Barrett esophagus are not detected because the recommendation is that these "normal" people should not be biopsied. We need to recognize that we cannot make Barrett esophagus disappear by the edict of definition. We made this mistake in patients with short-segment disease by the edict of an incorrect definition between 1961 and 1994. We are now doing the same thing by ignoring people with microscopic Barrett esophagus by declaring that it does not exist when we cannot see it with a technologically inferior instrument.

   This is proven by the fact that we are willing to accept that magnification endoscopy and chromoendoscopy can detect columnar-lined esophagus that is not seen by standard endoscopy. Better resolution cannot create Barrett esophagus; it just detects better what is already there and is being missed by standard endoscopy. The acceptance that better endoscopes can detect disease not visible by standard endoscopy is a statement that the present definition of Barrett esophagus is wrong. In the future, when the resolution of endoscopy becomes capable of identifying cells, we may be able to distinguish columnar-lined esophagus from normal gastric oxyntic mucosa. This capability may be in the horizon; we have tested acoustic microscopy on thin sections at a wavelength of 200 megahertz to 1 gigahertz; at this wavelength, the resolution is adequate to recognize cells and detect goblet cells, mucous cells, and parietal cells. Also, at this wavelength, the penetration of tissues is very small but within the 50-μm range where the surface epithelial cells can be visualized. When the practical reasons why this cannot be adapted to an endoscope (size of acoustic lens, stabilization, medium, etc.) can be worked out, *in vivo* endoscopic microscopy will become feasible and endoscopy will be able to diagnose Barrett esophagus. Until then, we need biopsy and microscopic study.
2. *That the present unproven definitions of the gastroesophageal junction are correct.*

   In Chapter 11 we showed that there is good evidence that esophageal tissue validated by the presence of submucosal glands is present distal to the end of the tubular esophagus and distal to the proximal limit of the rugal folds (the end-stage dilated esophagus). This results in an error where up to 2.5 to 3 cm of the reflux-damaged end-stage dilated esophagus is being called stomach. Cardiac and oxyntocardiac mucosa found in that region is called "normal gastric mucosa," and intestinal metaplasia found in that region is called "not Barrett esophagus" or "intestinal metaplasia of the gastric cardia."

   We are in the same place as Allison and Barrett were in the 1950s and Spechler and the medical community were in the 1990s. In both these situations, Barrett (in 1950) and the medical community (in 1961 through 1994) used an incorrect definition of the gastroesophageal junction or what was "normal" to declare by edict that Barrett esophagus did not exist. In the 1950s, Barrett used the incorrect definition of the gastroesophageal junction to declare that the entire columnar-lined esophagus was a tubular stomach. Allison and Johnstone (30) had to prove him wrong in 1953 by showing that Barrett's tubular stomach contained submucosal esophageal glands. In the 1961–1994 era, the medical community called short-segment Barrett esophagus "normal" and recommended that biopsies not be taken from the distal 2 to 3 cm of the tubular esophagus. It was up to Spechler et al. (39) to prove that intestinal metaplasia at risk for cancer was present in this "normal" mucosa.

   Today, we make all these mistakes of the past; we recommend that biopsies should not be taken when endoscopy is deemed normal. If it is biopsied, we declare by edict that what is distal to our incorrectly defined gastroesophageal junction is "normal" and "gastric." We become confused and nervous when intestinal metaplasia is found in those biopsies but refuse to call it Barrett esophagus because it is outside our definition. Just like Allison and Johnstone in 1953, we proved that this is wrong in Chapter 11 by demonstrating that esophageal submucosal glands are present distal to the end of the tubular esophagus and proximal limit of the rugal folds. What is now called "normal cardiac type gastric mucosa" is really reflux carditis, and what is now called "intestinal metaplasia of the gastric cardia" is Barrett esophagus (Fig. 13.19).

Just like when short-segment Barrett esophagus was born, recognition of this fact will produce the identical set of changes that occurred in 1994: (a) we will recognize Barrett esophagus involving the dilated end-stage esophagus; (b) we will change the definition of "normal" wherein any columnar epithelium proximal to gastric oxyntic mucosa will be considered abnormal, recognizing that endoscopy is not capable of defining the true gastroesophageal junction; (c) we will change the definition of Barrett esophagus to include patients with intestinal metaplasia involving the end-stage dilated esophagus; (d) we will recommend routine biopsy for patients who are endoscopically normal to detect Barrett esophagus involving the end-stage dilated esophagus; and (e) we will begin accumulating data to prove whether or not Barrett esophagus involving the end-stage dilated esophagus is a premalignant lesion.

This recognition will increase the length of the Barrett segment in all cases diagnosed before 2006 (or whenever this is accepted) by the part of the dilated end-stage esophageal length that has been regarded as normal. A Barrett segment that is now measured as 3 cm will need to be retrospectively revised as a segment that is 3 cm + the length of the end-stage dilated esophagus that is missed at present.

The new definition of Barrett esophagus: Barrett esophagus is intestinal metaplasia of the esophagus. This translates into Barrett esophagus occurring in metaplastic columnar epithelium of the esophagus. Because intestinal metaplasia does not occur in oxyntocardiac mucosa, Barrett esophagus = intestinal metaplasia in cardiac mucosa (Figs. 13.20 and 13.21).

According to this new definition, Barrett esophagus is a histologically defined disease that complicates reflux carditis. There is a lag phase, usually of many years, before intestinal metaplasia develops in reflux carditis. During this lag phase, increasing cardiac transformation of squamous epithelium occurs. The length of columnar transformation is dependent on the cumulative reflux exposure (severity × duration). Evidence suggests that when intestinal metaplasia develops in carditis, the process of columnar transformation stops and the columnar-lined esophagus remains the same length.

Barrett esophagus will therefore always be preceded by reflux carditis and always be recognizable as intestinal metaplasia of cardiac mucosa. When intestinal metaplasia arises, it will do so in the preexisting columnar-lined esophagus. Three different

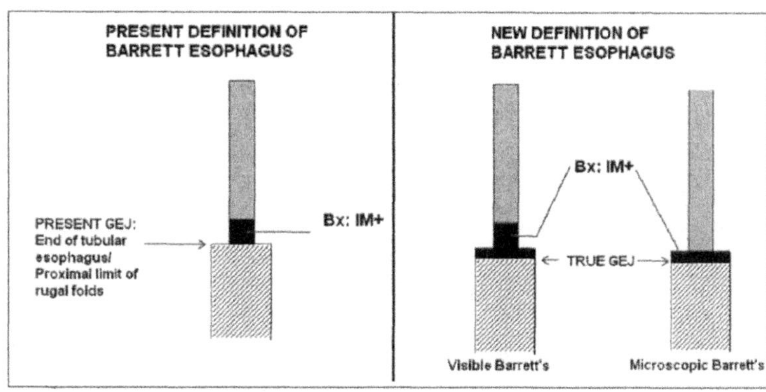

FIGURE 13.19 Present definition of Barrett esophagus requires a visible columnar-lined esophagus and intestinal metaplasia in a biopsy from this segment (*left*). The new proposed definition of Barrett esophagus (*right*). This Barrett esophagus includes the presently defined intestinal metaplasia in visible columnar-lined esophagus as well as microscopic Barrett esophagus where intestinal metaplasia is limited to the dilated end-stage esophagus.

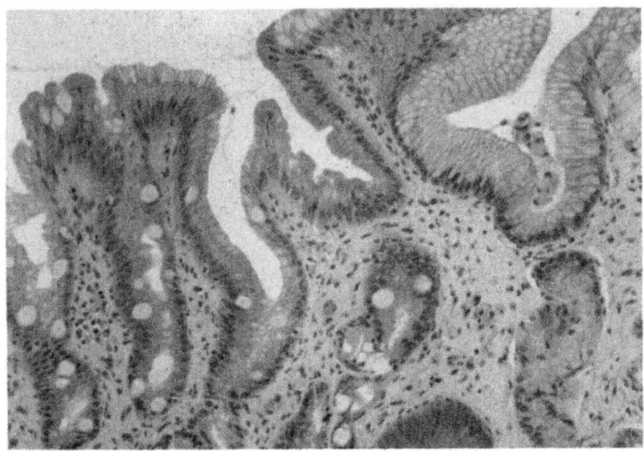

FIGURE 13.20 Barrett esophagus, defined as the occurrence of intestinal metaplasia in cardiac mucosa irrespective of its relationship to the endoscopic gastroesophageal junction; cardiac mucosa indicates esophageal location.

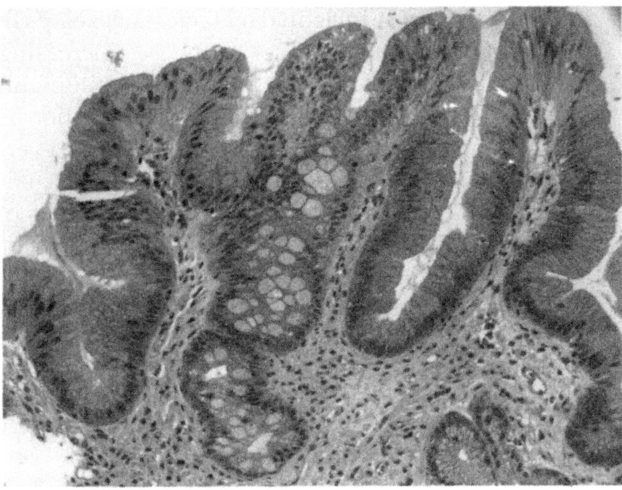

FIGURE 13.21 Barrett esophagus or intestinal metaplasia occurring in cardiac mucosa in a retrograde biopsy taken from the epithelium immediately distal to the squamocolumnar junction in an endoscopically normal person. (Cross reference: Color Plate 4.1)

types of Barrett esophagus can be artificially separated within the continuous spectrum: (a) Barrett esophagus occurring in a columnar-lined esophagus greater than 2 (or 3) cm above the end of the tubular esophagus, which is what we presently call long-segment Barrett esophagus; (b) Barrett esophagus occurring in a columnar-lined esophagus less than 2 (or 3) cm above the end of the distal esophagus, which is what we presently call short-segment Barrett esophagus; and (c) Barrett esophagus occurring distal to the end of the tubular esophagus, limited to the dilated end-stage esophagus of reflux disease. This entity is presently missed because of the incorrect definition of the gastroesophageal junction and the recommendation that patients with no visible endoscopic abnormality should not undergo biopsy.

References

1. Chandrasoma P. Controversies of the cardiac mucosa and Barrett's esophagus. Histopathol 2005;46:361–373.
2. An evidence-based appraisal of reflux disease management: the Genval Workshop Report. Gut 1999;44(Suppl 2):S1–S16.
3. Spechler SJ, Wang HH, Chen YY, Zeroogian JM, Antonioli DA, Goyal RK. GERD vs. *H. pylori* infections as potential causes of inflammation in the gastric cardia. Gastroenterology 1997;112: A297.
4. Kahrilas PJ, Shi G, Manka M, Joehl RJ. Increased frequency of transient lower esophageal sphincter relaxation induced by gastric distension in reflux patients with hiatal hernia. Gastroenterology 2000;118:688–695.
5. Goldblum JR, Vicari JJ, Falk GW, Rice TW, Peek RM, Easley K, Richter JE. Inflammation and intestinal metaplasia of the gastric cardia: The role of gastroesophageal reflux and *H. pylori* infection. Gastroenterology 1998;114:633–639.
6. Glickman JN, Fox V, Antonioli DA, Wang HH, Odze RD. Morphology of the cardia and significance of carditis in pediatric patients. Am J Surg Pathol 2002;26:1032–1039.
7. Lagergren J, Bergstrom R, Lindgren A, Nyren O. Symptomatic gastroesophageal reflux as a risk factor for esophageal adenocarcinoma. N Engl J Med 1999;340:825–831.
8. Arango L, Angel A, Molina RI, et al. Comparison between digestive endoscopy and 24-hour pH monitoring for the diagnosis of gastroesophageal reflux esophagitis: "Presentation of 100 cases." Hepatogastroenterology 2000;47:174–180.
9. Kiesslich R, Kanzler S, Vieth M, et al. Minimal change esophagitis: Prospective comparison of endoscopic and histological markers between patients with non-erosive reflux disease and normal controls using magnifying endoscopy. Dig Dis 2004;22: 221–227.
10. Riddell RH. The biopsy diagnosis of gastroesophageal reflux disease, "carditis," and Barrett's esophagus, and sequelae of therapy. Am J Surg Pathol 1996;20(Suppl 1):S31–S51.
11. Tummala V, Barwick KW, Sontag SJ, Vlahcevic RZ, McCallum RW. The significance of intraepithelial eosinophils in the histologic diagnosis of gastroesophageal reflux. Am J Clin Pathol 1987;87:43–48.
12. Takubo K, Honma N, Aryal G, Sawabe M, Arai T, Tanaka Y, Mafune K, Iwakiri K. Is there a set of histologic changes that are invariably reflux associated? Arch Pathol Lab Med 2005; 129(2):159–163.
13. Hirota WK, Loughney TM, Lazas DJ, Maydonovitch CL, Rholl V, Wong RKH. Specialized intestinal metaplasia, dysplasia, and cancer of the esophagus and esophagogastric junction: Prevalence and clinical data. Gastroenterology 1999;116:277–285.
14. Rex DK, Cummings OW, Shaw M, Cumings MD, Wong RKH, Vasudeva RS, Dunne D, Rahmani EY, Helper DJ. Screening for Barrett's esophagus in colonoscopy patients with and without heartburn. Gastroenterology 2003;125:1670–1677.
15. Haggitt RC. Adenocarcinoma in Barrett's esophagus: A new epidemic? Hum Pathol 1992:23:475–476.
16. Dulai GS, Guha S, Kahn KL, Gornbein J, Weinstein WM. Preoperative prevlance of Barrett's esophagus in esophageal adenocarcinoma: A systematic review. Gastroenterology 2002;122: 26–33.

17. DeMeester TR, Wang CI, Wernly JA, et al. Technique, indications and clinical use of 24-hour esophageal pH monitoring. J Thorac Cardiovasc Surg 1980;79:656–667.
18. Tharalson EF, Martinez SD, Garewal HS, Sampliner RE, Cui H, Pulliam G, Fass R. Relationship between rate of change in acid exposure along the esophagus and length of Barrett's epithelium. Am J Gastroenterol 2002;97:851–856.
19. Der R, Tsao-Wei DD, DeMeester T, et al. Carditis: A manifestation of gastroesophageal reflux disease. Am J Surg Pathol 2001;25:245–252.
20. Kauer WKH, Burdiles P, Ireland AP, et al. Does duodenal juice reflux into the esophagus of patients with complicated GERD? Evaluation of a fiberoptic sensor for bilirubin. Am J Surg 1995;169:98–104.
21. Tamhankar AP, Peters JH, Portale G, Hsieh C-C, Hagen JA, Bremner CG, DeMeester TR. Omeprazole does not reduce gastroesophageal reflux: New insights using multichannel intraluminal impedance technology. J Gastrointest Surg 2004;8: 890–898.
22. De Nardi FG, Riddell RH. Esophagus. *In*: Sternberg SS (ed), Histology for Pathologists. Raven Press, New York, 1992, pp 524–526.
23. Owen DA. Stomach. *In*: Sternberg SS (ed), Histology for Pathologists, second edition. Lippincott-Raven Publishers, Philadelphia, 1997. pp 481-493.
24. Hayward J. The lower end of the oesophagus. Thorax 1961;16: 36–41.
25. Sampliner RE. Practice guidelines on the diagnosis, surveillance, and therapy of Barrett's esophagus. Am J Gastroenterol 1998;93:1028–1031.
26. Clark GWB, Ireland AP, Chandrasoma P, DeMeester TR, Peters JH, Bremner CG. Inflammation and metaplasia in the transitional mucosa of the epithelium of the gastroesophageal junction: A new marker for gastroesophageal reflux disease. Gastroenterology 1994;106:A63.
27. Odze RD. Unraveling the mystery of the gastroesophageal junction: A pathologist's perspective. Am J Gastroenterol 2005; 100:1853–1867.
28. Chandrasoma PT, Der R, Ma Y, et al. Histology of the gastroesophageal junction: An autopsy study. Am J Surg Pathol 2000;24:402–409.
29. Jain R, Aquino D, Harford WV, Lee E, Spechler SJ. Cardiac epithelium is found infrequently in the gastric cardia. Gastroenterology 1998;114:A160 (Abstract).
30. Allison PR, Johnstone AS. The oesophagus lined with gastric mucous membrane. Thorax 1953;8:87–101.
31. Oberg S, Peters JH, DeMeester TR, et al. Inflammation and specialized intestinal metaplasia of cardiac mucosa is a manifestation of gastroesophageal reflux disease. Ann Surg 1997;226:522–532.
32. Paull A, Trier JS, Dalton MD, Camp RC, Loeb P, Goyal RK. The histologic spectrum of Barrett's esophagus. N Engl J Med 1976;295:476–480.
33. Chandrasoma PT, Der R, Dalton P, Kobayashi G, Ma Y, Peters JH, DeMeester TR. Distribution and significance of epithelial types in columnar lined esophagus. Am J Surg Pathol 2001;25: 1188–1193.
34. Chandrasoma PT, Der R, Ma Y, Peters J, DeMeester T. Histologic classification of patients based on mapping biopsies of the gastroesophageal junction. Am J Surg Pathol 2003;27: 929–936.
35. El-Serag HB, Sonnenberg A. Comorbid occurrence of laryngeal or pulmonary disease with esophagitis in the United States Military Veterans. Gastroenterology 1997;113:755–760.
36. Cameron AJ, Lomboy CT. Barrett's esophagus: Age, prevalence, and extent of columnar epithelium. Gastroenterology 1992;103: 1241–1245.
37. Jamieson JR, Stein HJ, DeMeester TR, et al. Ambulatory 24-hour esophageal pH monitoring: Normal values, optimal thresholds, specificity, sensitivity, and reproducibility. Am J Gastroenterol 1992;87:1102–1111.
38. Marsman WA, van Sandyck JW, Tytgat GNJ, ten Kate FJW, van Lanschot JJB. The presence and mucin histochemistry of cardiac type mucosa at the esophagogastric junction. Am J Gastroenterol 2004;99:212–217.
39. Spechler SJ, Zeroogian JM, Antonioli DA, Wang HH, Goyal RK. Prevalence of metaplasia at the gastroesophageal junction. Lancet 1994;344:1533–1536.
40. Genta RM, Huberman RM, Graham DY. The gastric cardia in *Helicobacter pylori* infection. Hum Pathol 1994;25:915–919.
41. Hackelsberger A, Gunther T, Schultze V, Labenz J, Roessner A, Malfertheiner P. Prevalance and pattern of *Helicobacter pylori* gastritis and the gastric cardia. Am J Gastroenterol 1997;92: 2220–2223.
42. Goldblum JR, Vicari JJ, Falk GW, Rice TW, Peek RM, Easley K, Richter JE. Inflammation and intestinal metaplasia of the gastric cardia: The role of gastroesophageal reflux and *Helicobacter pylori* infrection. Gastroenterology 1998;114: 633–639.
43. Kilgore SP, Ormsby AH, Gramlich TL, et al. The gastric cardia: Fact or fiction? Am J Gastroenterol 2000;95:921–924.
44. Wieczorek TJ, Wang HH, Antonioli DA, Glickman JN, Odze RD. Pathologic features of reflux and *Helicobacter pylori*–associated carditis: A comparative study. Am J Surg Pathol 2003;27: 960–968.
45. Lembo T, Ippoliti AF, Ramers C, Weinstein WM. Inflammation of the gastroesophageal junction (carditis) in patients with symptomatic gastro-esophageal reflux disease: A prospective study. Gut 1999;45:484–488.
46. Polkowski W, van Lanschot JJB, ten Kate FJW, Rolf TM, Polak M, Tytgat GNJ, Obertop H, Offerhaus GJA. Intestinal and pancreatic metaplasia at the esophagogastric junction in patients without Barrett's esophagus. Am J Gastroenterol 2000; 95:617–625.
47. Barrett NR. The lower esophagus lined by columnar epithelium. Surgery 1957;41:881–894.

CHAPTER

14

Diagnosis of Gastroesophageal Reflux Disease, Barrett Esophagus, and Dysplasia

In Chapter 13, we established that the present criteria used in the diagnosis of reflux disease have major problems (Table 14.1). They are not standardized, have low sensitivity and or specificity, and are generally worthless as diagnostic tests. Reflux disease is presently managed clinically without histopathologic support. Biopsies are not part of management, and there are no objective criteria to evaluate the progression or regression of the disease except clinical criteria such as relief of symptoms, healing of erosions, and changes in the endoscopic appearance of columnar-lined esophagus such as decreased length or the appearance of squamous islands. Biopsy only becomes necessary when a columnar-lined esophagus is endoscopically visualized. At this point, the disease is defined by histologic diagnostic criteria, which are necessary for a diagnosis of Barrett esophagus, dysplasia, and adenocarcinoma. Columnar-lined esophagus, even when endoscopically visible, is ignored if the biopsies do not contain intestinal metaplasia, dysplasia, or adenocarcinoma.

Symptomatic definitions of reflux disease based on the presence of heartburn, the presence of erosive esophagitis, and histologic changes in biopsies from the squamous epithelium are too insensitive to be of value. They provide guidance in the treatment of patients, but they do not satisfy the criteria for a scientifically valid diagnostic test. The 24-hour ambulatory pH test and impedance tests provider a highly accurate assessment of gastroesophageal reflux severity, but they do not assess cellular damage. These tests are rarely used except in academic centers and highly specialized swallowing centers.

In contrast to changes in squamous epithelium, which are not specific, columnar metaplasia of the esophagus is a highly specific change that is caused only by gastroesophageal reflux. It is the result of a combination of a complex injury of the squamous epithelium leading to increased permeability followed by a genetic switch that leads to metaplasia. It is unlikely that these events can be mimicked by any other disease. After the original thesis that this was a congenital abnormality was rejected in the 1960s, columnar metaplasia has never been even suggested as occurring with any other disease.

In Chapters 9 and 13, we discussed the evolution of columnar epithelia into its three main histologic types and their practical use in defining reflux disease (Fig. 14.1). Recognizing this evolution can provide highly specific information and a determination of what is the ideal diagnostic test for the diagnosis of reflux. Thus, virtually all patients with even the slightest cellular abnormality resulting from chronic reflux will have oxyntocardiac mucosa. The finding of oxyntocardiac mucosa in a biopsy is therefore 100% specific and nearly 100% sensitive. However, while it shows features of the ideal test, this finding is not practically useful. It is too sensitive (virtually every patient in the population will have it) and does not predict the end point of adenocarcinoma (oxyntocardiac mucosa does not progress to intestinal metaplasia and adenocarcinoma; Fig. 14.1). At the other extreme, intestinal metaplasia is 100% specific but has a very low sensitivity (only a small percentage of patients with reflux have intestinal metaplasia). It is not an adequately sensitive test for diagnosing reflux

TABLE 14.1 Specificity and Sensitivity of Criteria Used in the Diagnosis of Gastroesophageal Reflux Disease

Criterion	Specificity	Sensitivity
Classical symptoms	High	Intermediate
Erosive esophagitis	High	Low
24-hour ambulatory pH test	High	Intermediate
Multichannel impedance test	High	Intermediate
Histologic changes in squamous epithelium	Intermediate	Low
Barrett esophagus	100%	Low
Cardiac mucosal metaplasia (reflux carditis)	100%	High
Oxyntocardiac mucosal metaplasia	100%	Nearly 100%

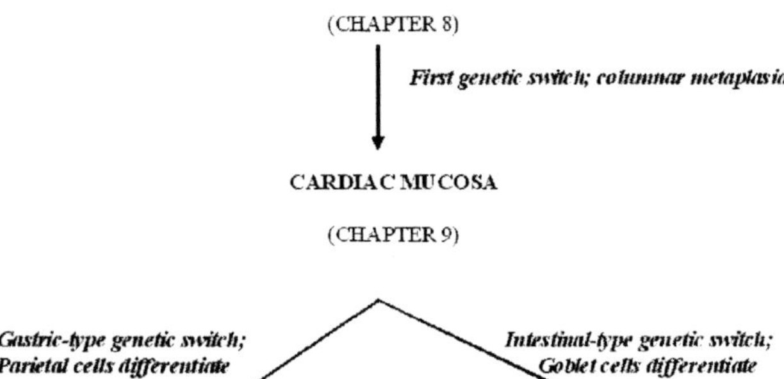

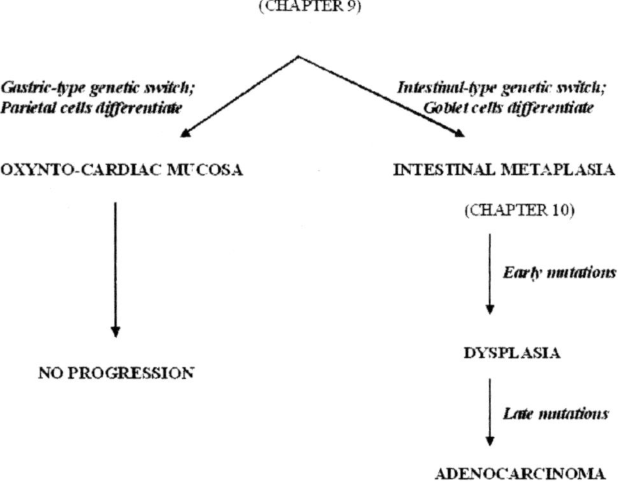

FIGURE 14.1 The reflux-adenocarcinoma sequence showing the progression of squamous epithelium through cardiac and intestinal metaplasia to adenocarcinoma. The conversion of cardiac to oxyntocardiac mucosa is an end point that is benign.

disease but is extremely useful as an indicator for the risk of adenocarcinoma; present belief, probably correct, is that esophageal adenocarcinoma develops only in intestinal metaplasia.

The presence of cardiac mucosa (= reflux carditis) on a biopsy is the best diagnostic criterion for reflux disease (Fig. 14.2). It is 100% specific; no disease other than gastroesophageal reflux causes the complex series

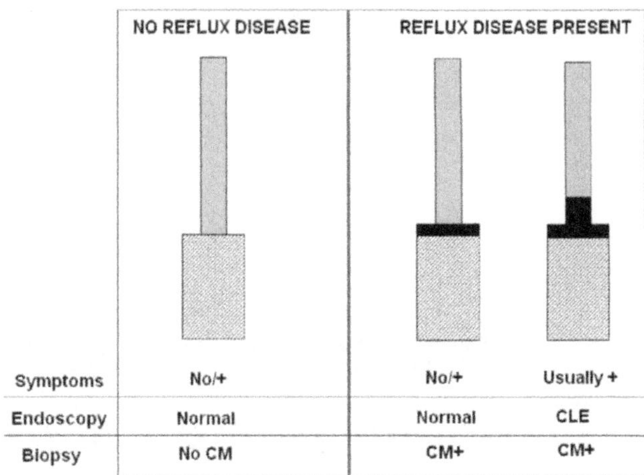

FIGURE 14.2 The diagnostic criterion for reflux disease is the presence of cardiac mucosa (= reflux carditis) in a biopsy. Patients who do not have cardiac mucosa do not have reflux disease; these patients are usually asymptomatic and are endoscopically normal. A clinically identical person (i.e., asymptomatic and endoscopically normal) who has cardiac mucosa immediately distal to the squamocolumnar junction has reflux disease. When a columnar-lined segment is endoscopically visible in the tubular esophagus, patients are usually symptomatic and almost always have cardiac mucosa in a biopsy taken immediately distal to the squamocolumnar junction. This is true irrespective of the endoscopic impression of location of biopsy in relation to the gastroesophageal junction.

of changes in the squamous epithelium that result in cardiac mucosa.

The sensitivity of reflux carditis in the diagnosis of gastroesophageal reflux is high but not 100%. False negatives occur for three reasons:

1. Sampling error, even four-quadrant endoscopic biopsy, usually samples less than 25% of the circumference of the junction and may miss small amounts of cardiac mucosa. Sampling error is much less at esophagectomy and autopsy where the entire circumference can be examined.
2. In a small number of patients, reflux damage of the squamous epithelium may not have caused cardiac metaplasia. This is very uncommon except in young children; most adult patients with significant reflux have columnar metaplasia of the esophagus even when the squamous epithelium is histologically normal.
3. All the metaplastic cardiac mucosa may have changed to oxyntocardiac mucosa. This is not uncommon in patients who have less than 1 cm of columnar metaplasia of the distal esophagus (1).

For these reasons, a suitable series of biopsy specimens will provide the diagnostic information presented in Table 14.2.

TABLE 14.2 Interpretation of Histologic Findings of Biopsies at the Squamocolumnar Junction

Finding	Interpretation
Gastric oxyntic mucosa and squamous epithelium only (Fig. 14.3)	No evidence of gastroesophageal reflux
Oxyntocardiac mucosa +/− squamous epithelium and gastric oxyntic mucosa (Fig. 14.4)	Evidence of mild gastroesophageal reflux (compensated reflux)
Cardiac mucosa +/− oxyntocardiac mucosa +/− squamous epithelium and gastric oxyntic mucosa (Fig. 14.5)	Gastroesophageal reflux disease
Intestinal metaplasia in cardiac mucosa +/− cardiac mucosa +/− oxyntocardiac mucosa +/− squamous epithelium and gastric oxyntic mucosa (Fig. 14.6)	Barrett esophagus

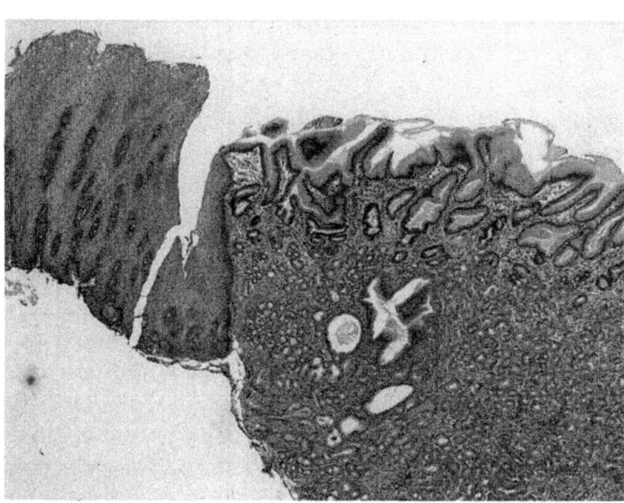

FIGURE 14.3 Section across the squamocolumnar junction showing squamous epithelium directly transitioning to gastric oxyntic mucosa. In this patient, there is no columnar-lined esophagus, and the squamocolumnar junction and true gastroesophageal junction are coincidental.

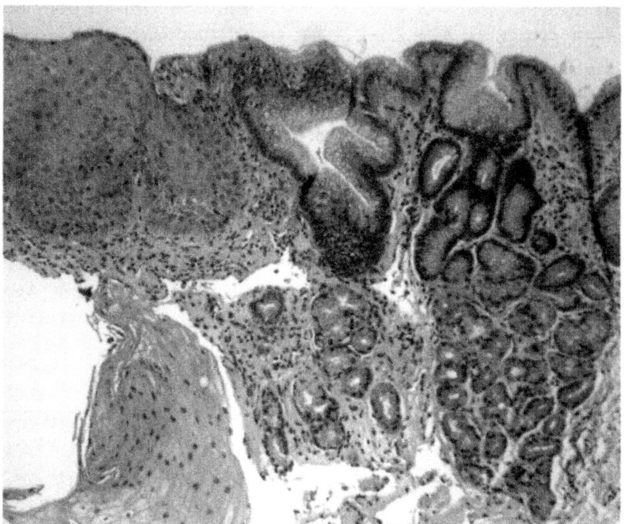

FIGURE 14.4 Squamous epithelium with oxyntocardiac mucosa immediately distal to it. Two foveolar-gland complexes are shown; the one on the right has numerous parietal cells; the one on the left has few parietal cells admixed with mucous cells. This patient has evidence of reflux-induced columnar metaplasia but has no cardiac mucosa. Because the defining criterion for reflux disease is cardiac mucosa, this patient has no reflux disease. We call this compensated reflux. (Cross reference: Color Plate 3.13)

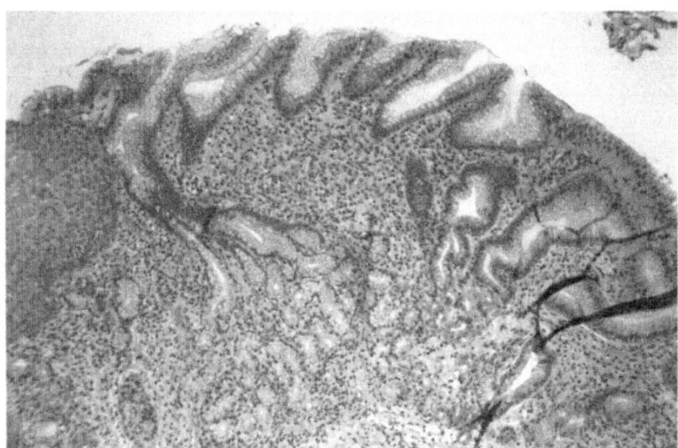

FIGURE 14.5 Biopsy straddling the squamocolumnar junction showing a small area of cardiac mucosa with irregular foveolar hyperplasia and chronic inflammation (reflux carditis) between the squamous epithelium (*left*) and parietal cells containing oxyntocardiac mucosa (*right*). This is evidence of reflux disease in this patient who was endoscopically normal. (Cross reference: Color Plate 3.8)

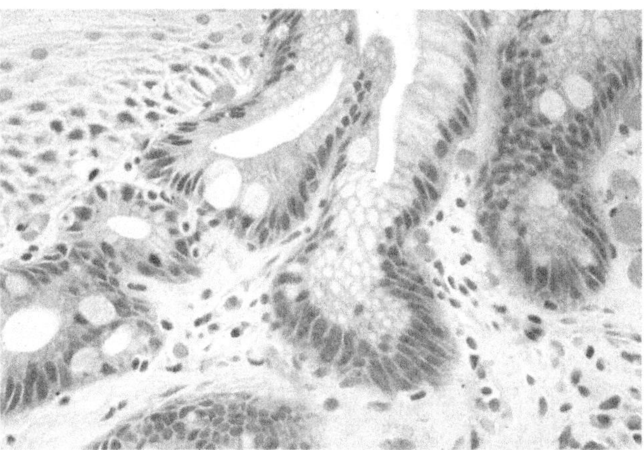

FIGURE 14.6 Biopsy straddling the squamocolumnar junction showing intestinal metaplasia (goblet cells in cardiac mucosa) adjacent to squamous epithelium. This represents Barrett esophagus, whatever the clinical features and endoscopic appearance.

A NEW DIAGNOSTIC METHOD FOR REFLUX DISEASE

The availability and recognition of a reliable and accurate histologic diagnostic test for reflux disease permit the development of a new method of assessing patients for gastroesophageal reflux (2, 3). This can provide scientifically validated information regarding the presence of reflux disease at a cellular level, the severity of reflux disease at a cellular level, and the risk of adenocarcinoma existing at the time of the biopsy.

It is only by having a standard validated system for diagnosis that we can convert the chaos and confusion that exists in the literature to something that has a semblance of scientific method. The method will provide standardized and reproducible data that can lead to an exact determination of etiology and risk for these patients upon which management protocols can be based.

The method of diagnosis of reflux disease is simple. It requires the endoscopist and the pathologist examining an esophagectomy specimen to recognize the following factors (Table 14.3): (a) that the determination of the Z line is a reliable marker of the proximal limit of any columnar-lined esophageal segment that may be present; (b) that we must recognize that the end of the tubular esophagus and the proximal limit of the rugal folds as endoscopic markers that are reliably assessed but do not define the true gastroesophageal junction (Fig. 14.7); (c) that there is no method other than histology to reliably differentiate columnar-lined esophagus from gastric mucosa; (d) that the true gastroesophageal junction can only be defined histologically by the proximal limit of gastric oxyntic mucosa, which is a reliable marker of the distal limit of the columnar-lined esophageal segment that may be present; (e) that the true gastroesophageal junction is 0 to 3 cm distal to the point of the end of the tubular esophagus or the proximal limit of the rugal folds, depending on the severity of reflux damage to the distal esophagus (0 = normal; increasing lengths are correlated with increasing reflux damage and end-stage dilated esophagus) (Fig. 14.8).

Once these issues are recognized, the diagnostic method aims at histologically defining the extent of the columnar-lined esophagus that may be present between the true gastroesophageal junction (the proximal limit of gastric oxyntic mucosa) and the squamocolumnar junction. This is achieved by a biopsy protocol (Fig. 14.9, Table 14.4). Ideally this begins 3 cm (2 cm is probably more than adequate in practice) distal to the end of the tubular esophagus or the proximal limit of the rugal folds, whichever appears to be the better defined. Four quadrant biopsies are taken at 1 to 2 cm intervals all the way up to the squamocolumnar junction. The decision between 1 and 2 cm intervals can be based on the length of visible columnar-lined esophagus. In patients without a visible columnar-lined esophagus, biopsies should be taken immediately distal to the squamocolumnar junction, attempting to straddle the actual junction, at 1 cm distal to this point and 2 cm distal to this point; for the most compulsive, an additional series of biopsies 3 cm distal will guarantee that the true gastroesophageal junction is reached. As shown by Jain et al. (4), who actually follow this suggested biopsy protocol to perfection in their study, the

TABLE 14.3 Endoscopic Landmarks and Their Meanings

Landmark	Precision	Normal	Abnormal	Significance
Squamocolumnar junction (= Z line)	High	Horizontal; at end of tubular esophagus	Serrated; migrated proximally into tubular esophagus	Defines the proximal limit of columnar lined esophagus
End of tubular esophagus	Moderate	Sharp; 2–3 cm below diaphragm	Ill-defined; flared; closer to diaphragm	None
Diaphragm	Low	Pinchcock appearance	Not detectable	Of value only in detecting hiatal hernia
Proximal limit of rugal folds	Moderate	Coincident with SCJ at end of tubular esophagus	Separated from SCJ by flat columnar epithelium	Best endoscopic estimate of GEJ; 0–3 cm proximal to true GEJ
Rugated mucosa	Moderate	Lines entire stomach	Lines entire stomach + end-stage dilated esophagus	Composed of oxyntic mucosa (stomach) or metaplastic esophageal columnar mucosa (proximal 0–3 cm)
Flat columnar mucosa	Moderate	Normally absent	Seen between SCJ and proximal limit of rugal folds	Composed of metaplastic esophageal columnar mucosa or atrophic gastric mucosa

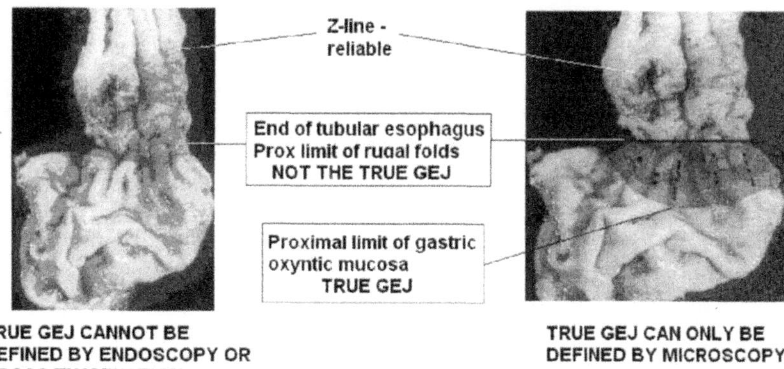

FIGURE 14.7 Esophagectomy specimen showing a long segment of columnar-lined esophagus. The recognizable gross pathologic (= endoscopic) landmarks are the squamocolumnar junction, the end of the tubular esophagus, and the proximal limit of rugal folds. In this specimen, the end of the tubular esophagus and proximal limit of rugal folds are coincidental. It is a mistake (made presently by definition) to designate the end of the tubular esophagus or the proximal limit of the rugal folds as the true gastroesophageal junction; these are simply endoscopic landmarks. The true gastroesophageal junction can be defined only by histologic mapping by the proximal limit of gastric oxyntic mucosa (*right*), which correlates with the distal limit of esophageal submucosal glands (black dots). (Cross references: Color Plates 3.5 and 3.6)

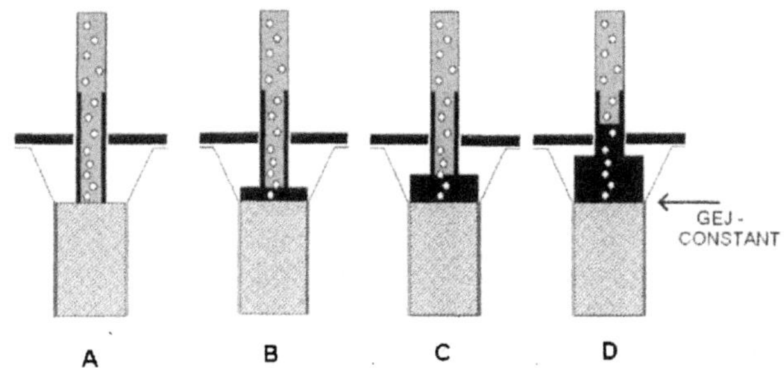

FIGURE 14.8 Reflux-induced cardiac mucosa. In the earliest stage, the cardiac mucosa is restricted to the dilated end-stage esophagus (b–c) with increasing lengths of cardiac mucosa separating the squamocolumnar junction from the gastric oxyntic mucosa. This early disease is presently regarded as normal. In the later stage, cardiac mucosa is visible in the tubular esophagus (d) in addition to that in the dilated end-stage esophagus. The true gastroesophageal junction (GEJ), which is the proximal limit of the oxyntic mucosa and the submucosal esophageal glands (white dots), remains constant. The reflux-damaged distal esophagus dilates.

probability of finding of cardiac mucosa 2 cm distal to the end of the tubular esophagus/proximal limit of rugal folds is very low (1% to 5%).

An *adequate biopsy series* is one that has squamous epithelium above the squamocolumnar junction proximally and gastric oxyntic mucosa in the most distal biopsy. Measured biopsies are always preferable; however, in the region beyond the end of the tubular esophagus, biopsies maybe difficult without retroflexing the endoscope. In these cases, a retroflex biopsy is acceptable but limits the ability to measure the columnar-lined segment because it is an unmeasured biopsy. The pathologic examination of this biopsy series consists of recording the epithelial types seen in each biopsy level. This permits mapping of the epithelial types that constitute the columnar-lined segment and their distribution within the segment (1) (Fig. 14.10).

The data recorded in this series of biopsies can now be analyzed to provide the answers to the following questions.

Does the Patient Have Evidence of Reflux-Induced Cellular Change?

In the normal person without any reflux damage, there will be no columnar-lined esophagus (i.e., cardiac mucosa with and without intestinal metaplasia or oxyntocardiac mucosa). The squamous epithelium of the esophagus transitions directly to gastric oxyntic mucosa (Figs. 14.2 and 14.3). This is designated as being a person without any evidence of reflux disease. Such patients are almost never encountered in clinical practice; they are asymptomatic, have no visible endoscopic abnormality, do not present to physicians, and can be seen only in autopsy studies that include young children or in clinical studies that include biopsy of large numbers of volunteers or asymptomatic patients with no endoscopic abnormality (5).

In most "normal" adults, the occurrence of chronic subclinical gastroesophageal reflux at a level below the threshold for abnormality in a 24-hour pH test will

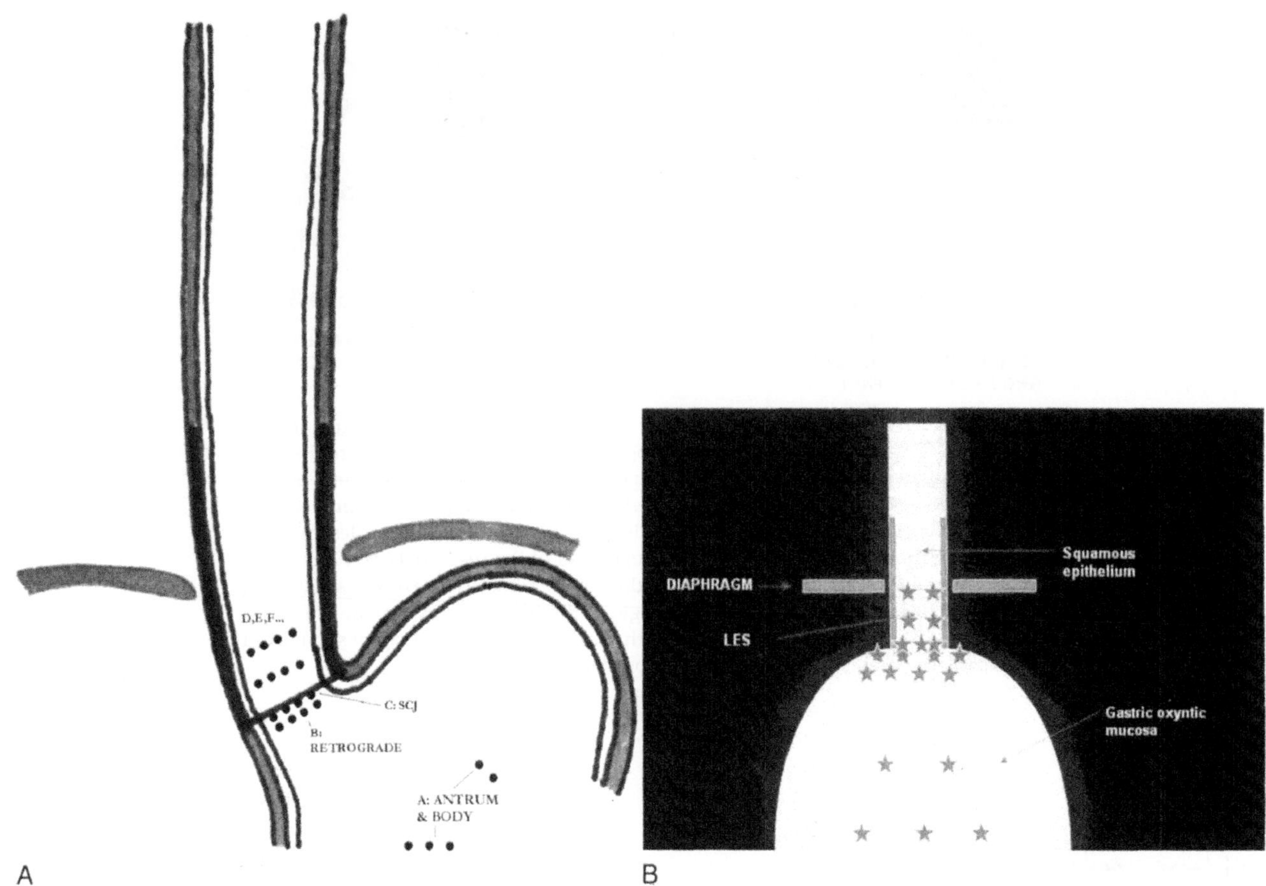

FIGURE 14.9 Biopsy protocol that is followed for complete evaluation of a patient with regard to detecting changes caused by gastroesophageal reflux; (a) shown in anatomic diagram; (b) shown diagrammatically. The aim of the biopsy series is to map the epithelium between the squamocolumnar junction and the proximal limit of gastric oxyntic mucosa, which is 0 to 3 cm distal to the proximal limit of rugal folds/end of tubular esophagus.

TABLE 14.4 Recommended Ideal Biopsy Protocol (Fig. 14.8)

Biopsy location	Circumstance	Purpose
A. At SCJ, attempting to straddle the junction	Always	Define the proximal limit of CLE; define epithelium at proximal limit of CLE; diagnose Barrett esophagus
B. 2–3 cm distal to end of tubular esophagus	Always	Find oxyntic mucosa to define distal limit of CLE and true GEJ
C. 1 cm distal to end of tubular esophagus	Always	Define reflux disease in the dilated end-stage esophagus; diagnose microscopic Barrett esophagus
D, E, F . . . Measured biopsies at 1–2 cm intervals between A and C	When CLE is seen endoscopically	Define length of CLE; map epithelial composition of CLE
X. Gastric antrum and body	Always	Define coexisting gastric pathology

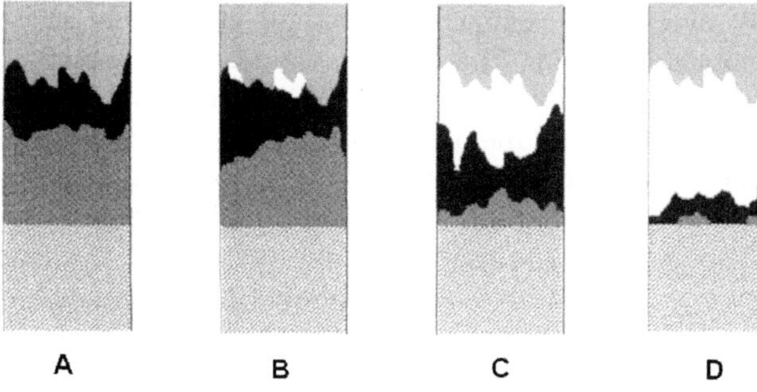

FIGURE 14.10 Vertical section of four examples of columnar metaplasia of equal length to show the possible variation in epithelial types (squamous epithelium—light gray; intestinal metaplasia—solid white; cardiac mucosa—solid black; oxyntocardiac mucosa—dark gray; gastric oxyntic mucosa—diagonal stripes). Patient A has no intestinal metaplasia and has a large amount of oxyntocardiac mucosa. Patients B, C, and D have increasing amounts of intestinal metaplasia and decreasing amounts of oxyntocardiac mucosa. These patterns are evaluable only by histologic mapping; by endoscopy, columnar epithelial types are indistinguishable from one another.

have produced very small amounts of columnar-lined esophagus (Fig. 14.8). When the high frequency of heartburn in the population is considered, it should not be surprising that minimal cellular changes produced by gastroesophageal reflux are present in almost everybody. The situation is analogous to finding anthracosis in the lung and atherosclerotic plaques in the aorta in urban American adults.

The distinction between reflux and reflux-induced cellular change is emphasized. Reflux is defined by a 24-hour pH test or impedance test and classified as normal or abnormal by quantitative criteria. While a good correlation exists between abnormal reflux in these tests and the presence of metaplastic columnar epithelium, these are independent characteristics.

Does the Patient Have Reflux Disease?

By our definition, patients with cardiac mucosa (= reflux carditis) have reflux disease. Patients without cardiac mucosa do not have reflux disease. Cardiac mucosa is not mimicked by any other epithelial type. If strict diagnostic criteria are adhered to (i.e., a columnar epithelium containing only mucous cells), the diagnosis is easy, precise, and not subject to any significant error. It is to be noted that cardiac mucosa may become intestinalized in a small number of patients; this defines Barrett esophagus and has the same significance as cardiac mucosa in defining the presence of reflux disease. In addition, Barrett esophagus indicates progression in the reflux-adenocarcinoma sequence (Fig. 14.1).

Because we have defined reflux disease as the presence of reflux carditis, patients who only have oxyntocardiac mucosa without cardiac or intestinal epithelium will be designated as not having reflux disease (Fig. 14.4). We refer to patients who have only oxyntocardiac mucosa in addition to squamous and oxyntic mucosa as having "compensated reflux" because, while they have evidence of reflux-induced cellular damage, they have no epithelia that are at risk to progress to intestinal metaplasia or adenocarcinoma (Fig. 14.1). Patients with only oxyntocardiac mucosa are rarely symptomatic, have a normal endoscopy, and have a histologically defined columnar metaplastic segment that is less than 1 cm in length. When the columnar metaplastic segment reaches 1 cm in length or when it is endoscopically visible, cardiac mucosa is almost always present (1).

When a patient has a very short segment of cardiac mucosa, it is located on either side of the end of the tubular esophagus or the proximal limit of rugal folds. In the less than 5-mm range, it is not usually visible by endoscopy and gross examination. Even with longer segments, cardiac mucosa may not be visible when it largely occupies the dilated end-stage reflux damaged esophagus. In this setting, cardiac mucosa is located distal to the end of the tubular esophagus and the mucosa may be rugated, depending on the degree of atrophy and inflammation in the metaplastic epithelium.

It must be recognized that the biopsies only give information at the point in time when the biopsy is performed. Patients with no evidence of reflux disease (i.e., reflux carditis) can develop cardiac mucosa in the future; patients with no intestinal metaplasia can develop intestinal metaplasia in the future. The significance of absence of reflux carditis is dependent on the age of the patient. If the patient is 60 years old and has no cardiac or intestinal mucosa, it is likely that the gastroesophageal reflux present is so minimal that it is unlikely that the patient will develop cardiac and intestinal mucosa in the future. This is different from how those conditions would effect a 10-year-old child; the risk of future cardiac and intestinal mucosa is much higher because the cellular changes may be evolving.

Reflux carditis establishes a cellular basis for the diagnosis of reflux disease. Patients who have reflux disease histologically can now be classified into various subgroups: (a) classically symptomatic, atypically symptomatic, mildly symptomatic, asymptomatic; (b) with or without an abnormal 24-hour pH test; or (c) with or without erosive esophagitis.

Reflux carditis also provides a method of evaluating progession and regression of the disease. Progression can be defined as follows (Fig. 14.11): (a) an

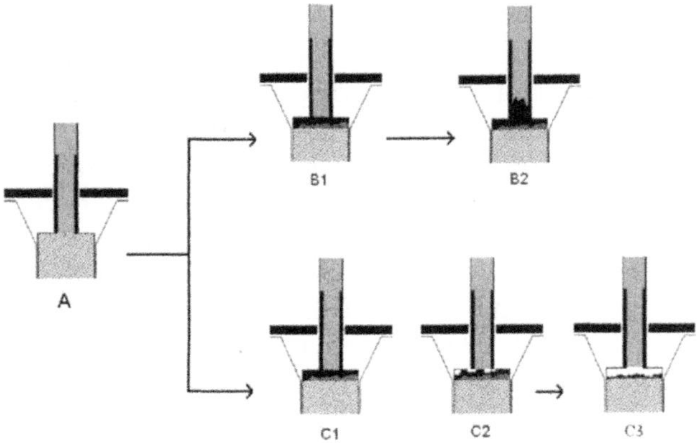

FIGURE 14.11 Progression of reflux disease from normal (a) to increasing columnar metaplasia with increasing cardiac mucosa (solid black; b1 and b2). A second type of progression results in the occurrence of intestinal metaplasia (solid white) of increasing amounts in the columnar-lined esophagus, which remains constant in length (c2 and c3). Combinations of the two methods is common.

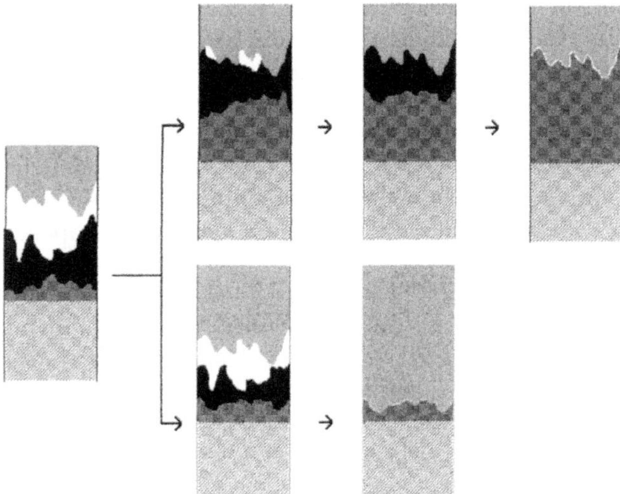

FIGURE 14.12 Regression of reflux disease from reflux carditis (solid black) with intestinal metaplasia (solid white) to reduction in the amount of intestinal metaplasia and increase in the amount of oxyntocardiac mucosa (stippled gray) without decrease in the length of columnar-lined esophagus (upper frames). In the lower frames, there is squamous replacement of intestinal and cardiac mucosa. In both series, the end result of regression (= cure) is the absence of cardiac and intestinal metaplasia, leaving only the benign squamous and oxyntocardiac epithelia in the esophagus.

increase in the amount of cardiac mucosa, by continuing metaplasia of squamous epithelium; or (b) development of intestinal metaplasia, dysplasia, or adenocarcinoma, which represent progression in the reflux-adencoarcinoma sequence, with and without an increase in the amount of cardiac mucosa.

Regression can be defined as follows (Fig. 14.12): (a) An increase in the amount of squamous epithelium, by reversal of cardiac or intestinal mucosa to squamous epithelium. This causes a decrease in the surface area of columnar-lined esophagus. Care must be taken here; squamous metaplasia occurring on the surface may contain metaplastic glandular epithelium in the lamina propria. This is invisible and may undergo dysplasia and malignancy while the surface appears normal. (b) An increase in the amount of oxyntocardiac mucosa, by conversion of cardiac mucosa to oxyntocardiac mucosa. (c) The reversal of intestinal metaplasia, by conversion to cardiac mucosa.

Cure of reflux disease is defined as the removal of all cardiac and intestinal metaplasia from the columnar-lined esophagus (Fig. 14.12). These are the only epithelial types that are in the reflux-carcinoma sequence, and their removal removes any risk of adenocarcinoma, which is the end point in our diagnostic scheme. The changes relating to regression of cardiac and intestinal epithelia and increase in the amount of oxyntocardiac mucosa cannot be detected endoscopically because the three columnar epithelial types are indistinguishable; only histologic mapping can provide this information. At present, regression to the extent of producing complete conversion to oxyntocardiac mucosa is rare and is seen only in patients with short segments of columnar-lined esophagus after successful antireflux surgery. However, this does not mean that regression is impossible; it is an excellent future goal to be aimed for.

Like any other common pathologic finding, careful evaluation is necessary before the pathologic change is ascribed as a cause of the patient's disease. For example, elderly patients who present with squamous carcinoma of the esophagus frequently have small amounts of cardiac mucosa at the junction (6). This represents an incidental finding; it does not mean that there is a relationship between squamous carcinoma and reflux disease. The situation is analogous to a patient with acute viral myocarditis who has coronary atherosclerosis; both are pathologic lesions, but they are unrelated to one another.

What Is the Severity of Reflux Disease in the Patient?

There are abundant data in the literature to show that the severity of reflux disease is related to the amount of squamous epithelium that has undergone columnar metaplasia (7, 8) (Fig. 14.13, Table 14.5). This is best assessed as the length of columnar metaplasia (see Chapter 8).

In the recommended biopsy series, the length of columnar-lined esophagus is determined by the maximum separation of measured biopsies that show

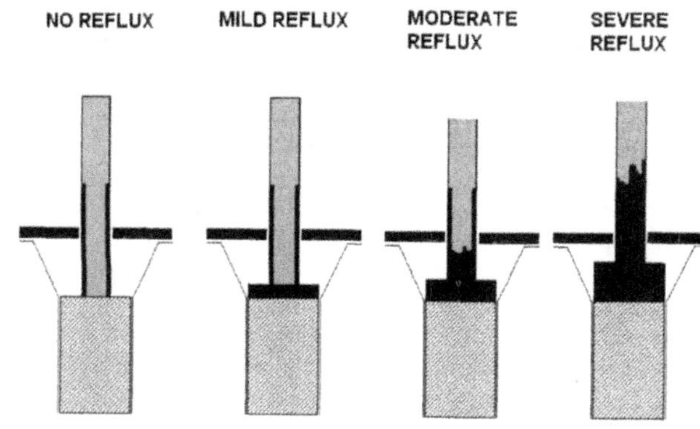

FIGURE 14.13 The severity of reflux disease is directly proportional to the length of columnar-lined esophagus (solid black area) that is interposed between the cephalad-displaced squamocolumnar junction and the proximal limit of gastric oxyntic mucosa (the true gastroesophageal junction).

TABLE 14.5 Assessment of Severity of Reflux Disease by the Measure of the Length of Columnar-Lined Esophagus Determined by Histologic Mapping

Criterion	Histologic length[b]	Interpretation
CLE[a] absent	0	No evidence of reflux
CLE[a] not visible endoscopically	<1 cm	Mild reflux disease
CLE[a] visible, short	1–2 cm	Moderate reflux disease
CLE[a] visible, long	>2 cm	Severe reflux disease

[a] Columnar-lined esophagus; this is defined histologically by the presence of oxyntocardiac and cardiac mucosa, the latter with and without intestinal metaplasia.

[b] Histologic length is defined by mapping biopsies taken at measured levels as the separation between squamous epithelium and gastric oxyntic mucosa; endoscopic correlations are secondary.

cardiac mucosa with or without intestinal metaplasia and oxyntocardiac mucosa (Fig. 14.8). In patients with intestinal metaplasia, this is the IM + CM + OCM length. In patients without intestinal metaplasia, this will be the CM + OCM length. In resection specimens and at autopsy, this length is determined by direct measurement in a vertical section taken from the squamocolumnar junction to oxyntic mucosa.

What Is the Risk of Cancer in These Patients?

If the presently accepted dogma that cancer arises only in intestinal metaplastic epithelium is correct, as we believe it is, the risk of cancer can be assessed by the presence of intestinal metaplasia. A patient with reflux disease who does not have intestinal metaplasia does not have any risk. However, if the patient has cardiac mucosa, this represents an epithelium that has the potential to undergo intestinal metaplasia in the future. As such, though cardiac mucosa has no cancer risk, there is the possibility that the patient will develop a risk in the future if intestinal metaplasia develops. This risk of transformation decreases with the increasing age of the patient because the more time that elapses without the development of intestinal metaplasia, the more likely is it that the milieu of the refluxate is one that has a low risk of intestinal metaplasia.

The assessment of cancer risk in patients with intestinal metaplasia is best achieved at the present time by the assessment of dysplasia. Other methods are not considered significant. Short-segment Barrett esophagus has a risk that is considered similar enough to long-segment disease to be managed by a similar surveillance protocol (9). The risk associated with Barrett esophagus restricted to reflux disease in the dilated end-stage esophagus is unknown because it has not been studied.

In Chapter 10, we suggested that new criteria to assess cancer risk by assessing volume of intestinal metaplasia are worthy of study. These include the following:

1. Assessment of the amount of intestinal metaplasia by the IM:OCM ratio (adjusted to the length of columnar-lined esophagus) or the IM number. The mapping of epithelial types in the different measured biopsies permits accurate quantitation of the amount of intestinal metaplasia within the columnar-lined segment. This will permit the evaluation of data that correlates the amount of intestinal metaplasia (the target epithelium for cancer) with future cancer risk. At present, Barrett esophagus is classified as short segment and long-segment based on the *endoscopic length* of columnar-lined esophagus irrespective of the amount of intestinal metaplasia that maybe present (Fig. 14.9). It is not surprising that studies based on this classification sometimes fail to find a difference in risk with different lengths of Barrett esophagus (6).
2. Flow cytometric study to evaluate aneuploidy and increased 4N fraction; and molecular markers (see criterion 3) (10). Flow cytometry presents technical difficulties and adds significant cost, which makes it unlikely to gain acceptance in standard practice despite its potential usefulness.
3. Molecular changes in intestinal metaplastic epithelium. At present, there are no accepted molecular tests that predict increased risk of carcinoma. Some suggestion exists for the possibility of using

expression of p53 and abnormalities in the proliferative patterns in the epithelium. An aberrant pattern of proliferation shown by expression of the proliferative marker Ki67 in the surface and superficial foveolar region was shown by Olvera et al. (11) to have a significant association with high-grade dysplasia. These molecular tests are expected to provide useful practical tests for predicting risk of carcinoma in the future, but none exists at the present time.
4. Dysplasia in intestinal metaplasia (discussion later).

A Classification of All People Based on Their Cellular Status as It Relates to Gastroesophageal Reflux

Combination of all these items of information permits the classification of the entire population into histologically defined groups/grades based on the presence of cellular changes of reflux, the severity of the cellular changes of reflux, and the cancer risk (3). The primary division is based on the risk of cancer, which must be the focus of any system. Placement of an individual within these groups provides a complete cumulative lifelong assessment of that person's cellular response to gastroesophageal reflux within the broad groups that are defined by the risk of adenocarcinoma at the time of the biopsy (Table 14.6).

This diagnostic method is recommended as the primary method for evaluating patients for reflux disease. It is the basis for all correlations with symptoms, endoscopic abnormalities, 24-hour pH testing, and impedance testing. For example, asymptomatic patients can belong to grade 0, 1a, 2a, and all subgroups of grade 3, but they are unlikely to belong to grades 1b and 2b and only very rarely belong to grades 1c and 2c. This type of classification is of critical importance because it provides the only method of stratifying asymptomatic patients with regard to their cancer risk. When it is appreciated that up to 95% of patients with adenocarcinoma present for the first time with symptoms related to the cancer, the ability of this method to recognize risk before cancer develops is extremely significant.

Also critically important is that the method allows the recognition of patients who are not at risk for cancer. There is no way, by present criteria, that the medical community can guarantee any patient that he or she is not at risk for reflux-induced cancer. Patients with no symptoms and no endoscopic abnormality are at lower risk that patients with severe reflux of long duration and those with long-segment Barrett esophagus, but they cannot be declared risk free. In Lagergren et al.'s (12) study, 40% of patients with esophageal adenocarcinoma and 71% of patients with adenocarcinoma of the gastric cardia (their definition includes patients with short-segment Barrett esophagus) did not have symptomatic reflux by their definition.

In this diagnostic method, the 55% to 65% of the population without cardiac mucosa and intestinal metaplasia can be declared free of risk at the point of the biopsy. If this patient is over 40 years old, the likelihood of developing cardiac mucosa that progresses to intestinal metaplasia and cancer during the natural lifetime is probably zero. This is because the failure of the gastric milieu to produce at-risk columnar epithelia for the age of the patient indicates a low risk; unless the milieu changes for some reason, it is unlikely to cause new changes. Such patients need never have another endoscopy for reflux; they can be managed symptomatically if they have symptoms without worry about cancer.

TABLE 14.6 Classification Based on Risk of Adenocarcinoma of the Esophagus (the percentage of the population within each group is in parentheses)

Grade 0: No risk of adenocarcinoma (55%–65%)

Definition: No cardiac mucosa or intestinal metaplasia in any biopsy.

Subgroups:

- Group 0a: *Normal*: Only squamous epithelium and gastric oxyntic mucosa
- Group 0b: *Compensated reflux*: Oxyntocardiac mucosa as the only epithelium in the metaplastic columnar esophagus.

Grade 1: Reflux disease (30%–45%)

Definition: Cardiac mucosa (reflux carditis) present; oxyntocardiac mucosa also present in most cases.

Subgroups:

- Group 1a: *Mild reflux disease*: CM + OCM length < 1 cm.
- Group 1b: *Moderate reflux disease*: CM + OCM length = 1–2 cm.
- Group 1c: *Severe reflux disease*: CM + OCM length > 2 cm.

Grade 2: Barrett esophagus (5%–15%)

Definition: Intestinal metaplasia (goblet cells) present in cardiac mucosa; nonintestinalized cardiac mucosa and oxyntocardiac mucosa also present in most cases.

Subgroups:

- Group 2a: *Barrett esophagus in reflux disease limited to dilated end-stage esophagus (microscopic Barrett esophagus)*: Endoscopy normal.
- Group 2b: *Short-segment Barrett esophagus*: Endoscopy shows a columnar lined esophagus < 2 cm long.
- Group 2c: *Long-segment Barrett esophagus*: Endoscopy shows a columnar lined esophagus > 2 cm long.

Grade 3: Neoplastic Barrett esophagus

Definition: Dysplasia or adenocarcinoma present.

Subgroups:

- Group 3a: Low-grade dysplasia.
- Group 3b: High-grade dysplasia.
- Group 3c: Adenocarcinoma

Application of the New Diagnostic Method in Practice

The application of this new diagnostic method into practice depends on the availability of resources. One of the problems about this diagnostic technique is that it may be too time-consuming for use in everyday practice. The requirement for multiple biopsies adds significant time to an upper endoscopy. This problem should not negate the development of the method. The present intent of the method is to standardize diagnosis in the academic setting so that accurate information can be developed. Once this has been done, practice guidelines can be determined based on available resources and cost-effectiveness. The potential practical problems caused by the enormity of the problem and the fact that it may stretch available resources must not deter the development of methods to determine the facts.

The great advantage with this system is that it uses time-honored diagnostic methods that are based on cellular pathology. In all human diseases, understanding dramatically improves when the diagnostic method shifts from noncellular to cellular criteria. This has happened with gastritis and colitis where histologic criteria permit a more precise assessment than clinical and endoscopic criteria. The use of biopsy as the primary diagnostic modality removes the need for 24-hour pH testing and impedance testing and the newer sophisticated endoscopic methods of chromoendoscopy and magnifying endoscopy. The use of biopsy as the primary diagnostic modality makes the diagnostic criteria available to every gastroenterologist because it is dependent on only endoscopy and biopsy. It removes the elitism of the swallowing center with the ability to perform specialized testing. Moving the diagnosis to the mainstream can only be positive.

We will consider in detail the application of the new diagnostic criteria in Chapter 17. In that discussion, we will modify the theoretically ideal biopsy protocol presented here to a more practical protocol that limits the required biopsies considerably and makes it more amenable to use in community practice.

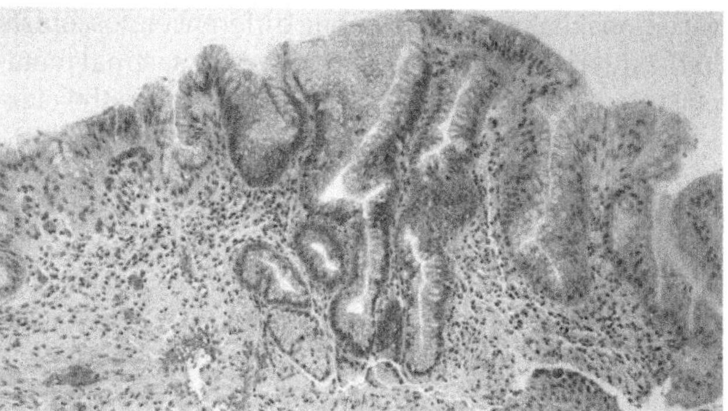

FIGURE 14.14 Biopsy showing cardiac mucosa with irregular foveolar hyperplasia and chronic inflammation (reflux carditis). This is the diagnostic criterion for reflux disease. Note that the mucosa on the left shows rudimentary foveolar pits, that on the right shows slightly more elongated foveolar pits, while the central region shows tortuous foveolar pits with possible glands. The epithelium contains only mucous cells and shows moderate chronic inflammation and mucin distension of some cells ("pseudo-goblet" cells). There are no goblet cells.

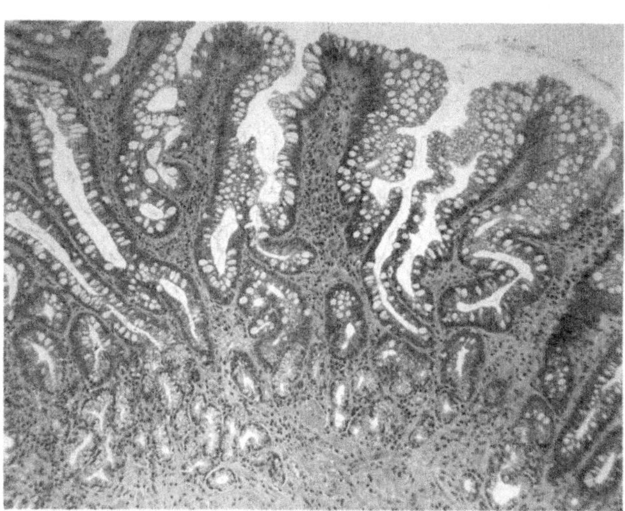

FIGURE 14.15 Cardiac mucosa showing marked hyperplasia of the foveolar region and extensive goblet cell change indicating intestinal metaplasia (Barrett esophagus). Note the presence of a small number of mucous glands below the intestinal foveolar region and the presence of irregular hyperplasia of the muscularis mucosae. (Cross reference: Color Plate 4.2)

THE DIAGNOSIS OF REFLUX CARDITIS AND BARRETT ESOPHAGUS

The diagnosis of reflux carditis is based on the accurate identification of cardiac mucosa in a biopsy (Fig. 14.14), which we considered in Chapter 5. The diagnosis of reflux disease by histology removes all possible errors and misconceptions that are derived from unproven endoscopic landmarks. Cardiac mucosa is always esophageal, irrespective of where the endoscopist believes is the location of the biopsy. There is no columnar mucosal type in this region that resembles cardiac mucosa; atrophic gastritis where there is loss of parietal cells (pseudopyloric metaplasia) is commonly associated with intestinal metaplasia.

The diagnosis of Barrett esophagus is based on the presence of intestinal metaplasia in cardiac mucosa (Fig. 14.15), also considered in Chapter 5. The recognition of histology as the mainstay of diagnosis removes variations that may exist among different endoscopists in the interpretation of landmarks. All intestinal metaplasia occurring in cardiac mucosa falls into the classification of Barrett esophagus without any regard to endoscopic landmarks.

When reflux carditis shows intestinal (Barrett-type) metaplasia, it can be difficult to differentiate it from chronic atrophic gastritis. This difficulty has been greatly exaggerated to the extent that this differentiation is believed impossible. This is far from true. Chronic atrophic gastritis involves large areas of the stomach in a multifocal fashion. The presence of a normal distal gastric biopsy in a patient with intestinal metaplasia in the gastroesophageal junctional region (i.e., the area within 3 cm from the proximal limit of rugal folds or the end of the tubular esophagus, which is the dilated end-stage esophagus) is diagnostic of Barrett esophagus. There is no "gastritis," atrophic or otherwise, restricted to the proximal 3 cm of the stomach. When chronic atrophic gastritis occurs in the proximal stomach, it is always associated with antral and corpus gastritis (13–15).

In rare cases where intestinalized cardiac mucosa (i.e., Barrett esophagus) coexists with atrophic gastritis, the distinction between the two can be difficult. In the vast majority of these cases, the following morphologic features permit their differentiation:

1. Atrophic gastritis occurs as multiple foci of involvement in gastric oxyntic mucosa, commonly with sharp demarcation between atrophic and non-atrophic oxyntic mucosa (Fig. 14.16).
2. Atrophic gastritis usually has a flat surface compared to the hyperplastic appearance of cardiac mucosa, which frequently has a villiform surface (Fig. 14.17).
3. Atrophic gastritis usually has flat, well-defined, and linear muscularis mucosae at the deep aspect of the foveolar region (Fig. 14.16), contrasting with reflux carditis where the muscularis mucosae are hyperplastic and have fibers extending into the mucosa in a vertical direction (Fig. 14.14).
4. It has frequent paneth cells in the foveolar pit (Figs. 14.16 and 14.17); though this can be seen in cardiac mucosa, it is relatively less common.
5. Atrophic gastritis is often associated with neuroendocrine cell hyperplasia including multiple micro-carcinoid tumors (Fig. 14.18). This results from the fact that atrophic gastritis is associated with hypochlorhydria resulting in elevated serum gastrin. The gastrin has a trophic effect on the neuroendocrine cells of the stomach, causing them to undergo hyperplasia. In very rare cases of persisting difficulty, the use of immunoperoxidase stains for cytokeratin 7 and 20 and DAS-1 have been suggested but are rarely of practical value. In our experience, cases of difficulty are uncommon.

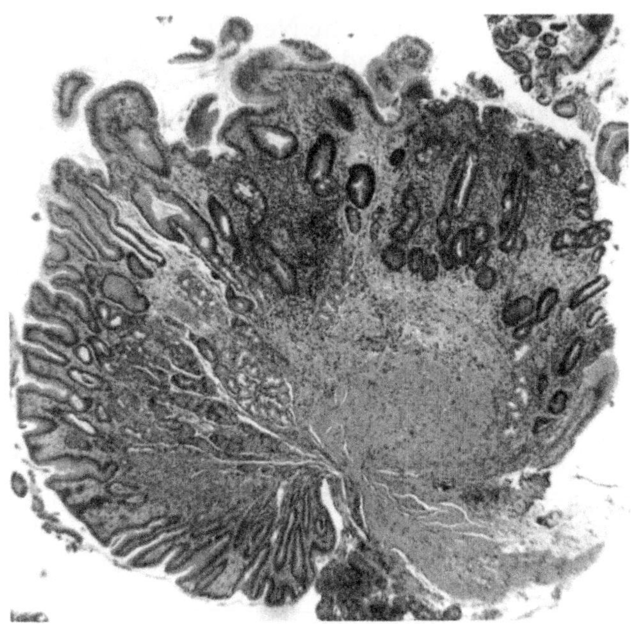

FIGURE 14.16 Chronic atrophic gastritis showing the sharp demarcation of the atrophic area from normal gastric oxyntic mucosa (*left*). The atrophic area is flat with an elongated foveolar region that shows goblet cells. Paneth cells are present in the basal region. The muscularis mucosae are flat. (Cross reference: Color Plate 4.6)

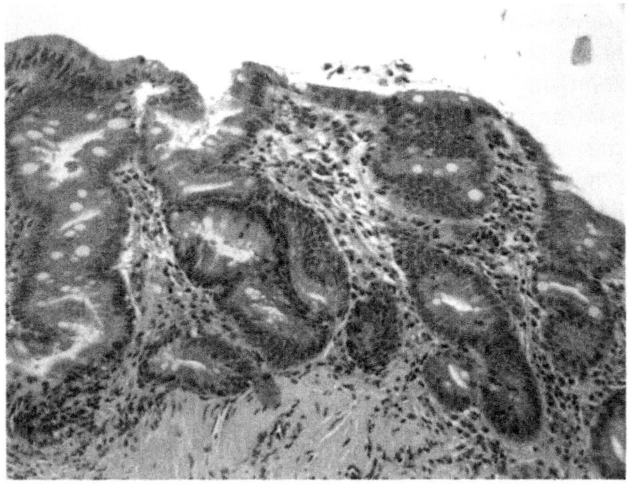

FIGURE 14.17 Chronic atrophic gastritis showing a flat epithelium with paneth cells at the base.

The supposed difficulty in differentiating between Barrett esophagus and atrophic gastritis has led some authorities to suggest that they are the same entity. This is absurd. It is often impossible by examination of a biopsy to differentiate a primary squamous carci-

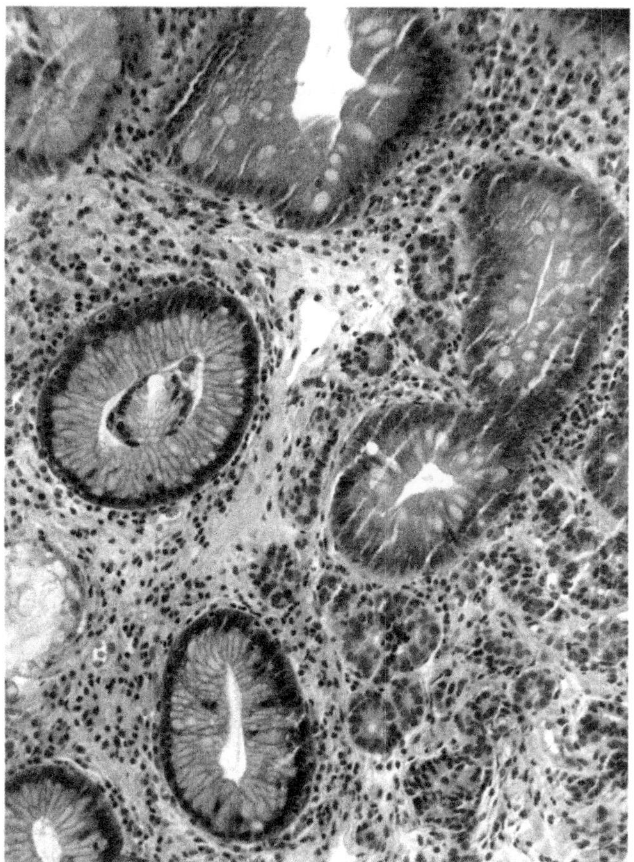

FIGURE 14.18 Neuroendocrine (enterochromaffin-like) cell hyperplasia associated with chronic atrophic gastritis. The ECL cells are seen as clusters of small cells with uniform round nuclei between the glands in the deep part of the mucosa.

noma of the lung from a metastatic squamous carcinoma to the lung; this does not mean that they are the same entity.

THE DIAGNOSIS OF DYSPLASIA IN BARRETT ESOPHAGUS

The aim of surveillance in patients with Barrett esophagus is the detection of dysplasia. The diagnosis of dysplasia in Barrett esophagus is an important clinical event because it indicates that the patient has progressed in the adenocarcinoma sequence. At present, the diagnosis of dysplasia is the best available evidence of progression in the reflux-adenocarcinoma sequence beyond the presence of intestinal metaplasia.

The diagnosis of dysplasia is entirely dependent on microscopic examination of biopsy samples from the segment of columnar-lined esophagus. Because dysplasia in Barrett esophagus occurs as a random process and because it occurs in a flat mucosa that is not distinguishable endocopically from nondysplastic columnar-lined esophagus, a biopsy protocol that ensures adequate sampling must be followed. The standard biopsy protocol in patients undergoing surveillance endoscopy after a diagnosis of Barrett esophagus requires the taking of four-quadrant biopsies at 1- to 2-cm intervals in the columnar-lined segment. With a 10-cm segment of columnar-lined esophagus, this amounts to a minimum of 24 biopsies.

This is not always easy to achieve in community practice where the taking of large numbers of biopsies adds significantly to the time and tedium of the procedure. As a result, biopsy protocols recommended by academic centers are commonly ignored in a busy community practice. In our experience, the number of biopsies taken during Barrett surveillance in many centers is significantly lower than that recommended.

The likelihood of detecting a random event by sampling is directly related to the amount of change present and the number of biopsies. Even with a stringent biopsy protocol, it is likely that less than 5% of the mucosal surface is subject to biopsy. With lesser sampling protocols, the sampled area is even less. The question as to whether this is adequate can only be determined by longitudinal studies.

Dysplastic change constitutes a spectrum of severity that is probably related to the degree of genetic abnormality in the cell. At the low end of the spectrum, the deviation from normal epithelium is difficult to discern. At the high end, the cells show changes that are identical with malignant cells and the only difference between the highest dysplasia and adenocarcinoma is the presence of invasion in the latter. Dysplasia occurring in the intestinal metaplastic epithelium is almost always associated with a failure of normal maturation of the cell. In intestinal epithelium, therefore, an early change of dysplasia is the failure to mature into goblet cells. The diagnosis of dysplasia in intestinal metaplasia can therefore be made in a columnar epithelium that does not contain goblet cells (16). This is still intestinal metaplasia as long as the diagnosis of dysplasia is correct.

The presence of dysplasia is an indication for some change in the regular management of the patient with Barrett esophagus. Because the degree of dysplasia generally shows a correlation with cancer risk and imminence, dysplasia is divided into low-grade and high-grade dysplasia (Fig. 14.19). This requires that a line is drawn between the two by developing criteria that can be reliably reproduced by pathologists and that, when applied, have clinical significance. As with all artificial separations, this division into low- and high-grade dysplasia has its problems, and there is a significant interobserver variation between pathologists.

Dysplasia is a cytologic abnormality in the cell that is characterized by nuclear enlargement, increased nuclear:cytoplasmic ratio, nuclear hyperchromasia with abnormalities in the distribution of chromatin, and decreased cytoplasm when compared with non-dysplastic Barrett epithelium (Figs. 14.20 to 14.28). Mitotic activity is present and often increased with the appearance of atypical mitotic figures (Figs. 14.23 and 14.25). The severity of the cytologic abnormality progressively increases, often with one criterion being more expressed than another, producing an infinite variation of histologic appearance. At some point in this process, the pathologist will decide that the cytologic abnormality is "severe." The exact point at which this occurs varies because definitive criteria cannot be stated that cover all the computations of change that are feasible.

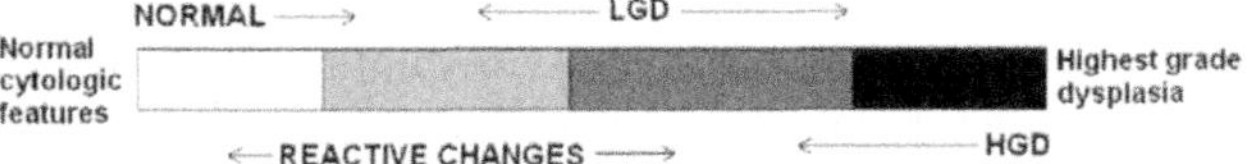

FIGURE 14.19 The spectrum cytologic changes in Barrett esophagus. At one end is absolutely normal intestinal metaplastic epithelium, and at the other end is the highest grade of dysplasia (to which the term "carcinoma *in situ*" is sometimes given). This is a continuous spectrum in which four diagnoses are possible: no dysplasia (normal), reactive changes, low-grade dysplasia, and high-grade dysplasia. There is some overlap in the cytologic features associated with these four entities, leading to significant interobserver variation because the lines drawn between the different diagnoses are not the same among every pathologist. The diagnosis "indefinite for dysplasia" includes some reactive cases and low-grade dysplasia where the associated changes overlap.

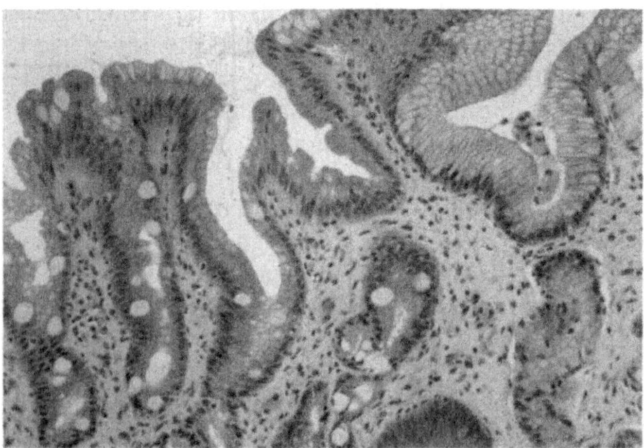

FIGURE 14.20 Intestinal metaplasia without dysplasia. The nuclei are flat and basal and resemble the nuclei of cardiac mucosa (*right*).

In the lesser forms of dysplasia, nuclear polarity is maintained. Normal nuclear polarity is defined as the presence of an elongated nucleus that is perpendicular to the basement membrane of the gland (Figs. 14.20 and 14.21). With increasing dysplasia, polarity is lost with the nuclei becoming more rounded and having their long axis becoming closer to parallel with the basement membrane (Figs. 14.22 to 14.27). At some

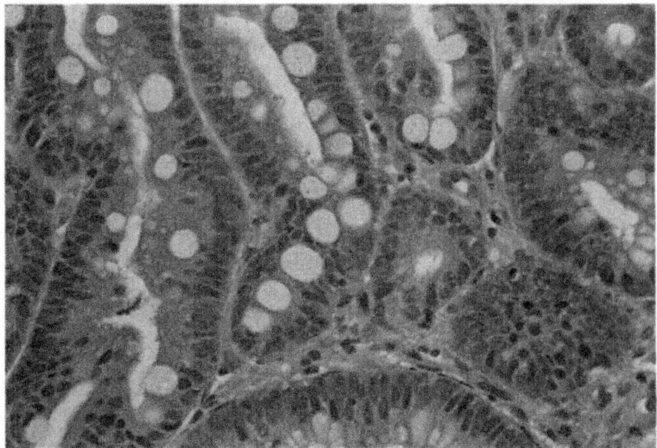

FIGURE 14.21 The cytologic changes of low-grade dysplasia include nuclear enlargement, stratification, and an increased nuclear:cytoplasmic ratio. In this example, goblet cells are still recognizable. These changes can be mimicked by cytologic changes associated with regeneration and active inflammation. (Cross reference: Color Plate 5.5)

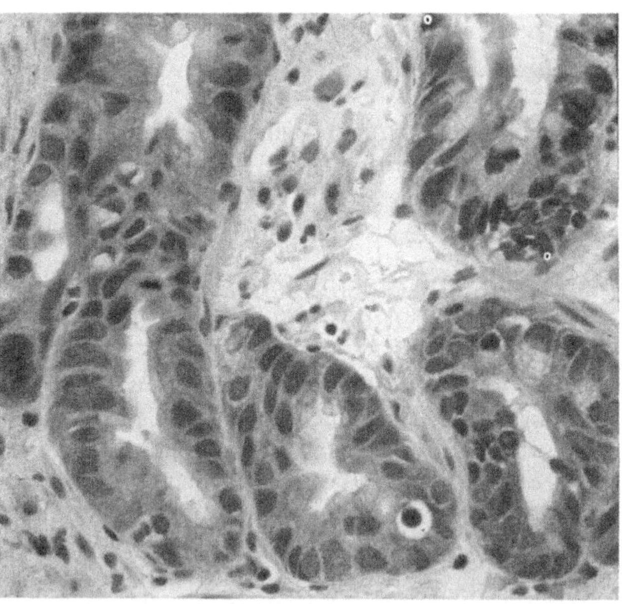

FIGURE 14.22 Mucin depletion in dysplasia resulting in a virtual absence of goblet cells. Dysplastic intestinal metaplastic epithelium may not have goblet cells. The glands retain their architecture, but nuclear polarity is altered (but not lost) as the nuclei begin to appear rounded.

point in this process, the pathologist will declare that nuclear polarity is "completely lost." Again, this point is not definable with precision.

In the lesser forms of dysplasia, the glands retain their simple round outline with a round lumen (Figs. 14.21 to 14.23). With increasing dysplasia, the glands tend to become complex with luminal bridges that split the lumen (Figs. 14.24 and 14.25). The best recognized form of this gland complexity is a cribriform architecture to the gland. At some point, the pathologist may decide that "gland complexity" is present. Again, this point is not precise and varies with different pathologists.

In the lesser degrees of dysplasia, the architecture of the epithelium retains its normal appearance. As dysplasia increases, the glandular architecture becomes altered by crowding (Fig. 14.26), microcystic

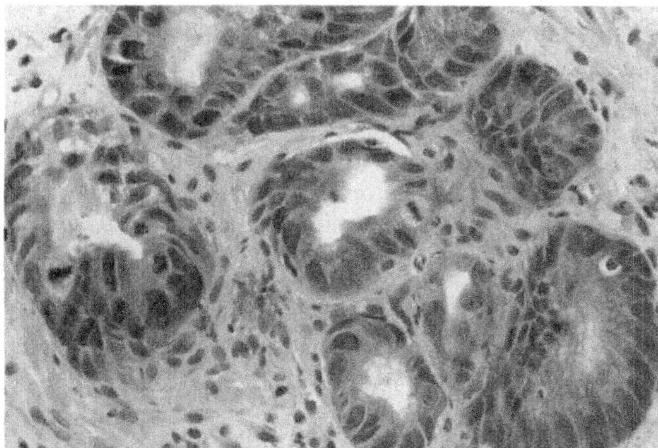

FIGURE 14.23 Another area in the same case as that in Figure 14.19 showing a slightly greater degree of cytologic abnormality and alteration of nuclear polarity. This is a borderline dysplasia where there will be considerable interobserver variation between low- and high-grade dysplasia if this was the worst or only area of dysplasia in the specimen. Our diagnosis was "low-grade dysplasia, recommend heightened surveillance."

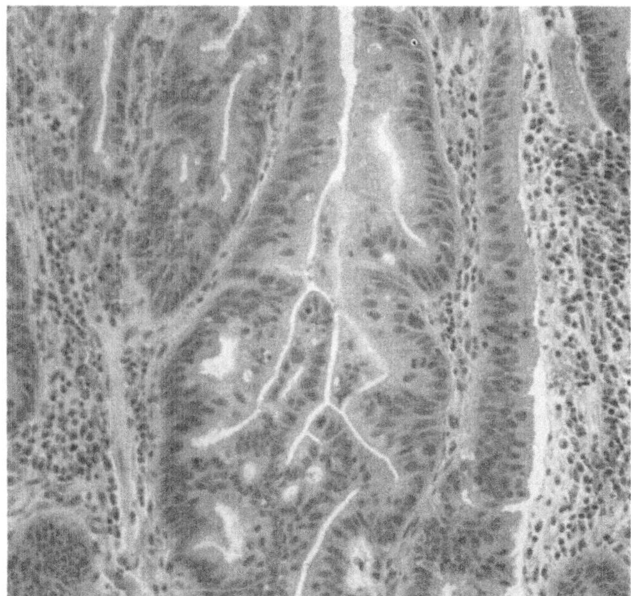

FIGURE 14.24 Barrett esophagus with cytologic changes within the range of low-grade dysplasia but with early gland complexity characterized by luminal bridging. In the absence of severe cytologic abnormality, a diagnosis of high-grade dysplasia is not made.

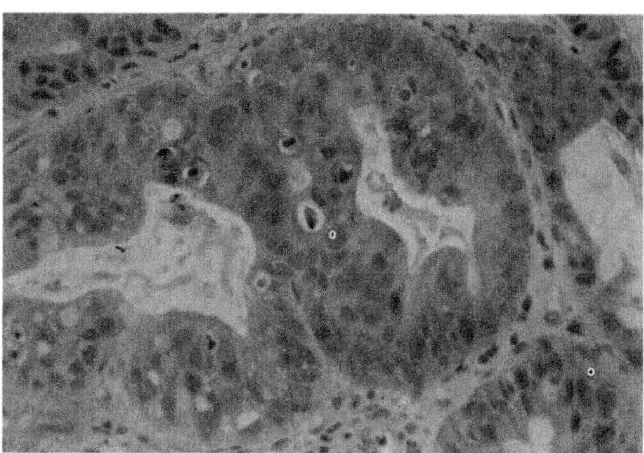

FIGURE 14.25 Diagnostic for high-grade dysplasia. The gland is enlarged and lined by cells that show severe cytologic abnormality. Nuclear polarity is completely lost with the nuclei being rounded. The gland is complex with a luminal bridge leading to early cribriform architecture. (Cross reference: Color Plate 5.7)

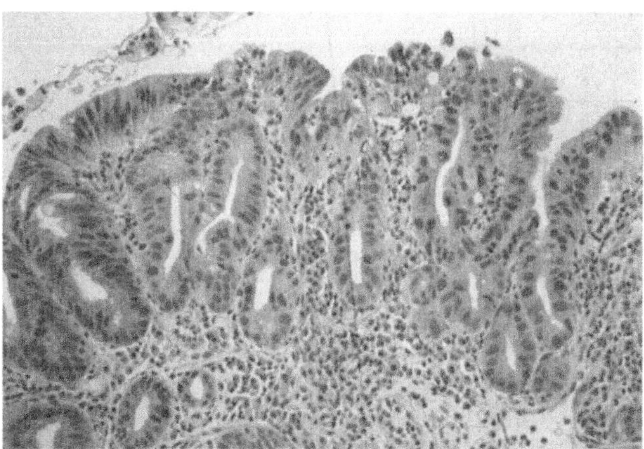

FIGURE 14.26 Dysplastic Barrett esophagus showing severe cytologic abnormality, gland crowding, and loss of nuclear polarity sufficient for a diagnosis of high-grade dysplasia. The dysplasia is clearly less severe than shown in Figure 14.22, and there is no gland complexity. (Cross reference: Color Plate 5.6)

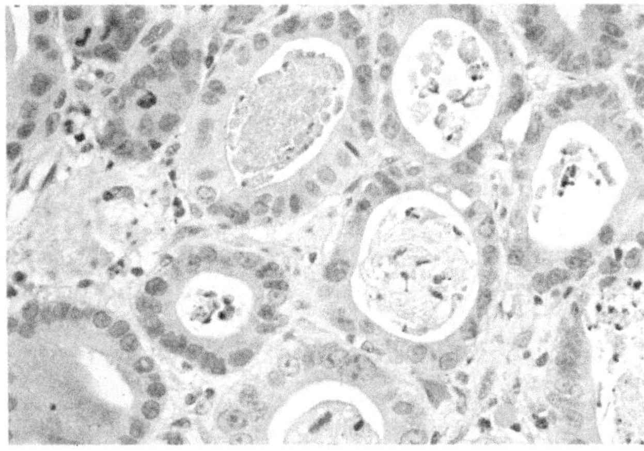

FIGURE 14.27 High-grade dysplasia characterized by microcystic dilatation of glands that contain necrotic debris in the lumen and are lined by a single layer of flattened severely dysplastic cells with complete loss of nuclear polarity. (Cross reference: Color Plate 5.8)

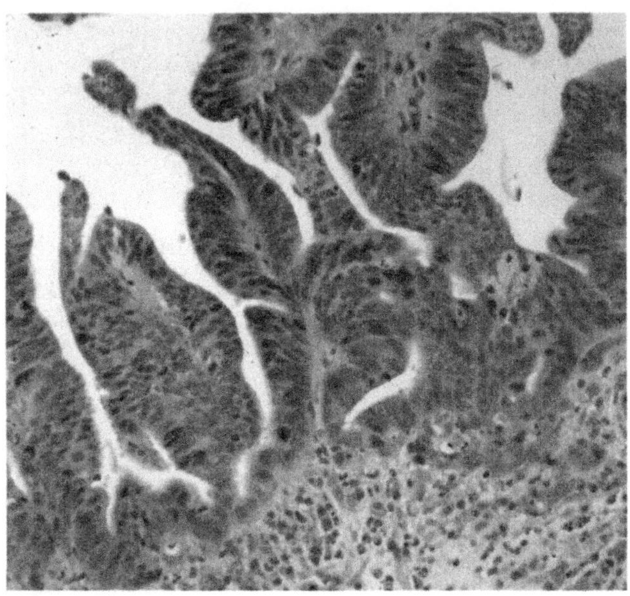

FIGURE 14.28 High-grade dysplasia. The architectural features of this biopsy show a villiform surface with stratified severely dysplastic cells. Despite the fact that an argument can be made that there is no complete loss of nuclear polarity, almost all pathologists will call this high-grade dysplasia.

change, often with necrotic debris in the lumen (Fig. 14.27), or surface villiform change (Fig. 14.28).

The criteria for the diagnosis of dysplasia in Barrett esophagus were initially published as a consensus statement in 1988 by the University of Washington group headed by Rodger Haggitt (17).

Literature Review

Reid BJ, Haggitt RC, Rubin CE, et al. Observer variation in the diagnosis of dysplasia in Barrett's esophagus. Hum Pathol 1988;19:166–178.

Negative for Dysplasia

The architecture is within normal limits. The nuclei do not vary greatly in size or shape and are located basally. The nuclear:cytoplasmic ratio is not increased. The nuclear envelope is generally smooth. Nucleoli are not markedly enlarged. Focal nuclear stratification is acceptable, as are small numbers of "dystrophic" goblet cells whose apical aspect does not communicate with the luminal surface. Greater nuclear alterations are acceptable when associated with evidence of inflammation, erosion, or ulceration. Numbers of abnormal-appearing mitoses are variable. Apical cytoplasmic mucin is usually present but may be reduced or absent in inflammation.

Indefinite for Dysplasia

The architecture may be moderately distorted. Nuclear abnormalities are less marked than those seen in dysplasia. Other features include more numerous dystrophic goblet cells, more extensive nuclear stratification, diminished or absent mucus production, increased cytoplasmic basophilia, and increased mitoses. The diagnosis of indefinite for dysplasia should be limited to cases in which the changes are too marked for negative but not sufficient for the diagnosis of dysplasia.

Positive for Dysplasia (LGD and HGD)

The diagnosis of low-grade dysplasia (LGD) or high-grade dysplasia (HGD) is based on the severity of both architectural and cytologic criteria that suggest a neoplastic transformation of the columnar epithelium. Although either architectural or cytologic abnormalities may predominate, high-grade dysplasia is diagnosed if either one is sufficiently prominent. Architectural abnormalities may include budded, branched, crowded, or irregularly shaped glands; papillary extensions into gland lumina; and villiform configuration of the mucosal surface. Nuclear features may include marked variation in size and shape, nuclear or nucleolar enlargement, increased nuclear:cytoplasmic ratio, hyperchromatism, and increased numbers of abnormal mitoses. Nuclear alterations are especially noteworthy if they involve the mucosal surface. Diagnostic features easily recognizable at lower power are cytoplasmic basophilia with loss of mucus and excessive nuclear stratification, often extending from the epithelial basement membrane to the luminal surface.

Intramucosal Carcinoma

Intramucosal carcinoma is defined as carcinoma that has penetrated through the basement membrane of the glands into the lamina propria but has not yet invaded through the muscularis mucosae into the submucosa.

This initial statement of criteria for recognizing dysplasia represented a great advance from the totally random pathologic diagnosis of dysplasia that existed at the time. However, the criteria are extremely complex and so vague that it is not likely that significant reproducibility would be achieved by their use. Over the years, the criteria have evolved, more by general usage than by any definitional studies.

The Diagnosis of High-Grade Dysplasia

High-grade dysplasia in Barrett esophagus is the only presently available criterion that indicates a high and imminent risk of adenocarcinoma in a patient with gastroesophageal reflux disease. It is the last point in the process before invasion occurs and the patient becomes at risk for metastatic disease. However, the natural history of high-grade dysplasia is not clearly defined. It is not known how many patients progress to adenocarcinoma or how quickly the transformation occurs in an individual patient. One of the complicating issues in evaluating the natural history is that the diagnosis of high-grade dysplasia is not a constant; it is only dysplasia above an arbitrarily drawn line within the spectrum of dysplasia (Fig. 14.19). The line drawn to separate high- from low-grade dysplasia is different between expert pathologists. There is no consideration given in the diagnostic criteria to the volume of high-grade dysplasia. In theory, a single gland showing the changes is sufficient for a diagnosis of high-grade dysplasia. It is not certain whether all pathologists will make a diagnosis of high-grade dysplasia based on changes in a single gland; if not, the absence of any volume consideration in the diagnostic criteria ensures a lack of uniformity. Studies of "high-grade dysplasia" are therefore not necessarily equivalent, making it difficult to evaluate data and compare different studies.

From a practical standpoint, the diagnosis of high-grade dysplasia is of great significance because its presence in a biopsy indicates two things: (a) there is a significant risk that the patient may already have invasive carcinoma in an unsampled area of the esophagus; and (b) it is a recognized indication for removing the area of dysplasia by either surgery or ablation.

This is not to suggest that surgery and ablation is always performed when high-grade dysplasia is diagnosed. Some authorities recommend careful follow-up using very strict endoscopy and biopsy protocols, suggesting that this protocol does not increase the danger of death from adenocarcinoma and permits retention of the esophagus. Careful follow-up is justified if the risk of adenocarcinoma actually developing in patients with high-grade dysplasia is low, if the time for progression from high-grade dysplasia is long, if the possibility of coexistent invasive carcinoma can be excluded, and if the patient prefers the strict follow-up regimen to esophagectomy. With the development of less radical removal methods such as endoscopic mucosal resection and ablation, the reason for not removing the area of high-grade dysplasia decreases.

Criteria for Diagnosis of High-Grade Dysplasia

The criteria for a diagnosis of high-grade dysplasia in Barrett esophagus is the presence of a severe cytologic abnormality and one of the two following features: complete loss of nuclear polarity and gland complexity (Figs. 14.25 to 14.28). Gland crowding is also taken into consideration. Loss of nuclear polarity consists of a rounding up of the nuclei, which cease to have their long axis perpendicular to the basement membrane. Gland complexity is characterized by luminal bridges and cribriform architecture.

The problems relating to the reliability of the diagnosis of high-grade dysplasia are that several of the criteria have subjective features: (a) What exactly is "severe cytologic abnormality," and does its determination vary between pathologists? (b) What exactly is "complete" loss of polarity, and does its determination vary between pathologists? (c) When faced with a biopsy, do all pathologists evaluate all criteria separately, or do they form an opinion based on a combination of the criteria? (d) What is the effect of the amount of high-grade dysplasia? Do pathologists hesitate to make a diagnosis of high-grade dysplasia when only one focus is affected compared to extensive high-grade dysplasia where multiple biopsies are positive? (e) Does the coexistence of extensive versus no low-grade dysplasia influence the facility in reaching a diagnosis of high-grade dysplasia? All of these questions are difficult to answer. Studies have shown that the natural history and progression to cancer is greater in patients who have multiple foci of high-grade dysplasia than in those who a single focus (18).

The most important variable, however, is the practice setting of the pathologist (Fig. 14.29). Assume that Dr. Chandrasoma works with Tom DeMeester, a foregut surgeon (this is true). If he makes a diagnosis of high-grade dysplasia in a patient with Barrett esophagus, he knows that the patient becomes a candidate for esophagectomy. It is by no means certain that the patient will have an esophagectomy, but he does not know that at the time he makes the diagnosis. In this setting, his diagnosis of high-grade dysplasia

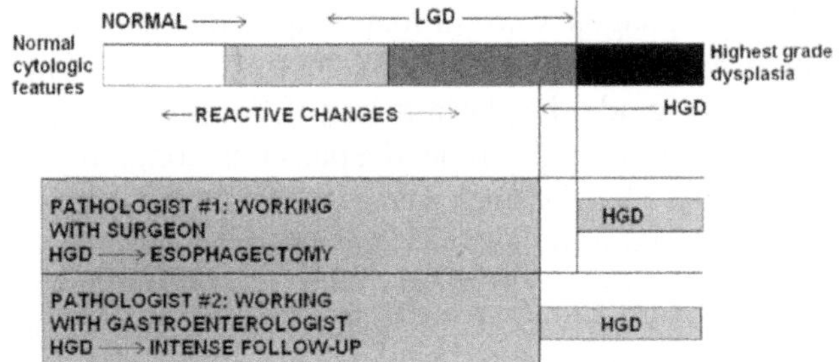

FIGURE 14.29 The influence of the practice setting of the pathologist making a diagnosis of high-grade dysplasia. Because there is no sharp line between low- and high-grade dysplasia, pathologist 1 who works with a surgeon who takes a diagnosis of high-grade dysplasia as an indication for esophagectomy will tend to err on the side of being more specific (i.e., line more toward the right). In contrast, pathologist 2 who works with a gastroenterologist whose protocol is intense surveillance for a diagnosis of high-grade dysplasia will tend to err on the side of greater sensitivity (i.e., line more toward the left). In this small number of cases, experts will disagree, not because either of them is wrong but because we are attempting to draw an artificial line between low- and high-grade dysplasia.

tends to have the highest level of specificity. Criteria have to be perfect. Any doubt will make him favor low-grade dysplasia with a recommendation for heightened surveillance. Overcalling high-grade dysplasia is a significant problem. The dictum in this setting is this: If uncertain, do not diagnose high-grade dysplasia.

Assume, on the other hand, that Dr. Chandrasoma works in a center where the accepted protocol for high-grade dysplasia is careful endoscopic surveillance with frequently repeated extensive biopsies. In this setting, it is very likely that he will call high-grade dysplasia at a slightly lower level of specificity (Fig. 14.29). The error here is to undercall high-grade dysplasia because failure to diagnose high-grade dysplasia results in a failure to increase the intensity of surveillance and places the patient at increased risk. The dictum is different: If in doubt, call it high-grade dysplasia.

The preceding example may sound absurd, but it is very true. This is because the distinction between low- and high-grade dysplasia is not absolutely clear cut in every case. There will always be cases in and around the borderline region of low- and high-grade dysplasia where the pathologist has to make a decision. These are the cases that will go one way or the other based on the practice setting of the pathologist. For this reason, it is difficult to compare the literature with regard to the management of high-grade dysplasia. In studies coming out of Dr. DeMeester's service at the University of Southern California (USC), the probability is high that there will be very few false positive diagnoses of high-grade dysplasia. Patients going on to esophagectomy for high-grade dysplasia will have a high probability of either having residual high-grade dysplasia or invasive carcinoma in their specimens; 35% of patients undergoing esophagectomy at USC for high-grade dysplasia have coexistent invasive carcinoma (19). In units where careful follow-up is the management for high-grade dysplasia, it is likely that the sensitivity of diagnosis of high-grade dysplasia is higher (and therefore the specificity lower); these studies will therefore have more patients who do not have invasive cancer at the time and are further removed temporally from invasive cancer. The difference is where the line is drawn between low- and high-grade dysplasia; this is not definable with complete reproducibility (Fig. 14.29).

Many patients are referred to Dr. DeMeester's service for surgery from all over the country after a diagnosis of high-grade dysplasia. In every referred case, the outside slides are reviewed and repeat biopsy performed before surgery is undertaken. In the majority of these cases, the diagnosis of high-grade dysplasia is confirmed by review, but a significant minority has a downgrading of the dysplasia to low grade.

Despite the problems and the lack of interobserver reproducibility, the criteria for the diagnosis of high-grade dysplasia are generally so well defined that a high level of accuracy of pathologic diagnosis is achieved. Despite this, it is the recommendation for patients who have a diagnosis of high-grade dysplasia that this diagnosis be confirmed by at least one independent expert pathologist.

The presence of invasion in high-grade dysplasia is recognized by the penetration of the basement membrane of the dysplastic glands and extension of malignant cells into the lamina propria (Figs. 14.30 to 14.32). When limited to the lamina propria without extension into the submucosa, this represents intramucosal adenocarcinoma. In a mucosal biopsy, the depth of invasion below the mucosa is difficult to assess with accuracy. However, mucosal biopsies of deeply invasive tumors show obvious invasive features, frequently with desmoplasia around the invasive glands.

Criteria of Diagnosis of Low-Grade Dysplasia

The criteria and reliability of diagnosis of low-grade dysplasia have problems that are infinitely greater than high grade dysplasia. Interobserver variation is so great that the diagnosis of low-grade dysplasia has little meaning. True low-grade dysplasia occurs as a result of an irreversible genetic change that is an early move toward the sequence necessary to produce adenocarcinoma. The change occurs in intact epithe-

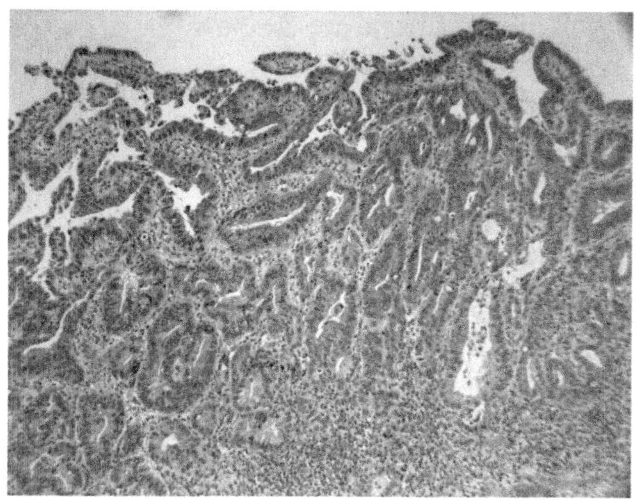

FIGURE 14.30 High-grade dysplasia with crowded small glands with irregularity of outlines sufficient to raise the probability of early lamina propria invasion (intramucosal adenocarcinoma).

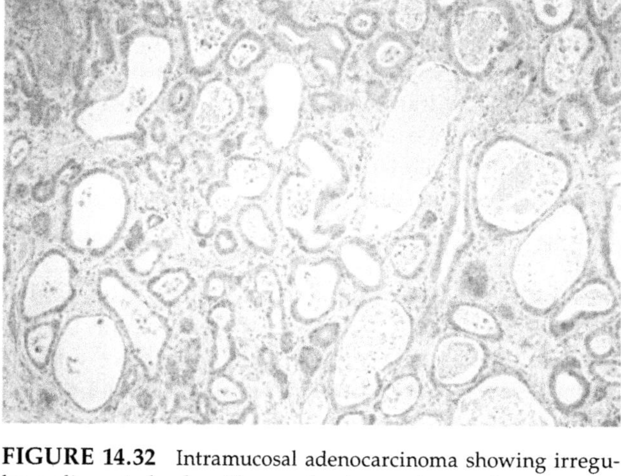

FIGURE 14.32 Intramucosal adenocarcinoma showing irregular malignant glands with some having a microcystic appearance involving the entire mucosa.

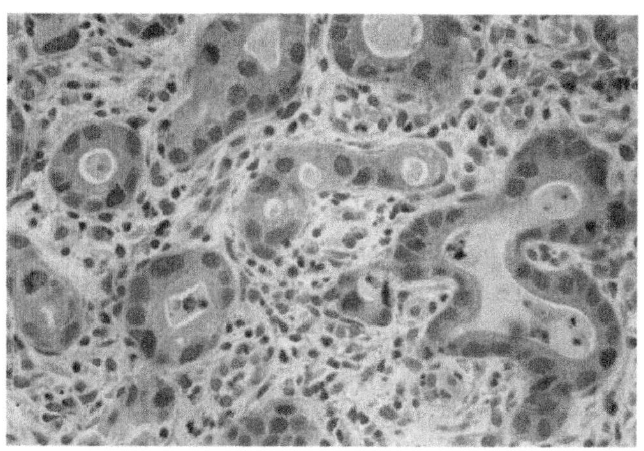

FIGURE 14.31 Early breach of the basement membrane in high-grade dysplasia, representing lamina propria invasion and early intramucosal adenocarcinoma. The irregularity of gland contour is typical. (Cross reference: Color Plate 5.9)

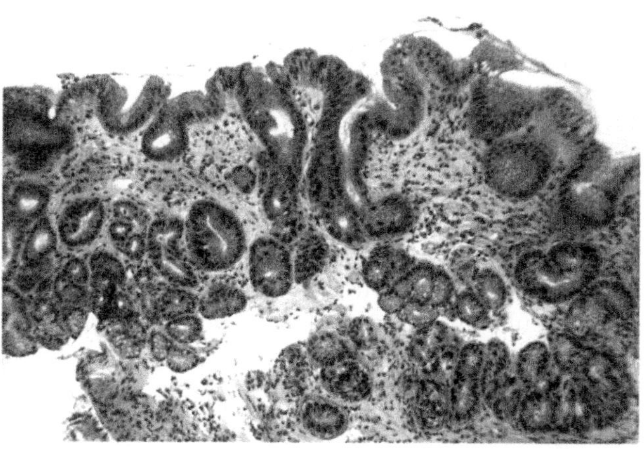

FIGURE 14.33 Relatively slight cytologic changes involving a deeper foveolar region with relatively normal surface maturation. Despite the absence of ulceration and active inflammation, this was diagnosed as a reactive change without evidence of dysplasia. There is likely to be significant interobserver variation for this diagnosis.

lium and is associated with a recognizable maturation abnormality. The typical low-grade dysplastic epithelium is therefore uneroded, without active inflammation, and shows the dysplastic cytologic change extending from the base of the foveolar pit (where the proliferating progenitor cell pool resides) to involve the surface (Figs. 14.21 and 14.22). The cytologic change includes nuclear enlargement, stratification, hyperchromasia with depletion of cytoplasm, and goblet-type mucin. The degree of change is less than that required to satisfy the criteria for high-grade dysplasia.

The cytologic abnormality of low-grade dysplasia is mimicked exactly by changes that are not associated with genetic abnormalities of the carcinogenetic pathway. These are known as reactive atypias; they are commonly seen in association with regenerative epithelial activity in areas of erosion and in areas of active inflammation (Fig. 14.33). Because of this, the diagnosis of low-grade dysplasia should not be made in areas of erosion or active inflammation. Most reactive atypias are associated with increased proliferation, but maturation of the cells remains normal; as such, the cytologic abnormality is greatest in the deep part of the foveolar pit with normal cytologic features at the surface. It is for this reason that surface epithelial abnormality is usually necessary for a diagnosis of low-grade dysplasia.

There is little practical difference between the diagnoses of intestinal metaplasia with low-grade dysplasia and reactive atypia; it is merely a matter of surveillance interval. The rate of transformation of low-grade dysplasia to adenocarcinoma is, with very rare exceptions, slow, and routine surveillance is probably adequate. We therefore restrict diagnosis of low-grade dysplasia to cases where the criteria of low-grade dysplasia are definite and well developed. In such cases, the shortening of the surveillance interval is probably justified. If there is any doubt about the criteria, we generally diagnose reactive change without dysplasia. We do not make a diagnosis of indefinite for dysplasia; because we only diagnose dysplasia when it is definite, "indefinite for dysplasia" is equivalent to "negative for dysplasia."

The use of the term "indefinite for dysplasia" is a source of confusion in the literature. Some authors take a diagnosis of "indefinite for dysplasia" to be equivalent to "low-grade dysplasia," while others take it to mean that dysplasia is absent. This makes comparison of data between research very difficult. The other problem about the use of the term "indefinite for dysplasia" is that it becomes a vehicle to express the pathologist's lack of ability to make a decision. It enables the pathologist to avoid a decision too easily; sitting on the fence has no penalty. We strongly encourage our trainees not to use the term. We have had no difficulty in diagnosis by restricting ourselves to the terms "low-grade dysplasia" and "reactive cytologic changes insufficient for a definite diagnosis of dysplasia."

Assessment of Reliability of Diagnosis of Dysplasia

There are two ways to assess the reliability of a diagnosis of dysplasia. The first is to see whether experts agree with the diagnosis. This is the usual method advocated for high-grade dysplasia where it is recommended that the diagnosis be reviewed by two experts and treatment undertaken only when there is concordance in the two opinions. It must be recognized that concordance and reproducibility of diagnosis are measures of precision; they are not the same as accuracy. Two pathologists can agree on a diagnosis; if they are both correct, the diagnosis is accurate; if they are both wrong, the patient will not receive the correct treatment.

Reproducibility

Montgomery et al. (20) presented an excellent study on the precision and reproducibility of the diagnosis of varying grades of dysplasia in Barrett esophagus.

Literature Review

Montgomery E, Bronner MP, Goldblum JR, Greenson JK, Haber MM, Hart J, Lamps LW, Lewin DN, Robert ME, Toledano AY, Shyr Y, Washington K. Reproducibility of the diagnosis of dysplasia in Barrett esophagus: A reaffirmation. Hum Pathol 2001;32:368–378.

This multinstitutional study assessed the reproducibility of criteria for diagnosing dysplasia with regard to interobserver and intraobserver variation. Ten participating institutions contributed 250 slides with the following submitting diagnoses: 86, no dysplasia; 44, indefinite for dysplasia; 42, low-grade dysplasia; 52, high-grade dysplasia; 15, intramucosal carcinoma; and 11, frankly invasive carcinoma. These were logged in again to blind the pathologists to the source and each of 12 pathologists were asked to assign a diagnosis by checking a box (Barrett esophagus without dysplasia, indefinite for dysplasia, low-grade dysplasia, high-grade dysplasia, intramucosal carcinoma, frankly invasive carcinoma). The pathologists were forced to commit to one of these diagnoses.

The method of selection of the pathologists is not stated. The pathologists were all from highly reputable institutions, and all the participants had subspecialty interest in gastrointestinal pathology (i.e., experts in the field). The total number of pathologists invited and the number accepting the invitation is not stated. No pathologists from outside the United States were among the participants.

The researchers circulated 125 odd-numbered slides to the 12 pathologists twice over a 12-month period without any discussion of diagnostic criteria. The 24 (2 per pathologist) diagnostic opinions were analyzed for interobserver and intraobserver agreement.

The participants then attended a consensus meeting. The statistical analysis of the first 125 slides was presented. After this, 53 cases representing a wide diagnostic range were reviewed microscopically and extensively discussed by the group at a multiheaded microscope. This led to a set of agreed-upon diagnostic criteria. The second set of slides (the 125 even-numbered slides) was then circulated twice to each of the 12 observers, again generating 24 sets of diagnoses.

The study presents the data in statistical terms. According to this, a mean k value greater than 0.80 = nearly perfect agreement; 0.60–0.80 = substantial agreement; 0.40 to 0.60 = moderate agreement within the group.

The results were as follows:

For the 125 Slides Evaluated before the Consensus Meeting

1. If the diagnoses were grouped into two categories—low-grade dysplasia and below versus high-grade dysplasia and above—the interobserver reproducibility was 0.66 ("substantial").
2. When a more clinically relevant grouping of diagnostic categories (negative, indefinite + low-grade dysplasia,

high-grade dysplasia, carcinoma) was used, the interobserver reproducibility was 0.43 ("moderate").

3. When the data were assessed in terms of diagnosis rather than reader, interobserver reproducibility was 0.67 ("substantial") for invasive carcinoma, "moderate" for Barrett esophagus without dysplasia (0.44) and high-grade dysplasia (0.40), "fair" for low-grade dysplasia (0.23), and "slight" for indefinite for dysplasia (0.14).

Consensus Meeting

The participants, evaluating 53 cases at a multiheaded microscope, developed an algorithm for grading of dysplasia that used previously established criteria and additional features developed at the meeting. This was based on four features as follows:

Feature 1: Surface maturation in comparison with the underlying glands
Feature 2: Architecture of the glands on the biopsy
Feature 3: Cytologic features of the proliferating cells
Feature 4: Inflammation and erosion/ulcers

Each feature might vary, and they were combined to arrive at a diagnosis. Each of these features was described in detail. Surface maturation depended on the presence of nuclei in the surface layer that were equal to or larger than those in the glands. Architectural abnormalities included crowding of glands with cribriform change, cystic dilation, and necrotic luminal debris considered to be severe abnormalities. Cytologic changes included a definition of polarity.

In applying these features, the expert pathologists developed an algorithm for the classification of dysplasia, as follows:

1. Barrett esophagus, negative for dysplasia: The surface appears more mature than the underlying glands, the architecture is normal, the cytologic features are normal including maintained normal polarity.
2. Barrett's esophagus, indefinite for dysplasia: This included cases that had "deeper cytologic changes suggestive of dysplasia but showed surface maturation." "Some degree of gland crowding" and cytologic abnormalities were permissible but not loss of nuclear polarity.
3. Low-grade dysplasia: This was characterized by absent or slight maturation at the surface, "at least focally." Architectural abnormalities varied from mild to marked distortion. Cytologic features of dysplasia were necessary; nuclear stratification could be present; "loss of nuclear polarity is not a feature." "Inflammation is typically minimal; cases with abundant inflammation and other features of low grade dysplasia are best classified in the indefinite category" (p. 374).
4. High-grade dydplasia: Surface maturation is lacking; "architecture may show crowding of cytologically abnormal glands or be markedly distorted with prominent glandular crowding and little intervening lamina propria. If the cytologic features are sufficiently dysplastic, lesser architectural distortion is acceptable. . . . Loss of polarity is seen" (p. 373).
5. Intramucosal carcinoma: Detection of the earliest lamina propria invasion remains difficult. The features are "an effacement of the lamina propria architecture and a syncytial growth pattern, extensive back-to-back microglands, and an intermingling of single cells and small clusters within the lamina propria" (p. 375). Frankly invasive carcinoma is easily recognized by desmoplasia and a clearly recognizable infiltrative pattern.

Slide Review after Consensus Meeting

1. If the diagnoses were grouped into two categories—low-grade dysplasia and below versus high-grade dysplasia and above—the interobserver reproducibility was 0.70 ("substantial"). This represented a significant ($p = 0.02$) improvement from the preconsensus meeting value.
2. When a more clinically relevant grouping of diagnostic categories (negative, indefinite + low-grade dysplasia, high-grade dysplasia, carcinoma) was used, the interobserver reproducibility was 0.46 ("moderate"). This was not significantly different to the preconsensus value of 0.43 ($p = 0.056$).
3. When the data were assessed in terms of diagnosis rather than reader, interobserver reproducibility was "substantial" for invasive carcinoma (0.64—less than the preconsensus value of 0.67) and high-grade dysplasia (0.64—improved from 0.40), "moderate" for Barrett esophagus without dysplasia (0.58—improved from 0.44), "fair" for low-grade dysplasia (0.32—improved from 0.23), and "slight" for indefinite for dysplasia (0.15—similar to the 0.14 value preconsensus).

An extremely interesting part of the results relates to the actual diagnoses of these 12 experts (actually multiplied by 2 because each reviewed the slides twice). Data relating to the number of cases in which 13 or more of the 24 opinions were the same (i.e., a simple majority, which was called "consensus diagnosis") showed the following: Of the 250 cases, 184 cases (70%) achieved a consensus. In the remaining 66 cases, there was no majority diagnosis. This means that these 12 expert pathologists reviewing each slide twice *failed to reach a majority opinion* in these 66 cases. Extrapolating from the numbers given for cases achieving consensus, these included significant numbers of cases of low- and high-grade dysplasia and even one case of frankly invasive carcinoma. The study emphasized that the number of cases that failed to reach a majority opinion decreased from 40 of 125 (32%) in the first set before the consensus meeting to 26 of 125 (21%) after the consensus meeting.

In their discussion, the authors appeared to be explaining why the results were so bad. One explanation was that the experts were not given the option of refusing to make a diagnosis because the material was suboptimal. They stated that many cases submitted were not suitable for interpretation and the pathologist reviewing the cases would have requested a repeat biopsy rather than rendering a diagnosis in clinical practice. This is not true; these were all cases submitted from parent institutions where a decision was made to make the diagnosis on this material. The authors also try to explain the failure of the consensus meeting to impact diagnostic reproducibility, but not with a great deal of conviction.

However, they concluded: "It is clear that the initially established criteria have served practicing pathologists well. The current data reaffirm that the existing criteria, further clarified herein, provide for reproducible diagnosis in the grading of dysplasia/neoplasia in Barrett esophagus" (p. 378). The word "reaffirmation" also appears in the title of the article.

A critical review of the data in this article suggests that the correct conclusion should be quite the opposite of that reached by the authors. The dismal results associated with the reproducibility of even the most critical diagnosis of low- and high-grade dysplasia among this group of selected experts in the United States should make one shudder and push for the development of better criteria. It is astonishing that of 42 cases of low-grade dysplasia and 52 cases of high-grade dysplasia (these were the diagnoses of the submitting institutions), only 27 of 42 cases of low-grade dysplasia and 27 of 52 cases of high-grade dysplasia reached a majority agreement. This means that more than 11 of 24 opinions given by these experts for these cases did not agree with the original diagnosis of low-grade dysplasia in 15 of 42 cases and high-grade dysplasia in 25 of 52 cases. When one considers that the institutions are the highest regarded in the United States and the pathologist participants were those with specialty interest in the field (and therefore those most likely to be sought for expert opinions), it should be a matter of grave concern how these patients were managed clinically. With these data for reproducibility of diagnosis among expert pathologists, the practice of confirming a diagnosis of high-grade dysplasia by an expert appears to be a hopelessly flawed recommendation. The data suggest that if a community pathologist sends the biopsy in question independently to two or three different experts among the group in the study, there is a high likelihood of different opinions.

Accuracy: Validation of the Diagnosis of Dysplasia

The second method of assessing the reliability of a diagnosis of dysplasia relates to accuracy as determined by outcome in the patient who has a diagnosis of dysplasia. This is only possible with longitudinal studies with follow-up over time to watch progression or reversal of disease in sequential biopsies or evaluation of the esophagus at esophagectomy and autopsy. For example, if a diagnosis of high-grade dysplasia is made on one biopsy and multiple subsequent biopsies show no dysplasia and the patient survives many years, it suggests that either high-grade dysplasia has reversed or the original diagnosis was incorrect. Logic would suggest the latter; practice and protocol would have multiple experts review the slide, reach agreement that it was indeed high-grade dysplasia, and declare the process as being reversed. Similarly, progression from low- to high-grade dysplasia in serial biopsies does not necessarily mean progression; it can mean diagnostic error or sampling error. These factors make it very difficult to assess the accuracy of a diagnosis of dysplasia.

Longitudinal studies that evaluate the outcomes in patients who have a diagnosis of dysplasia are also complicated by the fact that these patients often have treatment—acid suppression, antireflux surgery, endoscopic mucosal resection, ablation, and esophagectomy. The effect of such treatments must be included in the analysis.

If dysplasia is a highly significant clinical diagnosis associated with a significant risk of interobserver variation, it is important for the individual pathologist to continually assess his or her diagnosis of dysplasia against the clinical outcome in one's own practice setting. Dr. Chandrasoma works very closely with the foregut surgery team headed by Dr. DeMeester. When he makes a diagnosis of dysplasia, there is implicit trust in that diagnosis. This is true even when he contradicts a diagnosis made elsewhere. He is, for this group, the sole determinant of dysplasia and its grade. There is no second confirmation by another expert for any diagnosis. Review of the study of interobserver variation in the diagnosis of dysplasia by Montgomery et al. (20) supports this course of action; the study shows that if Dr. Chandrasoma were to send out any case that he diagnosed as high-grade dysplasia to these 12 foremost experts on two separate occasions, 13 or more of these 24 opinions would agree with his diagnosis only about half the time; in the other half, only 11 or fewer would agree. Would it be better for him to trust those who agreed or those who did not agree with his diagnosis? Chaos, not resolution, is the result of multiple disparate expert opinions.

The USC foregut surgery group clearly understands that when Dr. Chandrasoma makes a diagnosis of high-grade dysplasia, it is highly specific and based on the presence of definite criteria. If there is any doubt, he does not diagnose high-grade dysplasia. Cases where the features are in the gray area between low- and high-grade are expressed as "low-grade dysplasia with borderline features" and are treated as low-grade dysplasia with a recommendation for heightened surveillance. Time is used as the determinant in these cases with progression or reversal of changes in the rapid surveillance biopsies being used to guide man-

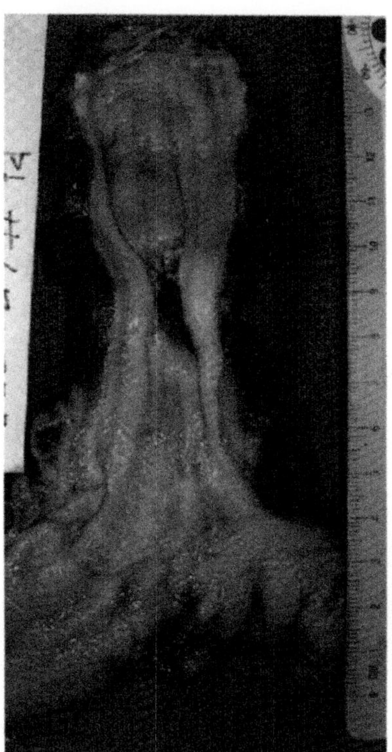

PLATE 1.1 Normal gross appearance of the gastroesophageal junctional region with a horizontal Z-line at end of tubular esophagus. Note the ulcerated squamous carcinoma in the mid-esophagus.
Cross references: Fig 4.1, 6.8, 7.1, 11.3, and 12.11

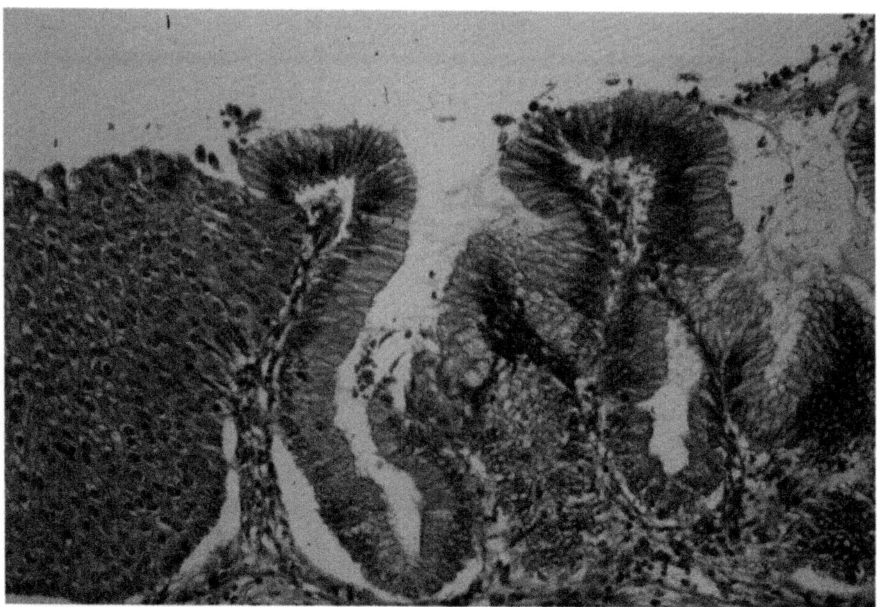

PLATE 1.2 Distal esophagus, third trimester fetus with fetal squamous and columnar epithelia.
Cross references: Fig 3.21c and 8.18

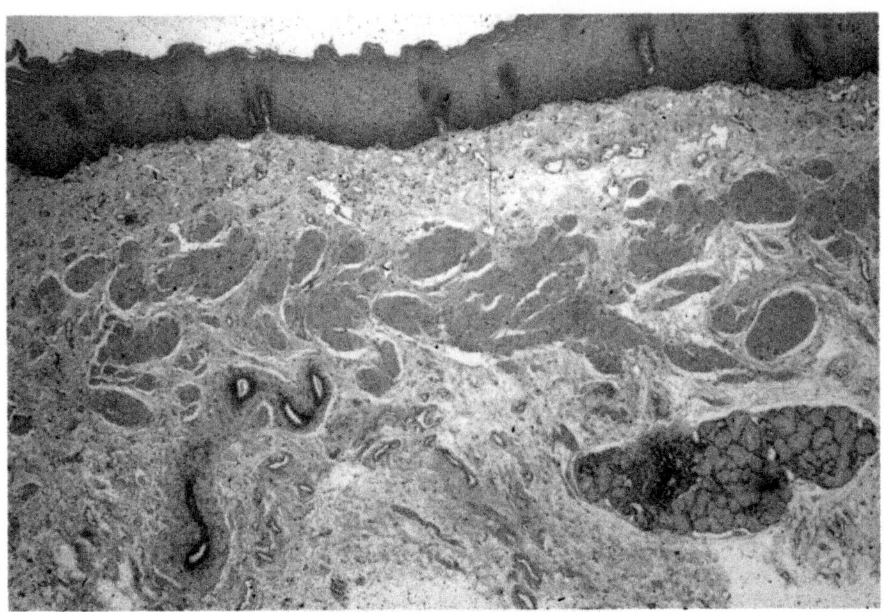

PLATE 1.3 Normal squamous lined esophagus with submucosal glands.
Cross reference: Fig 3.12 and 4.5

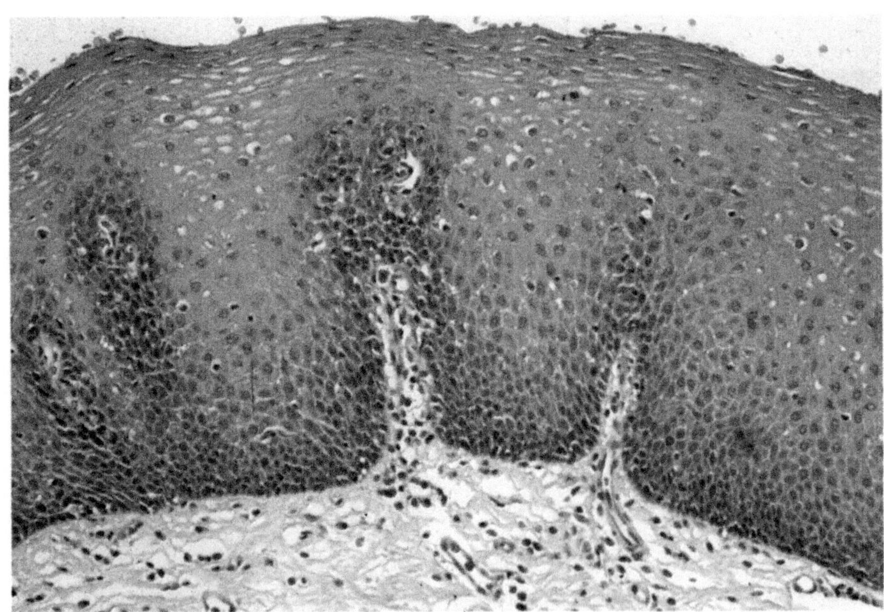

PLATE 1.4 Normal esophageal squamous epithelium.
Cross reference: Fig 5.1 and 8.3

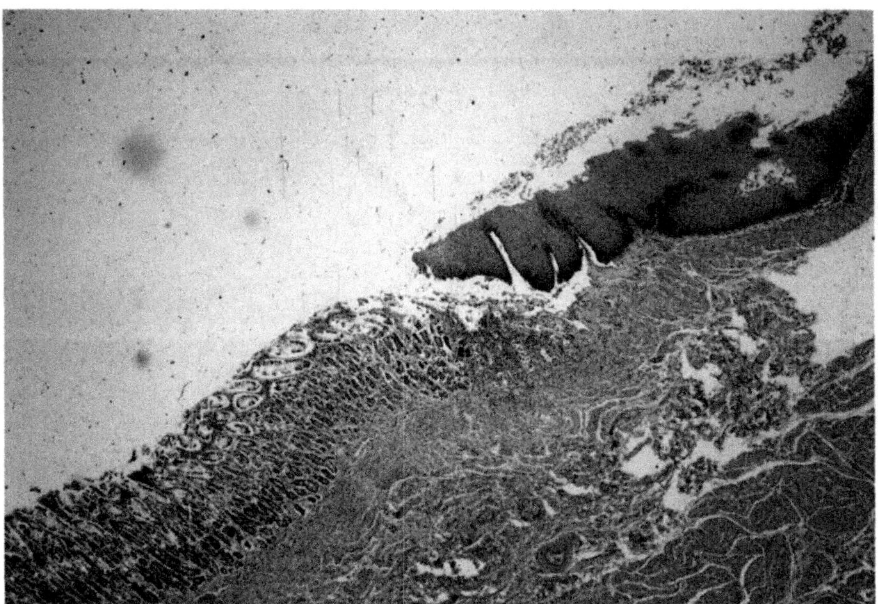

PLATE 1.5 Normal squamous to gastric oxyntic junction in autopsy specimen.
Cross references: Fig 3.23 and 6.11

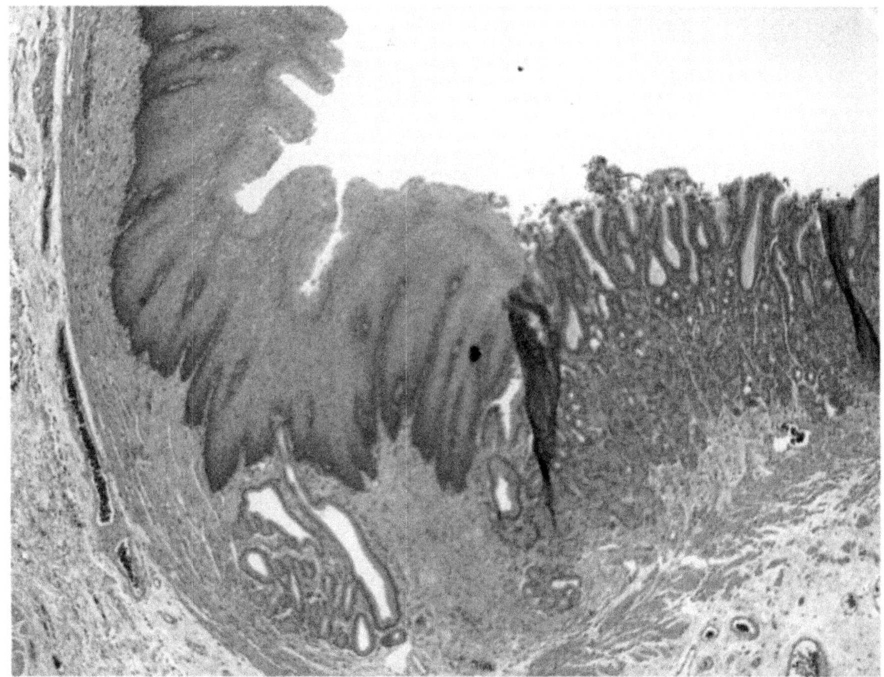

PLATE 1.6 Normal squamous to gastric oxyntic junction in esophagectomy specimen.
Cross references: Fig 7.7

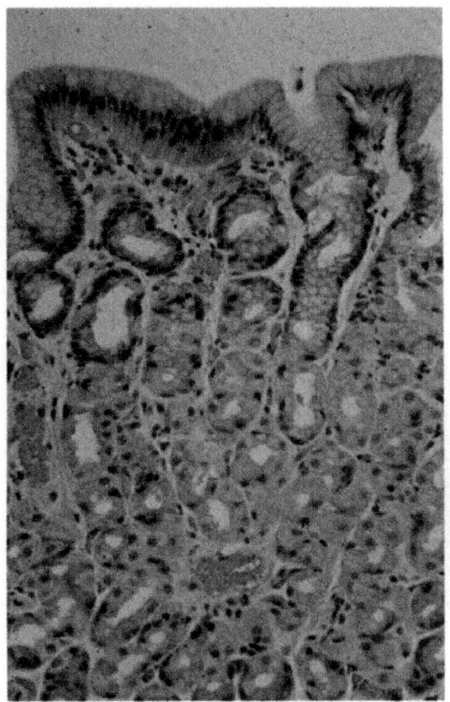

PLATE 1.7 Normal gastric oxyntic mucosa, low power.
Cross references: Fig 3.16a and 5.3

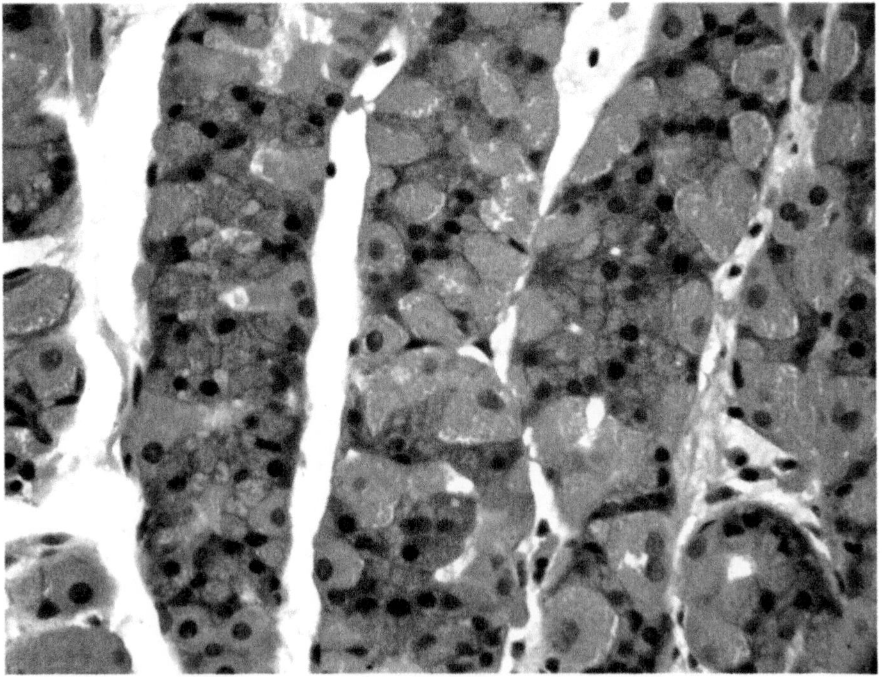

PLATE 1.8 Normal gastric oxyntic mucosa, high power.
Cross references: Fig 5.4

PLATE 2: REFLUX CHANGES IN SQUAMOUS EPITHELIUM

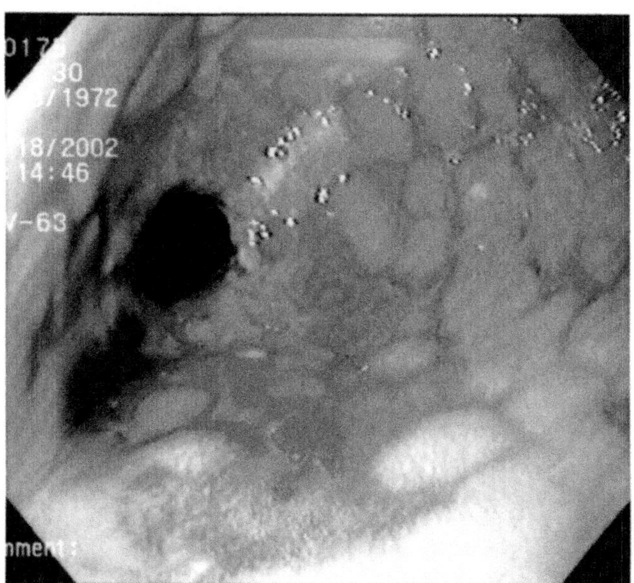

PLATE 2.1 Erosive esophagitis, endoscopy.
Cross references: Fig 8.8 and 13.1

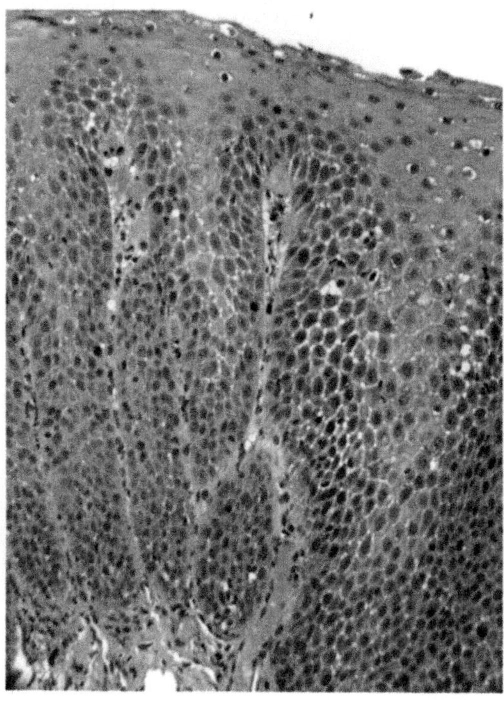

PLATE 2.2 Reflux esophagitis, squamous epithelium.
Cross references: Fig 8.5 and 13.2

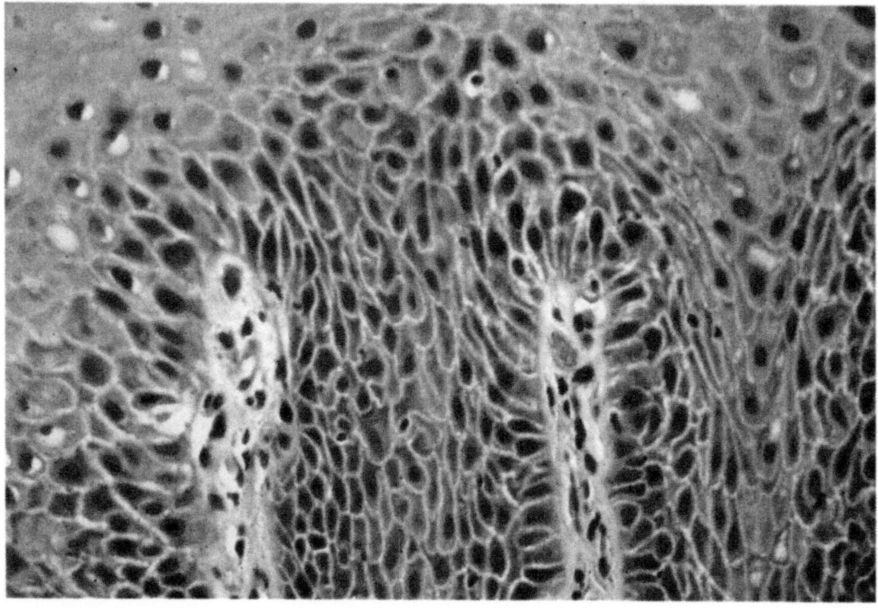

PLATE 2.3 Reflux esophagitis, squamous epithelium.
Cross references: Fig 8.10 and 13.3

PLATE 3: COLUMNAR METAPLASIA OF THE ESOPHAGUS (NON-INTESTINAL TYPES)

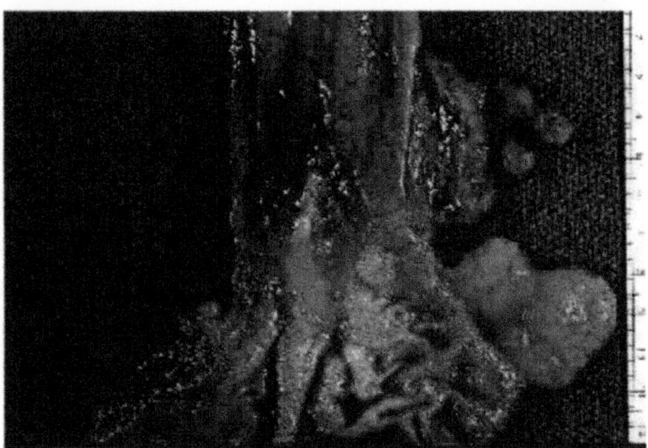

PLATE 3.1 Short segment of columnar lined esophagus, gross.
Cross references: Fig 4.15b and 11.29

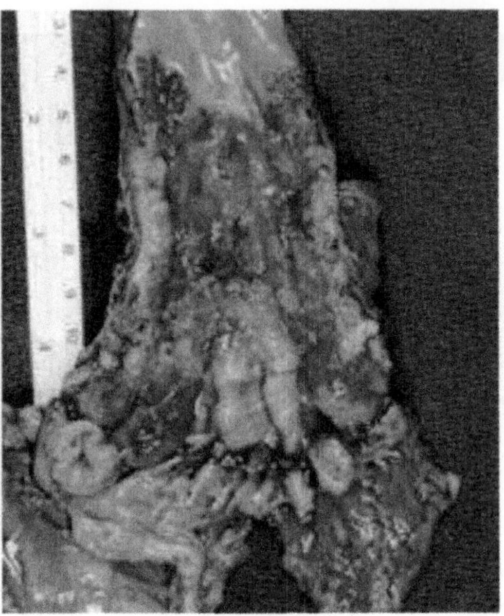

PLATE 3.2 Long segment of columnar lined esophagus, gross.
Cross references: Fig 4.10, 11.7, 12.20 and 13.4

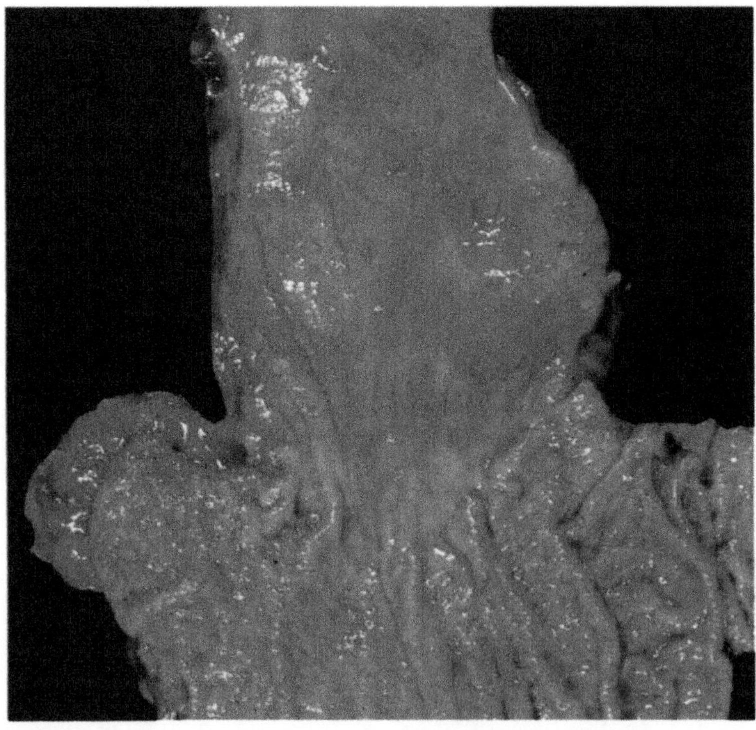

PLATE 3.3 Long segment of columnar lined esophagus, gross.
Cross references: Fig 4.14 and 12.24

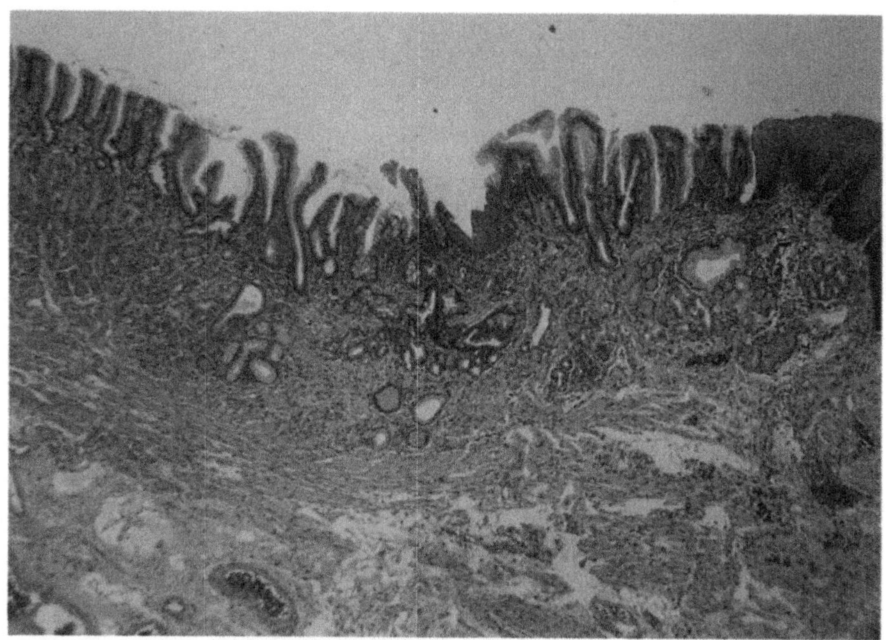

PLATE 3.4 Section of esophagus with oxyntocardiac, cardiac and intestinal types of metaplastic columnar epithelia.
Cross reference: 5.22 and 12.21

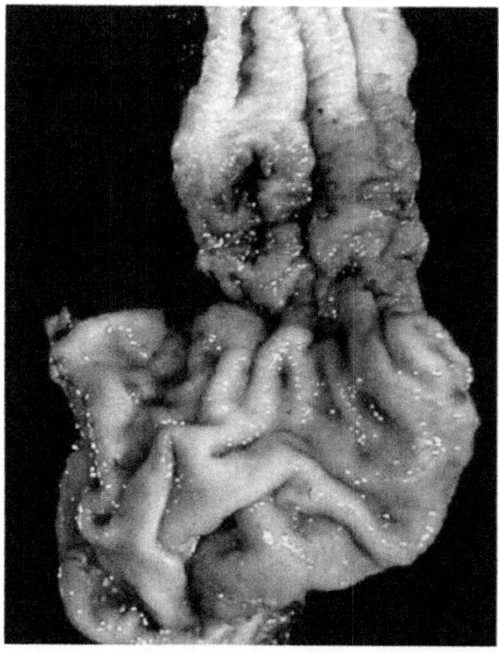

PLATE 3.5 Long segment columnar lined esophagus. Note coincidence of end of tubular esophagus and proximal limit of rugal folds.
Cross references: Fig 4.13b, 10.11, 11.8 and 14.7

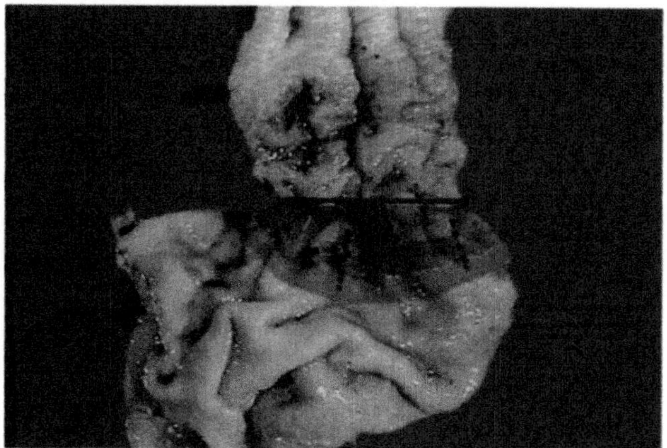

PLATE 3.6 Earlier figure with histologic mapping; yellow = intestinal metaplasia in cardiac mucosa, green = cardiac mucosa; pink = oxyntocardiac mucosa. Black dots represent submucosal esophageal glands in dilated end-stage esophagus.
Cross references: Fig 11.12, 12.22 and 14.7

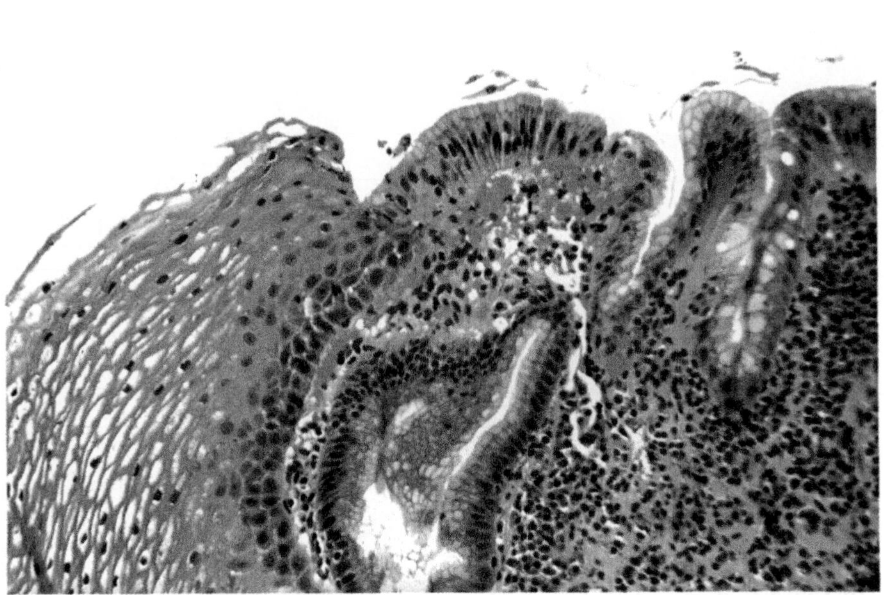

PLATE 3.7 Columnar metaplasia of squamous epithelium. Note line of suprabasal separation in yet intact squamous epithelium.
Cross references: Fig 8.15 and 13.12

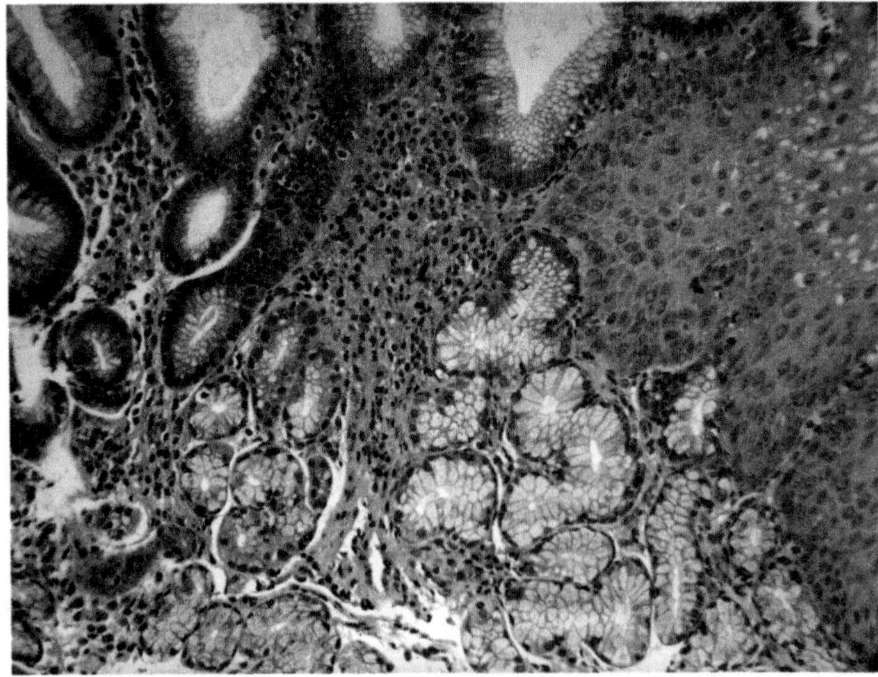

PLATE 3.8 Very short cardiac mucosa between squamous epithelkum and oxyntocardiac mucosa.
Cross references: Fig 5.23b, 6.14, 9.12, 12.14 and 14.5

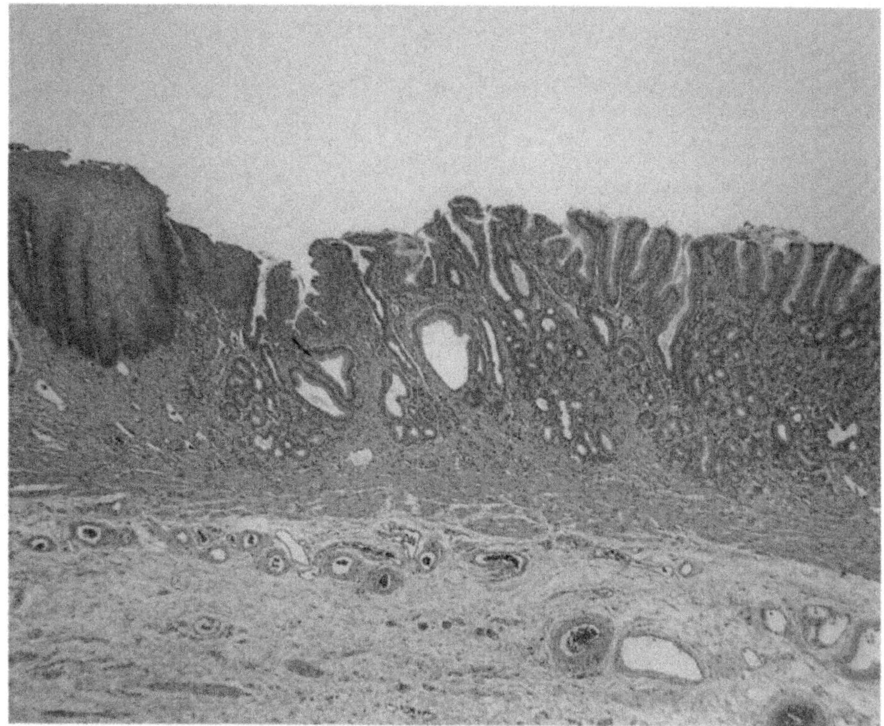

PLATE 3.9 Metaplastic cardiac mucosa between squamous epithelium and gastric oxyntic mucosa.
Cross references: Fig 7.9

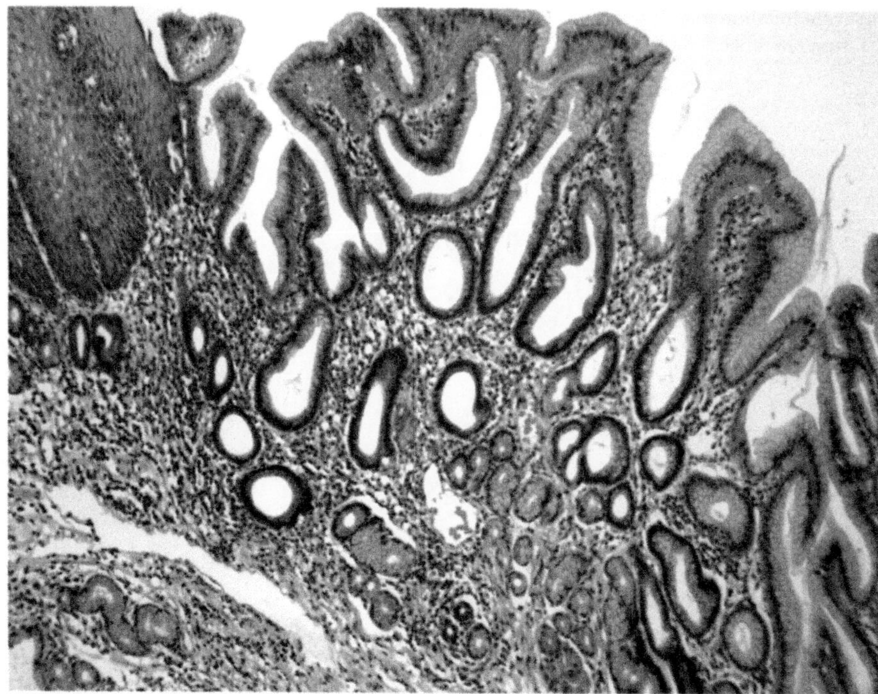

PLATE 3.10 Reflux carditis (metaplastic cardiac mucosa) showing foveolar hyperplasia, inflammation and lamina propria fibrosis.
Cross references: Fig 7.10 and 13.10

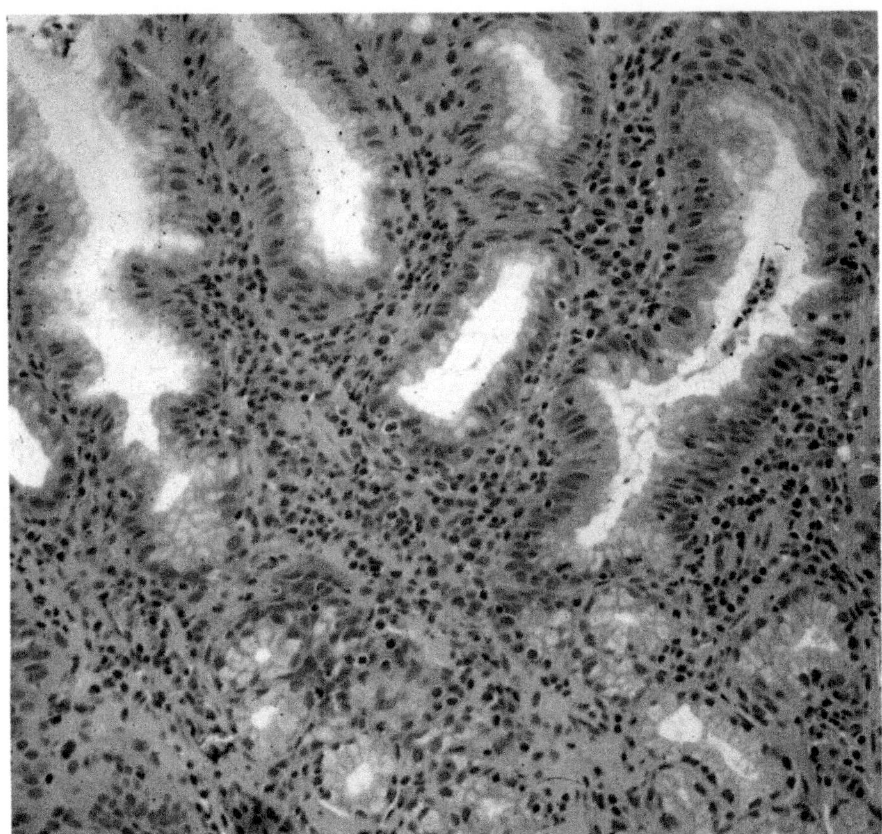

PLATE 3.11 Reflux carditis (metaplastic cardiac mucosa) showing foveolar hyperplasia and severe inflammation.
Cross references: Fig 8.19 and 13.7

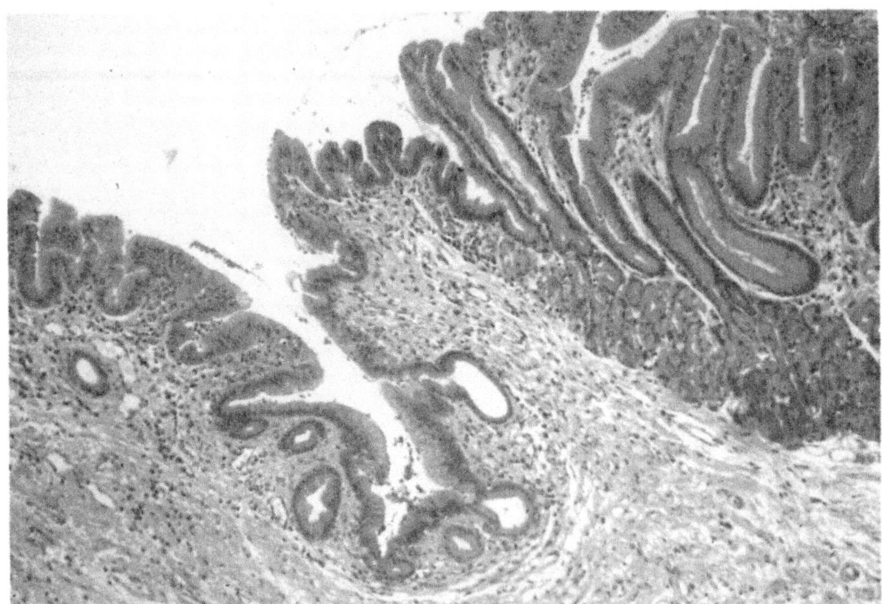

PLATE 3.12 Foveolar type cardiac mucosa (left) and oxyntocardiac mucosa with glands containing parietal cells (right) in metaplastic columnar esophagus.
Cross references: Fig 5.16 and 8.23

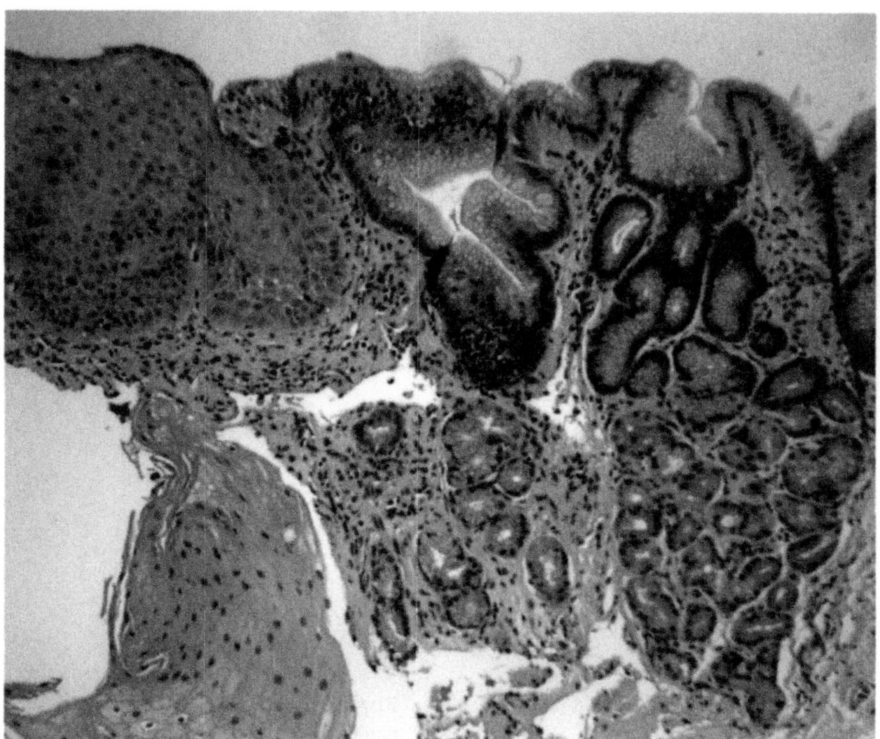

PLATE 3.13 Squamo columnar junction with squamous epithelium transitioning directly to oxyntocardiac mucosa. The foveolar complex on the right has numerous parietal cells; the one adjacent to the squamous epithelium has few.
Cross references: Fig 14.4

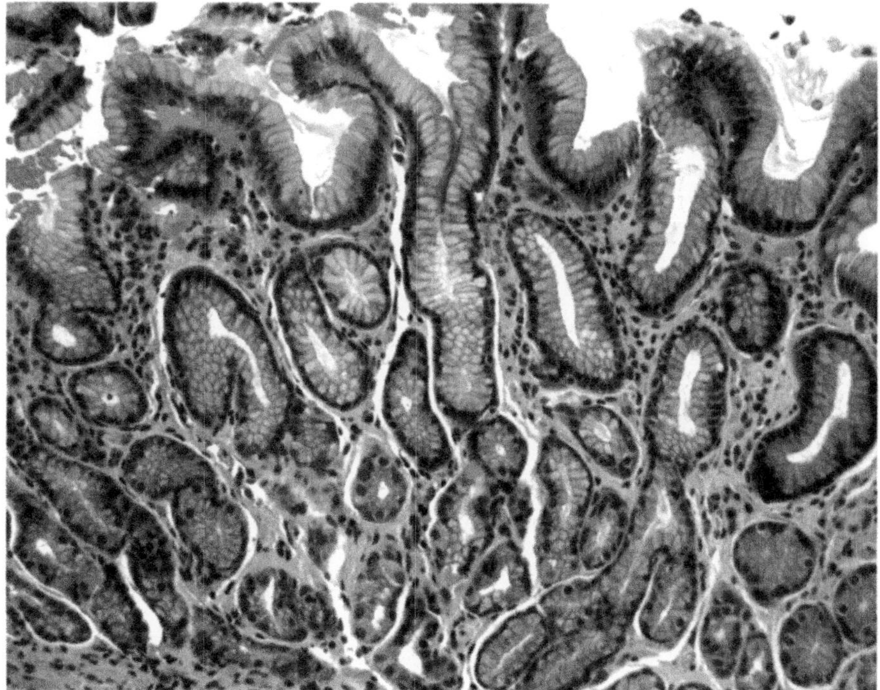

PLATE 3.14 Oxyntocardiac mucosa showing lobulated glands composed of parietal and mucous cells.
Cross references: Fig 5.17 and 9.19

PLATE 4: INTESTINAL METAPLASIA IN COLUMNAR LINED ESOPHAGUS (BARRETT ESOPHAGUS)

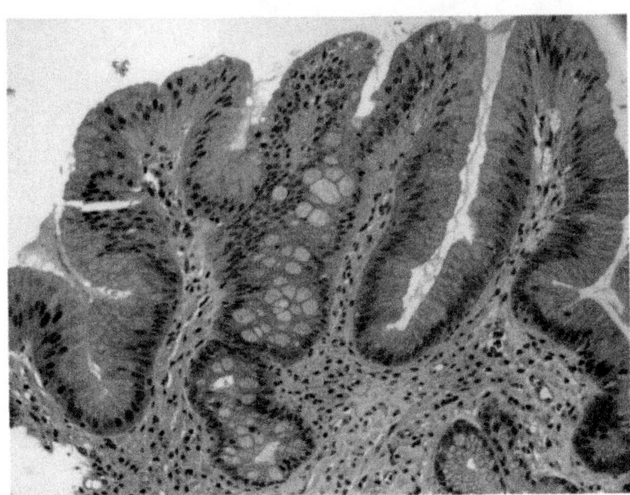

PLATE 4.1 Intestinal metaplasia with goblet cells in one central foveolar complex in cardiac mucosa.
Cross references: Fig 9.21, 13.4 and 13.21

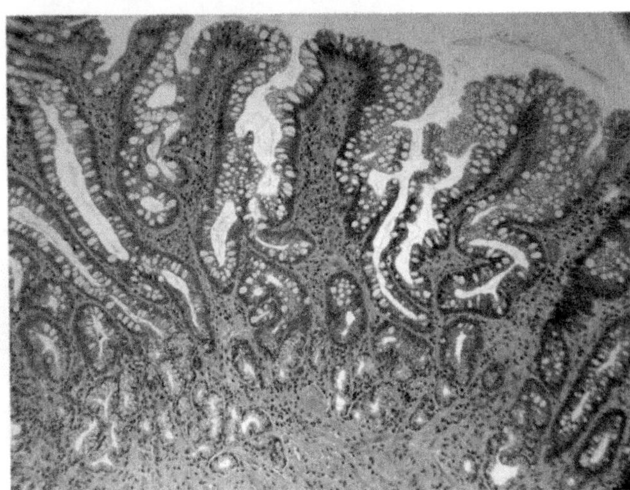

PLATE 4.2 Extensive intestinal metaplasia with numerous goblet cells throughout cardiac mucosa.
Cross references: Fig 9.26 and 14.15

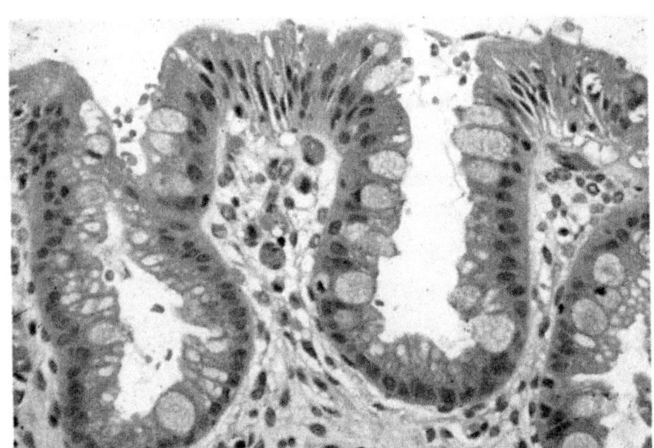

PLATE 4.3 Intestinal metaplasia showing goblet cells with basophilic cytoplasm in H&E stained section.
Cross references: Fig 5.20 and 9.22

PLATE 4: INTESTINAL METAPLASIA IN COLUMNAR LINED ESOPHAGUS (BARRETT ESOPHAGUS)

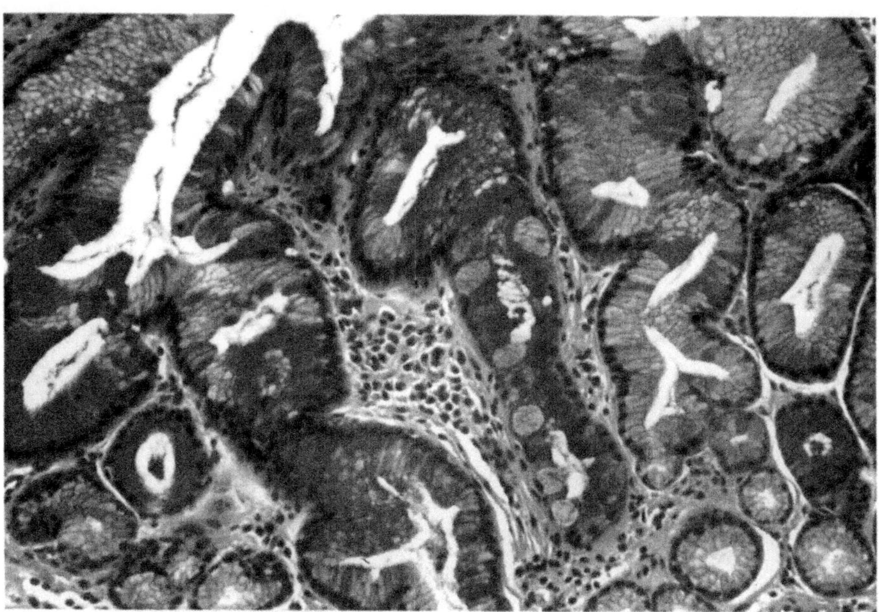

PLATE 4.4 Intestinal metaplasia in cardiac mucosa, Alcian blue stain. The goblet cells are seen as round blue vacuoles. Numerous non-goblet cardiac epithelial cells show intense blue cytoplasmic staining with alcian blue (columnar blue cells).
Cross references: Fig 5.29

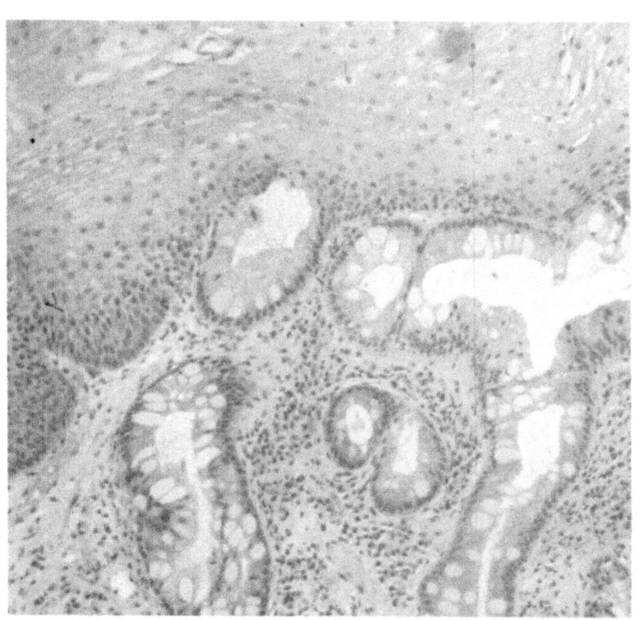

PLATE 4.5 Intestinal metaplasia under surface squamous epithelium.
Cross references: Fig 15.13 and 16.1

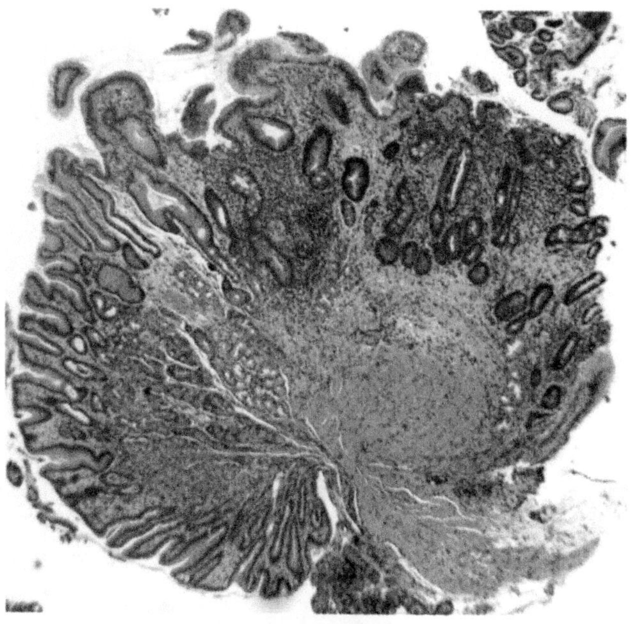

PLATE 4.6 Chronic atrophic gastritis showing atrophy of gastric oxyntic mucosa with gastric-type intestinal metaplasia.
Cross references: Fig 5.5 and 14.16

PLATE 5: NEOPLASTIC TRANSFORMATION OF BARRETT ESOPHAGUS

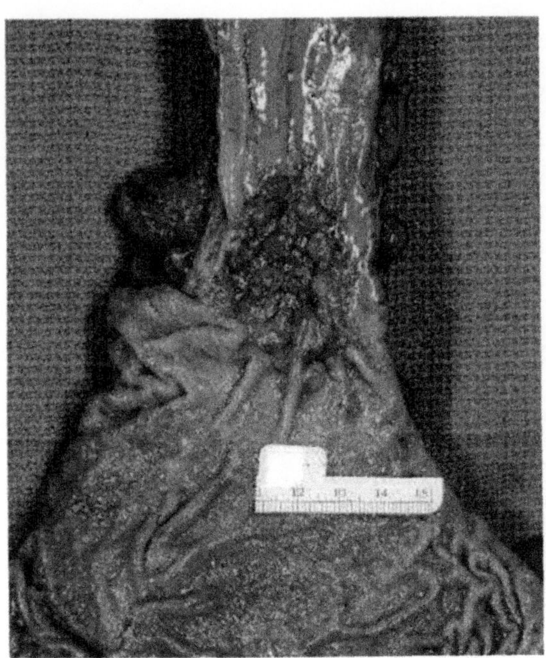

PLATE 5.1 Adenocarcinoma arising in columnar lined esophagus, gross.
Cross references: Fig 4.20 and 12.30

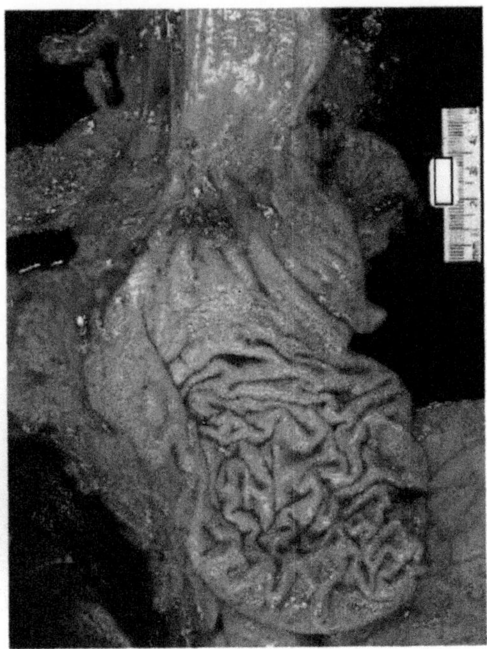

PLATE 5.2 Adencoarcinoma arising in columnar lined esophagus. Note dilated distal esophagus with flat, non-rugated columnar epithelium.
Cross references: Fig 4.21, 7.4 and 11.32

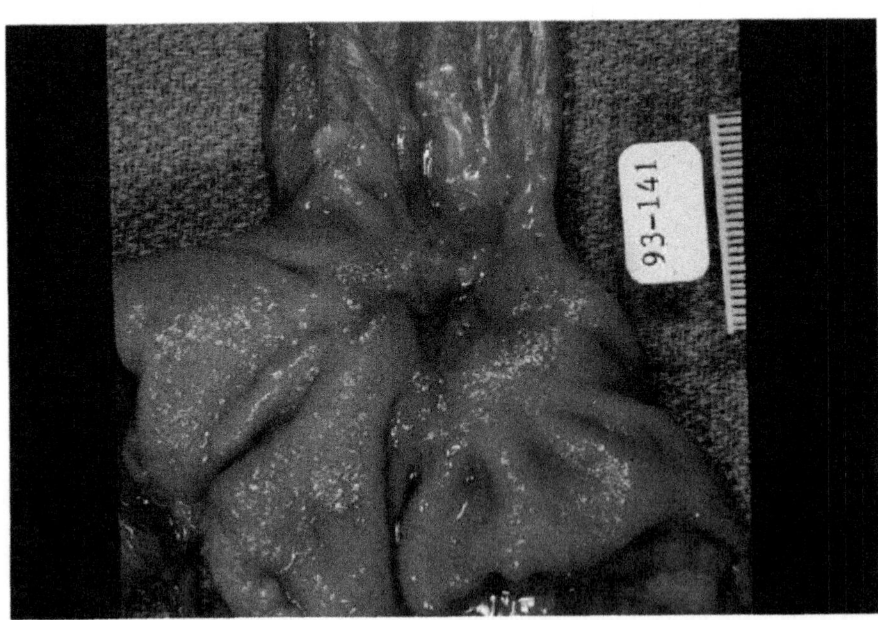

PLATE 5.3 Adenocarcinoma arising in dilated end-stage esophagus mimicking adenocarcinoma of proximal stomach.
Cross references: Fig 11.33 and 12.27

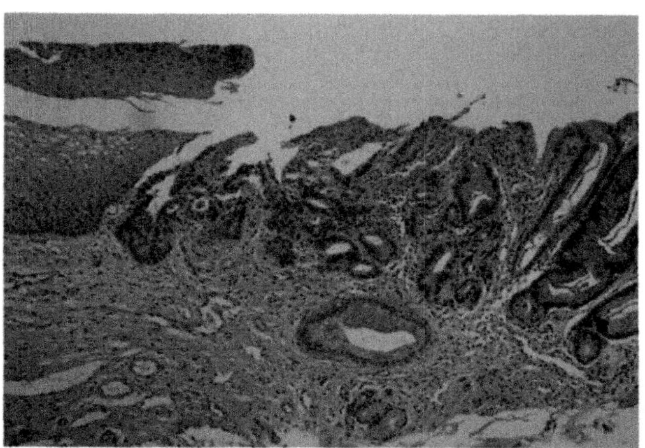

PLATE 5.4 Dysplasia in very short segment of columnar lined esophagus immediately distal to squamous epithelium.
Cross references: Fig 12.17

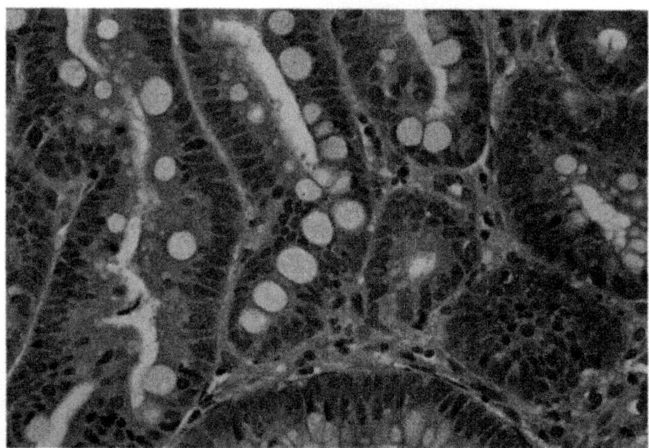

PLATE 5.5 Low grade dysplasia in intestinal (Barrett) metaplasia.
Cross references: Fig 14.21

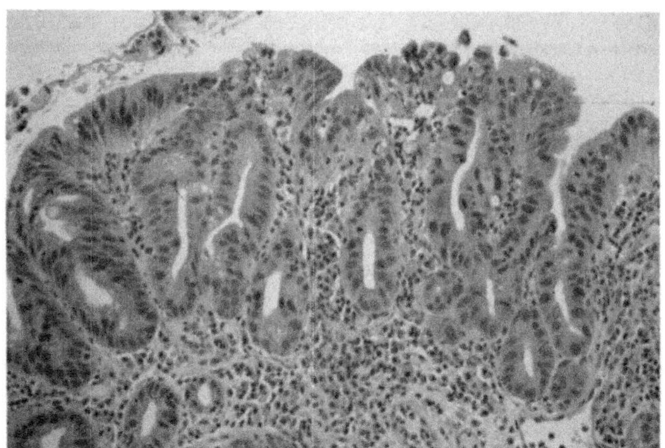

PLATE 5.6 High grade dysplasia in columnar lined esophagus.
Cross references: Fig 14.26

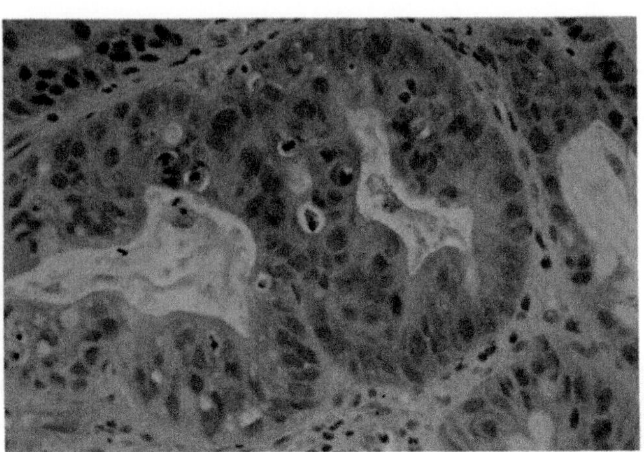

PLATE 5.7 High grade dysplasia. Note severe cytologic abnormality, loss of nuclear polarity and luminal bridging to form complex gland
Cross references: Fig 14.25

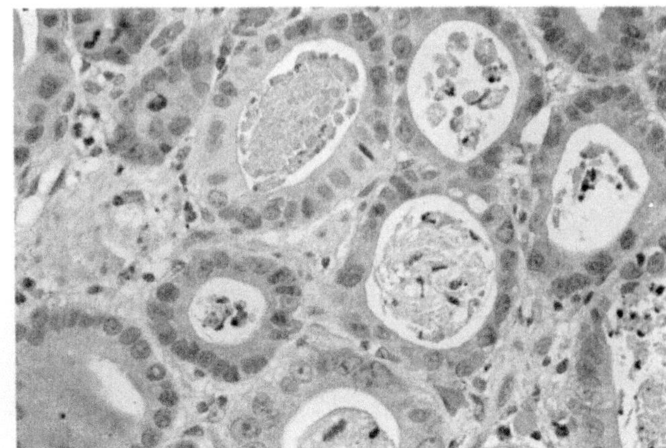

PLATE 5.8 Microcystic type of invasive adenocarcinoma of the esophagus.
Cross references: Fig 14.27

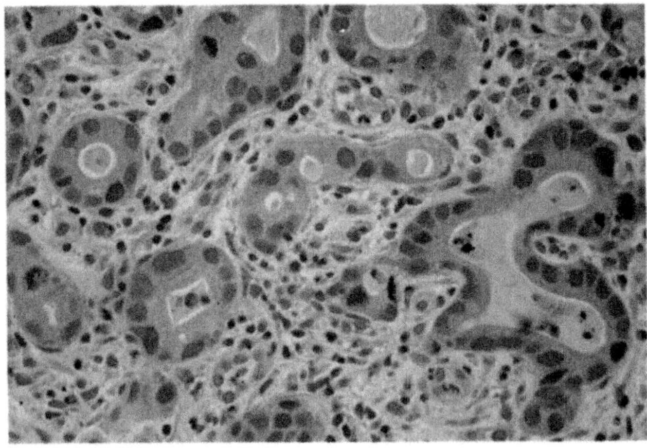

PLATE 5.9 Invasive adenocarcinoma of the esophagus.
Cross references: Fig 14.31

agement. No patient has a diagnosis of high-grade dysplasia until definitive criteria are present.

Cases that are diagnosed as low-grade dysplasia have definite criteria for the presence of a degree of dysplasia short of high-grade dysplasia (significant cytologic change with surface involvement) without active inflammation or erosion. When made in this manner, the diagnosis of low-grade dysplasia is reasonably specific although there are cases where "low-grade dysplasia" that Dr. Chandrasoma has diagnosed has disappeared in follow-up specimens; although this is cited as "reversal," it is more likely that the original diagnosis of low-grade dysplasia was suspect.

Any biopsy that does not have specific criteria for dysplasia is called negative for dysplasia. If there are cytologic changes, these are described and a statement made as to causation (regeneration in erosions, reactive, active inflammation, etc.). We do not use the diagnosis "indefinite for dysplasia." This has no meaning; dysplasia should be diagnosed only when it is definite. The likelihood is that even when we diagnose low-grade dysplasia with stringent criteria, we have a significant error rate in that dysplasia mimickers are included.

Over a period of 15 years, the reliability of Dr. Chandrasoma's diagnosis of dysplasia has been evaluated patient by patient on long-term follow-up, with and without treatment. The maintenance of trust in this setting is the real test of accuracy. Trust, however, must be verified by systematic study to see if that trust is justified. The following studies from our institutions have been published from the clinical perspective.

Literature Review

Peters JH, Clark GWB, Ireland AP, Chandrasoma P, Smyrk TC, DeMeester TR. Outcome of adenocarcinoma arising in Barrett's esophagus in endoscopically surveyed and nonsurveyed patients. J Thorac Cardiovasc Surg 1994;108:813–821.

This is a study of 17 patients who were referred from endoscopic surveillance programs for management of high-grade dysplasia or adenocarcinoma developing in Barrett esophagus (19). The referral diagnosis was adenocarcinoma in 6 of the patients and high-grade dysplasia in 11. After repeat endoscopy with aggressive biopsy protocols, 2 additional patients with adenocarcinoma were identified. Of the 9 patients who underwent esophagectomy for high-grade dysplasia, 5 had invasive adenocarcinoma in the esophagectomy specimen, which had been missed before the operation, despite the fact that the median number of biopsy specimens obtained per 2 cm of Barrett's mucosa was 7.8 (range, 1.5 to 15.0).

Literature Review

Tharavej C, Bremner CG, Chandrasoma P, DeMeester TR. Predictive factors of coexisting cancer in Barrett's high grade dysplasia. Submitted for publication.

The endoscopic, histologic, and demographic findings in 31 patients who underwent esophagectomy for high-grade dysplasia were reviewed (21). The presence of an ulcer, nodule, stricture, or raised area on preoperative endoscopy was noted. Results of endoscopic biopsies taken before resection every 1 to 2 cm along the Barrett's segment were reviewed. High-grade dysplasia was categorized as unilevel if the dysplasia was limited to a level of biopsy and as multilevel if more than one level was involved. Patients were divided into two groups according to the presence or absence of cancer in the resected specimens, and these variables were compared.

The prevalence of coexisting cancer in high-grade dysplasia patients was 45% (14 of 31). Of 31 patients, 9 had a visible lesion. Seven out of 9 patients (78%) with a visible lesion—compared to 7 out of 22 patients (32%) without a visible lesion—had cancer in the resected specimens ($p = 0.019$). Of 22 patients without a visible lesion, 10 had multilevel high-grade dysplasia and 12 had unilevel involvement. Six out of 10 patients with multilevel high-grade dysplasia compared to 1 out of 12 patients with unilevel high-grade dysplasia had cancer in the resected esophagus ($p = 0.009$). We concluded that in patients with high-grade dysplasia, the presence of a visible lesion on endoscopy or high-grade dysplasia in multiple biopsy levels are associated with an increased risk of the presence of occult cancer.

In this study, we validated the *high specificity* of a diagnosis of high-grade dysplasia at the University of Southern California. These patients do not include those in whom a diagnosis of intramucosal adenocarcinoma was made in any biopsy. We tend to be aggressive about the diagnosis of intramucosal adenocarcinoma in patients with high-grade dysplasia and call it when minimum criteria are satisfied; this is based on our belief and that of our group that these patients are at serious risk and that there should be a bias toward early surgery. The diagnosis of intramucosal adenocarcinoma in a biopsy would more certainly result in esophagectomy than one of high-grade dysplasia. These 31 patients therefore represent patients without any evidence of intramucosal adenocarcinoma.

At the time of biopsy diagnosis, which is made in my laboratory, we are blind to the presence or absence of a visible lesion. It is also relevant that the esophagectomy is performed at the USC University Hospital, which is not Dr. Chandrasoma's primary place of service. All esophagectomies are primarily evaluated by pathologists at this hospital; Dr. Chandrasoma does not review these cases until after they have been reported unless there is need for consultation. As such, the finding of invasive carcinoma is not made with any bias exerted by the pathologist reading the biopsies.

The finding of invasive carcinoma in 45% of this group of 31 patients who have high-grade dysplasia in a biopsy validated our diagnosis of high-grade dysplasia. When a visible lesion was present, 7 of 9 (78%) had invasive cancer in the specimen. In the 22 patients without a visible lesion, when high-grade dysplasia was present at multiple levels, the finding of invasive carcinoma in the resection specimen was 6 of 10 (60%) patients. The high risk of coexistent invasive cancer in the resection specimen leaves no doubt that patients with a diagnosis of high-grade dysplasia who have either a visible lesion or multilevel high-grade dysplasia must have resection because they are at imminent and serious risk.

The 12 of 22 patients without a visible lesion and a diagnosis of high-grade dysplasia at one level represented the group where surveillance may have been a feasible alternative to esophagectomy. In this group, 1 of 12 (8.3%) had invasive carcinoma, and the other 11 patients had residual high-grade dysplasia in the resected specimen. Because the natural history of high-grade dysplasia is not certain, it is not possible to determine the imminence or severity of risk in these 11 patients. There is a justification for limited esophagectomy in this group by virtue of providing the patient maximum safety in an uncertain setting, but only if the results of the esophagectomy are not associated with severe morbidity and loss of quality of life. At USC, the operation of choice for high-grade dysplasia is transhiatal and vagal sparing esophagectomy, which has a relatively low morbidity and excellent postsurgical quality of life.

The preceding study does not test or validate the *sensitivity* of our diagnosis of high-grade dysplasia. This is an important consideration, because it is possible that we have set the criteria at such a high level of specificity that there are a significant number of patients who really have high-grade dysplasia in a biopsy who are erroneously diagnosed as having no dysplasia or low-grade dysplasia. To validate the sensitivity of our diagnosis of high-grade dysplasia, we therefore have to evaluate the follow-up of patients without a diagnosis of high-grade dysplasia. This is a complicated issue. Because dysplasia is progressive, it is expected that some patients without dysplasia or with low-grade dysplasia will go on to develop high-grade dysplasia in a later biopsy.

The best validation of the sensitivity of high-grade dysplasia is to evaluate the follow-up of patients who have an antireflux surgical procedure after a biopsy diagnosis of Barrett esophagus, with and without low-grade dysplasia. These patients who undergo antireflux surgery will have a bias toward the most severe disease. Successful antireflux surgery stops gastroesophageal reflux, the natural progression of the disease is altered, and the follow-up permits testing of the diagnosis at the time of the prefundoplication biopsy. If there are patients with missed high-grade dysplasia in that biopsy, follow-up examinations after the fundoplication should reveal high-grade dysplasia and adenocarcinoma, assuming that at least some cases with high-grade dysplasia have the necessary mutations to progress to cancer at the time of diagnosis.

Literature Review

Hofstetter WL, Peters JH, DeMeester TR, Hagen JA, DeMeester SR, Crookes PF, Tsai P, Banki F, Bremner CG. Long-term outcome of antireflux surgery in patients with Barrett's esophagus. Ann Surg 2001;234:532–538.

This is a long-term follow-up study of 85 patients with Barrett esophagus treated by antireflux surgery at USC (22). Of these, 59 patients had long-segment Barrett esophagus and 26 patients had short-segment Barrett esophagus. After 408 patient-years of follow-up (with a mean follow-up per patient of 4.8 years after surgery), there were no patients with high-grade dysplasia or cancer.

This study validates the high sensitivity of our diagnosis of high-grade dysplasia. The fact that there were no patients with high-grade dysplasia or adenocarcinoma after a mean follow-up period of 4.8 years after antireflux surgery is strong evidence that there were no false negative diagnoses of high-grade dysplasia in the prefundoplication biopsy.

It is strongly recommended that pathologists establish criteria that validate their diagnosis of high-grade dysplasia in the manner that we have done. This is evidence of the accuracy of the diagnosis of high-grade dysplasia. If one can show by validation studies that one's diagnosis of high-grade dysplasia is accurate with a very high specificity and sensitivity, as we have done, reproducibility among experts becomes irrelevant. In the face of accuracy, expert disagreement only means that the expert who disagrees is wrong. However, if accuracy of diagnosis of high-grade dysplasia is not established, expert confirmation and reproducibility with all their inherent problems represent the only recourse.

References

1. Chandrasoma PT, Der R, Ma Y, Peters J, DeMeester T. Histologic classification of patients based on mapping biopsies of the gastroesophageal junction. Am J Surg Pathol 2003;27:929–936.
2. Chandrasoma P. Pathological basis of gastroesophageal reflux disease. World J Surg 2003;27:986–993.
3. Chandrasoma P. Controversies of the cardiac mucosa and Barrett's esophagus. Histopathol 2005;46:361–373.

4. Jain R, Aquino D, Harford WV, Lee E, Spechler SJ. Cardiac epithelium is found infrequently in the gastric cardia. Gastroenterology 1998;114:A160 (Abstract).
5. Oberg S, Peters JH, DeMeester TR, et al. Inflammation and specialized intestinal metaplasia of cardiac mucosa is a manifestation of gastroesophageal reflux disease. Ann Surg 1997;226: 522–532.
6. Sarbia M, Donner A, Gabbert HE. Histopathology of the gastroesophageal junction. A study on 36 operation specimens. Am J Surg Pathol 2002;26:1207–1212.
7. Csendes A, Maluenda F, Braghetto I, et al. Location of the lower esophageal sphincter and the squamocolumnar mucosal junction in 109 healthy controls and 778 patients with different degrees of endoscopic esophagitis. Gut 1993;34: 21–27.
8. Chandrasoma PT, Lokuhetty DM, DeMeester, TR, et al. Definition of histopathologic changes in gastroesophageal reflux disease. Am J Surg Pathol 2000;24:344–351.
9. Rudolph RE, Vaughan TL, Storer BE, et al. Effect of segment length on risk for neoplastic progression in patients with Barrett esophagus. Ann Intern Med 2000;132:612–620.
10. Reid BJ, Levine DS, Longton G, et al. Predictors of progression to cancer in Barrett's esophagus: Baseline histology and flow cytometry identify low- and high-risk patient subset. Am J Gastroenterol 2000;95:1669–1676.
11. Olvera M, Wickramasinghe K, Brynes R, Bu X, Ma Y, Chandrasoma P. Ki67 expression in different epithelial types in columnar lined esophagus indicates varying levels of expanded and aberrant proliferative patterns. Histopathology 2005;47:132–140.
12. Lagergren J, Bergstrom R, Lindgren A, Nyren O. Symptomatic gastroesophageal reflux as a risk factor for esophageal adenocarcinoma. N Engl J Med 1999;340:825–831.
13. Goldblum JR, Vicari JJ, Falk GW, et al. Inflammation and intestinal metaplasia of the gastric cardia: The role of gastroesophageal reflux and *H. pylori* infection. Gasteroenterology 1998;114: 633–639.
14. Genta RM, Huberman RM, Graham DY. The gastric cardia in *Helicobacter pylori* infection. Hum Pathol 1994;25:915–919.
15. Hackelsberger A, Gunther T, Schultze V, Labenz J, Roessner A, Malfertheiner P. Prevalance and pattern of *Helicobacter pylori* gastritis and the gastric cardia. Am J Gastroenterol 1997;92: 2220–2223.
16. Hamilton SR, Smith RRL, Camerson JL. Prevalence and characteristics of Barrett's esophagus in patients with adenocarcinoma of the esophagus or esophago-gastric junction. Hum Pathol 1988;19:942–948.
17. Reid BJ, Haggitt RC, Rubin CE, et al. Observer variation in the diagnosis of dysplasia in Barrett's esophagus. Hum Pathol 1988;19:166–178.
18. Buttar NS, Wang KK, Sebo TJ, et al. Extent of high grade dysplasia in Barrett's esophagus correlates with risk of adenocarcinoma. Gastroenterology 2001;120:1607–1619.
19. Peters JH, Clark GWB, Ireland AP, Chandrasoma P, Smyrk TC, DeMeester TR. Outcome of adenocarcinoma arising in Barrett's esophagus in endoscopically surveyed and nonsurveyed patients. J Thorac Cardiovasc Surg 1994;108:813–821.
20. Montgomery E, Bronner MP, Goldblum JR, Greenson JK, Haber MM, Hart J, Lamps LW, Lewin DN, Robert ME, Toledano AY, Shyr Y, Washington K. Reproducibility of the diagnosis of dysplasia in Barrett esophagus: A reaffirmation. Hum Pathol 2001;32:368–378.
21. Tharavej C, Bremner CG, Chandrasoma P, DeMeester TR. Predictive factors of coexisting cancer in Barrett's high grade dysplasia. (Submitted for publication.)
22. Hofstetter WL, Peters JH, DeMeester TR, Hagen JA, DeMeester SR, Crookes PF, Tsai P, Banki F, Bremner CG. Long-term outcome of antireflux surgery in patients with Barrett's esophagus. Ann Surg 2001;234:532–538.

CHAPTER

15

Research Strategies for Preventing Reflux-Induced Adenocarcinoma

One of the effects of delving deeply into any subject is that many new ideas arise in one's head. Many of these are fanciful and can never be expressed in any scientific forum. We have discussed many of these concepts with pathology residents that we train, and they have generally caused excitement, so we know there may be a glimmer of truth and future direction in at least some of them. One of the benefits of writing a book such as this is that one has the freedom to fantasize. Readers who are not interested in fantasy are advised to skip this chapter. This chapter is for the dreamers among us. For the dreamers that we are, this is the most important chapter in the book.

The failure in understanding and misdirection of research in gastroesophageal reflux disease ensures that the prevention of adenocarcinoma is not being achieved. At the present time, very little specific effort is being made to prevent death in the approximately 14,000 patients who die every year from esophageal adenocarcinoma. Only a small number of these patients are found during Barrett esophagus surveillance programs. We suggest that correction of understanding and attitude will change the direction of future research and result in new methods of preventing reflux-induced adenocarcinoma. This chapter outlines some of the dreams we have about how this will happen.

Part of the reason for this failure to understand the cellular basis of reflux disease is that even the most expert gastroenterologists and surgeons in the field have absolutely no training in microscopic pathology. When we present our histology-based data at conferences attended by gastroenterologists and surgeons, an inevitable barrier comes up. It is not that they do not believe; it is that they do not have the knowledge of cellular pathology needed to know whether to believe or not believe. The ignorance is profound.

The pathology establishment is not much better. It is slow to react to change, holds on to cherished dogmas even when evidence exists that they are wrong, and resists the emergence of new ideas. When we talk to groups of pathologists, we usually get an enthusiastic response at grass-roots level by pathologists who are sufficiently open-minded to understand and believe. Every pathology resident we have discussed our new ideas with has gone away convinced. However, these pathologist believers cannot put these concepts into practice because the pathology and gastroenterology establishments resist having these new concepts become recognized as standard practice (1). There is no doubt in our minds that the new diagnostic criteria that we outlined in Chapter 14 will become standard practice in the future. The trend toward acceptance is inexorable; it is the slow speed that is frustrating.

MECHANISMS OF CELLULAR CHANGES IN REFLUX DISEASE

The cellular events that occur when gastroesophageal reflux occurs result from the interaction of the cells that are exposed and the nature of the refluxate (Table 15.1). Ultimately, every change must be explained on the basis of a cellular interaction between a cell that is present in the patient and a molecule or combination of molecules in the refluxate. Such interactions may theoretically be of two kinds:

TABLE 15.1 Classification of Cellular Changes of Gastroesophageal Reflux Disease

Injurious/destructive phenomena	Agent involved
1. Increased squamous epithelial cell loss and reactive basal cell hyperplasia	Acid
2. Squamous epithelial inflammation (reflux esophagitis)	Acid
3. Chemoattraction of intraepithelial eosinophils	Acid
4. Squamous epithelial erosion/ulceration (erosive esophagitis)	Acid
5. Dilated intercellular spaces—increased squamous epithelial permeability	Acid
6. Cardiac epithelial inflammation, cell loss, and reactive proliferation	Acid; ?others
7. Intestinal epithelial inflammation, cell loss, and proliferation	Acid; ?others
Interactions between cells and molecules	
1. Gene (e.g., Cox-2) activation in squamous epithelium	Unknown
2. Cardiac metaplasia of squamous epithelium	Unknown
3. Parietal cell induction in cardiac mucosa (oxyntocardiac mucosa)	Unknown; ?acid
4. Goblet cell induction in cardiac mucosa (intestinal metaplasia)	Unknown; ?bile salts
5. Carcinogenesis in intestinal metaplasia	Unknown; ?bile salts

1. *Destructive or injurious.* This reaction occurs when a molecule causes cell damage by a chemical reaction with cell components. This is the obvious mechanism in high-dosage radiation and caustic damage where the physical agent directly causes cell necrosis on contact because of some lethal interaction between the cell (probably the cell wall) and the molecule. Acid in high enough concentration has the potential to induce direct cell damage in all types of esophageal epithelia, whether squamous or metaplastic columnar.
2. *Interactive.* This reaction occurs when the molecule in the refluxate combines with a cell component, usually a receptor, to induce a change in the structure or function of the cell. The esophageal epithelia, both squamous and metaplastic columnar, have a predominantly nonsecretory function, although mucous glands, pancreatic cells, paneth cells, and parietal cells are capable of secretion. As far as is known at present, secretion is not of great importance.

The important clinical result of molecule-cell interactions are alterations in genetic control of the cell that result in different metaplastic pathways and carcinogenesis. Interactive mechanisms generally preserve the altered cell in contrast to the destructive influences that damage and destroy the cells.

There is a broad difference in the type of agent most likely to produce these types of cellular changes. Destructive injuries tend to be the mechanism of physical agents and simple chemicals such as highly ionic compounds (the coagulative heat of a laser, the hydroxyl ions of lye, the free radicals generated by radiation, and the free hydrogen ions of acid). Ionic agents are harmless until they reach a sufficient concentration where they cause direct damage of the cell membrane or enter the cells, altering pH that may induce changes in chemical reactions in the cell. It is unlikely that any cell has receptors to simple ionic compounds. Cell receptors are generally complex and interact with larger molecules. It is therefore likely that larger molecules such as complex nitrogenous compounds and bile salt derivates are responsible for interactive changes.

Acid is very likely the main damage-inducing molecule in the refluxate, although evidence exists for significant damage-producing capability in other molecules in the refluxate such as bile acid components. Acid is responsible for numerous changes in the squamous epithelial cells, ultimately causing increased permeability that allows refluxate molecules of all types to gain access to the proliferative stem cell pool in the basal region of the epithelium (2, 3). Acid is therefore the key that unlocks the squamous epithelial barrier and permits other luminal molecules in the refluxate to gain access to the proliferative stem cell pool. Acid damage sets the stage for molecular interactions with cell surface receptors, causing metaplasia to different types of columnar epithelia and inducing genetic

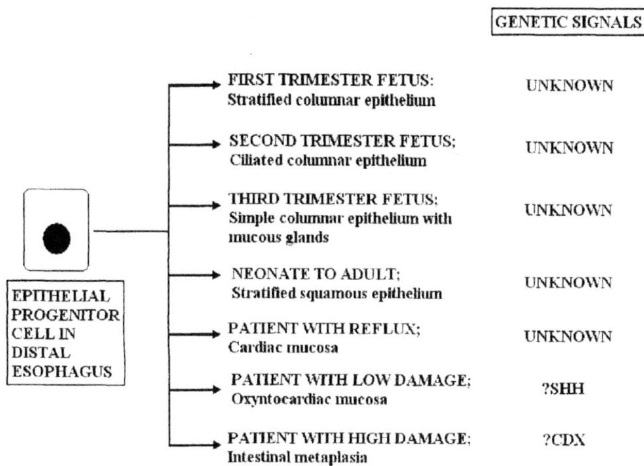

FIGURE 15.1 The range of expression of differentiation seen during life in a progenitor cell in the distal esophagus. In the normal fetus, this cell normally shows three types of columnar epithelial differentiations before finally suppressing all these differentiating signals and producing stratified squamous epithelium. During life, in patients who have reflux disease, other differentiating signals are activated, which result in different types of metaplastic columnar epithelia.

changes in the cells that cause progression in the reflux to adenocarcinoma sequence (see Chapter 8).

THE MECHANISM OF METAPLASIA: EPITHELIA DO NOT "MOVE"

The progenitor stem cell in any given location of the gastrointestinal tract is responsible for the type of epithelium that is present at that site. The type of epithelium that is generated depends on the genetic signals active in the cells that direct differentiation. In the esophagus, the differentiating signal develops via multiple fetal signals until the final adult genetic signal is established in early postnatal life that directs normal differentiation into stratified squamous epithelium (Fig. 15.1; see also Chapter 3). All fetal esophageal genetic signals that direct columnar epithelial cell differentiation are suppressed. The nature of the normal fetal and adult differentiating signals involved in the esophageal epithelial cell is unknown at this time. Genetic signals that direct epithelial differentiation in other parts of the gastrointestinal tract are not active at any point in normal development of the esophageal epithelial progenitor cell (except an intestinal signal, which may rarely be transiently expressed in the fetus). These signals remain within the genome of the esophageal epithelial cell in a suppressed state.

Metaplasia is a process that is simply a change in the genetic signal of the progenitor cell to direct the formation of a particular epithelium (Fig. 15.1). It does not involve "movement" of one epithelial type to another location. Thus, columnar metaplasia of the esophagus is caused by an alteration of the esophageal epithelial progenitor cell's genetic signal that causes it to differentiate into columnar rather than squamous epithelium. Different types of columnar epithelia are produced from activation of different genetic signals that direct cell differentiation. Columnar metaplasia is not the replacement of esophageal epithelium by "gastric epithelium moving up into the esophagus." The only situation where there is lateral movement of progenitor cells is when erosion and ulceration occur; in these cases, the progenitor cells move into the denuded area from the adjacent epithelium, from esophageal gland ducts, or even from circulating stem cells. It is likely that when the progenitor cell reaches its new location, it takes on the behavior and differentiation inherent to the new location. Circulating stem cells populating the esophagus do not differentiate into hepatocytes or hematopoietic cells. The misconception that columnar-lined esophagus is composed of gastric mucosa was introduced by Allison and Johnstone in the first description of the entity in 1953 (4). It still persists in the minds of many people.

Metaplasia may result from activation of suppressed genetic signals of two types:

1. *Fetal esophageal epithelial signals.* These will result in epithelia that were seen transiently during fetal life during development. Cardiac mucosa is one such epithelium. Cardiac mucosa resembles the nonciliated columnar epithelium that is present in the third trimester at the distal end of the esophagus (see Chapter 3). This may consist of a flat epithelium, a columnar epithelium with rudimentary foveolar pits, or one with mucosal glands composed only of mucous cells.
2. *Epithelial signals that are aberrant for the esophagus and normally seen in other parts of the gastrointestinal tract.* These produce columnar epithelia that are not normally seen in fetal esophagus during development. They include the gastric differentiating signal, which directs the formation of parietal cells; this is activated in certain conditions and only when cardiac mucosal metaplasia is present in the esophagus (see Chapter 9).

Oxyntocardiac mucosa does not arise directly from squamous epithelium. There is evidence that the Sonic Hedgehog gene system is involved in gastric-type differentiation. The other differentiating signal is the intestinal signal, which directs the formation of goblet cells similar to those in the intestine. This is activated in certain conditions and only when cardiac mucosal metaplasia is present in the esophagus (see Chapter 9). Intestinal metaplasia does not arise directly from squamous epithelium. There is evidence that the Cdx gene systems are involved in intestinal-type differentiation (Fig. 15.1).

In Chapter 10, we showed that the only esophageal epithelial type on which carcinogens in the refluxate act is the intestinal type (Fig. 15.2). This means that when the genetic switch that directs intestinal metaplasia occurs, there is some transformation of the esophageal epithelial cell that makes the cell susceptible to the action of carcinogens. This is most likely related to the alteration of the configuration of a cell receptor that makes the cell capable of a new interaction with a carcinogenic molecule (Fig. 15.3).

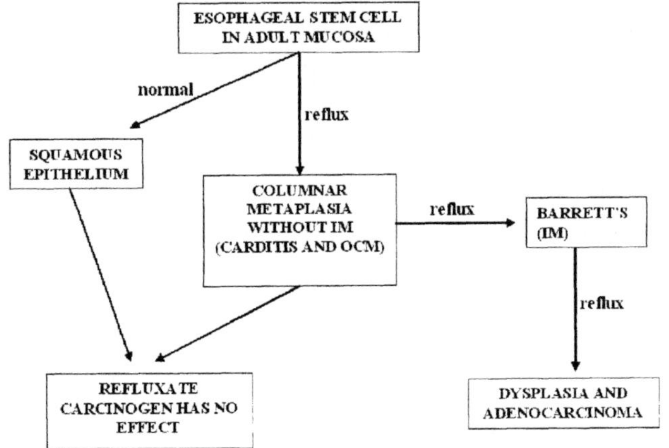

FIGURE 15.2 The relationship of different epithelial types in the esophagus to adenocarcinoma. Carcinogenesis occurs only in intestinal metaplastic (Barrett) epithelium.

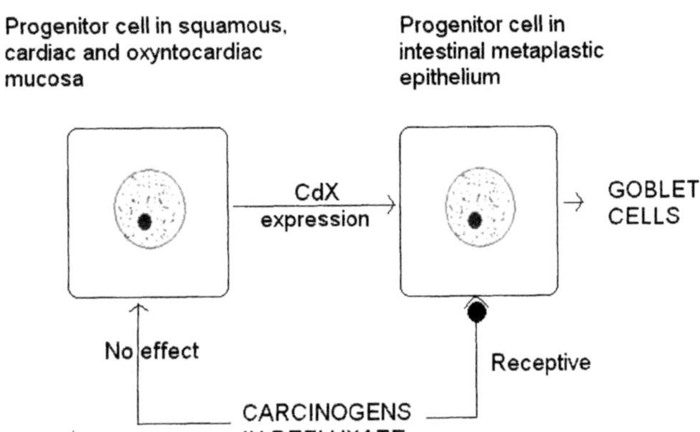

FIGURE 15.3 The progenitor cell in the esophagus becomes susceptible to carcinogenesis only when it undergoes metaplasia to intestinal metaplasia. In this diagram, this is shown as the emergence of a cell surface receptor consequent on a Cdx genetic activation that is hypothesized to be the differentiating genetic signal for intestinal metaplasia.

The early part of the reflux-adenocarcinoma sequence that takes squamous epithelium to intestinal metaplasia results from activation of normally suppressed genes that are involved in directing differentiation in the gastrointestinal tract ("genetic switches"). They are simple differentiating signals; they merely lead to the formation of an orderly adult epithelium that is different from the native squamous epithelium of the esophagus. These genetic switches are potentially easily reversible with appropriate manipulation and are likely to be easily amenable to therapeutic intervention.

In contrast, the genetic changes that convert intestinal metaplasia to dysplasia and adenocarcinoma are likely to be irreversible and abnormal genetic mutations. These mutations that are the essence of carcinogenesis give the affected cells a growth advantage, lack of orderliness in proliferation, and other traits that characterize the uncontrolled growth of neoplasia. They are difficult to reverse and less likely to be easily amenable to therapeutic intervention.

THE REFLUX-ADENOCARCINOMA SEQUENCE

In Chapters 8 to 10, we detailed the cellular changes in the reflux-adenocarcinoma sequence (Fig. 15.4). The

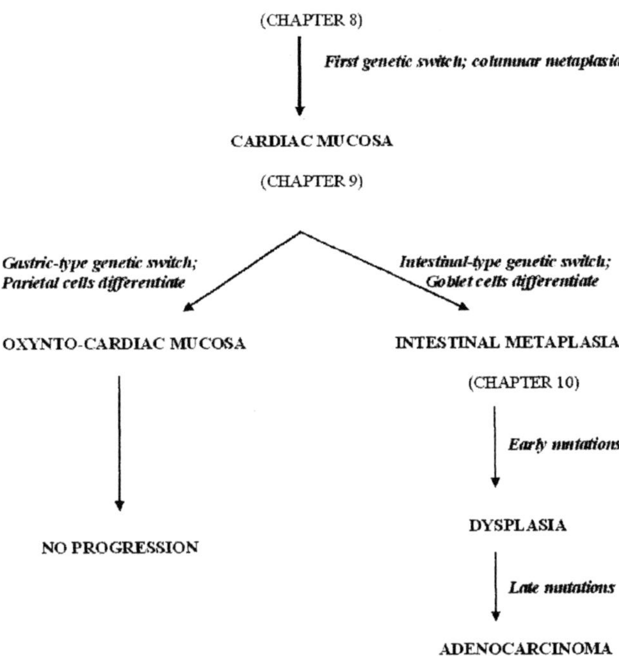

FIGURE 15.4 The reflux-adenocarcinoma sequence.

patient passes through this sequential change very slowly, over many years (Fig. 15.5). The actual speed and quantity of change in any patient as he or she progresses to higher risk stages in the reflux-carcinoma sequence depends on three critical and independent events (Fig. 15.6).

The Severity of Damage to Squamous Epithelium

This factor depends on the amount of reflux and the age of its onset and is acid-dependent; it is maximal at the gastroesophageal junction where acid exposure is greatest (Fig. 15.6). This converts squamous epithelium to increasing lengths of cardiac mucosa. Damage caused by reflux is a vicious cycle change. Squamous epithelial damage and columnar metaplasia is associated with shortening of the lower esophageal sphincter, which in turn aggravates the tendency to gastroesophageal reflux and causes further epithelial damage. The fact that this process is self-limited at a different degree of damage in different people suggests that there is a mechanism for a steady state to be reached.

Cameron et al. (5) showed that the final metaplastic length of squamous epithelium is established at a young age in most patients. One can reasonably assume that in most people, reflux begins fairly early in life and continues to cause increasing cardiac metaplasia of esophageal squamous epithelium for 10 to 20 years, it then reaches a state of equilibrium, where the separation of squamous epithelium from the point of reflux (the gastroesophageal junction) is adequate to prevent further damage. At this point, squamous epithelial damage stops sufficiently to prevent further cardiac metaplasia, and the columnar-lined esophagus becomes stable in length (Fig. 15.7). The length of columnar-lined esophagus is a testament to the cumulative lifetime of cellular damage that has occurred in the esophagus as a result of chronic gastroesophageal reflux. We have suggested that the component of reflux that focuses on cellular damage has remained unchanged since the 1950s as evidenced by the frequent presence of long segments of columnar-lined esophagus in the 1930–1960 era (see Chapter 2).

In many patients with mild reflux, all cardiac mucosa transforms to oxyntocardiac mucosa; this is a natural "cure" (= removal from the reflux-adenocarcinoma

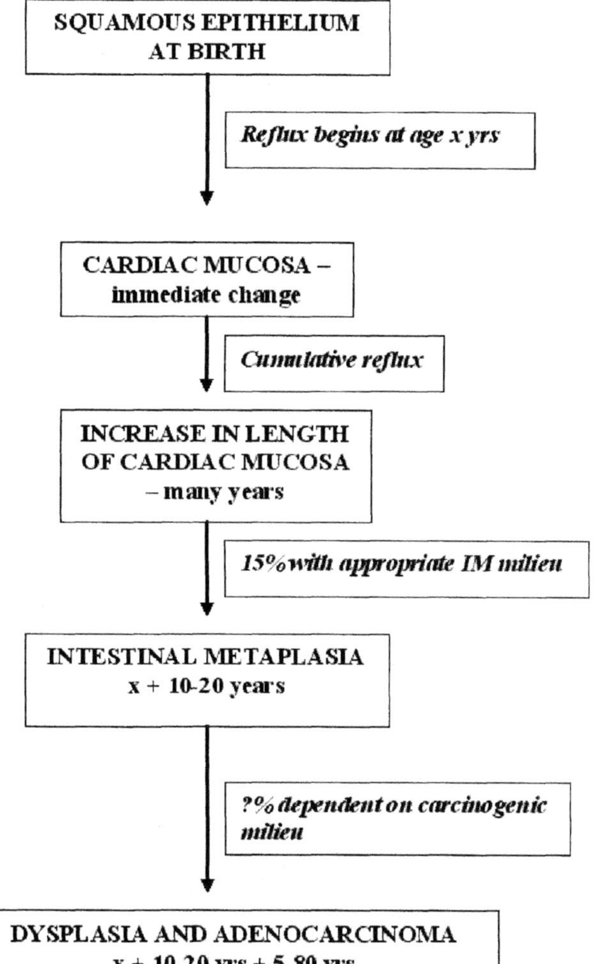

FIGURE 15.5 A typical time line for the changes in a patient with reflux. The onset of reflux is at *x* years. This time line varies infinitely in different patients, but it is likely that most patients have significant intervals between the onset of cardiac metaplasia and intestinal metaplasia and between intestinal metaplasia and adenocarcinoma.

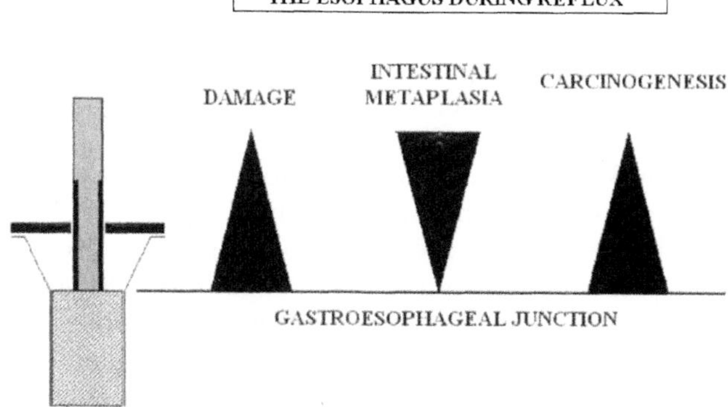

FIGURE 15.6 The three main events that are necessary for the development of adenocarcinoma. The elements that cause damage are all in the gastric juice and reach the esophagus by reflux. For damage (probably acid-induced) and carcinogenesis (dose dependent), there is a gradient that is maximal at the gastroesophageal junction and decreases more proximally. In contrast, the tendency to cause intestinal metaplasia is maximal in the more proximal esophagus and decreases toward the gastroesophageal junction.

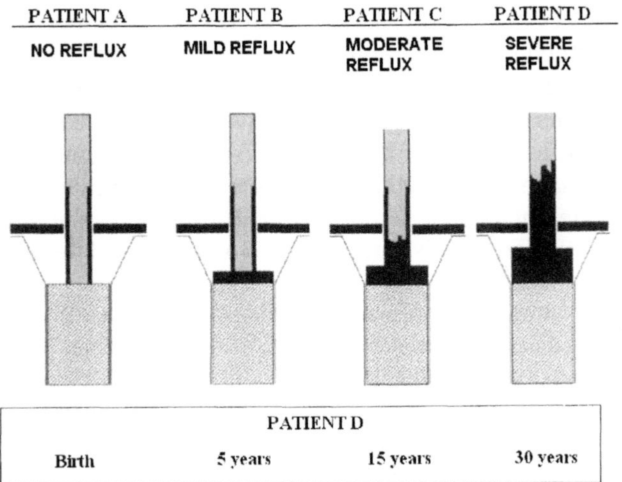

FIGURE 15.7 Four different patients at age 30 (A–D), showing varying lengths of cardiac transformation of the squamous esophagus resulting from four different levels of cumulative reflux damage (upper frame). This same figure can also show the evolution of cardiac metaplasia over the years in one patient with severe reflux (D). It is likely that by age 30 years, the final length of metaplastic columnar esophagus has been established.

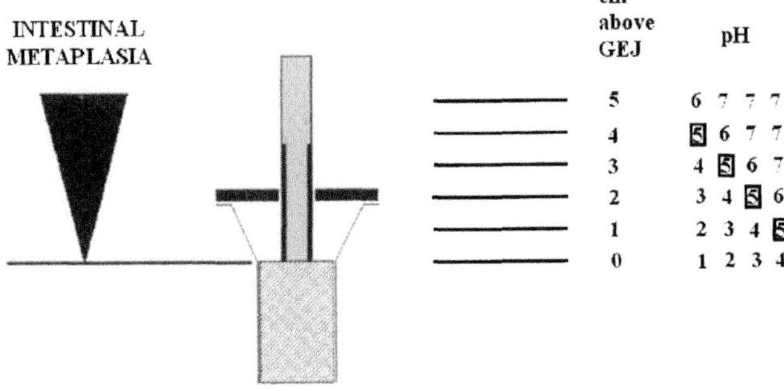

FIGURE 15.8 The effect of pH in the pathogenesis of intestinal metaplasia in cardiac mucosa. Four different baseline gastric pHs (1–4) are shown (*right*). With reflux of similar amount, the pH gradient varies from baseline gastric to neutral esophageal in the four instances. As shown, the actual pH at different levels of the esophagus will vary in these situations. If intestinal metaplasia is favored by a pH of 5 (highlighted in a box), the tendency to intestinal metaplasia of cardiac mucosa will increase in the more distal esophagus as the baseline gastric pH increases.

sequence) that we term compensated reflux because these patients are not at risk for intestinal metaplasia or carcinoma. These are the 55% to 65% of the population that have only oxyntocardiac mucosa between the squamous epithelium and gastric oxyntic mucosa when autopsy populations (6, 7) or clinical populations with many "normal" people are biopsied (8).

The Tendency to Intestinal Metaplasia

The cardiac mucosa generated from squamous epithelium by reflux evolves into oxyntocardiac and intestinal metaplasia based on the factors that we discussed in Chapter 9. This results in an infinitely variable mixture of the three epithelial types in columnar-lined esophagus.

Unlike damage, the tendency for intestinal metaplasia is greater in the more proximal esophagus than at the gastroesophageal junction (Fig. 15.6; also see Chapter 9). In a given segment of cardiac metaplasia, intestinal metaplasia almost always occurs in the most proximal region, immediately distal to the squamocolumnar junction. The prevalence of intestinal metaplasia in columnar-lined esophagus increases as the length of columnar-lined esophagus increases.

We have suggested that intestinal metaplasia of cardiac mucosa is a molecular change that is pH dependent; it is promoted by a reduction in acidity (i.e., an increase in pH). The pH at the gastroesophageal junction is the baseline gastric pH. In the normal patient without reflux, esophageal pH is neutral (i.e., 7). When reflux occurs, the pH in the esophagus has been shown to develop a gradient from low at the gastroesophageal junction to 7 in the proximal esophagus (Fig. 15.8). In patients who have alkaline (i.e., pH > 7) gastric juice, reflux can render the pH in the esophagus alkaline and the gradient is reversed, but this is a rare occurrence. In Figure 15.8, let us assume that intestinal metaplasia of cardiac mucosa occurs in a given patient at a pH of 5. The point in the esophagus where this pH is reached will vary with the baseline gastric pH. As the gastric juice becomes more alkaline, the critical point in the esophagus where intestinal metaplasia is generated becomes more distal. The natural alkalinizer of gastric juice is duodenogastric reflux. However, the use of acid suppressive drugs adds to this natural alkalinizing effect and tends to lower the critical point at which intestinal metaplasia occurs in the esophagus.

The molecular interactions in cardiac mucosa require not only the correct pH but also a minimum period of time exposure to effect the change. With small volume reflux and normal clearing of the refluxate, the exposure of the target cell to the correct alkaline environment may be too short. As the volume of reflux increases, it is likely that the time to which the target cell is exposed to the required pH also increases sufficiently to permit the molecular interactions to take place.

When one superimposes the damage element to this equation (Fig. 15.9), one can see that a given patient has cardiac mucosa in the esophagus to a point above the gastroesophageal junction that depends on acid

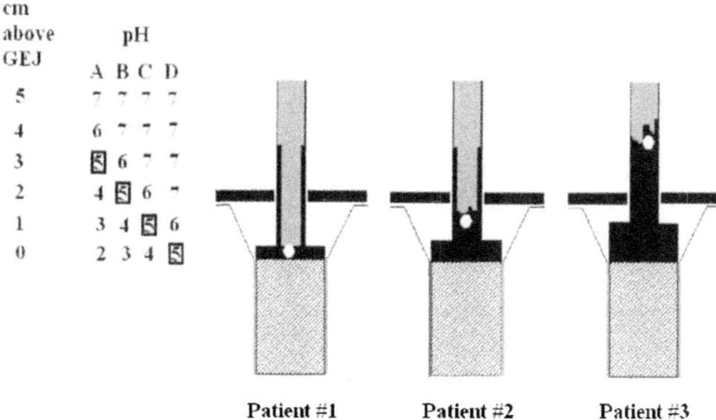

FIGURE 15.9 Intestinal metaplasia occurs only when cardiac mucosa (black) is present. Four different pH gradients in the esophagus (A–D) are shown (*left*). In patient 1, who has a very short-segment of cardiac mucosa, intestinal metaplasia (white circle) will occur only in the highest alkaline gradient (D). In patient 2, intestinal metaplasia will occur in gradients C and D only and will extend lower in gradient D than in gradient C. In patient 3, intestinal metaplasia will occur in all four gradients, but it will extend into increasingly distal regions of the esophagus from gradient A to D. Note that the extent of intestinal metaplasia in different pH gradient situations is not shown.

exposure of the squamous epithelium, which is almost opposite to that which causes intestinal metaplasia. Intestinal metaplasia will occur only if the patient has cardiac mucosa at the critical point where the pH is the correct level of alkalinity and the volume of reflux is sufficient to permit an adequate duration of exposure. The condition at highest risk for intestinal metaplasia is a long segment of cardiac mucosa with high volume reflux. This is associated with the maximum alkaline exposure in the most proximal region of the cardiac mucosa (Fig. 15.9). This explains why intestinal metaplasia occurs in the most proximal region of cardiac mucosa and why the prevalence of intestinal metaplasia increases as the length of cardiac mucosa increases.

We showed in Chapter 9 that there has been a historical increase in both the prevalence of intestinal metaplasia within columnar-lined esophagus and the amount of intestinal metaplasia in columnar-lined esophagus. This can be explained entirely by a decrease in the baseline acidity of gastric juice. It is probable that much of this alkalinity is natural, although it is difficult to find evidence for or a reason for a historic increase in duodenogastric reflux. However, there is no question that gastric acidity is reduced in patients who are treated with acid suppressive medications. With all other things being equal, acid suppression will tilt the equilibrium to (a) induce intestinal metaplasia in cardiac mucosa of shorter length and (b) increase the amount of intestinal metaplasia that occurs in columnar-lined esophagus (Fig. 15.9).

Intestinal metaplasia still occurs only in a minority of patients with cardiac mucosa, usually after many years following the initial cardiac metaplasia of squamous epithelium. It is uncommon for intestinal metaplasia to occur before 20 years of age, and the median age at which intestinal metaplasia (Barrett esophagus) is detected is around 40 years. The significant time interval between the occurrence of cardiac mucosa and intestinal metaplasia is a window of opportunity for therapeutic interventions to prevent intestinal metaplasia that has received scant attention because of the belief that cardiac mucosa is a normal epithelium.

Carcinogenesis in Intestinal Metaplasia

The risk of cancer begins only in those patients who develop intestinal metaplasia and at the time they develop intestinal metaplasia. Nonintestinalized columnar metaplastic epithelium and squamous epithelium of the esophagus are not believed to be directly susceptible to carcinogenesis (Fig. 15.2).

Carcinogenesis in intestinal metaplasia is likely caused by multiple interactions or "hits," whereby the susceptible target cell in intestinal metaplastic epithelium is acted upon by a carcinogenic molecule(s) in the gastric refluxate. This interaction depends on (a) the presence of the target cell, which is reliably indicated by the presence of intestinal metaplasia in the columnar-lined segment; (b) the presence of the carcinogen, which is unknown, but possibly a derivative of bile acid metabolism (see Chapter 10); and (c) the ability of the carcinogenic molecule to access the target cell. When these three essential elements are present, carcinogenesis is promoted by increased mitotic activity in the target cell, which increases the likelihood of mutagenic events and probably the concentration of the carcinogenic molecule, assuming that the molecule induces a dose-related effect, which is likely. Because the carcinogen is in gastric juice, the maximum concentration of carcinogen is at the gastroesophageal junction. Without reflux, the carcinogen never reaches the esophagus and cancer does not occur. With reflux, the carcinogen enters the esophagus. Based on the fact that reflux has an upward propulsion into the esophagus followed by a clearing mechanism, the exposure of the esophagus to carcinogen is maximal just above the gastroesophageal junction with a decreasing gradient as one moves proximally in the esophagus (Fig. 15.10). The exact height to which an effective carcinogen dose is delivered by reflux will depend on the severity of the reflux and the concentration of carcinogen (Fig. 15.10).

It is extremely interesting that the maximum carcinogenic tendency operates in the lowest regions of the

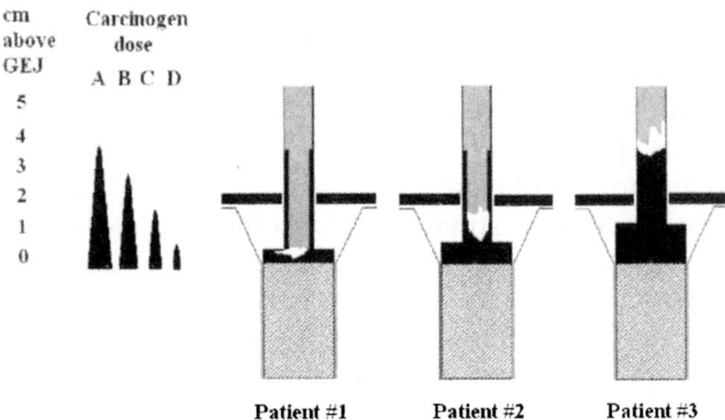

FIGURE 15.10 Interplay between different carcinogen doses (A–D) and the level of intestinal metaplasia in the esophagus. Carcinogenesis will not occur unless a sufficient carcinogenic dose reaches an area where there is intestinal metaplasia. In patient 1 (very short-segment Barrett esophagus limited to the dilated end-stage esophagus), carcinogenesis is possible in all four dose levels A to D. In patient 2, carcinogenesis will occur only in dose levels A, B, and C; it will not occur in the lowest level D. Patient 3 is at risk of cancer only with the highest carcinogen dose environment (A) because it is only at this level that carcinogenic effect reaches the area of intestinal metaplasia proximally in the esophagus.

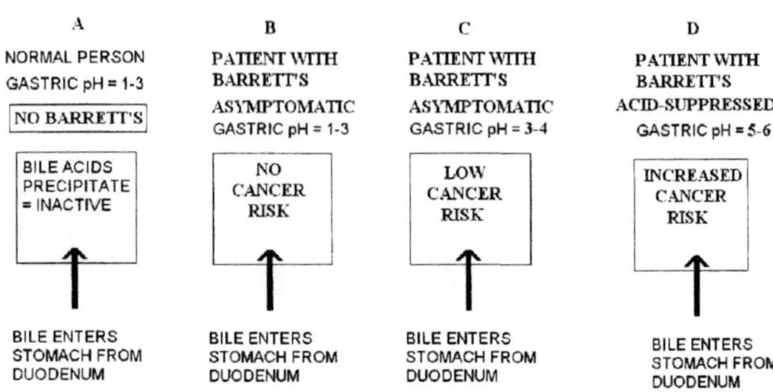

FIGURE 15.11 The effect of gastric pH on bile acids. Carcinogenic molecules are produced from bile acids at intermediate levels of gastric pH. Note that bile acids are derived from duodenogastric reflux; entry of bile acids is inevitably associated with alkalinization of gastric juice (B). This natural alkalinization is enhanced in patients receiving acid suppressive drugs (C), shown as increasing carcinogenicity.

esophagus. This is almost exactly opposite to the tendency to intestinal metaplasia. Patients with intestinal metaplasia limited to the most proximal region of a long segment of columnar-lined esophagus are protected because the likelihood that the carcinogen will reach an effective dose level so high in the esophagus is low (Fig. 15.10). The risk, however, increases as the intestinal metaplasia moves distally in such a patient. The patient who develops intestinal metaplasia in very short segments of columnar-lined esophagus has the highest theoretical exposure to the carcinogen because of the proximity to gastric juice (Fig. 15.10). This is counterbalanced by the fact the number of target cells in longer segments is greater, resulting in a complex interplay of factors. There is clinical evidence that provides supporting evidence for this hypothesis. Adenocarcinomas in Barrett esophagus much more commonly occur in the distal part of the esophagus than in the proximal esophagus, even when long segments of Barrett esophagus are present (see Chapter 10).

This interplay between the tendency to intestinal metaplasia in cardiac mucosa and carcinogenesis can also explain the historic increase in adenocarcinoma. There is evidence that the alkalinization of gastric juice by duodenogastric reflux promotes the conversion of bile acid to carcinogenic metabolites at pH ranges that are intermediate between normal gastric juice (pH 1 to 3) and normal duodenum (pH 6 to 7) (Fig. 15.11). Patients who have duodenogastric reflux may have a gastric juice in this dangerous intermediate pH range and will therefore generate a carcinogenic milieu for esophageal adenocarcinoma. For that person to actually develop adenocarcinoma requires that reflux results in effective doses of the carcinogen delivered to the target cell (i.e., intestinal metaplasia) (Fig. 15.10). If this does not happen, adenocarcinoma does not occur, *whatever the degree of carcinogenesis in the gastric juice.* If there has been a natural increase in the tendency to alkalinize gastric juice by increasing duodenogastric reflux, this can explain the historic increase in the incidence of adenocarcinoma. However, whatever carcinogenic tendency exists in the patient, acid suppressive drug therapy will cause a tilt in the equilibrium that promotes and increases the risk of carcinogenesis (Fig. 15.11). First, the increased alkalinization will tend to induce intestinal metaplasia of cardiac mucosa at increasingly distal regions of the esophagus (Fig. 15.9), bringing the target cell closer to the gastroesophageal junction and carcinogen. Second, the alkalinization of gastric juice will increase the generation of carcinogens from bile acids (Fig. 15.11), increasing the dose of carcinogen in gastric juice and driving the point of effective carcinogenesis in the esophagus more proximally. The combination of the two factors brings the target cell and carcinogen into increasingly greater contact in the patient using acid suppressive drugs.

RECOGNITION OF DIFFERENT RISK LEVELS IN THE EPITHELIA OF COLUMNAR-LINED ESOPHAGUS

Many experts in the field still believe that intestinal metaplasia arises directly from squamous epithelium and that the only columnar-lined esophagus that is abnormal is intestinal metaplasia. There is a general

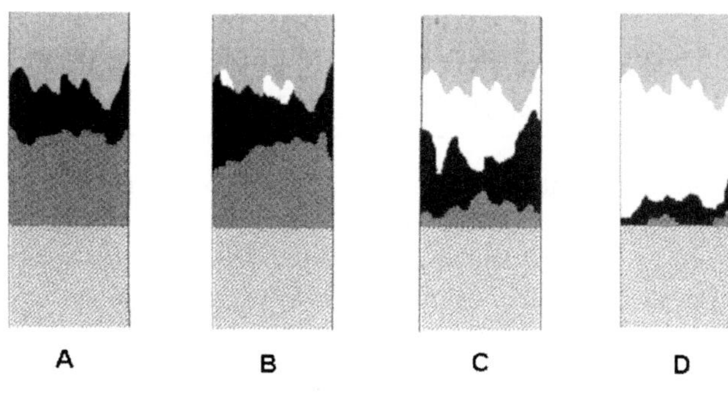

FIGURE 15.12 The distribution of columnar epithelial types in four patients (A, B, C, and D) with an identical length of columnar-lined esophagus. The amount of intestinal metaplasia (targets for carcinogensis) is greatest in patient D. Patient A has no intestinal metaplasia and is not at any risk. Patients B and C are at increasing risk (white = intestinal metaplasia; black = cardiac mucosa; stippled dark gray = oxyntocardiac mucosa).

tendency to treat a biopsy from columnar-lined esophagus in a manner similar to a biopsy from a mass lesion. If a biopsy from a mass lesion shows cancer, it can be assumed that the entire mass lesion is composed of that cancer. Gastroenterologists view a segment of columnar-lined esophagus and take multiple biopsies; if any one biopsy shows any amount of intestinal metaplasia, the assumption is made that the entire segment is lined by intestinal metaplasia. This becomes Barrett esophagus of a length that is determined by the endoscopically determined length of columnar-lined esophagus. This assumption fails to recognize that the amount of intestinal metaplasia varies greatly in different columnar-lined segments of equal or different lengths (Fig. 15.12). It is possible that a long segment has very few goblet cells while a short segment has more. To try to equate risk of cancer by the length of the entire epithelium rather than by the number of target cells is an error. This error results from the fact that most clinicians never look at histologic slides, pathologists fail to communicate, and cardiac and oxyntocardiac mucosa are simply ignored as being irrelevant. This misinformation creates havoc in the treatment and assessment of patients with Barrett esophagus.

Ablation of Barrett Esophagus

This misunderstanding results in potential overtreatment when Barrett esophagus is ablated. When a decision is taken to ablate Barrett esophagus, *the objective is to ablate the entire length of columnar-lined esophagus, not just the at-risk intestinal metaplastic epithelium.* To add risk and morbidity by ablating oxyntocardiac mucosa is unnecessary because the oxyntocardiac mucosa that ablators seek to replace with squamous

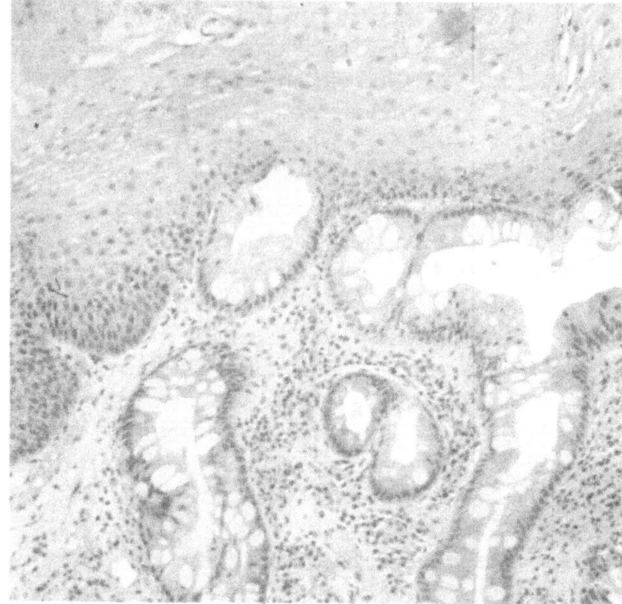

FIGURE 15.13 Biopsy of esophagus showing intestinal metaplastic epithelium in the lamina propria under a surface squamous epithelium. (Cross reference: Color Plate 4.5)

epithelium is not at risk to progress to adenocarcinoma. When a serious procedure such as ablation is planned, it is imperative to map the epithelia and ablate only the at-risk intestinal epithelium, which may actually be quite limited within the columnar-lined segment, and greatly decrease the amount of mucosa requiring ablation.

Assessment of Regression of Barrett Esophagus

Many studies have been conducted that look for evidence of regression of Barrett epithelium. The way these studies are designed is to define the study group by the presence of columnar-lined esophagus that has intestinal metaplasia, measure the endoscopic length of the columnar epithelium, and then evaluate whether the columnar epithelium has decreased in amount due to the growth of squamous epithelium in areas previously covered by columnar epithelium. It is clear that this is not regression in any sense of the word. It is well known, even to those conducting the studies, that the surface growth of squamous epithelium is often associated with columnar epithelium in the lamina propria under the squamous surface (Fig. 15.13). This columnar epithelium has been shown to undergo dysplasia and malignancy while still covered by squamous epithelium (Fig. 15.14). Therefore, a squamous surface is not evidence of regression.

The critical measure of cure or regression in Barrett esophagus is a reversal or decrease in the number of intestinal metaplastic cells, because only the proliferative cells in columnar epithelium marked by the presence of intestinal metaplasia are susceptible to carcinogenesis (Fig. 15.15). A columnar epithelium of equal length that has less intestinal metaplasia and more oxyntocardiac mucosa is an epithelium that has fewer target cells for carcinogenesis and is therefore less dangerous. The aim of treatment should be to reverse intestinal metaplasia to cardiac mucosa and to convert cardiac mucosa to oxyntocardiac mucosa. This is true regression of Barrett esophagus. A squamous replacement of columnar epithelium must be devoid of residual glands in the lamina propria to be true regression, and this is difficult to visualize and prove.

Studies that evaluate changes in different epithelial types resulting from various treatment modalities are urgently needed. We should not be looking at squamous replacement on the surface; we should be looking at changes in the density of goblet cells (any decrease is good) and parietal cells (any increase is good) within the segment of columnar-lined esophagus.

THE STATE OF PRESENT PHARMACEUTICAL INTERVENTION IN REFLUX DISEASE

Pharmaceutical research in reflux disease has been single-mindedly directed toward the development of ever-improving acid suppressive drugs used to treat symptoms and cause healing of erosive reflux disease. This research has been remarkably successful in achieving its goal. Acid suppression with drugs is now theoretically capable of producing almost complete suppression of acid secretion. Acid suppression has essentially reversed squamous epithelial damage, and the results have been the control of symptoms, an improvement in the patient's quality of life, and the prevention of complications such as complex strictures and severe uncontrolled ulceration.

Because of the "success" of acid suppressive drugs in treating reflux disease, the pharmaceutical drug industry has made almost no effort to develop alternate agents to impact reflux disease. These drugs are so profitable because of their widespread use that the thrust of competition has been merely to produce increasingly effective acid suppression. The medical community has placed all its pharmaceutical eggs for treating reflux disease in one basket.

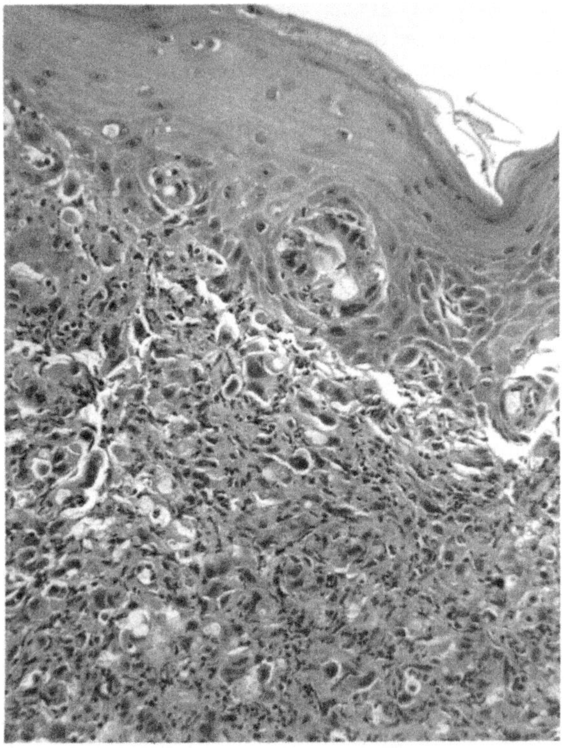

FIGURE 15.14 Poorly differentiated adenocarcinoma in the lamina propria under an intact surface squamous epithelium.

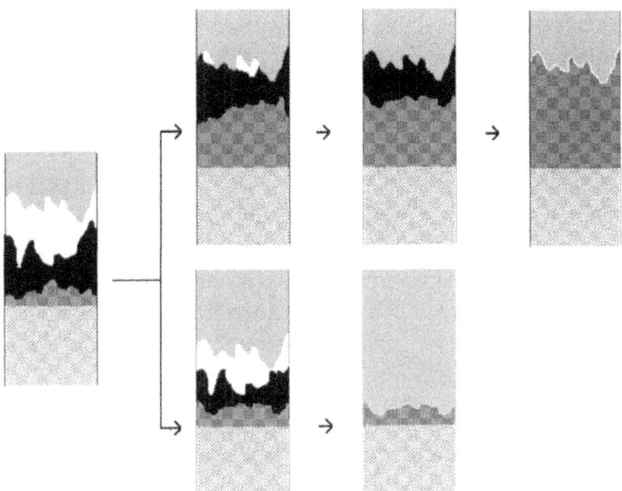

FIGURE 15.15 Criteria for "cure" (= removal from the reflux-adenocarcinoma sequence) in Barrett esophagus. This may result from conversion of all intestinal and cardiac mucosa to oxyntocardiac mucosa without any change in the length of columnar-lined esophagus (upper series) or replacement of columnar epithelium with squamous epithelium. In the latter event, it is difficult to detect the existence of columnar epithelial elements in the lamina propria under the squamous epithelium.

ACID SUPPRESSIVE DRUGS PROMOTE ESOPHAGEAL ADENOCARCINOMA

Even as acid suppression has effectively controlled the squamous epithelial damage and improved the life of many patients, there has been an alarming increase in the incidence of reflux-induced adenocarcinoma. In 1992, Rodger Haggitt suggested that this was an epidemic (9); Devesa et al. (10) reported a 350% increase in incidence in white males in the United States over the two decades preceding 1995. A more recent study by Pohl and Welch (11) showed that the overall esophageal adenocarcinoma incidence has continued to rise. In this report, the incidence increased nearly sixfold between 1975 and 2001 from 4 to 23 cases per million (Fig. 15.16). There is one glimmer of hope in this report; the incidence of adenocarcinoma of the "gastric cardia" has shown a flattening of the incidence curve in the past decade. It is still too early to know whether this represents a true flattening or just a temporary blip in an upward curve.

There has been vehement denial that there is any possible connection between the increasing usage of acid suppressive agents and the increasing incidence of adenocarcinoma. This is a highly profitable business for the pharmaceutical industry with annual worldwide revenue from this group of drugs exceeding $12 billion. There is a ubiquitous advertising blitz in all popular media that attest to the profitability of these drugs. The shelves of drugstores and supermarkets are

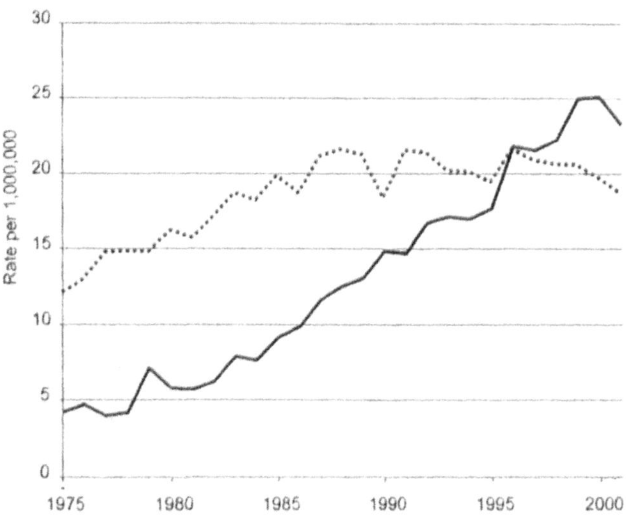

FIGURE 15.16 Incidence of adenocarcinoma of the distal esophagus (solid line) and adenocarcinoma of the gastric cardia (dotted line) between 1975 and 2000. Both show an increase; the incidence of adenocarcinoma of the esophagus has increased sixfold during this period.

replete with these agents (Fig. 15.17). If, as the epidemiologic data suggest (12; also see Chapter 16), they increase the risk of cancer, the emphasis on these drugs has been misplaced. It is based on a naive belief that acid damage of squamous epithelium causes intestinal metaplasia and then carcinoma. There is no evidence that acid suppression prevents cancer.

In the discussion of the three critical events that are necessary for carcinogenesis, presented earlier, we suggested that two of these events are promoted by the increased alkalinity of gastric juice. The occurrence and extent of intestinal metaplasia in esophageal cardiac mucosa increase as alkalinitiy of gastric juice increases (Fig. 15.9). The effective carcinogenic concentration in gastric juice and the height in the esophagus to which the carcinogenicity reaches is also increased by increasing gastric alkalinization (Figs. 15.10 and 15.11). Acid suppressive agents function by suppressing acid secretion in the stomach and *always* cause an increased alkalinity in gastric juice if they have any efficacy. As such, there is a strong theoretical basis for suggesting that acid suppressive agents promote carcinogenesis in Barrett esophagus.

It is to be emphasized that the use of acid suppressive drugs is unlikely to be the entire reason for the increasing incidence of esophageal adenocarcinoma over the past five decades. Most patients who present with esophageal adenocarcinoma have never taken acid suppressive medications. This suggests that there has been an increase in the yet unknown natural factors that are responsible for Barrett esophagus and esophageal adenocarcinoma. While acid suppressive drugs are not likely to be the main culprits, *they are clearly promoters, enhancing carcinogenesis and increasing the risk of cancer in patients taking these drugs for the treatment of reflux disease.*

The cancer-promoting effect of acid suppressive drugs need only be slight to have a significant numerical impact in the incidence of reflux-induced adenocarcinoma. Adenocarcinoma occurs only in a very small percentage of patients with reflux disease. However, the population at risk is large; 35% to 45% of the population has cardiac mucosa, which is our definition of reflux disease; 35% to 45% of the population of the United States is over 100,000 million people. A slight tilt in the balance to increase the conversion of cardiac mucosa to intestinal metaplasia will have the effect of greatly increasing the prevalence of Barrett esophagus. Acid suppression, by causing a slight shift of the pH in gastric juice, can have this effect; it can explain the increased prevalence of Barrett esophagus since the 1950s. A slight tilt in the balance to increase conversion of bile acid derivatives to carcinogenic agents (Fig. 15.11) will have the effect of significantly increasing the

FIGURE 15.17 Shelf at a local pharmacy showing the array of acid suppressive agents that are available over the counter without prescription. These include the most potent proton pump inhibitors.

absolute number of Barrett esophagus patients who progress to cancer. The sixfold increase in adenocarcinoma that has occurred in the United States since the 1970s requires only a minimal shift in the gastric pH resulting from acid suppression, assuming all other factors remain equal. The main factors causing Barrett esophagus and adenocarcinoma are the occurrence of adequate gastroesophageal reflux and a carcinogenic environment in the stomach. *Acid suppressive drugs, by causing a slight tilt in the gastric juice pH toward alkaline, enhances this natural carcinogenic effect. Patients with adequate carcinogenicity increase their risk and rate of progression; patients in the borderline range of carcinogenicity who would never have developed cancer receive an adequate nudge to tilt the balance. This explains how most reflux-induced cancers occur in asymptomatic people who have never taken acid suppressive agents develop cancer and how acid suppressive drugs increase the risk of cancer.*

There is potential danger here for patients suffering from reflux disease. If it becomes generally accepted that acid suppression increases the risk of cancer, there will be significant medicolegal risk to the pharmaceutical companies if they continue to market these drugs. This risk will lead to pressure to withdraw these drugs from the market; with the irrational panic associated with all such legally driven decisions, this withdrawal is likely to be abrupt and comprehensive. Withdrawal of acid suppressive drugs from the market removes the only effective medication for reflux disease. The only alternative to controlling symptoms of reflux disease, which can be debilitating to sufferers, is antireflux

Personal Health

THE JOURNAL REPORT

THE WALL STREET JOURNAL.

THE HIDDEN DANGERS *of* HEARTBURN

FIGURE 15.18 Headline of a two-page article in the Personal Health section of the *Wall Street Journal*, Monday, October 10, 2005. The hidden danger is adenocarcinoma, and the article suggests that acid suppressive drug use may be involved in the pathogenesis of cancer.

surgery. There are not enough trained surgeons at the present time to perform the required numbers of these operations. There are already rumblings in the educated media regarding the potential risks of heartburn and acid suppressive agents. In the October 10, 2005, issue of the *Wall Street Journal* (Fig. 15.18), a front-page article in the personal health section titled "The Hidden Dangers of Heartburn" highlighted the problem of reflux-induced adenocarcinoma with considerable acumen, even stating "the use of acid-suppressing drugs . . . may play a role in the rise of esophageal cancer." This isolated report was controlled by the

TABLE 15.2 Potential Therapeutic Targets That Have the Potential to Positively Impact the Incidence of Reflux-Induced Adenocarcinoma

Point of intervention	Possible agents
Inhibit cardiac mucosal metaplasia of squamous epithelium	(a) Acid suppression (b) Find new drugs that increase squamous epithelial resistance to acid damage (c) Find basis of genetic switch and develop an antagonistic drug
Reverse cardiac metaplasia and induce squamous epithelium	Find basis of genetic switch and develop drug to reverse it
Promote the conversion of cardiac mucosa to oxyntocardiac mucosa	(a) Find alternate new drugs to relieve reflux symptoms without acid suppression (b) Find drugs that simulate acid without causing squamous epithelial damage (c) Decrease the alkalinizing effect of duodenogastric reflux (d) Find basis of genetic switch and develop a drug that activates or simulates the causative agent
Inhibit the conversion of cardiac mucosa to intestinal metaplasia	(a) Find alternate new drugs to relieve reflux symptoms without acid suppression (b) Decrease the alkalinizing effect of duodenogastric reflux (c) Find basis of genetic switch and develop an antagonistic or inhibitory drug
Reverse intestinal metaplasia to cardiac mucosa	Find basis of genetic switch and develop a drug to reverse the change

pharmaceutical industry's vehement denial of any association between acid suppression and cancer, but the fact that this article appeared in such a prominent media outlet represents a warning of things to come.

PROPOSED NEW PHARMACEUTICAL RESEARCH TARGETS AIMED AT PREVENTING ADENOCARCINOMA

It is time to direct the pharmaceutical and basic research into molecular interactions that are actually responsible for the cellular transformations that take squamous epithelium through cardiac mucosa into intestinal metaplasia (Fig. 15.4, Table 15.1; also see Chapter 9). There is fertile ground for meaningful research into these areas, which we will explore (Table 15.2).

The technology for cell culture has reached an advanced stage. Fitzgerald et al. (13) showed the feasibility of growing human cells in culture and seeing the effects of exposure to acid and bile salt derivates. They measured the response of cells grown from human Barrett epithelium in terms of proliferative rate and differentiation by evaluating the expression of proliferative cell nuclear antigens and villin. More recently, Kazumori et al. (14) tested rat keratinocytes in culture against different bile salt derivatives, showing that cholic acid and dehydrocholic acid increased Cdx2 promoter activity in keratinocytes and induced the production of intestinal type mucin, suggesting that these bile salt derivatives may be involved in columnar metaplasia. Their technology permitted them to demonstrate that Cdx2 promotion was induced via NF-kB binding sites on the surface of the cell. The focus is again on squamous epithelium and intestinal metaplasia; the knowledge does not exist to test cardiac mucosa because of the false belief that cardiac mucosa is not a relevant epithelium in the pathway.

The ultimate goal in the prevention of adenocarcinoma is the recognition of the pathogenesis of each cellular event in terms of a molecular event. This will permit a specific intervention at a critical step in the carcinogenic sequence that will prevent the progression to adenocarcinoma. Let us look at the sequential events that form targets for attack and suggest research that may be directed at these points.

Squamous Epithelial Transformation to Cardiac Mucosa

The first change that takes the esophageal epithelial cells toward susceptibility to carcinogenesis (i.e., intestinal metaplasia) is the conversion of squamous epithelium to cardiac mucosa (see Chapter 8). This is an almost invariable change when there is reflux as it occurs in virtually all people with reflux as a very early cellular event. The presence of metaplastic columnar epithelium is by far the most sensitive indicator of reflux disease.

Cardiac mucosa is commonly seen in children with reflux (15, 16) and occurs above the anastomotic line in patients who have had an esophagectomy within a short period after surgery (17, 18). Cameron et al. (5) produced data that suggested that the length of columnar-lined esophagus increases to its maximum at a relatively early age and then remains static.

These data suggest that the transformation of squamous epithelium to cardiac mucosa is a relatively rapid and essentially invariable change in patients with significant reflux. Acid is necessary for the process because it is the agent that damages the squamous cells, increasing the spaces between them and permitting refluxate molecules to enter the epithelium and access the pool of proliferating progenitor cells in the basal region (Fig. 15.19). Because cardiac mucosal metaplasia occurs very early, acid suppression can only be effective if it is used in everyone in the population from an early stage in life. This is obviously not feasible.

Cardiac metaplasia is likely to be caused by an interaction between a larger refluxate molecule and a cell receptor. Elucidating the actual cell interaction that activates the genetic switch that converts squamous epithelium to cardiac mucosa is extremely important from a basic science angle because it will involve identification of one of the basic differentiating genetic signals in the normal fetal esophagus. This would involve exposing the esophageal squamous epithelial cells in tissue culture to various molecular components in gastric juice and recognizing the occurrence of columnar (mucous) cell transformation. The fact that squamous to cardiac metaplasia is inevitable in patients with reflux means that the molecule causing this change is ubiquitous in gastric juice. Examining differences between the activated genes in the original squamous and transformed columnar epithelial cell can lead to the detection of the actual gene activated in the metaplastic switch.

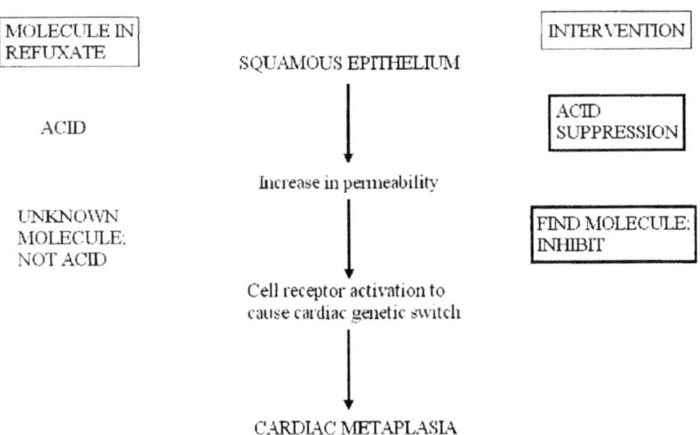

FIGURE 15.19 The basic mechanism whereby squamous epithelium is damaged by acid to cause an increased permeability of the epithelium. Other nonacid molecules in the refluxate can then enter the squamous epithelium to interact with the proliferating progenitor cells in the basal region, causing the genetic switch that results in cardiac metaplasia. Points of attack of this process are shown on the right.

However, despite its theoretical interest, it is unlikely that elucidation of this molecular reaction and genetic switch and developing a method to inhibit it will be practical because cardiac mucosal metaplasia occurs so early and with great rapidity.

Once cardiac metaplasia occurs in a patient with reflux disease, it is unlikely that reversal to squamous epithelium occurs naturally. Interventions that change the milieu, however, can potentially reverse this reaction. This is one mechanism of "cure" for reflux disease. Squamous islands appear in columnar-lined esophagus after both acid suppressive therapy and antireflux surgery, suggesting that reversal of metaplasia is feasible. Elucidating the molecular mechanism of cardiac metaplasia may provide more effective methods of reversal to squamous epithelium. One problem with such reversal is that surface squamous transformation may result in the entrapment of metaplastic glandular elements in the lamina propria. These are hidden under the squamous surface at endoscopy and gross examination but may retain the potential to progress to intestinal metaplasia, dysplasia, and adenocarcinoma (Figs. 15.13 and 15.14).

One interesting and counterintuitive mechanism of treating symptomatic reflux without the use of acid suppressive agents is to promote the conversion of reflux-damaged squamous epithelium to cardiac mucosa by a drug that stimulates this conversion. Esophageal pain is likely to be the result of stimulation of nerve endings in squamous epithelium; columnar epithelia are believed to be much less sensitive (see Chapter 8). As a result, the conversion of pain-generating acid-damaged squamous epithelium to cardiac mucosa can have the effect of decreasing heartburn. If mechanisms are also available to prevent intestinal metaplasia in cardiac mucosa and promote conversion of cardiac mucosa to oxyntocardiac mucosa, they can be highly effective and practical pain-relieving tools in reflux disease. From the point of view of pain, columnar epithelia are better than squamous epithelium.

Cardiac Mucosal Transformation to Oxyntocardiac Mucosa

The conversion of cardiac mucosa to oxyntocardiac mucosa represents one mechanism of "cure" for reflux disease because oxyntocardiac mucosa is a stable epithelium that does not progress to intestinal metaplasia. The patient is removed from the carcinoma sequence (Fig. 15.4). We have provided evidence that the amount of oxyntocardiac mucosa in columnar-lined esophagus has decreased since the 1970s (see Chapter 9) and suggested that this may be a reason for the increase in esophageal adenocarcinoma. There is evidence that oxyntocardiac mucosal transformation of cardiac mucosa is the result of the following conditions being present: (a) a low-damage environment; this exists in patients with mild reflux, in the distal part of the esophagus when there is continuous rather than pulse (or intermittent) exposure of the epithelium to the refluxate (13) and after successful antireflux surgery when reflux has ceased; and (b) when the pH is low; parietal cell generation in cardiac mucosa appears to be promoted by an acid environment (see Chapter 9). There is also evidence that the genetic system whose activation leads to conversion of cardiac to oxyntocardiac mucosa is the Sonic Hedgehog gene system (Fig. 15.20; also see Chapter 9).

If the cardiac to oxyntocardiac conversion is pH based and promoted by low pH, acid suppression will be expected to have an inhibitory effect on this change. We have suggested that there is historic evidence for this (Chapter 9) and that this is one reason why acid suppression has caused an increase in cancer. If found to be true by appropriate research, many improvements can be made: (a) Alternate drugs can be found to relieve symptoms and complications of reflux like erosions, ulcers, and strictures that do not work by suppressing acid; the disease can then be controlled without increasing the gastric pH. (b) Drugs that simulate the action of acid on cardiac mucosa that promotes its conversion to oxyntocardiac mucosa without causing squamous epithelial damage would also serve as an agonist to this change. (c) Because the natural alkalinizing agent of gastric juice is duodenogastric reflux, drugs that are designed to prevent duodenogastric reflux or antagonize the alkalinity of duodenogastric refluxate need to be developed and tested. (d) Finally, acid suppressive drug use should be used with appropriate caution in the management of symptomatic reflux disease because they can theoretically prevent the cardiac to oxyntocardiac transformation.

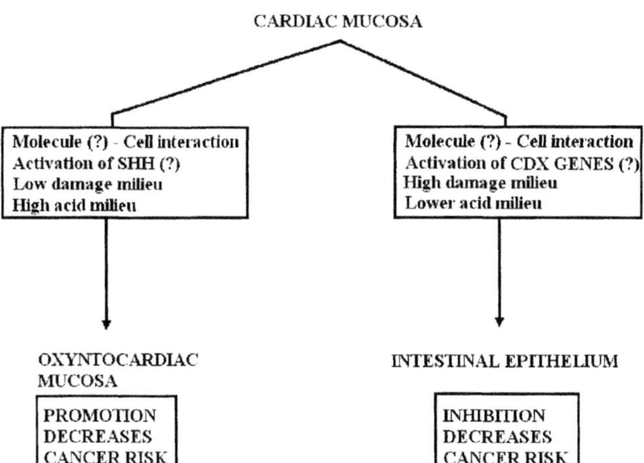

FIGURE 15.20 The evolution of cardiac mucosa in one of two different pathways resulting from the presence of exactly opposite factors. Oxyntocardiac mucosa is generated in a high-acid, low-damage environment; intestinal metaplasia is generated in a less acid (i.e., more alkaline), high-damage environment. Only the intestinal metaplastic pathway is susceptible to carcinogenesis.

Research into the molecular mechanism and identification of the exact molecular agent that causes the transformation of cardiac to oxyntocardiac mucosa has great potential value. The research will involve growing esophageal cardiac epithelial cells in tissue culture and exposing these cells to a variety of conditions and different molecules normally present in the gastric juice. The end point of the experiment is the detection of parietal cell differentiation or Sonic Hedgehog gene activation. Once the exact agent is identified, drugs can be developed that can replicate the action of this molecule and promote this transformation. This will have the effect of converting cardiac mucosa to oxyntocardiac mucosa and preventing adenocarcinoma.

Cardiac Mucosal Transformation to Intestinal Epithelium

The conversion of cardiac mucosa to intestinal epithelium represents progression of reflux disease in the adenocarcinoma sequence. Unlike cardiac mucosa, intestinal metaplastic epithelium is the target epithelium for carcinogens (Fig. 15.4). We have provided evidence that the amount of intestinal epithelium in columnar-lined esophagus has increased since the 1970s (see Chapter 9) and suggested that this may be a reason for the increase in esophageal adenocarcinoma. There is evidence that the intestinal transformation of cardiac mucosa is the result of the following conditions being present (Fig. 15.20): (a) a high-damage environment; this exists in patients with severe reflux, in the more proximal part of the esophagus when there is pulse (or intermittent) rather than continuous exposure of the epithelium to the refluxate (13); (b) when the pH is high; goblet cell generation in cardiac mucosa appears to be promoted by a less acid environment (see Chapter 9). This condition is seen in patients who have acid suppressive drug treatment, providing the basis for the observed association between acid suppressive drug use and the increased incidence of adenocarcinoma. There is also evidence that the genetic system whose activation leads to the conversion of cardiac to intestinal epithelium is the Cdx gene system (Fig. 15.4; also see Chapter 9).

If the cardiac to intestinal conversion is pH based and promoted by a high pH, acid suppression will be expected to have a stimulating effect on this change. We have suggested that there is historic evidence for this (Chapter 9) and that it is one reason why acid suppression has caused an increase in cancer. Note that this is opposite to the cardiac to oxyntocardiac conversion. If found to be true by appropriate research, many of the suggestions for ways to promote the cardiac to oxyntocardiac transformation will also act to inhibit the cardiac to intestinal conversion: (a) new alternatives to the present acid suppressive drugs to treat reflux symptoms and complications, (b) drugs that inhibit or antagonize the alkalinizing effect of duodenogastric refluxate, and (c) cautious and limited use of acid suppressive drugs in the treatment of reflux disease. Acid suppressive agents should be regarded as serious drugs with significant adverse effects including the promotion of intestinal metaplasia in cardiac mucosa. They should be taken away from the realm of over-the-counter drugs that are marketed directly to the public by intense advertising and made prescription-only with adequate warnings regarding their use.

Research into the exact molecular mechanism that causes the transformation of cardiac to intestinal epithelium has great potential value. Any drug that can inhibit or reverse this reaction or antagonize the agent involved will tend to decrease the incidence of adenocarcinoma. The research will involve growing esophageal cardiac epithelial cells in tissue culture and exposing them to a variety of conditions and different molecules normally present in the gastric juice. The end point of the experiment is the detection of goblet cell differentiation or Cdx1 and Cdx2 gene activation. Once these are identified, drugs can be developed that can either inhibit the agent that causes the transformation or effectively reverse the metaplastic molecular reaction, thereby converting intestinal epithelium to cardiac mucosa and preventing adenocarcinoma.

We are not saying that any of the proposed research will produce significant new drugs or an effective method of preventing cancer; they only have a theoretical possibility of doing so. However, failing to recognize the potential avenues of research that are available while continuing the single-minded development of acid suppressive agents guarantees that there will be no new drugs. Many of the suggested research avenues deal with simple molecular reactions and simple molecules that the pharmaceutical industry has considerable proven success in dealing with. It is uncommon in medicine to have an opportunity such as this to prevent cancer.

RESEARCH INTO SURGICAL METHODS OF PREVENTING ADENOCARCINOMA

Many different surgical methods into preventing adenocarcinoma are progressing from the research arena into clinical application. The most advanced technique is antireflux surgery, usually a Nissen fundoplication where a valve effect is created at the end of the esophagus by wrapping the fundus of the stomach around the tubular esophagus. This procedure creates a new sphincter action that, when it is successful, prevents gastroesophageal reflux. The surgery is based on the principle that preventing reflux sequesters the esophageal epithelium from all molecules in gastric juice and prevents injury. A few controlled studies are emerging to suggest that successful antireflux surgery is effective in preventing the progression to adenocarcinoma. The most commonly performed antireflux procedure is surgical, usually laparoscopic. More recently, endoscopic techniques to create a barrier to reflux are also under study; these have limited proven success and acceptance at this time.

Methods of ablating Barrett epithelium are also gaining favor. When combined with effective control of reflux, the ablated surface can be induced to regenerate as squamous epithelium, decreasing cancer risk. Endoscopic mucosal resection and ablation with physical methods such as laser, electrocoagulation, and photodynamic therapy are increasing in effectiveness and acceptance. In general, these techniques are presently used mainly for Barrett esophagus complicated by high-grade dysplasia or early adenocarcinoma in patients who are not surgical candidates. With increasing success, these techniques can be extended to uncomplicated Barrett esophagus.

The final target of surgical treatment is the control of duodenogastric reflux which is widely believed to be involved in carcinogenesis. Csendes et al. have devised and reported a radical operation for the treatment of Barrett esophagus, which includes separating the stomach from the duodenum and preventing gastroduodenal reflux by creating a Roux-en-Y loop. This procedure is still in an experimental phase and is not performed with any frequency. It provides a logical basis for preventing adenocarcinoma. The idea that it is too radical a surgery is counteracted by the serious nature of the adenocarcinoma it aims to prevent. Dr. Csendes correctly states that the surgery, though it appears radical, is similar to bariatric surgical procedures very commonly performed to control obesity.

IDENTIFYING FACTORS IN GASTRIC JUICE RESPONSIBLE FOR MOLECULAR EVENTS

Every molecular event that occurs in reflux disease, including reflux-induced adenocarcinoma, is induced by a molecule in gastric refluxate acting upon an esophageal epithelium. The gastric refluxate consists of gastric juice. Assuming that all the changes of reflux disease are caused by a luminally acting agent, every injury-producing molecule must be present in the gastric juice. There is no evidence that any event in reflux disease is caused by a blood-borne agent.

Research into identification of the molecules responsible for molecular events is complicated because there is the possibility that the agent is not constantly present. If the agent causes a molecular event and disappears, no amount of searching in gastric juice can identify the agent. As such, this type of research is often of limited value. However, within broad limits, the probability of identifying any causative agent is greatest when a comparison is made between the gastric juice of those at highest and lowest risk for that event. Careful examination of clinical groups can be used to identify the types of patients at highest and lowest risk for a given event.

Squamous Epithelial Damage

There is excellent evidence that the early squamous epithelial changes are the result of acid exposure. Tobey et al. (3) have shown that exposure squamous epithelial cells *in vitro* to acid causes dilatation of intercellular spaces and increased permeability. Acute and severe injury that leads to erosion of the squamous epithelium (*erosive esophagitis*) is also certainly caused by acid, as evidenced by the rapid healing that is induced by acid suppressive drugs. It is not certain that other molecules in the refluxate are synergistic with acid in causing squamous epithelial damage. However, the practical importance of acid as the agent that causes squamous epithelial damage is strongly suggested by the historical decline in severe ulceration and strictures of squamous epithelium that has occurred with improving ability to pharmacologically suppress acid.

Squamous to Cardiac Transformation

The specific interactive agent that transforms squamous epithelium to cardiac mucosa is unknown. This is the chronic manifestation of *nonerosive reflux disease*. The proven relationship between the amount of cardiac mucosa present and the severity of cumulative chronic reflux permits the practical recognition of patients at highest risk for cardiac metaplasia (see Chapter 6). *The longer the cardiac mucosa and the younger the patient, the greater the likelihood that the environment favors the squamous to cardiac mucosal transformation.*

There is excellent correlation between the length of cardiac mucosa present in the esophagus (i.e., nonerosive reflux disease) and the 24-hour pH test. However, a correlation of a change with the 24-hour test does not prove that the change is caused by acid; even if the change is caused by another molecule that accompanies acid, the change will show a correlation with the 24-hour pH test. However, the fact that there is such a good correlation suggests that the agent that actually causes the squamous to cardiac transformation is *ubiquitous in gastric juice.* The transformation is the result of amount of reflux rather than any differences in the concentration of the agent.

Cardiac mucosa is a common finding in patients who have been on long-term acid suppressive drug therapy. This does not prove anything because of the fact that the squamous to cardiac mucosal metaplasia occurs early in life; the start of acid suppressive therapy may have been after the transformation had already occurred. This finding only shows that acid suppression is likely incapable of reversing the cardiac mucosa to squamous epithelium.

Cardiac to Oxyntocardiac Transformation

The interaction that causes the development of parietal cells in cardiac mucosa is likely the result of activation of a gastric-type genetic signal in the progenitor cell of cardiac mucosa. Like the squamous to cardiac mucosal transformation, it is likely that the molecule responsible in *ubiquitous in gastric juice.* Oxyntocardiac mucosa appears to result from the existence of a milieu created by the amount and type of reflux rather than changes in the concentration of any molecule in gastric juice. Oxyntocardiac mucosa is frequently the only epithelial type seen in patients with reflux disease limited to end-stage dilated esophagus. In patients with visible columnar-lined esophagus, oxyntocardiac mucosa is present largely in the more distal region, suggesting that the cardiac to oxyntocardiac transformation is the result of factors or molecules present at highest levels closest to the gastroesophageal junction. The possibilities include (a) continuous exposure to gastric juice, as occurs in the dilated end-stage esophagus where oxyntocardiac mucosa dominates; (b) any molecule in gastric juice, because this will tend to have the highest concentration at the gastroesophageal junction; and (c) low pH; high acidity tends to favor the cardiac to oxyntocardiac transformation. This is likely to be due to the fact that the interaction is pH dependent; searching for molecules in gastric juice that are more active at low pH is likely to detect the molecule responsible for this change.

Cardiac to Intestinal Metaplasia Transformation

In complete contrast to oxyntocardiac mucosa, intestinal metaplasia as well as goblet cell density is greatest in the most proximal part of the columnar-lined segment immediately distal to the squamocolumnar junction (19). The fact that intestinal metaplasia has increased in prevalence since the 1970s suggests that this interaction is not caused by acid; the change is either not influenced by acid or potentiated by acid suppression. There is a suspicion that bile acid derivatives produced in the refluxate as a result of changes in pH caused by incomplete acid suppression may be responsible for this event (14).

The fact that intestinal metaplasia increases in prevalence as the length of columnar-lined esophagus increases, and becomes invariable at a given length, suggests that the molecule responsible for this change is *ubiquitous in gastric juice.* In our study, 100% of patients with a columnar-lined esophagus greater than 5 cm had intestinal metaplasia. If the molecule causing the change was only present in some people, the prevalence of intestinal metaplasia would not reach 100% at any length of columnar-lined esophagus. There is evidence in the literature that the length at which the prevalence of intestinal metaplasia reaches 100% has decreased since the 1950s. In Allison and Johnstone (4), many patients with columnar-lined esophagus reaching the arch of the aorta did not have intestinal metaplasia. By 1976 (20), it seems that intestinal metaplasia was seen in the majority of patients with a 10-cm length of columnar-lined esophagus. We have suggested that this increased tendency to develop intestinal metaplasia at increasingly lower levels is the result of alkalinization of gastric juice.

The fact that intestinal metaplasia has a maximal likelihood of occurring at a point farthest away from the gastroesophageal junction suggests that the factors favoring this transformation are dependent on the molecule interacting with the progenitor cell at a given pH. This could be the result of two independent circumstances: (a) the concentration of the factor in gastric juice is increased so that it reaches a higher point in the esophagus during reflux, and (b) the pH of the gastric juice is increased so that the correct pH is present at a lower level in the esophagus during reflux. There is also the possibility that the transformation is favored by the higher damage level associated with the intermittent, pulse-type exposure of the epithelium to the refluxate that is likely in the higher regions of the esophagus.

If patients are classified into those with low and high tendency to cause the cardiac to intestinal mucosal transformation, we can predict what will happen to patients in terms of the likelihood of Barrett esophagus at different lengths of cardiac mucosa in the esophagus (Table 15.3).

By this reasoning, patients with *intestinal metaplasia in the shortest segments of columnar-lined esophagus are those in whom the factors favoring the cardiac to intestinal transformation are greatest* (Fig. 15.21). Comparing the group at highest risk for intestinal metaplasia with the group that has the least risk (i.e., patients with the longest segments of columnar-lined mucosa that do not have intestinal metaplasia) is likely to detect which factors and molecules in the gastric refluxate are responsible for intestinal metaplasia.

This is counterintuitive; while damage by reflux produces increasing lengths of columnar-lined esophagus, the factors causing intestinal metaplasia are maximal in those patients who have the shortest segments of Barrett esophagus. In these patients, the intestinal metaplasia has occurred in cardiac mucosa present at a more distal point in the esophagus at a more acidic pH and with a lesser pulse effect than patients with long-segment Barrett esophagus (Fig. 15.21). This line of reasoning produces an expectation that is very different from the one presently believed.

This line of reasoning can explain the differences in gender distribution that is seen with different lengths

TABLE 15.3 Clinical Effect of Different Damage Levels in Conjunction with Low and High Tendency to Develop Intestinal Metaplasia (IM)

Damage level	IM tendency	Clinical effect
Low	Low	Microscopic CLE with CM and OCM
Intermediate	Low	<2 cm CLE with CM and OCM
High	Low	2–5 cm CLE with CM and OCM
Very high	Low	>5 cm CLE with IM restricted to proximal end of CLE
Low	High	Microscopic CLE with IM
Intermediate	High	Short-segment CLE with IM (SSBE)
High	High	Long-segment CLE with IM (LSBE)
Very high	High	>5 cm CLE with IM involving most of CLE

Note: The damage level determines the length of columnar-lined esophagus (CLE) while the tendency to intestinal metaplasia determines the occurrence and amount of intestinal metaplasia.

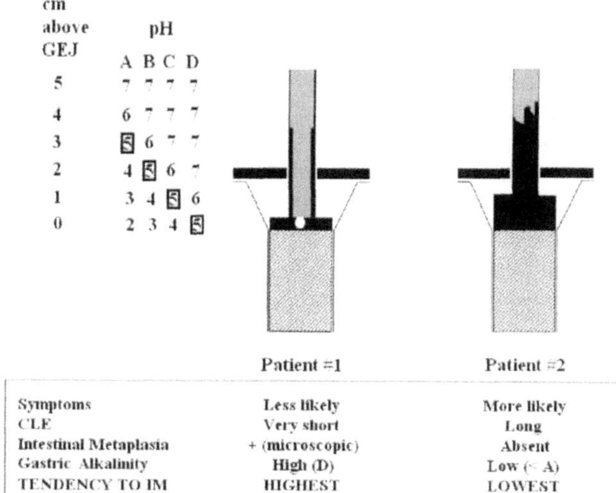

cm above GEJ
pH
A B C D
5 7 7 7 7
4 6 7 7 7
3 5 6 7 7
2 4 5 6 7
1 3 4 5 6
0 2 3 4 5
Patient #1
Patient #2
Symptoms Less likely More likely
CLE Very short Long
Intestinal Metaplasia + (microscopic) Absent
Gastric Alkalinity High (D) Low (< A)
TENDENCY TO IM HIGHEST LOWEST

FIGURE 15.21 Characterization of patients in terms of the highest and lowest tendency to result in the cardiac to intestinal transformation. Patient 1, with intestinal metaplasia in the shortest segment of columnar-lined esophagus, has the highest tendency because this requires the greatest alkalinity of gastric juice (D). Patient 2, with a long segment of columnar-lined esophagus devoid of intestinal metaplasia, has the lowest tendency because intestinal metaplasia has not occurred in the higher, more alkaline region of the esophagus. (*left*) The pH at different levels in the esophagus; if intestinal metaplasia occurs above a pH of 6, the reason for this becomes evident.

of Barrett esophagus. If females have a greater tendency to convert the cardiac mucosa to intestinal metaplasia than do males, females will develop intestinal metaplasia with shorter segments of cardiac metaplasia than males. This assumes that once intestinal metaplasia has occurred in a segment of columnar-lined esophagus, it ceases to increase in length, which we have suggested is the case (see Chapter 9).

The critical factor to assess when trying to evaluate the tendency to intestinal metaplasia is not the overall prevalence of intestinal metaplasia or the overall length of Barrett esophagus in males and females. Both of these are much greater in males. The critical measure is the ratio between patients with long and short Barrett esophagus. Short-segment Barrett esophagus is much more frequently seen than long-segment Barrett esophagus in women; long-segment Barrett esophagus is much more common in men. By having a SSBE : LSBE ratio that is much higher, women would be identified as the gender with the greater tendency to develop intestinal metaplasia.

Studying the differences in gastric contents between women who have microscopic segments of Barrett esophagus (i.e., presently diagnosed as intestinal metaplasia of the gastric cardia) and males with long segments of columnar-lined esophagus who do not have intestinal metaplasia will be valuable because these two groups represent the two extremes in terms of tendency to induce intestinal metaplasia (Fig. 15.21, Table 15.3). Identification of molecules in the gastric juice that show the greatest difference in concentration in these two groups is most likely to lead to the agent causing intestinal metaplasia in cardiac mucosa.

Carcinogenesis in Intestinal Metaplasia

Many factors are involved in carcinogenesis:

1. *The presence of carcinogen(s).* The very low incidence of adenocarcinoma in reflux disease suggests that the carcinogen in *not ubiquitous in gastric juice.* The carcinogen(s) may be endogenous secretions of the foregut or exogenous molecules in food. If the patient has duodenogastric reflux, the endogenous compounds available greatly increase. All these molecules interact in the stomach to produce new molecules that may be transient, disappearing after they have induced carcinogenesis. Whatever the carcinogenic molecule(s), it is self-evident that they are present at highest concentration in the stomach, reach the esophagus by reflux, and decrease in concentration from the gastroesophageal junction to the more proximal esophagus (Fig. 15.22). The effective carcinogen concentration is always greatest in the most distal esophagus and reaches a variable height in the esophagus (A–D in Fig. 15.22).
2. *The presence or absence of the target cell.* The target cell for reflux-induced adenocarcinoma is the actively proliferating cell in intestinal metaplastic epithelium. If there is no intestinal metaplasia, there is no risk of malignancy irrespective of the length of columnar-lined esophagus. Of course, there is the possibility that the target cell may appear at a point in the future in a given patient. The absence of the target cell in a 50-year-old patient is much more highly predictive of a lack of cancer risk than the absence of intestinal metaplasia in a child. The presence of the target cell indicates risk of cancer, irrespective of length of columnar-lined esophagus.
3. *The number of target cells present.* This is an unevaluated risk, but it is logical to believe that the cancer risk will be directly proportional to the number of target cells present because the likelihood of a mutational event will be proportional to the number of targets. We have suggested that reclassification of Barrett esophagus by some measure of the number of target cells (see Chapter 10) will provide better risk delineation than the present classification into long- and short-segment Barrett esophagus.

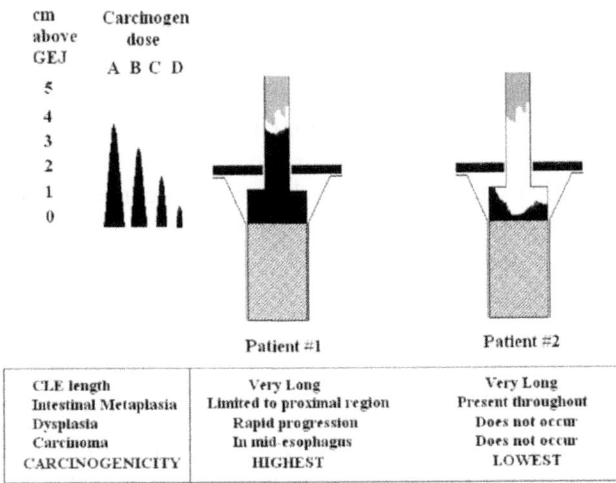

	Patient #1	Patient #2
CLE length	Very Long	Very Long
Intestinal Metaplasia	Limited to proximal region	Present throughout
Dysplasia	Rapid progression	Does not occur
Carcinoma	In mid-esophagus	Does not occur
CARCINOGENICITY	HIGHEST	LOWEST

FIGURE 15.22 Characterization of patients in terms of the carcinogenicity of gastric juice. The highest carcinogenicity is in patient 1, who develops cancer in a small area of intestinal metaplasia in the most proximal part of the columnar-lined esophagus. The concentration of carcinogen in gastric juice must be very high to reach this target cell. In contrast, patient 2, who has intestinal metaplasia extending all the way down to the gastroesophageal junction, is exposed to the highest concentration of carcinogen. If this patient's condition does not progress to cancer, carcinogenicity must be very low.

4. *The location of the target cells.* Because the concentration of the carcinogen is highest distally and the occurrence of intestinal metaplasia is maximal in the most proximal region of the segment of columnar-lined esophagus, it is essential that the intestinal metaplasia reach low enough to encounter sufficient carcinogen (Figs. 15.9 and 15.10).
5. *The proliferative rate of the target cell.* Intestinal epithelium has the highest proliferative rate among the different epithelial types in columnar-lined esophagus (21). It is logical that cancer risk will correlate with the proliferative rate because the probability of mutations increases with an increasing proliferative rate.

The exact molecular mechanism(s) of carcinogenesis is unknown, though there is suspicion that bile salt metabolites may be involved (see Chapter 10). Clinical observations on patients with reflux-induced adenocarcinoma can provide hints about the carcinogenic process. The observed rates of progression to dysplasia and cancer are variable in patients with Barrett esophagus. At the time of diagnosis, these patients are placed on acid suppression (if symptomatic) and regular surveillance. At the time they are placed on surveillance, the duration for which intestinal metaplasia has been present is unknown. Some of these patients develop cancer within 1 year of the index endoscopy. This is a high carcinogenic group. The risk of cancer in the remainder of the patients is 0.5% per year and is greatest in patients who have evidence of progression in the carcinogenic pathway at the index biopsy—that is, have low-grade dysplasia (22) or aneuploidy (23). The majority of patients who are on surveillance for Barrett esophagus have very stable disease without progression to increasing dysplasia or cancer over many years. This clinical course identifies a low carcinogenic environment; there is no progression to cancer despite the presence of large numbers of target cells.

Most patients who develop reflux-induced adenocarcinoma have been completely asymptomatic or take over-the-counter acid suppressive agents for mild symptoms. They present for the first time with symptoms related to the cancer. The rate of progression in this group is unknown because it is not known when they developed intestinal metaplasia in the absence of surveillance.

Based on these observations, it is possible to predict the likelihood of carcinogenesis based on differences in the elements of damage (which controls length of columnar-lined esophagus), the tendency to intestinal metaplasia (which controls the number and location of target cells), and the carcinogenicity that we have discussed here (Table 15.4):

Table 15.4 shows that the length of columnar-lined esophagus and the presence of symptoms is a function of the damage environment; the age of onset of intestinal metaplasia is a function of the tendency to produce intestinal metaplasia; and the progression to cancer from the time of onset of intestinal metaplasia is a function of the carcinogenic environment. These are all likely to be independent mechanisms with at least the tendency to intestinal metaplasia and carcinogenicity being unrelated to acid.

This construct fits well with observed patterns of reflux-induced adenocarcinoma if the dominant factor for the development of cancer is the carcinogenicity of the refluxate. In a given individual with a highly carcinogenic refluxate, the following events may happen:

1. If the patient has a low-damage environment and a low tendency to develop intestinal metaplasia, that patient may never develop the critical length of columnar-lined esophagus needed to develop intestinal metaplasia. Without intestinal metaplasia, the patient will not develop cancer. This individual will not have symptoms of reflux and at the time of natural death will have a very short segment of cardiac or oxyntocardiac mucosa.

TABLE 15.4 Patterns of Adenocarcinoma Seen in Patients Related to Their Damage, Intestinal Metaplasia, and Carcinogenic Environments (Only Patients Who Develop Intestinal Metaplasia Are Considered; Many Patients with a Low Tendency to Intestinal Metaplasia Will Never Develop Intestinal Metaplasia)

Damage environment	Intestinal metaplasia tendency	Carcinogenic environment	Characteristics in patient
Low	Low	Low	Short CLE; asymptomatic; Late age onset of IM; No/slow progression to cancer
Low	Low	High	Short CLE; asymptomatic; Late age onset of IM; Rapid progression to cancer
Low	High	Low	Short CLE; asymptomatic Early age onset of IM; No/slow progression to cancer
Low	High	High	Short CLE; asymptomatic; Early age onset of IM; Rapid progression to cancer
High	Low	Low	Long CLE; symptomatic; Late age onset of IM; No/slow progression to cancer
High	Low	High	Long CLE; symptomatic; Late age onset of IM; Rapid progression to cancer
High	High	Low	Long CLE; symptomatic Early age onset of IM; No/slow progression to cancer
High	High	High	Long CLE; symptomatic Early age onset of IM; Rapid progression to cancer

2. If the patient has a low-damage environment with a strong tendency to develop intestinal metaplasia, the patient will not have symptoms but will develop very short-segment Barrett esophagus at a relatively young age. Because the carcinogenicity is high, this patient represents the common patient who develops adenocarcinoma in the end-stage dilated esophagus without developing symptoms and without detection of Barrett esophagus even if an endoscopy is done.

In contrast, the patient with a low carcinogenic environment will progress to cancer very slowly or not at all. These patients will have a length of columnar-lined esophagus and symptoms based on their damage environment. If they develop intestinal metaplasia and are so diagnosed, they will be the population of patients with Barrett esophagus who remain stable on surveillance over long periods without progressing to cancer.

Research toward recognizing the factors that induce carcinoma in intestinal metaplasia is best undertaken by evaluating differences in the composition of gastric juice in those at lowest and highest risk for cancer. Unlike the tendency to intestinal metaplasia, which is highest in patients who develop intestinal metaplasia in the shortest segments of cardiac mucosa, the tendency to carcinogenesis in intestinal metaplasia is greatest in patients who develop cancer in the most proximal esophagus at the youngest age (Fig. 15.22). The patient who has a high-damage element and minimal tendency to intestinal metaplasia will develop intestinal metaplasia restricted to the most proximal region of a long segment of columnar-lined esophagus (patient 1 in Fig. 15.22). This patient can develop cancer only in the most proximal segment of the columnar-lined esophagus where target cells are present, and only if there is a high concentration of carcinogen (A in Fig. 15.22). The patient with the highest concentration of carcinogenic molecules in the gastric juice is therefore identified as the youngest patient who develops the most proximal adenocarcinoma in a very long segment of columnar-lined esophagus with intestinal metaplasia limited to its most proximal region (patient 1 in Fig. 15.22).

In contrast, the lowest concentration of carcinogen is most likely in the oldest patient who has a long segment of columnar-lined esophagus with intestinal metaplasia reaching the most distal esophagus who has remained stable for long periods without developing dysplasia or cancer during surveillance. This patient develops intestinal metaplasia in the distal esophagus and has the highest number of target cells for the carcinogen to act at a point that is close to the highest carcinogen concentration (patient 2 in Fig. 15.22). The fact that this patient does not develop dysplasia and cancer over a long period can only be explained by the presence of a low carcinogenic environment.

It should be realized that the two patients with the highest and lowest carcinogenicity will both be classified by present criteria as having long-segment Barrett esophagus. It is only when we recognize that patients with long-segment Barrett esophagus have infinitely variable amounts of intestinal metaplasia with consequent differences in the risk of cancer that we will make any sense of the cancer risk in Barrett esophagus.

Directed research into the gastric juice composition of patient groups who demonstrate characteristics of a high and low carcinogenic tendency is more likely to detect the actual molecules involved in carcinogenesis. Molecules present at higher concentration in the gastric juice of patients with a high compared to low carcinogenic tendency will be more likely to be the true carcinogen(s). Once these are characterized, their ability to induce carcinogenesis can be tested in experimental cell culture systems. This method is likely to be far more effective in characterizing the actual carcinogenic molecule(s) than the present shotgun approach to studying gastric juice molecules.

MOLECULAR RESEARCH INTO CARCINOGENESIS

Extensive research is being conducted at present is an attempt to determine the molecular pathway of carcinogenesis in Barrett esophagus. Many molecular abnormalities have been identified in Barrett esophagus as well as dysplasia and adenocarcinoma. Putting these molecular changes together to work out the exact molecular/genetic pathway of carcinogenesis in Barrett esophagus is on the near horizon.

Unfortunately, it is not likely that understanding the molecular pathway of Barrett esophagus carcinogenesis will have practical value any time soon. We have understood the molecular/genetic pathway in colorectal cancer for more than a decade now, but this knowledge has not significantly affected the management and prevention of colorectal cancer. The reversal of those molecular events responsible for carcinogenesis awaits the development of genetic manipulation and the delivery of gene therapy agents. As more time passes without success in these areas, the probability that researchers will not succeed increases.

Little research has been directed at the molecular events in the pre-Barrett metaplastic stage of the disease. This is because of the false dogma that cardiac mucosa is a normal gastric epithelium, and it is impossible to study cardiac mucosa without being able to distinguish normal from abnormal cardiac mucosa. Until we accept that *all cardiac mucosa is abnormal and that "normal" cardiac mucosa is a myth, we will not research this mucosa.*

Recognizing that cardiac mucosa is always abnormal will provide an impetus to molecular research on this epithelial type. Research should address the following questions:

1. What is the genotype of cardiac epithelial cells and how does this differ from squamous epithelium? This research may give insight to the mechanism by which squamous epithelium transforms into cardiac mucosa and the genetic signal that causes the progenitor cell to differentiate into squamous or cardiac epithelium.
2. What differences exist in the genetic makeup of cardiac and intestinal epithelium? There is evidence that Cdx genes may be important in the transformation of cardiac to intestinal epithelium. If so, what is the exact mechanism of this process? What interaction between the target cell in cardiac epithelium and molecule in gastric refluxate is involved? What is the receptor on the cell? What is the molecule in the refluxate? What are the steps between receptor binding with the molecule and the genetic change? Can any of these factors be altered by new agents (drugs)?
3. What differences exist in the genetic makeup of cardiac and oxyntocardiac epithelium? There is evidence that the Sonic Hedgehog gene may be important in the transformation of cardiac to oxyntocardiac epithelium. A similar series of questions can be developed. Is this transformation promoted by acid, as we have suggested is theoretically possible?

The first crucial step to successful research is to develop an accurate understanding of the cellular processes involved. The failure to do this in reflux disease has resulted in much misdirected research. The second step is to make careful observations on patients, trying to divide the patients by various characteristics. The third step is to develop logical and reasonable hypotheses. This chapter has tried to achieve these first steps, which are crucial in directing research in the correct path. It is up to people with training in research techniques to actually complete the experiments that will prove or disprove these hypotheses.

Our dream is that the identification of a molecular mechanism that converts cardiac mucosa to oxyntocardiac mucosa will result in a drug that can induce this change. Once a drug is developed, all that will need to be done to prevent intestinal metaplasia will be to screen people for cardiac mucosa and remove them from the reflux-adenocarcinoma sequence by inducing oxyntocardiac mucosa. Or perhaps a drug can be developed to inhibit or reverse the molecular step involved in the conversion of cardiac mucosa to intestinal metaplasia. This will have a similar effect in preventing esophageal adenocarcinoma and appears much more feasible than gene therapy directed at esophageal dysplasia or cancer. There is precedent for such success: identification of the molecular mechanism associated with CD117 and tyrosine kinase receptors permitted the development of Gleevac, a drug directed at this molecular interaction that effectively controls the growth of CD117+ gastrointestinal stromal tumors. Our hope is for a similar drug that can be used at an early metaplastic stage of reflux-induced esophageal carcinogenesis.

References

1. Odze RD. Unraveling the mystery of the gastroesophageal junction: A pathologist's perspective. Am J Gastroenterol 2005;100:1853–1867.
2. Tobey NA, Carson JL, Alkiek RA, et al. Dilated intercellular spaces: A morphological feature of acid reflux-damaged human esophageal epithelium. Gastroenterology 1996;111:1200–1205.
3. Tobey NA, Hosseini SS, Argore CM, Dobrucali AM, Awayda MS, Orlando RC. Dilated intercellular spaces and shunt permeability in non-erosive acid-damaged esophageal epithelium. Am J Gastroenterol 2004;99:13–22.
4. Allison PR, Johnstone AS. The oesophagus lined with gastric mucous membrane. Thorax 1953;8:87–101.
5. Cameron AJ, Lomboy CT. Barrett's esophagus: Age, prevalence, and extent of columnar epithelium. Gastroenterology 1992;103: 1241–1245.
6. Chandrasoma PT, Der R, Ma Y, et al. Histology of the gastroesophageal junction: An autopsy study. Am J Surg Pathol 2000;24:402–409.
7. Zhou H, Greco MA, Daum F, et al. Origin of cardiac mucosa: Ontogenic considerations. Ped Dev Pathol 2001;4:358–363.
8. Marsman WA, van Sandyck JW, Tytgat GNJ, ten Kate FJW, van Lanschot JJB. The presence and mucin histochemistry of cardiac type mucosa at the esophagogastric junction. Am J Gastroenterol 2004;99:212–217.
9. Haggitt RC. Adenocarcinoma in Barrett's esophagus: A new epidemic? Hum Pathol 1992:23:475–476.
10. Devesa SS, Blot WJ, Fraumeni JF. Changing patterns in the incidence of esophageal and gastric carcinoma in the United States. Cancer 1998;83:2049–2053.
11. Pohl H, Welch HG. The role of overdiagnosis and reclassification in the marked increase of esophageal adenocarcinoma incidence. J Natl Cancer Inst 2005;97:142–146.
12. Lagergren J, Bergstrom R, Lindgren A, Nyren O. Symptomatic gastroesophageal reflux as a risk factor for esophageal adenocarcinoma. N Engl J Med 1999;340:825–831.
13. Fitzgerald RC, Omary MB, Triadafilopoulos G. Dynamic effects of acid on Barrett's esophagus: An ex-vivo proliferation and differentiation model. J Clin Invest 1996;98:2120–2128.
14. Kazumori H, Ishihara S, Rumi MA, Kadowaki Y, Kinoshita Y. Bile acids directly augment caudal-related homeobox gene Cdx2 expression in esophageal keratinocytes in Barrett's epithelium. Abstract ID: S1674, DDW, 2005.
15. Hassall E. Columnar lined esophagus in children. Gastroenterol Clin N Amer 1997;26:533–548.
16. Glickman JN, Fox V, Antonioli DA, Wang HH, Odze RD. Morphology of the cardia and significance of carditis in pediatric patients. Am J Surg Pathol 2002;26:1032–1039.
17. Dresner SM, Griffin SM, Wayman J, Bennett MK, Hayes N, Raimes SA. Human model of duodenogastro-oesophageal reflux in the development of Barrett's metaplasia. Br J Surg 2003;90:1120–1128.
18. Lord RVN, Wickramasinghe K, Johansson JJ, DeMeester SR, Brabender J, DeMeester TR. Cardiac mucosa in the remnant esophagus after esophagectomy is an acquired epithelium with Barrett's like features. Surgery 2004;136:633–640.
19. Chandrasoma PT, Der R, Dalton P, Kobayashi G, Ma Y, Peters JH, DeMeester TR. Distribution and significance of epithelial types in columnar lined esophagus. Am J Surg Pathol 2001; 25:1188–1193.
20. Paull A, Trier JS, Dalton MD, Camp RC, Loeb P, Goyal RK. The histologic spectrum of Barrett's esophagus. N Engl J Med 1976;295:476–480.
21. Olvera M, Wickramasinghe K, Brynes R, Bu X, Ma Y, Chandrasoma P. Ki67 expression in different epithelial types in columnar lined esophagus indicates varying levels of expanded and aberrant proliferative patterns. Histopathology 2005;47:132–140.
22. Weston AP, Badr AS, Hassanein RS. Prospective multivariate analysis of clinical, endoscopic, and histological factors predictive of the development of Barrett's multifocal high grade dysplasia or adenocarcinoma. Am J Gastroenterol 1999;94: 3413–3419.
23. Reid BJ, Levine DS, Longton G, et al. Predictors of progression to cancer in Barrett's esophagus: Baseline histology and flow cytometry identify low- and high-risk patient subset. Am J Gastroenterol 2000;95:1669–1676.

CHAPTER

16

Rationale for Treatment of Reflux Disease and Barrett Esophagus

The present treatment of patients with gastroesophageal reflux disease is highly successful in controlling pain, healing erosive esophagitis, preventing severe ulceration and stricture formation, and generally improving the quality of life for the vast majority of patients. This has been achieved as a result of pharmacological advances and the development of drugs that are increasingly effective in suppressing acid. The newest proton pump inhibitors, when used in adequate dosage, can effectively produce a level of acid suppression that is close to complete. In the few (approximately 5% to 10%) patients in whom acid suppression cannot be controlled medically, new surgical antireflux procedures are highly effective. These antireflux surgeries are now performed laparoscopically with a high level of success, extremely low mortality, and low post-surgical morbidity. Few patients today remain significantly symptomatic with treatment.

Clouding this rosy assessment is the rapid increase in the incidence of Barrett esophagus and reflux-induced esophageal adenocarcinoma since the 1970s. The trend continues and a plateau is not yet in sight for esophageal adenocarcinoma, although the curve for adenocarcinoma of the gastric cardia has flattened (1). Whereas adenocarcinoma of the esophagus was considered a rare cancer type in the 1960s, it now has an annual incidence of more than 8000 cases in the United States. When one adds the more than 8000 to 10,000 patients who develop adenocarcinoma of the gastric cardia (which we believe is largely adenocarcinoma arising in Barrett esophagus occurring in a reflux-damaged dilated end-stage esophageal segment), reflux-induced adenocarcinoma accounts for 16,000 to 18,000 cases per year. With an overall mortality in the range of 80%, the number of deaths is around 14,000 per year and increasing. When one adds to this equation the extreme suffering associated with advanced esophageal adenocarcinoma, this disease is becoming a medical nightmare.

THE PRESENT RATIONALE FOR TREATMENT OF REFLUX DISEASE

At present, two major methods of treatment are used for patients with precancerous stages of reflux disease: (a) acid suppression and (b) antireflux surgery (Table 16.1). Both are highly effective in controlling symptoms of reflux such as heartburn and healing erosive esophagitis. Newer nonsurgical (endoscopic) antireflux procedures are available, but their long-term efficacy is unproven.

When Barrett esophagus is found by the demonstration of intestinal metaplasia in an endoscopic biopsy taken from a columnar-lined esophagus, these treatment options remain unchanged. However, a third option arises because of the risk of malignancy in Barrett esophagus. This is to remove the pathologic area either by surgery (endoscopic mucosal resection, some type of esophagectomy) or with ablative techniques that destroy the mucosa (photodynamic therapy or some type of physical agent that destroys cells on contact).

At present, this third option of ablation is used clinically only in patients who have a very high risk of malignancy (either high-grade dysplasia or early

TABLE 16.1 Comparison of Acid Suppressive Drugs and Antireflux Surgery as the Primary Treatment Modality of Gastroesophageal Reflux Disease

	Acid suppressive drugs	Antireflux surgery
Availability	Unrestricted; over the counter	Surgical expertise is limited
Ease of use	High	Surgery is not simple
Risk to life	Presently believed to be near zero	Surgical mortality near zero
Side effects	Presently believed to be minimal	Significant incidence of dysphagia
Immediate cost	Low; no pretreatment testing is essential	High; requires preop testing, surgery, and hospital stay
Lifelong cost	High; drugs expensive and require to be taken lifelong	Lower than drugs if surgery is successful
Failure of symptom control	Symptoms are not controlled in <5% with combined/high-dose therapy	Significant (20%) risk of failed surgery
Treatment of failures	Antireflux surgery	Redo surgery; acid suppressive drugs

adenocarcinoma). This is largely because these methods have significant morbidity, complications (mainly strictures and perforation), and cost. As newer methods of removing the pathologic lesion are developed that have lower morbidity, complications, and cost, there is every possibility that ablation and mucosal resection can be used in patients with lower risk and at earlier stages in their cancer pathway.

In the absence of an accepted cellular definition of reflux disease, the effect of the treatment of reflux disease is difficult to assess in any objective or rational manner. At the present time, the effect of treatment is assessed as follows:

1. *Control of symptoms.* To refer to reflux disease as "cured" when symptoms are controlled is appropriate only if the end point of treatment is the control of symptoms. When one recognizes, however, that symptom control does not prevent the progression in the reflux–adenocarcinoma sequence, symptom control has no value in assessing treatment if the end point of the disease is adenocarcinoma.
2. *Healing of erosive esophagitis.* Again, while this is a laudable goal in treatment, there is no evidence that healing erosive esophagitis has any effect in decreasing the progression to adenocarcinoma. In fact, most patients who develop adenocarcinoma have never had erosive esophagitis. Using the healing of erosive esophagitis as the end point of treatment is an admission that there is no attempt to address the risk of adenocarcinoma in the patient.
3. *Control of acidity and maintaining the pH of gastric contents greater than 4.* This is a laudable goal for acid suppressive drug therapy because it shows that the drug is effective in the manner it has been prescribed and used by the patient. There is good evidence that maintaining the pH in the greater than 4 range is effective in controlling symptoms and healing erosive esophagitis, but there is no evidence that it does anything to prevent the progression to cancer (2). Maintaining gastric pH above 4 will normalize the 24-hour pH test.

Impedance Studies

Impedance-measuring devices in the lower esophagus detect the retrograde entry of air and liquids into the esophagus from the stomach and quantitate the height to which the refluxate passes in the esophagus within the range of placement of the probes. They are useful in showing that reflux continues to occur in patients who are on acid suppressive medication despite the fact that the 24-hour pH test is normalized (3).

ACID SUPPRESSION WITH DRUGS AS TREATMENT FOR REFLUX DISEASE

Acid suppression with drugs is achieved in two ways. Many patients with heartburn are controlled by simple antacids, which are alkaline agents that neutralize secreted acid in the stomach when taken orally. These drugs have been popular for a long time and are used when the patient experiences heartburn to provide immediate relief. These drugs are taken episodically and have limited efficacy in the severely symptomatic patient and the patient who has erosive esophagitis. However, sales of these drugs are prolific, suggesting that a significant number of patients with mild reflux control their symptoms with these agents alone without needing to seek medical care. Direct marketing of these drugs encourages patients to pop a Tums, Rolaids, Mylanta, or Maalox either when they get heartburn or before they eat a heavy meal.

Much more effective are drugs that suppress the secretion of acid by the stomach. The first of these suppressors of acid secretion were the histamine-2 receptor blockers (cimetidine; ranitidine sold as Pepcid and Zantac), which came on the market in the late 1960s. These were followed in the late 1980s by the very powerful proton pump inhibitors (omeprazole, lansoprazole, esomeprazole sold as Prilosec, Prevacid, and Nexium). Acid suppression with H2-receptor blockers and proton pump inhibitors requires chronic and often lifelong treatment with these drugs. Despite the fact that these powerful acid suppressive agents are available without prescription, chronic therapy is associated with high cost.

Most patients with severe symptoms of reflux disease can be controlled with proton pump inhibitors. This treatment sometimes requires increasing the dose to maximum and combining this high dose with an H2-receptor blocker. Less than 5% of patients do not achieve symptomatic control and require antireflux surgery (Table 16.1). Proton pump inhibitors are generally used without evaluating the level of acid suppression by 24-hour pH testing; the drug dosage is adjusted to produce symptom control. Acid suppression has no valid basis for use in anything other than the control of symptoms of reflux. There is no claim for any efficacy in reversing Barrett esophagus or preventing adenocarcinoma. However, it is not unusual to find asymptomatic patients diagnosed as having Barrett esophagus on lifelong acid suppressive drug therapy, probably based on the vain hope that acid suppression in some way will reverse the Barrett esophagus or prevent cancer.

It is worthwhile to examine in detail the efficacy of acid suppressive drug therapy in terms of the different pathophysiologic events that are responsible for reflux disease with a view to seeing whether they prevent, reverse, or promote each pathologic change and what effect they produce when used in practice. In the next section, we will ask these identical questions as they relate to antireflux surgery.

Is Acid Suppressive Drug Therapy Successful in Preventing Reflux?

The cause of reflux is an abnormality in the lower esophageal sphincter, which results in an exaggeration of sphincter relaxations caused by gastric distension. With the increasing severity of the sphincter abnormality, the factors needed to cause sphincter relaxation become less and the severity of reflux increases (4). Increased reflux, which is at first postprandial and intermittent, becomes frequent and occurs at all times and postures, greatly increasing the exposure of the distal esophagus to gastric refluxate.

There has never been any evidence, either experimental or in human studies, that has suggested that acid suppressive agents improve the functional characteristics of the defective lower esophageal sphincter. The expectation, therefore, is that the occurrence of reflux will not be significantly changed in patients who are on acid suppressive medications. The control of symptoms and healing of erosions is the result of the absence of acid in the refluxate; it is not the result of an absence of reflux.

At one time, it was believed that acid suppression in some way decreased the volume of reflux because the 24-hour pH test was frequently normalized after treatment with proton pump inhibitors. Peters et al. (5) showed that the acid exposure decreased to 0.1% of the 24-hour period in patients who were on omeprazole at a dose of 40 mg BID. They reported that: "Omeprazole reduced reflux to 0.1%" (p. 490). It was believed that because acid secretion was a major component of gastric volume, acid suppression decreased gastric secretion, decreased gastric distension, and prevented reflux.

With new impedance technology, this has proven to be largely false. Patients who are effectively acid suppressed still continue to reflux gastric contents into the esophagus at a frequency and volume that are only slightly decreased from pretreatment levels, with the majority of patients continuing to have abnormal levels of reflux.

Literature Review

Tamhankar AP, Peters JH, Portale G, Hsieh C-C, Hagen JA, Bremner CG, DeMeester TR. Omeprazole does not reduce gastroesophageal reflux: new insights using multichannel intraluminal impedance technology. J Gastrointest Surg 2004;8:890–898.

This is a study of 6 asymptomatic volunteers who had combined 24-hour impedance pH monitoring before and after 7 days of 20 mg BID omeprazole. Reflux episodes were identified by multichannel intraluminal impedance and classified by pH measurements 5 cm above the lower esophageal sphincter as acid (pH < 4), weak acid (pH > 4 but associated with a 1 pH unit decrease during the episode), and nonacid (pH > 4 without a 1 pH unit decrease during the episode). A pH electrode was placed in the stomach to measure gastric pH.

Impedance detected 116 reflux episodes before and 96 reflux episodes after omeprazole treatment. The median number of reflux episodes (18 versus 16, $p = 0.4$), median duration of reflux episodes (4.7 versus 3.6 minutes, $p = 0.5$), and total duration of reflux episodes per subject (27.2 versus 42.4 minutes, $p = 0.5$) were similar before and after the omeprazole treatment. Acid reflux episodes reduced from 63% before to 2.1% after omeprazole ($p \leq 0.0001$), whereas nonacid reflux episodes increased from 15% to 76% ($p \leq 0.0001$). The authors concluded that, in asymptomatic subjects, omeprazole at a dose of 20 mg BID does not affect the number or duration of reflux episodes. It has a strong effect in converting acid reflux to less acid reflux, thus exposing the esophagus to altered gastric juice at a higher pH.

This study proves that the 24-hour pH test cannot be used to assess the presence of gastroesophageal reflux while patients are on acid suppressive therapy. Effective acid suppression results in an increase of gastric pH above 4, making acid a useless detector of reflux. One requirement of the 24-hour pH test is that patients be off acid suppressive drugs for 2 weeks before the test can be performed.

When patients who are on acid suppressive therapy are tested with a marker of reflux that is different from acid (the 24-hour pH test) such as the Bilitec test, which uses bile pigments as the marker, the presence of abnormal reflux is shown even in patients who have normalization of the 24-hour pH test (6).

Literature Review

Parrilla P, deHaro LFM, Ortiz A, Munitiz V, Molina J, Bermejo J, Canteras M. Long term results of a randomized prospective study comparing medical and surgical treatment of Barrett's esophagus. Ann Surg 2003;237:291–298.

In a very small part of this study, 12 of 43 patients who were in the acid suppressive therapy arm for 1 year had 24-hour and Bilitec 2000 tests while on their medications. In these 12 patients, who were all asymptomatic, there was a significant decrease in the 24-hour pH test from 16% (median percentage time $pH < 4$; range, 6% to 43%) to 0.9% (range, 0 to 9%) ($p < 0.05$). Nine of these patients had normalization of the test, the other 3 patients had a percentage time $pH < 4$ between 4% and 9%, indicating a mildly abnormal acid exposure. The Bilitec test in these 12 patients also showed a decrease from 31% (range, 6% to 66%) to 15.4% (range, 2% to 32%); this decrease was not statistically significant. Nine of these 12 patients (75%) had an abnormal Bilitec test, showing continued abnormal reflux.

This study shows that patients continue to have their target cells exposed to refluxate molecules other than H^+ ions while being acid suppressed with omeprazole. Only the damaging effect of acid on target cells is prevented by acid suppression. All the cellular changes caused by molecules other than acid continue unabated in patients who are on acid suppression treatment.

Conclusion: Acid suppressive drug therapy does not decrease the volume or frequency of reflux; it simply makes the refluxate nonacid or less acid.

Is Acid Suppressive Drug Therapy Successful in Treating Reflux Disease?

Acid suppression has proven efficacy in controlling symptoms and healing erosive esophagitis. This is to be expected because acid is a powerful injurious agent to squamous epithelium, and pain and erosion are the direct results of acid on squamous epithelium. The control of symptoms often occurs with even incomplete acid suppression (6). This may actually be a curse rather than a blessing because it prevents people with reflux disease from seeking care. It converts the symptomatic person to an asymptomatic person because of the ready availability of even the strongest acid suppressive drugs without prescription. It hinders the early detection of Barrett esophagus in this group of patients by preventing the initial patient-physician interaction.

The overwhelming importance given by primary care physicians to controlling symptoms, which is equated with curing reflux disease, is another problem. The vast majority of primary care physicians and even gastroenterologists do not appreciate the fact that this patient-physician encounter should be regarded as a cancer-preventing opportunity. A proton pump inhibiting pill and a smile, and the patient is on his or her way to an outcome defined as being "successful" when the pain has been controlled. No attempt is made to prevent the cancer that will inevitably occur in some of these patients; the risk of cancer is considered too small to even think about it. The risk of cancer, however, is 100% for the patients so treated who ultimately develop cancer. The equivalent situations in medicine are the failure to consider malignancy when evaluating a patient with a breast lump, chronic hoarseness, and rectal bleeding. All of these clinical situations are taken extremely seriously despite a low cancer risk.

Most physicians will answer the question that heads this section with an enthusiastically positive response. They will say that acid suppressive agents are miracle drugs that permit the cure of patients with reflux disease. This is absolutely wrong. Many more people die of reflux disease today than ever before in history.

At present, the number of deaths from reflux-induced adenocarcinoma is approximately 14,000 per year; this number is continuing to increase in almost epidemic fashion (1, 7, 8). The medical community is missing the forest (deaths caused by reflux-induced adenocarcinoma) for the trees (symptoms of reflux); we are just making people comfortable without addressing the fact that they die from this disease. The success of treatment of a disease must be primarily determined by mortality rates; our number 1 job is to prevent death. We are failing dismally when we provide symptom relief and completely ignore the increasing incidence of cancer.

Conclusion: Acid suppressive drugs have been an abject failure in treating reflux disease if one uses the benchmark of number of deaths from reflux disease per year. It continues to get worse rapidly.

Is Acid Suppressive Drug Therapy Successful in Preventing Barrett Esophagus?

There are no studies to evaluate this question, largely because there is no recognition of the sequence of changes leading to Barrett esophagus. Without recognizing that cardiac mucosa is a precursor stage of Barrett esophagus, studies will not be designed to evaluate this change. Most gastroenterologists begin the discussion of Barrett esophagus at the point at which the patient develops intestinal metaplasia.

Many studies in the literature assess the prevalence of intestinal metaplasia. The prevalence of intestinal metaplasia varies greatly in different studies and depends largely on the population that is under study. The prevalence of intestinal metaplasia is well known to increase as the length of columnar-lined esophagus increases. Chandrasoma et al. (9) reported that the incidence of intestinal metaplasia increased from 15% with a columnar-lined esophagus of less than 1 cm, to 70% at 1 to 2 cm, 90% at 3 to 4 cm, and 100% at greater than 5 cm. In this study, the columnar esophageal length was determined by histologic mapping; the lengths are greater than the endoscopically visible columnar-lined esophagus. Most other recent studies are concordant with these numbers. Spechler et al. (10) showed that 18% of patients without endoscopically apparent Barrett esophagus had intestinal metaplasia. At the time, endoscopic Barrett esophagus required greater than 2 cm of columnar-lined esophagus. In a screening study, Rex et al. (11), studying a population of 961 patients, reported that 12 (1.2%) had a columnar-lined segment that was greater than 3 cm; all of these patients had intestinal metaplasia; 164 of 961 (17%) had a visible columnar-lined esophagus measuring 0.5 to 2.9 cm; the prevalence of intestinal metaplasia was 53 of 164 of these patients (32%).

The prevalence of intestinal metaplasia is also related to the severity of reflux disease as assessed by symptoms. In Rex et al.'s study, the prevalence of Barrett esophagus and long-segment Barrett esophagus in patients who did not have heartburn was 5.6% and 0.36%, respectively, compared to 8.3% and 2.6% in patients with heartburn. There are no data relating to acid suppressive drug use in these patients, but it is reasonable to assume that patients who had heartburn had acid suppressive medications more often than those who did not. It is amazing that the use of acid suppressive drugs is not reported in these studies. Any explanation for the difference in the incidence of Barrett esophagus among patients with and without heartburn must take into account the possibility that different usage of acid suppressive agents may be the cause. It is almost as if there is a subliminal fear about investigating this possibility.

There are no large-scale controlled studies designed to evaluate whether symptomatic patients without Barrett esophagus have an increased risk of developing Barrett esophagus if they use acid suppressive drugs. Such a study would be difficult to execute because there is no alternate form of drug treatment for symptoms of reflux disease and it is difficult to have a no-treatment arm in a symptomatic patient population. The fact that Barrett esophagus has increased in prevalence since the 1970s (see Chapter 9) strongly suggests that the use of acid suppressive drugs to treat symptomatic reflux does not prevent the esophagus from developing Barrett esophagus. We have provided theoretical reasons why acid suppressive drugs actually promote the conversion of cardiac mucosa in columnar-lined esophagus to intestinal metaplasia (Chapters 9, 14). This would suggest that it is reasonable to at least partially ascribe the increased incidence of Barrett esophagus since the 1970s to the increasing use and effectiveness of acid suppressive drugs used to treat symptomatic reflux disease.

Conclusion: There is no evidence that acid suppressive drugs can prevent intestinal metaplasia (Barrett esophagus). There is a probability that the use of acid suppressive drugs promote the development of Barrett esophagus when used to treat symptomatic reflux.

Is Acid Suppressive Drug Therapy Successful in Treating Barrett Esophagus?

During the American Gastroenterological Association Consensus Workshop in Chicago, 18 experts

(15 gastroenterologists, 2 surgeons, and a pathologist) critically evaluated the data available regarding the efficacy of treatment of patients with Barrett esophagus with acid suppressive agents (12). Statement 32, considered by the experts, was formulated as follows: "Acid suppressive therapy has been shown to improve symptoms and to heal and prevent relapse of erosive esophagitis in patients with Barrett esophagus" (p. 322).

The experts concluded that symptom relief and healing of erosive esophagitis was achieved by acid suppression in patients with Barrett esophagus. Because proton pump inhibition was effective in preventing relapse of symptoms, all the experts felt that long-term drug therapy is effective in patients with Barrett esophagus. This is not a correct conclusion. The correct conclusion is that long-term drug therapy is *only* effective in healing and preventing squamous epithelial damage in patients with Barrett esophagus. *There is no evidence whatsoever that intestinal metaplasia is reversed by acid suppression.* In fact, in Chapter 15 we provided a theoretical basis that suggests that acid suppression may potentiate the conversion of cardiac mucosa in the esophagus to intestinal metaplasia. It is highly unlikely that the interactive molecular phenomena that cause the changes in intestinal metaplasia to progress to cancer are prevented or slowed by acid suppression.

Our response to this conclusion is not very complimentary. We see this as equivalent to a patient with angina pectoris being treated with drugs to control the angina. If the coronary artery problem is not addressed, the patient will progress to death when coronary thrombosis causes complete obstruction. The recognition of this fact led to intensive research to find ways to reverse the coronary disease; coronary artery bypass surgery and angioplasty emerged. These procedures were initially used without good evidence of their efficacy in prolonging survival because the disease, if left untreated, was devastating. At the present time, gastroenterologists are treating Barrett esophagus by controlling pain (= angina in the preceding example) *without attempting to do anything about preventing cancer, which is the devastating killer in this disease* (= *coronary thrombosis in the preceding example*). They are not doing anything to address Barrett esophagus, which is the precursor lesion for adenocarcinoma (= *coronary atherosclerosis in the above example*).

The attitude among gastroenterologists is that the problem of adenocarcinoma in Barrett esophagus can be ignored because cancer is a very uncommon event. Epidemiologic data suggest this is not true; approximately 14,000 patients die every year in the United States from reflux-induced adenocarcinoma. This is equal to 38 people every day and 1.5 people every hour. The only reason why gastroenterologists think this is rare is that the number of gastroenterologists has increased so much that these patients are distributed thinly within this specialty.

It is also difficult to understand why asymptomatic patients who have Barrett esophagus are treated with acid suppressive drugs. The only benefit of acid suppression in reflux disease is the relief of symptoms and healing of erosions. In the absence of any evidence that these drugs reverse intestinal metaplasia, there is absolutely no reason to use them in the absence of symptoms. However, many gastroenterologists continue to treat asymptomatic Barrett esophagus in the vain hope that acid suppression has a magically beneficial effect.

Conclusion: Acid suppressive drugs are ineffective in preventing or reversing molecular interactions occurring in Barrett esophagus. Treating patients who have Barrett esophagus with acid suppressive drugs is at best like not treating the disease at all.

Does Acid Suppressive Drug Therapy Cause a Regression of Barrett Esophagus?

The claim within the discussion relating to statement 35 in the AGA Chicago workshop, quoted earlier, that "observations that partial regression of metaplastic mucosa may be induced by suppression of acid . . . with proton pump inhibitors (PPIs) . . . provide some plausibility for the notion that PPIs may be able to reduce the cancer risk" is worthy of closer examination. Among several studies that show slight "regression" of Barrett esophagus after long-term PPI treatment, the article from the Netherlands has been chosen for review because it is the best designed.

Literature Review

Peters FT, Ganesh S, Kuipers EJ, Sluiter WJ, Klinkenberg-Knol EC, Lamers CB, Kleibeuker JH. Endoscopic regression of Barrett's oesophagus during omeprazole treatment: A randomized double blind study. Gut 1999;45:489–494.

This is a study of 68 patients with endoscopically and histologically proven Barrett esophagus. The authors used the original definition of Barrett esophagus, which includes all three epithelial types as long as the endoscopic length of the columnar mucosa was greater than 3cm. However, all but 3 patients had intestinal metaplasia evident in at least one biopsy specimen; 75% of the patients revealed no dysplasia at inclusion, and 25% had indefinite/low-grade dysplasia.

The selected patients were allocated to treatment with either omeprazole, 40 mg twice daily, or ranitidine, 150 mg twice daily (as a control group), according to a computer-generated randomization list; drugs were given for a total duration of 2 years. Of the 68 patients, 33 were assigned to omeprazole and 35 to ranitidine; 26 in the omeprazole group and 27 in the ranitidine group completed the study with the assigned drug and were suitable for statistical analysis at 24 months.

Endoscopy was performed at baseline and after 3, 9, 15, and 24 months, while the patients were on medication. The esophagogastric junction was defined to be at the upper border of the gastric folds as viewed after desufflation. For the measurement of the length of Barrett esophagus, the highest point of the squamocolumnar junction was taken. All distances were measured in relation to the incisors. To determine the total area of Barrett's mucosa, the endoscopist made a drawing on graph paper of the mucosal landmarks, the course of the Z line, and of islands of squamous and metaplastic epithelium during or immediately after each endoscopy. The area of Barrett's mucosa was expressed as the number of complete squares within the borders of the metaplastic epithelium, including islands of metaplastic epithelium proximal to the Z line and excluding islands of squamous epithelium distal to the Z line.

Ambulatory 24-hour esophageal pH-metry was performed at baseline, while the patient took no acid-reducing drugs or an H_2 receptor antagonist in a dose not exceeding the equivalent of ranitidine 150 mg twice daily. pH-metry was repeated three months after the start of the study while the patient was on study medication.

For both the length and the area of Barrett esophagus, the measurements as obtained at the five subsequent endoscopies were combined into one value, that being the area under the curve (AUC). This AUC represents in each individual the change, either absolute or percentage, from a state of no change. Negative values represent a decrease in length or area, whereas positive values represent an increase. The intraobserver coefficient of variability was 5% for measurement of the length and 4.5% for the area of Barrett esophagus.

Symptoms were ameliorated in both groups. The authors evaluated the presence of reflux by a 24-hour pH test while the patients were on the acid suppressive medication. Omeprazole reduced acid reflux to 0.1%, ranitidine to 9.4%, per 24 hours.

Elimination of acid reflux by omeprazole resulted in a significant reduction of both the length and the total area of Barrett esophagus, whereas no change was noted during persistent reflux in the ranitidine group. When the results in the omeprazole group and the ranitidine group were compared, a statistically significant difference was found between the effects on total area of Barrett esophagus; the difference between the effects on length showed a trend in the same direction. When only the lengths and the areas as measured at baseline and after 2 years were compared with each other, the changes in both length and area in the omeprazole group were statistically significant (−6.4% [C.I. −3.2 to −9.5%] and −7.9% [C.I. −3.0 to −12.9%], respectively), but in the ranitidine group they were not (+0.4% [C.I. 4.3 to −3.5%] and −0.6% [C.I. 3.3% to −2.1%], respectively). Both the reduction of the length and that of the total area occurred gradually during the 2-year treatment period.

The authors concluded that this randomized, double-blind study showed that the elimination of acid gastroesophageal reflux, as accomplished by omeprazole 40 mg twice daily, results in statistically significant regression of Barrett esophagus, whereas persistent pathological reflux during ranitidine 150 mg twice daily does not. The regression is quantitatively small, however: approximately 8% in 2 years. Therefore, according to the authors, the clinical significance of this regression is to be qualified as modest and the findings do not obviate the search for other ways to induce regression.

In fact, it is important to recognize that this "modest 8% regression" was determined by a complicated and nonstandard and nonduplicated method devised by these authors where the intraobserver coefficient of variability was 5% for measurement of the length and 4.5% for the area of Barrett esophagus. This further adds to the suggestion that "regression" is at most modest and probably less.

This study is one of the more optimistic. In contrast, Sharma et al. (12) studied 27 patients on 60 mg lansoprazole. Of these, 13 underwent ambulatory 24-hour esophageal pH monitoring while on therapy. Symptoms were recorded, and the length of Barrett's epithelium was measured, photographed, and biopsied every 6 months over an average of 5.7 years. They reported that 8 of 13 patients had a normal 24-hour pH (group I, mean pH < 4, 0.8%) and 5 patients had abnormal results (group II, mean pH < 4, 10.6%). Symptoms improved in all patients, and there was complete healing of erosive esophagitis in all patients. An increase in the number of squamous islands was noted in 62.5% of patients in group I and in 80% of patients in group II. The mean length of Barrett's epithelium at baseline and study completion in group I was 5.6 and 5.0 cm, respectively (mean decrease, 0.6 cm), and for group II it was 4.2 and 4.2 cm, respectively (mean decrease, 0 cm). There was no significant difference in the change in length between the two groups ($p = 0.494$). These authors concluded that although symptoms improved, erosive esophagitis healed, and squamous islands increased, there was no significant decrease in the length of Barrett esophagus. Their findings indicate that control of esophageal pH alone was not sufficient for the reversal of Barrett esophagus.

Wilkinson et al. (13) reported that the regression was observed in the first 2 years of treatment and remained stable in the 2 to 5-year period. They also, like many other workers, showed the frequent presence of surface overgrowth of squamous epithelium as squamous islands.

These studies indicate that what is called regression of Barrett's epithelium either does not occur or occurs to a very limited extent even with the highly effective and well-controlled acid suppression that is achieved in these academic centers with strict regimens for acid suppression.

A more basic question about these studies is whether the definition that is used for regression of Barrett esophagus is correct. There are many reasons to reject the basic premise of these studies that a decrease in the endoscopic surface area or length of the columnar-lined esophagus represents regression. The only thing that is being tested is the conversion of the surface epithelium from columnar to squamous. This occurs as a surface overgrowth. In columnar-lined esophagus, such surface overgrowth occurs over retained columnar epithelial elements under the squamous surface (Fig. 16.1). Squamous surface overgrowth may cause a decrease of the surface area of endoscopic columnar-lined esophagus without actually decreasing the amount of columnar-lined epithelium. Endoscopic regression of Barrett esophagus is almost always a myth, and a dangerous one at that because dysplasia and adenocarcinoma can occur in the glands under the squamous epithelium (Fig. 16.2).

These studies would have had no value even if they had shown that the area of *surface* columnar epithelium is reduced significantly by reflux. Surface columnar epithelial cells are terminally differentiated nonmitotic, Ki67-negative cells that are not susceptible to carcinogenesis. The target cell for carcinogenesis in metaplastic columnar epithelium is the proliferating stem cell pool at the base of the foveolar pit; if these are retained, the conversion of the surface from columnar to squamous will have no bearing on the risk of adenocarcinoma. The historical data clearly show that the incidence of adenocarcinoma has continued to increase despite increasing acid suppression. These powerful data will never be negated by studies that demonstrate an irrelevant "regression of Barrett esophagus" defined in a meaningless manner.

As we have discussed before, *true regression of Barrett esophagus* must be defined by some measure showing a decrease in the number of target cells for carcinogenesis. This is shown by a reversal of intestinal metaplasia to cardiac mucosa and a conversion of cardiac mucosa to oxyntocardiac mucosa (Fig. 16.3). It is not necessary for squamous regeneration to occur; in fact, it is probably better if squamous regeneration does not occur because this masks underlying columnar epithelium at risk. Establishing this type of regression requires histologic mapping of the area of

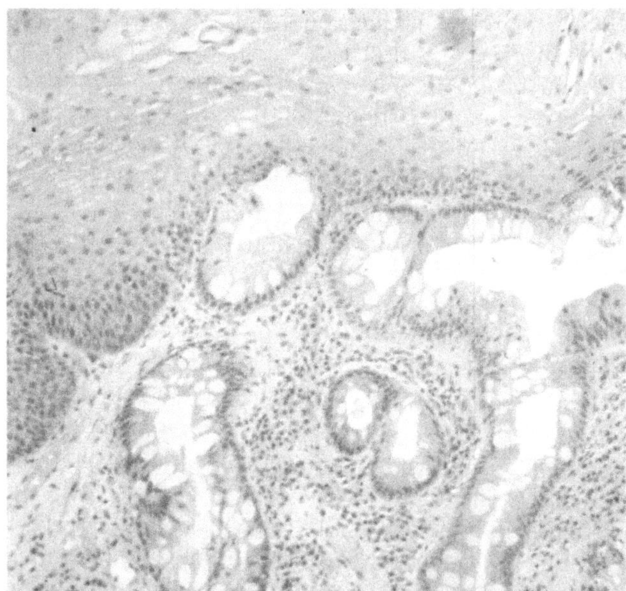

FIGURE 16.1 Squamous epithelial surface overgrowth over Barrett epithelium in the lamina propria. This patient will have endoscopic "regression" but not histologic regression. (Cross reference: Color Plate 4.5)

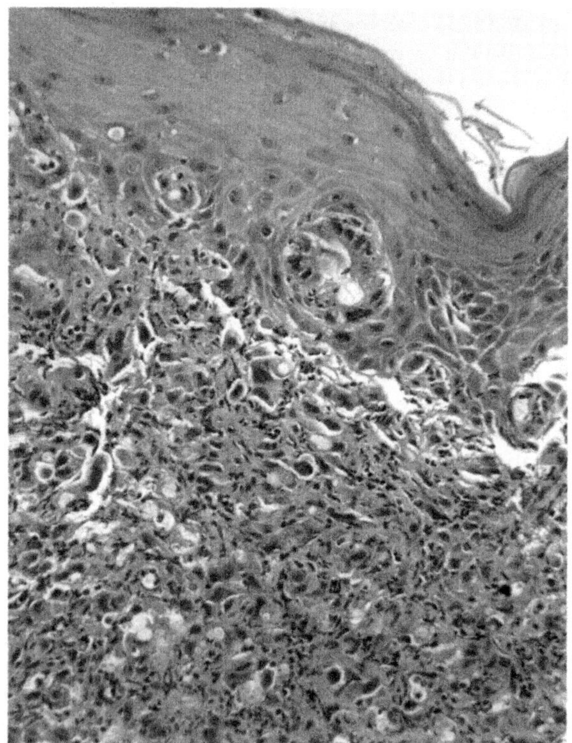

FIGURE 16.2 Poorly differentiated adenocarcinoma occurring under a squamous epithelial surface. The lamina propria under the normal squamous epithelium shows malignant cells.

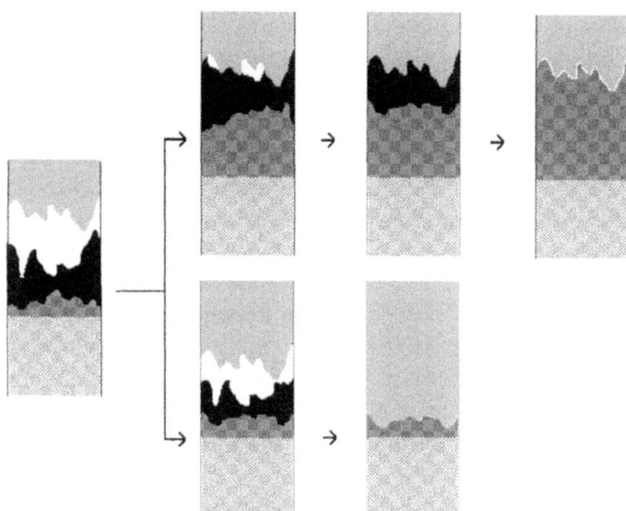

FIGURE 16.3 True regression of Barrett esophagus. In the upper frame, the amount of intestinal metaplasia (white) is seen to decrease and the amount of oxyntocardiac mucosa (dark gray stippled) increases. In this patient, regression has occurred without any decrease in the extent of columnar-lined esophagus. In the lower frame, the amount of intestinal metaplasia as well as cardiac mucosa (black) has decreased, resulting in a shorter columnar-lined esophagus. In these cases, it is difficult to know whether residual glandular epithelium is lurking invisibly under the normal squamous epithelium.

columnar epithelium after treatment aimed at detecting changes in the target cell population (i.e., intestinal metaplasia). Such studies have not been done largely because the level of understanding of cellular pathology, even among the foremost physicians in the field, is at a very low level. There is not even the recognition that "Barrett esophagus (= intestinal metaplasia)" is not the same as "columnar-lined esophagus (= all metaplastic columnar epithelial types)" in these studies. Until we recognize that every cell in columnar-lined esophagus is not at risk for cancer and we understand that histologic assessment is the only way to evaluate the epithelium in any meaningful manner, we will swim in a sea of confusing data and meaningless studies.

Conclusion: Acid suppression with drugs, even when highly effective, has little or no effect in decreasing the amount of intestinal metaplasia in columnar-lined esophagus.

Does Acid Suppressive Drug Therapy Decrease the Risk of Adenocarcinoma?

In the AGA Chicago workshop (12), the 18 experts dealt with the question of whether acid suppression had an effect on the risk of adenocarcinoma in patients with Barrett esophagus. Unfortunately, they combined the acid suppression of drug therapy and antireflux surgery in this question without considering the two separately.

Statement 35 considered by the experts was as follows: "Normalization of esophageal acid exposure by acid suppression (in patients with Barrett esophagus) reduces the risk for development of esophageal adenocarcinoma" (p. 322).

Seven of 18 experts rejected this statement completely, 9 rejected it with reservation, and 2 accepted it with major reservation. In the discussion of this statement:

> The group acknowledged that there are no clinical data that indicate that normalization of esophageal acid exposure by acid suppression reduces the risk of development of esophageal adenocarcinoma in patients with Barrett esophagus. Observations that partial regression of metaplastic mucosa may be induced by suppression of acid . . . with proton pump inhibitors (PPIs) . . . provide some plausibility for the notion that PPIs may be able to reduce the cancer risk. In the shorter term, intermediate markers of biopsy specimens from patients with Barrett esophagus in whom intra-esophageal pH had been normalized on PPI therapy showed decreased cell proliferation and improved differentiation. In contrast, incomplete acid suppression may allow short episodes of acid reflux that may lead to epithelial changes and could select for poorly differentiated cells with increased proliferative potential. . . . The majority of workshop members felt that, in the absence of clinical data, the statement should be rejected. (p. 322)

Let us consider the molecular changes that the statement alludes to as possibly decreasing the risk of cancer with effective acid suppression. Fitzgerald et al. (14) exposed cultured cells from biopsies of endoscopically defined Barrett esophagus (BE) and showed that continuous exposure to acid increased cell differentiation (defined as increased expression of villin, which correlated ultrastructurally with brush border maturation) and blocked cell proliferation in Barrett's epithelium. In contrast, pulse-type acid exposure did not alter villin expression but caused enhanced cell proliferation.

In a follow-up study by this group, Ouatu-Lascar et al. (15) evaluated these same elements 6 months after treatment with the proton pump inhibitor lansoprazole, evaluated for control of intraesophageal pH by 24-hour pH testing. Compared to baseline, the proliferation index had decreased and villin expression had increased in 24 patients whose esophageal pH had normalized; 15 patients who had persistent abnormal intraesophageal pH by the 24-hour pH test showed no significant change in the level of proliferative activity or villin expression. Histologic examination showed that there were no differences in the occurrence of dysplasia after 6 months in either group. The authors concluded that there is an inverse relationship between villin and proliferative cell nuclear antigen in Barrett epithelium. Dysplasia was unrelated to villin expression and well correlated with PCNA expression. Effective intraesophageal acid suppression favors differentiation and decreases proliferation. They commented that the intriguing possibility that acid suppression can be used to prevent dysplasia remains to be explored.

The decrease in the proliferative rate in columnar-lined esophagus that accompanies acid suppression suggests only that acid is an important factor in causing cell damage in columnar-lined esophagus. Any decrease in the proliferative rate will tend to decrease the probability of mutations and is a theoretical basis by which acid suppression can decrease the risk of and progression to cancer. Because this is likely to be a relatively minor factor compared with continued exposure of the target cells to mutation-causing refluxate molecules and the possibility that carcinogens are promoted by a more alkaline environment, the overall effect of acid suppression is not likely to reduce the risk of cancer. The exact effect of these factors on cancer risk can only be determined by evaluating the practical effect of treatment on the risk.

It is difficult to find accurate data addressing the question of the impact of acid suppression on the occurrence of dysplasia or cancer. The reason for this is that it would be difficult to randomize patients with reflux disease or Barrett esophagus into groups that do and do not receive acid suppressive treatment. The only patients who may not receive acid suppression are those who never had symptoms. These patients will not be found without screening for Barrett esophagus (11).

In a somewhat convoluted study attempting to evaluate this question, El-Serag et al. (16) reported that treatment with proton pump inhibitors significantly decreased the incidence of dysplasia in patients with Barrett esophagus. This study determined treatment received by 236 patients over a 20-year period (1981 to 2000) by an analysis of pharmacy information and research files. The receipt and duration of H2RA or PPI use was compared between patients with and without dysplasia. During 1170 patient-years of follow-up, 56 patients developed dysplasia, 14 of whom had high-grade dysplasia. The cumulative incidence of dysplasia was significantly lower among patients who received PPI after BE diagnosis than in those who received no therapy or H2RA. This study is difficult to understand for many reasons. The study accrued fewer than 12 patients per year, it had a very low overall incidence of high-grade dysplasia, and the method of establishing therapy is unusual and unreliable.

The theoretical basis for any effect of acid suppression in impacting progression of Barrett esophagus to cancer would depend on three possible mechanisms. First, removal of acid may decrease the damage factor and decrease the proliferative rate of Barrett epithelial cells. Any reduction in proliferative rate decreases the likelihood of mutation and would tend to decrease carcinogenesis. Second, the removal of acid would decrease cancer if acid were the actual carcinogen or if the alkalinization of gastric juice decreased the concentration or inhibited the effectiveness of the carcinogen. It is highly unlikely that a simple molecule such as an H^+ ion is carcinogenic. Evidence suggests that the alkalinization of gastric juice actually increases the concentration and effectiveness of carcinogenesis rather than inhibiting it (Fig. 16.4; also see Chapter 10; 15).

The practical reality of the relationship between acid suppression and cancer is self-evident. The present treatment of reflux disease and Barrett esophagus with acid suppression is associated with a rapidly increasing incidence of esophageal adenocarcinoma even as symptoms and erosive esophagitis are well controlled. Weston et al. (17), in a prospective study of 108 newly diagnosed patients with Barrett esophagus that was followed with a standard endoscopic and bioptic surveillance protocol for a period of 12 to 101 months (mean 39.9 months) for a total of 361.8 patient-years, reported that 5 patients developed adenocarcinoma and 5 patients developed multifocal high-grade dysplasia. This translates to an incidence of 1 per 71.9 patient-years for both adenocarcinoma and multifocal high-grade dysplasia. A simpler way of expressing this result is that 10 of 108 patients (9.3%) developed adenocarcinoma or high-grade dysplasia that precipitated esophagectomy within 8.4 years. This is the fate of a

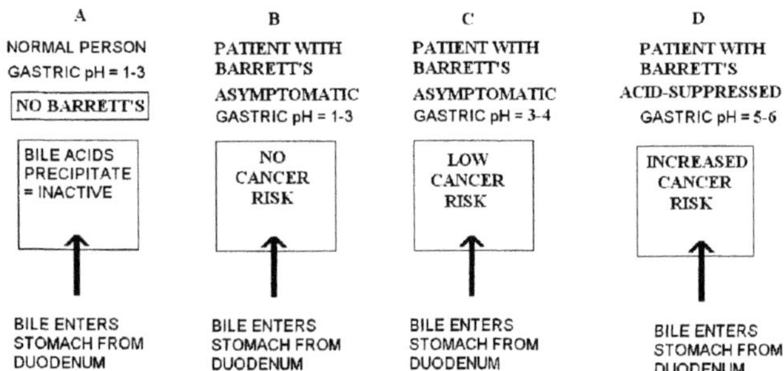

FIGURE 16.4 The gastric juice composition of four patients. Patient A has no Barrett esophagus and has a normal gastric pH. Patients B and C are patients with asymptomatic Barrett esophagus. Patient C, who has a gastric pH of 3 to 4, has a low cancer risk compared to patient B who has no risk because no carcinogens are generated at normal pH from bile acids. Patient D is a symptomatic patient with Barrett esophagus who is on acid suppressive drugs. The increased pH favors increased carcinogen production from bile acids and increases the risk of adenocarcinoma of the esophagus.

patient diagnosed with Barrett esophagus and treated by the best medical treatment that is available.

Esophageal adenocarcinoma develops in approximately 0.5% of patients with Barrett esophagus per year (18, 19) and has increased by 350% since the 1970s (1). It would be a stretch to say that this dramatic increase in esophageal adenocarcinoma would have been even more rapid if there was no acid suppression, which is the requirement if acid suppression had an inhibiting influence on cancer progression.

Conclusion: The present practice of using acid suppressive agents to treat patients with Barrett esophagus is a failure in terms of preventing cancer.

DOES ACID SUPPRESSIVE DRUG THERAPY INCREASE THE RISK OF ADENOCARCINOMA?

In the AGA Chicago Consensus Workshop (12), the question of whether acid suppressive drugs increased the likelihood of cancer was not directly addressed. However, one phrase in the experts' discussion of statement 35, quoted earlier, raises an interesting and troublesome question. They stated that "incomplete acid suppression may allow short episodes of acid reflux that *may lead to epithelial changes and could select for poorly differentiated cells with increased proliferative potential*" (p. 322). Are the experts saying that incomplete acid suppression may increase the risk of progression to cancer?

If the AGA workshop formulated a statement that evaluated the possibility that acid suppressive drugs increased the risk of esophageal adenocarcinoma, they would have found strong evidence in the most influential study published on the epidemiology of esophageal and gastric cardiac adenocarcinoma.

Literature Review

Lagergren J, Bergstrom R, Lindgren A, Nyren O. Symptomatic gastroesophageal reflux as a risk factor for esophageal adenocarcinoma. N Engl J Med 1999;340:825–831.

Lagergren et al.'s epidemiologic study, which established that patients with symptomatic reflux had an increased risk of esophageal adenocarcinoma, has a significant statement hidden in the report without much emphasis. They reported:

> We compared the risk of esophageal adenocarcinoma among persons who used medication for symptoms of reflux at least five years before the interview with that among symptomatic patients who did not use such medication. The odds ratio was 3.0 (95 per cent confidence interval, 2.0 to 4.6) without adjustment for the severity of symptoms and 2.9 (95 per cent confidence interval, 1.9 to 4.6) with this adjustment. (pp. 827–829)

This shows that there is a three times increased risk among symptomatic patients if they had taken acid suppressive medication than if they had not taken such medication. The magnitude of the risk does not change significantly when an adjustment is made for the severity of symptoms. This suggests that the increased risk in this group is primarily due to the use of acid suppressive drugs in symptomatic patients.

Lagergren et al.'s study (20) is remarkable for its attention to detail. It is amazing that we have never heard this item of data, which has such incredible significance presented by anyone at any meeting that we have attended. In effect, *this item of data provides epidemiologic proof that when a symptomatic patient uses acid suppressive medications, he or she has a three times greater risk of developing esophageal adenocarcinoma than a similarly symptomatic person who did not use acid suppressive drugs.*

The way this item of data is presented in Lagergren et al.'s study is unusual. To make this statement, the authors must have collected detailed data on the medications used by the patients. The methods section of the study provides a detailed account of information collected about "several potential confounding factors" (p. 826). These include age, sex, body mass index, smoking history, alcohol history, socioeconomic status, dietary intake of fruit and vegetables, energy intake, whether the patients worked in a stooped posture, physical activity at work, and physical activity during leisure time. The way each of these factors was assessed in the questionnaire is given in minute detail (e.g., energy intake was estimated by the amounts of seven kinds of dishes eaten at each meal!). These factors were subject to multivariate analysis. (The authors' Table 3 in their article details the results of this analysis, essentially showing no impact on the odds ratio for each of these factors.) There is absolutely no mention of medication history in either the methods or this table. The ability of the authors to make a statement on the effect of medication use must mean that this information was sought in the questionnaire. Why is this fact not noted in their methods section? Medication use should have been detailed into type of medication and duration of therapy. It is unfortunate that data are not provided about the effect of acid suppressive medication use in the 40% of asymptomatic patients who developed esophageal adenocarcinoma. It is an editorial lapse that the authors were not asked to provide these extremely important data. The fact that the data were present for the symptomatic group must mean that they were also available for the asymptomatic group, unless the questionnaire loop skipped these questions if the patient was not symptomatic.

The importance of the finding that the use of acid suppressive medication is associated with an increased risk of esophageal adenocarcinoma in this study is that it potentially negates the main finding in this study that symptomatic reflux is associated with esophageal adenocarcinoma. If one assumes that the likelihood of a person taking acid suppressive medication increases with the increasing severity of reflux symptoms, the relationship between symptomatic reflux and adenocarcinoma may be more the result of acid suppressive medications than the reflux.

In fact, logic suggests that it is very unlikely that symptomatic reflux bears a strong relationship to the incidence of esophageal adenocarcinoma (Table 16.2). Severe reflux was common in historic times. In 1929, Chevalier Jackson (21) claimed to have seen 88 cases of esophagitis in 4000 consecutive endoscopies (he must have been seeing erosive esophagitis); Allison and Johnstone's original report on columnar-lined esophagus in 1953 (22) is based on the findings in 125 patients with esophageal ulcers and stenosis. Despite the frequent occurrence of severe reflux in those times, the incidence of adenocarcinoma was extremely rare; even if symptomatic reflux was less frequent then than now (which is unlikely), the frequency at that time was sufficient to make esophageal adenocarcinoma fairly common. This suggests that some factor other than symptomatic reflux is the cause of the increased incidence of esophageal adenocarcinoma.

Lagergren et al.'s study looked at a large variety of comorbid factors and concluded that none of these had any significant impact on the odds ratios that existed between symptomatic reflux and adenocarcinoma. When taken in conjunction with historical data, Lagergren et al.'s data strongly suggest that acid suppressive drug usage may be a part of the explanation for the increasing incidence of esophageal adenocarcinoma and its association with symptomatic reflux.

We have discussed a mechanism whereby incomplete acid suppression can theoretically increase the carcinogenic environment in the stomach. Incomplete acid suppression is associated gastric pH levels that are intermediate between normal (<4) and complete acid suppression (>6). At pH levels between 4 and 6, bile salt metabolism generates molecules that have been shown to have carcinogenic effects in animals (see Fig. 16.4 and Chapter 10).

It is quite clear that the use of acid suppressive drugs is not the primary reason for the increase in carcinogenesis in reflux disease. Most patients who

TABLE 16.2 Historical Relationship Between Symptomatic Reflux, Use of Acid Suppressive Drugs, and the Incidence of Columnar-Lined Esophagus, Barrett Esophagus (= Intestinal Metaplasia), and Reflux-Induced Adenocarcinoma

Characteristic	1929–1953	1980–1997	Conclusion
Symptomatic reflux disease	Common	Common	Similar in the two periods
Acid suppressive medication	Did not exist	Increasingly prevalent	Marked increased in usage of acid suppressive drugs
Columnar-lined esophagus	Common	Common	Similar prevalence in the two periods
Intestinal metaplasia	Rare	Common	Marked increase in prevalence
Esophageal adenocarcinoma	Very rare	More common	Marked increase in prevalence

Lagergren et al. study finding	Comment
Symptomatic reflux is causally associated with esophageal adenocarcinoma	Conclusion is highly unlikely to be correct because it cannot explain why this causal relationship did not exist in historical times when symptomatic reflux was common
Risk of esophageal adenocarcinoma increased in patients using acid suppressive medications	It is highly likely this relationship is the primary association rather than symptomatic reflux
Risk of adenocarcinoma increases with severity and duration of symptoms	The threefold increase in adenocarcinoma remained when adjusted for severity of symptoms; use of acid suppressive drugs is more likely as reflux symptoms increase in severity

Note: Careful study of the table shows that it is highly unlikely that symptomatic reflux disease is primarily associated with reflux-induced adenocarcinoma because of the rarity of cancer in historic times when symptomatic reflux was common. It is much more likely that the relationship between symptomatic reflux and adenocarcinoma operated through a promoter action of acid suppressive drugs on reflux-induced carcinogenesis.

develop reflux-induced adenocarcinoma have never had symptomatic reflux and have never taken acid suppressive drugs. In some unknown way, the carcinogenic milieu in reflux disease has increased. The role of acid suppression is to *promote* this natural carcinogenic mechanism (Chapters 9, 10, 15; Table 16.3).

The amount of promotion of carcinogenesis by acid suppressive drugs that is required to cause a large increase in the absolute number of cases of reflux-induced adenocarcinoma is very small. This is because the population at risk (patients with reflux disease who take acid suppressive drugs) is very large. It is probable that 30% to 40% of people in the United States take acid suppressive drugs to account for the approximately $12 billion sales per year of these drugs. This amounts to more than 100 million people. A very slight tilt in the balance toward adenocarcinoma (the threefold increase shown in Lagergren et al.'s study) would be expected to cause a large increase in the absolute number of patients developing cancer.

If it is possible that incomplete acid suppression can promote the progression to cancer, and because most acid suppression practiced in the general community is incomplete, should the experts at the AGA Chicago Consensus Workshop not have considered the next logical question: "Because we know that acid suppression does not reduce the risk for adenocarcinoma, and because there is evidence that acid suppression may promote cancer, should we consider not using acid suppressive agents to treat Barrett esophagus?" This question, however, was not considered. Of course, the answer would be that there is no other way to control the patients' symptoms; but this answer is wrong—antireflux surgery is available.

TABLE 16.3 Pathophysiologic Effects Caused by Acid Suppressive Drugs That May Contribute to Their Action as Promoters of Carcinogenesis

1. Increase in gastric pH—objective of acid suppression and the mechanism of action of these drugs in controlling symptoms, squamous epithelial erosions.
2. Increase in alkalinity of the refluxate inhibits the conversion of cardiac mucosa in the esophagus to oxyntocardiac mucosa (Chapter 9).
3. Increase in alkalinity of the refluxate promotes the conversion of cardiac mucosa in the esophagus to intestinal metaplasia, increasing the likelihood of Barrett esophagus in columnar-lined esophagus (Chapter 9).
4. Alkalinity increases the amount of intestinal metaplasia occurring in columnar-lined esophagus, increasing the number of target cells available for carcinogenesis (Chapter 9).
5. Alkalinity promotes the occurrence of intestinal metaplasia at increasingly distal regions of the esophagus, increasing the access of intestinal metaplasia to the carcinogens in gastric juice (Chapter 9).
6. Alkalinity promotes the conversion of bile acid derivatives to carcinogenic molecules in the stomach, increasing carcinogen concentration in the refluxate (Chapters 10 and 15).

There are no randomized clinical studies that have established that cancer incidence is increased in patients who are being treated with acid suppressive treatment. There can be no such studies because of the difficulty in randomizing symptomatic patients to acid suppression and no treatment. This may become possible in the future if screening for Barrett esophagus is undertaken; in this setting, it is very feasible to randomize asymptomatic patients to acid suppression and no treatment to test the question of whether acid suppression results in a decrease, increase, or has no effect on the incidence of cancer in Barrett esophagus.

The increasing incidence of esophageal adenocarcinoma that is well documented by powerful epidemiologic data has occurred coincidentally with an increasing use and effectiveness of acid suppressive agents. While this is not proof that the two are related, it is certainly not proof that the two are unrelated. In fact, the safest course is to assume that acid suppressive drugs are responsible for the increasing incidence of esophageal adenocarcinoma until a reasonable alternative explanation for the increase in cancer is found. The minimum action that is needed is to recognize that the safety of these drugs is sufficiently unproven that they be withdrawn as over-the-counter drugs and come into the category of potentially dangerous drugs that must be used with caution and by prescription only. This is a failure of public health as well as a failure within the pharmaceutical industry. The governmental agencies that approved these drugs for sale directly to patients without prescription share the blame if these drugs are ultimately proven to promote cancer.

Conclusion: There is significant evidence suggesting that acid suppressive drug use increases the incidence of cancer progression in patients with symptomatic reflux disease.

ANTIREFLUX SURGERY AS TREATMENT FOR REFLUX DISEASE

Antireflux surgery is an effective method of treating gastroesophageal reflux disease. There are many types of operations, the selection and indications of which are complex surgical issues. In this section, we will concentrate on the rationale and the theoretical and practical effects antireflux surgery has on the cellular events of gastroesophageal reflux disease.

In the past, antireflux surgery was largely restricted to patients with severe reflux who are referred to a surgical unit when medical therapy with acid suppressive drug therapy in maximum dosage has failed to control symptoms. Recently, antireflux surgery has become increasingly used as a primary treatment for reflux disease that represents a reasonable alternative to lifelong acid suppressive drug therapy.

The main difference between acid suppressive drugs and antireflux surgery is that while acid suppressive drugs aim at specifically removing H^+ ions from the gastric refluxate, antireflux surgery aims at preventing reflux completely (Fig. 16.5). While a patient who is on acid suppressive drugs will continue to have all of his or her target cells exposed to all the molecules in the refluxate except acid, successful antireflux surgery prevents all reflux and stops all interactions between refluxate molecules and target cells.

It should be very clear from this that the acid suppressive drugs have a theoretical limit to their effectiveness that is determined by an ability to reverse and prevent only acid-induced cellular events. They have no capability to prevent or control cellular changes that are caused by refluxate molecules other than acid. While acid suppressive agents are highly effective in controlling symptoms and healing erosive esophagitis and preventing strictures, they can have only a secondary change on the cellular interactions that convert cardiac mucosa to intestinal metaplasia and carcinogenesis in intestinal metaplasia by altering the pH. We have provided evidence that suggests that alkalinizing gastric refluxate actually promotes the generation of intestinal metaplasia in cardiac mucosa and increases carcinogen concentration in gastric juice (Fig. 16.4).

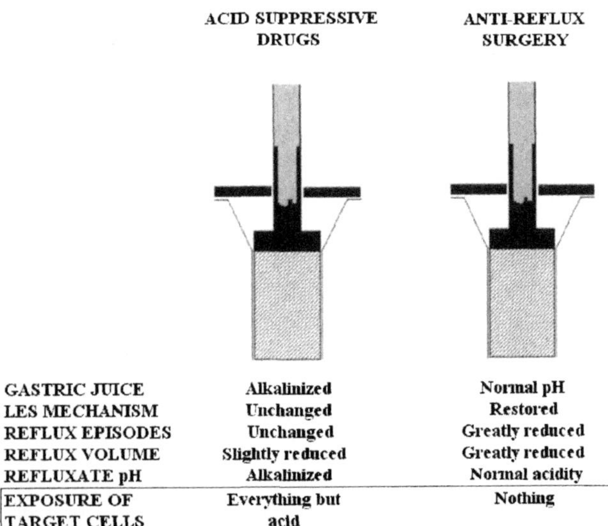

ACID SUPPRESSIVE DRUGS
ANTI-REFLUX SURGERY
GASTRIC JUICE
LES MECHANISM
REFLUX EPISODES
REFLUX VOLUME
REFLUXATE pH
EXPOSURE OF TARGET CELLS
Alkalinized
Unchanged
Unchanged
Slightly reduced
Alkalinized
Everything but acid
Normal pH
Restored
Greatly reduced
Greatly reduced
Normal acidity
Nothing

FIGURE 16.5 Fundamental difference between the effect of acid suppressive drug therapy and antireflux surgery. Acid suppressive drugs simply change acid reflux to nonacid reflux of similar volume. The target cells in the esophagus continue to be bombarded by all refluxate molecules except H^+. Antireflux surgery eradicates all reflux when it is successful.

Antireflux surgery, on the other hand, by stopping reflux has a theoretical effectiveness that is unlimited. It completely removes the target cell from the molecules that cause all the cellular changes that result in progression of the disease to adenocarcinoma. Theoretically, if a perfect antireflux procedure can be developed and all patients with reflux disease have this operation at an early stage of disease before the required genetic changes required for adenocarcinoma have occurred, esophageal adenocarcinoma can be prevented. This is not practically feasible, but *it is rare to encounter a human cancer where a theoretically certain basis for prevention exists.* The gap between theoretical absolute cancer prevention and the practical effect of antireflux surgery in preventing reflux-induced adenocarcinoma is a function of the lack of technical perfection in preventing reflux and the performance of the surgery after genetic events required for adenocarcinoma have already been established in the target cells. The latter is not a problem; genetic problems required for adenocarcinoma are slow to develop, and there is a long window of opportunity for effective antireflux surgery in the vast majority of patients. It is really only the technical imperfection, lack of availability, and high cost of antireflux surgery that is keeping us from having an effective cancer prevention method.

The arguments against antireflux surgery are as follows:

1. *That antireflux surgery is frequently not successful in stopping reflux.* This is true; there is a significant failure rate for antireflux surgery. However, it is likely that this is a function of experience. As the frequency of antireflux surgery increases and surgical expertise widens, there is a likelihood that the success rates will increase. In a study of inpatient surgeries in 503 nonfederal acute care hospitals in California, Liu et al. (23) showed that the number of surgical procedures increased 20.4% in the state of California, from 135,795 in 1990 to 163,468 in 2000. During this same period, the number of antireflux surgeries increased by 258%. As with any other surgery, success is more likely in the hands of surgeons who perform a large number of these procedures. When evaluating the effectiveness of antireflux surgery, therefore, it is important to use data from centers performing large numbers of antireflux operations. The probability is that these

numbers will be closer to the ultimate success rates when the number of antireflux procedures performed permit surgeons to specialize in this procedure. The situation is not unlike coronary artery bypass surgery, which has shifted from being a dangerous procedure limited to surgeons with expertise to a routine and highly successful procedure in the hands of virtually every cardiothoracic surgeon. The technical requirement for antireflux surgery is much less than coronary artery bypass surgery.

2. *That antireflux surgery has significant side effects.* This is also true; patients have a significant risk of dysphagia and slipped fundoplications. But it is also true that these side effects will very likely decrease as the procedure is used more. The main side effect of antireflux surgery is the occurrence of dysphagia resulting from a wrap that is too tight. The necessity to prevent dysphagia makes it impossible to create an antireflux procedure that completely eliminates reflux. The aim of the surgery is to establish a balance between substantially decreasing reflux and ensuring that swallowing is unimpeded. As the use of this technique increases, it is also very likely that technical advances will occur that increase the effectiveness and reduce the complications of antireflux surgery. Even today, the surgery is highly effective; the advances necessary for perfection are relatively small.
3. *It is more expensive.* This is arguable. Theoretically, a successful antireflux operation will require no acid suppressive medication and potentially limited surveillance. This is not true in practice where approximately 10% to 20% of patients are placed on acid suppression and kept under surveillance. Postoperative acid suppression is overused after antireflux at the present time. These drugs should only be used after it has been established by 24-hour pH testing that any postsurgery symptoms are actually caused by continuing acid reflux. The exact cost will depend on the cost of surgery versus the reduction in the need for acid suppressive drugs and surveillance. This is not easy to assess and is likely to change continuously as the success rate of antireflux surgery improves.

Although antireflux surgery is increasing in acceptance and use in a rapid manner, it is still largely reserved for patients who have either severe complicated disease or have failed medical therapy. As such, antireflux surgery in general treats a population that has more severe disease than most populations who are maintained on acid suppressive drug therapy. As the number of antireflux surgeries increases and technical ability with the procedure becomes more widespread, the procedure will become available as a treatment option for patients who are diagnosed earlier or who have fewer complications. If one extrapolates the concept of dilated end-stage esophagus to this situation, it is likely that antireflux surgery is performed today largely in the setting of a severely damaged and dilated esophagus that is not recognized at surgery. This has the potential for maximum complications and failures. It is very likely that the failure rate will decrease when the procedure is performed in patients with less damaged esophagi.

We will assess antireflux surgery as a treatment for reflux disease using the same series of questions that we used for acid suppressive drug therapy (Table 16.4).

Is Antireflux Surgery Successful in Preventing Reflux?

Antireflux surgery when it is successful, prevents reflux. In these patients, reflux can be assessed by the standard 24-hour pH test if the patients are not on acid suppressive medications postoperatively. If the patients are on acid suppressive drugs, these must be

TABLE 16.4 Comparison of Acid Suppressive Drug Treatment and Antireflux Surgery in Their Ability to Influence Different Elements of Reflux Disease

	Acid suppressive drugs	Antireflux surgery
Does it improve lower esophageal sphincter function?	No	Yes
Does it prevent reflux?	No	Yes
Does it control symptoms?	Yes	Yes
Does it heal erosions/ulcers?	Yes	Yes
Does it normalize the 24-hour pH test?	Yes	Yes
Does it normalize the impedance test?	No	Yes
Does it normalize the Bilitec test?	No	Yes
Does nonacid reflux continue?	Yes	No
Does it decrease columnar-lined esophagus length	Maybe slightly	Maybe slightly
Does it reverse intestinal metaplasia?	No evidence	Yes; mainly short segments
Does it decrease intestinal metaplasia?	Unknown	Unknown
Does it promote cancer?	Yes	No
Does it prevent cancer?	No	Yes

withdrawn for 2 weeks before the test is done. Because patients have normal intrinsic acid secretion, pH measurement in the esophagus is a sensitive marker for reflux.

Literature Review

Eubanks TR, Omelanczuk P, Richards C, Pohl D, Pellegrini CA. Outcomes of laparoscopic anti-reflux procedures. Am J Surg 2000;179:391–395.

This is a study of 640 patients who had antireflux surgery between 1993 and 1999 at the University of Washington. Of these, 228 (36%) agreed to repeat manometry and 24-hour pH monitoring 8 to 12 weeks postoperatively and are the subject of this study. Symptom resolution was defined as a frequency of symptoms less than once per week. Normal acid exposure consisted of a distal esophageal pH below 4 less than 4% of the time and a DeMeester composite score of less than 14.7. The accuracy of symptom scoring was calculated using acid exposure as the standard.

The primary symptom was improved in 93% of the 228 patients. Acid exposure was reduced from a preoperative DeMeester score of 71 to 16 ($p < 0.05$). Eighty percent of patients had normalization of acid exposure postoperatively. Heartburn was the only symptom to have a significant correlation with acid exposure in the postoperative period ($p < 0.05$). Heartburn resolved in 181 patients, 168 of whom had normal acid exposure (true negative). Thirty-eight patients without symptoms had abnormal acid exposure (false negative). Nine patients had persistent heartburn with abnormal acid exposure (true positive), whereas 13 patients had persistent heartburn with normal acid exposure (false positive). Thus, the positive predictive value of heartburn was 43%, the negative predictive value was 82%, and the overall accuracy was 78%.

The authors concluded that antireflux surgery improves both the symptoms of reflux disease and the degree of acid exposure as measured by pH monitoring. The most accurate symptom for predicting acid exposure in the postoperative period is heartburn. Although the absence of heartburn postoperatively is fairly reliable at predicting normal acid exposure on pH testing, the presence of heartburn warrants postoperative pH monitoring, as more than half of these patients will have normal acid exposure.

This study (24) shows that reflux is controlled to levels that are normal by criteria of normalcy of the 24-hour pH test in 80% of patients. Even those patients who continued to have an abnormal 24-hour pH test frequently showed a reduction of acid exposure from the preoperative level. In a given patient, the reduction of acid exposure in the 24-hour pH study after surgery is a measure of the effectiveness of the antireflux procedure because it can be assumed that gastric acid secretion did not change significantly during this short period of time between the two tests. Antireflux surgery is therefore effective in decreasing the amount of reflux in nearly all patients and bringing the reflux amount to less than what is regarded as normal by the 24-hour test criteria in 80%. Booth et al. (25) and Anvari et al. (26) have shown by 24-hour pH testing that reflux remains controlled after a median follow-up of 5 years, showing that the reflux control is long-lasting and probably permanent.

Mason et al. (27), in an animal experiment, provided insight to the mechanism by which antireflux surgery is controlled. In Chapter 11, the importance of sphincter shortening secondary to gastric distension in causing transient lower esophageal sphincter relaxations and reflux episodes was discussed. This is believed to be the major mechanism of reflux.

Literature Review

Mason RJ, DeMeester TR, Lund RJ, Peters JH, Crookes P, Ritter M, Gadenstatter M, Hagen JA. Nissen fundoplication prevents shortening of the sphincter during gastric distention. Arch Surg 1997;132:719–724.

This study tested the manometric changes in the lower esophageal sphincter caused by fluid distension of the stomach before and after a Nissen fundoplication in 10 baboons. The experiment included continuous manometric evaluation of the esophagus, cardia, and stomach during distention of the stomach with water. The sphincter characteristics were assessed by a slow motorized pull-through of the lower esophageal sphincter after each successive intragastric increment of 50 mL of water.

The outcomes that were evaluated before and after Nissen fundoplication were (a) the lower esophageal sphincter length and (b) the frequency of reflux episodes after each volume increment. The pressure and intragastric volume at the yield point are defined as the point of permanent loss of the gastroesophageal pressure gradient.

Gastric distention of the stomach with water resulted in a progressive decrease in lower esophageal sphincter length. Before fundoplication, the median length (+/− interquartile range) of the lower esophageal sphincter decreased by 1.5 mm (+/− 0.3 mm) for every 1 mmHg increase in gastric pressure. After fundoplication, this shortening decreased dramatically to 0.2 mm (+/− 0.1 mm) for every 1 mmHg increase in gastric pressure ($p < .02$).

The number of reflux episodes per minute produced by gastric distension also decreased after fundoplication from 2.19 (median; interquartile range [IQR]: 1.36, 3.11) to 0 (IQR: 0, 0.59). This was related to a higher intragastric pressure required to produce reflux, which was 3.60 mmHg (median; IQR: 2.23, 6.64) before fundoplication compared with 8.09 mmHg (IQR: 4.07, 13.70) after fundoplication. The pressure and volume at which the sphincter yielded were also significantly increased after fundoplication indicating an increased ability of the sphincter to withstand gastric distension. The yield intragastric pressure was 13 mmHg (IQR: 5, 14) and 39 mmHg (IQR: 29, 65) before and after fundoplication, and the yield intragastric volume was 825 ml (IQR: 350, 1205) and 1250 (IQR: 1100, 1850) before and after fundoplication.

The authors concluded that a Nissen fundoplication improves the competency of the lower esophageal sphincter to progressive degrees of gastric distention by its effect in preventing sphincter shortening when the stomach distends.

Parrilla et al. (5) demonstrated significant improvement in sphincter characteristics in 58 patients after antireflux surgery. Their results are as follows:

	Preop	Postop	p Value
Total length of LES (cm)	2.5 (1.0–4.2)	3.5 (2.0–5.1)	<0.001
Abdominal length of LES (cm)	0.5 (0.0–2.0)	2.4 (0.0–3.8)	<0.001
Resting pressure of LES (mmHg)	7 (0–21)	15.5 (4–31.3)	<0.001
Percentage time pH <4	19 (1.8–83)	0.6 (0–46)	<0.001

We have used the analogy of a balloon to describe how gastric distension causes sphincter shortening (Fig. 16.6). As a standard balloon is blown up beyond its normal filling capacity, the balloon overdistends, pressure increases, and the neck of the balloon shortens as it is "taken up" into the body of the balloon (Fig. 16.6B). If one now repeats blowing up the balloon after placing a constriction at the junction of the neck and body of the balloon (e.g., a wedding ring or a tight string loop), one can see the effect of a Nissen fundoplication. As the balloon distends, the shortening of the neck is prevented by a constriction at the neck in a manner that is similar to the effect of a fundoplication (Fig. 16.6C).

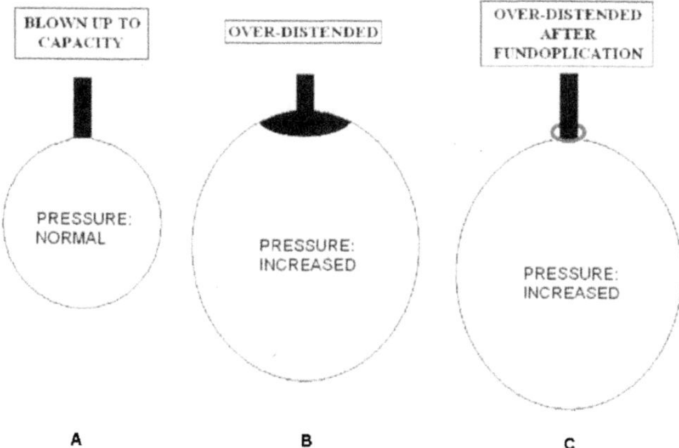

FIGURE 16.6 The balloon experiment. When the balloon is inflated beyond its normal capacity, the intraluminal pressure increases and the neck of the balloon shortens. This is the mechanism of sphincter shortening with gastric distension that is responsible for transient sphincter relaxation and reflux. If the same balloon is inflated after a constriction (e.g., a wedding band) is placed around the base of the neck, the neck does not shorten. This is the basic mechanism whereby antireflux surgery prevents reflux.

Conclusion: Antireflux surgery almost always decreases and frequently completely abolishes gastroesophageal reflux by improving the mechanism of the lower esophageal sphincter.

Is Antireflux Surgery Successful in Treating Reflux Disease?

Antireflux surgery is highly effective in controlling symptoms of reflux disease. Most studies report symptom relief in approximately 90% of patients, which is similar to that achieved with acid suppression with drugs (Table 16.5).

The studies listed in Table 16.5 show that the control of reflux symptoms occurs immediately after surgery, as shown by Eubanks et al. (24) where the symptoms and acid exposure were shown to be controlled 2 to 3 months after surgery, and is long-lasting, as shown by Bammer et al. (28) who showed control of symptoms persisted at a median follow-up of 6.4 years in 96% of patients. Bammer et al.'s series from the Mayo Clinic

TABLE 16.5 Single Institution Experiences with Antireflux Surgery

Reference	Number of patients	Median follow-up (years)	Percentage with control of symptoms
Eubanks et al. (24)	228	2–3 months	93[a]
Bammer et al. (28)	171	6.4	96
Lafullarde et al. (29)	178	6	90
Beldi and Glattli (30)	60	3.6	100
Kamolz and Pointner (31)	104	5	98
Booth et al. (25)	48	5	93[a]
Bloomston et al. (32)	100	5	90
Anvari and Allen (26)	181	5	88[a]

[a] pH testing showed decreased acid exposure.

in Florida focuses on the first 171 patients who underwent laparoscopic fundoplications at that facility and represents excellent results at the beginning of a surgical experience with a procedure. Similarly Lafullarde et al. (29), reporting from Australia on their first 178 laparoscopic Nissen fundoplications, showed good or excellent long-term results in 90% of patients after 5 years. This suggests that laparoscopic antireflux surgery is a type of operation that is relatively easy to do with success, requiring a relatively short learning curve. Excellent results have been reported from all over the world; the studies listed in the table are from the United States, the United Kingdom (25), Australia (29), Switzerland (30), and Canada (26), again emphasizing the potential for a wide application of this surgical technique.

The control of symptoms does not necessarily correlate with absence of reflux. In Eubanks et al.'s (24) study, 38 patients without reflux symptoms had an abnormal 24-hour pH test. Khajanchee et al. (33) reported that out of the 155 of 209 patients who were asymptomatic after fundoplication, 12% had an abnormal 24-hour pH test. This finding is not surprising; it is similar to the control of symptoms that occurs in patients who have incomplete acid suppression with acid suppressant drugs. In Peters et al. (6), there was good symptom control with ranitidine despite that fact that the 24-hour pH test was frequently abnormal in these patients. This group of patients who have persistent reflux despite symptom control should be considered failures of antireflux therapy because the target cells in their esophagus continue to be exposed to refluxate molecules, albeit at a lower level than before the fundoplication.

Care has to be taken when evaluating the success or failure of antireflux surgery. While symptoms caused by reflux have a high rate of control, patients not infrequently develop heartburn and many other abdominal complaints after surgery. There is a tendency to assume that these are due to recurrent reflux and precipitate the use of proton pump inhibitors. Many of these patients with postoperative symptoms do not have reflux as shown by a negative 24-hour pH study. In Eubanks et al. (24), of 22 patients who complained of heartburn, only 9 had an abnormal 24-hour pH test; the others continued to have heartburn despite a normal test. Bammer et al. (28) reported that 14% of their 171 patients were on continuous proton pump inhibitor therapy for a variety of symptoms (bloating, diarrhea, regurgitation, heartburn, and chest pain); of these, 79% were treated for vague abdominal or chest symptoms unrelated to reflux, which calls into question the indications for this therapy. Booth et al. (25) reported on 175 patients treated by fundoplication. Of the 19 patients (14%) taking regular acid suppressive medication after surgery, 8 used it for nonreflux symptoms and 12 had normal postoperative pH tests. For these reasons, studies that use the frequency of proton pump inhibitor use after antireflux surgery as an outcome of success are unreliable.

Desai et al. (34) from the University of Washington in St. Louis reported on the clinical outcome following laparoscopic Nissen fundoplication of 84 patients with nonerosive esophagitis (group 1) and 330 patients with preoperative endoscopic erosive esophagitis (group 2). The results of their study are summarized in the following table (note that all differences between preop and postop percentages are statistically significant):

	Group 1 (n = 84)		Group 2 (n = 330)	
	Preop	Postop	Preop	Postop
Medication use	98%	18%	99%	16%
Proton pump inhibitor use	86%	14%	86%	11%
Heartburn	98%	15%	93%	15%
Regurgitation	85%	7%	85%	7%
Dysphagia	45%	25%	48%	14%
Chest pain	65%	23%	54%	14%
Abdominal bloating	12%	35%	14%	32%

These authors concluded that antireflux surgery was an effective treatment in patients with nonerosive as well as erosive reflux disease. It is interesting that the number of patients in this study with erosive disease was much greater, despite the fact that nonerosive reflux disease is much more common. This is due to the fact that it is the patients with more severe reflux disease who tend to have antireflux surgery.

Antireflux surgery has been shown to be at least as good as or better than long-term proton pump inhibitor therapy in healing erosive esophagitis and preventing its recurrence.

Literature Review

Lundell L, Miettinen P, Myrvold HE, Pedersen SA, Liedman B, Hatlebakk JG, Julkonen R, Levander K, Carlsson J, Lamm M, Wiklund I. Continued (5-year) followup of a randomized clinical study comparing antireflux surgery and omeprazole in gastroesophageal reflux disease. J Am Coll Surg 2001;192:172–179.

(*Note:* The abstract of this study is reproduced almost verbatim).

This study (35) consists of 310 patients with erosive esophagitis enrolled in a prospective randomized clinical trial. There were 155 patients randomized to continuous omeprazole therapy and 155 to open antireflux surgery, of whom 144 later had an operation. Because of various withdrawals during the study course, 122 patients originally having an antireflux operation completed the 5-year follow-up; the corresponding figure in the omeprazole group was 133. Symptoms, endoscopy, and quality-of-life questionnaires were used to document clinical outcomes.

Treatment failure was defined to occur if at least one of the following criteria were fulfilled: moderate or severe heartburn or acid regurgitation during the last 7 days before the respective visit; esophagitis of at least grade 2; moderate or severe dysphagia or odynophagia symptoms reported in combination with mild heartburn or regurgitation; if randomized to surgery and subsequently required omeprazole for more than 8 weeks to control symptoms, or having a reoperation; if randomized to omeprazole and considered by the responsible physician to require antireflux surgery to control symptoms; if randomized to omeprazole and the patient, for any reason, preferred antireflux surgery during the course of the study. Treatment failure was the primary outcomes variable.

Results

When the time to treatment failure was analyzed by use of the intention to treat approach, applying the life table analysis technique, a highly significant difference between the two strategies was revealed ($p < 0.001$), with more treatment failures in patients who originally were randomized to omeprazole treatment. The protocol also allowed dose adjustment in patients allocated to omeprazole therapy to either 40 or 60 mg daily in case of symptom recurrence. The curves subsequently describing the failure rates still remained separated in favor of surgery, although the difference did not reach statistical significance ($p = 0.088$). Quality of life assessment revealed values within normal ranges in both therapy arms during the 5 years.

Conclusions

In this randomized multicenter trial with a 5-year follow-up, we found antireflux surgery to be more effective than omeprazole in controlling gastroesophageal reflux disease as measured by the treatment failure rates. But if the dose of omeprazole were adjusted in case of relapse, the two therapeutic strategies reached levels of efficacy that were not statistically different.

There is much argument about the relative efficacy of proton pump inhibitors and antireflux surgery between gastroenterologists and surgeons in meetings that we attend (where frequently Dr. Chandrasoma is the only pathologist in the audience). These arguments border on the ludicrous. Both these treatment modalities are extremely effective in controlling symptoms and healing erosive esophagitis. The difference is that proton pump inhibitor therapy, by not reducing any component of reflux except acid, does not have any impact on cellular changes that occur as a result of interactions between target cells in the esophagus and nonacid refluxate molecules. Antireflux surgery, on the other hand, will decrease all interactions between target cells in the esophagus and refluxate molecules in virtually all patients whose 24-hour pH test shows a decrease from preoperative levels. The comparison between proton pump inhibitors and antireflux surgery that is argued about only relates to the effect of acid suppression by these two techniques. Acid suppression only heals the squamous epithelium; it does not impact the progression of change in columnar metaplastic epithelia that results in Barrett esophagus and cancer.

Conclusion: Antireflux surgery is at least as good as and probably better than long-term acid suppressive therapy in controlling symptoms and healing erosive esophagitis.

Is Antireflux Surgery Successful in Preventing Barrett Esophagus?

Unlike acid suppressive drugs, there is evidence that antireflux surgery can prevent the development of intestinal metaplasia in columnar-lined esophagus. Cardiac-type mucosa in columnar-lined esophagus evolves into either oxyntocardiac mucosa or intestinal (Barrett-type) epithelium (Chapter 9). We have suggested that this is related to interactions that occur between the target cell in cardiac mucosa and molecules in the refluxate. We have suggested that the occurrence of intestinal metaplasia is favored by the following factors: (a) a high-damage environment where proliferative activity in cardiac mucosa is greatest, (b) a more alkaline (higher) pH, and (c) intermittent rather than continuous exposure to refluxate. We have also suggested that the conversion of cardiac mucosa to oxyntocardiac mucosa is favored by (a) a low-damage environment, (b) a lower (more acid) pH, and (c) a more continuous exposure to refluxate.

If we examine the state of affairs after antireflux surgery, we will see that the target cell in the esophagus is more likely to be exposed to the following conditions: (a) there is less total exposure to refluxate molecules, which decreases the likelihood of damage; (b) the refluxate is more likely to have a lower pH in those patients who are not on acid suppressive medication after surgery; and (c) the reflux is likely to be more intermittent. Two of these conditions, the decreased total exposure and low pH, would tend to cause cardiac mucosa to switch to oxyntocardiac mucosa; the more intermittent nature of the reflux would tend to favor intestinal metaplasia.

A study from Sweden by Oberg et al. (36) shows that the tendency of cardiac mucosa in the esophagus to undergo intestinal metaplasia is significantly less in patients who have antireflux surgery than in patients who are maintained on acid suppressive drugs.

Literature Review

Oberg S, Johansson J, Wenner J, Johnsson F, Zilling T, von Holstein CS, Nilsson J, Walther B. Endoscopic surveillance of columnar-lined esophagus: Frequency of intestinal metaplasia detection and impact of antireflux surgery. Ann Surg 2001;234:619–626.

This is a study of 177 patients enrolled in a surveillance program for columnar-lined esophagus. This study is unique because the indication for surveillance was simply the presence of an endoscopic columnar-lined esophagus without the requirement of intestinal metaplasia. To be included in the study, the patients needed to have had 3 or more surveillance endoscopies over a minimum of 2 years. Intestinal metaplasia was present in 108 of 177 patients in the first two endoscopies; 69 patients did not have intestinal metaplasia. Intestinal metaplasia was more prevalent in those patients with a longer columnar-lined segment; 88.9% of patients with long (>3 cm) segments had intestinal metaplasia compared with 30.5% of patients with segments shorter than 3 cm.

The incidence of intestinal metaplasia in patients with no evidence of intestinal metaplasia on the two first endoscopies was evaluated on the subsequent endoscopies and compared in patients with medically and surgically treated gastroesophageal reflux disease. The authors pointed out that sampling error may account for the absence of intestinal metaplasia and attempted to decrease this possibility by requiring two successive biopsies that are negative for intestinal metaplasia.

Patients with surgically treated reflux disease were 10.3 times less likely to develop intestinal metaplasia compared with a group receiving standard medical therapy in surveillance biopsies after the first two biopsies that were negative for intestinal metaplasia. In the medically treated group of 49 patients, there was a progressive increase in the number of patients developing intestinal metaplasia, with 80% of the patients progressing to intestinal metaplasia at 8 years. In the antireflux surgery group of 20 patients, 12 (60%) remained without intestinal metaplasia at 16 years.

This is an extremely important study because it strongly supports our theoretical construct of the molecular interaction that predicts the occurrence of intestinal metaplasia in cardiac mucosa. The fact that acid suppressive drugs were completely ineffective in preventing the development of intestinal metaplasia in cardiac mucosa strongly supports the hypothesis that this is not an acid-induced change. The fact that antireflux surgery was able to prevent this interaction in 60% of patients again fits well with the hypothesis that antireflux surgery creates a milieu around the target cell (the progenitor cell in cardiac mucosa) that decreases the likelihood of conversion to intestinal metaplasia. The fact that intestinal metaplasia is not completely prevented is explained easily by the fact that surgery does not completely abolish reflux; normalization of the 24-hour pH test is not equivalent to abolition of reflux. At the upper limit of what is considered normal in the 24-hour pH test, there is exposure of the target cell to refluxate for 4% of the 24-hour period, which is longer than 1 hour per day.

Literature Review

Tharavej C, Bremner CG, Hagen JA, DeMeester SR, Chandrasoma PT, Lipham JC, Peters JH, DeMeester TR. Transformation of cardiac mucosa after antireflux surgery. Unpublished Data.

This is a study of 61 consecutive patients who underwent antireflux surgery and who had pre- and postoperative biopsies from the gastroesophageal junction. Patients with intestinal metaplasia were excluded; all patients had cardiac mucosa in these biopsies. Matched pre- and postoperative biopsies were evaluated for the type of columnar epithelium present at the gastroesophageal junction in each patient. Regression of cardiac mucosa was defined when conversion from cardiac mucosa to oxyntocardiac or oxyntic mucosa occurred (i.e., when parietal cells were present). Nonregression was defined as the persistence of cardiac mucosa and progression as the occurrence of intestinal metaplasia (i.e., when goblet cells were present).

At a median follow-up of 2 years after surgery, regression of cardiac mucosa was demonstrated in 18 (29%) of patients; 17 to oxyntocardiac mucosa and 1 to oxyntic mucosa. Cardiac mucosa was unchanged in 39 (64%) and progressed to intestinal metaplasia in 4 (6%). The decrease in prevalence of cardiac mucosa before and after antireflux surgery was significant ($p = 0.001$).

Thirty-six of these patients had postoperative pH studies. Of these, 28 had a normal 24-hour pH test. Of these, 12 (43%) had regression to oxyntocardiac mucosa, 16 (57%) had unchanged cardiac mucosa, and none progressed to intestinal metaplasia. Of the 8 patients who had an abnormal 24-hour pH test, cardiac mucosa was unchanged in 6 and progressed to intestinal metaplasia in 2. The frequency of regression of cardiac mucosa was significantly higher when the 24-hour pH test was normalized (12 of 28 versus 0 of 8, p = 0.02).

This study (37) shows that antireflux surgery, when it successfully normalizes the 24-hour pH test, creates a reflux milieu around the target cell (progenitor cell in cardiac mucosa) above the gastroesophageal junction that causes it to evolve in the direction of oxyntocardiac mucosa. Oxyntocardiac mucosa is a stable epithelium not at risk for developing intestinal metaplasia or progressing to cancer. As such, this is a highly beneficial change in terms of decreasing the risk of progression to cancer. We have suggested that the best estimate of cancer risk in columnar-lined esophagus is the IM:OCM ratio. Any increase in OCM would decrease this ratio and decrease the cancer risk.

Conclusion: Antireflux surgery, when done in patients with intestinal-metaplasia-negative columnar-lined esophagus, is effective in preventing Barrett esophagus.

Does Antireflux Surgery Cause Regression of Barrett Esophagus?

In the discussion of the effect of acid suppressive drug therapy in treating Barrett esophagus, we made the point that it is worthless to use control of symptoms as an outcome indicator because this very likely tests the effect on the squamous epithelium and not the columnar epithelium. We also made the point that assessment of regression by evaluating the surface overgrowth of squamous epithelium is meaningless because columnar epithelium frequently persists under the squamous epithelium. True regression of Barrett esophagus is a decrease in the absolute number of cells at risk for carcinogenesis (i.e., an absolute decrease in the amount of intestinal metaplasia in the columnar-lined segment).

True regression of Barrett esophagus (i.e., an absolute decrease in the number of target cells susceptible to carcinogenesis) is best assessed by the amount of intestinal metaplasia that is present after antireflux surgery. This is difficult to assess accurately in biopsies because of sampling error even with extensive biopsy protocols because it requires quantitation of goblet cells. The only reasonable outcome measure that can be used in the absence of data from extensive and standardized biopsy protocols is the frequency with which intestinal metaplasia that was present in a preoperative biopsy disappears postoperatively.

Literature Review

Hofstetter WL, Peters JH, DeMeester TR, Hagen JA, DeMeester SR, Crookes PF, Tsai P, Banki F, Bremner CG. Long-term outcome of anti-reflux surgery in patients with Barrett's esophagus. Ann Surg 2001;234:532–538.

This is a study of 97 patients with Barrett's esophagus treated with antireflux surgery. Follow-up was complete in 88% (85 of 97) at a median of 5 years; these 85 patients represented the study group. Barrett esophagus was defined as the presence of intestinal metaplasia in a biopsy taken from an endoscopically visualized columnar-lined segment. Patients with intestinal metaplasia of the cardia were excluded. Fifty-nine patients had long-segment and 26 had short-segment Barrett esophagus. Fifty patients underwent a laparoscopic Nissen procedure, 20 a transthoracic procedure, and 3 abdominal Nissen operations. Nine had a Collis-Belsey procedure, and 3 had other kinds of partial wraps.

At a median follow-up of 5 years, 9 of 63 patients (14%) with nondysplastic Barrett esophagus regressed to cardiac mucosa. Seven of 16 patients who had Barrett esophagus with low-grade dysplasia regressed to nondysplastic intestinal epithelium. No patient developed high-grade dysplasia or cancer in 410 patient-years of follow-up.

This study shows that in conventional Barrett esophagus, both long and short segment, antireflux surgery is relatively ineffective in reversing intestinal metaplasia. In another study from the University of Southern California (USC), patients with intestinal metaplasia of the gastric cardia (which we believe is largely microscopic Barrett esophagus occurring in end-stage dilated esophagus) were included.

Literature Review

DeMeester SR, Campos GM, DeMeester TR, Bremner CG, Hagen JA, Peters JH, Crookes PF. The impact of an antireflux procedure on intestinal metaplasia of the cardia. Ann Surg 1998;228:547–558.

This is a study of 60 patients with intestinal metaplasia of the esophagus or cardia treated with antireflux surgery. Patients in the intestinal metaplasia of the cardia (CIM) group (n = 15) had no endoscopically visible segment of columnar epithelium. Patients in the Barrett group (n = 45) had columnar epithelium visible within the esophagus with intestinal metaplasia present in a biopsy. Median follow-up was 25 months in each group. Postoperative biopsies showed complete loss of intestinal metaplasia in 73% of the patients with CIM compared with 4.4% of the patients in the Barrett group. Low-grade dysplasia, present in 10 patients preoperatively, regressed in 7 patients (70%). No patient progressed to high-grade dysplasia or cancer. The authors concluded that small microscopic areas of intestinal metaplasia are able to regress much more frequently than longer, visible segments of intestinal metaplasia.

These studies show that the microscopic amounts of intestinal metaplasia that occur in reflux disease limited to the end-stage dilated esophagus are more difficult to find in postoperative biopsies. Because the amount of intestinal metaplasia is so small to begin with, it is difficult to know in an individual patient whether the absence of intestinal metaplasia in postoperative biopsies is a sampling problem or true regression. As the numbers of these patients studied increases, the likelihood is that there is indeed a true reversal of short segments of intestinal metaplasia after antireflux surgery. With longer segments of Barrett esophagus, it is clear that intestinal metaplasia only rarely reverses after antireflux surgery.

It must be appreciated that a total reversal of intestinal metaplasia is not necessary to decrease cancer risk. The critical factors are a decrease in the absolute number of target cells (i.e., the amount of intestinal metaplasia, best assessed as the IM:OCM ratio) and possibly the rate of proliferation in residual intestinal metaplasia.

The observations of histologic changes in these three studies are (a) Tharavej et al. (37): cardiac mucosa at the gastroesophageal junction either remains as such or transforms into oxyntocardiac mucosa after successful antireflux surgery; (b) Hofstetter et al. (38): there is no significant reversal of intestinal metaplasia in short- and long-segment Barrett esophagus after antireflux surgery; and (c) DeMeester et al. (39): 73% of patients with intestinal metaplasia of the cardia lost their intestinal metaplasia after antireflux surgery while there was no significant reversal in those patients who had longer segments of Barrett esophagus. Is there any explanation for these findings?

The gastric wrap during the antireflux procedure is positioned at the end of the tubular esophagus in the standard Nissen operation. This separates the columnar-lined esophageal segment into two regions: one above the point of the wrap's effectiveness and one below the point of the wrap's effectiveness (Fig. 16.7).

Patients with traditional short- and long-segment Barrett esophagus have their intestinal metaplasia above the point of the wrap's effectiveness (Fig. 16.7A and b). An effective operation prevents reflux and all interaction between the epithelium and refluxate molecules. The intestinal metaplasia undergoes no change because there are no interactions. The only change it will undergo will be the phenotypic expression of genetic changes that already exist, that is, if cancer mutations are complete at the time of surgery, and the epithelium is in the lag phase, the intestinal metaplasia can progress to cancer.

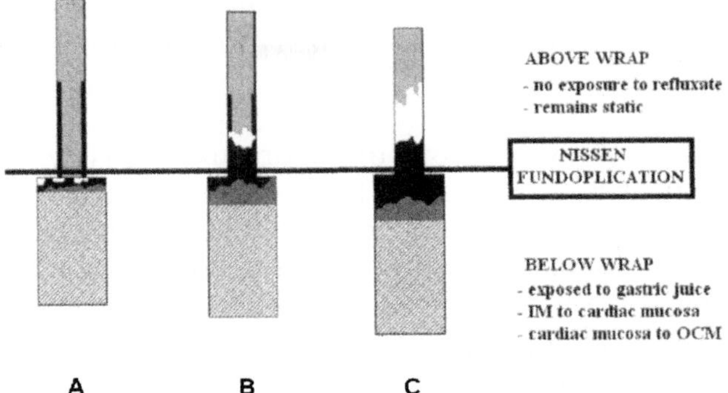

FIGURE 16.7 Effect of a Nissen fundoplication. The effective point of the wrap is designed to be at the end of the tubular esophagus. In patients with severe reflux, this point is *above the dilated end-stage esophagus.* The area below the wrap is exposed to the normalized gastric pH, which induces reversal of intestinal metaplasia and promotes the conversion of cardiac mucosa to oxyntocardiac mucosa. The columnar-lined esophagus above the wrap is no longer exposed to refluxate and remains static.

In contrast, in patients with intestinal metaplasia restricted to the end-stage dilated esophagus who did not have any visible columnar epithelium in their tubular esophagus before surgery (= microscopic Barrett esophagus, which will be misinterpreted by present criteria as intestinal metaplasia of the cardia), *the intestinal metaplasia is in the dilated esophageal segment distal to the end of the tubular esophagus and therefore below the point of effectiveness of the fundoplication.* This segment of columnar-lined esophagus continues to be exposed to gastric contents. With successful antireflux surgery, the patients no longer receive acid suppressive drugs; the pH of the gastric contents becomes more acid as acid suppression is removed and exposure remains continuous. This is the milieu that favors conversion of cardiac mucosa to oxyntocardiac mucosa.

If this hypothesis is correct, it would mean that what has happened to this segment below the point of effectiveness of the fundoplication is that it has reverted to the historic state of columnar-lined esophagus that existed before acid suppressive drugs became available where the low pH favored the evolution of cardiac mucosa to oxyntocardiac mucosa over intestinal metaplasia. The antireflux surgery has created an experimental model where the effectiveness of the surgery in preventing reflux controls the squamous epithelial changes above the point of the fundoplication (and therefore controls the patient's symptoms) and permits the natural forces to shift the columnar epithelia in a benign direction away from intestinal metaplasia and toward oxyntocardiac mucosa in the dilated end-stage reflux-damaged esophagus.

Conclusion: Antireflux surgery causes significant histologic regression of intestinal metaplasia in patients with short segments of Barrett esophagus; regression frequency decreased with the increasing length of Barrett esophagus.

Does Antireflux Surgery Decrease the Risk of Adenocarcinoma?

This is an emotional issue among gastroenterologists. If it is ever shown that antireflux surgery decreases the risk of adenocarcinoma when used to treat reflux disease, the basic control of treating patients with reflux disease will shift from gastroenterologists to surgeons. Patients will demand surgery if they know it will decrease the risk of cancer, particularly if they also become aware that acid suppressive drugs are suspected of promoting adenocarcinoma. Gastroenterologists will be required to send their patients to surgeons for this treatment; failure to do so without the informed consent of patients will amount to malpractice. Matters of control and finance are those that evoke the most severe emotional battles. As a pathologist, Dr. Chandrasoma has no vested interest in this argument; he does not have a financial interest in who treats these patients. If anything, it is possible that successful surgery will decrease the number of biopsies that he will receive from these patients and decrease his income. His hope is that his income will decline by the large number of esophagectomies for adenocarcinoma that will no longer come his way.

If nothing else, there must be a theoretical presumption that successful antireflux surgery *must* prevent esophageal adenocarcinoma. Adenocarcinoma results from an unknown molecule in the refluxate interacting with a susceptible target cell (the progenitor cell in intestinal metaplastic epithelium in the esophagus) to produce a series of genetic alterations that give this cell the biological properties of a malignant neoplasm.

There are only two theoretical ways by which successful antireflux surgery will not decrease the incidence of adenocarcinoma: (a) if the surgery is performed after all the genetic events necessary for neoplasia have already taken place—that is, the target cell is in the lag phase between acquiring the genetic basis of cancer and the phenotypic expression of cancer. Carlson et al. (40) showed that acid suppression fails to prevent the progression to dysplasia and cancer in patients who have p53 gene expression in intestinal metaplasia. (b) If the carcinogen is not luminally acting. Antireflux surgery will not prevent cancer if the carcinogen reaches the target cells by a route that is not transluminal (e.g., blood-borne). There is no evidence that reflux-induced esophageal adenocarcinoma is caused by a blood-borne carcinogen, but this is a possible unknown.

In the discussion on the effect of antireflux surgery, we have shown that there is strong evidence that when surgery is performed in patients without Barrett esophagus, the surgery is effective in preventing the development of intestinal metaplasia in cardiac mucosa and effective in converting cardiac mucosa to oxyntocardiac mucosa. Both events will have a strong impact in decreasing cancer incidence *if antireflux surgery is performed before Barrett esophagus has developed.* At the present time, antireflux surgery is performed for the more severe and complicated cases. A shift in this attitude toward antireflux surgery early in the course of the disease can be expected to be more effective in decreasing cancer incidence.

We have also provided evidence that with successful antireflux surgery, intestinal metaplasia can reverse, particularly in patients with very short segments of intestinal metaplasia. In patients who have a reversal of intestinal metaplasia, this must result from either (a) exposure to a milieu that favors conversion of intestinal to cardiac epithelium such as increased acidity of the refluxate or (b) removal of a factor that is necessary to maintain the genetic signal required for intestinal metaplasia. There are no data on the amount of intestinal metaplasia after antireflux surgery; regression is assessed simply by complete absence. If the patients who had persistent intestinal metaplasia had a decrease in the amount of intestinal metaplasia, it would suggest decreased risk. The fact that intestinal metaplasia disappears in at least some patients suggests a likelihood of a decrease in the amount of intestinal metaplasia in more patients. In patients who do not have a reversal of intestinal metaplasia after antireflux surgery, the absence of further interactions between refluxate molecules and the target cells can be assumed to prevent the progression of intestinal metaplasia to dysplasia and cancer.

The Bias among Gastroenterologists

It is very interesting to look at the AGA Chicago consensus workshop analysis of this question. This workshop consisted of 18 experts, of whom 15 were gastroenterologists, 2 were surgeons, and 1 was a pathologist. This subject was addressed by statement 39: "In patients with Barrett esophagus, anti-reflux surgery has not proven to have a major protective effect against the development of esophageal adenocarcinoma" (p. 323).

The reader will notice that when we framed the questions in this chapter to address the efficacy of acid suppression treatment and antireflux surgery, we used the identical questions with only the words "acid suppressive drugs" and "antireflux surgery" interposed. If one does the same for the statement in the AGA workshop that evaluated this question for acid suppression, one would end up with a completely different statement. This would read as follows (this is statement 35 with the words "antireflux surgery" substituted for "acid suppression"): "Normalization of esophageal acid exposure by antireflux surgery reduces the risk for development of esophageal adenocarcinoma." This statement has a much lower burden of evidence than the actual statement 39 that was formulated; no longer would there be a need for "proof" or "a major protective effect." The duplicity is compounded by the fact that "acid suppression" in statement 35 actually includes antireflux surgery and does not restrict itself to acid suppressive drugs. It almost appears that the objective of the AGA workshop was to avoid reaching the correct conclusion that antireflux surgery is clearly superior to acid suppressive drug therapy in preventing cancer. While the former has been shown to be of definite benefit in reducing the risk, there is evidence that the latter actually increases the risk of cancer.

Few people will agree with the statement that there is *proof* that antireflux surgery has *a major protective effect* against the development of adenocarcinoma. There is, however, strong evidence in the literature that antireflux surgery decreases the incidence of adenocarcinoma when it is successful. The restrictive nature of the question caused even the two surgeons among the expert panel to accept this statement with some reservation (we are assuming that the two surgeons were among the three experts who did not accept this statement completely). Unfortunately, this consensus statement is often quoted as evidence that experts in the field are of the opinion that antireflux surgery does not do anything to prevent esophageal adenocarcinoma. This is the opinion of most gastroenterologists. It is biased opinion that is not based on scientific evidence.

As we will show, the conclusion reached by the experts based on the five references they quote in their statement as their sources of evidence is probably incorrect.

Studies from Nonsurgical Sources

Most gastroenterologists are not qualified to make the assessment of whether antireflux surgery decreases the risk of adenocarcinoma. They have no substantial firsthand experience with patients who have had antireflux surgery that permit them to evaluate this question. Patients who have antireflux surgery are followed systematically by the surgical program that performed the surgery. Gastroenterologists only encounter those patients who have dropped out of follow-up programs. These are patients who develop problems after their surgery that make them sufficiently dissatisfied to prevent them from returning to their surgical institution. This is a highly skewed population whose numbers tend to be too small for statistical analysis.

In the assessment of the workshop statement on the effect of antireflux surgery, five studies were quoted. The two studies originating from nonsurgical sources were as follows:

1. Spechler et al. (41). This was a randomized controlled study of 247 patients randomized to acid suppressive drugs (165 patients on two different regimens) or antireflux surgery (82 patients); 129 of these patients (91 in the drug treatment group and 38 in the surgery group) participated in the follow-up. One of the outcome measures was the incidence of esophageal adenocarcinoma.

   Quoting from the research: "Five patients (all white men; 4 medical patients and 1 surgical patient) developed esophageal adenocarcinoma after a mean follow-up of 7.1 years (range: 4–12 years). There was no significant difference in the rate of esophageal cancer development between the medical and surgical groups ($p = 1.0$ by the Fisher exact test). However, we calculated that a study designed to demonstrate a 50% relative reduction in the risk of esophageal cancer during this period would have required a sample size of 460 patients in each treatment group (assuming a baseline cancer incidence of 0.5% per year; power = 80%; $p = 0.05$). *Therefore, our study did not have sufficient statistical power to detect potentially important differences between groups in the rate of cancer development.*" The experts should have heeded the conclusion of the authors and not included this article in their discussion.
2. Ye et al. (42). This was an epidemiologic study from the Swedish Inpatient Register, which identified a cohort of 35,274 male and 31,691 female patients with a discharge diagnosis of gastroesophageal reflux disease, and another cohort of 6406 male and 4671 female patients who underwent antireflux surgery. Follow-up was attained through record linkage with several nationwide registers. The standardized incidence ratio (SIR) was used to estimate the relative risk of upper gastrointestinal cancers,

using the general Swedish population as reference.

After exclusion of the first-year follow-up, 37 esophageal and 36 gastric cardia adenocarcinomas were observed among male patients who did not have surgery (SIR, 6.3, 95% confidence interval [C.I.], 4.5 to 8.7; SIR, 2.4, 95% C.I., 1.7 to 3.3, respectively). Among male patients who had undergone antireflux surgeries, risks were also elevated (16 esophageal adenocarcinoma, SIR, 14.1, 95% C.I., 8.0 to 22.8; 15 gastric cardia adenocarcinomas, SIR, 5.3, 95% C.I., 3.0 to 8.7) and remained elevated with time after surgery. The cancer risk pattern in women was similar to that for men, but the number of cases was much smaller. The authors concluded that gastroesophageal reflux is strongly associated with the risk of esophageal adenocarcinoma and, to a lesser extent, with gastric cardia adenocarcinoma. The risk of developing adenocarcinomas of the esophagus and gastric cardia remains increased after antireflux surgery.

This population-based study only informs us that 16 esophageal and 15 cardiac cancers developed in patients who had antireflux surgery. It gives no information on whether this represents a change from the expected risk if the patients had not had antireflux surgery. Comparisons are impossible in studies like this because of the reasonable presumption that the group who had antireflux surgery is likely to have selected those patients with the most severe reflux disease. There is also no information on the type or success of the antireflux operation.

The preceding two studies confirm that the population of patients who have had antireflux surgery is not available for scientific review by gastroenterologists. The Ye et al. study (42) is an epidemiologic study. The Spechler et al. study (41), which accessed patients from 10 VA hospitals in the United States, had only 82 patients undergoing antireflux surgery over the 2-year period and a self-admitted lack of statistical ability to address the question because of the small numbers.

Studies from Surgical Sources

The incidence of dysplasia and reversal of dysplasia in patients who have antireflux surgery can be seen in the following study from the University of Washington. Unfortunately, as with most surgical studies, the number of cases is small and follow-up is not sufficient to draw definitive conclusions.

Literature Review

Oelschlager BK, Barreca M, Chang L, Oleynikov D, Pellegrini CA. Clinical and pathologic response of Barrett's esophagus to laparoscopic anti-reflux surgery. Ann Surg 2003;238:458–464.

This is a study (43) of 90 patients with Barrett esophagus who underwent laparoscopic antireflux surgery between 1994 and 2000 and agreed to have a postoperative endoscopy in 2002 after a median postoperative interval of 40 months (range: 12 to 95 months). The change in the preoperative and postoperative biopsy findings are as follows:

Histology	Preop biopsy (number of patients)	Postop biopsy finding (number of patients)
IM without dysplasia	75	Remained unchanged = 48 Intestinal metaplasia not present = 26 Developed indefinite for dysplasia = 1
Indefinite for dysplasia	12	Remained unchanged = 4 Intestinal metaplasia not present = 3 IM without dysplasia = 4 Progressed to low-grade dysplasia = 1
Low-grade dysplasia	3	Intestinal metaplasia not present = 1 Progressed to high-grade dysplasia = 1 Progressed to adenocarcinoma = 1

Fifty-four of the 90 patients with endoscopic follow-up had short-segment BE (<3cm), and 36 had long-segment BE (>3cm) preoperatively. Postoperatively, endoscopy and pathology revealed complete regression of intestinal metaplasia in 30 (55%) of 54 patients with short-segment BE but in 0 of 36 of those with long-segment BE. Among patients with complete regression, 89% of those tested with pH monitoring had normal esophageal acid exposure. A normal 24-hour pH test was seen in 69% of those who failed to have complete regression of intestinal metaplasia. One patient developed adenocarcinoma within 10 months of surgery; this patient probably had the genetic changes for carcinoma already present at the time of the surgery. The numbers in this study are too small for definite conclusions about the incidence of dysplasia but confirm the significant reversal of intestinal metaplasia in short-segment Barrett esophagus. The fact that dysplasia (low grade) developed in only 1 of 87 patients who did not have dysplasia in the preoperative biopsy in this series is encouraging.

Several studies from single surgical centers are available with well-studied patient groups that provide data relating to the incidence of adenocarcinoma in patients who have had antireflux surgery. In evaluating these studies it is important to recognize the following things:

1. In patients with Barrett esophagus, antireflux surgery will not prevent cancer in those patients who already have the necessary mutations and are in the lag phase. A certain number of cancers are to be expected. If this is the explanation for cancer occurring after antireflux surgery, the cancers should cluster in the first year and then tail off in number, becoming extremely rare after about 5 years. This assumes a lag phase of 5 years. A careful review of cancers occurring after antireflux surgery indicates that it is difficult to find cases where cancer has developed more than 5 years after a successful antireflux procedure.
2. Antireflux surgery cannot abolish reflux completely. In 80% of patients, reflux is decreased to normal levels by the 24-hour pH test. A normal 24-hour pH test, however, means exposure of the target cell to refluxate for up to 60 minutes per day (4.5% of the 24-hour period). Therefore, it should not be expected that antireflux surgery will completely eradicate the possibility of cancer. One must look for a decreased incidence from what is expected if the patient did not have surgery, not a complete absence of cancer.

In the AGA workshop analysis of whether antireflux surgery has a protective effect against the development of esophageal adenocarcinoma, three surgical studies were selected. These studies are as follows:

1. McDonald et al. (44) reported on 112 patients who survived antireflux surgery (there was one operative death) in a series that extends from 1975 to 1994 with a mean follow-up of 6.5 years (with a range of 4 months to 18.2 years). Of 99 surviving patients (the other 13 died of unrelated causes), 3 patients developed adenocarcinoma 13, 25, and 39 months after the antireflux procedure. It should be recognized that the cancers arose within the 5-year period that is generally considered the lag phase in patients who have prevalent cancers at the time of surgery. The incidence of adenocarcinoma in this select group of patients was 1 in 273.8 patient-years of follow-up. Though this cancer incidence is lower than that in Weston et al. (17), which was 1 cancer per 71.9 patient-years, it is not possible to compare the two because Weston et al. included 20 patients who had low-grade dysplasia/indefinite for dysplasia and 8 patients who had unifocal high-grade dysplasia in the entry biopsy.
2. This contrasts with a study from Chile by Csendes et al. (45) who reported on 152 patients who had antireflux surgery (either Nissen fundoplication or posterior gastropexy) studied prospectively for a mean of 100 months. There was one death (0.7%) after operation. The late follow-up at 100 months demonstrated a high percentage of failures among patients with uncomplicated Barrett esophagus (54%) and an even higher figure in patients with complicated Barrett esophagus (64%). Twenty-four-hour pH monitoring demonstrated a decrease in acid reflux into the esophagus, and Bilitec studies also demonstrated a decrease of duodenoesophageal reflux, but in all cases with a higher value than the normal limit. In 15 patients, low-grade dysplasia appeared at 8 years of follow-up and an adenocarcinoma appeared in 4 patients. This study is unusual in that no patient maintained a normalized 24-hour pH study after antireflux surgery. The main conclusion to be drawn from this study is that unsuccessful antireflux surgery is not effective in preventing the progression from Barrett esophagus to dysplasia and adenocarcinoma.
3. In the study by Hofstetter et al. from the USC group, which we reviewed earlier, 85 patients with Barrett's esophagus who had been treated with antireflux surgery and had complete follow-up at a median of 5 years were evaluated. Fifty-nine patients had long-segment and 26 had short-segment Barrett's. Fifty patients underwent a laparoscopic procedure, 20 a transthoracic procedure, and 3 abdominal Nissen operations. Nine had a Collis-Belsey procedure, and 3 had other partial wraps. At a median follow-up of 5 years, 9 of 63 patients (14%) with nondysplastic Barrett esophagus regressed to cardiac mucosa, and 7 of 16 patients who had Barrett esophagus with low-grade dysplasia regressed to nondysplastic intestinal epithelium. No patient developed high-grade dysplasia or cancer in 410 patient-years of follow-up.

An objective analysis of these three studies shows the following: (a) Two of the studies from the United States can be assumed to have had high success rates in terms of normalizing reflux by the 24-hour pH test. In Hofstetter et al., 0 of 85 patients progressed to high-grade dysplasia or adenocarcinoma after a median follow-up of 5 years and a cumulative 410 patient-years of follow-up. In McDonald et al., 3 of 112 patients followed for a mean of 6.5 years developed cancer. All three cancers arose within 40 months after the surgery; no cancers arose after 40 months in any patient. In

Csendes et al. (45), the incidence of cancer was 5 out of 152 patients after a mean follow-up of 100 months, but there was an almost universal lack of success with the antireflux procedure in terms of normalizing the 24-hour pH test.

Literature Review

Parrilla P, deHaro LFM, Ortiz A, Munitiz V, Molina J, Bermejo J, Canteras M. Long term results of a randomized prospective study comparing medical and surgical treatment of Barrett's esophagus. Ann Surg 2003;237:291–298.

This is a randomized prospective clinical trial and has the best evidence regarding the effect of treatment on Barrett esophagus. This is a follow-up to a prior study by this group (46) with increased patient numbers and follow-up duration.

This study shows that the value of antireflux surgery in decreasing the incidence of adenocarcinoma in patients with Barrett esophagus is related to the effectiveness of the antireflux procedure. Of the 58 patients who underwent antireflux surgery, 56 patients had a Nissen fundoplication and 2 patients with significant esophageal shortening had a Collis-Nissen procedure. The postoperative 24-hour pH test was normalized (<4% with pH < 4) in 49 patients; in the other 9 patients, reflux was decreased from the preoperative levels but remained at abnormal levels (median: 19.5%, range: 5% to 46%). These 9 patients included the 5 with clinical recurrence.

In the 58 patients who had antireflux surgery, preoperative biopsies showed nondysplastic intestinal metaplasia in 53 patients and low-grade dysplasia in 5 patients. After a mean follow-up of 7 years (median: 6 years; range: 1 to 18 years), 50 of the 53 patients with no dysplasia and 5 of the 5 patients with low-grade dysplasia had no evidence of dysplasia. Three patients with no dysplasia preoperatively developed low-grade dysplasia after surgery, and 2 of these went on to develop high-grade dysplasia on biopsy; both of these patients had adenocarcinoma in the esophagectomy specimen.

If the 49 patients who had a successful antireflux operation were considered after the 9 patients whose surgery was not successful in normalizing the 24-hour pH test were excluded (same follow-up as the entire group), 44 had no dysplasia in the preoperative biopsies and 5 had low-grade dysplasia. After surgery, no dysplasia was seen in the 5 patients with preoperative low-grade dysplasia. One of 44 patients with no dysplasia preoperatively developed de novo low-grade dysplasia after surgery; no patient developed high-grade dysplasia or adenocarcinoma for the duration of the follow-up period.

The 2 patients who developed adenocarcinoma after antireflux surgery showed clinical and pH-metric recurrence of gastroesophageal reflux 1 year postoperatively; the percentage of total time with pH below 4 was 12% and 37%. Medical treatment was given to these 2 patients and high-grade dysplasia appeared after a follow-up of 4 and 6 years postoperatively.

Any objective analysis of these studies should suggest that there is strong evidence that antireflux surgery, *when it is successful as measured by a normalization of the 24-hour pH test*, has a significant protective effect against the development of esophageal adenocarcinoma. When one considers that most studies report that antireflux surgery has a rate of successful normalization of the 24-hour pH test in the 80% to 90% range (24), antireflux surgery can be expected to have a significant effect in reducing overall adenocarcinoma rates if this procedure is performed in an adequate number of patients.

The management of the 15% of patients whose antireflux surgery is not successful as determined by a failure of normalization of the 24-hour pH test should be addressed. There is a strong reason to move away from the passive course of using acid suppressive medication to a more aggressive control of reflux, even with a complicated reoperation, because these failed patients continue to be at significant risk for adenocarcinoma.

Conclusion: Antireflux surgery has a significant protective effect against the development of adenocarcinoma in patients. When performed in patients without Barrett esophagus, it prevents intestinal metaplasia. When performed in patients with Barrett esophagus, it significantly reduces the progression to high-grade dysplasia and adenocarcinoma.

Does Antireflux Surgery Increase the Risk of Adenocarcinoma?

The answer to this question is that there is absolutely no evidence that cancer risk increases after antireflux surgery. Because it has the effect of reducing exposure of the target cell to gastric refluxate molecules even in patients whose surgery is deemed unsuccessful by some outcome measures, antireflux surgery can only do some good and certainly no harm. A completely failed antireflux procedure simply restores the patient to the preoperative cancer risk.

It is interesting to compare the study by Csendes et al. (45) with that of Weston et al. (17). Csendes et al. reported that their patients had a decrease in the amount of reflux without normalization in the 24-hour pH test or Bilitec test. From a theoretical standpoint, these patients have a decreased exposure of the target cells to all elements of the refluxate. In contrast, patients who are on acid suppressive drugs have suppression of acid without any decrease in other elements of the refluxate. Csendes et al. reported low-grade dysplasia in 15 of 152 patients (9.9%) and cancer in 4 of 152 patients (2.6%) with a mean follow-up of 100 months. The correct comparison in Weston et al.'s series would be the 80 patients who did not have any dysplasia (the 20 patients with low-grade dysplasia and indefinite for dysplasia and the 8 patients with unifocal high-grade dysplasia are excluded). Of these 80 patients, 23 of 80 (28.8%) developed low-grade dysplasia or were indefinite for dysplasia, and 3 of 80 (3.8%) developed cancer (2) or multifocal high-grade dysplasia, which were the end points of the study that led to a recommendation of surgery. We will resist the temptation to make a statistical evaluation between these sets of data because it would not be meaningful. There are many unknowns that may influence these data and the numbers are very small. However, at worst, the failed antireflux surgery series appears to have an incidence of adenocarcinoma that is similar to the medically treated series after exclusion of all patients with any semblance of dysplasia in the initial biopsy. It should be stated that the same consideration was not given to Csendes et al.'s series; it is not known how many of these patients had low-grade dysplasia or were indefinite for dysplasia in the index preoperative biopsy. In Weston et al.'s series, if the 20 patients with low-grade dysplasia were included (still excluding 8 patients with unifocal high-grade dysplasia), the incidence of adenocarcinoma/multifocal high-grade dysplasia was 5 of 100 (5%), and if the 8 unifocal high-grade dysplasia patients were included, the incidence rises to 10 of 108 (9.3%).

Conclusion: There is no theoretical possibility that antireflux surgery, even when unsuccessful, can increase the risk of cancer in Barrett esophagus. Practically, even failed antireflux surgery has an incidence of dysplasia and cancer that is either similar or better than the best medical therapy.

ABLATION AS TREATMENT FOR BARRETT ESOPHAGUS

Ablation of the target epithelium, either surgically by endoscopic mucosal resection or by various physical modalities like laser coagulation and photodynamic therapy, is theoretically extremely attractive as a method of treating columnar-lined esophagus. Removal of the target lesion that has premalignant potential is the standard for many other body sites (e.g., cone biopsy of the cervix for cervical dysplasia, prophylactic colectomy in chronic ulcerative colitis, or removal of adenomas in colorectal screening).

In a patient whose mucosa has been removed, either by ablation or endoscopic mucosal resection, the type of epithelium that regenerates in the process of healing varies in different circumstances. As shown in the early animal experiments of Bremner et al. (47), when the epithelium regenerates in the absence of reflux, it tends to regenerate as squamous epithelium. In the presence of reflux, it tends to regenerate as columnar epithelium. If this is true, if complete acid suppression can be maintained after ablation of the offending mucosa, healing can be induced to develop a not-at-risk squamous epithelium to replace the offending columnar epithelium. Cure is achieved.

Ablation by laser and photodynamic therapy is presently limited to late-stage reflux disease with high-grade dysplasia or adenocarcinoma to control disease in patients who either refuse esophagectomy as an option or are not physically fit to withstand major surgery. Ablation is not a first-line treatment for uncomplicated Barrett esophagus, mainly because it is considered to be a procedure with high-morbidity and risk. It is frequently complicated by recurrence of intestinal metaplastic epithelium, even with high-dosage acid suppressive drug treatment and has a significant risk of stricture formation. Endoscopic mucosal resection has similar problems, and its use is also generally restricted to the treatment of high-grade dysplasia and carcinoma in patients who are not fit for surgery.

As the safety and efficacy of ablation and endoscopic mucosal resection improve and methods of acid suppression that are reliable and effective enough to prevent the recurrence of Barrett esophagus are found, these may ultimately be the methods of choice in treating Barrett esophagus. The management of Barrett esophagus then moves from the type of regimen recommended for chronic ulcerative colitis (surveillance with random biopsies for detection of dysplasia followed by esophagectomy at a designated end point—multifocal high-grade dysplasia or early adenocarcinoma) to the regimen for colonic polyps or cervical dysplasia where the offending epithelium is removed.

The ideal ablative technique must remove the entire mucosa all the way down to the muscularis mucosae without leaving any surface epithelium or lamina propria glandular elements behind. This is not easy because the folds of the esophageal epithelium preclude the administration of a consistent dosage of power through the mucosa, and islands of epithelium may be left untreated. Care is necessary when one evaluates the epithelium after ablation. When ablation has failed to remove the entire thickness of the mucosa, columnar epithelium may remain under the surface squamous epithelium, representing lack of true regression. These patients will appear to have squamous epithelialization on the surface at endoscopy but still have at risk epithelium below the surface, which has malignant potential.

References

1. Pohl H, Welch HG. The role of overdiagnosis and reclassification in the marked increase of esophageal adenocarcinoma incidence. J Natl Cancer Inst 2005;97:142–146.
2. Sharma P, McQuaid K, Dent J, Fennerty B, Sampliner R, Spechler S, Cameron A, Corley D, Falk G, Goldblum J, Hunter J, Jankowski J, Lundell L, Reid B, Shaheen N, Sonnenberg A, Wang K, Weinstein W. A critical review of the diagnosis and management of Barrett's esophagus: The AGA Chicago Workshop. Gastroenterology 2004;127:310–330.
3. Tamhankar AP, Peters JH, Portale G, Hsieh C-C, Hagen JA, Bremner CG, DeMeester TR. Omeprazole does not reduce gastroesophageal reflux: New insights using multichannel intraluminal impedance technology. J Gastrointest Surg 2004;8:890–898.
4. Kahrilas PJ, Shi G, Manka M, Joehl RJ. Increased frequency of transient lower esophageal sphincter relaxation induced by gastric distension in reflux patients with hiatal hernia. Gastroenterology 2000;118:688–695.
5. Peters FT, Ganesh S, Kuipers EJ, Sluiter WJ, Klinkenberg-Knol EC, Lamers CB, Kleibeuker JH. Endoscopic regression of Barrett's oesophagus during omeprazole treatment: A randomized double blind study. Gut 1999;45:489–494.
6. Parrilla P, deHaro LFM, Ortiz A, Munitiz V, Molina J, Bermejo J, Canteras M. Long term results of a randomized prospective study comparing medical and surgical treatment of Barrett's esophagus. Ann Surg 2003;237:291–298.
7. Haggitt RC. Adenocarcinoma in Barrett's esophagus: A new epidemic? Hum Pathol 1992:23:475–476.
8. Devesa SS, Blot WJ, Fraumeni JF. Changing patterns in the incidence of esophageal and gastric carcinoma in the United States. Cancer 1998;83:2049–2053.
9. Chandrasoma PT, Der R, Ma Y, Peters J, DeMeester T. Histologic classification of patients based on mapping biopsies of the gastroesophageal junction. Am J Surg Pathol 2003;27:929–936.
10. Spechler SJ, Zeroogian JM, Antonioli DA, Wang HH, Goyal RK. Prevalence of metaplasia at the gastroesophageal junction. Lancet 1994;344:1533–1536.
11. Rex DK, Cummings OW, Shaw M, Cumings MD, Wong RKH, Vasudeva RS, Dunne D, Rahmani EY, Helper DJ. Screening for Barrett's esophagus in colonoscopy patients with and without heartburn. Gastroenterology 2003;125:1670–1677.
12. Sharma P, Sampliner RE, Camargo E. Normalization of esophageal pH with high dose proton pump inhibitor therapy does not result in regression of Barrett's esophagus. Am J Gastroenterol 1997;92:582–585.
13. Wilkinson SP, Biddlestone L, Gore S, Shepherd NA. Regression of columnar lined (Barrett's) oesophagus with omeprazole 40mg daily: Results of 5 years of continuous therapy. Aliment Pharmacol Ther 1999;13:1205–1209.
14. Fitzgerald RC, Omary MB, Triadafilopoulos G. Dynamic effects of acid on Barrett's esophagus: An ex-vivo proliferation and differentiation model. J Clin Invest 1996;98:2120–2128.
15. Ouatu-Lascar R, Fitzgerald RC, Triadafilopoulos G. Differentiation and proliferation in Barrett's esophagus and the effects of acid suppression. Gastroenterology 1999;117:327–335.
16. El-Serag HB, Aguirre TV, Davis S, Kuebeler M, Bhattacharyya A, Sampliner RE. Proton pump inhibitors are associated with reduced incidence of dysplasia in Barrett's esophagus. Am J Gastroenterol 2004;99:1884–1886.
17. Weston AP, Badr AS, Hassanein RS. Prospective multivariate analysis of clinical, endoscopic, and histological factors predictive of the development of Barrett's multifocal high grade dysplasia or adenocarcinoma. Am J Gastroenterol 1999;94: 3413–3419.
18. Spechler SJ. Barrett's esophagus. N Engl J Med 2002;346: 836–842.
19. Shaheen NJ, Crosby MA, Bozymski EM, Sandler RS. Is there a publication bias in the reporting of cancer risk in Barrett's esophagus? Gastroenterology 2000;119:333–338.
20. Lagergren J, Bergstrom R, Lindgren A, Nyren O. Symptomatic gastroesophageal reflux as a risk factor for esophageal adenocarcinoma. N Engl J Med 1999;340:825–831.
21. Jackson C. Peptic ulcer of the esophagus. JAMA 1929;92: 369–372.
22. Allison PR, Johnstone AS. The oesophagus lined with gastric mucous membrane. Thorax 1953;8:87–101.
23. Liu JH, Etzioni DA, O'Connell JB, Maggard MA, Hiyama DT, Ko CY. Inpatient surgery in California: 1990–2000. Arch Surg 2003;138:1106–1111.
24. Eubanks TR, Omelanczuk P, Richards C, Pohl D, Pellegrini CA. Outcomes of laparoscopic antireflux procedures. Am J Surg 2000;179:391–395.
25. Booth MI, Jones L, Stratford J, Dehn TC. Results of laparoscopic Nissen fundoplication at 2–8 years after surgery. Br J Surg 2002;89:476–481.
26. Anvari M, Allen C. Five-year comprehensive outcomes evaluation in 181 patients after laparoscopic Nissen fundoplication. J Am Coll Surg 2003;196:51–57.
27. Mason RJ, DeMeester TR, Lund RJ, Peters JH, Crookes P, Ritter M, Gadenstatter M, Hagen JA. Nissen fundoplication prevents shortening of the sphincter during gastric distention. Arch Surg 1997;132:719–724.
28. Bammer T, Hinder RA, Klaus A, Klingler PJ. Five- to eight-year outcome of the first laparoscopic Nissen fundoplications. J Gastrointest Surg 2001;5:42–48.
29. Lafullarde T, Watson DI, Jamieson GG, Myers JC, Game PA, Devitt PG. Laparoscopic Nissen fundoplication: Five-year results and beyond. Arch Surg 2001;136:180–184.
30. Beldi G, Glattli A. Long-term gastrointestinal symptoms after laparoscopic Nissen fundoplication. Surg Laparosc Endosc Percutan Tech 2002;12:316–319.
31. Kamolz T, Pointner R. Expectations of patients with gastroesophageal reflux disease for the outcome of laparoscopic antireflux surgery. Surg Laparosc Endosc Percutan Tech 2002;12: 389–392.
32. Bloomston M, Nields W, Rosemurgy AS. Symptoms and antireflux medication use following laparoscopic Nissen fundoplication: Outcome at 1 and 4 years. JSLS 2003;7:211–218.
33. Khajanchee YS, O'Rourke RW, Lockhart B, Patterson EJ, Hansen PD, Swanstrom LL. Postoperative symptoms and failure after antireflux surgery. Arch Surg 2002;137:1008–1013.
34. Desai KM, Frisella MM, Soper NJ. Clinical outcomes after laparoscopic antireflux surgery in patients with and without preoperative endoscopic esophagitis. J Gastrointest Surg 2003;7:44–55.
35. Lundell L, Miettinen P, Myrvold HE, Pedersen SA, Liedman B, Hatlebakk JG, Julkonen R, Levander K, Carlsson J, Lamm M, Wiklund I. Continued (5-year) followup of a randomized clinical study comparing antireflux surgery and omeprazole in gastroesophageal reflux disease. J Am Coll Surg 2001;192:172–179.

36. Oberg S, Johansson J, Wenner J, Johnsson F, Zilling T, von Holstein CS, Nilsson J, Walther B. Endoscopic surveillance of columnar-lined esophagus: Frequency of intestinal metaplasia detection and impact of antireflux surgery. Ann Surg 2001;234: 619–626.
37. Tharavej C, Bremner CG, Hagen JA, DeMeester SR, Chandrasoma PT, Lipham JC, Peters JH, DeMeester TR. Transformation of cardiac mucosa after antireflux surgery. Abstract presented at Digestive Diseases Week, Chicago, IL 2005.
38. Hofstetter WL, Peters JH, DeMeester TR, Hagen JA, DeMeester SR, Crookes PF, Tsai P, Banki F, Bremner CG. Long-term outcome of antireflux surgery in patients with Barrett's esophagus. Ann Surg 2001;234:532–538.
39. DeMeester SR, Campos GMR, DeMeester TR, Bremner CG, Hagen JA, Peters JH, Crookes PF. The impact of antireflux procedure on intestinal metaplasia of the cardia. Ann Surg 1998;228:547–556.
40. Carlson N, Lechago J, Richter J, Sampliner RE, Peterson L, Santella RM, Goldblum JR, Falk GW, Ertan A, Younes M. Acid suppression therapy may not alter malignant progression in Barrett's metaplasia showing p53 protein accumulation. Am J Gastroenterol 2003;98:213–214.
41. Spechler SJ, Lee E, Ahnen D, Goyal RK, Hirano I, Ramirez F, Raufman JP, Sampliner R, Schnell T, Sontag S, Vlahcevic ZR, Young R, Williford W. Long term outcome of medical and surgical therapies for gastroesophageal reflux disease: Follow-up of a randomized controlled trial. JAMA 2001;285: 2331–2338.
42. Ye W, Chow WH, Lagergren J, Yin L, Nyren O. Risk of adenocarcinomas of the esophagus and gastric cardia in patients with gastroesophageal reflux diseases and after antireflux surgery. Gastroenterology 2001;121:1286–1293.
43. Oelschlager BK, Barreca M, Chang L, Oleynikov D, Pellegrini CA. Clinical and pathologic response of Barrett's esophagus to laparoscopic antireflux surgery. Ann Surg 2003;238:458–464.
44. McDonald ML, Trastek VF, Allen MS, Deschamps C, Pairolero PC, Pairolero PC. Barrett's esophagus: Does an antireflux procedure reduce the need for endoscopic surveillance? J Thorac Cardiovasc Surg 1996;111:1135–1138.
45. Csendes A, Braghetto I, Burdiles P, Puente G, Korn O, Diaz JC, Maluenda F. Long-term results of classic antireflux surgery in 152 patients with Barrett's esophagus: Clinical, radiologic, endoscopic, manometric, and acid reflux test analysis before and late after operation. Surgery 1998;123:645–657.
46. Ortiz A, Martinez de Haro LF, Parrilla P, Morales G, Molina J, Bermejo J, Liron R, Aguilar J. Conservative treatment versus antireflux surgery in Barrett's oesophagus: Long-term results of a prospective study. Br J Surg 1996;83:274–278.
47. Bremner CG, Lynch VP, Ellis FH. Barrett's esophagus: Congenital or acquired? An experimental study of esophageal mucosal regeneration in the dog. Surgery 1970;68:209–216.

CHAPTER

17

Treatment Strategies for Preventing Reflux-Induced Adenocarcinoma

The ultimate solution for the treatment of gastroesophageal reflux disease is not only the reversal of symptoms; it is preventing Barrett esophagus and treating it in a manner that will prevent reflux-induced adenocarcinoma. Ideally, these will be in the form of simple drugs that will impact the disease at a molecular level, causing reversal of intestinal metaplasia, prevention of the conversion of cardiac mucosa to intestinal metaplasia, or the promotion of generation of oxyntocardiac mucosa from cardiac mucosa. When these new molecular treatments are developed, they can be put into effect quickly. For example, the availability of Gleevac as an effective drug to control CD117-positive gastrointestinal stromal tumors and Herceptin in breast cancers that overexpress HER2/neu oncoprotein has transformed the outlook for survival in these diseases.

While we await a new miracle drug that will specifically control the progression to cancer, treatment strategies need to be developed that will minimize mortality from reflux disease. It is important for the medical community to assess how we are presently trying to achieve this goal, recognize that we are failing abjectly, and see if we can change treatment strategies and attitudes to control the increasing death rate from reflux disease with treatment modalities that are presently available.

The first steps whereby squamous epithelium is damaged and transforms into cardiac mucosa occur quickly and are not likely to be preventable. Most patients destined to develop adenocarcinoma very likely develop cardiac mucosa early in life, often during childhood, and long before they ever present to a physician. It is unlikely that complete acid suppression can be practically achieved before cardiac metaplasia becomes established. Once established, it is unlikely that acid suppression has any positive impact on the transformation of cardiac mucosa to intestinal metaplasia and progression to cancer. It is therefore unlikely that acid suppression alone can impact the progression to cancer once cardiac metaplasia has occurred.

This is borne out by the evidence. With acid suppression as the cornerstone of treatment, incidence rates of reflux-induced adenocarcinoma have increased 350% since the 1970s (1) and continue to increase (2). The medical community has been slow to adapt to this change. We treat reflux disease as if cancer is the rare complication that it was in the 1970s and ignore the fact that approximately 14,000 patients die every year from it in the United States.

As physicians, we must stop to evaluate whether we are truly incapable of doing anything about preventing this cancer and descend into futility or whether a different approach to the treatment of this disease can alter the course in a positive manner. We suggest the latter is true and will try to present both a new approach to this disease as well as practical recommendations for treatment changes that are aimed at preventing reflux-induced adenocarcinoma. It is time to take this disease seriously; it is time to graduate from the present management of easing patients' symptoms by having them swallow pills to doing something that will prevent cancer and death.

ASYMPTOMATIC VERSUS SYMPTOMATIC PATIENTS

The present treatment of reflux disease is only directed at patients who are symptomatic with reflux disease. The truly asymptomatic patient cannot be identified without a mass-screening program. It is therefore important to accurately quantitate the prevalence of the asymptomatic patient who develops adenocarcinoma. The higher this number, the greater is the futility in trying to control reflux-induced adenocarcinoma by treating only symptomatic patients. If it is true that the majority of patients who develop cancer never have any symptoms of reflux, mass population screening becomes necessary to control the disease in any significant way.

For this reason, extreme care is needed when making statements about the prevalence of the asymptomatic patient in the population. Spechler (3), in a statement regarding the feasibility of using symptoms as a criterion for screening for Barrett esophagus, stated: "Such an approach will have a limited impact on the rates of death from cancer, because *up to 40% of patients with esophageal adenocarcinoma have no history of gastroesophageal reflux disease*" (p. 837). Spechler's number of 40% is derived from Lagergren et al.'s (4) epidemiologic study, which he cited. Let us carefully examine Lagergren et al.'s study to see if this statement is accurate.

Literature Review

Lagergren J, Bergstrom R, Lindgren A, Nyren O. Symptomatic gastroesophageal reflux as a risk factor for esophageal adenocarcinoma. N Engl J Med 1999;340:825–831.

In the methods section, the authors defined symptomatic reflux for the purpose of this study. Each patient with a diagnosis of adenocarcinoma of the esophagus or gastric cardia was given an 80-minute interview (169 to 553 questions depending on question loops that were present in the protocol) conducted by a trained Swedish government agency interviewer; controls were treated the same. The patients were asked about recurrent heartburn and regurgitation only. To avoid reverse causality—that is, to avoid the possibility that the symptoms were due to the adenocarcinoma—the authors disregarded symptoms that had occurred less than 5 years before the interview. Their Table 4 defines "symptomatic" reflux as the presence of heartburn or regurgitation occurring 1 time per week, which was the definition of the first level of "symptomatic."

The statement that Spechler (3) should have used for his argument is that *40% of patients with esophageal adenocarcinoma did not have heartburn or regurgitation at a frequency of more that once per week more than 5 years before the time of presentation with adenocarcinoma.* It is highly likely that many more patients would have had lesser degrees of reflux symptoms that were recognizable but did not fall under the rather strict definition of the study. It is highly appropriate for Lagergren et al. to use a strict definition, but we must not take their data to suggest that patients who did not reach the criteria of their definition were "asymptomatic."

If 60% of patients with esophageal adenocarcinoma had symptomatic reflux by Lagergren et al.'s (4) criteria, it is certain that this percentage would have been lower if the criteria used to define symptomatic reflux were more stringent (e.g., heartburn or regurgitation at least once per day) and higher if the criteria were less stringent (e.g., heartburn or regurgitation at least once per month or any heartburn or regurgitation at all). In the screening study of Rex et al. (5), 40% of patients gave a history of heartburn, many at a frequency of less than once per week (see later in this chapter for a review of this study).

There is no doubt that some patients who present for the first time with symptoms of cancer give no history suggestive of reflux episodes, even with vigorous inquiry. It is important to know how frequently adenocarcinoma occurs in truly asymptomatic people; these data are unknown. Exaggeration of an asymptomatic population who develop reflux-induced adenocarcinoma creates a sense of futility about our ability to detect and prevent esophageal adenocarcinoma because it suggests that this is not possible without mass screening of the population. This pessimistic attitude is in large part responsible for the inertia and sense of futility that exists at the present time. Though Lagergren et al.'s (4) number of 40% is often used, it is likely that a more sensitive definition of symptomatic reflux would reduce this number considerably.

PRESENT TREATMENT OF REFLUX DISEASE

At the present time, the treatment of reflux occurs at a variety of levels. In this section, we will carefully define these various patient groups and describe how they are presently managed. In the next section, we will show that the present treatment regimen addresses prevention of reflux-induced adenocarcinoma in only a small number of patients. Finally, we will try to show what changes in attitude and treatment regimens are needed to impact the incidence of reflux-induced adenocarcinoma.

Completely Asymptomatic Patients Who Never Present to a Physician

Patients in this group never have any symptoms of reflux, never take any acid controlling medications, never complain to any physician that they have any symptoms of reflux (typical or atypical), and never have any indication for having an upper endoscopy. The exact risk of adenocarcinoma in this group is unknown. However, the likelihood is that it is extremely small. Lagergren et al. (4) showed that the risk of adenocarcinoma progressively increased with severity and duration of symptoms. The odds ratio for esophageal adenocarcinoma for a patient with symptoms (defined as heartburn or regurgitation at least once a week) compared to persons who did not reach this symptom threshold was 7.7 (95% C.I. 5.3 to 11.4). The risk increased progressively with increasing severity and duration of symptoms to reach an odds ratio that was 43.5 (95% C.I. 18.3 to 103.5) with the highest symptom scores and duration greater than 20 years. When one recognizes that 60% of patients with esophageal adenocarcinoma had heartburn or regurgitation at least once a week, extrapolation of this positive relationship between adenocarcinoma risk and symptom severity and duration would suggest that the vast majority of patients who develop adenocarcinoma are likely to have had some symptomatic reflux. This may be classical heartburn or regurgitation at a lower frequency than once a week or one of the myriad atypical symptoms of reflux (chest pain, chronic cough, hoarseness, dysphagia, asthma, etc.) that did not qualify as symptomatic reflux in Lagergren et al.'s study (4).

The likelihood of symptoms in patients who develop adenocarcinoma of the gastric cardia is lower. In Lagergren et al.'s study, the odds ratio for symptomatic reflux in patients with adenocarcinoma of the cardia was 2.0 (95% C.I. 1.4 to 2.9), increasing to 4.4 (95% C.I. 1.7 to 11.0) when reflux was most severe and of long duration. Most patients who develop adenocarcinoma of the cardia (really adenocarcinoma of the dilated end-stage esophagus) do so in very short segments of Barrett esophagus. These occur in patients with low-damage environments with a lower likelihood of symptomatic reflux than patients with longer segments of columnar-lined esophagus that develop esophageal adenocarcinoma. The detection of adenocarcinoma occurring in very short segments of columnar-lined esophagus are likely to prove much more difficult than those associated with longer segments because they are much less likely to have typical reflux symptoms.

The optimistic viewpoint is that esophageal adenocarcinoma is rare in the truly asymptomatic patient. If this is true, and we believe there is some evidence for it, then we can control this disease effectively by addressing the symptomatic patient without the need for mass population screening. Specificity of reflux symptoms is important in the diagnosis of reflux disease. However, the critical factor in screening for reflux disease is the sensitivity of symptoms as an indicator of reflux, not specificity. We need to screen the patient for the presence of *any possible symptom that may indicate the presence of any reflux* before concluding that a patient is asymptomatic.

On a personal note, if any interviewer were to ask Dr. Chandrasoma whether he has had symptoms of reflux and if that interviewer defined it as classical heartburn or regurgitation at any frequency, he would answer that he had no such symptoms. However, we know that he probably does have reflux. This is manifested as a "weird feeling" in his chest that comes on every once in a while over many years. This is quite uncomfortable for a short period and almost immediately eased by a sip of milk. The fact that he takes milk (or swallows excessively) to relieve some symptom in his chest makes him symptomatic. He does not have symptomatic reflux to fit any diagnostic criteria that will ever be used in any study of reflux disease, but he considers himself at risk for reflux-induced adenocarcinoma. Because he is not Caucasian and because his symptoms are mild and infrequent, he believes (optimistically, probably incorrectly, and because he dodges medical procedures) that his risk is very small.

The completely asymptomatic person who develops adenocarcinoma and presents for the first time with symptoms related to the cancer is beyond the scope of any preventive method short of some form of screening protocol.

Completely Asymptomatic Patients Who Have an Upper Endoscopy

Upper endoscopy is performed for many reasons with increasing frequency in the Western world. Many of these patients will be asymptomatic for reflux disease. When the upper endoscopy is performed for vague symptoms such as chronic dyspepsia, the possibility that this is the result of an atypical manifestation of reflux disease is sometimes considered.

At present, these patients receive a passing glance at their distal esophagus and gastroesophageal junction. If an obviously visible columnar-lined esophagus is seen, it is biopsied, and if the pathologist reports intestinal metaplasia, the patient is placed on a Barrett esophagus surveillance protocol. If no intestinal metaplasia is present in the biopsy, the patient is essentially ignored despite the endoscopic abnormality that was present. They are simply classified as "not Barrett's" as if that is a legitimate diagnosis.

If the person appears normal endoscopically, nothing is done if one follows the American Gastroenterological Association practice guidelines (6). With present criteria for diagnosis, these patients cannot have Barrett esophagus unless there is a visible endoscopic abnormality. We hope we convinced you in Chapter 11 that this is not correct. These endoscopically "normal" persons frequently have reflux disease limited to end-stage dilated esophagus, and a significant number will have microscopic Barrett esophagus distal to the presently (and incorrectly) defined endoscopic gastroesophageal junction. Present practice misses this group because of a failure of definition.

Improving understanding of the disease, recognizing the existence of reflux disease limited to end-stage dilated esophagus, and recognizing the existence of microscopic Barrett esophagus by biopsy of endoscopically normal patients are prerequisites to developing a strategy to prevent adenocarcinoma in these patients. Without these, those patients in this group who are destined to develop adenocarcinoma are doomed. The risk of adenocarcinoma in this asymptomatic group is likely to be small, but the cost of a screening biopsy is negligible in a patient already undergoing endoscopy.

Minimally Symptomatic Patients Who Never Seek Medical Care

This is probably the largest population of patients with reflux disease. They can be best identified by the fact that they use some kind of acid controlling device to relieve their minor reflux symptoms. This may be something as simple as the antacid action achieved by increasing swallowing of saliva or sipping milk. These patients will be found among the large population who use over-the-counter antacids to control their symptoms, taking a Tums, Maalox, Mylanta, or Rolaids before or after a meal. They are encouraged to do this by massive advertising campaigns by the manufacturers of these drugs. As over-the-counter drug availability has extended to H2-receptor blockers (Pepcid, Zantac) and proton pump inhibitors (Prilosec), these patients have increasingly moved into the group that needs never seek medical care because of the increasing effectiveness of these drugs in controlling symptoms. The use of these drugs partially suppresses acid, in essence titrating their gastric pH to produce an increase that is just adequate to control symptoms. Peters et al.'s study (7) showed that ranitidine was highly effective in controlling symptoms with partial acid suppression. There is a strong theoretical reason to believe that the resulting increase in gastric pH into the 3 to 5 range may create a milieu that increases the carcinogenicity of the refluxate.

These people do not know that they have a risk of future adenocarcinoma or even recognize that their symptoms are caused by gastroesophageal reflux. At present, there is no effort to recognize these patients and bring them into the care of a physician. Those in this group who are destined to develop adenocarcinoma are doomed. The exact number in this group is unknown but probably significant. They can be identified by careful inquiry into past acid suppressive drug use when they present with cancer.

Symptomatic Patients Who Seek Care and Do Not Have an Endoscopic Examination

Many patients with classical symptoms of reflux disease such as heartburn or regurgitation seek the care of a primary care physician or gastroenterologist and are placed on acid suppressive medications, often long term, without any endoscopic examination or confirmation of the presence of abnormal reflux by a 24-hour pH test. While many of these patients have reflux disease, a significant number of patients whose symptoms are likely not caused by reflux are treated with chronic acid suppressive drugs. In a study of pharmacy billing data from two insurers in Eastern Massachusetts, Jacobson et al. (8) reported that 4684 of 168,727 patients (2%) were prescribed chronic (>3 months) acid suppressive medications. A relevant gastrointestinal diagnosis was found in only 61% of these patients (GERD in 38% and dyspepsia in 42%). Diagnostic testing was uncommon with only 19% having undergone an upper endoscopy within the prior 2 years.

The more sophisticated of these physicians use the proton pump inhibitor as a therapeutic test, withdrawing the drug if it does not control the symptoms and confirming the diagnosis if symptoms are completely controlled. Confirmation of the diagnosis of reflux disease by the proton pump therapeutic test is an indication for long-term acid suppressive therapy. An endoscopy is not considered to be indicated for these patients because the diagnosis of reflux disease has been confirmed and the disease "cured." Those among this group who are destined to develop adenocarcinoma are doomed. In effect, this treatment regimen completely ignores the cancer risk.

At present, this is probably the largest group of patients with recognized reflux disease. In general, they are contented with their visit to the physician because their symptoms have been effectively controlled. Most of these patients are subject to long-term drug therapy at considerable cost, which they generally consider a worthwhile price to pay for the improvement in their quality of life. Their physicians do not inform these patients that they are at risk for cancer. The patients are not sufficiently informed to question the treating physician as to what effort is being taken to decrease the risk of cancer. This is inappropriate even though the risk is very small. There is excellent epidemiologic evidence that the majority of patients who develop reflux-induced adenocarcinoma will be the patients in this group. In Lagergren et al.'s study (4), 60% of patients who developed esophageal adenocarcinoma gave a history of heartburn or regurgitation at a frequency of at least once a week more than 5 years preceding the diagnosis of adenocarcinoma. Even more stunning, Lagergren et al. (4) showed that when this group was compared with regard to their medication history, patients who had taken acid suppressive drugs had an odds ratio of 3.0 for developing esophageal adenocarcinoma compared to symptomatic patients not taking acid suppressive drugs.

The patient-physician interaction in this group is one where the patient has been treated entirely to control symptoms and improve quality of life. However, a known cancer risk has been hidden from the patient and a treatment administered without informed consent that is known not to decrease and probably increase the cancer risk. This is close to the definition of the worst possible patient-physician interaction that one can find in medicine. An opportunity for cancer prevention has been missed without the patient's ever being informed of the cancer risk.

Another group of patients who seek the care of a physician have symptoms that are frequently atypical. They are not correctly diagnosed as having reflux disease when they do seek medical help, and they fall into "control" populations in studies that define reflux by a specific set of symptoms. These "control" populations vary in different studies. In Lagergren et al. (4), the control group is defined by the absence of heartburn or regurgitation at least once a week. Spechler et al. (9) from Harvard defined symptomatic GERD as the presence of heartburn equal to or greater than 1 day per week without mention of regurgitation. At the Cleveland Clinic (10), reflux disease ("classic GERD") was defined by the presence of heartburn and/or acid regurgitation at least twice per week for at least 6 months. This is different from the definition used at the University of Chicago, where Kahrilas et al. (11) defined symptomatic reflux as heartburn equal to or greater than three times per week controlled by acid suppression. These strict definitions of reflux are appropriate for these studies as long as it is clearly recognized that the different studies do not produce comparable data *and that the control group almost certainly includes persons who suffer from symptomatic reflux that is less than defined by the criteria used.* We must not be deluded into believing that we can define away reflux symptoms. Certainly, many of Spechler et al.'s (9) "symptomatic reflux patients" would have fallen into Kahrilas et al.'s (11) "asymptomatic controls" with Cleveland Clinic controls being somewhere in the middle.

The correct diagnosis of patients with atypical symptoms requires a high degree of suspicion on the part of the physician. Any heartburn or regurgitation (irrespective of frequency), or any unexplained chest pain, chronic cough, asthma, or hoarseness must be considered a potential symptom of reflux. Unlike a diagnostic test that should have no false positives, a screening method should have no false negatives. Anyone with even the remotest possibility of symptomatic reflux must be included. The literature has informed physicians about these atypical symptoms of reflux, and physicians tend to consider reflux as a possible diagnosis in these patients with atypical symptoms. Despite this, endoscopy is rarely recommended; rather, these patients are managed with a trial of acid suppressive drugs.

Symptomatic Patients Who Seek Care and Have an Upper Endoscopic Examination

Some patients with reflux symptoms seek the care of a gastroenterologist who will perform an upper endoscopy to evaluate the epithelium and seek an endoscopic confirmation of the diagnosis. Most gastroenterologists will use standard endoscopy. The more sophisticated will use newer techniques such as chromoendoscopy and magnifying endoscopy. Because the patient is having the endoscopy primarily to investigate symptoms of reflux, attention is focused to detect the most subtle changes in the epithelium that are believed to be at risk, which is the tubular esophagus above the endoscopically defined gastroesophageal junction.

With presently accepted criteria for the diagnosis of reflux disease, upper endoscopy is of value only in the approximately 30% of patients who have erosive esophagitis and 5% to 10% of patients who have a columnar-lined esophagus. If there is no erosive esophagitis, a biopsy of the squamous epithelium has such low sensitivity that it has almost no practical value in the diagnosis of reflux disease and is correctly not done in most cases. If there is no visible columnar-lined esophagus, the American Gastroenterological Association guidelines recommend that biopsy not be done (6).

In most patients with reflux symptoms who do not have erosive esophagitis or columnar-lined epithelium, a biopsy is not performed. A diagnosis of nonerosive reflux disease (NERD) is made and the patients placed on acid suppressive drugs with the aim of controlling symptoms. Acid suppressive drugs have been shown to be highly effective in controlling symptoms in nonerosive reflux disease; these patients are "cured."

In the patient who is shown to have erosive esophagitis without columnar-lined esophagus, acid suppressive drugs are used with the aim of not only controlling symptoms but also to effectively heal the eroded squamous epithelium. A biopsy may be taken from the eroded squamous epithelium, but it has little practical value because the end point of treatment success is the healing of endoscopic erosive esophagitis.

In patients who are seen to have a columnar-lined distal esophagus at upper endoscopy done for any reason, biopsies are taken in an attempt to diagnose Barrett esophagus. If there is no intestinal metaplasia, these patients have no diagnosis; the fact of the endoscopic abnormality is ignored. The only exception to this is in the rare patient who has a greater than 3 cm length of columnar-lined esophagus and does not have intestinal metaplasia in the biopsy. These patients are either subject to repeat biopsy because there is evidence that intestinal metaplasia may have been missed or they are classified as "Barrett esophagus" by those who do not believe that intestinal metaplasia is necessary for a diagnosis of Barrett esophagus (12). The latter practice is more prevalent in Europe than in the United States.

In the United States, patients who do not have intestinal metaplasia in the biopsied columnar-lined esophagus are diagnosed as having "no Barrett esophagus." There is no attempt to explain the significance of cardiac mucosa in the abnormal columnar-lined segment in the esophagus from which the biopsy is taken. The level of understanding varies among physicians. The better informed physicians will recognize that this is metaplastic columnar epithelium without intestinal metaplasia, but they may fail to realize its significance as an indicator of reflux disease. The less informed physicians will believe that this is gastric epithelium that has "grown into" the esophagus in some mysterious fashion. Few pathologists will correctly call this reflux carditis and use it as an absolutely reliable criterion for the diagnosis of reflux disease. Most people will regard this patient as not being at risk for adenocarcinoma, although some experts wonder whether adenocarcinoma can arise from cardiac mucosa without an intervening phase of intestinal metaplasia (see the discussion of statetments 1 and 4 in the AGA Chicago workshop of experts; [12]). In general, these patients are treated in a manner identical to the way those who did not have a columnar-lined esophagus at endoscopy are treated. The finding of cardiac mucosa is ignored and the patient placed on long-term acid suppressive therapy. The trigger for surveillance of any kind is not present without intestinal metaplasia.

In all of these situations, the thought that the patient without intestinal metaplasia is at risk for cancer never enters the consciousness of either the patient (who does not know and is not informed that there is a cancer risk) or the physician (who does not know, does not believe, or ignores the existence of a cancer risk). The patients who are destined to develop adenocarcinoma in this group are doomed, because the treatment they are receiving at best has no effect on preventing progression to cancer and at worst may actually increase the cancer risk.

Patients Diagnosed as Having Barrett Esophagus

If the pathologist reports intestinal metaplasia in a biopsy taken from a visible segment of columnar-lined esophagus, the patient is designated as having Barrett esophagus. This raises awareness in the physician (and, with the proliferation of the Internet, also in many patients) that there is a risk of future cancer. These patients are still treated with chronic acid suppressive drugs (often even when they do not have symptoms that need to be controlled). They are also placed under a surveillance protocol with regular endoscopy and biopsy to detect dysplasia and early adenocarcinoma.

On an individual basis, there is evidence that surveillance is capable of detecting a lesion, either early cancer or high-grade dysplasia that precipitates aggressive surgery to remove the lesion before it reaches an advanced stage (13). Patients who have surgical intervention for cancer in the population under surveillance have a much higher chance of survival than patients presenting for the first time with symptoms of the cancer (14).

Unfortunately, the number of patients with a known diagnosis of Barrett esophagus prior to developing cancer is less than 5% of the total number of patients with reflux-induced adenocarcinoma (15). The present treatment regimens are therefore not likely to have any significant impact on the incidence of and mortality from reflux-induced adenocarcinoma.

OUTCOME OF PRESENT TREATMENT REGIMENS

Present treatment strategies for treating reflux disease have had the practical outcome of causing a dramatic increase in the number of deaths from reflux disease since the 1970s. Before the 1960s, deaths from reflux disease were rare and were largely caused by strictures and severe ulceration of the esophagus. While these causes of death have come under control, deaths from reflux-induced adenocarcinoma have increased greatly. In the United States, approximately 8000 people will develop adenocarcinoma of the esophagus and at least a similar number will develop adenocarcinoma in the gastric cardia. Approximately 80% of these patients will die of their disease; this means that around 14,000 people will die every year from reflux-induced adenocarcinoma.

Available evidence suggests that the baseline risk of cancer for a patient with asymptomatic reflux has increased since the 1970s (Fig. 17.1). The risk increases from this baseline as the severity and duration of symptomatic reflux increase (4). For a given level of risk in a symptomatic patient, the odds ratio is 3.0 for persons treated with acid suppressive drugs compared to those without such medications (4).

The response of the medical community to this rapid increase in cancer has been less than stellar. It has not resulted in a change in the attitudes of most primary care physicians and gastroenterologists who continue to treat patients with all stages of reflux disease with acid suppressive drugs, essentially hoping that their patients will not develop cancer. This hope is in vain for 95% of those who are destined to develop cancer; only the 5% of patients who have a diagnosis of Barrett esophagus will be helped by present regimens of treatment where early detection of cancer by surveillance protocols will improve their survival. Treatment has only changed in the direction of increased use of increasingly effective acid suppression. Continuing to treat reflux disease with the sole aim of controlling symptoms and improving the quality of life of patients without recognizing that the emphasis of the disease has shifted and become oncologic is an unforgivable inertia.

If the rate of increase continues unabated and the next two decades bring a further 350% increase in incidence, mortality from reflux-induced adenocarcinoma will approach the approximately 55,000 annual deaths from colorectal cancer. If the present inertia and the rate of increase of this cancer continue, a public health problem of considerable magnitude will soon be upon us. Whether this will happen or not is not predictable.

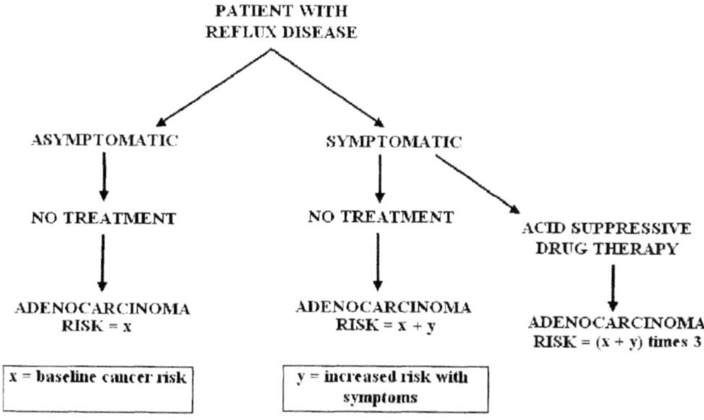

FIGURE 17.1 Factors affecting cancer risk in patients with reflux disease. The asymptomatic patient (*left*) has the baseline risk (x). As the severity of symptoms increases, the risk of cancer increases by a factor y (middle patient). In Lagergren et al.'s study, the risk for esophageal adenocarcinoma is 7.7× for a patient at the lowest severity of symptoms and 43.5× for a patient with the highest severity of symptoms. The symptomatic patient who is on acid suppressive drugs (*right*) has a three times greater risk of esophageal adenocarcinoma than a similarly symptomatic patient who has not taken acid suppressive drugs (Lagergren et al. [4]).

REFLUX DISEASE IS THE PREMALIGNANT PHASE OF CANCER

Gastroesophageal reflux disease, including asymptomatic reflux disease, is a premalignant disease. It is similar to chronic ulcerative colitis and colonic adenomas. This is not presently recognized despite the fact that the data are overwhelming that it is true. At present, only Barrett esophagus is considered to be premalignant.

The treatment strategy for reflux disease without Barrett esophagus must shift from the present attitude of treating it with the aim of pain relief and improvement of quality of life. Oncologic principles must be invoked with aggressive attempts to prevent cancer. We must treat reflux disease with the same attitude that we adopt for preventing death from breast cancer, with aggressive early detection techniques and aggressive eradication of premalignant disease or risk factors. Oncologic principles demand that we err on the side of aggression in managing the premalignant state and do so at the earliest point of prevention. Only 30% of abnormal mammographic lesions that are excised turn out to be malignant; unnecessary surgery in the 70% of patients who have benign causes for their mammographic abnormality is considered a worthwhile price to pay for detecting and treating the 30% of patients who will have early cancer. This is a sea change from the present attitude of treating reflux disease where we essentially do nothing for most patients until cancer develops.

It is crucial to look at reflux disease in this new light. We must concentrate on the approximately 14,000 people in the United States who will die of this disease every year rather than the millions of people with reflux disease who will never get cancer. Preventing cancer must take precedence over everything else. If this means that some patients are overtreated, so be it, as long as the consequences of overtreatment are not associated with significant morbidity. A balance needs to be established whereby the maximum number of cancers are prevented at the minimum cost to those patients who are overtreated and who would never have developed cancer. This balance is not easy. The present treatment has no balance at all; we simply sacrifice the more than 90% of patients who present for the first time with cancer in order that we not overtreat patients with reflux disease who are destined not to get cancer.

The preventive protocol in place for breast cancer requires that every woman undergo a screening mammography at age 40. It requires every mammographic abnormality that is found to be aggressively followed to a specific diagnosis by biopsy. The 70% of patients who have benign causes for their mammographic abnormality are subject to the trouble of getting regular mammograms, the significant psychologic stress of knowing they have an abnormal mammogram, and a needle biopsy often followed by a large excisional biopsy that leaves a scar and disfigurement of the breast. This is all for naught because if the screening mammogram was not done, these patients would never have had any need for surgery and would probably never have developed cancer. However, if this scenario is painted to the women who are subject to this exercise, there will be overwhelming agreement that it is worthwhile to detect early cancer in the 30%. Mammographic screening is extremely costly; it has been estimated that it costs $55,000 for each breast cancer detected (3). Mammographic screening was recommended without any evidence that it would be capable of decreasing the mortality from breast cancer; in fact, a Danish study questioned the effectiveness of mammography in improving overall survival. This is the typical aggressive response to an oncologic disease with a significant risk of death.

If we are to look at reflux disease as an oncologic disease in a manner similar to breast cancer, we have to ask a whole new series of questions to see if a change in the attitude from the present passive stance to an aggressive onslaught has the ability to impact the incidence of reflux-induced adenocarcinoma. If we do nothing, we commit to doing increasing numbers of radical esophagectomies as the number of cancers increase.

ANTIREFLUX SURGERY CAN PREVENT PROGRESSION IN THE REFLUX TO ADENOCARCINOMA SEQUENCE; CANCER CAN BE PREVENTED

The present treatment of reflux disease is based on the absolute belief that while reflux diseases with or without Barrett esophagus are premalignant diseases by epidemiologic data, it is pointless to attempt to identify these patients because there is nothing that can be done about them. This sense of futility is clearly evident when one reviews the consensus of experts at the AGA Chicago Workshop on Barrett esophagus (12). The opinion of these 18 experts (15 gastroenterologists, 2 surgeons, and 1 pathologist) regarding those issues that may impact on the ability to prevent adenocarcinoma are as follows:

Statement	Number of experts (n = 18) accepting statement[a]
11—Endoscopic screening for BE and dysplasia has been shown to improve mortality from esophageal adenocarcinoma.	1 (C)
16—Endoscopic screening for BE and dysplasia should be performed in all adults >50 years of age with heartburn.	0
22—Endoscopic surveillance in patients with BE has been shown to prolong survival.	5 (B) + 4 (C)
35—Normalization of esophageal acid exposure by acid suppression reduces the risk for development of esophageal adenocarcinoma. (Note: Includes drugs and antireflux surgery.)	2 (C)
39—In patients with BE, antireflux surgery has not proven to have a major effect against the development of esophageal adenocarcinoma.	15 (A) + 3 (B)
41—Mucosal ablation with intensive acid suppression or antireflux surgery prevents adenocarcinoma in patients with BE without dysplasia.	1 (A)

[a] A = accept completely; B = accept with some reservation; C = accept with major reservation; D = reject with reservation; E = reject completely.

The opinion of this expert group, which is largely made up of gastroenterologists, is that screening for Barrett esophagus is worthless in terms of improving mortality and therefore should not be considered, surveillance of patients with Barrett esophagus does not prolong survival even though it is the standard of practice, and all treatments for Barrett esophagus including acid suppressive drug therapy, antireflux surgery, and ablation have no ability to prevent or decrease the incidence of adenocarcinoma in patients with Barrett esophagus. The sense of futility is profound. These experts have nothing to offer the 14,000 people annually in the United States who are destined to develop reflux-induced adenocarcinoma and die of their disease. One of the reasons for the pessimism in the results of this workshop is the way the statements were framed; it almost seemed as if those who were framing the questions had a desire to establish complete futility.

Such passivity and pessimism are reflections of a disease whose patients are managed primarily by physicians who are not adequately trained to deal with an oncologic disease (internists and gastroenterologists). The passive approach to reflux disease contrasts with the aggressive approach to breast cancer, at least partly because breast cancer is primarily handled by physicians trained in oncology (surgeons).

There is no reason for this futility. Reflux-induced adenocarcinoma is theoretically an eminently preventable disease. The only cancer-preventive protocol today is long-term surveillance with random biopsy in patients with Barrett esophagus to detect early and noninvasive cancer. There is strong evidence that surveillance improves survival in patients with Barrett esophagus (Peters et al.). Surveillance with extensive biopsy is a tedious and time-consuming protocol. Unless there is an absolute commitment to the value of surveillance, there is a possibility that surveillance protocols will be less than adequate. In the University of Washington, Seattle, Barrett esophagus project, the actual median surveillance interval in 309 patients in one study during the 1983–1988 period was 25 months for patients with a baseline diagnosis of nondysplastic Barrett esophagus, 18 months for patients with a baseline diagnosis of indefinite for dysplasia or low-grade dysplasia, and 5 months for patients with a baseline diagnosis of high-grade dysplasia (16). If this is the norm at the University of Washington, which is the home of the Seattle biopsy protocol, the probability is that the surveillance interval is even longer in institutions that are less academically focused on this issue. Surveillance will not work if it is not done properly.

In Chapter 16, we analyzed the data that show that considerable evidence exists that antireflux surgery decreases the risk of reflux-induced adenocarcinoma (Table 17.1). Theoretically, successful antireflux surgery performed early in the course of the disease *must halt the progression to cancer because such progression is dependent on the interaction of refluxate molecules and target cells.* Evidence indicates that this is true. If performed before intestinal metaplasia develops in columnar-lined esophagus, antireflux surgery is effective in preventing the occurrence of intestinal metaplasia and encourages the conversion of cardiac mucosa to oxyntocardiac mucosa. If performed in patients with Barrett esophagus before all the genetic mutations required for cancer have occurred, it prevents the occurrence of cancer.

TABLE 17.1 The Choice between Acid Suppressive Therapy and Antireflux Surgery in the Treatment of Patients with Reflux Disease

	Acid suppressive drug therapy	Antireflux surgery
Number of patients treated	Millions	Thousands
Control of symptoms	90%+	90%+
Healing erosive esophagitis	90%+	90%+
Stopping acid exposure of esophagus (pH test)	Effective	Effective
Stopping bile exposure of esophagus (Bilitec)	Not effective	Effective
Decreasing total reflux	Not effective	Effective
Evidence that it causes regression of intestinal metaplasia	Maybe	Yes, only with short lengths
Evidence that it increases conversion of cardiac mucosa to oxyntocardiac mucosa	No	Yes
Evidence that it decreases proliferative activity in Barrett esophagus	Yes	Yes
Theoretical likelihood of preventing cancer	No, unless acid is the only agent	Yes, all agents removed
Evidence that it may increase cancer incidence	Yes	No
Evidence that it decreases incidence of cancer	No	Yes, not strong
Effect on decreasing incidence of cancer adequately tested	Yes	No

If it is accepted that antireflux surgery is capable of decreasing the progression to carcinoma in patients with target cells in the esophagus that are susceptible to carcinogenesis, the prevention of reflux-induced adenocarcinoma becomes theoretically possible (Fig. 17.2). All that needs to happen is to identify patients who are at risk and treat them with antireflux surgery, assuming that this procedure has a high success rate and that it does not cause severe morbidity.

It is inevitable that this course of action will result in a large number of antireflux procedures in patients who will never develop adenocarcinoma. This is the equivalent of breast resections for mammographic abnormalities when it is known that 70% will be benign. The emphasis of attack has shifted from making people who will never develop cancer comfortable to focusing on preventing cancer in those who are destined to do so.

In actuality, antireflux surgery has far greater value for a person with reflux disease than breast resection for a patient with a benign mammographic lesion. For the patient who has a benign breast lesion excised simply because it was detected at screening, there is absolutely no benefit to compensate for the unnecessary surgery and consequent disfigurement. For the patient with reflux disease, antireflux surgery represents an excellent method of controlling reflux symptoms in addition to its role in preventing cancer. Around 90% to 95% of patients achieve complete symptom control and healing of erosive esophagitis, and over 80% of patients do not require long-term acid suppressive drug therapy. The lifelong cost of antireflux surgery is less than the cost of long-term acid suppressive drug therapy when surgery is successful.

Antireflux surgery truly cures patients with reflux disease in all senses of the word. It controls symptoms, heals erosive esophagitis, prevents intestinal metaplasia, reverses some cases of intestinal metaplasia, increases the conversion of cardiac mucosa to the more benign oxyntocardiac mucosa, and prevents adenocarcinoma in Barrett esophagus (Fig. 17.2). This is aggressive cancer prevention in the true oncologic sense.

There is no proof as yet that antireflux surgery will reduce esophageal adenocarcinoma. There is only the theoretical expectation that it will and some evidence that it does. The proof must come after large numbers of cases have been explored in a correctly designed study setting. There is, however, a strong rationale for making antireflux surgery the primary treatment for reflux disease. The alternative, which is acid suppressive drug therapy, has been used in millions of people with general agreement that it is a futile exercise if one's aim is to prevent cancer.

Passivity and inertia will maintain the present guidelines of management of reflux disease. There is a certainty that this will not decrease the probability of death in 95% of the 16,000 to 18,000 patients in the United States who are destined annually to develop reflux-induced adenocarcinoma. In fact, this number is likely to continue to increase, either because acid suppressive drug therapy promotes carcinogenesis or because there are unknown factors causing an increase in the incidence of reflux-induced adenocarcinoma. To continue to use a treatment regimen that has no chance without looking for an alternative is the ultimate inertia. Requiring absolute proof that antireflux surgery will decrease the incidence before recommending its use in the treatment of reflux is a classical catch-22 situation; the proof cannot come until large numbers of patients are treated and followed in appropriately designed study settings. The fact that there are indications of a positive effect in decreasing cancer with the small series of patients treated with antireflux surgery to date points strongly to a high probability that antireflux surgery is likely to have a profound cancer-preventing effect. *The choice is clear; retain present treatment regimens and ensure continued failure or change the treatment and expect control of reflux-induced adenocarcinoma.*

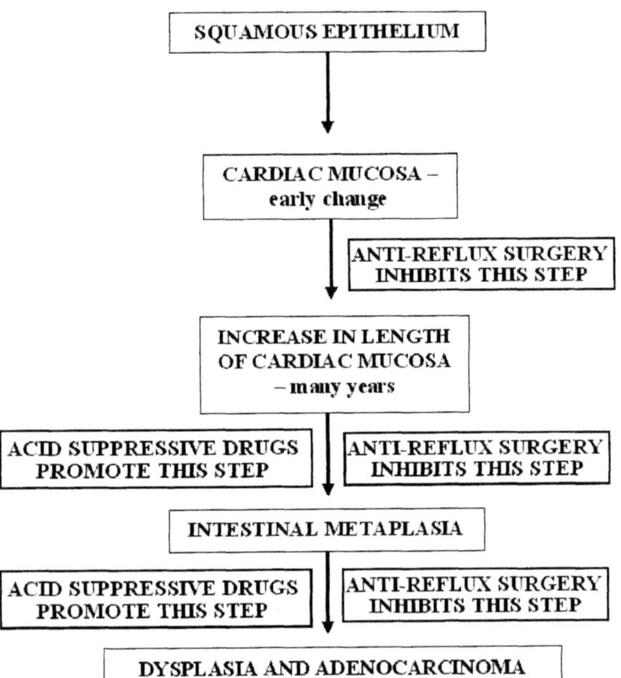

FIGURE 17.2 Steps in the reflux-adenocarcinoma sequence. The conversion of cardiac mucosa to intestinal metaplasia and carcinogenesis in intestinal metaplasia are promoted by acid suppressive drug therapy and inhibited by antireflux surgery.

DEFINITION OF THE BIOPSY PROTOCOL FOR INDEX DIAGNOSIS

The availability of a method that can impact the cellular events of the reflux-adenocarcinoma sequence makes it imperative that we become aggressive in identifying risk accurately in these patients. This requires a radical change from the present passivity to an aggressive seeking of information related to the cellular events in these patients by dramatically increasing the indications for endoscopy in these patients, using an aggressive biopsy protocol in all patients who undergo endoscopy, and then making logical treatment decisions based on the cellular changes that are present.

All the data in our studies are based on a standard biopsy protocol that systematically samples the area between the squamocolumnar junction and gastric oxyntic mucosa (the true gastroesophageal junction) (Fig. 17.3, Table 17.2). The protocol defines the adequacy of a biopsy series when the most proximal level contains squamous epithelium above the squamocolumnar junction and the most distal level contains gastric oxyntic mucosa (17). The location of the true gastroesophageal junction cannot be determined by endoscopy; it is 0 to 3 cm distal to the end of the tubular esophagus and the proximal limit of the rugal folds, depending on the length of end-stage dilated esophagus present in the patient. By this protocol, the length of columnar-lined esophagus is defined by the maximum separation of the squamocolumnar junction from gastric oxyntic mucosa in the measured biopsies. If the length of columnar-lined esophagus is greater than 1 cm, there will be at least two measured biopsies containing metaplastic esophageal columnar epithelia (cardiac mucosa with and without intestinal metaplasia and oxyntocardiac mucosa). Any patient who has metaplastic epithelia in biopsies that do not represent two measured levels greater than 1 cm apart is classified as having a columnar-lined segment less than 1 cm long. In patients who have a segment that is longer than 1 cm, the maximum distance between two measured biopsies that contain metaplastic epithelia defines the length.

In practice, this biopsy protocol means that four sets of biopsies are taken in patients who have a normal

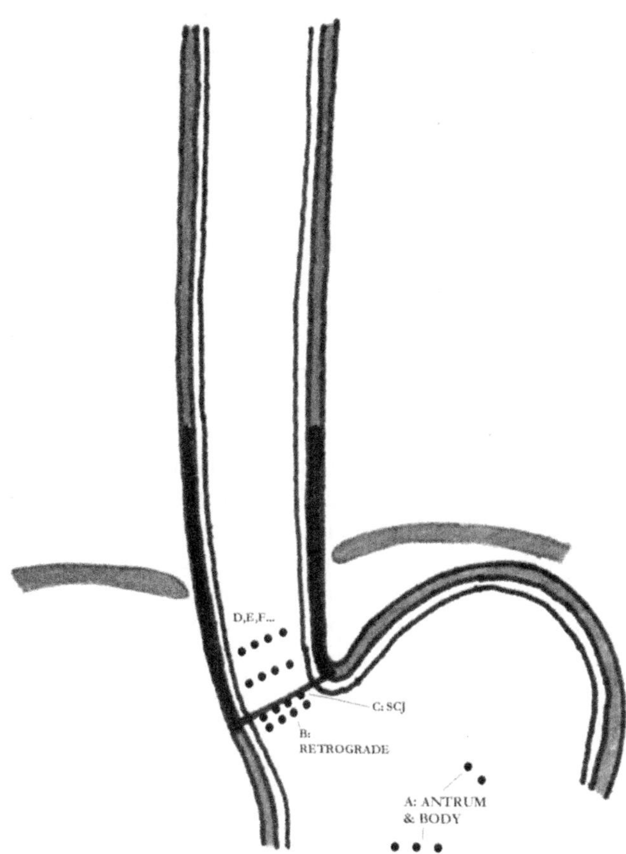

FIGURE 17.3 Ideal biopsy protocol for a patient with reflux disease. Biopsies are taken from the squamocolumnar junction, the proximal limit of rugal folds, and 1 and 2 cm distal to the proximal limit of rugal folds and distal stomach. In a patient who has a visible columnar-lined esophagus between the proximal limit of rugal folds and the squamocolumnar junction, additional biopsies are taken at 1- to 2-cm intervals.

TABLE 17.2 The Ideal Biopsy Protocol for Diagnosis

Biopsy location	Circumstance	Purpose
A. At SCJ, attempting to straddle the junction	Always	Define the presence and proximal limit of CLE; define columnar epithelium at SCJ
B. 2 cm distal to end of tubular esophagus	Always	Find oxyntic mucosa to define distal limit of CLE
C. 1 cm distal to end of tubular esophagus	Always	Define reflux disease in end-stage dilated esophagus
Measured biopsies at 1–2-cm intervals between A and C	When CLE is seen endoscopically	Define length of CLE
D. Gastric antrum and body	Recommended	Define coexisting gastric pathology

endoscopic appearance: (a) A four-quadrant biopsy from the squamocolumnar junction that attempts to straddle the actual junction so that the epithelium immediately distal to the squamous epithelium can be seen, (b) a four-quadrant biopsy from the columnar epithelium 2 cm distal to the squamocolumnar junction under the gastric rugal folds and distal to the end of the tubular esophagus, and (c) a four-quadrant biopsy from the columnar epithelium within 1 cm from the squamocolumnar junction under the gastric rugal folds and distal to the end of the tubular esophagus; (b) and (c) can be combined. (d) Biopsies from the antrum and body of stomach.

In a patient who has a visible columnar-lined segment, four-quadrant biopsies are taken at 1 to 2 cm intervals beginning 2 cm distal to the end of the tubular esophagus to the squamocolumnar junction. The number of levels biopsied depends on the length of the columnar-lined esophagus present. Again, the gastric antrum and body are also sampled.

This is the perfect biopsy protocol. This ideal protocol is cumbersome and is unlikely to be used routinely at upper endoscopy by gastroenterologists in a busy community practice. As such, recommending this protocol for general use has a high likelihood of failure in practice. Extensive biopsies are essential during surveillance of patients with known Barrett esophagus where the aim is to find dysplasia, which is a random event that can occur anywhere in the columnar lined segment (18). Extensive biopsy is not essential at the index endoscopy.

At the first or index endoscopy, the intent is to identify the presence of intestinal, cardiac, oxyntocardiac and gastric oxyntic mucosa distal to the squamocolumnar junction. The presence of a regular and consistent zonation of epithelial types in any columnar-lined segment of esophagus, short or long, means that the index biopsy needs to sample only the area immediately distal to the squamocolumnar junction (19, 60; Table 17.3). *All that is needed in this index endoscopy is a single four-quadrant biopsy sampling of the columnar epithelium immediately distal to the end of the squamous epithelium*; ideally, the biopsy should attempt to straddle the junction so that the transition from squamous to columnar epithelium is present in the biopsy. If intestinal metaplasia is present, this will be its location in the vast majority of cases; its absence in this biopsy reliably excludes Barrett esophagus. In patients without intestinal metaplasia, the presence of cardiac mucosa will define the presence of reflux disease; the absence of cardiac mucosa in this biopsy reliably excludes the presence of reflux disease. In patients who have neither intestinal epithelium nor cardiac mucosa at the junction, the columnar epithelium will be either oxyntocardiac or gastric oxyntic. Patients with only oxyntocardiac mucosa as a metaplastic esophageal columnar epithelium have no risk of cancer; we call this group "compensated reflux." Patients with a squamous-oxyntic mucosal junction are normal because they have no evidence of columnar-lined esophagus; this is a rare finding at endoscopy in an adult population.

The taking of a single four-quadrant biopsy at the index endoscopy from the columnar epithelium immediately distal to the squamocolumnar junction is an effective strategy whether or not the patient has a visible segment of columnar-lined esophagus. Even with long-segment columnar-lined esophagus, the zonation is maintained, and the highest yield of critical information will result from a single four-quadrant biopsy taken in the manner described. In patients with a visible segment of columnar lined esophagus, the probability of finding intestinal metaplasia is sufficiently high that it is probably worthwhile to go with the standard protocol of four-quadrant biopsies at 1- to 2-cm intervals at the index endoscopy. The probability of finding intestinal metaplasia increases with increasing length of columnar lined esophagus.

This recommended new biopsy protocol for an index endoscopy is a significant change from standard biopsy practice at present. The present recommendation is to not take biopsies in patients whose endoscopy appears normal. Upon encountering a patient with a visible segment of columnar-lined esophagus, most gastroenterologists will usually take multiple random biopsies from the entire segment. This has the highest probability of sampling error (i.e., missing intestinal metaplasia) for a given number of biopsies.

TABLE 17.3 Adjusted Practical Biopsy Protocol for Index Biopsy

Biopsy location	Circumstance	Purpose
A. At SCJ, attempting to straddle the junction	Always	Define presence and type of columnar epithelium at SCJ
B. Gastric antrum and body	Recommended	Define coexisting gastric pathology

If a limited number of biopsies is to be taken, they should all be concentrated in the area immediately distal to the squamocolumnar junction rather than randomly distributed. This will minimize a false negative diagnosis of intestinal metaplasia based on the regular zonation of intestinal epithelium to the most proximal part of the columnar-lined segment.

The intent of biopsy at the index endoscopy is to maximize the probability of identifying the presence of epithelia that are at highest risk (i.e., primarily intestinal and then cardiac). The intent is not to search for dysplasia or cancer unless a visible abnormality that is suspicious for cancer is seen at endoscopy. While the ideal biopsy protocol is the one described at the top, the objective of diagnosis is met by the limited, less time-consuming biopsy protocol, balancing the practical need for efficiency with retention of diagnostic accuracy that is essential for a busy community practice.

CLASSIFICATION OF RISK BASED ON BIOPSY RESULTS

In Chapter 14, we developed a system to stratify all persons in the population into logical risk groups based on the recognition of the reflux-adenocarcinoma sequence and significance of different epithelial types (Fig. 17.4). This recognizes (a) that the esophageal target cell for carcinogenesis is found only in intestinal metaplastic epithelium; (b) that the only esophageal target cell that is at risk for undergoing intestinal metaplasia resides in cardiac mucosa; and (c) that squamous, oxyntocardiac, and gastric oxyntic mucosa are outside the sequence of cellular events that lead to adenocarcinoma in reflux disease (Fig. 17.4).

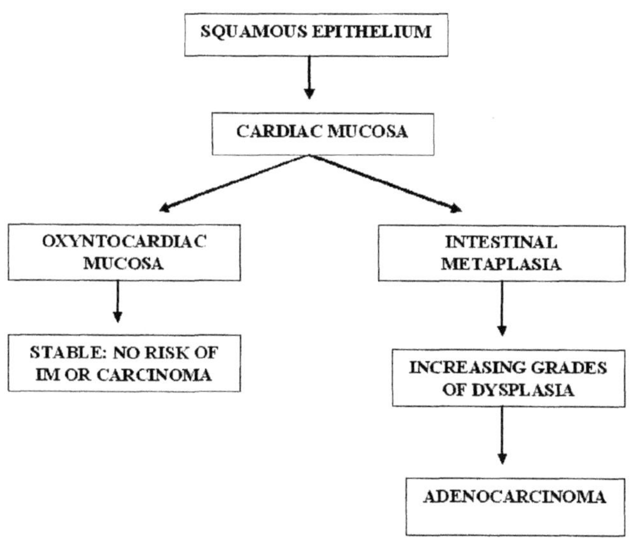

FIGURE 17.4 The steps in the reflux-adenocarcinoma sequence, which forms the basis for the histologic grading system recommended for the assignation of risk level to patients.

Based on these facts regarding the sequence of events that occur as reflux damages esophageal squamous epithelium and induces the development of different types of columnar epithelia, we derived the classification of patients shown in Table 17.4 (17).

The practical index biopsy protocol recommended has a significant risk of missing dysplasia because this is a random event that occurs anywhere in the columnar-lined segment. This is not a serious problem because dysplasia is limited to patients who have intestinal metaplasia. Patients who have intestinal metaplasia detected at the index endoscopy fall into a high-risk group that will need a repeat endoscopy with extensive sampling to evaluate whether dysplasia is present before a decision is made regarding further management.

The application of the biopsy protocol permits the assessment of risk in these patients in terms of how far they have progressed in the sequence of the cellular event that ultimately result in adenocarcinoma. Appropriate treatment strategies can be formulated based on this risk.

RECOMMENDATIONS FOR TREATMENT OF BIOPSY-DEFINED GROUPS

At the index endoscopy, which is now done in the vastly increased pool of people defined as being at risk for cancer, all patients will be biopsied according to the recommended new biopsy protocol. The limited biopsy protocol recommended has a likelihood of general application in community practice because it is easy and not time-consuming. If a reimbursement code is attached to ensure that an endoscopy with biopsy is valued higher than one without biopsy, successful implementation can be further assured.

Based on the results of the biopsies done according to protocol, patients will be classified by the grading system that we have suggested (Table 17.4). Patients who are in grade 0 (i.e., without cardiac and oxyntocardiac mucosa) can be reassured that they are at very low risk for cancer. If these patients are asymptomatic, they need no further consideration unless they are very young when the possibility that their reflux disease may significantly evolve in the future must be considered (Table 17.5). If they are symptomatic, a decision can be taken between acid suppressive drug

TABLE 17.4 Classification of the Population into Groups Based on Their Risk of Developing Adenocarcinoma of the Esophagus (the percentage of the population that is probably within each group is given in parentheses)

Grade 0 (55% to 65%)

Definition: No cardiac or intestinal metaplasia in any biopsy.
Histology: Squamous, oxyntocardiac, and oxyntic mucosa present.
Inference: No target cells directly at risk for intestinal metaplasia or cancer.
Risk of intestinal metaplasia: Zero.
Risk of cancer progression: Zero.

Grade 1: Reflux disease (30% to 45%)

Definition: Cardiac mucosa (reflux carditis) present; intestinal metaplasia absent.
Histology: Squamous, oxyntocardiac, and oxyntic mucosa can be present in addition to cardiac mucosa.
Inference: No target cells directly at risk for cancer; target cells directly at risk for intestinal metaplasia present.
Risk of intestinal metaplasia: Present.
Risk of cancer progression: Zero.

Grade 2: Barrett esophagus (5% to 15%)

Definition: Intestinal metaplasia (goblet cells) present in cardiac mucosa.
Histology: Squamous, oxyntocardiac, nonintestinalized cardiac mucosa and oxyntic mucosa can be present in addition to intestinal metaplasia.
Inference: Target cells directly at risk for cancer present.
Risk of cancer progression: Present.

Grade 3: Neoplastic Barrett esophagus

Definition: Dysplasia or adenocarcinoma present.
Histology: Squamous epithelium, nondysplastic intestinal metaplasia, cardiac, oxyntocardiac, and oxyntic mucosa can be present.
Inference: Progression toward cancer is present.
Risk of invasive cancer: Greater with increasing degrees of dysplasia.

therapy and antireflux surgery based on cost and patient preference without any regard to cancer risk. In these patients without cardiac and intestinal epithelia, acid suppression is likely to carry no risk of promoting cancer.

Patients with symptomatic reflux carditis (grade I) should be offered immediate antireflux surgery as an alternative to acid suppressive drug treatment as the primary treatment modality. These patients are in the earliest point in the reflux to adenocarcinoma sequence. The earlier the reflux is stopped, the more effective is antireflux surgery as a cancer preventive measure. Cardiac mucosa will be induced by antireflux surgery to switch to oxyntocardiac mucosa, and intestinal metaplasia will often be prevented. The target cell for carcinogenesis never arises in the esophagus, and cancer is prevented. Acid suppressive drugs, while they effectively control symptoms, cause the cardiac mucosa to switch to intestinal metaplasia and increase the likelihood of cancer. There is no logical justification for choosing life-long acid suppressive drug therapy over antireflux surgery in the symptomatic patient with cardiac mucosa. Surgery is at least as effective and probably not more costly on a lifetime basis.

Asymptomatic patients with reflux carditis (grade I) are common. These patients have no immediate risk of carcinogenesis and require no treatment. However, they should have a follow-up endoscopy (after an interval to be determined, e.g., in 5 years) to ensure that they have not progressed to grade II.

Patients with intestinal metaplasia (Barrett esophagus, grade II) should be urged to have an antireflux operation whether or not they are symptomatic. The primary objective of this is to prevent cancer. A secondary benefit is the control of symptoms in the symptomatic patient. Antireflux surgery in patients with Barrett esophagus must have the proviso that this surgery will prevent cancer *only if the genetic changes required for cancer have not already occurred.* These patients need surveillance after surgery to ensure they do not have prevalent cancer at the time of the antireflux surgery and to ensure, with 24-hour pH testing, that the surgery has successfully abolished reflux.

TABLE 17.5 Comparison of the Present Approach to Management of Different Patient Groups with a Suggested Aggressive Approach Designed to Prevent Adenocarcinoma

Patient group	Present management	Suggested approach
Completely asymptomatic patients who never present to a physician	Ignore	Screening of high-risk groups; mass population screening
Completely asymptomatic patients who have an upper endoscopy	Ignore if gastroesophageal junction is normal; biopsy if there is a visible columnar-lined esophagus	Mandate biopsy according to protocol to assess risk
Minimally symptomatic patients who never seek care	Self-medication with over-the-counter acid suppressive drugs to control symptoms	Make these drugs prescription only; warning labels if over the counter
Symptomatic patients who seek care and do not have endoscopy	Treated with acid suppressive drugs to control symptoms	Mandate endoscopy with biopsy to assess risk
Symptomatic patients who have endoscopy and have a visible columnar lined esophagus	Biopsy to diagnose intestinal metaplasia; if negative, treat with acid suppressive drugs	Mandate biopsy according to protocol to assess risk
Symptomatic patients who have a normal endoscopy	Treat with acid suppressive drugs	Mandate biopsy according to protocol to assess risk
Patient group after endoscopy and biopsy	**Present management**	**Suggested approach**
Grade 0: Asymptomatic patients with no carditis or intestinal metaplasia	Ignore	Ignore; reassure there is no risk of cancer
Grade 0: Symptomatic patients with no carditis or intestinal metaplasia	Treat with acid suppressive drugs	Treat with acid suppressive drugs; reassure there is no risk of cancer
Grade 1: Asymptomatic patients with carditis	Ignore	Repeat endoscopy in 5 years
Grade 1: Symptomatic patients with carditis	Treat with acid suppressive drugs	Antireflux surgery
Grade 2: Asymptomatic patients with Barrett esophagus	Surveillance; ?acid suppressive drugs	Antireflux surgery
Grade 2: Symptomatic patients with Barrett esophagus	Surveillance; acid suppressive drugs	Antireflux surgery
Grade 3a (low-grade dysplasia)	Treat with long-term acid suppressive drugs; long-term surveillance	Antireflux surgery; surveillance for 5 years if pH normalized and dysplasia disappears
Grade 3b (high-grade dysplasia)	Intense surveillance; esophagectomy if there is a visible lesion or multifocal high-grade dysplasia	Transhiatal vagal sparing esophagectomy
Grade 3c (invasive adenocarcinoma)	Esophagectomy	Esophagectomy

Patients with dysplasia and cancer (grade III) are presently managed reasonably well and require only a slight shift in recommendations. Patients with low-grade dysplasia should be encouraged to have antireflux surgery before being placed on a surveillance protocol to detect cancer. If cancer has not occurred 5 years after a proven (by 24-hour pH test) successful antireflux operation, surveillance can be stopped. Patients with proven multifocal high-grade dysplasia and cancer should be managed by surgery. The present suggestion that unifocal high-grade dysplasia can be treated with careful observation will appear meaningless in an aggressive cancer-preventive approach to surgery. High-grade dysplasia is equivalent to carcinoma in situ and must be managed aggressively as a high-risk lesion.

We have only considered acid suppressive drug treatment and antireflux surgery as treatment options. At present, ablative methods including endoscopic mucosal resection are considered appropriate only for patients with high-grade dysplasia and adenocarcinoma who are poor surgical candidates. As these ablative techniques improve and become more available, they have the potential to replace surgery in patients with Barrett esophagus without dysplasia and with low-grade dysplasia. Ablative methods must be combined with some method of effective acid suppression; as such the cost of these methods is likely to be additive to the method of acid suppression selected (drugs or antireflux surgery).

Refinement of this basic strategy will become necessary as new molecular predictors of risk and new molecular treatment modalities become available. There is a suggestion that this is already possible with special procedures such as flow cytometric data (see Chapter 10) and the presence of molecular markers such as p53 or Cox-2, but this is of secondary importance to histology at this time and not indicated for routine patient management because the data lack specificity.

RECOMMENDED CHANGES IN THE TREATMENT OF DEFINED PATIENT GROUPS

The exact application of this system and the way it changes treatment will now be considered in the various patient groups that we have identified, stressing the manner in which the approach to these patients must change from the present passivity in order that an aggressive cancer preventing approach can be taken (Table 17.5).

Completely Asymptomatic Patients Who Never Present to a Physician

We have emphasized that the number of truly asymptomatic patients may be greatly exaggerated because of the fact that restrictive definitions of symptomatic reflux are used. If symptomatic reflux is defined in a more sensitive manner to include heartburn and regurgitation at low frequency (less often than once per week) and atypical symptoms (unexplained chest pain, "dyspepsia," chronic or recurrent cough, hoarseness, asthma, dysphagia, etc.), the incidence of the truly asymptomatic patient may actually be low. The best test to define a truly asymptomatic patient maybe to ask the question: "Have you ever taken any over-the-counter antacid medications?" If the answer to this question is affirmative, the patient is very likely to have symptomatic reflux, sometimes in denial, and sometimes not even aware that he or she is taking these medications for reflux.

In Rex et al. (5), where upper endoscopic screening was offered to patients presenting for colonoscopy (reviewed fully later in the chapter), the patients were carefully questioned for symptoms of reflux. The results show that 25.1% reported heartburn less than once per week; 6.4% had heartburn once per week; 6.5% had heartburn several times per week; and 2.7% had heartburn daily. This shows that 15.6% would be considered "symptomatic" by the standard definition used for symptomatic reflux in most studies, but almost double this number would have heartburn at a frequency that would be less than the standard definition. These patients have the classical reflux symptom and are certainly not truly asymptomatic. If reflux symptoms other than heartburn are included, the numbers that could be considered as symptomatic reflux disease will increase still further.

The truly asymptomatic patients are completely outside the control of the medical establishment. Only some type of screening measure can detect them. This may represent (a) mass screening of the population, which could be either the entire population or one with a defined demographic risk factor for reflux-induced adenocarcinoma such as age greater than 50 years, male, of the Caucasian race; or (b) upper endoscopy offered to patients as an additional test when they present for colorectal cancer screening.

At present, there is no mass screening protocol aimed at detecting the precursor lesions of esophageal adenocarcinoma in the general asymptomatic population. This is appropriate. Recommending mass screening for the entire population (e.g., at age 40) is unlikely to be cost-effective because the risk of adenocarcinoma in this group is likely to be extremely small. As we will show, the prevalence of long-segment Barrett esophagus in this population is less than 0.5%, the prevalence of short-segment Barrett esophagus is around 5%, and microscopic intestinal metaplasia in the distal end-stage dilated esophagus is around 10%. Lagergren et al. (4) showed an increasing risk of adenocarcinoma with increasing severity of symptoms; this suggests that the risk in the truly asymptomatic is likely to be too low to justify mass screening of patients at this time. Mass screening may become cost-effective in the future if the incidence of reflux-associated adenocarcinoma continues to increase.

It has been suggested that upper endoscopic screening be offered to patients presenting for screening colonoscopy. This represents a prime group for considering upper endoscopic screening for esophageal adenocarcinoma; the upper endoscopy is done at the same sitting with the same sedation and adds little to the screening colonoscopy. This group of patients is sufficiently concerned about cancer to have screening colonoscopy and usually have health care benefits that cover the cost of this screening. This affluent group is also at highest risk for esophageal adenocarcinoma. It is the logical first group to evaluate the cost-effectiveness of screening because the procedure is additive to an already planned procedure and likely to be the least expensive in terms of screening.

Literature Review

Rex DK, Cummings OW, Shaw M, Cumings MD, Wong RKH, Vasudeva RS, Dunne D, Rahmani EY, Helper DJ. Screening for Barrett's esophagus in colonoscopy patients with and without heartburn. Gastroenterology 2003;125:1670–1677.

The study group consisted of 961 outpatients older than 40 years who were scheduled for colonoscopy and accepted an invitation to have a screening upper endoscopy. They had no clinical indication for and had not had a previous upper endoscopy. Patients completed two validated heartburn questionnaires prior to their procedure.

At endoscopy, the following landmarks were identified: (1) the squamocolumnar junction (SCJ); (2) the esophagogastric junction (EGJ), defined as the proximal end of the gastric folds where the tubular esophagus meets the stomach; and (3) the diaphragmatic hiatus. Long-segment Barrett esophagus (LSBE) was defined as columnar epithelium with specialized intestinal metaplasia (SIM) of greater than or equal to 3 cm proximal to the EGJ, whereas short-segment Barrett esophagus (SSBE) was defined as columnar epithelium with SIM less than 3 cm above the EGJ.

Four-quadrant biopsy samples were taken from the tubular esophagus at least every 2 cm in the case of circumferential segments and at least one sample from each tongue in the case of columnar epithelial tongues extending into the squamous epithelium. Two samples were taken from the gastric cardia (defined as the proximal edge of the gastric folds, just distal to the end of the tubular esophagus). Beginning with the 150th patient, biopsy samples were taken from the gastric antrum and body.

The results of this study can be summarized as follows:

1. Prevalence of heartburn: 59.1% of patients had never experienced heartburn; 25.1% reported heartburn less than once per week; 6.4% had heartburn once per week; 6.5% several times per week; and 2.7% had heartburn daily.
2. Barrett esophagus: The total number of patients with Barrett esophagus was 65 of 961 (6.8%). Twelve of 961 (1.2%) had long-segment Barrett esophagus (9 with lengths between 30 and 39 mm; 1 each with 40-mm, 50-mm, and 80-mm lengths); 164 patients had visible columnar epithelium in the tubular esophagus measuring 5 to 29 mm. Of these, 53 had specialized intestinal metaplasia (BE) identified by histology. The length of the columnar segment in these 53 patients with short-segment Barrett esophagus was 5 to 9 mm in 33 patients, 10 to 14 mm in 13 patients, 15 to 19 mm in 2 patients, 20 to 24 mm in 3 patients, and 25 to 29 mm in 2 patients. Of these 53 patients with SSBE, only 5 had circumferential segments; in the remainder the BE was seen as tongues of columnar-lined esophagus. The histology in the 111 patients with abnormal columnar-lined esophagus without intestinal metaplasia was not given.
3. Association of Barrett esophagus with symptoms: Among 556 subjects who had never had heartburn, the prevalence of Barrett esophagus and long-segment Barrett esophagus was 5.6% and 0.36%, respectively. Among 384 subjects with a history of any heartburn, the prevalence of Barrett esophagus and long-segment Barrett esophagus was 8.3% and 2.6%, respectively.
4. *Helicobacter pylori*: The Pyloritek was positive for *H. pylori* in 194 of 812 patients. The prevalence of *H. pylori* was higher in blacks than in whites.
5. Erosive reflux esophagitis was present in 155 of 961 patients (16.1%). Erosive esophagitis was more common in patients with Barrett esophagus (31%) than in those without (15%, $p = 0.005$). There was no association between erosive esophagitis and gender or *H. pylori* status.
6. Intestinal metaplasia of the cardia: Of the 940 patients with evaluable cardia tissue, intestinal metaplasia (IM-cardia) was identified in 122 (12.9%). There were 7 patients in whom intestinal metaplasia was identified in both tubular esophagus (BE) and in the gastric cardia.
7. Dysplasia: Of the 12 patients with long-segment Barrett esophagus, 1 had low-grade dysplasia and 1 was diagnosed as "indefinite for dysplasia." One patient with short-segment Barrett esophagus was diagnosed as "indefinite for dysplasia." No patient had high-grade dysplasia or cancer. The patients with low-grade dysplasia and "indefinite for dysplasia" had three follow-up biopsies at 6-month intervals; in all three, no dysplasia was found on two occasions and low-grade dysplasia was seen on one occasion.

This study shows an almost perfect endoscopic biopsy protocol in a screening setting except for the lack of measured biopsies of what they call the gastric cardia (this is a minor problem). Unfortunately, there is a complete lack of histologic data. The only thing of relevance to these authors from a histologic standpoint is whether or not intestinal metaplasia is present. If it is present and the patient had a visible columnar-lined esophagus, it is Barrett esophagus. If there is no intestinal metaplasia, the patients are not further considered even if they had a visible segment of columnar-lined esophagus.

The data show that in this total population, the prevalence of Barrett esophagus and long-segment Barrett esophagus was 6.8% and 1.2%. The prevalence of short-segment Barrett esophagus (5.5%) outnumbered that of long-segment Barrett esophagus (1.2%) by a factor of over 4 times (53 versus 12 patients). Also, 122 (12.9%) of patients had intestinal metaplasia of the cardia, nearly double the number with Barrett esophagus.

The study also shows that patients with any heartburn had a higher prevalence of Barrett esophagus (8.3% versus 5.6% of patients without any heartburn) and long-segment Barrett esophagus (2.6% versus 0.36% of patients without any heartburn). There are no data relating to possible reflux symptoms other than heartburn. In fact, the study specifically excluded patients with any symptom except heartburn and regurgitation that may have remotely been an indication for an upper endoscopy; these included vomiting, dysphagia, upper abdominal pain, weight loss, and anemia.

The fact that the study does not provide information of epithelial type other than the presence of intestinal metaplasia precludes grading the patients within our system below-grade 2 (Barrett esophagus). These are important data that would be readily available if only the authors recognized the significance of cardiac and oxyntocardiac mucosa. They were, however, following the presently accepted incorrect dogma that these were normal gastric epithelia or irrelevant even when these epithelia were present in an endoscopically defined columnar-lined esophagus. The study shows, however, that 111 of 961 patients (11.6%) had a visible columnar-lined esophagus without intestinal metaplasia. In our mapping biopsy study, we found that those patients who had a visible columnar lining (almost always a total histologic metaplastic epithelial length of at least 1 cm) almost always had cardiac mucosa. It therefore appears reasonable to suggest that these 111 patients (11.6%) had cardiac mucosa (therefore, classified as grade 1 in our system).

Unfortunately, the data provided regarding intestinal metaplasia of the cardia do not lend themselves to meaningful evaluation. The authors provide no information as to what kind of epithelium this intestinal metaplasia was found in: was it cardiac mucosa (and therefore Barrett esophagus) or was it in gastric oxyntic mucosa (and therefore atrophic gastritis)? There is no information that this intestinal metaplasia of the cardia was localized to this region (and therefore likely to be Barrett esophagus) or whether it was part of an atrophic pan-gastritis. They reported that 41 of 109 patients (37.6%) with intestinal metaplasia of the cardia had evidence of *H. pylori* infection of the stomach. This compared with a prevalence of *H. pylori* of 194 out of 812 (23.9%) in the total population who had *H. pylori* testing. This suggests that at least some of their cases of intestinal metaplasia of the cardia were *H. pylori*–related atrophic pan-gastritis involving gastric oxyntic mucosa. If one assumes that all of their *H. pylori*–negative patients who had intestinal metaplasia of the cardia were reflux related, this would suggest a 68 of 812 (8.4%) prevalence of Barrett esophagus in end-stage esophagus. Though the numbers are not assessable with total accuracy because of the imperfect data reporting, it is likely that there is a progressive increase in the prevalence of intestinal metaplasia in the general population presenting as long-segment Barrett esophagus (12 of 961, or 1.2%), short-segment Barrett esophagus (53 of 961, or 5.5%), and Barrett esophagus restricted to an end-stage dilated esophagus (a minimum of 68 of 940, or 8.4%). This will match perfectly with the fact that reflux disease limited to end-stage dilated esophagus is the most common type of reflux disease and the likelihood that intestinal metaplasia in columnar-lined esophagus increases as its length increases. If Barrett esophagus is at risk for adenocarcinoma irrespective of its length, this is the population at risk for adenocarcinoma. The actual risk can only be determined by follow-up; these data are available for short- and long-segment Barrett esophagus (see Chapter 10) but are not even considered for microscopic Barrett esophagus limited to dilated end-stage esophagus, which is the largest group.

The data in this study show that significant screening is possible in patients presenting for screening colonoscopy. Among the first 3130 consecutive persons undergoing outpatient colonoscopy in whom reasons for exclusion were recorded, only 10% (313) refused the invitation to participate. Large numbers of those accepting the invitation were excluded for a variety of reasons. This suggests that there will be a high acceptance rate of an added upper endoscopy in this group.

The following table presents the results of this screening in terms of placing these patients within the risk categories that we have defined in our classification:

Grade	Number of patients	Comment
Zero	Unknown	Histologic data inadequate
1 (reflux carditis)	Total unknown	Histologic data inadequate
1a (<1 cm)	Unknown	Histologic data inadequate
1b (1 to 2 cm)	111/961 (11.6%)	Histology presumed
2 (Barrett esophagus)	65/961 (6.8%) in all persons; 8.3% of 384 when heartburn is +; 5.6% of 556 when heartburn is –	
Long-segment BE	12/961 (1.2%) in all persons; 2.6% of 384 when heartburn is +; 0.36% of 556 when heartburn is –	
Short-segment BE	53/961 (5.5%) in all persons	
IM limited to end-stage dilated esophagus	At least 68/812(8.4%)	Histologic data inadequate; based on assumptions
3a (low-grade dysplasia)	1/961 at index biopsy; 3/961 at index biopsy; 3/961 on follow-up	Not counting "indefinite"; including "indefinite"
3b (high-grade dysplasia)	Zero at index; zero on follow-up	
3c (adenocarcinoma)	Zero at index; zero on follow-up	

The yield of grade 2 disease (Barrett esophagus) in all patients was 65 of 961 (6.8%) by the presently used definition. Let us assume that the population over 40 years in the United States is 150 million people. The data would mean that approximately 10 million people greater than 40 years old in the United States will have Barrett esophagus by the present definition. If the estimated 8.4% of people who have microscopic Barrett esophagus in dilated end-stage esophagus (those that are missed by present definition criteria) are included, the total reaches 22.5 million. This assumes, probably incorrectly because of demographic selection, that the risk is the same in the population not seeking colonoscopy screening. If Barrett esophagus is the only precursor for adenocarcinoma, the 8000 annual incidence of esophageal adenocarcinoma and the 8000 to 10,000 annual incidence of adenocarcinoma of the "gastric cardiac" must occur in this population at risk. These figures are generally in line with the data in epidemiologic studies, which indicate that the incidence of adenocarcinoma of the esophagus is approximately 23 per million population and adenocarcinoma of the gastric cardia is approximately 18 per million in the United States (2).

Arguing purely from a theoretical viewpoint, if a mass screening of the population is performed, 22.5 million more persons with Barrett esophagus will be detected and need a strategy:

1. The present strategy is long-term surveillance with or without acid suppression. The aim of surveillance is to detect early cancer (including multifocal high-grade dysplasia); surveillance is a cancer detection method, not a cancer prevention method. Acid suppressive drugs either do nothing to prevent cancer or actually increase the incidence of cancer. If this is the plan of attack for newly detected Barrett esophagus patients, screening is not worthwhile. We are back to the status quo of futility.
2. On the other hand, if the detection of Barrett esophagus is immediately followed by antireflux surgery, data suggest that reflux will be normalized in approximately 85% of these patients and that the incidence of adenocarcinoma will be greatly reduced in this group. Because this is the entire population at risk for reflux-induced adenocarcinoma, the incidence of this disease can be expected to decline from the present 18,000 per year (8000 for esophageal adenocarcinoma + 10,000 for adenocarcinoma of the "gastric cardia") to 2700 per year (patients whose antireflux surgery is unsuccessful). To do this will require 150 million endoscopies and 22.5 million antireflux surgeries, which are beyond the capability of the medical establishment. *However, the fact that such a reduction is theoretically possible should be a source of tremendous encouragement and optimism.* There is no other cancer type in the human for which there is such a powerful theoretical basis for effective cancer prevention, however impractical it may be.

Using this theoretical model, one can adjust downward to reach a practical level of screening of asymptomatic persons. This can be done in the following ways:

1. *Make a general recommendation for screening for Barrett esophagus.* The experience with the similar recommendation for colonoscopic screening for colorectal cancer suggests that a downward adjustment would happen automatically. While colonoscopic screening is recommended in all persons over 40 years, the number of screening colonoscopies that are done is only a small fraction of the approximately 150 million people over the age of 40 years. If upper endoscopic screening were recommended, the demand for this procedure would be determined by availability and the ability to afford the screening procedure, thereby limiting the numbers to an extent that is practically within the resources. It would also tend to select patients at highest risk when one considers that reflux-induced adenocarcinoma is primarily a disease of affluent Caucasians.
2. *Limit screening to patients willing to have antireflux surgery if Barrett esophagus is found by screening.* Screening for Barrett esophagus is futile if the result of detection is to do nothing with the information or place the patients on acid suppressive drugs. There will be no expectation of a decreased risk. Surveillance is not a cancer prevention method; it is a cancer detection method. If this is explained to the patient and a commitment to antireflux surgery mandated before screening, the number of patients willing to undergo screening will probably decline.
3. *Recommend upper endoscopic screening as an additive procedure for patients presenting for colonoscopy screening.* Rex et al.'s study suggests that 80% to 90% of persons coming in for colonoscopic screening will agree to the added procedure. It will be done in one sitting and not add greatly to the cost. Again, this can be offered only to patients who agree to an antireflux procedure if Barrett esophagus is detected.
4. *Limit screening to patients who are symptomatic for reflux disease* (discussed later). In Rex et al., 59.1% of patients never had heartburn but may have had other less specific symptoms indicative of reflux.

The other advantage of screening upper endoscopy is that it identifies the 55% to 60% of the population that falls into grade 0 disease (no cardiac or intestinal epithelium). If a person has neither of these epithelial types at age 40 years, it indicates a low-damage environment with a low tendency to convert cardiac mucosa to intestinal metaplasia. Because refluxate characteristics do not change greatly during life (as long as acid suppressive drugs are not used when significant pH change is inevitable), the theoretical chance that this person will develop cardiac mucosa followed by intestinal metaplasia and cancer within his or her natural lifetime is extremely small. These patients can be reassured that they have little risk.

Screening also identifies a significant number of patients with grade 1 disease (reflux carditis characterized by the presence of nonintestinalized cardiac mucosa). These patients may have a visible columnar-lined esophagus; in Rex et al. (2003), 111 of 961 patients (11.4%) belonged to this group. These patients, while they did not have a target cell for cancer, were at risk of developing intestinal metaplasia in the future. This risk is proportional to the length of metaplastic epithelium. Some kind of repeat examination after a several-year interval is necessary to assure these patients that they are not progressing in the cancer sequence.

Completely Asymptomatic Patients Who Have an Upper Endoscopy

Upper endoscopy has become a common procedure in gastroenterology practice and is done for a variety of indications other than reflux disease. If it is appreciated that reflux-induced adenocarcinoma can be prevented and occurs in asymptomatic patients, every upper endoscopy becomes a potential screening procedure for cancer prevention. All patients should have one set of four-quadrant biopsies that aims to sample the columnar epithelium immediately distal to the squamocolumnar junction. Patients will be graded with regard to the reflux disease status. Patients in grade 0 can be reassured; those in grade 1 can be reevaluated in 5 years for progression to grade 2; and those in grade 2 (Barrett esophagus) can be offered antireflux surgery.

This type of screening examination will have an effect only if the patient is sufficiently educated to recognize that the finding of Barrett esophagus is a serious cancer risk and demands prevention. Antireflux surgery is preventive; long-term surveillance is protective by detecting early cancer.

Minimally Symptomatic Patients Who Never Seek Medical Care

These are the patients who are induced by the advertisements of the pharmaceutical industry to take over-the-counter antacids, histamine-2-receptor antagonists, or proton pump inhibitors for any symptoms in the upper abdomen. As these drugs have become increasingly effective, they "cure" in a shotgun manner many symptoms like heartburn and dyspepsia. Many patients with classical symptoms of reflux are converted to asymptomatic or mildly symptomatic patients with these drugs, removing the need and desire to seek medical care. At present, the majority of these people do not know that they have a risk of future adenocarcinoma.

The medical establishment has a responsibility to inform the population at risk about the fact that symptomatic reflux is a proven risk indicator for esophageal adenocarcinoma. This can be done at three levels depending on what the medical establishment perceives is the true risk of acid suppressive drug treatment:

1. *If we believe (probably incorrectly) that acid suppressive drug use has absolutely no effect in promoting reflux-induced adenocarcinoma, we should still inform the public that having symptoms of reflux disease is a risk indicator for esophageal cancer.* The most effective method of doing this is to recognize that the first response of the symptomatic person is to go to a pharmacy and buy an acid suppressive drug. The most efficient method of public education is to print a warning label on the stronger of these drugs—the H2RAs and PPIs. The label can state that the symptom of heartburn, which is the most common reason for the use of these drugs, indicate that the person taking the medications is at risk for developing esophageal adenocarcinoma. The warning can include a suggestion that the person should seek medical care. This is not a condemnation of acid suppressive drugs; it is simply a recognition that the symptom for which the drug is used is a known risk indicator for a cancer type that will affect approximately 18,000 people annually in the United States. An increase in the public's awareness that reflux disease is precancerous is the first step toward developing public support for any future measures that may become necessary if the cancer rates keep increasing. It is never too early to start this kind of public education because of the significant lag phase that exists with such efforts.
2. *If there is a consensus that the safety of acid suppressive agents in terms of their potential to promote esophageal*

adenocarcinoma is unproven, there is a strong case to be made for removing H2-receptor antagonists and proton pump inhibitors from the realm of over-the-counter drugs that are directly marketed to the population. Recognizing that these drugs possibly promote cancer pushes them to the realm of serious drugs that must be used by physicians with caution, rather than the cavalier attitude that governs their present over-the-counter use. Removing acid suppressive drugs as over-the-counter medications will bring these patients into the hands of the medical community where their risk of cancer can be addressed.

3. *If Lagergren et al.'s (1999) data are correct and there is a threefold increased risk of adenocarcinoma among symptomatic patients who take acid suppressive medications compared to those who are similarly symptomatic but have not taken such drugs, there is an argument for severely restricting their use.* These drugs then become agents that promote human cancer, and the use of such drugs is not usually considered acceptable. To those who will say that a threefold increased risk is not sufficient to remove a drug from the shelves, one needs only to realize that Merck voluntarily withdrew Vioxx from the market when it was shown to have a 1.9 time relative risk for producing strokes among users.

All of these suggested new actions will have the effect of bringing this group of mildly symptomatic reflux patients who presently self-medicate their disease into the care of physicians where their disease can be addressed before these patients develop cancer. These self-medicated patients may actually be a significant part of the population presently considered "asymptomatic."

Symptomatic Patients Who Seek Care and Do Not Have an Endoscopic Examination

This is the group of patients who are most likely to develop adenocarcinoma in the future. The present treatment regimen of acid suppressive drugs without endoscopy dooms the patient destined to develop adenocarcinoma; he or she presents for the first time with new symptoms caused by the cancer, with its usually advanced stage and poor prognosis.

Treating these patients merely to control symptoms without informing them of the risk involved is a serious failure of education. These patients must be told that they have an increased risk, that the magnitude of the risk cannot be assessed without endoscopy, and that acid suppressive drug therapy, while effective in controlling symptoms, does not decrease the cancer risk in any way. Only when a sufficient number of people in this highest risk symptomatic group are educated in this manner will there be public pressure to do something about reflux-induced adenocarcinoma.

In the more aggressive approach to reflux disease that we are recommending, patients who present with symptoms even remotely suggestive of reflux will be subject to an upper endoscopy with biopsy to assess their level of risk by histologic examination (Table 17.4). Antireflux surgery is recommended for patients with symptomatic carditis (as an alternative to acid suppressive drugs) and Barrett esophagus to both treat symptoms effectively and prevent cancer.

Symptomatic Patients Who Seek Care and Have an Upper Endoscopic Examination

The failure to perform an adequate series of biopsies in symptomatic patients who have an upper endoscopy represents a profound failure by the medical establishment to understand this disease. Present recommendations are to detect erosive esophagitis and take biopsies only if there is a visible segment of columnar-lined esophagus. This practice catches only the tip of the iceberg of reflux disease. Patients who have a "normal" examination cannot be reassured that they have no risk of future cancer.

The newly mandated biopsy protocol will stratify all patients according to their risk (Table 17.4). Older patients in grade 0 (i.e., without carditis and intestinal metaplasia) can be reassured that they have no future risk of cancer; the number in this group will be smaller than in the asymptomatic population. Patients with grade 1 (i.e., carditis, the most common group) can be offered antireflux surgery with the knowledge that this will prevent intestinal metaplasia and cancer. This is the earliest and therefore most effective act of cancer prevention. Patients in grade 2 (i.e., Barrett esophagus) should be urged to have antireflux surgery with the proviso that this surgery will prevent cancer *only if the genetic changes required for cancer have not already occurred.* These patients need at least 5 years of surveillance to ensure they do not have prevalent cancer at the time of the antireflux surgery.

Patients Diagnosed as Having Barrett Esophagus

Patients diagnosed as having Barrett esophagus are presently treated with acid suppression (without any evidence that it decreases cancer risk) and surveillance to detect early cancer (or multifocal high-grade dysplasia), at which point an esophagectomy is recommended. This has no hope of decreasing cancer incidence, although by detecting cancer at an early stage, this regimen may decrease mortality in this patient group. At present rates of endoscopy for patients with reflux, less that 5% of patients who develop adenocarcinoma are diagnosed with Barrett esophagus. This is an expression of the rarity with which endoscopy is performed in patients with reflux disease and the failure of diagnosis because of a failure to biopsy.

In the more aggressive stance that we have recommended, the aim is to identify all patients in the population with Barrett esophagus. The highest yield of Barrett esophagus will be in patients with symptoms suggestive of reflux (5). The presence of any symptoms suggestive of reflux will trigger endoscopy immediately (the aim is to develop a rule similar to the necessity of a colonoscopy for every patient who presents with rectal bleeding), and biopsies will be done according to the recommended protocol. The only patients with Barrett esophagus not detected by this protocol will be the completely asymptomatic patients who need a public health mandate for endoscopic screening if they are to be found. If endoscopy is limited to the symptomatic patient who is most at risk for cancer, this newly diagnosed pool of Barrett esophagus patients is likely to contain the majority of patients destined to develop adenocarcinoma in the future.

If this aggressive protocol is followed by immediate antireflux surgery as soon as the diagnosis of Barrett esophagus is made, adenocarcinoma can be expected to decrease in incidence. Barrett esophagus is an indication for antireflux surgery purely by virtue of the cancer risk; it is irrelevant whether or not the patient is symptomatic. This is an appropriately aggressive approach to a serious premalignant disease. Just as many patients with benign mammographic lesions will have unnecessary breast excisions to detect and treat those that develop breast cancer at an *in-situ* stage, many patients with asymptomatic Barrett esophagus who are not destined to develop cancer will have antireflux surgery. However, in patients who are symptomatic, the surgery is highly beneficial because it controls symptoms without the need for lifelong drug treatment in addition to reducing cancer risk.

In the approximately 85% of patients who have successful antireflux surgery, the disease will cease to progress because the surgery effectively stops the interaction between the target cell in the esophagus and whatever unknown molecule in the refluxate causes the carcinogenic genetic events.

The ultimate proof that antireflux surgery is effective in preventing cancer and is superior to acid suppressive drug therapy will require well-designed randomized prospective clinical trials that compare medical and surgical treatment. These studies will be slow in coming because of the relative rarity of antireflux surgery in any given institution. The following study from Murcia, Spain, is an excellent example of such a study. It is perfectly designed and well controlled. The first report by Ortiz et al. in 1996 (21) was followed by a 2003 report (22) with more patients and longer follow-up. The trend in this study as patient numbers increase is toward proving that successful antireflux surgery is effective in preventing adenocarcinoma. The numbers are still very small, making it difficult to prove differences in treatment types that are valid.

Literature Review

Parrilla P, deHaro LFM, Ortiz A, Munitiz V, Molina J, Bermejo J, Canteras M. Long term results of a randomized prospective study comparing medical and surgical treatment of Barrett's esophagus. Ann Surg 2003;237:291–298.

This is a prospective study of 146 patients with Barrett esophagus that began in 1982 and is ongoing; this study summarized data to the year 2000. Patient accrual up to 1993, reported by Ortiz et al. (21), was 59 patients, indicating increased accrual in the last 7 years of the study. Until 1997, only patients with long-segment Barrett esophagus were entered; after 1997, patients with short-segment columnar-lined esophagus with intestinal metaplasia were entered. Twenty-five patients with significant stenosis and 8 patients who refused randomization were excluded. The remaining 113 patients were randomized to medical treatment (51) and antireflux surgery (62). During follow-up, 12 patients were excluded because they rejected endoscopic surveillance, leaving 101 study patients (43 undergoing medical treatment and 58 undergoing antireflux surgery).

The medical treatment consisted of H2 antagonists (150 mg twice daily) initially and omeprazole (20 mg twice daily) from 1992 onward for all patients. Of the 58 patients in the surgical treatment arm, 56 had a short Nissen fundoplication; the other 2 had a Collis-Nissen procedure because of significant esophageal shortening. The effectiveness of surgery was checked by pH-metric and manometric studies performed in all patients 1 year postoperatively and then every 5 years.

Results

After a median follow-up of 5 years, clinical control was similar in the two groups; 91% of patients had a satisfactory outcome. Four patients in the medical group and 5 patients in the surgery group had a clinical outcome less than satisfactory. There were no operative deaths, and the only surgical complication was the need for splenectomy in 1 patient.

The 24-hour pH studies showed decreased postoperative reflux in all patients with normalization of the 24-hour pH test in 49 of 58 patients. Nine patients had an abnormal 24-hour pH study postoperatively (median 19.5%; range, 5% to 46%); these included all 5 patients with a less than satisfactory clinical outcome.

The dysplasia status of the patients was considered in three groups (the medical group, the total surgery group, and the successful surgery group):

1. *Medical group.* Pretreatment, 40 patients had no dysplasia and 3 patients had low-grade dysplasia. In 2 patients with low-grade dysplasia, no dysplasia was seen posttreatment. Low-grade dysplasia persisted in 1 of the 3 patients with low-grade dysplasia and developed in 8 of 40 patients who did not have dysplasia pretreatment. Two patients went on to develop high-grade dysplasia and had an esophagectomy in which adenocarcinoma was present.
2. *Total surgery group.* Preoperatively, 5 patients had low-grade dysplasia and 53 had no dysplasia. All 5 patients with low-grade dysplasia reversed to no dysplasia after surgery. De novo low-grade dysplasia developed in 3 patients; 2 of these progressed to high-grade dysplasia and had an esophagectomy in which adenocarcinoma was present. Both patients who progressed to cancer had a failed antireflux surgery (defined by failure of normalization of the 24-hour pH test).
3. *Successful surgery group (excluding the 9 failures).* Preoperatively, 5 patients had low-grade dysplasia and 44 had no dysplasia. Postoperatively, all 5 patients with low-grade dysplasia reversed to no dysplasia. De novo low-grade dysplasia developed in 1 patient; no patient developed high-grade dysplasia or adenocarcinoma.
4. *Rates of malignancy.* The rate of malignancy was 1 in 129 patient-years (0.8% per year) for the medical treatment group and 1 in 203 patient-years (0.5% per year) for the surgical treatment group. When surgery was successful, the rate was zero.

Conclusions

The authors concluded that there is no statistical difference between medical and surgical treatment with respect to preventing Barrett esophagus from progressing to dysplasia and adenocarcinoma. They found a statistical difference between the rate of de novo dysplasia in the medical and successful surgery groups (8 of 40 versus 1 of 44). This is not a valid difference because the authors used the criterion of de novo dysplasia for statistical comparisons. This includes low-grade dysplasia. Low-grade dysplasia is not an adequately reliable diagnosis to permit comparisons with regard to neoplastic progression as shown by the fact that 2 out of 3 patients who had low-grade dysplasia in the medical group and 5 of 5 in the surgical group reversed. The comparisons should be made regarding incidences of high-grade dysplasia and adenocarcinoma, which are much more reliable diagnoses. The number of patients in this series who developed high-grade dysplasia is too small. As such, comparisons the authors make between the medical and surgical groups should be rejected at this time. However, the lack of statistical significance is clearly a function of small numbers, which make the study lack the statistical power to make this determination. There is a clear trend that suggests that patients in the medical group have an incidence of adenocarcinoma (1 per 129 patient-years, or 0.8% per year) that is close to established rates while the successful surgery group has no patients progressing to high-grade dysplasia. With a median follow-up of 6 years, the expectation is that more than half of the patients in the antireflux surgery group have now lost their cancer risk (cancer very rarely develops more than 5 years after successful antireflux surgery [23]). In contrast, the medical group will continue to show the approximately 0.5% per year cancer risk, which is a lifetime phenomenon. If these assumptions, which are based on strong evidence in the literature, are true, this study is likely to show a divergence of cancer incidence in the medical and successful surgery groups that will ultimately reach statistical significance.

It should also be remembered that this study dates back to 1982 and that accrual rates were slow initially because antireflux surgery was a rare procedure at the time. As accrual rates increase, it is likely that the failure rate from surgery will decline, making the divergence in results between the total surgical group and successful surgical group less apparent.

Parrilla et al.'s study has a head start of more than 20 years and shows the type of data that must be accumulated to prove the cancer reducing power of antireflux surgery. This will not happen easily as long as antireflux surgery remains a relatively rare secondary treatment for reflux disease. What is important is that the theoretical presumption that isolating the target cells of the esophagus from the carcinogens in the gastric refluxate will prevent cancer is confirmed by data from studies that test this hypothesis. This should produce great hope that we can prevent reflux-induced cancer and move the medical community from its present regimen of futile acid suppressive drugs to an aggressive and optimistic attack using antireflux surgery as the mainstay of treatment for reflux disease.

Follow-up of a patient after antireflux surgery for Barrett esophagus is essential for two reasons:

1. *To ensure that the patient did not have prevalent cancer at the time of surgery.* There is a time lag between the occurrence of the entire complement of genetic changes needed for cancer and the phenotypic expression of invasive cancer. Annual surveillance endoscopy with biopsies to ensure the absence of

dysplasia is necessary. If there is no progression to dysplasia in 5 years in a patient who has had a successful antireflux operation, it is unlikely that cancer will occur and surveillance can be made infrequent or even discontinued.
2. *To ensure that the surgery has been successful in normalizing reflux.* This is best done by a 24-hour pH study 1 year after surgery in patients who do not have symptoms. In patients who develop symptoms, a 24-hour pH test is essential. The failure of antireflux surgery must be defined as an abnormal 24-hour pH test. Many patients have a variety of symptoms after antireflux surgery and are declared to be failures and placed on acid suppressive medication without confirming failure by pH studies. This is not appropriate.

The rate of success in most centers that use the Nissen fundoplication (with a Collis-type procedure when the esophagus is significantly shortened) is presently around 85%. It is likely to improve as the number of antireflux procedures increases.

The failure of antireflux surgery is a serious problem because it brings the risk of cancer toward pretreatment levels. The risk does not decrease with acid suppressive drug therapy. These patients should be aggressively evaluated to see if corrective surgery can reestablish the control of reflux.

Conclusion: Antireflux surgery, when successful, decreases the incidence of cancer and should be recommended as the primary treatment whenever Barrett esophagus is diagnosed.

COST OF ANTIREFLUX SURGERY

Antireflux surgery has a significant cost associated with the preoperative testing, the actual surgery (usually laparoscopic), the short hospital stay, management of complications, and any surveillance and postoperative testing that may be necessary. The cost of antireflux surgery in the asymptomatic patient who is found to have Barrett esophagus during a screening procedure can be measured only against the number of future cancers that are prevented. As new methods are developed to better identify cancer risk in the asymptomatic Barrett patient, the number of antireflux surgeries needed will decrease and the yield of prevented cancers will increase, making this a cost-effective exercise.

In contrast, an aggressive recommendation of antireflux surgery is feasible in symptomatic patients because the alternative is long-term acid suppressive drug treatment, which has a significant lifetime cost. After antireflux surgery, 85% of patients do not need acid suppressive medication. Over the long term, this is a significant cost savings that compensates for the initial capital expense of the surgery.

We will now review two studies that evaluated the costs involved in the treatment of symptomatic reflux patients with a long-term drug regimen and antireflux surgery.

Literature Review

Van Den Boom G, Go PM, Hameeteman W, Dallemagne B, Ament AJ. Cost effectiveness of medical versus surgical treatment in patients with severe or refractory gastroesophageal reflux disease in the Netherlands. Scand J Gastroenterol 1996;31:1–9.

This study from the Netherlands (24) compares maintenance treatment with omeprazole with open and laparoscopic Nissen fundoplication from a health-economic perspective. This is a meta-analysis of published articles to assess effectiveness, and simple decision-analytic techniques to combine costs and effects are used. The findings and assumptions were submitted to sensitivity analysis.

It was estimated that it costs approximately 1880 Dutch guilders to initially heal a patient with severe or refractory esophagitis with 40 mg omeprazole daily. When medical maintenance therapy was compared with surgery, it appeared that medical maintenance therapy with omeprazole (20 to 40 mg daily) for a prolonged period of time (more than 4 years) was less cost-effective than an open Nissen procedure. It was estimated that a laparoscopic Nissen will shift this so-called break-even point toward 1.4 years, mainly due to a shorter hospital stay.

Although the authors recommended caution in drawing conclusions, it appears that replacing long-term omeprazole treatment with (laparoscopic) Nissen fundoplications in these patients might lead to substantial savings.

Literature Review

Viljakka M, Nevalainen J, Isolauri J. Lifetime costs of surgical versus medical treatment of severe gastro-oesophageal reflux disease in Finland. Scand J Gastroenterol 1997;32:766–772.

This similar study from Finland (25) compared the long-term costs of medical and surgical management. The medical regimens were ranitidine (150 or 300 mg per day), omeprazole (20 or 40 mg per day), and lansoprazole (30 mg per day), with costs calculated for total life expectancy after diagnosis and for one third of that time. Costs for open or laparoscopic surgery (Nissen fundoplication) included pre- and postoperative investigations, sick leave, and calculated financial loss due to a fatal outcome.

Costs were lowest with ranitidine, 150 mg per day, for one third of the patient's lifetime, and costs were highest with lifelong omeprazole, 40 mg daily. The cost of an open or laparoscopic operation was less than that for lifelong daily treatment with proton pump inhibitors or ranitidine at 300 mg daily. The authors concluded that in Finland, antireflux surgery for gastroesophageal reflux disease is cheaper than lifetime treatment with proton pump inhibitors.

The data from these European studies cannot be used to study cost-effectiveness in the United States where cost structures are very different. However, they lead to the conclusion that antireflux surgery is at least not more expensive than long-term acid suppressive therapy.

If the total cost of antireflux surgery is equivalent to or less than the cost of acid suppressive drug therapy, it is impossible to continue to justify the use of long-term acid suppressive treatment over antireflux surgery as the primary treatment modality in any patient with symptomatic reflux disease.

RESOURCES NEEDED TO IMPLEMENT RECOMMENDED CHANGES

In Chapter 15, we outlined the future research that can provide permanent solutions to prevent Barrett esophagus and reflux-induced adenocarcinoma. These are dreams. While we await the fulfillment of these dreams, we must do the best we can with the resources that are immediately available.

Education and Correct Understanding of the Pathology of Reflux Disease

The main purpose of this book is to provide a step-by-step, cell-by-cell, millimeter-by-millimeter explanation of the cellular events caused by gastroesophageal reflux. The book recommends clear, long-accepted, and reproducible definitions of histology as the mainstay of defining normal and abnormal states. It provides a method to classify all people into risk categories based on the solid foundation of histology. All this is achieved with simple technology (routine histologic examination of hematoxylin and eosin-stained biopsies) that is available to every physician in all corners of the world.

The concepts detailed in this book are not generally accepted. However, they have become increasingly known to exist and have piqued the interest of serious authorities. There is considerable resistance to many of these new ideas, *but the resistance is not accompanied by any alternative reasonable explanation or scientific data.* This trend of increased interest, resistance, and lack of alternative explanations is one that is highly predictive of success.

The pathologic basis of reflux disease is out in the open and available for full comprehension. We are willing to come to any forum to openly debate any part of these ideas with any person. We are willing to hold training sessions for anyone who wants to learn the pathologic system that is being suggested. The sad truth, however, is that the speed of change of the medical establishment is a reflection of inertia and establishment resistance that is ingrained. It took the self-administration of a culture of *H. pylori* by Dr. Marshall to begin to move the establishment. We hope this book will be a rational alternative to destroying our lower esophageal sphincters for this concept to be believed.

Change in Attitude

The medical establishment has to recognize that its inertia toward the rapid increase in reflux-induced adenocarcinoma is inappropriate. The attitude that this is a disease we can adequately treat by controlling symptoms and ignoring the cancers is not appropriate. We urgently need to recognize that acid suppressive drugs are dangerous and should not be used in the present cavalier over-the-counter manner. We desperately need to recognize that we have an excellent theoretical basis of cancer prevention in antireflux surgery and that initial studies show that it has great promise as a cancer prevention method.

The present attitude of total pessimism and futility that was expressed by the 18 experts at the AGA workshop in Chicago (12) must change. We need to urgently convene a workshop of optimistic experts who will suggest that reflux is a disease that is a serious cancer risk, that we have at hand a method to prevent cancer, and that we should aggressively develop resources to implement a plan that will successfully decrease the incidence of cancer. This requires only the education and change in attitude of a few leaders in the field. Most changes in medicine are made not at a grass-roots level but by the proclamations of our leaders. We need to find optimistic leaders to replace the present pessimism.

Recognition of the Dangers of Acid Suppressive Drugs

This is primarily a governmental function. There are sufficient data in the literature to remove the stronger acid suppressive drugs from the realm of the supermarket shelf and make them available by prescription only. The pharmaceutical companies that market these drugs should make themselves aware of the research that suggests they are promoters of cancer and take appropriate action.

Resource Development

The plan suggested to prevent adenocarcinoma requires the performance of a greatly increased number of endoscopic examinations with a limited biopsy protocol. If done over a period of time and limited in some way, this will probably not overwhelm the system. The plan also calls for the performance of a greatly increased number of antireflux operations. An adequate number of surgeons must be trained in this procedure to satisfy the demand. Laparoscopic Nissen fundoplication has been shown to have a rapid learning curve for surgeons, with a high success rate even in the first patients in many series. As the numbers of these surgeries increase, it can be anticipated that technical skills will increase, failure rates decrease, and surgical innovations to improve results will emerge.

Endoscopy, pathologic examination of biopsies, and antireflux surgery are readily available at present. The only resource development that is needed is the ability to handle an increased volume. It is anticipated that these changes will increase the cost of care of reflux disease, but this cost is not as high as one is led to believe. On the side of increasing cost is the cost of endoscopy, biopsy interpretation, and antireflux surgery. On the side of cost saving is the potential decrease in cost of acid suppressive drugs and the cost of caring for patients who develop adenocarcinoma. The suggestion from European studies that the lifetime cost of treating a patient with reflux disease with antireflux surgery is less than the costs of acid suppressive drugs suggests that the overall cost will not increase by the new plan as long as the plan is limited to symptomatic patients. Only when we start screening for asymptomatic patients will the cost increase; this can be a second phase of prevention. The human suffering that is saved by preventing adenocarcinoma of the esophagus is the reason for the plan. This is arguably one of the worst cancers in humans, and to prevent even one cancer in one patient is worth a substantial financial cost.

SUMMARY OF RECOMMENDED CHANGES

Recommendations to Governmental Agencies

1. Review acid suppressive drug safety and consider removing these from over-the-counter availability and making them available through prescription only.
2. Demand pharmaceutical companies that market these drugs to provide safety data that prove there is no increased risk of cancer associated with their use.
3. Demand that pharmaceutical companies place warning labels on acid suppressive drugs to educate the public about the association between reflux disease and cancer.
4. Consider suitable advertising to the public that symptomatic reflux disease (heartburn) is a risk factor for esophageal cancer and that rates of this cancer type are increasing.

Recommendations to Primary Care Physicians

1. Inform physicians that gastroesophageal reflux disease should be regarded as a premalignant disease in a manner that is similar to ulcerative colitis and that treatment with acid suppressive drugs, even when highly effective in controlling symptoms, does not reduce cancer risk.
2. Educate primary care physicians that upper endoscopy is indicated in all patients with a diagnosis of reflux disease. Reflux symptoms should precipitate upper endoscopy in a manner that is similar to rectal bleeding precipitating colonoscopy.
3. Educate primary care physicians that using acid suppressive medication for more than 3 months without endoscopy is inappropriate.

Recommendations to Gastroenterologists

1. Recommend upper endoscopy in all patients with reflux symptoms. If this requires added resources, develop these resources.
2. Educate gastroenterologists on the correct biopsy protocol aimed to stratify patients by their stage in the reflux-adenocarcinoma sequence and maximize the detection of Barrett esophagus. This requires new practice guidelines that recommend that endoscopically normal patients should be biopsied.
3. Educate gastroenterologists about the danger of acid suppressive drugs in promoting reflux-induced adenocarcinoma.
4. Recommend that antireflux surgery be used as the primary treatment modality for all patients with symptomatic reflux who have reflux carditis in their biopsies.
5. Recommend that Barrett esophagus be recognized as a premalignant disease that demands antireflux surgery immediately upon diagnosis, whether or not the patient is symptomatic.

Recommendations to Pathologists

1. Understand the cellular pathology of reflux disease, get rid of false dogmas, and use reproducible histologic definitions for diagnosis.
2. Educate clinical colleagues about the pathological basis of reflux disease.

Recommendations to Surgeons

1. Develop resources to permit the performance of a substantially increased number of laparoscopic Nissen fundoplications.
2. Improve training in and methodology of the laparoscopic antireflux procedure to improve the present success rate of 85% to 100%.

References

1. Devesa SS, Blot WJ, Fraumeni JF. Changing patterns in the incidence of esophageal and gastric carcinoma in the United States. Cancer 1998;83:2049–2053
2. Pohl H, Welch HG. The role of overdiagnosis and reclassification in the marked increase of esophageal adenocarcinoma incidence. J Natl Cancer Inst 2005;97:142–146.
3. Spechler SJ. Barrett's esophagus. N Engl J Med 2002;346: 836–842.
4. Lagergren J, Bergstrom R, Lindgren A, Nyren O. Symptomatic gastroesophageal reflux as a risk factor for esophageal adenocarcinoma. N Engl J Med 1999;340:825–831.
5. Rex DK, Cummings OW, Shaw M, Cumings MD, Wong RKH, Vasudeva RS, Dunne D, Rahmani EY, Helper DJ. Screening for Barrett's esophagus in colonoscopy patients with and without heartburn. Gastroenterology 2003;125:1670–1677.
6. Sampliner RE. Practice guidelines on the diagnosis, surveillance, and therapy of Barrett's esophagus. Am J Gastroenterol 1998;93:1028–1031.
7. Peters FT, Ganesh S, Kuipers EJ, Sluiter WJ, Klinkenberg-Knol EC, Lamers CB, Kleibeuker JH. Endoscopic regression of Barrett's oesophagus during omeprazole treatment: A randomized double blind study. Gut 1999;45:489–494.
8. Jacobson BC, Ferris TG, Shea TL, Mahlis EM, Lee TH, Wang TC. Who is using chronic acid suppression therapy and why? Am J Gastroenterol 2003;98:51–58.
9. Spechler SJ. Esophageal columnar metaplasia (Barrett's esophagus). Gastrointest Endosc Clin North Am 1997;7:1–18.
10. Goldblum JR, Vicari JJ, Falk GW, et al. Inflammation and intestinal metaplasia of the gastric cardia: The role of gastroesophageal reflux and *H. pylori* infection. Gasteroenterology 1998; 114:633–639.
11. Kahrilas PJ, Shi G, Manka M, Joehl RJ. Increased frequency of transient lower esophageal sphincter relaxation induced by gastric distension in reflux patients with hiatal hernia. Gastroenterology 2000;118:688–695.
12. Sharma P, McQuaid K, Dent J, Fennerty B, Sampliner R, Spechler S, Cameron A, Corley D, Falk G, Goldblum J, Hunter J, Jankowski J, Lundell L, Reid B, Shaheen N, Sonnenberg A, Wang K, Weinstein W. A critical review of the diagnosis and management of Barrett's esophagus: The AGA Chicago Workshop. Gastroenterology 2004;127:310–330.
13. Weston AP, Badr AS, Hassanein RS. Prospective multivariate analysis of clinical, endoscopic, and histological factors predictive of the development of Barrett's multifocal high-grade dysplasia or adenocarcinoma. Am J Gastroenterol 1999;94: 3413–3419.
14. Peters JH, Clark GWB, Ireland AP, Chandrasoma P, Smyrk TC, DeMeester TR. Outcome of adenocarcinoma arising in Barrett's esophagus in endoscopically surveyed and nonsurveyed patients. J Thorac Cardiovasc Surg 1994;108:813–821.
15. Dulai GS, Guha S, Kahn KL, Gornbein J, Weinstein WM. Preoperative prevalence of Barrett's esophagus in esophageal adenocarcinoma: A systematic review. Gastroenterology. 2002; 122:26–33.
16. Rudolph RE, Vaughan TL, Storer BE, et al. Effect of segment length on risk for neoplastic progression in patients with Barrett esophagus. Ann Intern Med 2000;132:612–620.
17. Chandrasoma P. Controversies of the cardiac mucosa and Barrett's esophagus. Histopathol 2005;46:361–373.
18. McArdle JE, Lewin KJ, Randall G, et al. Distribution of dysplasias and early invasive carcinoma in Barrett's esophagus. Hum Pathol 1992;23:479–482.
19. Paull A, Trier JS, Dalton MD, Camp RC, Loeb P, Goyal RK. The histologic spectrum of Barrett's esophagus. N Engl J Med 1976; 295:476–480.
20. Chandrasoma PT, Der R, Dalton P, Kobayashi G, Ma Y, Peters JH, DeMeester TR. Distribution and significance of epithelial types in columnar lined esophagus. Am J Surg Pathol 2001;25:1188–1193.
21. Ortiz A, Martinez de Haro LF, Parrilla P, Morales G, Molina J, Bermejo J, Liron R, Aguilar J. Conservative treatment versus antireflux surgery in Barrett's oesophagus: long-term results of a prospective study. Br J Surg 1996;83:274–278.
22. Parrilla P, deHaro LFM, Ortiz A, Munitiz V, Molina J, Bermejo J, Canteras M. Long term results of a randomized prospective study comparing medical and surgical treatment of Barrett's esophagus. Ann Surg 2003;237:291–298.
23. McDonald ML, Trastek VF, Allen MS, Deschamps C, Pairolero PC, Pairolero PC. Barrett's esophagus: Does an antireflux procedure reduce the need for endoscopic surveillance? J Thorac Cardiovasc Surg 1996;111:1135–1138.
24. Van Den Boom G, Go PM, Hameeteman W, Dallemagne B, Ament AJ. Cost effectiveness of medical versus surgical treatment in patients with severe or refractory gastroesophageal reflux disease in the Netherlands. Scand J Gastroenterol 1996;31:1–9.
25. Viljakka M, Nevalainen J, Isolauri J. Lifetime costs of surgical versus medical treatment of severe gastro-oesophageal reflux disease in Finland. Scand J Gastroenterol 1997;32:766–772.

Index

D

G

H

T

U

V

Printed and bound by CPI Group (UK) Ltd, Croydon, CR0 4YY
08/05/2025
01864915-0001